# BOTANY

## PRINCIPLES AND APPLICATIONS

# BOTANY

## PRINCIPLES AND APPLICATIONS

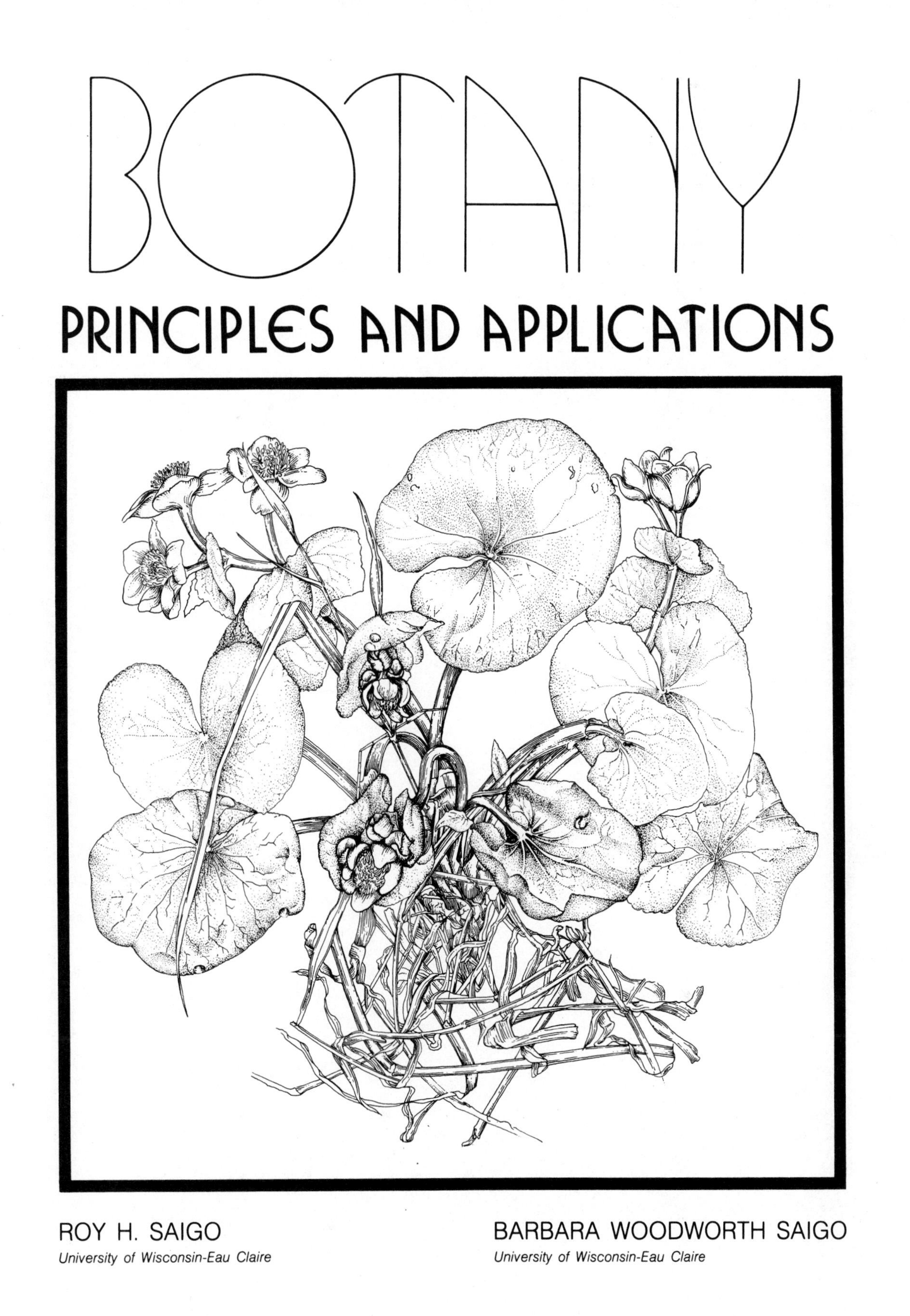

ROY H. SAIGO
*University of Wisconsin-Eau Claire*

BARBARA WOODWORTH SAIGO
*University of Wisconsin-Eau Claire*

Prentice-Hall, Inc., Englewood Cliffs, New Jersey 07632

*Library of Congress Cataloging in Publication Data*

Saigo, Roy H.
Botany, principles and applications.

Bibliography: p.
Includes index.
1. Botany. 2. Botany, Economic. I. Saigo, Barbara
Woodworth. II. Title.
QK47.S25 1983 581 82-15093
ISBN 0-13-080234-4



Editorial/production supervision by Maria McKinnon
Interior design by Lee Cohen and Maria McKinnon
Page layout by Diane Heckler-Koromhas
Cover design by Lee Cohen
Manufacturing buyer: John Hall
All photos not otherwise credited are by the authors.

Cover: This brilliant Indian paintbrush (*Castilleja*) was photographed by the authors at approximately 7500 feet elevation on Mount Jefferson in the Cascade Mountains in Central Oregon.

ISBN 0-13-080234-4

Printed in the United States of America
10 9 8 7 6 5 4 3 2 1

Prentice-Hall International, Inc., *London*
Prentice-Hall of Australia Pty. Limited, *Sydney*
Editora Prentice-Hall do Brasil, Ltda., *Rio de Janeiro*
Prentice-Hall of Canada Inc., *Toronto*
Prentice-Hall of India Private Limited, *New Delhi*
Prentice-Hall of Japan, Inc., *Tokyo*
Prentice-Hall of Southeast Asia Pte. Ltd., *Singapore*
Whitehall Books Limited, *Wellington, New Zealand*

To our families, with love and appreciation, and to the teachers and advisors who made a difference in our lives.

# SUMMARY CONTENTS

# COMPLETE CONTENTS

## four
## CELLULAR ENERGETICS: THE FIRES OF LIFE 50



## five
## NEW GENERATIONS OF CELLS AND ORGANISMS 68



## six
## INHERITANCE AND GENETIC CHANGE 107

## seven
## PLANT TISSUES 121



## eight
## PLANT ARCHITECTURE: ROOTS, STEMS, AND LEAVES 137



## nine
## PLANT GROWTH REGULATION 174



## ten
## FROM SOIL TO SEEDLING 193

## eleven
## CULTIVATED GROWTH: FROM SEEDLING TO HARVEST 213



## twelve
## ALTERED GROWTH: DISEASE MECHANISMS AND RESPONSES 233



## thirteen
## DIVERSITY: PROKARYOTES AND NONVASCULAR PLANTS 257

## fourteen
## DIVERSITY: VASCULAR PLANTS 301



## fifteen
## BIOMES OF NORTH AMERICA 322



## sixteen
## USEFUL PLANTS AND THEIR PRODUCTS 371

## seventeen
## FOOD AND OTHER USES OF WILD PLANTS 407



## eighteen
## HAZARDOUS PLANTS: INJURY, ALLERGY, MEDICINE, POISON 437



## nineteen
## ENVIRONMENTAL PRINCIPLES IN PRACTICE: INDOOR PLANTS 468

# PREFACE

## REASON FOR THIS TEXT

Why another botany book? This was the first question raised by some of our reviewers, as there are already several good texts on the market. After reviewing the manuscript, however, they encouraged Prentice-Hall and us to continue. They commented on such things as enthusiasm, readability, scientific level, sense of involvement with the student, and the linking of science with everyday life.

We initiated the text after 12 years of working with undergraduate students. We found that they assimilated botanical information better and were more stimulated to become involved with botany when they could connect course information to their daily contacts with plants, plant materials, and the natural environment.

Thus we evolved a presentation of basic botany in a context that literally "spills over" out of the classroom and laboratory into students' daily lives. This approach seems especially appropriate as the percentage (now 35+) of older (over 23) and part-time students enrolled in colleges and universities increases.

Our text incorporates not only scientific, technological, and agricultural aspects of botany, but also aesthetic and recreational uses of plants. Where appropriate, human relationships to the biosphere are used as examples, and the interrelatedness of all life on earth is consistently emphasized. Thus the approach is holistic.

## OBJECTIVES

The text has three objectives:

1. To teach basic botanical facts and principles.
2. To relate these facts and principles to familiar examples and human activities.
3. To enhance environmental understanding and awareness.

These objectives have been accomplished in two ways—by the selection of topics and by the integration of facts and principles with examples of their practical applications. The topics presented have a strong basic botany core, and are varied enough to be used in entirety or to be selected from by individual instructors.

## TEXT CHARACTERISTICS

### Content

All basic topics in most introductory botany courses are covered. Where feasible and logical, facts and principles are related to common examples that are likely to be familiar to students, including human uses of plants. For example, discussions of apical dominance refer to the practices of pinching back house and garden plants and shearing Christmas trees. Properties of the vascular cambium and secondary growth processes are identified as the basis for grafting and budding. Xylem anatomy is related to properties and uses of woods.

Throughout, the text emphasizes the continuous interaction of the living plant with its environment, and how environmental factors affect plant growth and survival.

Special chapters are provided on such applied topics as plant pathology, horticulture (indoor plants, landscaping), and human uses of plants (economic botany, using wild plants, medicinal/poisonous plants). These special chapters are also written to emphasize basic botanical facts and principles. The plant pathology chapter, for example, emphasizes the biology of disease-causing organisms (mainly viruses, bacteria, and fungi), and the nature of host responses (how pathogens disrupt normal host-plant growth and metabolism as it has been studied in earlier chapters) rather than emphasizing disease diagnosis and control.

Thus the applied chapters are fully integrated with the more traditional general botany topics. They have been designed especially to enhance and emphasize, rather than to stand alone as accessory topics. The Text Overview section of this preface demonstrates how the chapters are interrelated.

### Accuracy

General texts always lag behind research discoveries and usually must be restricted to introductory rather than definitive detail. Throughout this project each chapter has been read by one or more experts in the appropriate fields to reach a realistic consensus of what could be included. If errors or misconceptions are present, however, we hope they will be reported to us for future correction.

### Readability

We have attempted to capture some of the excitement and flow of a true narrative, hoping to convey botany to the student as the interesting and dynamic science it truly is! We have tried to write in a clear, straightforward manner, using appropriate, noncondescending language.

With the assistance of many expert reviewers, we have concentrated on developing a text that covers topics with the appropriate terminology and concise language required in any introductory botany course. The rigor of science is maintained without introducing excessive terms and details that would be appropriate for a more advanced course. Many discussions are crossreferenced.

Special attention has been devoted to avoiding the sexist language and connotations that often unnecessarily burden scientific writing, including some existing texts.

### Illustrations

Most of the photographs and line drawings are original. We have tried to avoid excessive use of previously published illustrations to give the text a fresh look and have chosen familiar, rather than exotic, examples when possible.

### Organization

The chapter sequence has been carefully analyzed and reviewed. Its logic is elaborated in the Text Overview that follows. We are aware of several different arrangements an instructor may wish to choose, however, so we have written the chapters to be somewhat independent of one another. This style and the crossreferencing should permit flexible use of the text.

## TEXT OVERVIEW

Chapter 1 presents some reasons for studying botany and briefly introduces the scientific method.

Chapter 2 introduces characteristics of living things and describes how specific environmental factors affect plant growth and survival. There is a brief introduction to plant classification.

Chapter 3 explores cell structure and function, introducing a basic vocabulary essential to understanding modern botany.

Chapter 4 discusses the life-sustaining processes of photosynthesis and respiration. This chapter has a bilevel organization so it can be used to provide an overview of cellular metabolism or to provide greater depth.

Chapter 5 explains how new cells and organisms are produced. It presents details of sexual and asexual reproduction, including specific methods for propagating ornamental plants.

Chapter 6 relates cellular reproductive processes to genetic mechanisms. Although briefly treated, basic genetics is explained and genetic processes are related to adaptation and improvement of plant species through natural and artificial selection.

Chapter 7 examines the origin, structure, and functions of plant tissues, emphasizing how the tissue types represent adaptations to plant needs. This chapter reinforces and builds on preceding chapters and introduces the vocabulary necessary for learning about plant organs and growth processes.

Chapter 8 describes roots, stems, and leaves—their functional and structural diversity—as well as vascular transport, building the discussion on previous chapters.

Chapter 9 reviews plant growth, including the influences of growth regulators, phytochrome-mediated phenomena, and circadian rhythms. Common horticultural uses of growth regulators are introduced.

Chapter 10 takes plants from a self-contained existence in the seed to a dependence on the external environment. Soils, nutrients, seed structure, and germination are discussed.

Chapter 11 discusses plants from germination through maturation. Growth characteristics and requirements are related to home gardening practices, providing a practical context for the discussion.

Chapter 12 examines some principles of plant pathology. The chapter is *not* a disease and pest manual. Instead, disease is discussed as a biological interaction between the disease-causing organism or condition and the afflicted plant. The tremendous human consequence of plant diseases is noted.

Chapters 13 and 14 discuss kinds of plants (past and present), their environmental requirements and life cycle patterns, and some human factors. Structural and reproductive details of seed plants are covered in early topical chapters. The sequence through these two chapters emphasizes progressive degrees of adaptation, culminating in the flowering plants.

Chapter 15 synthesizes much of what has gone before, presenting various North American biomes as a traveler might encounter them, and explaining the physical and biological factors that characterize each kind of environment. There is special emphasis on plant adaptations to harsh or unusual environments—such as arctic, alpine, and desert areas. Characteristic plant and animal species are cited.

Chapter 16 describes some of the many plants and plant products humans use, and is organized to remind students of facts learned earlier about plant tissues, organs, and products.

Chapter 17 introduces foraging and other uses of wild plants, presenting guidelines for collecting and preparing wild foods and author-tested suggestions. It reinforces several skills and concepts—the ability to make careful observations, to identify plants, and to appreciate the natural environment.

Chapter 18 is related to the two previous chapters, as it discusses plants that may have both useful and harmful aspects—those that contain toxic principles or that otherwise affect the health and well-being of humans.

Chapter 19 explains houseplant culture in terms of matching growing conditions to species requirements. Facts and principles about plant structure, growth, reproduction, and ecology are reinforced.

Chapter 20 focuses on adapting the home landscape to meet both aesthetic and functional needs, taking into account the natural growth characteristics and environmental requirements of plant species that are used.

## ACKNOWLEDGMENTS

The acknowledgments represent the building blocks and evolution of a text. Individuals cited range from those who provided encouragement, assistance, and support throughout the entire project to those who assisted with specific portions; from those who guided the direction and accuracy of the scientific information and its presentation to the students who used the manuscript and made suggestions.

Our sincere gratitude is extended to each person named below, as well as to others whose assistance was perhaps less direct but nevertheless appreciated.

For their assistance throughout the project: Dr. Jerry D. Davis, University of Wisconsin-LaCrosse; Dr. Robert A. Cecich, U.S. Dept. of Agriculture Forest Genetics Laboratory; and Dr. Fred Rose, Idaho State University.

For reading and evaluating the entire manuscript at one or more of its various stages: Dr. Jerry D. Davis; Dr. Eva Estermann, San Francisco State University; Dr. William C. Steere, New York Botanical Garden; Dr. Richard M. Klein, University of Vermont; Dr. Wilford M. Hess, Brigham Young University; Dr. Charles R. Curtis, University of Delaware; Dr. David Webb, University of Puerto Rico, Rio Piedras; Dr. Robert S. Mellor, University of Arizona; Dr. Edward P. Klucking, Central Washington University; and Dr. Glenn W. Patterson, University of Maryland.

For reading and assisting with portions of the manuscript: Dr. Robert A. Cecich; Dr. Marcus J. Fay, Dr. Lloyd E. Ohl, Mr. Walter Schaffer, and Mr. Gilbert Tanner, University of Wisconsin-Eau Claire; Dr. David Koranski, University of Minnesota-St. Paul; Dr. Douglas Maxwell, Dr. Darrel G. Morrison, Dr. David M. Peterson, and Dr. Peter Quail, University of Wisconsin-Madison; Dr. Jerome P. Miksche, University of North Carolina-Raleigh; Dr. Thomas Moore and Dr. Fred Rickson, Oregon State University; Ms. Deon Nontelle, University of Wisconsin-LaCrosse; Dr. Lee H. Pratt, University of Georgia; Dr. Fred L. Rose and Dr. Jay Anderson, Idaho State University; Dr. Robert J. Tomesh, University of Wisconsin-River Falls; and Dr. Mary Hotze Witt, University of Kentucky-Lexington.

For assisting with specific information on special topics: Dr. Arthur Anderson, (Pioneer) Hi-bred Seeds; Dr. Gary Breckon, University of Puerto Rico, Mayagüez; Dr. William Barnes, Dr. Victor Cvancara, Dr. Thomas Jewell, Dr. Joel Klink, Dr.

Johng Lim, Dr. Thomas Rouse, Dr. Birdell Snudden, Dr. Marshall Sundberg, and Dr. John Woodruff, University of Wisconsin-Eau Claire; Dr. Ray F. Evert and Dr. John Thomson, University of Wisconsin-Madison; Mrs. Olive Thomson, Madison, Wisconsin; and Mr. William W. Woodworth, Sr., Crown Zellerbach Corp.

To students who assisted in various ways: Mary Bartholomew Binkley, Nancy Knutson Hoeser, Jayne Reiche Kievet, Rose Forster Knutson, Annette Proehl, Sharon Smith, Beth Murel Stroede, and Martha Torrey. We also thank those students who tested the manuscript in class.

For illustrations: We thank the many generous individuals who are credited in the captions for making photographs and line drawings available to us. In addition, we express special gratitude to Dr. Harry T. Horner, Dept. of Botany, Iowa State University, and Ms. Candice Elliott, Dept. of Plant Pathology, University of Wisconsin-Madison, for unselfishly coordinating the acquisition of many photographs from others. The exquisite life drawings of plants are the work of Ellen Dudley, Madison, Wisconsin; most of the remaining art is by Sara Rogers, Eau Claire, Wisconsin. We are grateful to the Media Development Center, University of Wisconsin-Eau Claire, for their advice and assistance.

For typing: Linda Glenna, Barbara McKnight Highley, Rose Forster Knutson, Susan Madden, Yvonne Plomdahl, and Rhonda Sander.

For personal and family assistance and support: We give special thanks to Mr. and Mrs. William Woodworth, Sr., and the Toshiaki Saigo family for support and assistance in many ways during the project. We also express tremendous gratitude to Kären Hegge Passow, Patricia R. Rouse, the Orville and Marilyn Lee family, and Mary Breckon for loving child care while mom and dad "worked on the book."

Last, but not least, our appreciation is expressed to Heather (10), Holly (7½), and Dustin (2), who provided needed laughter, joy, encouragement, and understanding while sharing their lives with this demanding but creative and exciting task.

R.H.S.

B.W.S.

# BOTANY
## PRINCIPLES AND APPLICATIONS

# chapter one WHY STUDY BOTANY?

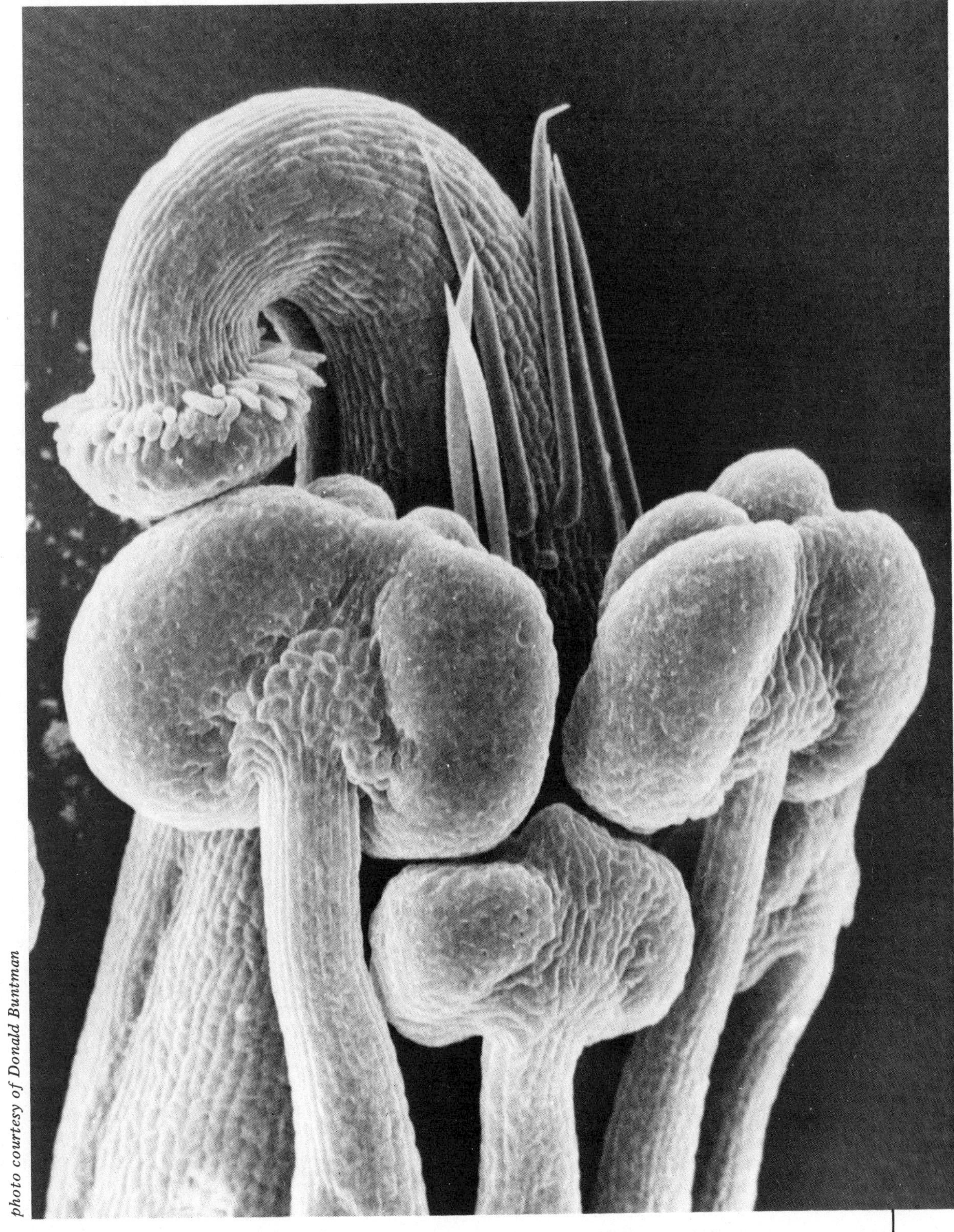

*photo courtesy of Donald Buntman*

## WHY CARE ABOUT PLANTS?

Humans have many reasons for studying and caring about plants. For one thing, plants are interesting! They are marvels of intricate biological adaptations for survival under diverse environmental conditions and they determine the animal life that is present in a place. Plants are beautiful! In the process of living their own lives, they produce aesthetic and colorful functional forms that give humans visual, tactile, and olfactory pleasure. Their striking sculptural beauty is especially apparent when they are viewed microscopically, as in the chapter opening art, a scanning electron microscopic photograph of the reproductive structures of a soybean (*Glycine max*) flower.

But a more significant reason is that we *need* plants. Throughout the long development of the human species people have been surrounded by and dependent on a world defined by vegetation. Prehistoric humans foraged for greens and grains; fruits and roots; healing and soothing herbs, barks, and berries; timbers, thatch, and fibers to protect their naked and vulnerable bodies from the elements.

In existing primitive societies there is still a very direct relationship between humans and the plants in their environments. In industrial societies like ours we also rely on plants in innumerable although often unrecognized ways. We take much for granted. We assume that we will be able to buy food and clothing, building materials, books, medicines. Yet how often do we think about the sources of these items? Table 1-1 provides a small sample of how plants touch our lives, the subject to which Chapter 16 is devoted.

**TABLE 1-1 A morning in your life with plants.**

| | |
|---|---|
| 6:30 A.M. | Wake up, put on slippers and robe* (fibers from cotton seed) |
| | Blow sniffly nose* (facial tissue made from wood pulp) |
| | Make bed* (cotton sheets, wooden headboard and frame) |
| 6:45 A.M. | Turn up heat* (fueled by wood, coal, oil, or natural gas—all derived from plant matter) |
| | Start the coffee* (berries of coffee plant) |
| | Brush teeth* (mint oil for flavor, powdered diatom shells for abrasion) |
| | Shower* and shampoo* (fats, waxes, solvents, and fragrances derived from plants in soap, shampoo, and hair rinse; cotton fibers in towel) |
| 7:00 A.M. | Breakfast |
| | orange juice* (from citrus berry) |
| | coffee, with nondairy creamer* (contains soybean solids and other plant derivatives) |
| | corn flakes*, milk, and sugar* |
| | vitamin*/mineral supplement (natural vitamins synthesized by plants, including microorganisms) |
| | read back and side panels of cereal box* (wood pulp, starches, and other plant products in pasteboard; wax in inner liner) |
| 7:20 A.M. | Get dressed |
| | Final "fixing" |
| | lotions*, cosmetics*, hair care products*, perfumes*/colognes* (including such plant products as cocoa, coconut, carnauba, candelilla, and other fats and waxes; aloe gel; olive, arnica, calendula, sesame, and other oils; fragrances; alcohols; fat-soluble vitamins; balsam; herbal extracts; cellulose compounds; etc.!) |
| 7:40 A.M. | Take cold medication* and aspirin* (belladonna alkaloids, caffeine; acetyl salicytic acid) |
| | Apply mentholated oil* to cold sore (menthol, camphor, and eucalyptol oils) |
| 7:45 A.M. | Put on rainboots* and coat (rubber made from plant latex) |
| | Leave in car* (rubber tires; gas and lubricants from plant matter; upholstery padding of foam rubber and miscellaneous plant and synthetic fibers) |
| 11:50 A.M. | Lunch |
| | taco* (lettuce and onion leaves; tomato and pepper fruits; sauce of vinegar fermented by bacteria, plus oil from corn, sunflower, or soybean seeds) |
| | cola drink* (cola flavor from cola plant; sugars; other plant extracts for flavor) |
| | vanilla ice cream* (thickened with stabilizers derived from seaweeds; flavor from seed pod of vanilla orchid) |

Plants are with you at every asterisk (*).

Even more basically, humans and all other life forms on Earth depend on plant life for *survival.* The life processes of complex plants and animals require a constant source of energy. Earth's constant source of energy is the sun, but **solar energy** cannot be used directly by living things, or **orga-**

**nisms.** First, it must be converted to **food energy**, the energy stored in the molecular bonds of certain chemical compounds. Green plants are the only organisms that can convert solar energy to food energy. All animals, including humans, therefore rely directly on green plants for the energy to conduct their life processes.

In addition, however, we depend on green plants for the continuing ability to USE that food energy. Our body processes require oxygen to release food energy. Where does the oxygen come from? Earth's atmosphere is a vast storehouse of available oxygen. Much if not all of that oxygen originally came from green plants, and is replenished by them. In the same process by which they bind the sun's energy, green plants release oxygen. The process is called **photosynthesis** and it is the basis for Earth's life systems.

So it would seem that as long as there are green plants to carry out photosynthesis, the human species will have its minimum requirements for survival. But is it so simple? We have always taken the availability of plants for granted. We have learned many ways to use them. In this process, however, we have not paid equal attention to what *plants* need to survive. In past years we stripped the land of its complex native organic cover, covering it with cultivated crops, pavement, and cities or leaving nothing but bare soil to be washed and blown away. In the Middle East the fertile lands that cradled civilization before the Pharoahs are now desert. The forested lands referred to in the Bible have been reduced to scrub vegetation. The vast fertile prairies of the Oklahoma territory and regions farther west were turned into a gigantic "Dust Bowl" in less than a hundred years.

And "we," collectively, are still doing it. There is a major project in South America to make the Amazon wilderness accessible. The native vegetation is cut and bulldozed away, leaving the thin, infertile soil exposed to torrential rains and subsequent erosion instead of being protected from the onslaught of pelting raindrops by many layers of leaves. Even if this effort were to be abandoned, the complex mature rain forest would take untold years to reproduce itself—if it could, in fact, do so. It is feared that loss of the tropical rain forests would alter global climatic patterns. In the Philippine Islands unwise logging practices convert steep-sided verdant hillsides into raw, unvegetated, and gullied wastelands. In Europe and North America pollution of waterways contaminates the estuaries where hundreds of edible shellfish live and where many ocean fish begin their lives. We bury chemical and radioactive wastes in the ground instead of detoxifying them. In these and many other ways, the human species continues to "soil its own nest" and that of other life forms as well.

The United States and many other nations are highly industrialized. We manufacture all conceivable necessities, luxuries, and frivolities. Our factories are powered by burning enormous amounts of combustible fuels. The heat and electricity for most of our homes and the power to operate our forms of transportation come from the oxidation of combustible fuels. Burning is oxidation—it consumes oxygen. Some violent acts of nature also use oxygen, such as forest fires and volcanic explosions.

In a real sense, human activities involving combustion compete for oxygen with the life processes of humans and all other organisms. What would be the consequences if oxygen-using processes on Earth exceeded the total capacity of green plants for oxygen producing?

So now we are back to the human association with plant life on Earth. Nearly all the factors that threaten the existence of plant life, worldwide or on an acre-by-acre basis, are human activities. If we accept the assumption that in order for humans to survive we must keep the rest of life on Earth healthy, then it is in our own deepest self-interest to be informed about the nature of plant life, its diversity, the requirements of plants, and their ecological interactions.

Knowledge about plants is the foundation for a personal and collective strategy for maintaining our planet. How is this so? One person does not contaminate an environment and one person alone is rarely able to save an environment from contamination or destruction. People, however, determine the legal and political provisions of land-use policies, long-range planning, and international economics.

To bring it closer to home, it depends on how municipalities obtain their power, dispose of their solid and liquid wastes, control pollution, plan their housing, transportation, green areas. It depends on how resources and agriculture are managed. And it depends on treaties and legislation to prevent existing natural environments, such as oceans, waterways, and wilderness areas, from being irreversibly altered. The tragedy occurs when an environment is disrupted or destroyed before we fully understand its intricate interactions and operation in relationship to the balance of life on Earth.

People have the power to learn, to know, to save or destroy. Public policies are determined by people. People *need* plants.

## LEARNING TO OBSERVE

Over the centuries thousands of observers have added their many separate observations to the growing body of human knowledge about living things. Today it is possible to explain and understand many natural phenomena in causal rather than mystical terms. For the truly aware person, this understanding is complementary to aesthetic stimuli. Autumn leaves, varying from purple to palest yellow, for example, are beautiful, especially when backlit by the afternoon sun. Yet our appreciation can be heightened further by an understanding of the natural seasonal rhythms of physical and chemical changes occurring within the dying leaf cells.

Observation is the source of both questions and answers. Scientific observation differs from casual observation in that it follows a logical sequence called the **scientific method.** The scientific method involves identification of a specific question; proposal of tentative explanations or hypotheses (sing., **hypothesis**); design of precise, controlled tests to yield quantitative measurements; comparison of the hypotheses to the test results; refinement, rejection, or acceptance of each hypothesis as the explanation or answer to the posed question. Table 1-2 presents an example of how we might apply the scientific method to a problem.

When you observe from now on, try to go beyond merely registering sensory impressions of sight, sound, touch, taste, and smell to a more purposeful, sensitive, educated type of observation.

**TABLE 1-2 Application of the scientific method.**

1. **Observation** of a natural event: My favorite houseplant wilted.
2. Formulation of a **question:** Why did it wilt?
3. Formulation of tentative explanations, **hypotheses:** Perhaps the problem was
   (a) lack of watei in the soil.
   (b) root disease caused by too much water in the soil.
   (c) inadequate nutrients.
   (d) inadequate sunlight.
   (e) excessive temperature.
   (f) insect pests.
   (g) unknown causes.
4. Making **organized observations** to test the hypotheses: The plant has been maintained on a regular recommended fertilizing schedule. It is hanging in a drainless container in a bright, south-facing window. The soil is wet. No insects are present.
5. **Evaluating** the hypotheses: Hypotheses (a), (c), (d), and (f) are rejected; (b), (e), and (g) are still viable.
6. **Further observations:** The soil is warm, soggy, and has a putrid odor.
7. Generalization of a **more acceptable hypothesis:** Although unknown causes may still be involved, the evidence is in favor of the explanation that overwatering and inadequate drainage caused the root system to rot.

The scientific method involves a series of logical processes, as in the imaginary example presented here. This example demonstrates the applicability of a scientific approach to an ordinary situation.

# chapter two EARTH LIFE FORMS AND ENVIRONMENT

## INTRODUCTION

A universe, a solar system, a planet. A sheltering atmosphere with hospitable gases and temperatures . . . water . . . LIFE! Earth, of the planets we know, provides an environment that supports living entities, or organisms, of great variety and complexity. Earth and its atmosphere constitute an enormous "solar vivarium," a life-nurturing contained system that is driven by the radiant energy blasted into space by solar fires.

Green plants are intermediaries between the sun and other life forms, transforming solar energy into food and other complex molecules. As animals, humans depend on plants (directly and indirectly) for food. In addition, humans have learned to use plants in thousands of non-food-related ways.

Part of our purpose here is to introduce briefly the nature of plants and how they are studied. Primarily, however, we emphasize the general characteristics of Earth's environments that determine the nature of life on Earth.

## WHAT IS LIFE?

What distinguishes the living from the nonliving? Through the ages humans have interpreted the essence called "life" mystically, metaphysically, and scientifically. To this day that essence remains elusive, as shown by attempts to refine the definitions of when human "life" begins and when it ends. Medical, moral, and legal issues are involved in these attempts. However, several interdependent characteristics that are common among organisms can be identified objectively. Considered together, they separate living from nonliving things.

1. Typical organisms are composed of **cells** and cell products; cells originate from preexisting cells; and each cell is capable of maintaining its own existence. These are the three basic tenets of the **cell theory** or cell doctrine. Cells are the basic units of life and their structure and life-sustaining activities are discussed in detail in the next few chapters.
2. Organisms consist mainly of **organic compounds,** of which the carbon atom is the basic structural entity. Oxygen, hydrogen, and nitrogen are the principal components of organic compounds. Organic compounds are the basis of both structural and functional molecules within living organisms. The three major types of organic compounds are **carbohydrates** (e.g., sugars, starches, cellulose), **proteins** (composed of amino acids), and **lipids** (e.g., fats, oils, waxes).
3. Two complex information molecules called **nucleic acids** have a primary role in regulating cell activities. They are deoxyribonucleic acid (DNA) and ribonucleic acid (RNA), and are discussed in detail in Chapter 3. They are called nucleic acids because they were first studied in cellular structures called nuclei, even though RNA is abundant throughout the cell.
4. An organism is not a random assemblage of molecules but is highly organized, both structurally and functionally. This **organization** is obvious at all levels—biochemical, subcellular, cellular, tissue, organ, and the whole organism. Illness or disruption of any sort interrupts the organized activities and structure of an organism. Organization is maintained by constant checks and balances and loss of internal organization results in death.
5. Organisms experience **growth,** by total enlargement, production of new cells, and increase in complexity. These three criteria distinguish "true" growth from increase in size by simple addition of material (accretion), such as in the formation of plant crystals (Fig. 2-1), mineral deposits in a steam iron, pearls, gall and kidney stones, and some geological formations.
6. Organisms **reproduce** their own kind, passing on inherited "instructions" in the form of specific DNA molecules.
7. Organisms are capable of **change.**
8. Organisms demonstrate **irritability**, the ability to respond dynamically to stimuli. The response may be internal and chemical in nature or may involve an external or mechanical response.
9. **Movement** is a characteristic of organisms. Many organisms and reproductive cells are capable of locomotion, the ability to propel themselves from one place to another. Others, such as the sensitive plant (*Mimosa*)(Fig. 2-2) and the Venus flytrap (*Dionaea*), are able to

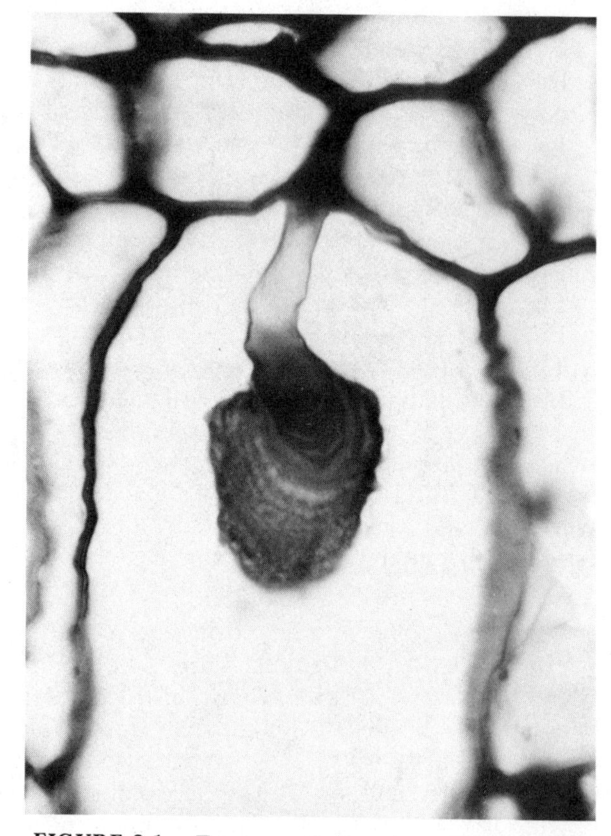

**FIGURE 2-1** True growth is more complex than increase in size by accretion, the manner by which crystals "grow," as this calcium carbonate crystal in the leaf of a rubber plant (*Ficus*).

**FIGURE 2-2**
Sensitive plant (*Mimosa*) leaflets fold toward the leaf midrib in response to being touched.

move individual parts of the body. In addition, the internal cell contents are constantly circulating, a phenomenon called cyclosis.

10. Organisms actively convert energy and matter to other forms and the overall term for an organism's biochemical activities is **metabolism.** Among the metabolic activities of an organism are digestion and assimilation of food, respiration, the conversion of food energy to chemical energy to "drive" cellular activities, formation of structural compounds, and elimination of harmful byproducts. In addition, green plants are able to synthesize food by trapping light energy during a process known as **photosynthesis.**
11. Finally, living organisms experience **death**, the natural conclusion of a living condition.

These characteristics are easily applied to most organisms. The viruses and their relatives, however, possess some, but not all, of the characteristics (see Chapter 13).

Complex organisms are generally **multicellular**, composed of many cells. Multicellularity allows and necessitates internal specialization. **Tissues** are groups of cells specialized to perform specific functions, such as food production, internal transport, support, reproduction, storage, protection, locomotion, and even growth.

In complex or "higher" plants and animals tissues are organized for increased efficiency into distinct structures called **organs.** Plant organs are leaves, stems, roots, and reproductive structures. Cones and flowers are assemblages of reproductive organs.

## WHAT IS A PLANT?

The answer to this question would seem fairly straightforward! In thinking of a plant, you probably envision aquatic algae, pondweeds and seaweeds, mosses and lichens, ferns, grasses, shrubs, trees, and smaller flowering plants. All these (and more) are indeed "plants."

Chapters 13 and 14 describe how botanists attempt to sort out the various kinds of "plant" life on Earth into logical categories, based on complexity of structure and metabolism, life histories, and evolutionary relationships, as indicated by many kinds of evidence. As a result, there has been a recent trend to define the term "plant" more narrowly as those forms that are eukaryotic (see below), photosynthetic, and multicellular. The merits of various classification schemes are discussed in greater detail in the beginning of Chapter 13.

For present purposes, we will use the more general "plants" concept to refer to all major categories of organisms listed on Table 2-1 unless otherwise specified. Note that a distinction is made between "plants" that are prokaryotic and those that are eukaryotic. **Prokaryotic** refers to organisms with characteristics that are considered more primitive, such as the absence of an enclosed nucleus and certain other membrane-bounded cellular structures. In contrast, **eukaryotic** organisms possess a membrane-bounded nucleus that contains chromosomes, bearers of the inherited information of previous generations, as well as other complex cellular structures. (These distinctions are explored further in Chapter 13.)

TABLE 2-1 Organisms commonly referred to as "plants."[a]

| | | |
|---|---|---|
| **Prokaryotic groups** | | |
| Bacteria | Blue-green "algae" | |
| **Eukaryotic groups** | | |
| Green algae | Water molds | Mosses, hornworts, liverworts |
| Brown algae | Chytrids | Whisk fern |
| Red algae | Slime molds | Club mosses |
| Diatoms | Bread molds | Horsetails, scouring rushes |
| Golden-brown algae | Sac fungi | Ferns |
| Dinoflagellates | Club fungi | Gymnosperms: Cycads, *Ginkgo*, gnetophytes |
| Euglenoids | "Imperfect" fungi[b] | Conifers |
| | Lichens | Flowering plants (Angiosperms) |

[a]The sorting out of these groups of organisms into taxonomic categories is detailed in Chapters 13 and 14.
[b]Refers to fungi for which no sexual stage exists or has been discovered.

Plants are sometimes subdivided into "lower" and "higher" groups. In general, higher plants are those that possess specialized tissues for conducting water and nutrients *and* that are able to produce seeds—the gymnosperms and angiosperms (see Table 2-1).

## PLANTS AND ANIMALS: SIMILARITIES AND DIFFERENCES

Higher plants and higher animals differ greatly from one another in overall structure and appearance, but they are amazingly similar at the cellular level, as detailed in Chapter 3. Their basic biochemical and metabolic activities are also similar. These similarities pose a strong argument for a common origin of life on Earth. Certain distinctive differences do exist, however.

Some special problems confront plants. Confined to one spot, they must absorb water and nutrients from their immediate surroundings. They must conserve water and require a plumbing system to allow for internal transport. They must compete with adjacent plants for solar energy and extend themselves upward, which requires a complex support system. They cannot escape fires, floods, insects, diseases, and changes in their environment. Therefore they must have innate defenses or the ability to adapt; otherwise they will die.

Cells reproduce by division but there are some differences in the ways that plant and animal cells divide, as detailed in Chapter 5. Higher plants and animals differ also in the manner of tissue support. Both plant and animal cells possess a thin outer membrane, the **plasma membrane**, as an outer boundary. In addition, plant cells are contained within a flexible, secreted "box," the **cell wall** (Chapter 3), which is composed mainly of cellulose fibrils and binding substances called pectins. In some tissues the cell wall becomes impregnated with hardening or waterproofing substances. Thus the "skeleton" of a plant is made up of the cell walls of many individual cells.

A major difference between higher plants and animals concerns their nutrition. Green plants are primarily **autotrophic** (self-feeding), producing food by photosynthesis. (More will be said later of the role of the green pigment chlorophyll in photosynthesis.) Animals are all **heterotrophic** (other-feeding), acquiring food by ingestion or absorption. Fungi, most bacteria, and some higher plant species are also heterotrophic.

Higher plants generally display indefinite growth throughout the lifetime rather than attaining a fixed maximum size. This is especially true of woody **perennial** plants, those that live for several to many years. Juvenile growth is the most rapid (Chapter 9). Unlimited growth is provided by localized growing points or **meristems** (Chapter 7).

Woody perennials are also distinctive for having an indefinite life span whereas higher animals have relatively fixed life expectancies.

If part of a plant dies or is removed, the plant can survive—continuing to live and grow—and has a capacity for **regeneration** of lost parts. Grass that is grazed or mowed is an example, as is the dandelion that regenerates even though its top is removed repeatedly! Higher animals, however, must maintain the wholeness of their physiological systems for full function and, with few exceptions, do not regenerate lost parts.

Most plants must rely on such external forces as wind, water, and animals to bring together sperm and egg cells and to disperse seeds to sites away from the parent plant. Vegetative plant parts may act as asexual propagating agents (Chapter 5).

## FOR EACH KIND OF ORGANISM, ITS OWN NAME

Every kind of plant has a Latin or latinized **scientific name**, and most plants have several familiar **common names**. Lambsquarters, a common wild plant and garden weed, for example, is also known (in the English language) as wild spinach, goosefoot, pigweed, white goosefoot, and smooth pigweed. Several other related and unrelated plants are also called pigweeds. As you can see, common names are not a reliable means of identification.

The confusing array of common names led a Swedish naturalist, Carl von Linné (latinized to Carolus Linnaeus), to a monumental lifework that included sorting out and classifying plants and animals in a systematic way. Linnaeus assigned each different kind of organism a Latin or latinized binomial (double name), consisting of a group, or **genus** name, followed by an exclusive **species** name. The full binomial is called the scientific name of a species. Linnaeus' work, *System Naturae,* was published in 1735 and his system of **binomial nomenclature** is the basis for all subsequent botanical naming and classification, the science of **taxonomy**. Unlike common names, the scientific name is internationally consistent.

In a scientific name the genus name is always capitalized and the species name begins with a lowercase letter. Sometimes subcategories are included as a part of the scientific name, as a **subspecies** or **variety** name. The scientific name is always set apart from other writing either by italics or underlining. The capital letter L. behind a scientific name means that it is a name that was given by Linnaeus. If the scientific name was given by another botanist, that person's name, or a standardized abbreviation of it, follows the species name. The concept of the scientific name cannot be overstressed, for it is the only wholly accurate way to identify a specific organism.

So whatever common name is applied and in whatever language it is described, the common eastern North American species of lambsquarters is always correctly identified by the scientific name *Chenopodium album;* a similar western species is *Chenopodium fremontii.* Although similar enough to be classified in the same genus, they are sufficiently distinct to have different species names.

Before beginning to study plants, it is helpful to know a little about how plants are classified into categories, based on increasing degrees of similarity. The largest categories are the major *divisions,* which are subdivided into *classes,* which are subdivided into *orders,* which are subdivided into *families,* which are subdivided into *genera* (the plural for genus). Each genus may contain only a single species or many closely related species. The **biological species** is a concept applied to organisms that are physically and genetically similar to the extent that they may interbreed to produce viable offspring. Because many qualifications and exceptions to this definition exist (especially among plants), there is no universally agreed-upon exact definition of species. This definition, however, is generally acceptable.

In this classification scheme each category is more exclusive than its predecessor. Table 2-2 shows how a flowering dogwood (*Cornus florida* L.)

**TABLE 2-2 An example of taxonomic classification. Family, genus, and species are the "working level" of plant classification and identification.**

Division: Anthophyta (flowering seed plants, the angiosperms)
Class: Dicotyledones (two-cotyledon flowering seed plants)
Order: Apiales
Family: Cornaceae (dogwood family)
Genus: *Cornus* (dogwood genus)
Species: *florida*

Scientific Name = *Cornus florida* L.
Common Name = Flowering dogwood
Some related species = *Cornus alternifolia* L.: Alternate-leaved, pagoda, or blue dogwood
*Cornus canadensis* L.: Bunchberry or dwarf dogwood
*Cornus nuttalli* Aud.: Pacific or western flowering dogwood
*Cornus stolonifera* Michx.: Red-osier dogwood
*Cornus stricta* Lam.: Stiff dogwood

When more than one variation of the species is known, a **subspecies** name is added to the scientific name:
*Cornus florida rubra:* Pink-flowering dogwood

Cultivated varieties (**cultivars**) are designated by the addition of the cultivar name to the scientific name:
*Cornus florida* Fragrant Cloud: a scented flowering dogwood

When discussing several species within a single genus, it is permissible to abbreviate the scientific name by using the initial of the genus name for successive references:
*Cornus florida, C. nuttalli, C. stolonifera*

**FIGURE 2-3** Flowering dogwood, *Cornus florida*. (Photo courtesy of Carolina Biological Supply Co.)

(Fig. 2-3) is classified. The family level of classification is an important "working level," for each family contains plants that are closely similar in basic structure and reproductive characteristics. Therefore plants are often identified by both the scientific name and the family name. Plant family names are capitalized and end with -aceae (pronounced ay-see-ay). The first part of the modern family name refers to a common genus in the family; thus the dogwood family name is Cornaceae.

## THE ENVIRONMENT FOR LIFE ON EARTH

An organism's **environment** is the sum of all the physical and biological factors that affect its existence. The lifetime of an individual plant is shaped and limited by environmental forces and the plant's innate ability to respond to them. In addition, all the diverse plant species that we know are the results of genetically accumulated responses to the environment over many generations. Environment determines not only whether an organism can survive but also what form it will take.

Therefore, an understanding of the forces of the environment is fundamental to understanding the structure of plants, their metabolic functions, their diversity, *and* the manner in which botanical principles can be applied to such human activities as gardening, growing houseplants, and economic uses of plants.

When a single factor determines the presence or absence of a particular organism more than any other factor, that factor is said to be the **limiting factor.** Limiting factors include humidity, temperature, pH, light, and competition, as for growth sites and nutrients. Water availability is most frequently the limiting factor for plant and animal distribution on land. Light is an important limiting factor in aquatic environments.

**Ecology** (derived from the Greek *oikos,* meaning house) is the science that seeks to understand and interpret the interactions of organisms with their environments.

Much has been written in recent years about ecology, not all of it based on scientific understanding, however. The very term ecology has been thrown into our contemporary jargon without regard for its meaning so that it frequently emerges as an erroneous substitute term for environment. "It's good for the ecology" or "We must be concerned about the ecology," we are told. There have been "ecology" shirts, shoes, patches, and notebooks, for advertisers have taken advantage of popular environmental awareness.

It would be extremely unfortunate if a science that has struggled to establish its scientific "respectability" for the past five decades were to suffer the fate of lesser fads and go out of fashion. In view of the tremendous human capabilities for altering all aspects of the natural environment, it would be doubly tragic for future generations to ignore ecological considerations, both in their lives as individuals and as citizens within a world society.

Plants live in a variety of environments, some of which are so unusual and harsh that we would expect them to be devoid of life, such as icebergs, hot springs, salt flats, and geyser pools. Most environments, however, are of three major types: marine, freshwater, and terrestrial. Each type presents its own problems for life, requiring special plant adaptations. (See also Chapter 15.)

The **marine** environment is that of the world's seas and oceans (chapter opening art). As a place to live, it offers many advantages to organisms because it is the most stable and least varying type of environment. **Freshwater** environments provide more variable physical conditions than the oceans, often due to dramatic seasonal variations. Of the three basic types, the **terrestrial** (land) environment is the harshest. Its basic characteristics require plants to have adaptations that are not necessary to an aquatic way of life. In addition, environmental changes occur *rapidly* and with great *amplitude* in the land environment.

## Physical Factors

The physical components of the environment are radiation (light and heat), water, atmosphere, and soil. They are commonly grouped into two broad categories: **climatic factors** (relating to climate) and **edaphic factors** (relating to soil). Each organism can survive within a certain **tolerance range** for *each* physical factor of its environment. Within this broader range of extreme upper and lower limits is an **optimum** point or range at which the condition is most favorable to that organism.

### Radiation

Earth's life systems are based on metabolic reactions that require energy. Virtually all this energy originates as solar radiation, so that life on Earth is truly dependent on our sun, the middle-aged star at the center of our solar system. Green plants collect solar energy and convert it into the energy of chemical bonds (bonds that form between atoms) through the process of photosynthesis (*photo* = light; *synthesis* = putting together), as detailed in Chapter 4. All life forms are fed by the organic compounds produced by green plants. Through agriculture humans have taken advantage of the energy-trapping ability of green plants, intensively cultivating a relatively few species that are efficient food sources for humans and domestic animals. If our star, the sun, were to fade out, however, the primary basis for Earth's life systems would be lost. Complex plants and animals and eventually all organisms would die off as the planetary food supply became exhausted and temperatures dropped below life-tolerating levels.

Solar radiation includes visible light, ultraviolet, infrared, and cosmic (nuclear) types. Much solar radiation never reaches Earth's surface but is reflected or radiated back into space or absorbed in bands of matter encircling Earth. Earth's ozone gas layer, for instance, filters out large amounts of ultraviolet radiation.

Because photosynthesis utilizes visible-light wavelengths, **visible light** is the radiation spectrum most significant to the daily maintenance of life. Plants are affected by specific variables of visible-light **intensity** (whether exposure is direct, indirect, or filtered), **quality** (wavelengths or colors present), and daily duration or **photoperiod.**

**Ultraviolet (UV) radiation** is not involved in photosynthesis and actually damages living cells at higher levels of exposure, a characteristic that has made it useful for sterilizing everything from scientific equipment to bowling shoes. Pigments in plant and animal outer tissues protect internal tissues from UV radiation.

We perceive **infrared radiation** as heat. Each organism has a temperature optimum for its life processes. Excessive heat destroys protein compounds within cells, causing death, whereas excessive cold slows life processes to a state of virtual suspension. Cold temperatures can kill organisms that lack cold-survival adaptations.

### Water

Water ($H_2O$) is essential to life and most organisms are unable to cope with low water availability. The living substance of the cell, **protoplasm,** is more than 90% water.

Water plays many important roles in plants.

1. Water is the medium in which all the chemical interactions of metabolism occur. Many of these interactions may be dependent upon proper water concentration.
2. Water is the solvent in which substances, including food, are dissolved for transport within organisms.
3. Water participates directly in cellular chemical reactions, contributing hydrogen and oxygen atoms necessary for molecular synthesis.
4. Water provides form and support for plant tissues.
5. Because of its own molecular properties, water provides a buffer against rapid and excessive temperature changes within cells. Water cools and heats at a slower rate than air.
6. Water molecules contribute to the acid/base balance of cells.
7. Water molecules also interact in the uptake and utilization of mineral nutrients.

Besides light, water availability may be the single most limiting factor for plants, and plants can be grouped according to their relationships to water. Those that live in water are called **hydrophytes** ("water plants"); those adapted to live in dry environments are called **xerophytes** ("dry plants"); those adapted to terrestrial environments that have abundant available moisture are called **mesophytes** ("middle plants").

Water is continuously recycled between the atmosphere and the earth through the processes of precipitation and evaporation. In the **water cycle** most of the exchange occurs between the atmosphere, the land, and surface waters. Water is also taken in by organisms; some is used and some is lost by evaporation from body surfaces and with products of elimination. Evaporation of water from plant surfaces is called **transpiration.**

#### Substrate

**Substrate** is the base or surface upon which an organism lives or to which it is attached. The physical composition and chemical properties of the substrate influence plant life in many important aspects, including moisture availability, nutrition, root penetration, and anchorage or attachment. Because living cells must have an immediate aqueous environment, even the roots of terrestrial plants are surrounded by a film of moisture between soil particles. The nature of the substrate surface has a direct effect on the ability of organisms to exist on or in it. Aquatic substrate surfaces, for instance, fit into general categories of rocky, gravelly, sandy, or muddy. The amount of organic litter or detritus on the substrate surface is also an important characteristic. Some organisms use other organisms as their substrate—for example, moss and lichens growing on tree bark.

Minerals are an important part of the substrate and their availability to plant and animal life is closely linked to the evaporation-precipitation cycle of water. The **mineral cycle** begins and ends with rock and is an extremely long-term phenomenon. Weathering continually releases minerals from the substrate. Plants take in minerals by absorption and the minerals are passed on to other organisms by feeding. Most minerals are subsequently released by decomposition that follows death of an organism, are returned to the substrate, and may be converted to rock once again by such physical processes as sedimentation and metamorphosis. Some minerals are assimilated into plant and animal structures that do not deteriorate rapidly, such as wood, teeth, shells, and bones, and a long erosive process is required to release them.

If plant and animal remains are protected from decay and erosion, they may be preserved as **fossils,** remains of prehistoric organisms. There are many kinds of fossils and ways for plant matter to become fossilized. Among the most common sources of fossils are peat and coal—the organic remains of plants that lived millions of years ago. Some plant fossils are found in layers of sedimentary rock, originating when they were buried by the sand and silt of bodies of water. Petrified wood was formed by the gradual infiltration and replacement of wood tissues by crystallizing minerals, forming a literal "tree of stone."

Relative acidity or alkalinity, as measured in **pH** (proportion of $H^+$ ions present), is another substrate characteristic (Fig. 2-4). (See Chapter 10 for a more complete explanation of pH.) Bogs and forest floors are acidic because of organic acids released by decomposition. The soil of many desert areas is alkaline. When large amounts of precipitation or irrigation water soak into desert soil, soil minerals dissolve. Then rapid evaporation of water from the soil surface causes the mineral-laden deeper soil solution to move upward. As water continues to evaporate, the minerals accumulate at the soil surface. In some desert areas these white, crusty "alkali" deposits are extensive.

**Salt** concentration in the substrate also limits plant life because salt has a tendency to draw water from living cells. Plants adapted to survival in such unusually salty conditions as salt ponds, mineral springs, salt flats, and saltwater marshes are called **halophytes** ("salt plants"). Irrigation can contribute to soil salinity by bringing salt to the soil surface in the same way that alkaline minerals are raised. Salinization of agricultural soil claimed from the desert is a severe problem in the vegetable-growing Imperial Valley of California.

Soil is "living" in a sense understood and

**FIGURE 2-4** Soil **pH** ranges in various habitats.

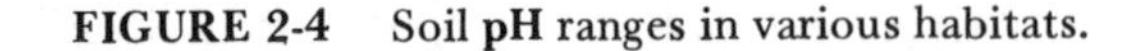


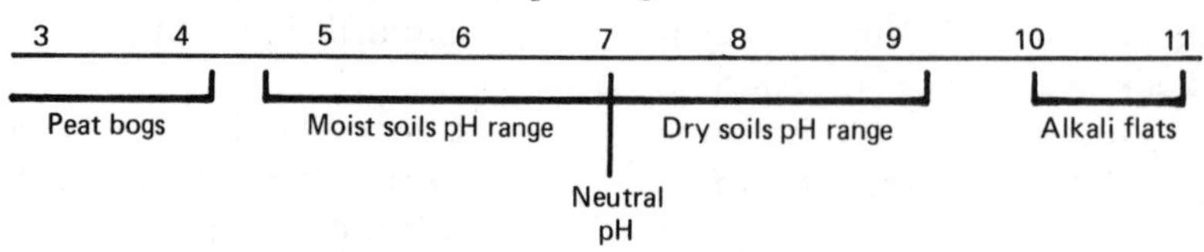

utilized by organic gardeners, supporting such **microorganisms** (microscopic organisms) as bacteria, fungi, and invertebrate animals. Certain bacteria are essential to soil fertility because of their ability to decompose organic matter and transform nitrogen into a form usable by green plants, as noted below.

### Atmosphere

A final physical component of the environment is the atmosphere. The major gases of Earth's atmosphere are **nitrogen** ($N_2$), **oxygen** ($O_2$), **carbon dioxide** ($CO_2$), and **water vapor** ($H_2O$). All four are essential to the life processes of organisms, providing the atomic "building blocks" of organic compounds. The continued availability of oxygen, carbon dioxide, and nitrogen is maintained by complex recycling patterns (Figs. 2-5, 2-6, and 2-7).

Many rarer gases are also present in the atmosphere and some of these "minor" gases may become biologically significant as a result of human activities. There are numerous examples: the air near intense automobile and aircraft traffic is high in toxic carbon monoxide gas and particulate matter; various industries and coal-burning power plants discharge large amounts of sulfur dioxide gas; some types of sewage treatment increase atmospheric concentrations of "sewer gases," foul-smelling hydrogen sulfide and combustible methane; ozone concentrations are higher near high-tension electrical lines; fluorocarbon gases used as aerosol propellants accumulate in the atmosphere where they may damage Earth's protective ozone layer.

A particularly vivid example of atmospheric pollution involves sulfur dioxide ($SO_2$), which has a distinctive, pungent odor often associated with emissions from pulp and paper mills. Susceptible plants, such as most lichens, many mosses, and Eastern white pine (*Pinus strobus*), are killed outright by concentrations now existing in industrial areas. Nonlethal concentrations damage leaves of pines and many broad-leaved plants.

Sulfur dioxide combines with atmospheric moisture to form sulfuric acid ($H_2SO_4$). When high atmospheric levels of $SO_2$ accompany precipitation (fog, mist, rain, snow), an **acid rain** of sulfuric acid occurs. Acid rain decreases soil and stream pH, etches statuary, masonry, grave markers, automobile finishes, and other exposed surfaces, and causes additional well-documented environmental effects.

The acid rain phenomenon was first noticed and extensively studied in parts of Scandinavia

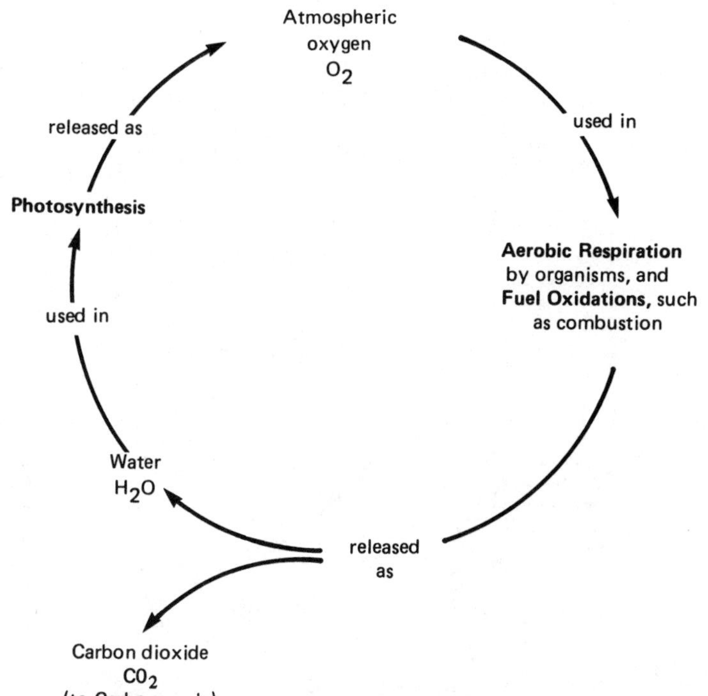


**FIGURE 2-5** The **oxygen cycle.** Atmospheric oxygen is used to liberate energy from organic molecules, whether by **combustion** or by **aerobic respiration** in the cells of plants and animals. Water ($H_2O$) and carbon dioxide ($CO_2$) are byproducts of these oxidations. When water and carbon dioxide are utilized by green plants for **photosynthesis,** carbon (C), hydrogen (H), and oxygen (O) atoms are combined to form carbohydrates. During photosynthesis, however, "extra" oxygen is liberated when water molecules are split and is released into the atmosphere. Thus green plants are Earth's "oxygen factories."

downwind from industrial areas of Britain and northern Europe. Following acid rainfall, streams many miles away from the pollution sources became more acidic, causing significant changes in stream plant and animal life.

Acid rainfall is now a significant problem in the United States, especially in the Northeast and Upper Midwest, areas that have coal-fired power plants. There is great concern that Great Lakes' fish populations may be affected by the higher acidity (lower pH) of streams that empty into the lakes. Lower pH in streams is known to affect adversely the migration of salmon upstream to reproduce.

The atmosphere also contains **particulate mat-**

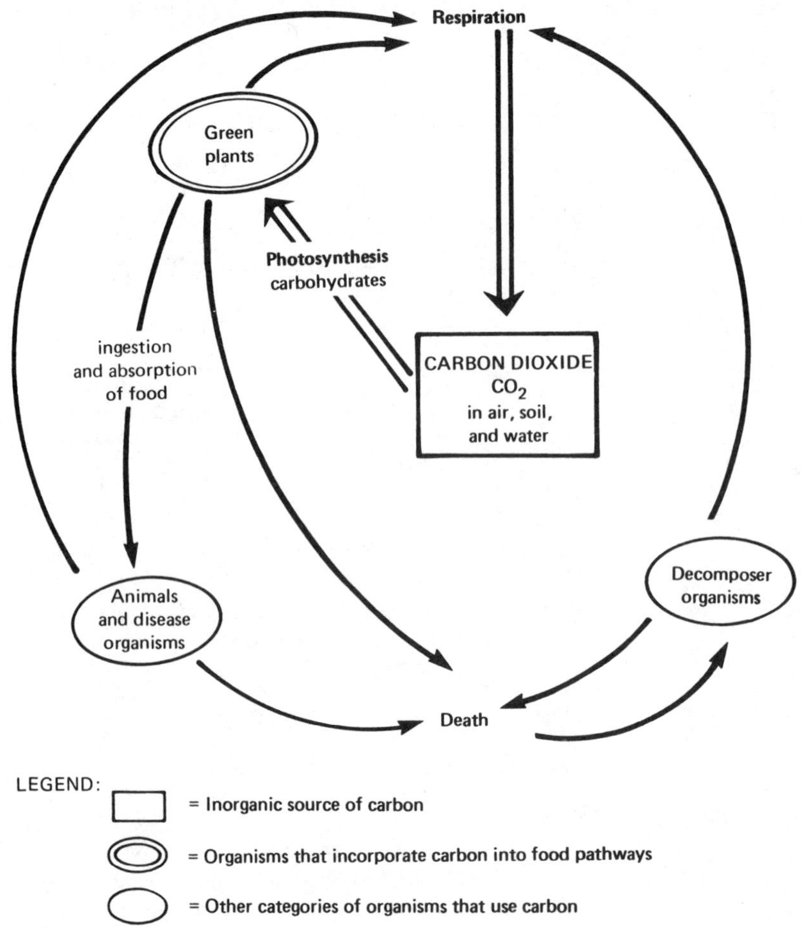


**FIGURE 2-6**
The **carbon cycle.** As noted in Fig. 2-5, green plants use carbon dioxide and water during **photosynthesis,** forming carbohydrates—food to nourish and build the bodies of the plants themselves as well as all other organisms that rely on them (directly and indirectly) for food. Plants and the organisms that feed on them return carbon dioxide to the atmosphere by their cellular **respiration** processes (Chapter 4), and **combustion** is an increasingly significant source of atmospheric $CO_2$. All organisms die, and when they do the organic (carbon-containing) substances locked up in their bodies provide food for decomposer organisms. Decomposers liberate much of the carbon by their cellular respiration. Upon their death, even decomposer organisms decompose! (*Note:* Only those aspects of the carbon cycle that are involved in the daily dynamics of an ecosystem are included in this diagram, even though other physical and geological factors are important.)

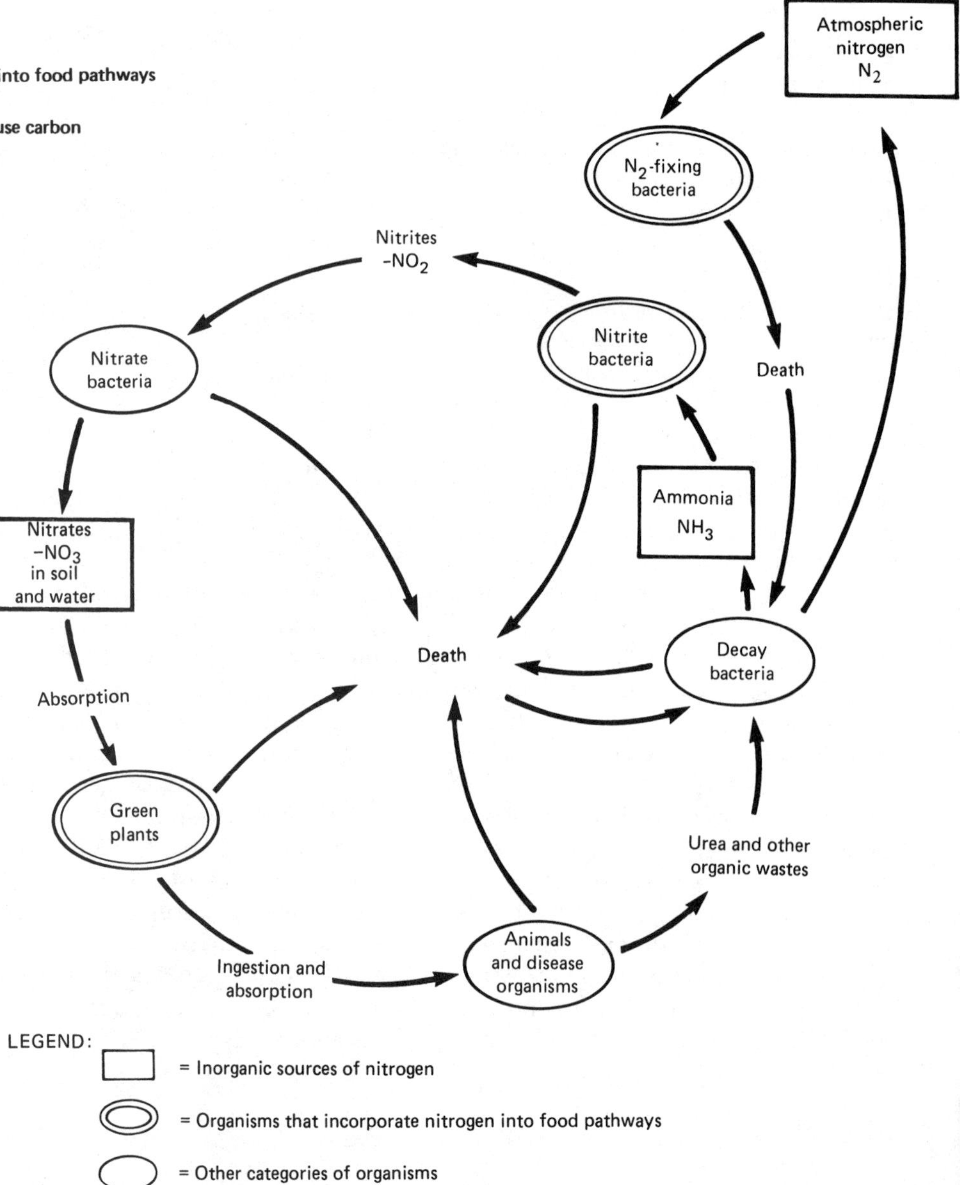


**FIGURE 2-7**
The **nitrogen cycle** in an ecosystem. Nitrogen is subject to complex recycling patterns, *some* aspects of which are presented here. **Soil** nitrogen is present as nitrogen gas ($N_2$), nitrates, nitrites, ammonia, ammonium, and in organic compounds. It is in the **atmosphere** as nitrogen and other gases. Green plants assimilate nitrogen only as **nitrates,** so they depend on the soil nitrogen supply. Various bacteria species replenish soil nitrates through direct **fixation** of nitrogen, **transformation** of nitrogen-containing compounds (nitrification), and **decomposition.** Nitrogen-fixing bacteria incorporate gaseous nitrogen into their amino acids and proteins, a food source for other soil organisms. Decay bacteria release ammonia, which is converted to nitrites by certain bacteria, then to nitrates by other bacteria. Green plants use nitrates to build proteins. In animal bodies, proteins are digested, and nitrogen-containing wastes are released. These wastes and dead organisms are then decomposed, continuing the cycle!

ter—solid rather than gaseous components—including dust. soot, radioactive fallout, lead, asbestos fibers, pollen, spores, and even bacteria and viruses. Particles block out sunlight in proportion to their size and concentration and some cause human health problems when inhaled or deposited on edible plant surfaces.

Atmospheric temperatures and currents are responsible for major climatic regions as well as local "microclimates." Air movement across land, water, and plant and animal surfaces increases evaporation. In the mountains dehydration plus **wind chill** (excessive temperature reduction accompanying evaporation) kills buds on the windward side of subalpine trees and shrubs, creating characteristic "flagged" and matted shapes. A similar effect occurs when wind whips sand and salt spray against trees on oceanic bluffs, killing buds on the windward side. Winds may actually have a pruning as well as shaping effect, physically breaking off windward-growing limbs.

## Biological Factors

An individual organism is rarely, if ever, truly solitary. It lives in an environment that is influenced or even determined by other organisms. Biological factors include all direct and indirect effects upon the life of an organism by members of its own species (**intraspecific interactions**) and other species (**interspecific interactions**).

### Intraspecific Interactions

Individual interactions between members of the same species are usually competitive, because members of the same species growing side by side compete exactly for such environmental essentials as specific nutrients, light, moisture, and space.

### Interspecific Interactions

Interspecific interactions are frequently antagonistic in nature, such as competition, feeding, parasitism, and pathogenicity. **Interspecific competition** affects daily survival of individuals and shapes species evolution through a selection process that includes survival to reproduce (Chapter 6). Each organism shares its immediate environment with members of other species, competing for environmental essentials. Competition may result in gradually improved species adaptation to local environmental conditions; development of slightly different requirements by the various competing species, thus sidestepping head-on competition; and/or gradual changes in the species composition of an area as some species flourish and others are eliminated.

Plants are fed upon by many animals but usually survive the feeding if most of the plant body is left. Another type of feeding is **predation**, whereby one organism kills and eats another. Predation usually refers to animals killing other animals. Some plant species, however, are predators in a sense, capturing and consuming insects or aquatic animals.

During **parasitism** one organism (the parasite) lives attached to or within the body of another organism (the host), obtaining its nutrition from the tissues of the host but not necessarily causing death of the host. Parasitism borders on and intergrades with **pathogenicity**, whereby a pathogenic organism (usually a virus, bacterium, or fungus) causes **disease**, a degenerative condition often leading to death.

**Symbiosis** ("together-living") is an interspecific interaction in which a member of one species lives intimately with a member of another species. Symbiosis usually involves mutual adaptation or accommodation and each participant in a symbiotic relationship is called a **symbiont**. Some scientists classify parasitism as a form of symbiosis while others don't.

If both symbionts benefit, the relationship is called **mutualism**, and there are many plant examples. Certain bacteria, for instance, that are capable of converting nitrogen to plant nutrients live within nodules (small nodes or bumps) on the roots of legumes, plants belonging to the pea family (Fabaceae) (Figs. 2-8 and 2-9). These bacteria benefit the legumes nutritionally and are provided a protected and nourishing environment for their own growth and reproduction. Because of the nodule bacteria, legumes (e.g., soybeans, peas, clover, alfalfa, vetch) increase and restore soil fertility; therefore they are rotated with other crops from season to season. Legume foliage and seeds are high in food value, specifically proteins, of which nitrogen is a major component.

Certain fungi and the roots of many woody plants form mutualistic associations called **mycorrhizae** (Gr., *mykes* = fungus, *rhiza* = root). Some mycorrhizae consist of a dense fungal mat covering root surfaces. Most mycorrhizae, however, are formed by penetration of the root itself by fungal filaments. In pines fungal penetration stimulates root branching. Thus many short lateral roots, plus the absorptive surfaces of the fungal filaments, provide greater functional surface area for water and

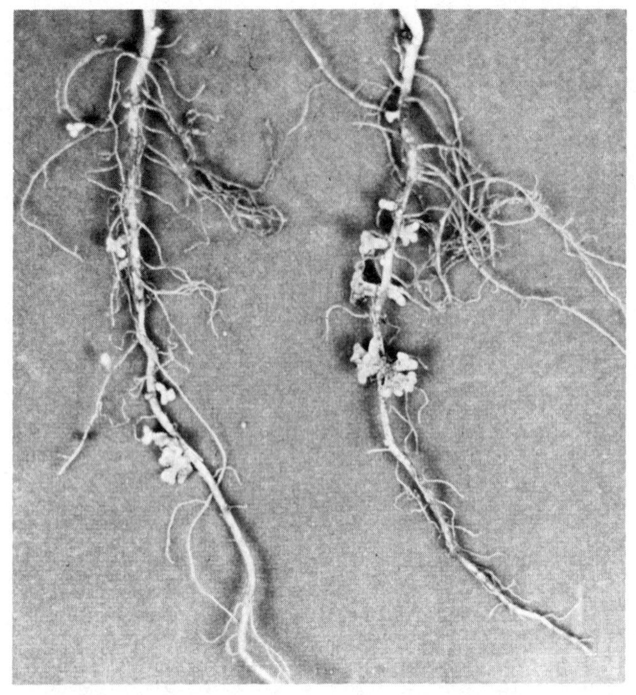

FIGURE 2-8 Nodules containing nitrogen-fixing bacteria on legume roots. (Photo courtesy of Arthur Anderson.)

nutrient absorption. Mycorrhizal formation is greatest in soils deficient in nitrogen, phosphorus, potassium, or calcium, essential plant nutrients; on the other hand, mycorrhizae are not as likely to form in fertile soils. In pines the relationship is significant to the point of complete mutual interdependence under extreme conditions.

Mycorrhizae are also significant in orchids and members of the heath family (Ericaceae), which includes manzanita, heather, blueberries, huckleberries, cranberries, rhododendrons, and azaleas, for example. In some plants the mycorrhizal fungus seems to be more of a parasite than a symbiont.

Mutualistic relationships between plant and animal species are common, especially in marine environments. Microscopic, single-celled algae often live within the tissues of such invertebrates as protozoans, flatworms, coral polyps, sea anemones, and giant clams. There the algae carry out photosynthesis, presumably exchanging oxygen and some food molecules as "rent" for the benefits of their residence.

**Commensalism** is a symbiotic relationship in which one member benefits and the other neither benefits nor suffers. At times it is not possible to determine clearly if a relationship is commensal or mutualistic, especially where plants are involved.

FIGURE 2-9 Nitrogen-fixing bacteria within the cells of a legume root nodule. (Photo courtesy of Carolina Biological Supply Co.)

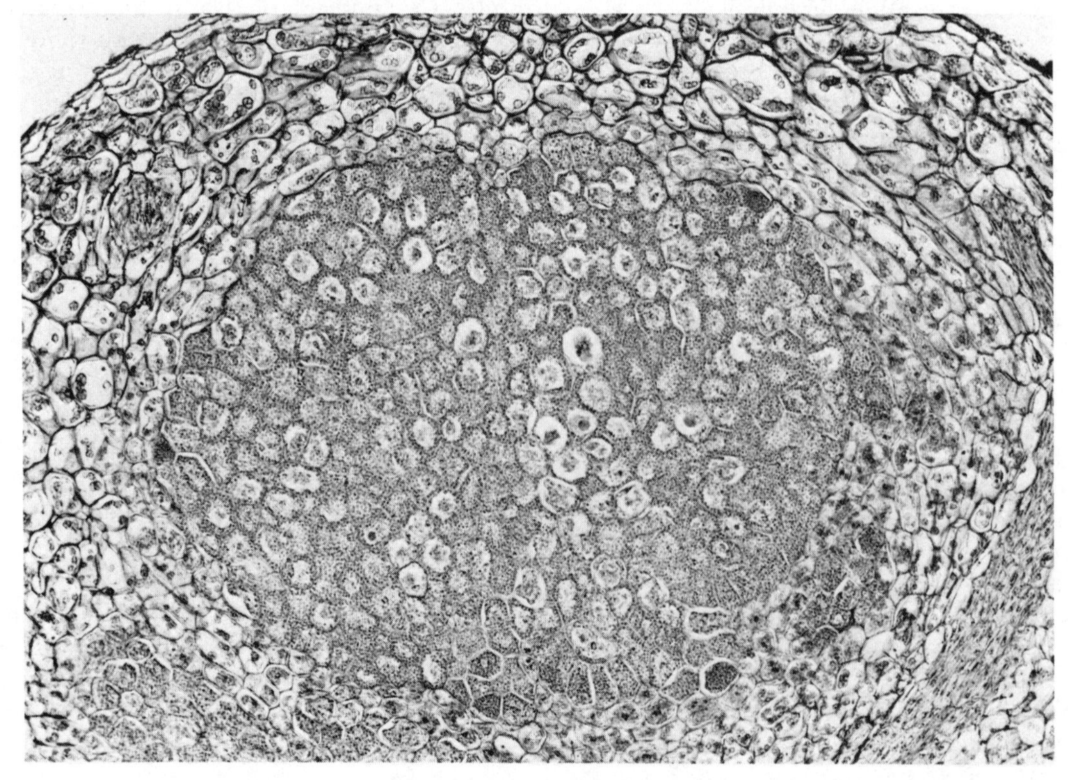

Some algae living within animal cells, for example, may actually be commensal, benefiting only the animal. There are numerous excellent examples of commensalism between animal species, particularly in the marine environment.

A plant that uses another plant as its substrate is called an **epiphyte.** Many mosses, lichens, and ferns, for instance, are epiphytic on trees. The most striking epiphytes grow *only* as epiphytes (Fig. 2-10), such as many tropical bromeliads and orchids and some cactus species. Debate exists as to whether epiphytism is truly symbiosis, for it frequently seems to be a circumstantial matter. In the tropics, for example, certain bromeliads also live on electric lines! The degree of adaptation of the epiphyte and the degree of specificity of the relationship are used to interpret each situation.

## ENERGY TRANSFERS WITHIN ECOSYSTEMS

Species survival, at the most basic level, relies on fulfillment of the nutritional and energy requirements of individuals. To meet these needs, an individual organism interacts with the whole array of physical and biological factors that make up its environment. In other words, any given organism is part of a larger, functional environmental system or **ecosystem.**

Ecosystem energetics begin with the sun. Unlike matter, energy cannot be contained indefinitely (cycled) in an ecosystem but must be continually replenished. Therefore there is limited value in emphasizing the *total amount* of energy present in an ecosystem. Instead the amount of energy *avail-*

**FIGURE 2-10**
Tropical epiphytes. (Photo courtesy of James T. Brock.)

*able* to various organisms is significant. The first and second **Laws of Thermodynamics** summarize the limitations of energy utilization, as in an ecosystem.

1. *Energy can neither be created nor destroyed, but it can be transferred or transformed from one state to another.*
2. *At each successive transfer or transformation there is a net loss of available energy.*

These important principles are demonstrated in the following discussion.

Only a small fraction of the light energy that enters an ecosystem ends up as food energy. Much of the energy that strikes inanimate and living surfaces is reflected back into the atmosphere (from outer space Earth shines like Venus and Mars) and some is absorbed as heat. Photosynthetic organisms (mainly green plants) absorb some light wavelengths, trapping and converting some of the light energy into chemical energy—the energy that holds the atoms of molecules together. The energy is used to form carbohydrates, which are then converted to food, structural, and other metabolic substances. Because green plants produce their own food in this manner, they are called **producers.** All other organisms utilize the energy trapped by the producers and so are called **consumers.** Consumers are categorized on the basis of how they obtain their food.

Of the little light energy actually utilized by green plants, even less is made available to each successive consumer in a **food chain** (linked feeding series)—for example, a short food chain in which a rabbit eats grass and is, in turn, eaten by a fox or hawk. Some energy is *used* during metabolism, releasing heat, most of which is *dissipated* into the environment. Some energy is *locked up* in structural compounds that are indigestible to the next consumer. Diagrammatically represented, the flow of *available* energy through successive nutritional or **trophic levels** (consumer feeding levels) in an ecosystem assumes a pyramid shape, the wider base representing energy available to green plants. Each successive trophic level is represented by a smaller block of available energy.

The amount of available energy determines the actual proportions of **biomass,** or total living mass by weight, of each species in an ecosystem. Producers are the most abundant, supporting a lesser biomass of **herbivores** (plant-eaters), which support, in turn, a lesser biomass of **carnivores** (flesh-eaters).

In addition to consumers that are clearly herbivores or carnivores, some animals (e.g., humans, bears, raccoons) are **omnivores,** eating both plant and animal matter. Living within or upon the bodies of organisms at all feeding levels are the **parasites.**

Carnivorous plants are interesting in that they are both producers and consumers. The Venus flytrap (*Dionaea*), butterwort (*Pinguicula*), pitcher plant (*Sarracenia*), Cobra lily (*Darlingtonia*) [Fig. 2-11 (a) and (b)], and sundew (*Drosera*) (Fig. 2-12) carry out photosynthesis. The acidic bog environments they inhabit, however, are low in available nitrogen, a necessary nutrient. So the carnivorous plants have adaptations enabling them to trap and digest insects, whose bodies are high in nitrogen-containing proteins!

Scavengers, decomposers (Fig. 2-13), reducers, and transformers are important consumer categories within an ecosystem. Discarded plant parts, such as leaves, twigs, blossoms and fruits, the feces and excreted metabolic wastes of animals, and the dead bodies of all organisms return large amounts of complex organic matter to the ecosystem (Fig. 2-14). Consider what happens to a carcass or a "cowpie." Such **scavengers** as crows, rats, dung beetles, scavenger beetles, and ants come to feed directly. Flies lay eggs so that their wormlike larvae (maggots) will hatch in contact with an abundant food supply. After scavengers expose internal tissues, odor becomes obvious to the human observer, the result of digestion by **decomposers,** primarily fungi and bacteria. The decomposers are **saprophytes,** releasing digestive enzymes into the organic substrate and then absorbing the products of digestion. Finally, **reducers** and **transformers,** primarily bacteria, reduce the simple organic compounds left by the decomposers into inorganic components and then transform or convert them into forms that can be utilized by green plants. By this stage most of the initial *energy* brought into the ecosystem by the original producers has been dissipated, but much of the *matter* has been made available for another cycle. Such is the economy of nature.

Decomposers are not always free-living. The fact that termites can eat wood and cows can digest hay depends on symbiotic bacteria and protozoans that live within their digestive systems. Termites, cows, and other ruminants (e.g., sheep, deer) cannot produce enzymes that digest cellulose and would literally starve to death with full stomachs without symbiotic microorganisms. Occasionally deer that have depleted browse in their winter "yards" are suddenly given hay by well-meaning

**FIGURE 2-11** The cobra lily (*Darlingtonia*) is a carnivorous plant of Pacific coastal bogs. (a) Light passes through translucent "windows" of the hood and insects enter the hood from beneath, near the winglike appendages. (b) Downward-pointing epidermal hairs in the hollow stalk permit insects to move down but not up, until they die in a pool of digestive fluid in the base of the stalk.

**FIGURE 2-12**
Insects are trapped on the sticky secretions of the sundew (*Drosera*) plant's leaves. (Photo by Mark R. Fay.)

**FIGURE 2-13**
Fungi are major decomposers of dead plant matter.

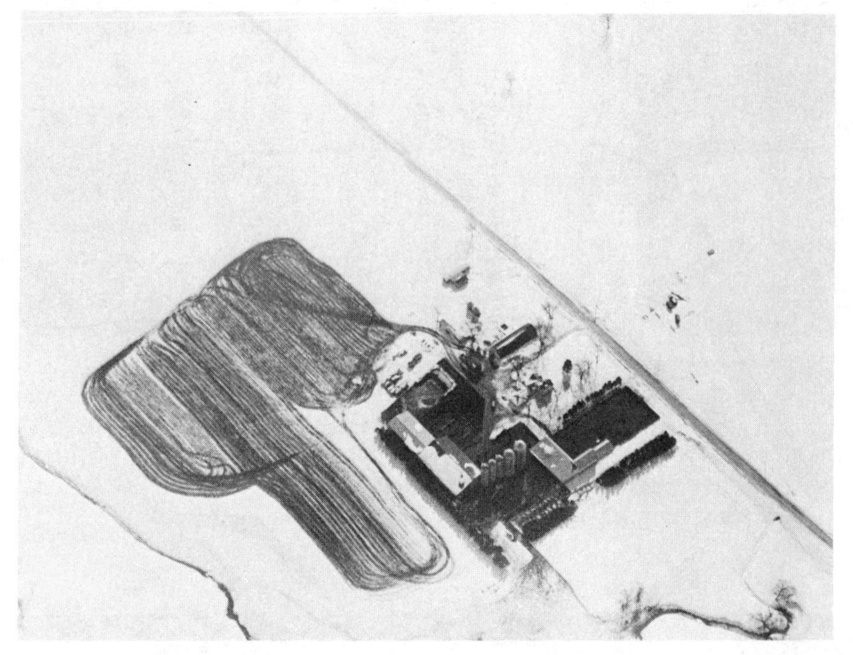

**FIGURE 2-14**
Farmers participate in recycling nutrients by returning to the soil the organic wastes from their livestock.

humans. If the balance of digestive tract symbiotic species does not have time to adjust to the abruptly different diet, normal digestion is disrupted and deer may die in spite of the presence of food. This situation is especially true with high-protein hay, such as alfalfa.

## ECOLOGICAL SUCCESSION

Change is constant. On the surface this statement appears contradictory, but maintenance of living systems at all levels relies on endless influx, output, and recycling. Some changes within ecosystems are sudden and disruptive, such as fire, flood, volcanic eruption, earthquake, strip mining, and real estate development (Fig. 2-15). Some natural changes occur as regular cycles, such as day and night, lunar months, and the seasons. Some of the most significant ecosystem changes, however, occur as *gradual* shifts in both physical and biological factors. This long-term, gradual change is called **ecological succession**, a sequential replacement (Fig. 2-16) of intermediate **communities** (plant and animal associations) leading to a more stable, longer-lasting **climax community**.

**FIGURE 2-15**
Intensive real estate development completely alters the natural ecosystem.

**FIGURE 2-16** Vegetation around this pond shows clearly the successional stages of open water to emergent plants to shoreline plants, eventually blending into shrubs and forest. In time, as succession continues, there will be no pond.

Each organism changes its own environment, often in ways that make the environment less suitable for it and other members of its species. Eventually (perhaps in a single year or over many years) the accumulative activities of all organisms (plant and animal) in a community alter the environment sufficiently so that it is more suitable to a different assortment of species, which replace, one by one, the original species. The collective result is succession, as defined earlier.

Succession that occurs on a previously unoccupied site is called **primary succession.** The earliest stages of primary succession on newly-formed bodies of water, sandbars, lava flows, volcanic islands, sand dunes, and similar sites involve colonization by hardy "pioneer plants" that have low nutrient requirements (Fig. 2-17). Over a period of time small amounts of organic debris are built up, supplying nutrients and a substrate for other plants, and so on. For each immigrant into the developing community, conditions are slightly different than for its predecessors and it, in turn, creates changes that must be met by organisms yet to come as well as for those already there.

As conditions become unsuitable for a species, it is gradually outcompeted and replaced by others that are adapted for survival under the new conditions. All physical aspects of the environment may be altered by the plants and the animal composition of a community is determined by conditions created by the plants.

If an existing community is destroyed or disrupted, **secondary succession** will occur, such as after a fire (Fig. 2-18) or flood, abandonment of a garden plot or field, and following logging (Figs. 2-19 and 2-20).

**FIGURE 2-17**
Lichens are pioneer species that colonize bare rock, a process that may take centuries, as this Indian petroglyph in Wyoming dramatically demonstrates.

**FIGURE 2-18**
Secondary succession occurring after a forest fire.

**FIGURE 2-19**
Clear-cutting removes all standing timber on a site. (Photo courtesy of Fred L. Rose.)

**FIGURE 2-20**
Secondary succession on a site logged by clear-cutting. (Photo courtesy of Fred L. Rose.)

In the past the climax community has been defined as self-maintaining, with a species composition that exists in a state of dynamic (rather than static) equilibrium. Theoretically, the plants and animals in a climax community have stable relationships and few further significant shifts in species composition occur by replacement.

The validity of the climax community concept as described above has been increasingly questioned. Environments are not static and unchanging (Fig. 2-21); genetic variation and natural selection continue to occur. Physical factors of the environment are altered by changing global climates and geological events. In addition, the community continues to alter its own environment. These continuing processes suggest that even though a climax community may be stable over a long period of time, its composition is not ultimate. With these limitations in mind, the modified concept of the climax community remains useful.

Within a given community at a given time, a particular plant species or group of species exerts a major modifying influence. Therefore these species are referred to as the **dominant species** of the community. They may be the largest individuals and occupy the most space or they may be the most numerous. Dominant species are usually those that successfully outcompete other species with similar environmental requirements. The dominants most successfully exploit the full range of opportunities of the present environment so that species that are subdominant to them generally have more specialized, limited roles within the community.

Many general types of communities are present in North America's mountains, plains, deserts, seacoasts, wetlands, waters, and arctic region. They are discussed in detail in Chapter 15.

## INTRODUCED SPECIES

Environment, as we have discussed it, shapes each species, adapting species to survive where they are and permitting the development of stable communities. When alien species are introduced into a community, however, the existing competitive

**FIGURE 2-21** Sand dunes invading a coastal forest, which was originally formed by colonization of and succession on an older dune. (Photo courtesy of Fred L. Rose.)

balance may be disrupted and the introduced species may proliferate at the expense of native species.

Early European explorers and settlers of the North American continent often regarded what they saw not as magnificent, beautiful, and benevolent but as overgrown, savage, and intimidating—something to be "tamed" with reciprocal ferocity. Part of this taming was the clearing of forests and undergrowth and part was the introduction of familiar and therefore "desirable" plants and animals. Many introductions were motivated by a psychological need to bring a "civilized" reminder of the homeland to a strange new country. To this day, for instance, most of the popular cultivated flowers, trees, and shrubs are regarded more highly than are our native species, although this attitude is beginning to change somewhat. In addition, many species were accidentally introduced, such as by seeds "hitchhiking" among plant products and containers of desired seeds.

In the New World the many Old World plant species had no natural pests or diseases. More significantly, they had no natural competitors, especially on the freshly disturbed soils of habitation, and they flourished. When settlers moved westward, so did the introduced plants. In the tradition of Johnny Appleseed, the settlers left an identifiable trail of imported plant life, both desirable and undesirable. Some of our most persistent weeds came to America as imports and grew with rampant success. Two such examples are dandelion (*Taraxacum*) and Russian thistle (*Salsola*), the large western "tumbleweed."

## SUMMARY

The planet Earth and its atmosphere may be considered a **solar vivarium**—a life-sustaining system supported by energy from the sun. Green plants convert solar energy to food, a form in which energy is useful to all living things or **organisms.**

Organisms share certain common characteristics. Typical organisms are composed of **cells** and cell products. They contain carbon-based **organic compounds**, the most prominent of which are carbohydrates, proteins, and lipids. Nucleic acids **DNA** and **RNA** regulate cell functions. Organisms are highly **organized**, both in form and function, and **grow** with increasing complexity. Living things also **reproduce** their kind, **react** to stimuli in their environments, and exhibit **movement.** They are capable of **change.** Organisms exhibit **metabolism,** converting available energy and matter into usable forms. Organisms also **die.**

Conglomerates of cells, called **tissues,** perform specialized functions. Tissues are organized into **organs,** such as leaves, stems, roots, and reproductive structures. Many kinds of organisms are known as plants.

The cells of higher plants and higher animals are amazingly similar in structure and metabolism. There are, however, certain distinct differences between plants and animals.

Green plants are **autotrophic** whereas animals and nonphotosynthetic plants are **heterotrophic.**

Every organism has a **scientific name,** a binomial consisting of a genus name and a species name. The major taxonomic categories for plants are division, class, order, family, genus, and species.

Environmental forces shape the growth of individual plants as well as the evolution of species. A **limiting factor** is the environmental factor that most directly limits or determines the survival of an organism in a given environment. **Ecology** is the science that attempts to understand interactions of living things with their environments.

Of the three major kinds of environments, **terrestrial** living, with its constant change, is by far the harshest environment for organisms. **Fresh water** varies more than the **marine** environment, which presents the least obstacles for life.

Physical and biological factors make up an organism's **environment. Physical factors** include radiation, water, atmosphere, and soil. Complex recycling patterns assure continued availability of essential elements. **Biological factors** are all the direct and indirect effects on an organism by other organisms, including intraspecific and interspecific interactions.

**Symbiosis** is an intimate living-together association of members of different species, including relationships of mutualism, commensalism, and, at times, epiphytism.

An **ecosystem** is a functional ecological system in nature, including interactions of all the physical and biological factors within an area.

Energy cannot be created or destroyed, but it can be transferred or transformed from one state to another. Available energy diminishes, however, with each successive transfer or transformation. These **first and second Laws of Thermodynamics** describe the energy limitations that affect organisms and ecosystems.

**Producers**, mostly green plants, transfer solar light energy into chemical energy (food) for themselves and **consumers.** In each step along the food chain some energy is used, some is dissipated as heat, and some is locked up in compounds that are indigestible to the next consumer.

Producers represent the greatest **biomass** in a given ecosystem. Herbivores, carnivores, omnivores, parasites, scavengers, decomposers, reducers, and transformers are all types of consumers.

**Ecological succession** is the gradual replacement of one community by another on a site. Primary succession occurs on previously barren sites. Secondary succession occurs when an existing community has been disrupted or destroyed. Succession tends to lead to a complex long-lasting **climax community.**

Within a community, one plant species or group of species exerts a **dominant** influence on the community.

**Introduced species** are often particularly successful in a new environment to the detriment of competing native species.

## SOME SUGGESTED READINGS


Billings, W.D. *Plants, Man and the Ecosystem.* Belmont, CA.: Wadsworth Publishing Co., 1970. A brief, readable introduction to ecology with an emphasis on the role of humans in modifying the environment.

Carson, R. *Silent Spring.* Boston: Houghton Mifflin Co., 1962. The popular bestseller that depicted the effect of humans on the environment and, in so doing, stimulated the ecology movement.

Colinvaux, P. *Introduction to Ecology.* New York: John Wiley & Sons, 1973. An ecology text that emphasizes the interrelationships between evolution and ecology.

Daubenmire, R.F. *Plants and Environment,* 2nd ed. New York: John Wiley & Sons, 1959. A good introduction to the factors affecting plant growth.

Gabriel, M.L., and S. Foge. *Great Experiments in Biology.* Englewood Cliffs, N.J.: Prentice-Hall, 1955. An extensive collection of excerpts from important original papers in biology.

Leopold, A. *A Sand County Almanac.* Oxford: Oxford University Press, 1949. An environmental classic that concludes with a conservation ethic more timely today than when it was written.

Odum, E. P. *Fundamentals of Ecology,* 3rd ed. Philadelphia: W.B. Saunders Co., 1971. A comprehensive ecology text; the "standard" textbook.

Sears, P.B. *Deserts on the March.* Norman, Oklahoma: University of Oklahoma Press, 1935. An early call to ecology with an emphasis on land utilization and its misuse.

Smith, R.L. *Ecology and Field Biology,* 2nd ed. New York: Harper and Row, 1974. A well-written ecology textbook that emphasizes many applied aspects of the science.

Thomas, L. *The Lives of a Cell: Notes of a Biology Watcher.* New York: The Viking Press, 1974. A collection of stimulating and sensitive essays dealing with science and humanity.

Treshow, M. *Environment and Plant Response.* New York: McGraw-Hill Book Co., 1970. A textbook of environmental pathology.

# chapter three THE CELL AS THE BASIS FOR LIFE

## INTRODUCTION

What could differ more in appearance and activity than a redwood tree (*Sequoia sempervirens*) and a human being (*Homo sapiens*) (chapter opening art)? Yet if we move from the gross structure, the cells of these organisms are amazingly similar. The similarity in the cellular structure of such different organisms may seem a paradox, but these cellular similarities support the theory of the commonality of origin of primitive cells.

Because of smallness, the cell is a frontier that has been difficult to explore. Our visual tools, both light and electron microscopes, deprive us of most of the benefit of our natural stereoscopic vision. We have had to rely on flat images, thin slices of severely preserved and stained cells, to study inner cellular detail. The relatively recent development of the scanning electron microscope has restored our three-dimensional vision to some extent. We are still unable, however, to "put our eyes" directly into the living, throbbing cell to watch its components interact. Thus our understanding remains limited to what we can deduce from magnified images and biochemical tests.

In this chapter we explore our current knowledge of cell structure. You may find some of the terminology intimidating, although we have eliminated many of the more specialized terms. This unfamiliarity is natural, for you are entering the internal world of the cell, a world that is foreign to you even though it is a part of you. This vocabulary is essential to biological literacy, however, and *must* be learned. We hope that Table 3-1 will help you to learn the information that is presented.

## CELLULAR STRUCTURE

For many years observation of plant cells with the light microscope revealed a cellular structure consisting of **cell wall, nucleus, plasma membrane, central vacuole,** some **plastids,** granular "stuff"

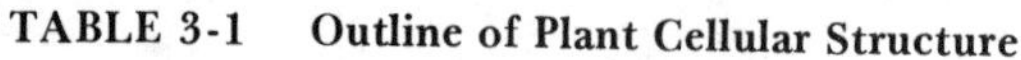
**TABLE 3-1 Outline of Plant Cellular Structure**

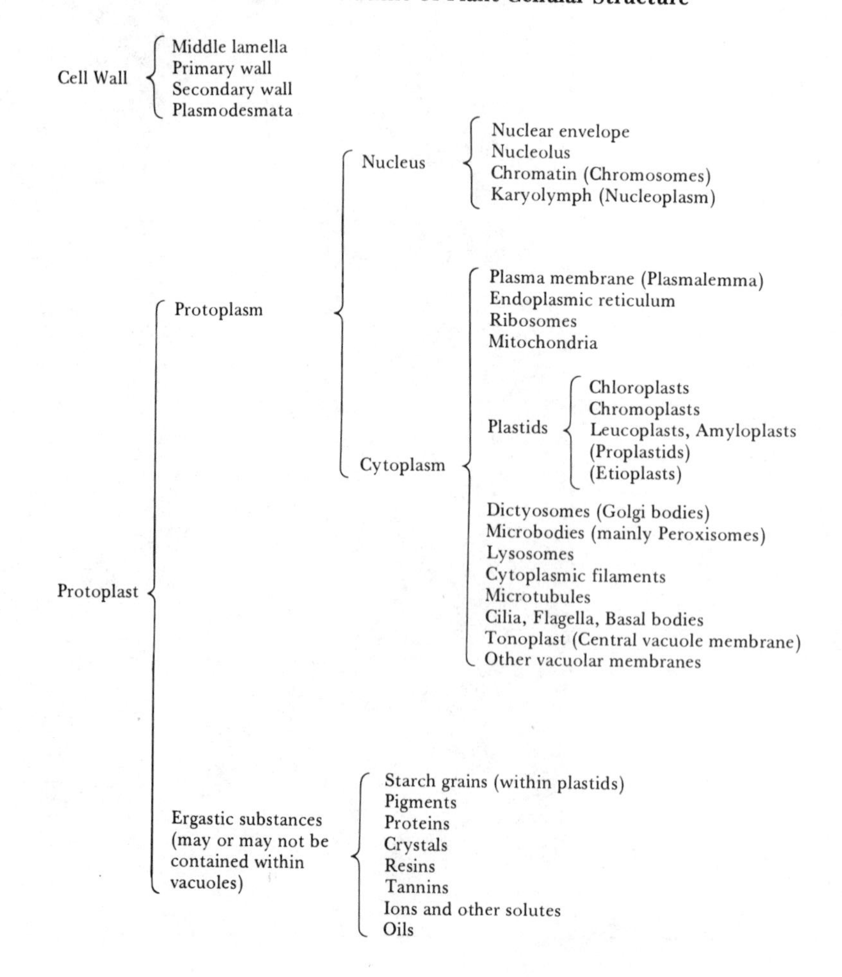

- Cell Wall
  - Middle lamella
  - Primary wall
  - Secondary wall
  - Plasmodesmata
- Protoplast
  - Protoplasm
    - Nucleus
      - Nuclear envelope
      - Nucleolus
      - Chromatin (Chromosomes)
      - Karyolymph (Nucleoplasm)
    - Cytoplasm
      - Plasma membrane (Plasmalemma)
      - Endoplasmic reticulum
      - Ribosomes
      - Mitochondria
      - Plastids
        - Chloroplasts
        - Chromoplasts
        - Leucoplasts, Amyloplasts
        - (Proplastids)
        - (Etioplasts)
      - Dictyosomes (Golgi bodies)
      - Microbodies (mainly Peroxisomes)
      - Lysosomes
      - Cytoplasmic filaments
      - Microtubules
      - Cilia, Flagella, Basal bodies
      - Tonoplast (Central vacuole membrane)
      - Other vacuolar membranes
  - Ergastic substances (may or may not be contained within vacuoles)
    - Starch grains (within plastids)
    - Pigments
    - Proteins
    - Crystals
    - Resins
    - Tannins
    - Ions and other solutes
    - Oils

called **cytoplasm,** and such nonliving cell components (**ergastic substances**) as starch grains and crystals. Use of special stains indicated some greater detail. The development of the transmission and scanning electron microscopes, however, has pushed our observation of the physical structure of the cell to increasingly intimate detail (Figs. 3-1 and 3-2).

Table 3-1 categorizes cellular structures, as revealed by electron microscope study. Such a list confounds any notion of a "simple" cell. For many species, a single cell is the entire organism! Some structures listed on this chart are typical of plant but not animal cells (the cell wall, pigment-containing plastids, central vacuole and its membrane, and various ergastic substances). Otherwise the basic components of the protoplasm are the same. The

**FIGURE 3-1** This is an electron microscope view of a meristematic cell (one that is capable of division) in a cattail (*Typha*) leaf. Many **cellular organelles** and other **cytoplasmic structures** are visible, some of which are labelled. The many tiny dots are **ribosomes.** This is an example of a eukaryotic plant cell. (Photo courtesy of Albert P. Kausch.)

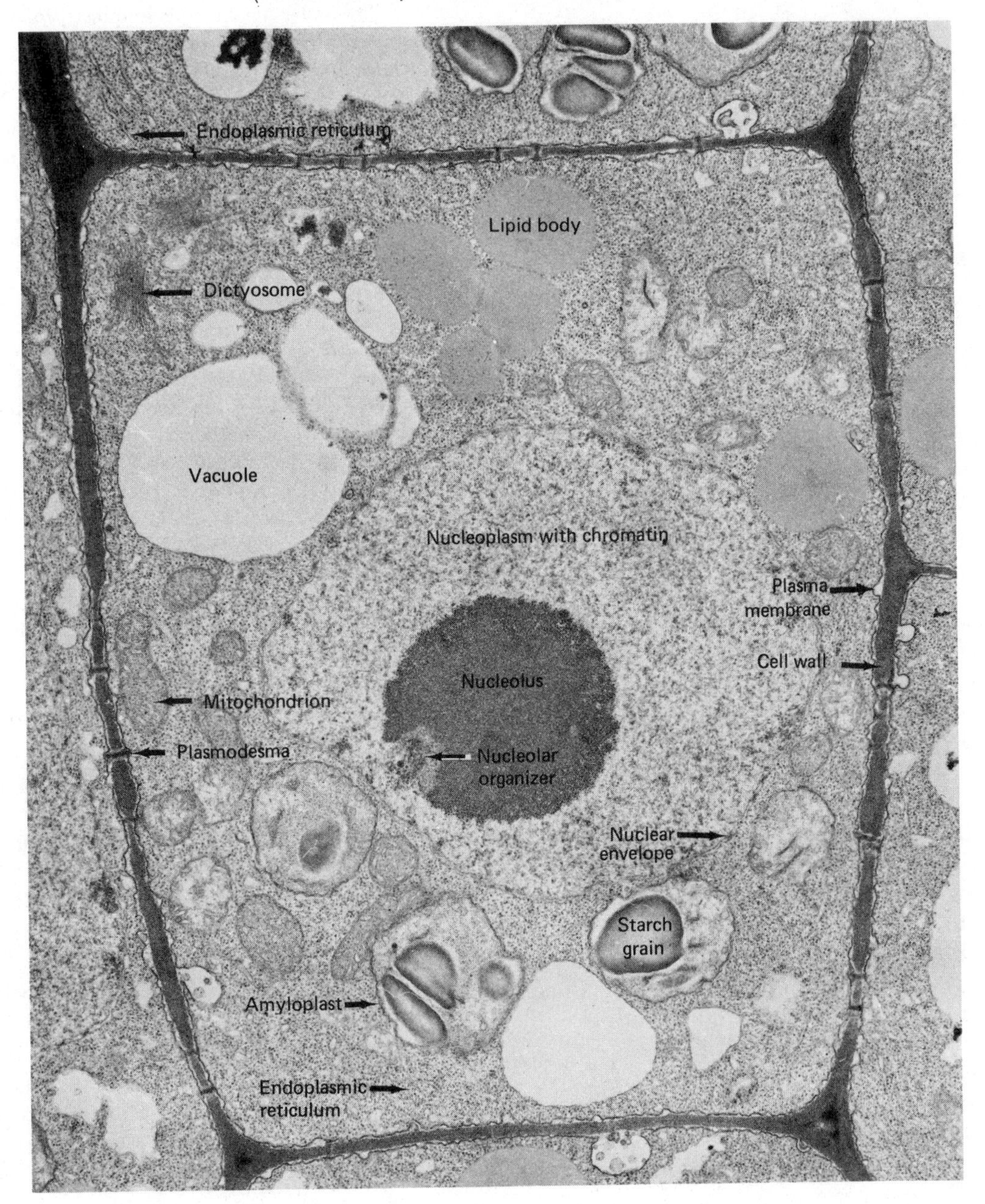

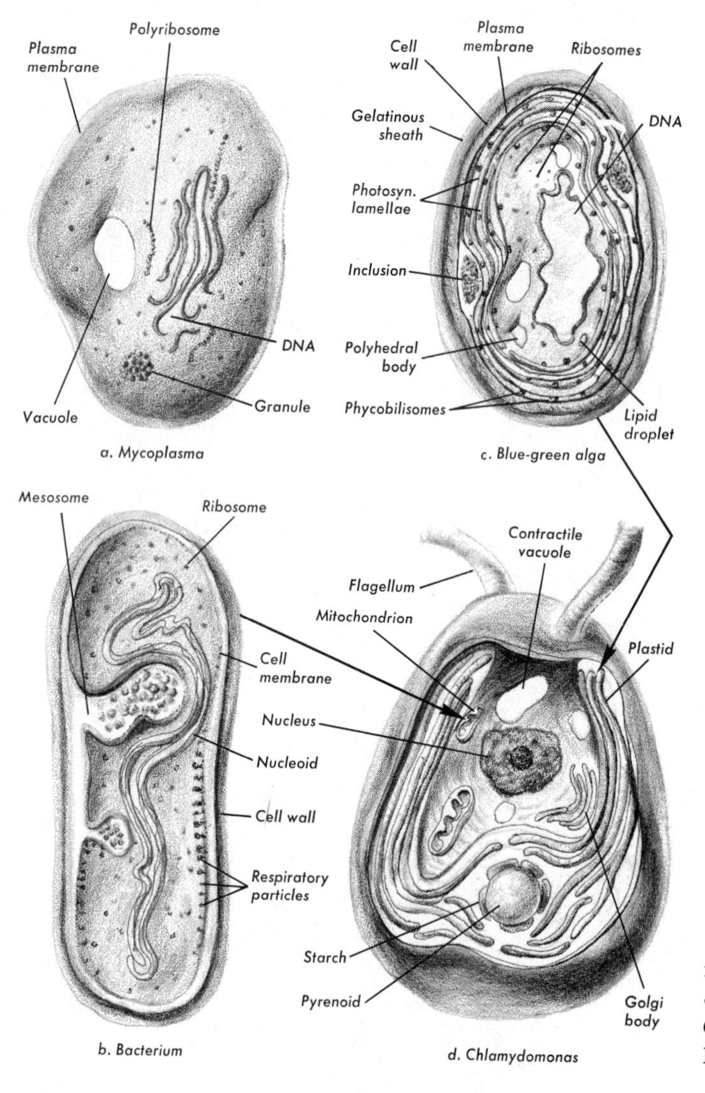


**FIGURE 3-2**
Diagrams of major cell types (not drawn to scale). The two heavier arrows call attention to the current hypothesis that primitive bacteria and blue-green algae cells invaded or were engulfed by primitive eukaryotic cells, initiating a symbiotic relationship that led to the permanent development of the mitochondria and chloroplasts of eukaryotic cells (Chapter 13). (a) **Mycoplasmas** are the simplest **prokaryotes** and are the smallest known cells, ranging in size from 0.2 to 1.0 nm. (b) **Bacteria** are the largest, most successful, and diverse prokaryotes. As in other prokaryotes, the DNA is not contained within an organized nucleus. The DNA molecules are arranged in rings, and are attached to infoldings of the plasma membrane that are called mesosomes. (c) **Blue-green algae** cells greatly resemble those of bacteria in several significant ways (Chapter 13). These prokaryotes contain photosynthetic layers of membrane (thykaloids) as well as special bodies (-somes) or granules containing the pigment phycobilin. (d) Cell of the single-celled photosynthetic **eukaryote** *Chlamydomonas*. Depending on the classification scheme employed, *Chlamydomonas* is considered to be a "freshwater alga" or a "photosynthetic freshwater protist" (Chapter 13). Like other eukaryotes, it has complex, membrane-bounded organelles, including nucleus, mitochondria, and chloroplasts. (From A.M. Elliott and D.E. Outka, *Zoology*, 5th ed., 1976. Reprinted by permission of Prentice-Hall, Inc., Englewood Cliffs, N.J.)

following discussion describes cellular components listed in Table 3-1, most of which can be seen in Figs. 3-1 and 3-2.

### The Cell Wall

The outermost structure of a typical plant cell is the **cell wall**, which supports and protects the cell. In tissues the cell wall that separates adjacent cells is a composite of cell walls formed by both cells, cemented together by the **middle lamella**, a layer of jellylike substances called **pectins.** Fibers of **cellulose**, a complex carbohydrate, are the main structural components of a cell wall. Cellulose and pectin make up the **primary cell wall.**

The presence of a well-developed cell wall is usually a barrier to cell division and expansion. Therefore cells specialized for growth are very thin walled. The cell wall does not, however, completely isolate cells from one another. It is perforated by numerous membrane-lined channels called **plasmodesmata** (meaning plasma bonds). The plasmodesmata provide cytoplasmic continuity between cells so that it is possible for some substances to be transported between cells without having to pass through a membrane barrier (Figs. 3-3 and 3-4).

Water, in which chemical substances are dissolved, is held in the spaces between the cellulose fibers of the cell wall. Thus the cell wall provides an immediate aqueous environment for the cell it surrounds.

Some cells become specialized for protective, supportive, or transport functions. Their cell walls become "customized" to these special functions by the addition of one or more substances (cutin, suberin, lignins) to the primary cell wall. Such a reinforced cell wall is called a **secondary cell wall.**

Deposition of waxy substances called cutin

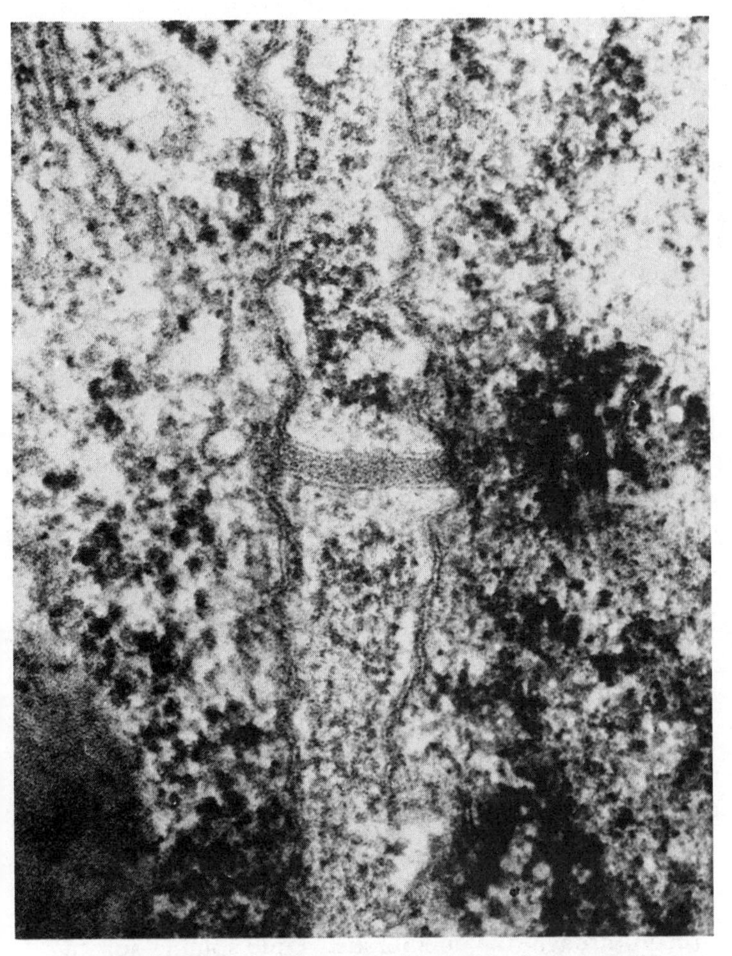

**FIGURE 3-3** Electron micrograph of adjacent plant cells, showing a **plasmodesma** penetrating the cell wall. (Photo courtesy of Robert A. Cecich.)

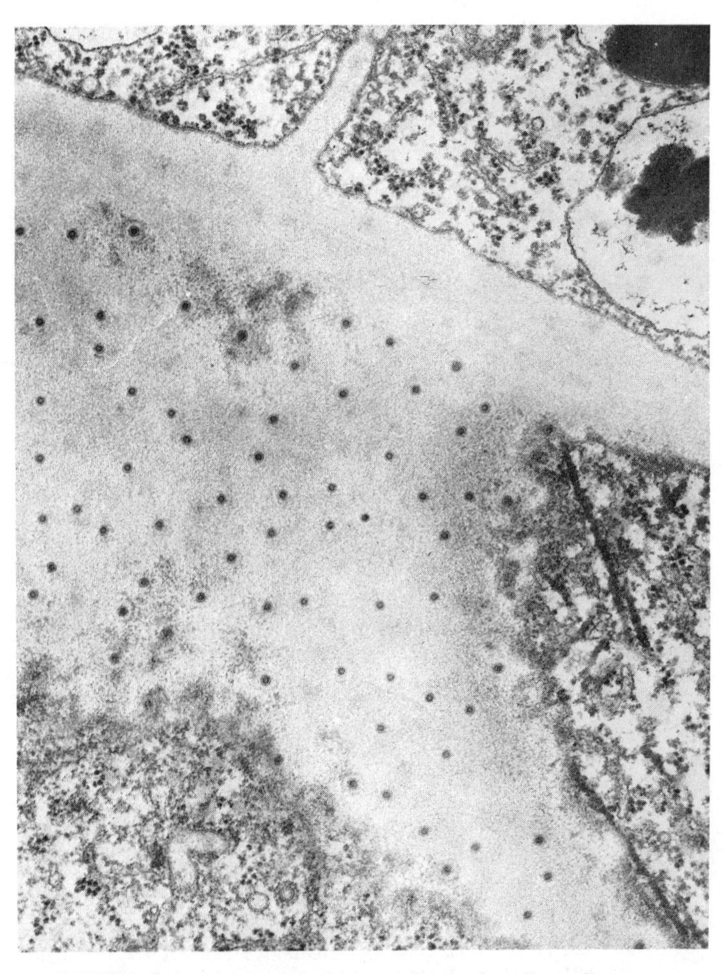

**FIGURE 3-4** A lengthwise section of cell wall, showing **plasmodesmata** in cross section. (Photo courtesy of Fred Rickson.)

and suberin causes the cell wall to become impervious to water. **Cutin** usually is found on outer walls of epidermal cells. It creates the gloss on leaves and fruits, for example, and reduces water loss from these exposed surfaces. **Suberin** is found in **cork** tissue and effectively seals off dying cells, converting them to air chambers. Cork tissue insulates and protects the tissues it surrounds; the bark of a tree contains cork tissue.

The walls of cells that are specialized for water transport or support become impregnated with hardening substances, the **lignins.** The hardness of wood and nut shells is due to lignified cell walls.

## The Protoplast

Everything contained within the cell wall is the **protoplast.** The protoplast is made up of the **protoplasm** (the vital or living part) and **ergastic substances,** nonliving inclusions of the cell. The protoplasm may be further broken down into the cytoplasm and the nucleus. The **cytoplasm** consists of a viscous fluid matrix (ground substance) that contains water, many organic and inorganic substances, and a variety of tiny, specialized functional structures. Cytoplasmic structures range from simple to complex, and the simpler ones can be formed as needed by the cell. The more complex, membrane-bounded structures are called **organelles.** The **nucleus** is a large organelle that contains the cell's inherited information and hence serves to control and coordinate cell metabolism. The cell is enclosed by a single living cytoplasmic membrane (the **plasma membrane** or **plasmalemma**) and the organelles are bounded by either a single or double membrane, depending on the organelle.

Perhaps the most important concept in understanding the cell is that of **cytoplasmic membranes.** A cell is not just a hollow balloon, full of ground substance with the other "stuff" floating around in the ground substance! It is composed of many compartments and functional surfaces, all defined by membranes. The actual structure of cytoplasmic membrane has been an enigma for many years. Far from being a firm enclosure, it is a flexible, dynamic, self-healing, fluidlike lipid-and-protein complex.

Membrane structure is now believed to be a "fluid mosaic" with large protein molecules floating in a "sea" of smaller lipid molecules (Fig. 3-5). Some of the protein molecules are found on the inner surface of the membrane, others on the outer surface. Some protein molecules extend from one side of the membrane to the other. Protein molecules located on either the inside or outside of the membrane apparently are able to drift laterally through the lipid but not from inside to outside or outside to inside the membrane. They seem to be "anchored" to microtubules and microfilaments (see p. 35) that underly the membrane. Specific carbohydrates are often attached to proteins on the outer surface of the plasmalemma.

When examined with the electron microscope, cytoplasmic membrane appears to be layered, which is why it was previously likened to a "sandwich" of lipid between two layers of protein. One complete thickness (protein-lipid-protein) of cytoplasmic membrane is often referred to as a **unit membrane.**

One of the major characteristics of cytoplasmic membrane is its **semipermeability** or **differential permeability.** This means that the membrane selectively allows certain molecules to permeate or pass through while restricting the passage of others.

#### Plasma Membrane

The plasma membrane (plasmalemma or cell membrane) is a single unit membrane forming the outer boundary of a cell. It is continuous through the plasmodesmata of the cell wall and appears to be continuous (at least in some cells) with the endoplasmic reticulum, which is a folded membrane network within the cell (see following section). The cell membrane is in direct contact with the cell wall; cellulose is apparently assembled at the cell membrane surface and then incorporated into the cell wall.

The cell membrane maintains the chemical composition of the cell by regulating the passage of substances in and out of the cell. Water, dissolved gases, and many other small molecules travel freely through the membrane. Lipid-soluble substances that become dissolved in the lipid of the

**FIGURE 3-5** The **fluid mosaic model** of cell membrane structure, with protein molecules suspended in a double layer of lipid. Carbohydrates are sometimes attached to the outer surface of the proteins. (Adapted from "The Fluid Mosaic Model of the Structure of Cell Membranes," by S.J. Singer and G.L. Nicolson. *Science* 175: 720–731, February 1972. Copyright 1972 by the American Association for the Advancement of Science.)

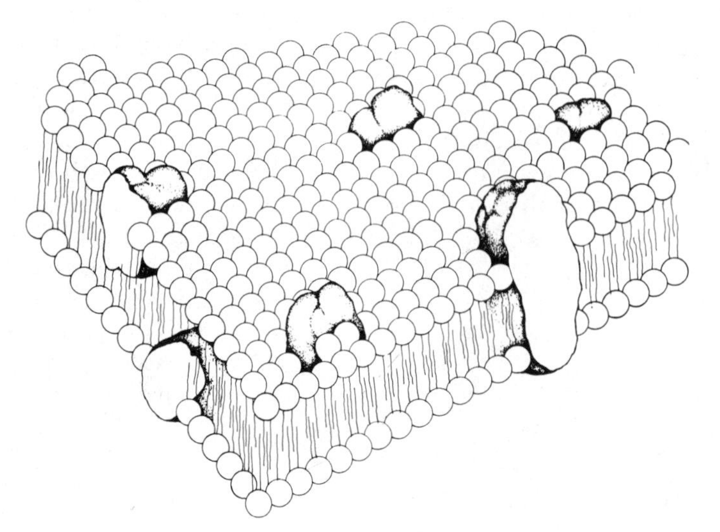

membrane penetrate with particular ease. It is thought that very small molecules, such as water, pass through actual pores in the membrane. Because the membrane is semipermeable, such large and complex molecules as proteins and complex carbohydrates synthesized inside the cell are not allowed to escape.

Yet membrane regulation of cell contents cannot be explained adequately in terms of simple permeability. The cell membrane also maintains a dynamic imbalance of small molecules, especially such ions as sodium, potassium, magnesium, and calcium, on its surface. In addition, the membrane can actively resist or facilitate the passage of specific molecules. Molecular "traffic" in and out of cells is discussed in detail later in this chapter.

#### Endoplasmic Reticulum and Ribosomes

**Endoplasmic reticulum (ER)**, literally translated, means "within the plasma network." The ER [Fig. 3-6 (a) and (b)] is a continuous network of unit membrane folds, forming flattened pouches and channels that function in intracellular transport, storage, and protein synthesis. The convolutions of the membrane provide a great surface area for these activities. The ER provides functional compartmentalization of the cell interior, creating an abundance of "work areas" analogous to the various rooms of an office building. There is evidence that the ER may be continuous with both the plasma membrane and the outer membrane that surrounds the nucleus, at least in some cells.

The ER is somewhat difficult to visualize three dimensionally. In the thin slices necessary for electron microscopic observation it appears as two more or less parallel unit membranes, with a space in between. In some preparations, this space appears clear, but in others it appears granular because of the presence of such substances as proteins that have absorbed the stain.

The ER is often studded with dense bodies,

(a)

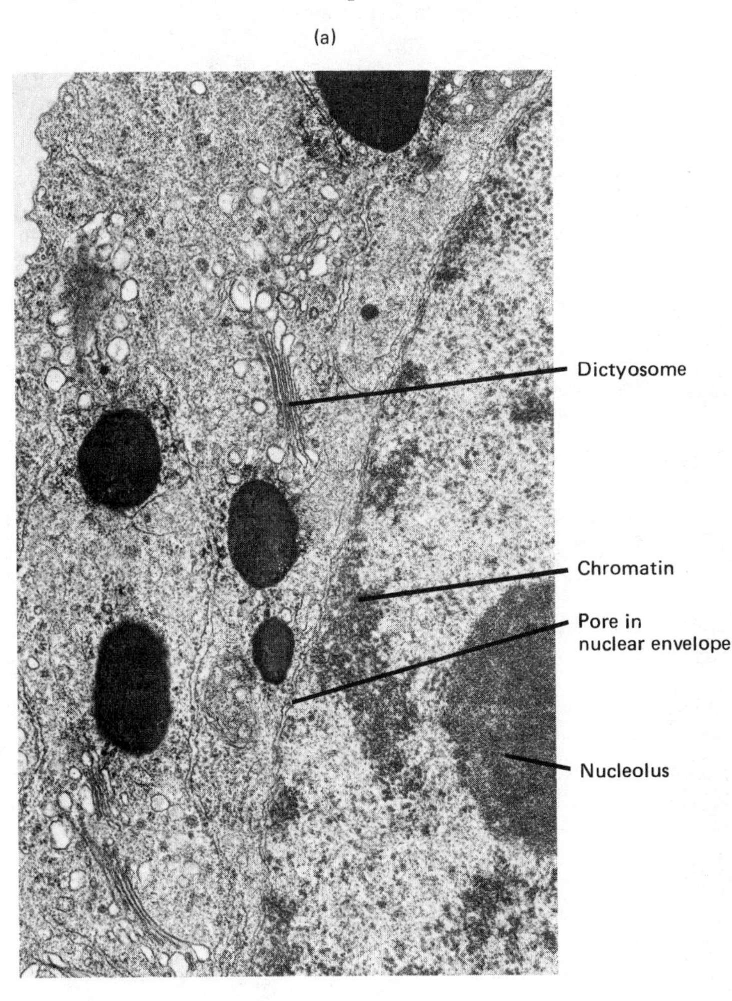


**FIGURE 3-6 (a) and (b)**
Electron micrographs of plant cells, showing **dictyosomes**, the **nuclear envelope** with its pores, **chromatin**, the **nucleolus**, mitochondria, and rough **endoplasmic reticulum** with **ribosomes**. (Photos courtesy of Robert A. Cecich.)

(b)

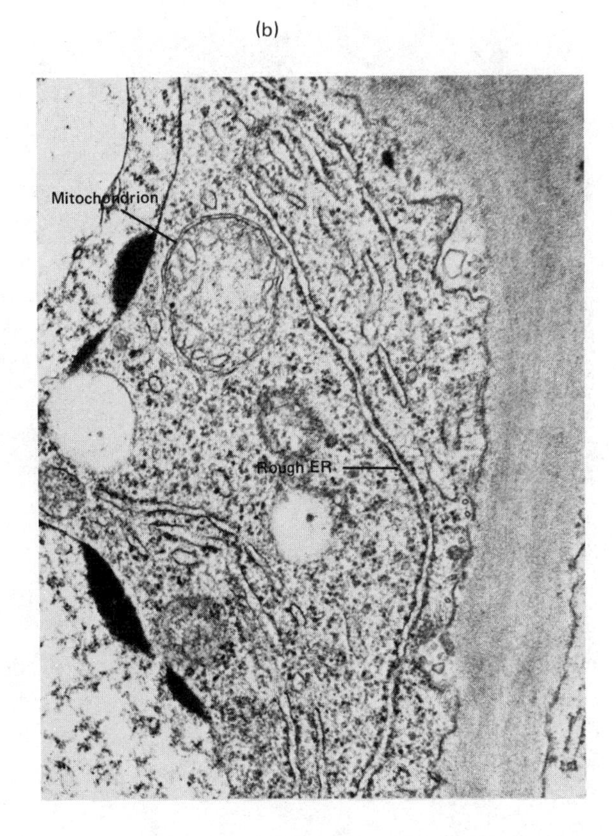

**ribosomes,** which are composed of the nucleic acid ribonucleic acid (RNA) and associated proteins. Ribosomes are sites of protein synthesis, described subsequently. ER having associated ribosomes is called rough ER in contrast to smooth ER, which lacks ribosomes. The function of the ER as a means of communication for chemical substances within and between cells is probably related to the role of ribosomes as sites of protein synthesis. The ER is particularly well developed in cells in which there is abundant protein synthesis.

Just as certain areas of the ER lack ribosomes, ribosomes are not always associated with ER but are common in the cytoplasmic matrix as well. These "free" ribosomes seem to produce proteins that are utilized within the cell, whereas those associated with ER synthesize proteins for secretion to the outside of the cell. A mechanism for this secretion of ER products is that the ER pinches off small vesicles, which move toward and fuse with the plasma membrane, thus releasing their contents to the outside of the cell. Ribosomes are also found within mitochondria and chloroplasts.

#### Dictyosomes

Imagine looking at a stack of three to seven frilly margined pancakes and you can mentally approximate the microscopic appearance of a **dictyosome.** Dictyosomes were first observed in animal cells and named Golgi bodies or Golgi apparatus in honor of the scientist who initially described them, but they exhibit their greatest complexity in higher plants. These stacks of flattened, unit membrane-bounded saccules have fingerlike, interconnected projections around the margins. The tips of the projections often pinch off to form tiny vesicles that then move around in the cytoplasm. The vesicles are "packages" of substances from within the membranes of the dictyosome. Some substances are entirely synthesized within the dictyosomes.

It appears that, at least in some cells, saccules are added *to* a dictyosome by fusion with vesicles that have been pinched off the ER. In this way, substances produced by the ER may be modified within the dictyosome and the final product can be distributed through the cytoplasm within dictyosome vesicles.

#### Microbodies and Lysosomes

**Microbodies** and **lysosomes** (*lyse* = to break, *soma* = body) are spherical, enzyme-containing organelles bounded by a single unit membrane and typically lack internal membrane structures (Fig. 3-7). At least some microbodies and lysosomes originate as dictyosome vesicles.

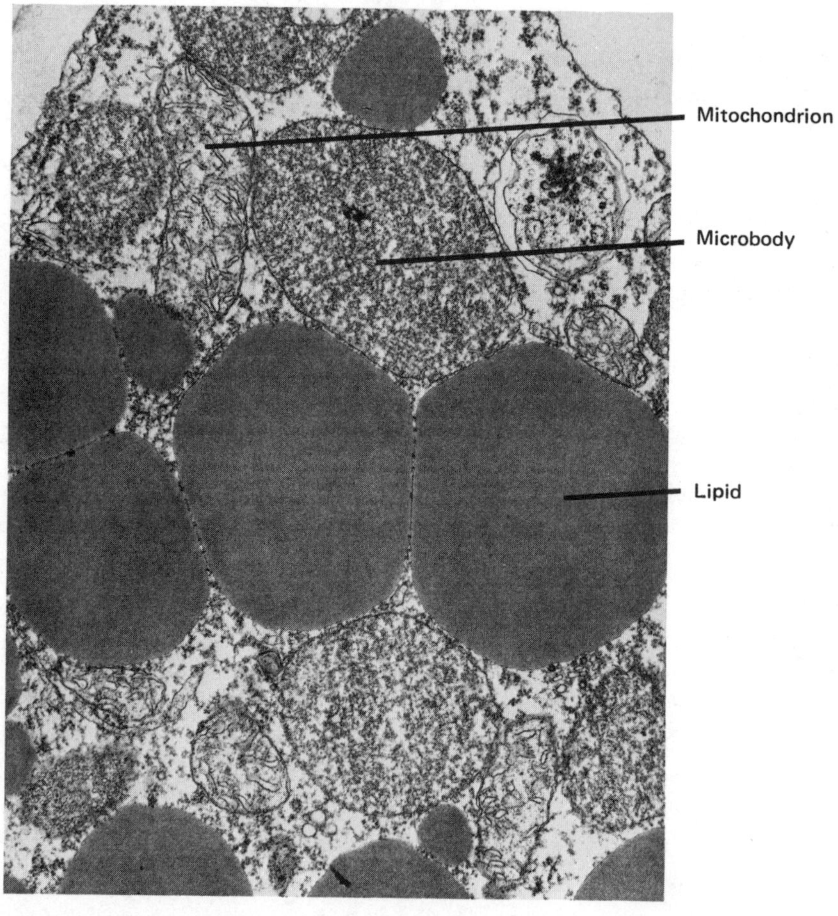

FIGURE 3-7
Electron micrograph of a plant cell, showing **mitochondria, microbodies,** and **lipid droplets.** (Photo courtesy of Robert A. Cecich.)

Microbodies in different plants or different tissues of the same plant may contain different enzymes and hence perform different functions. Thus various specific terms are used to distinguish different microbodies by their functions. Most microbodies are **peroxisomes,** so named because they contain enzymes that degrade hydrogen peroxide ($H_2O_2$) to water ($H_2O$) and oxygen ($O_2$). This is a significant function because frequently the highly reactive (and therefore dangerous) hydrogen peroxide is formed instead of water as the final product in oxygen-using (oxidation) metabolic reactions. In leaf tissue peroxisomes have a key role in photorespiration (Chapter 4). The storage tissue of seeds has specialized microbodies called glyoxysomes that aid in the breakdown of lipids into sucrose during early seedling development.

**Lysosomes** are present in some plant cells but are especially characteristic of animal cells. They contain a variety of digestive enzymes. In plants lysosomes fuse with vacuoles containing stored food and their enzymes digest the food to make its components available to the cell. They also break down other organelles, such as in damaged or starved cells, and digest molecules taken into the cell by endocytosis, discussed later.

#### Cytoplasmic Filaments and Microtubules

Increasingly sophisticated means of observation have revealed the presence within cells of many tiny filaments and tubules whose total functions have yet to be revealed but that form a fine **cytoskeleton** (cell skeleton). **Microfilaments** are the smallest (4 to 6 nm) and often form a weblike network just inside the cell membrane. (Refer to Appendix 1 for an explanation of metric units.) Some plasma membrane proteins are anchored to microfilaments to achieve contraction and thus are related to cell movements. A less well-defined group of filaments is identified by its size—**10 nm (or 100 Å) filaments.** They are common and at times seem to function in intracellular movements, but their function(s) is not clearly known.

**Microtubules,** in comparison, are larger (25 nm) and more complex. Each microtubule is a flexible, hollow rod, usually composed of 13 individual filaments, and is similar to a straw in that its ends are open to the cytoplasmic matrix. Microtubules apparently have many functions and are common in cells. They are thought to function in movement of components within the cell. They also make up the **spindle fibers** that appear during nuclear divisions. Locomotory structures, the hairlike **cilia** and whiplike **flagella** of both plant and animal cells, are precisely arranged bundles of microtubules, as are the **basal bodies** to which they are attached. Because some microtubules lie parallel to the cellulose fibrils of the cell wall, it is thought that they organize cellulose microfibril orientation. Centrioles (in animal cells) have the same organization as basal bodies.

#### Mitochondria

**Mitochondria** are highly complex organelles. They are sites of cellular respiration, the biochemical "fire" that releases energy from food molecules within the cell. As such, they are the main sites of oxygen use in the cell by far. Mitochondria are usually ovoid or oblong with a smooth outer unit membrane, separated from an inner membrane by a thin intermembrane space. The inner unit membrane is folded inward into numerous flattened or fingerlike projections (**cristae**), which increase surface area for chemical activity within the organelle and partially compartmentalize the interior (Fig. 3-8). Special techniques in preparing mitochondrial membranes for electron microscope observation show that the inner membrane has tiny mushroomlike projections into the fluid matrix of the mitochondrion interior. The projections are called **respiratory assemblies.**

FIGURE 3-8 Structure of a **mitochondrion.** The inner and outer membranes create separate compartments in which the various chemical reactions of aerobic respiration occur (Chapter 4). Infoldings of the inner membrane are called **cristae.** The knob-like **respiratory assemblies** on the cristae seem to be primary sites of oxygen utilization in the cell.

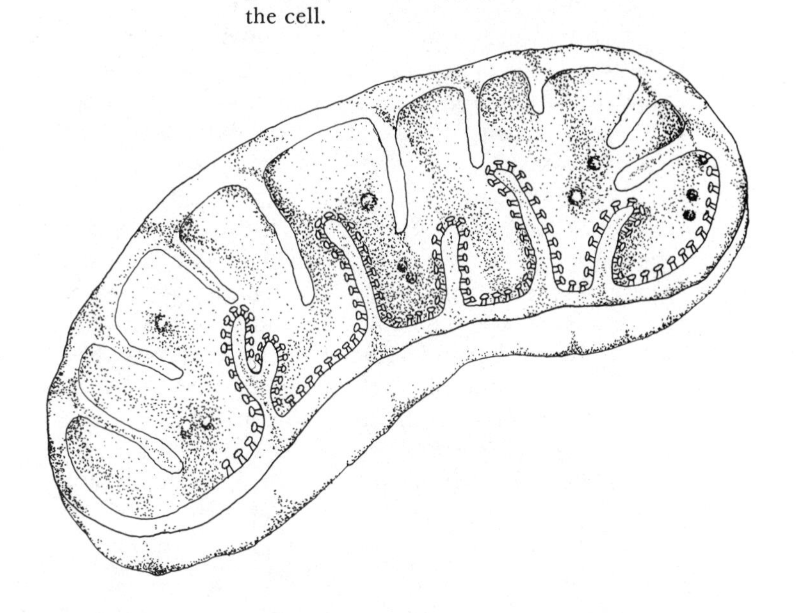

Although relying on the rest of the cell to supply food energy, each mitochondrion is an independent functional unit in several respects. The activities of respiration provide the mitochondria with energy for its own synthetic activities as well as for the rest of the cell. Mitochondria contain ribosomes and some of their own genes (in the form of deoxyribonucleic acid or DNA) and so are able to carry out some of their own protein synthesis. Mitochondria grow and even duplicate themselves by division.

### Plastids

Plastids are very complex organelles found only in plant cells, although not all plant cells have them. Plastids often contain lamellae, layers of membranes.

**Chloroplasts** are plastids that contain the green pigment **chlorophyll**, which is essential to photosynthesis. Each chloroplast is bounded by two unit membranes and shows a high degree of internal organization (Fig. 3-9). The chlorophyll molecules are bound to proteins in the membranes of stacked, flattened sacs called **thykaloids.** The chlorophyll-protein complexes of the thykaloid membrane are sites of the initial, light-using phase of photosynthesis (Chapter 4). In addition to several kinds of chlorophyll, orange **carotene** and yellow **xanthophyll** pigments are usually present, assisting in light absorption. Some double layers of membrane (lamellae) extend between thykaloid stacks. The internal membranes are surrounded by a liquid matrix called the **stroma.** The stroma is the site of carbon dioxide fixation, the second phase of photosynthesis. During active photosynthesis starch grains may form within the chloroplasts.

Like mitochondria, chloroplasts contain ribosomes and DNA strands and are apparently capable of synthesizing some of their own proteins by utilizing the energy captured by photosynthesis. Electron micrographs (photographs taken through the electron microscope) have demonstrated that chloroplasts duplicate themselves by dividing.

**Chromoplasts** are brightly colored plastids that may also have interior membrane layers, but they are not as highly organized as chloroplasts. Their pigments are primarily carotenes and xanthophylls. These pigments are lipid soluble (dissolved in lipids) and are thus present as droplets within the chromoplasts. Chromoplasts can be seen by light microscopic examination of scraped tissue from such plants as red peppers and carrots.

Carotenes from plants are an important source of vitamin A for animal nutrition. In animals, for example, vitamin A is utilized in the chemical reactions of "night vision," the accommodation of the eye to light and dark—hence the old saying about eating carrots to improve vision! Animals have varying abilities to digest and assimilate carotenoid pigments. The yellow of egg yolks is due to carotenoids ingested by the hens. Among cattle, beef breeds can break down carotenoids

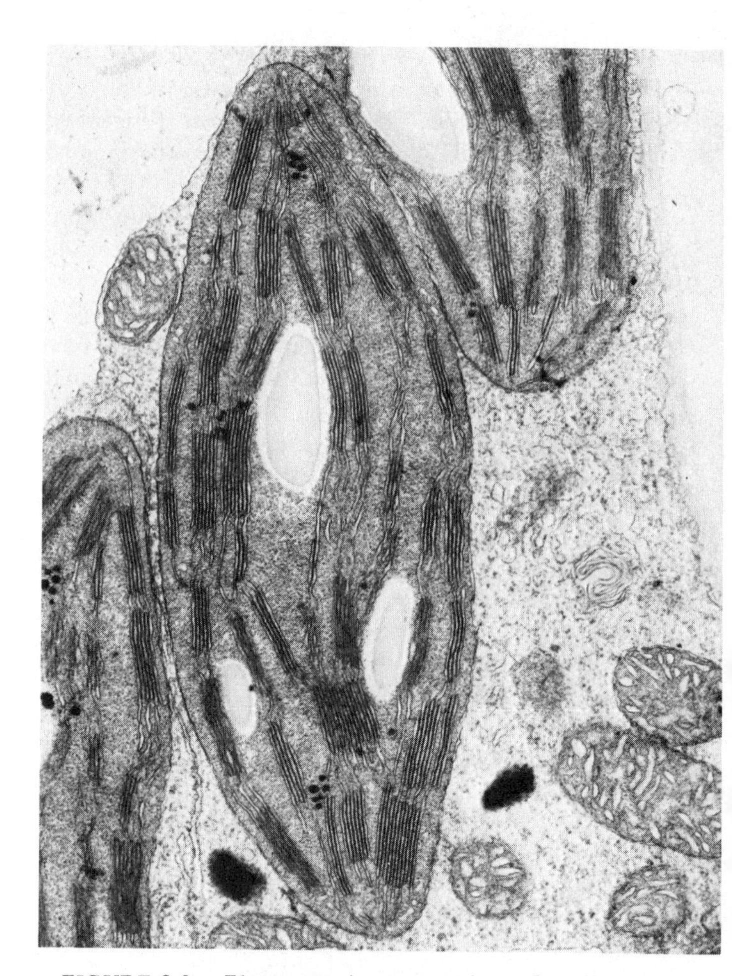

**FIGURE 3-9** Electron microscope view of an oat **chloroplast.** The granular-appearing substance that makes up the chloroplast matrix is the **stroma,** which contains DNA, RNA, and many other substances. The stacked structures are called **grana** (sing., granum) and consist of photosynthetic units called **thykaloids.** Each granum thykaloid is a coinlike closed disk, formed by a pair of chlorophyll- and carotenoid-containing membranes. In addition, thykaloids in the form of flattened sacs extend through the stroma between grana. Clear-appearing **starch grains** are also present.

more effectively than can dairy animals. Because carotenoids have an affinity for lipids, the fat of certain dairy breeds is therefore quite yellow rather than whitish. Corn grains contain carotenoids, and the regional notion that yellow-fatted animals were corn-fed crudely recognizes this relationship; however, white-fatted meat can also come from corn-fed animals. Yellow-fatted meat is probably an indication of dairy breed ancestry!

**Leucoplasts** are colorless plastids found in white tissues, such as inner lettuce leaves, onion bulbs, and potato tubers. Leucoplasts may contain lamellae and are associated with starch and oil metabolism and storage. Starch-containing leucoplasts are called **amyloplasts.**

Two additional kinds of plastids seem to be structural or developmental variants of other plastids. **Etioplasts** are apparently chloroplasts that have degenerated because of a lack of light. **Proplastids** seem to be immature plastids. Both etioplasts and proplastids can develop into normal-appearing chloroplasts after exposure to light. Proplastids may also mature into plastids other than chloroplasts.

### Vacuoles and Ergastic Substances

Most mature living plant cells have a large **central vacuole,** a membrane-enclosed sac that occupies up to 80 or 90% of the cell volume. The central vacuole content, called the **cell sap,** provides a ready supply of water and inorganic and organic substances. It is also a repository for crystals, oils, and other ergastic substances excreted or secreted by the cytoplasm. Some substances are actively accumulated within the vacuole. The central vacuole is bounded by a single unit membrane, the **tonoplast,** which maintains a barrier between vacuole contents and cytoplasm. The solute concentration in the central vacuole and cytoplasm is greater than in the water surrounding the cell. This concentration imbalance is necessary to maintain cell turgidity, as discussed subsequently.

The red-to-blue, water-soluble **anthocyanin** pigments are contained within the central vacuole and provide an example that demonstrates the importance of the living vacuolar membrane in containing ergastic substances. Freshly cut red cabbage, after being rinsed thoroughly, does not "bleed" into the water. Cooking destroys the membranes, however, and the pigment readily seeps into the cooking water. Anthocyanins also react to pH conditions. Red cabbage cooked in alkaline, mineral-containing ("hard") water turns dark purple or even blue. This is a reversible reaction and addition of vinegar (acetic acid) restores the red color.

Other ergastic substances in the central vacuole are certain proteins, tannins, resins, oils, inorganic ions and solutes, and a variety of **crystals.** Some crystals appear to serve as the plant's way of taking excess inorganic substances out of solution. This step prevents the accumulation of harmful concentrations, which would interfere in cellular water balance and other cytoplasmic activities. The two most common crystals are calcium carbonate (Fig. 2-1) and calcium oxalate. Calcium carbonate crystals (chalk or shell-like) are generally smooth surfaced, unlike calcium oxalate crystals, which are sharply pointed (Fig. 3-10). Smaller vacuoles containing a variety of substances are commonly present in the cytoplasm.

**FIGURE 3-10** Fresh section of parenchyma tissue of spiderwort (*Tradescantia*), showing calcium oxalate **crystals.** (Photo by Jeff P. Carlson.)

### Nucleus

The cell **nucleus** (pronounced new-klee-us, not new-cue-luss) is the largest cellular organelle and is usually spherical to oval (Fig. 3-11). The nucleus controls the metabolic activities of a cell. This "master control" function is due to the presence of DNA strands, which carry the cell's inherited information for regulating cellular activities. Although the strands are small and scattered throughout the nucleus as lightly staining **chromatin** ("colored matter") during active metabolism, they shorten and condense for nuclear (pronounced new-klee-are) division, becoming apparent as darkly staining "colored bodies," the **chromosomes** (Fig. 3-12). Chromatin also contains RNA and proteins. DNA and RNA interact with cellular components and the cellular environment to regulate and determine both the form and activities of an organism. (The mechanisms of this interaction are detailed subsequently.)

The contents of the nucleus are separated from the cytoplasm by a double unit membrane, the **nuclear envelope.** The nuclear envelope can be recognized easily in electron micrographs, for it resembles the pitted surface of a golf ball. The **nuclear pores** account for up to a third of the surface and seem to allow certain substances (e.g., ions, nucleotides, nuclear proteins, messenger RNA, and immature ribosomes) to pass between the nuclear matrix (nucleoplasm or karyolymph) and the cytoplasm. Ribosomes are often attached to the outer nuclear membrane. The latter is directly attached to the ER in many cells. The inner nuclear membrane lacks ribosomes and is attached to chromatin fibers, especially near the nuclear pores.

Plant cell nuclei commonly have several nucleoli. A **nucleolus** is a small, dense, round or doughnut-shaped body composed of granules of protein and RNA. The granules are ribosome precursors. Nucleoli disassemble just before a nucleus begins to divide, and then reassemble in the newly formed nuclei after division (Chapter 5). A nucleolus is formed in contact with the DNA of the chromosomes and the specific sites on chromo-

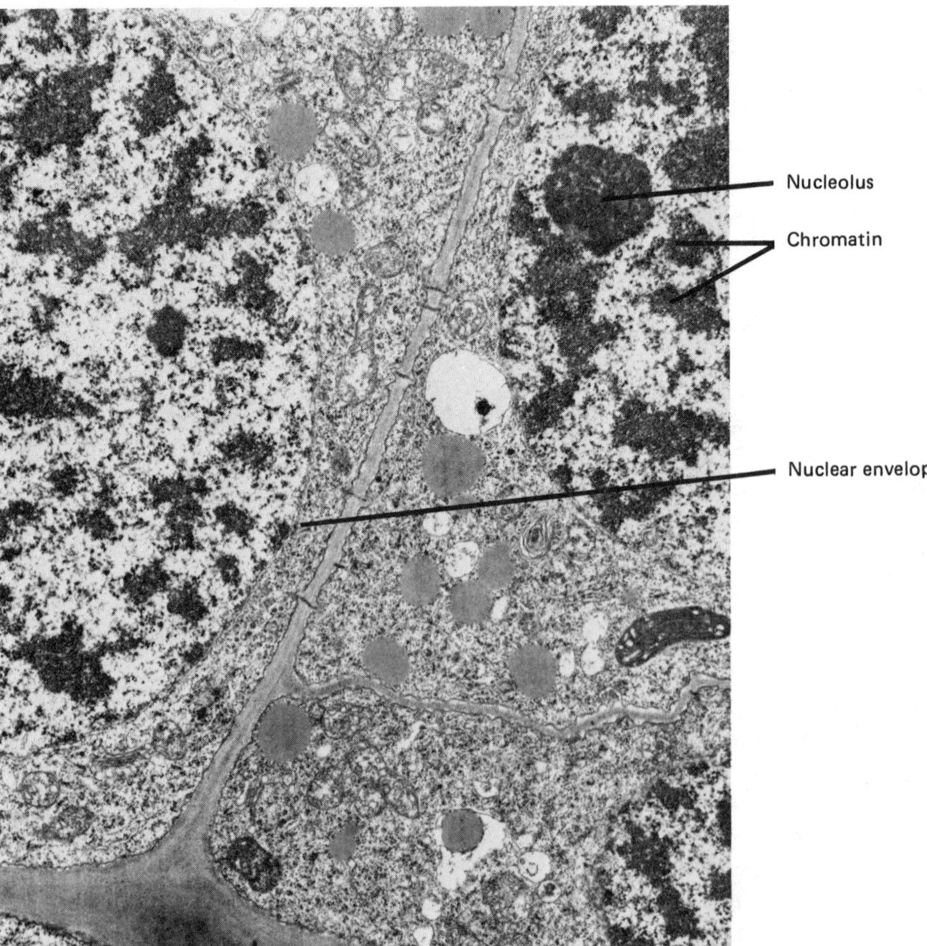


**FIGURE 3-11**
Several adjacent cells, showing details of the **cell wall,** cytoplasm, and nuclei. Note **nuclear envelope, chromatin,** and **nucleolus.** (Photo courtesy of Robert A. Cecich.)

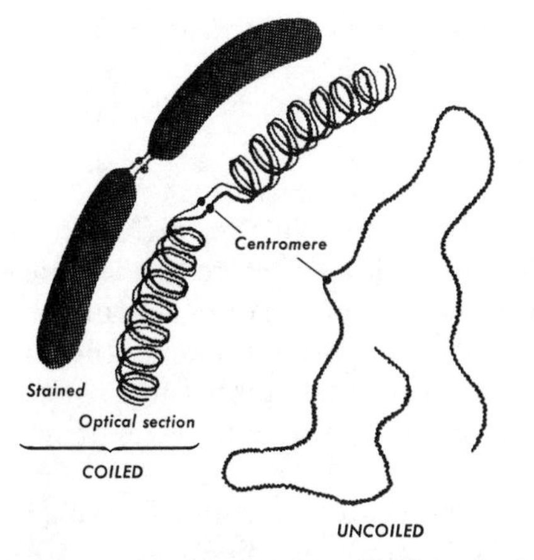


**FIGURE 3-12** Structure of a duplicated **chromosome.** DNA molecules are present in an uncoiled form during the active, metabolic state of a cell—that is, when the cell is not preparing for or involved in nuclear division. Prior to nuclear division, the long, threadlike DNA molecules condense by coiling, which can be seen in an unstained or optical section. When stained, the chromosomes ("colored bodies") appear as solid masses because proteins and other associated substances that surround the coiled DNA also absorb the stain. (From A.M. Elliott and D.E. Outka, *Zoology*, 5th ed., 1976. Reprinted by permission of Prentice-Hall, Inc., Englewood Cliffs, N.J.)

somes where this process occurs are called the **nucleolar organizers** (Fig. 3-1).

The cells of some organisms do not have an organized nucleus but have their DNA and RNA present in the cytoplasm. As noted in Chapter 2, organisms that lack a true nucleus are called prokaryotes, distinguishing them from organisms that do have nuclei, the eukaryotes.

#### Development of Cytoplasmic Structures

Most cytoplasmic structures are formed and destroyed as needed by the cell. The dynamics of their origins and fates differ for three general categories of structures. Those composed mainly of protein subunits (e.g., ribosomes, microfilaments, and microtubules) self-assemble in response to favorable conditions within the cell. When conditions are otherwise, the structures disassemble into the protein subunits.

Of structures composed mainly of membrane, only the endoplasmic reticulum seems able to grow by direct incorporation of lipid and protein molecules. The other membrane structures (e.g., plasma membrane, dictyosomes, nuclear envelope, and microbodies) form directly or indirectly from the endoplasmic reticulum, as diagrammed here.

nuclear envelope ← ENDOPLASMIC RETICULUM → dictyosomes (Golgi bodies) → plasma membrane
nuclear envelope → ENDOPLASMIC RETICULUM
lipids and proteins ↖ (to ENDOPLASMIC RETICULUM)
ENDOPLASMIC RETICULUM ↓ peroxisomes → lysosomes ← vacuoles formed during endocytosis
plasma membrane ↓ vacuoles formed during endocytosis
lipids and proteins ↖ (from lysosomes)

Our information about these relationships is still incomplete and this scheme may be altered as our understanding increases. Note that the relationship between nuclear envelope and ER is bidirectional. Prior to nuclear division (mitosis and meiosis, Chapter 5) the nuclear envelope disintegrates and the fragments are incorporated into the ER. The reverse process occurs to form new nuclear envelopes when division is completed.

Other membranes seem to be formed by small vesicles that pinch off the ER and then fuse to become part of the developing structure. The plasma membrane is increased by fusion with secretion vesicles from the dictyosomes; it is decreased by the fusion of endocytic channels and vesicles (p. 42) with the lysosomes. Digestion of membranes by lysosome enzymes releases membrane components that can then be used to synthesize new membrane. This is another example of the recycling that occurs in natural systems.

The complex organelles (plastids and mitochondria) are formed by growth and division of existing plastids and mitochondria.

## MOLECULAR TRAFFIC

The environment of a cell is constantly changing and yet it is crucial for the cell to maintain a relatively constant internal state. Many changes result from the cell's own activities. Cellular metabolism, for example, produces organic acids, alcohols, and other compounds that can be harmful if allowed to accumulate. On the other hand, certain necessary metabolic "ingredients" are used up and must be constantly resupplied to the cell. Finally, food from photosynthetic cells must be moved from the site of production to other parts of the plant for utilization and storage.

Because higher plants are immobile, they depend on the immediate external environment to supply them with water and mineral nutrients. The water and mineral solution is absorbed through the roots and then transported throughout the plant by specialized tissues, ultimately becoming the fluid environment surrounding each cell. The interaction of cells with this environment is one of constant molecular traffic, involving processes of **diffusion, osmosis, imbibition, transpiration,** and others. The balanced internal composition is therefore maintained by constant change and adjustment, a dynamic rather than static equilibrium. This condition of dynamic equilibrium in living systems is called **homeostasis.**

### Diffusion

Molecules are in constant motion. Where packed tightly together, as in solids, there is relatively little motion; more movement is possible in liquids, the most in gases. A moving molecule tends to go in the same direction unless it bumps into something else, like another molecule, which usually happens. Such constant molecular jostling is the basis of a process called diffusion.

**Diffusion** is usually defined as the tendency for molecules to move in response to a concentration gradient. In other words, they move from an area of greater concentration to one of lesser concentration. By thinking of many molecules bumping together, it is easy to visualize how they tend to become evenly dispersed. There are many examples: odoriferous molecules from a solid air-freshener perfume an entire room; smoke from a pipe or cigarette disperses; sugar added to a beverage sweetens the contents of the entire container even if left unstirred for a long-enough time.

There are many diffusable substances in and around cells: water, oxygen, carbon dioxide, simple sugars, amino acids, minerals, and growth substances, for instance. Most molecules that enter and leave the cell do so by free diffusion. Lipid-soluble substances (e.g., oxygen and carbon dioxide) diffuse through the lipid-matrix cell membranes with particular ease.

To many molecules, however, the semipermeable cell membrane presents a barrier to diffusion. In this way, important larger molecules are retained within the cell; others are kept out.

The rate of diffusion is affected by pressure, concentration, and temperature. Perfume applied to warm human "pulse points," for example, releases fragrance molecules more rapidly than from cooler skin surfaces.

### Osmosis

**Osmosis** refers to the diffusion of solvent molecules through a differentially permeable membrane (whether the membrane is living or artificial). In living cells the predominant solvent is water. Therefore a practical discussion of osmosis in organisms refers to water movement into and out of cells.

Water molecules continually pass in and out of cells. An **osmotic gradient** is established, however, when there is a higher concentration of solute molecules on one side of the membrane than on the other. A container of 30% sucrose solution (a nonpermeating sugar), for instance, contains more sucrose molecules and fewer water molecules than an equal volume of a 10% sucrose solution. If these two solutions are on opposite sides of a differentially permeable membrane, more water molecules from the 10% sucrose solution tend to enter the 30% sucrose solution than move the other way. The result is a *net* influx of water molecules into the 30% solution.

Three terms are used to describe the osmotic relationship of a living cell to its aqueous environment. If the total solute concentration outside a cell is greater than the concentration inside—for example, in mineral waters—there is a greater tendency for water to leave the cell, producing a net outflux. Such an environmental solution is **hypertonic.** If the inside and outside solute concentrations are the same so that influx and outflux of water are balanced, the cellular environment is **isotonic.** If there is a lower concentration of solute molecules outside the cell than inside, there is a net influx of water. Such an environment is **hypotonic.**

Most plant cells require a hypotonic environment. Imagine a differentially permeable water-filled bag (the cell membrane and contents) within a cardboard box (the cell wall). As more and more water enters the cell in response to the osmotic gradient, the cell becomes swollen and engorged with water or **turgid.** Most of the water is contained in the central vacuole. An animal cell would eventually burst, but in plants the cells are kept from bursting by their cell walls.

The internal pressure of water against the cell wall (**turgor pressure**) creates great architectural strength, especially when many cells are cemented together to form tissues. In nonwoody plants turgidity is a structural necessity, for the stiff, crisp tissues provide support for the shoot. When turgidity is lost, nonwoody tissues become soft or **flaccid** and the plant wilts. Flaccidity is caused by excessive transpiration or by exposure to a hypertonic solution. If water loss continues and the central vacuole shrinks, water is drawn from the cytoplasm. Finally, the cytoplasm shrivels, a condition known as **plasmolysis** (Fig. 3-13). Most plants are able to recover from plasmolysis if it is not too severe or prolonged. Some plants, such as succulents and cacti, have gelatinous sap that holds water in the tissues during dry periods, thereby resisting desiccation.

## Other Membrane Phenomena

Both plant and animal cells are able to concentrate (accumulate) some substances within the cell *against* a diffusion gradient. This act requires work (energy expenditure) on the part of the cell membrane and is called **accumulation.** The process also works in reverse and the cell membrane may specifically remove or exclude substances against a diffusion gradient.

In addition to free diffusion, many molecules enter or leave cells by **carrier-assisted transport,** whereby specific carrier molecules "ferry" other molecules through the plasma membrane. The carriers are usually proteins, which have very

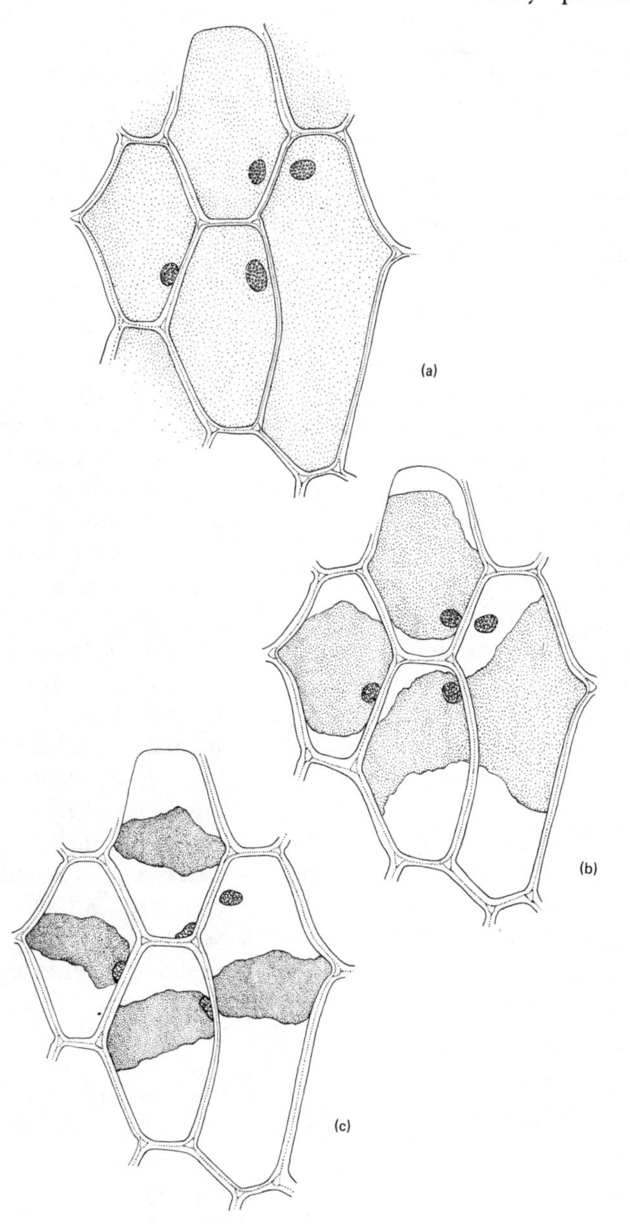


**FIGURE 3-13**
Changes in **central vacuole** volume during **plasmolysis.** (a) In a living plant cell viewed with the light microscope, the full central vacuole (stippled) presses the cytoplasm into a hard-to-see thin layer against the cell walls. (b) When the cell is immersed in a sucrose solution water passes from the cell into the surrounding solution and plasmolysis occurs. As the central vacuole shrinks, the vacuolar membrane (tonoplast) becomes visible. (c) The central vacuole is greatly shrunken. Part of the plasma membrane and cytoplasm (marked by the position of the nucleus) remains attached to the cell walls, and part may be pulled away, forming strands stretching between the cell walls and tonoplast. After a time, the cytoplasm and nucleus may become so dehydrated that their metabolic functions are irreversibly disrupted.

specific surface configurations, enabling specific carrier proteins to match up and form a reversible combination with specific molecules that are transported. A fully acceptable model has yet to be derived to explain the exact mechanism of carrier-assisted transport, but Fig. 3-14 presents several possibilities. When this kind of membrane transport occurs in response to (in the direction of) a diffusion gradient, it is called **facilitated diffusion.** Movement against a diffusion gradient, however, requires energy expenditure and is called **active transport.**

**FIGURE 3-14** Three hypothetical models for **carrier-assisted transport** involving membrane-bound carrier molecules. Each model shows a molecule (ball) being transported from outside to inside the cell membrane. The upper model is simple, but seems unlikely because membrane proteins may be "anchored" by cytoplasmic filaments and do not readily flip, nor do they move between the inner and outer lipid layers of the membrane. The center model is of a two-unit, complex carrier, which would enable carrier molecules to retain their positions. The bottom model represents a substrate-specific channel, which could be opened and closed by subtle changes in the shape of the protein.

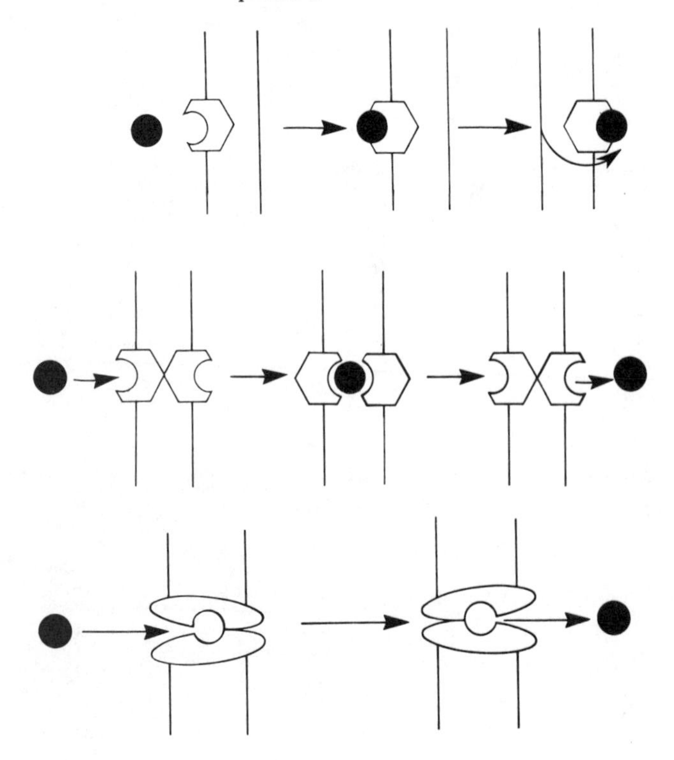

Although accumulation of substances against a diffusion gradient is often accomplished by active transport, in other situations the unequal distribution of diffusable substances results from a phenomenon called a **Donnan (or Gibbs–Donnan) equilibrium,** named by the English chemist Frederick Donnan, who described it in 1927. In a Donnan equilibrium large molecules (macromolecules) within the cell, such as proteins and nucleic acids, attract small positively charged ions to their negatively charged sites. As a result, the concentration of specific positive ions inside the cell may be greater than outside, even though the overall condition is net electroneutral equilibrium. The higher solute ion concentration within the cell however in turn creates an osmotic gradient, leading to a net influx of water. This osmotic gradient affects all living cells. In a hypotonic solution, as described previously, water would continually enter a cell. Plant cells are protected from bursting by the cell wall, but animal cells must rely on an active transport process to reduce osmotic pressure. This mechanism is called a **sodium pump** because sodium ions ($Na^+$) are actively pushed out of the cell, thus reducing the total solute concentration within the cell.

In certain lower plants and in animals some cells take external fluids or food particles into the cell by vesicle or vacuole formation instead of by passage through the cell membrane, a process called **endocytosis.** During endocytosis the membrane actually flows around, or engulfs, a portion of fluid or food particle and then pinches off inside the cell, forming a vesicle or vacuole. The result is that a small portion of the external fluid environment is brought into contact with the cytoplasm. In plant cells the cell membrane may form narrow invaginations, or channels, from the tips of which endocytic vesicles are pinched off into the cytoplasm. (Imagine the kind of surface invaginations created by pushing your fingers into a ball of dough.) Endocytic vesicles often fuse with lysosomes, the enzymes of which digest both the vesicle contents and surrounding membrane.

#### Imbibition, Transpiration, and Guttation

Several special types of diffusion occur in plants. **Imbibition** (L., *imbibere,* to drink in) is the diffusion of gas or liquid molecules (usually water) between molecules of a solid, causing the solid to swell. Imbibition requires a strong intermolecular attraction. In plants the most common partners are water and the cellulose of cell walls. One example of imbibition is the absorbency of paper towels and tissues. The swelling of door jambs and window frames during humid weather is also caused by imbibition.

Evaporation involves diffusion of molecules into the air during a change from a liquid to gaseous state. As noted in Chapter 2, evaporation of water from plant surfaces is called **transpiration.**

Water is also lost from plants by **guttation.** Guttation occurs when the rate of osmotic uptake of water by plant roots exceeds the rate of transpiration, causing water droplets to ooze out along leaf margins or tips. Special openings called **hydathodes** may be present for guttation. Cool, still, humid nights when there is plenty of soil moisture favor guttation.

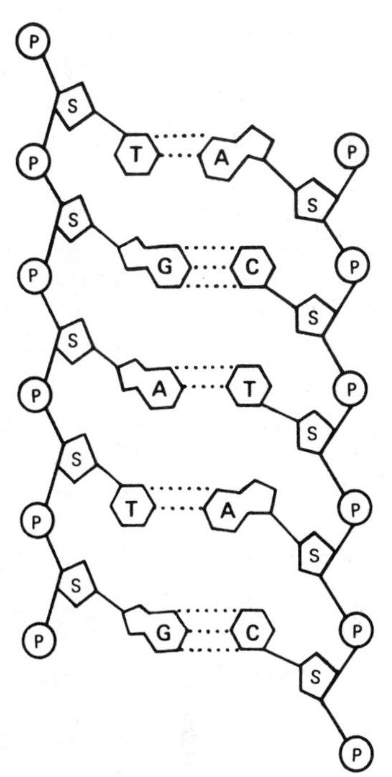


**FIGURE 3-15**
A diagrammatic view of the **nucleotide** arrangement that makes up a **DNA** molecule. The outer framework is made of molecules of the sugar (**S**) deoxyribose and phosphate (**P**) units. The nucleotide bases adenine (**A**), thymine (**T**), cytosine (**C**), and guanine (**G**) pair according to the configuration of their molecules. Two **hydrogen bonds** hold thymine and adenine together. Three hydrogen bonds hold cytosine to guanine.

## DNA, RNA, AND PROTEIN SYNTHESIS

### DNA: The Double Helix

#### DNA Structure

The great significance of deoxyribonucleic acid (DNA) to the living world has already been mentioned. An organism's DNA carries the coded instructions for all the life processes of the organism. Alteration of the heritable DNA can change some of these processes (see genetic engineering, Chapter 6). True appreciation of its importance requires an explanation of what DNA is and how it works.

The DNA molecule is composed of structural units called **nucleotides.** Each nucleotide is composed of a five-carbon sugar (deoxyribose), a nitrogen-containing base (adenine, guanine, cytosine, or thymine), and a phosphate unit ($PO_4$) (Fig. 3-15).

The DNA molecule is organized as a **double helix** composed of precisely matched nucleotide pairs. An imperfect but helpful way to visualize a double helix is to think of a rope ladder with wooden rungs. Twisting the ladder causes each rope on the side to form a helix through space and the ropes are held equidistant by the rungs (Fig. 3-16).

In each nucleotide the sugar and phosphate contribute to the "rope" on one side of the structure and the nitrogenous base forms half a "rung." Nucleotides are connected along the outside by the

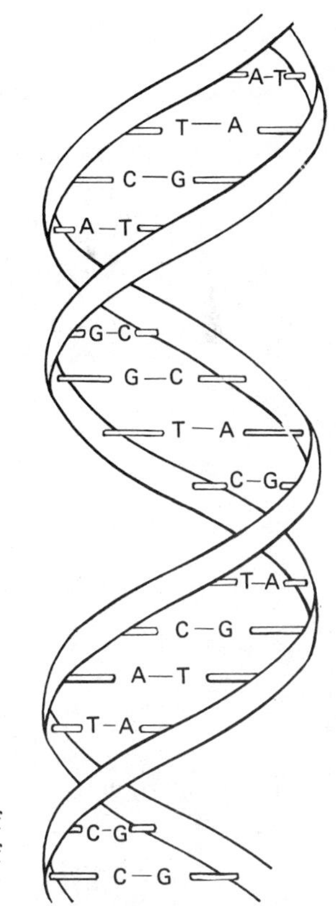


**FIGURE 3-16**
A simplified visualization of the **double helix** structure of DNA.

linking of phosphates to sugars. The nucleotides are connected in the middle (to form complete "rungs") by the linking of complementary nitrogenous bases by the formation of weak hydrogen (H) bonds. The four nitrogenous bases are of two types: the smaller bases (pyrimidines) are **thymine** and **cytosine**; the larger (purines) are **adenine** and **guanine.** A small base can pair only with a large base, resulting in uniformly sized "rungs." The pairing is even more specific than that, however, for adenine will pair only with thymine and cytosine only with guanine. (One way to remember the complementary base pairs is by alphabetical position—the two farthest from one another link, A–T, as do the closest, C–G.) The specific pairing of bases is the key to DNA coding of genetic information, as discussed subsequently.

#### DNA Replication (Duplication)

One of the criteria that established DNA as the genetic material passed from generation to generation is its ability to duplicate itself in an exact manner. Therefore both nuclei resulting from a mitotic division (Chapter 5) are theoretically identical to each other and the parent nucleus.

To accomplish duplication, the double helix of DNA separates along its midline by the breaking of the hydrogen bonds connecting the base pairs. By analogy, it could be said that the molecule "unzips," leaving nucleotide bases exposed on each single strand (Fig. 3-17). With the assistance of enzymes, the open sites are matched up with new complementary nucleotides that have been previously synthesized and are present in the nucleus. Gradually each single strand forms its own double helix. Both new double helices are nearly always identical to the parent molecule. In fact, however, slight changes occasionally occur, either spontaneously or because of the influence of some environmental factor. Changes in the DNA structure are **mutations** (see also Chapter 6).

**FIGURE 3-17** A model of the process of **DNA replication** (duplication). The smaller strands are new, formed by the attachment of free nucleotides to exposed nucleotide bases on the "unzipping" DNA molecule. The replication process works its way along the length of the DNA molecule, eventually yielding two double helices identical to the original helix.

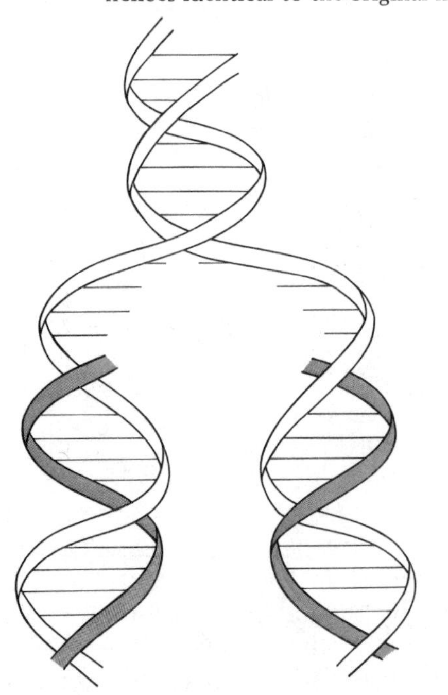

#### The Search for the Secret of Life

"New" discoveries in science are initially subject to much skepticism within the scientific community and such was the attitude toward the hypothesis that DNA was "the secret of life." Groups of scientists had attacked the problem of identifying the structural pattern of the molecule by different methods. Biochemical analysis indicated that certain components were present in certain ratios. X-ray diffraction studies indicated a regularly repetitive pattern. Mathematical and stereochemical applications attempted to put the data together into a logical geometric form. Much of this research occurred in the King's College laboratory of Maurice Wilkins in London, where Rosalind Franklin assembled the definitive x-ray diffraction information. In 1953 Francis Crick (Great Britain) and James Watson (the United States), working at Cambridge University, finally managed to synthesize the available information into a successful model. For their work, Watson, Crick, and Wilkins were awarded a Nobel Prize in 1961. The double helix model they hypothesized was immediately subjected to cautious and intensive scientific scrutiny and is now regarded as an established fact.

### RNA: Complementary Molecules

The genetic information on the DNA molecule is in a coded form and would be of no use to a cell unless a decoding mechanism existed. As we now understand it, the decoding, or **transcription**, is accomplished by ribonucleic acid (RNA). Furthermore, RNA is also responsible for the direct **translation** of the decoded information into the synthesis of protein, as summarized here:

$$\text{DNA} \xrightarrow{\text{transcription}} \text{RNA} \xrightarrow{\text{translation}} \text{protein.}$$

### RNA Structure and Formation

Ribonucleic acid (RNA) is similar to DNA in structure but with significant differences.

1. The five-carbon sugar is ribose instead of deoxyribose.
2. Instead of thymine, **uracil** is present and pairs with adenine.
3. Typically the molecule is composed of a single strand that may fold back on itself in various places to form a modified double helix.

RNA is synthesized directly from DNA, which temporarily "unzips" to expose base sequences. Assisted by enzymes, RNA nucleotides are attracted to the DNA strands and are connected in a precise sequence to form an RNA strand that is an exact complement to the segment of DNA it forms on.

#### Types of RNA

Three general types of RNA are involved in protein synthesis—messenger RNA (*m*RNA), transfer RNA (*t*RNA), and ribosomal RNA (*r*RNA). Messenger and transfer RNA are named for their roles in protein synthesis and ribosomal RNA for its location in the ribosomes.

Each **messenger RNA** molecule carries a sequence of bases that is complementary to the base sequence present on the section of DNA from which it was formed. *m*RNA usually does not fold back on itself but travels as a single coded "tape" from the nucleus to the cytoplasm, presumably through the nuclear pores.

**Transfer RNA** molecules are the smallest of the three RNA types. Each *t*RNA molecule is a single strand folded back on itself, forming a double helical region and three loops. The loop on the end consists of three unpaired nitrogenous bases. Each set of three unpaired bases is called an **anticodon.** In addition, each *t*RNA molecule carries an amino acid. *t*RNA molecules with specific anticodons carry specific amino acids, a situation analogous to owner-specific vehicles.

**Ribosomal RNA,** like *m*RNA and *t*RNA, is formed from DNA. The physical structure of *r*RNA is not known, although it seems to be a strand composed of both double and single helices. Bacterial ribosomes have been studied extensively and are apparently composed of two subunits, each consisting of RNA and associated proteins. Ribosomes provide the sites for the actual synthesis of proteins.

## The Genetic Code

The inherited messages for protein formation are found on DNA molecules in coded form. The portion of a DNA molecule that specifies one complete protein molecule is called a **gene.** The DNA code is transferred from the DNA to specific strips of *m*RNA in a special code language that consists of "three-letter words" called **codons.** Each codon is a sequence of three nucleotides on an *m*RNA molecule. A specific series of codons forms a code "sentence" that is translated by complementary anticodons on *t*RNA molecules into a specific protein, usually an enzyme. (This process is detailed in the next section.)

In the genetic code 64 ($4^3$) three-letter nucleotide combinations (codons) are mathematically possible and, in fact, 61 of these combinations correspond to specific amino acids (Table 3-2). The remaining three (UAA, UAG, and UGA) act as "stop" signals, terminating the formation of protein along a given strip of *m*RNA.

The code is not "letter perfect." Several codons may correspond to the same amino acid. In addition, one *t*RNA may accept several codons due to slight differences in the anticodon structure. Although the code is specific for the 20 commonly occurring amino acids in Table 3-2, at least one other amino acid (*N*-formyl methionine) can be coded for by the codons AUG and GUG, which usually specify the amino acids methionine and valine. The dual attachment capacity of AUG and GUG may be related to a role as a "start" signal for chain initiation.

Accurate translation of the genetic code is also subject to inherent variations present because of inherited mutations, including amino acid replacement, chain termination, and frame-shift (code sequence altered by insertion of a nucleotide). New gene mutations are a further cause of differences.

Perhaps the most spectacular single statement that can be made about the genetic code is, in the words of James Watson, that it "appears to be essentially the same in all organisms." The maintenance of nearly identical innate control mechanisms is strong evidence for the common origin of Earth's organisms.

TABLE 3-2 The genetic code.

| First Nucleotide Base | Middle Nucleotide Base |  |  |  |  |  |  |  | Third Nucleotide Base |
|---|---|---|---|---|---|---|---|---|---|
|  | U |  | C |  | A |  | G |  |  |
|  | UUU |  | UCU |  | UAU |  | UGU |  | U |
| U | UUC | Phe | UCC |  | UAC | Tyr | UGC | Cys | C |
|  | UUA |  | UCA | Ser | UAA[b] |  | UGA[b] |  | A |
|  | UUG | Leu | UCG |  | UAG[b] |  | UGG | Try | G |
|  | CUU |  | CCU |  | CAU |  | CGU |  | U |
| C | CUC |  | CCC |  | CAC | His | CGC |  | C |
|  | CUA | Leu | CCA | Pro | CAA |  | CGA | Arg | A |
|  | CUG |  | CCG |  | CAG | Gln | CGG |  | G |
|  | AUU |  | ACU |  | AAU |  | AGU |  | U |
| A | AUC | Ileu | ACC |  | AAC | Asn | AGC | Ser | C |
|  | AUA |  | ACA | Thr | AAA |  | AGA |  | A |
|  | AUG[a] | Met | ACG |  | AAG | Lys | AGG | Arg | G |
|  | GUU |  | GCU |  | GAU |  | GGU |  | U |
| G | GUC |  | GCC |  | GAC | Asp | GGC |  | C |
|  | GUA | Val | GCA | Ala | GAA |  | GGA | Gly | A |
|  | GUG[a] |  | GCG |  | GAG | Glu | GGG |  | G |

[a] AUG and GUG are chain-initiation codons.

[b] UAA, UAG, and UGA are chain-termination codons. When one of these codons is reached, synthesis of the specific protein chain (polypeptide) is finished and the polypeptide breaks away from the ribosome.

Abbreviations for the 20 commonly occurring natural amino acids, as presented above, are:

| | | | |
|---|---|---|---|
| Phe = phenylalanine | Ser = serine | Tyr = tyrosine | Cys = cysteine |
| Leu = leucine | Pro = proline | His = histidine | Try = tryptophan |
| Ileu = isoleucine | Thr = threonine | Gln = glutamine | Arg = arginine |
| Met = methionine | Ala = alanine | Asn = asparagine | Ser = serine |
| Val = valine | | Lys = lysine | Gly = glycine |
| | | Asp = aspartic acid | |
| | | Glu = glutamic acid | |

### Protein Synthesis

An analogy may be useful to assist you in understanding how protein synthesis occurs, so we will liken the process to a stage production. The performance location is a ribosome. The script is written by nuclear DNA. The performance is directed by *m*RNA, which is also a member of the cast. Other cast members are *t*RNA molecules with their attached amino acids. The stage crew consists of various specific enzymes. Props include inorganic molecules to be used for chemical bonding. Financial backing (energy) is provided in the form of adenosine triphosphate (ATP) molecules (Chapter 4). The plot is summarized in Fig. 3-18(a)-(d).

An *m*RNA molecule assumes position on the ribosome, attaching to the two adjacent synthesizing "active" sites. Each synthesizing site accommodates a single *m*RNA codon. The *m*RNA/*r*RNA combination attracts two *t*RNA molecules with anticodons (a) corresponding to the two *m*RNA codons. With the two *t*RNA molecules thus parked in close proximity, a bonding reaction (formation of "peptide" bond) occurs between their attached amino acids (b). Next, the *m*RNA strip slides over by one codon (c), allowing the end amino acid to release from the ribosome (d). The "spent" *t*RNA leaves. A new "loaded" *t*RNA attaches to the exposed active site. Another bonding reaction occurs so that now three amino acids are linked. This process occurs repeatedly, forming a protein, a polypeptide chain in which amino acids are the links. Each specific protein differs from other proteins in amino acid sequence. The outcome of each synthesis is precise because the sequence of specific amino acids has been determined by the *m*RNA

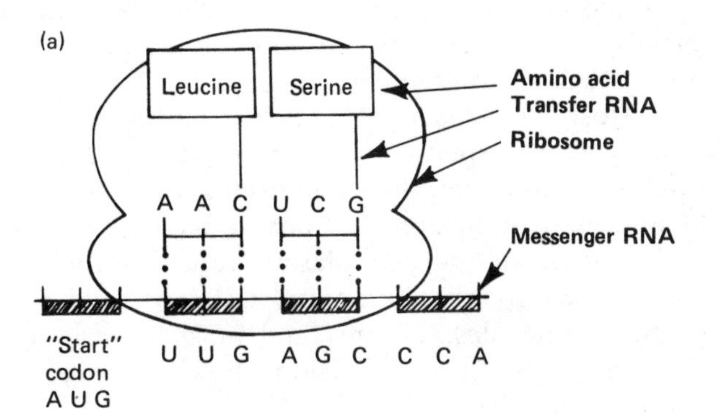


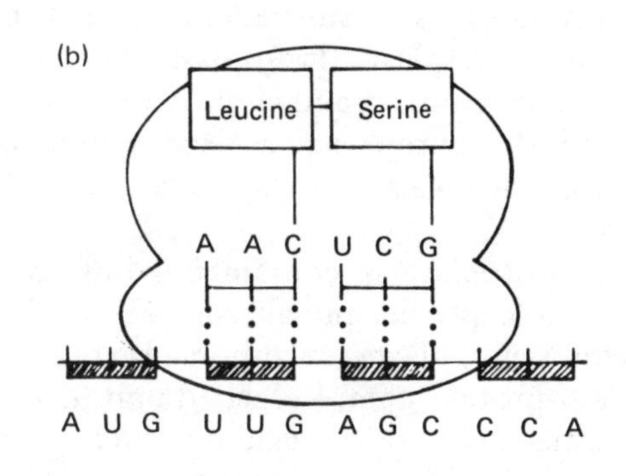


(c)
Proline
Leucine
Serine
G G U
A A C
U C G
A U G U U G A G C C C A

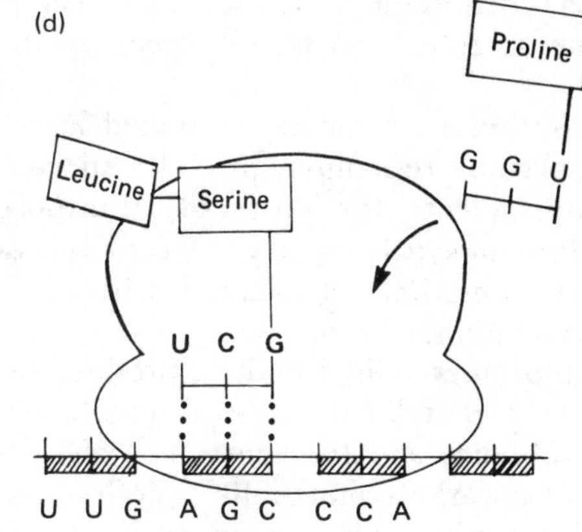


**FIGURE 3-18** A model for **protein synthesis,** showing interactions of **messenger RNA** (*m*RNA) and **transfer RNA** (*t*RNA) at the two active sites on a **ribosome.** Only a small portion of a messenger RNA molecule is diagrammed. In reality, an individual *m*RNA molecule may consist of dozens to hundreds of three-unit **codons.** See text for a detailed description of the sequence from (a) through (d).

codons as per instructions from the DNA. Thus the DNA has controlled the entire process without ever leaving the nucleus or being used up itself in any way!

Unlike its explanation, protein synthesis occurs rapidly, enabling the cell to respond quickly to the demands of chemical circumstances. More than one ribosome may become associated with a particular *m*RNA molecule at a single time, forming a veritable assembly line for the rapid formation of many molecules of the desired protein.

Protein synthesis is apparently controlled by a feedback mechanism whereby the cell responds to needs and surfeits of specific proteins. Therefore an *m*RNA molecule may continue its activity until it is inactivated by enzymes produced according to the information on subsequent *m*RNA molecules. In this way, a dynamically changing and yet stable system of checks and balances prevails.

Meanwhile the unburdened *t*RNA molecules are recycled, replacing their lost amino acids with identical counterparts for continued participation in protein synthesis.

By now you may be feeling overwhelmed by the complexity of a single "simple" cell! The important thing to do, however, is to connect each part and particle to its role in the cell's existence. The definitive identification of these roles will require continuous future scientific research, just as our current understanding is the result of past and present scientific research projects.

## SUMMARY

Typical plant cells are protected by **cell walls** consisting primarily of **cellulose** and **pectins.** Membrane-lined **plasmodesmata** perforate the cell walls, allowing transport between cells. Waxy **cutin** and **suberin** in cell walls retard dehydration. **Lignins** harden cell walls, such as those of wood.

Within the cell wall, the living **protoplasm** and

the nonliving **ergastic substances** are collectively called the **protoplast.** Protoplasm includes the **nucleus,** which controls and coordinates cell metabolism, and the **cytoplasm,** a viscous fluid matrix filled with chemicals and specialized cytoplasmic structures.

The protoplast is contained within a living, semipermeable **plasma membrane.** Like other cellular membranes, plasma membrane is composed of a lipid and protein complex called **unit membrane.** The plasma membrane regulates and maintains cytoplasmic composition. Substances may cross the plasma membrane in several ways. The plasma membrane secretes and is reinforced by the cell wall.

A continuous network of folded membrane, the **endoplasmic reticulum,** provides surface areas and compartments for synthesis, transport, and storage. Protein synthesis takes place on **ribosomes,** which are sometimes associated with the endoplasmic reticulum.

**Dictyosomes** (Golgi bodies) produce or accumulate enzymes and other secretory materials.

**Microbodies** and **lysosomes** contain enzymes and perform various metabolic functions in cells. Most microbodies are peroxisomes. Lysosomes contain digestive enzymes.

Tiny filaments and tubules form a fine skeletal network within cells and perform many functions. Functions of complex **microtubules** include movement of substances within the cell, cell division, locomotion, and cell wall synthesis.

**Mitochondria** are sites of cellular respiration, have DNA and ribosomes, grow, and reproduce by dividing.

**Chloroplasts** contain **chlorophyll** and are sites of photosynthesis. Chloroplasts also have DNA and ribosomes, grow, and reproduce by dividing.

**Chromoplasts** are less complex plastids, containing mostly orange **carotenes** and yellow **xanthophylls. Leucoplasts** are colorless plastids. **Etioplasts** are apparently degenerate, light-starved chloroplasts. **Proplastids** seem to be immature chloroplasts, although they also may develop into other types of plastids.

The large **central vacuole** occupies 80 to 90% of cell volume within most mature plant cells. Its contents, the **cell sap,** provide a repository of dissolved nutrients, water, and ergastic substances. The central vacuole aids in maintaining cell **turgidity.**

The **nucleus** is the largest organelle and contains the chromosomes. Nuclear contents are separated from the cytoplasm by a **nuclear envelope,** which is perforated to allow larger molecules to pass into the cytoplasm. Plant cell nuclei usually contain several **nucleoli,** dense bodies of protein and RNA granules.

Most cytoplasmic structures are formed and destroyed in response to cell needs. Protein structures self-assemble when conditions are favorable. Membrane structures seem to be derived directly or indirectly from the endoplasmic reticulum, which itself grows by direct incorporation of lipid and protein molecules. Plastids and mitochondria are formed by growth and division of existing plastids and mitochondria.

Cells perform many activities to maintain **homeostasis.** Many substances move in and out of cells by **diffusion. Osmosis** is the diffusion of water through a semipermeable membrane. Most plant cells require a **hypotonic** environment to maintain the turgidity that supports nonwoody tissues. If too much water is lost from a cell, it becomes **flaccid.** Severe water loss causes **plasmolysis.**

Cells can actively **accumulate** or expel substances against a diffusion gradient, an energy-using activity that is also called **active transport. Carrier-assisted transport** and **facilitated diffusion** are means by which specific molecules are ferried through the plasma membrane by special carrier molecules. A **Donnan equilibrium** is caused by the higher macromolecule concentration within the cell, attracting small positive ions and hence also water. During **endocytosis** the plasma membrane engulfs external fluids or particles.

Primary cell walls absorb and retain water by **imbibition.** Plants lose water by **transpiration** and **guttation.**

**Deoxyribonucleic acid** (DNA) is composed of **nucleotides,** each of which contains a sugar (deoxyribose), a nitrogen-containing base (adenine, guanine, cytosine, or thymine), and a phosphate unit. DNA is organized as a **double helix,** with complementary base pairs held together by hydrogen bonds.

DNA can duplicate itself exactly and yet has the potential for alteration. Changes that do occur are called **mutations.** DNA was established as the genetic material only after rigorous research.

The structure of **ribonucleic acid** (RNA) resembles that of DNA, being composed of nucleotides formed of a sugar (ribose), a nitrogen-containing base (adenine, guanine, cytosine, or uracil), and a phosphate unit.

Three types of RNA interact to decode the genetic information on DNA and translate the information into protein synthesis. A **messenger RNA** molecule has a base sequence that is complementary to the base sequence on the section of DNA from which it was formed. Each **transfer RNA** molecule loops back on itself, leaving three unpaired bases called an **anticodon.** Each transfer RNA molecule carries a specific amino acid. The physical structure of **ribosomal RNA** is not completely known.

Coded genetic information is transferred from DNA to strips of messenger RNA in "three-letter words" called **codons.** A series of translated codons creates a specific amino acid sequence—a protein, usually an enzyme.

The act of protein synthesis takes place on ribosomes. A messenger RNA codon attaches to synthesizing sites on a ribosome, attracting transfer RNA molecules with complementary anticodons. A bonding reaction occurs between the amino acids, the messenger RNA strip moves over by one codon, and so on, until a protein chain has been completed.

## SOME SUGGESTED READINGS

Brown, W.V., and E.M. Bertke. *Textbook of Cytology.* St. Louis: The C.V. Mosby Co., 1969. A textbook of cytology emphasizing the structure and variation of cells.

Dyson, R.D. *Cell Biology: A Molecular Approach.* 2nd ed. Boston: Allyn & Bacon, 1978. A clearly written and up-to-date textbook that emphasizes our emerging knowledge of the dynamic functions and interactions of cell components.

—. *Essentials of Cell Biology.* Boston: Allyn & Bacon, 1975. As the title implies, a less comprehensive treatment of cell biology that emphasizes the most significant details.

Jensen, W.A., and R.B. Park. *Cell Ultrastructure.* Belmont, CA.: Wadsworth Publishing Co., 1967. Electron micrographs of many cells and organisms.

Ledbetter, M.C., and K.R. Porter. *Introduction to the Fine Structure of Plant Cells.* Berlin and New York: Springer-Verlag, 1970. A profusely illustrated introduction to plant ultrastructure, with excellent electron micrographs.

Loewy, A.G., and P. Siekevitz. *Cell Structure and Function.* 2nd ed. New York: Holt, Rinehart and Winston, 1969. A good introduction to the cell and how it works.

Watson, J.D. *The Double Helix.* New York: Atheneum Publishers, 1968. A lively, popular account of the work and the people that led to the discovery of DNA as the hereditary material.

—. *Molecular Biology of the Gene,* 2nd ed. Menlo Park, CA.: W.A. Benjamin, 1970. An excellent monograph on the gene written for the beginning student.

# chapter four CELLULAR ENERGETICS THE FIRES OF LIFE

## INTRODUCTION

Being alive uses energy. All organisms require food to live because food molecules are held together by energy. That energy is released when food molecules are broken apart, as during digestion. Where do food molecules come from, and specifically how do living cells harness the energy released from food?

Green plants absorb solar energy and use some of it to build food molecules by the process of **photosynthesis.** Plants, and organisms that feed upon them, use that "food from the sun" to fuel their own internal "fires" via the process of **cellular respiration.** Rice is an important cultivated food plant. The chapter opening art shows women in the Tohoku province of northern Japan shocking and stacking cut rice stalks to dry.

Because it is impossible to share all that is known about metabolism in a few pages, in this chapter we focus primarily on the two great life-sustaining energy transfer processes, photosynthesis and respiration. Anyone desiring different perspectives and greater detail is encouraged to investigate the Suggested Readings at the end of the chapter.

The Laws of Thermodynamics introduced in Chapter 2 also apply to the energy transfers and transformations that occur within cells. In Chapter 2 they were discussed in the context of energy flow through ecosystems. Brought to the cellular level, their context is the flow of energy (electrons) through the metabolic biochemical pathways of the living cell.

## MOLECULAR ENERGY TRANSFERS

### Role of Electrons

The energy exchange reactions that occur within metabolic systems depend on the ability of **electrons** (negatively charged atomic particles) to react to energy changes. Figure 4-1 illustrates a change in energy states by a single electron. When the electron ($e^-$) is energized in some way, the additional energy it absorbs displaces it from its normal orbital level into an orbital level farther from the atom's nucleus. In such a condition, the electron is at a higher energy level because additional energy holds the electron outward against the naturally attractive force between it and the positively charged nucleus. When this extra energy is released, the electron resumes its stable orbital level. The released energy can be released as heat or light or it can be transferred to form chemical bonds, as during the initial stages of photosynthesis.

Many of the chemical reactions of photosynthesis and respiration involve electron transfers, called reduction and oxidation. A **reduction** involves the *addition* of electrons ($e^-$), which may then attract positively charged protons or hydrogen ions ($H^+$). Electrons (and hence also protons) are *removed* during **oxidation**. So for practical purposes, it is customary to describe reduction as the addition of $H^+$ and oxidation as the removal of $H^+$—the net results. Oxygen is usually the eventual electron acceptor, although many oxidation reactions are intermediate, not involving oxygen directly.

### Role of ATP

Some of the energy released during metabolic processes is captured in the bonding that converts **adenosine diphosphate (ADP)** to **adenosine triphosphate (ATP)** molecules. Structural and functional aspects of ADP and ATP are summarized in Fig. 4-2. ATP is a "high-energy" molecule because it is highly reactive and readily transfers its terminal phosphate-containing group to other molecules, releasing energy in the process. By analogy, you could consider each ATP molecule to be an energy "bank," holding a "currency unit" of energy to be transferred during cellular metabolism.

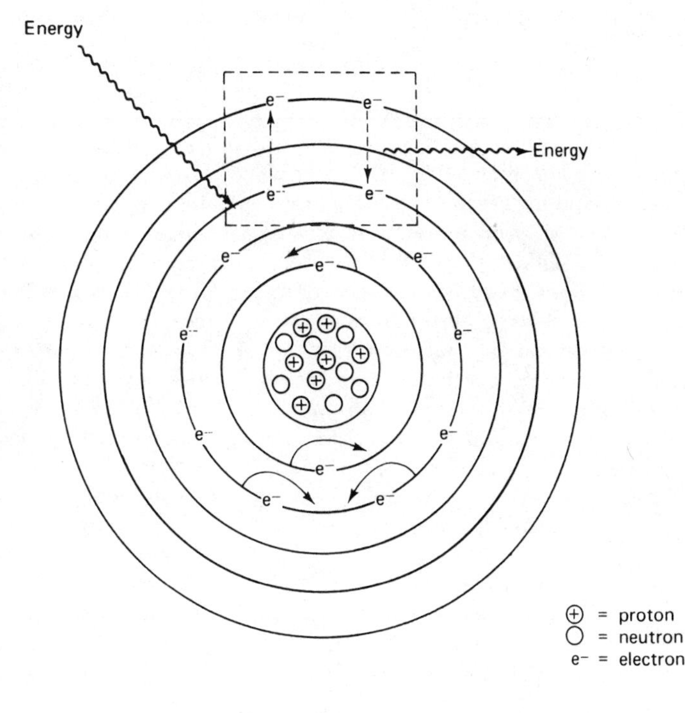


**FIGURE 4-1**
An atomic model, showing **electrons** in orbitals around the atom nucleus, which is composed of **protons** and **neutrons.** The outlined box demonstrates changes in energy states by an electron ($e^-$). An electron is in a normal stable state in an inner orbital. When excited by the addition of energy it moves to an outer orbital. This is a higher energy level, with a stronger repelling force between the electron and the positively charged atom nucleus. Eventually the electron releases energy as it returns to its stable state. The released energy may be in the form of heat, light, or electrical energy; or, as in many metabolic reactions, it may be used to bind atoms together to form molecules. (From G.R. Noggle and G.J. Fritz, *Introductory Plant Physiology*, 1976. Reprinted by permission of Prentice-Hall, Inc., Englewood Cliffs, N.J.)

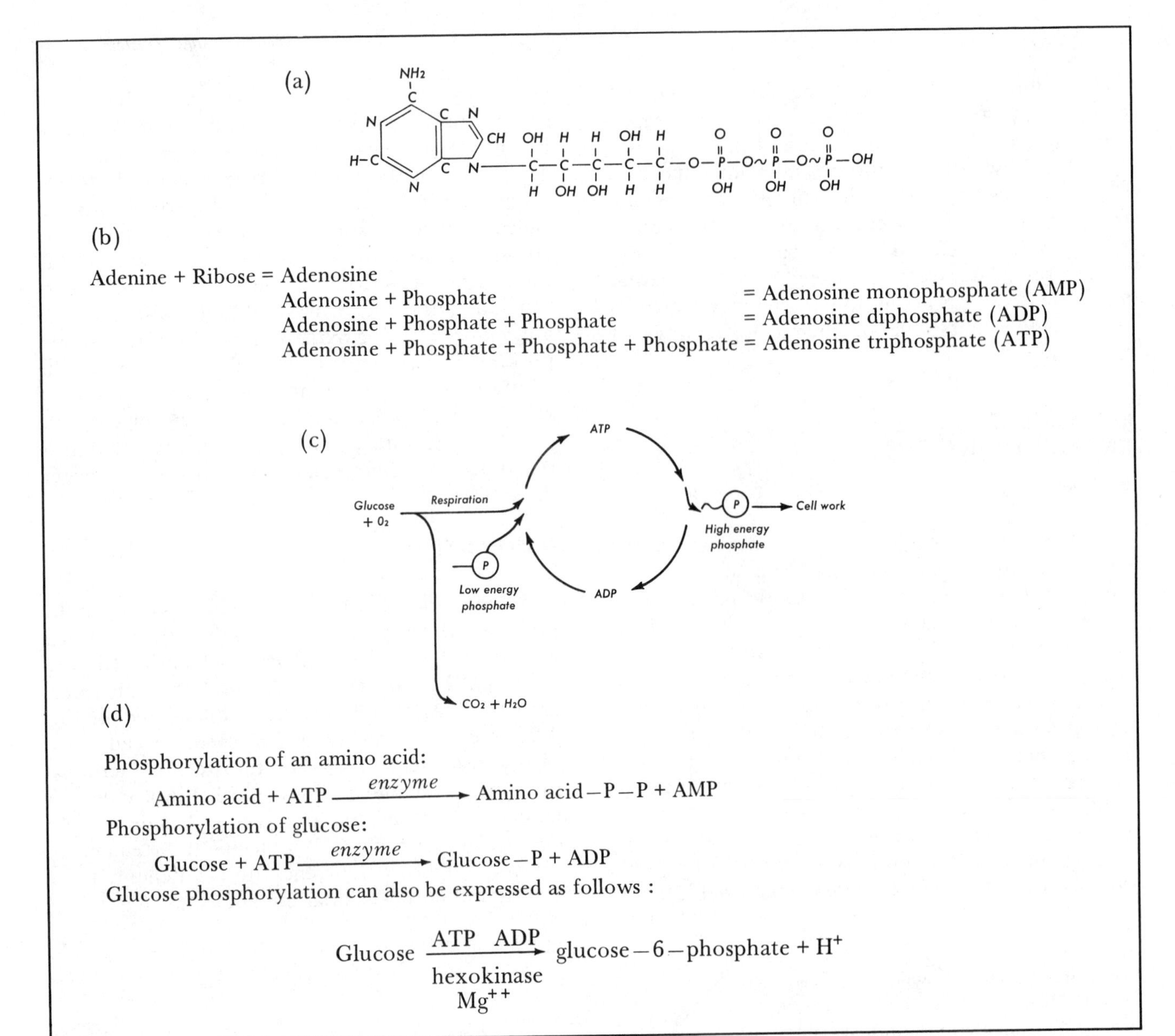


**FIGURE 4-2**
The compound **adenosine triphosphate** (ATP) is of central importance to the energy-requiring reactions of organisms. (a) Chemical structure of ATP, showing the base, sugar, and phosphate components. The linked rings form the base **adenine**, which is also a component of DNA and RNA. The five-carbon chain is the sugar **ribose**. Abbreviations are for **carbon** (C), **oxygen** (O), **hydrogen** (H), **nitrogen** (N), and **phosphorus** (P). Lines between letters indicate chemical bonds. The wavy bonds connecting the terminal phosphate units indicate a high-energy condition. When such bonds are broken, relatively large amounts of energy are released. Because of their high energy content, such bonds are less stable and more easily (spontaneously) broken. (See also Fig. 4-1.)

(b) Word formulas summarize the relationship of ATP to AMP and ADP. All three compounds play roles in metabolic pathways, accepting and contributing phosphate groups ($P_i$) during energy transfers called **phosphorylations.**

(c) Under the influence of specific enzymes, ATP is reduced to ADP by the loss of a phosphate ion, releasing energy to be utilized directly by organisms for chemical processes of maintenance and activity (cell work). The reaction is easily reversible ($ADP + P_i + energy \rightleftharpoons ATP$) allowing great conservation of materials in the energy traffic of cells via continuous recycling, as shown here.

(d) Other examples of phosphorylation. Phosphorylation involves an energy transfer that can activate an otherwise stable molecule—such as glucose or an amino acid—making it more reactive. A word written above or below the arrow of a reaction indicates that the reaction requires the presence of what is named, usually an enzyme. Sometimes inorganic **cofactors** (or organic cofactors called **coenzymes**) also must be present. As an example, glucose phosphorylation is an essential reaction at the beginning of **glycolysis**, the breakdown of glucose during cellular respiration. Hexokinase is the specific enzyme involved. Magnesium ($Mg^{++}$) is a necessary cofactor.

## Role of Enzymes

Some chemical reactions happen spontaneously; that is, whenever sufficient quantities of the reacting substances are present a reaction will take place. Usually, however, even spontaneous reactions occur too slowly to be of use to living systems. So almost all reactions of the metabolic processes of plants and animals occur under the control of specific **enzymes.** Enzymes are proteins that catalyze (Gr., *kata* = down; *lyein* = to loosen) reactions by reducing the amount of energy that must be present to cause the reactions to happen (activation energy). Enzymes function in all manner of reactions, whether the process is putting together, taking apart, or rearranging components.

Enzymes have certain functional characteristics.

1. They are specific, each catalyzing a particular chemical reaction.
2. An enzyme is not destroyed by the reaction; so it can be used repeatedly.
3. Enzymes do not alter the products of reactions, only the rates.
4. Each enzyme functions best within certain ranges of environmental conditions, having an optimum temperature and pH. Enzymatic reactions are slowed by low temperature, but the enzymes themselves are not destroyed. High temperatures and pH extremes do destroy (denature) enzymes.
5. Finally and significantly, all enzymes are proteins, composed of polypeptide chains.

Proteins are among the longest molecules, their structure is complicated by twisting and folding, and there are cross-connections between portions of the molecule (Fig. 4-3). Therefore each specific protein has a particular shape and surface configuration, a key to enzyme specificity. The structure of some enzymes includes a small, nonprotein part (prosthetic group). Others consist of two distinct parts that can separate from one another—the protein part (apoenzyme) and a smaller inorganic or organic part (cofactor or coenzyme). Both parts, however, must be present for the enzyme to function.

**FIGURE 4-3**
Molecular configuration of a hypothetical **complex protein molecule.** The folding and formation of cross-bridges gives such molecules distinctive topographic properties.

The mechanism of enzyme function is based on the physical attachment of each enzyme molecule to its substrate, the substance on which it acts. An enzyme may be many times larger than its substrate molecule. Enzyme activity is inhibited by anything that blocks the formation of the enzyme-substrate complex. Each substrate requires the action of a specific enzyme, a specificity caused by the fitting together of enzyme and substrate molecule surfaces. By analogy, their surfaces must fit together like adjacent (but three-dimensional) jigsaw puzzle pieces.

Often more than one enzyme is necessary to complete a particular reaction. For example, when a large and complex protein molecule is dismantled, specific enzymes work only on the ends of protein chains; others work only on internal chemical bonds.

Enzyme synthesis and enzyme activation are the keys to the cell's selective control of chemical reaction pathways. You will recall that a cell's DNA controls enzyme synthesis (Chapter 3). In this way DNA controls which of the many potential chemical reactions that could occur within cells actually do occur and when.

Enzymes are named according to their functions. The name may indicate the specific substrate on which the enzyme acts, the general category of function, or the type of chemical bond on which the enzyme works. Enzyme names usually end in the suffix *-ase.* Proteinase, for instance, is a general term for an enzyme that breaks down proteins into smaller protein fragments or (eventually) amino acids; lipases work on lipids (fats and oils); enzymes that break down carbohydrates are called carbohydrases; and phosphatases remove phosphate from various molecules. Beyond such general terminology, each enzyme is named specifically according to its function. DNA synthesis, for example, requires the presence of the enzyme DNA polymerase; RNA synthesis requires the presence of RNA polymerase; the release of energy from ATP requires the presence of ATPase.

### Cofactors and Coenzymes

Certain enzymes will not function unless a particular loosely associated but necessary nonprotein part is present. Such a part is called a **cofactor** or enzyme activator. The metals magnesium, zinc, iron, and copper are some inorganic cofactors required by various oxidative enzymes during the reactions of photosynthesis and respiration.

Organic cofactors are called **coenzymes.** Many vitamins function as coenzymes, including thiamin, riboflavin, nicotinic acid, and pyridoxine. Plants produce a wide range of vitamins whereas animals possess a more limited ability to synthesize vitamins. So animals must depend on vitamins formed by the plants on which they feed, and vitamins produced by microorganisms that live in their digestive tracts. Some enzymes require that both inorganic and organic cofactors be present at the correct sites on the protein before the enzymes can be active.

During photosynthesis and respiration two coenzymes have a major role in reduction and oxidation reactions. These two substances are nicotinamide adenine dinucleotide (**NAD** or **coenzyme I**) and nicotinamide adenine dinucleotide phosphate (**NADP** or **coenzyme II**). Even a "bare bones" discussion of photosynthesis and respiration must include mention of these coenzymes. NAD functions primarily in degradative removal of hydrogen ions, as during respiration. NADP functions most frequently in synthetic reactions, such as photosynthesis.

## FOOD MANUFACTURE: PHOTOSYNTHESIS

$$\underset{\text{carbon dioxide}}{6\ CO_2} + \underset{\text{water}}{6\ H_2O} + \text{light energy} \xrightarrow[\text{and enzymes}]{\text{chlorophyll}} \underset{\text{glucose}}{C_6H_{12}O_6} + \underset{\text{oxygen}}{6\ O_2\uparrow}$$

The sun's radiant energy, via green plants, is the main source of energy to operate life systems on Earth. Only chlorophyll-containing organisms can absorb light radiation and convert it into chemical energy to be used by other living organisms. In its simplest terms, this capture and conversion of energy involves the *absorption of light energy and its use to combine carbon dioxide and water to form the simple sugar glucose.* The foregoing formula is balanced to show the *net* results of photosynthesis. (Actually, 12 molecules of $H_2O$ are used, while 6 are generated; thus the simplified equation shows a *net* use of 6 $H_2O$.)

This simplistic view of photosynthesis belies the innate complexity of the process. Hundreds of individual enzymatically controlled chemical interactions occur in a precise and orderly sequence from the moment light energy bombards and is absorbed into the photosynthetic pigments of green plants until glucose is eventually generated. The entire process takes place extremely rapidly.

Other byproducts of photosynthesis include organic phosphates, some amino acids, and organic acids. Sugar-phosphate combinations are early products of photosynthesis, are often converted to sugar-nucleotide combinations (e.g., ADP-glucose), and are then used to produce sucrose, starch, cellulose, and other organic compounds. Figure 4-4

**FIGURE 4-4** This *highly* simplified diagram shows the flow of energy (thicker line) during **photosynthesis,** from solar units to carbohydrate chemical bonds. Note that the hydrogen (H), carbon (C), and oxygen (O) atoms in the carbohydrates originate as parts of water ($H_2O$) and carbon dioxide ($CO_2$) molecules. The oxygen ($O_2$) that is liberated comes from water molecules. See Supplements 4-1 and 4-2 for more detailed diagrams of the actual reactions involved.

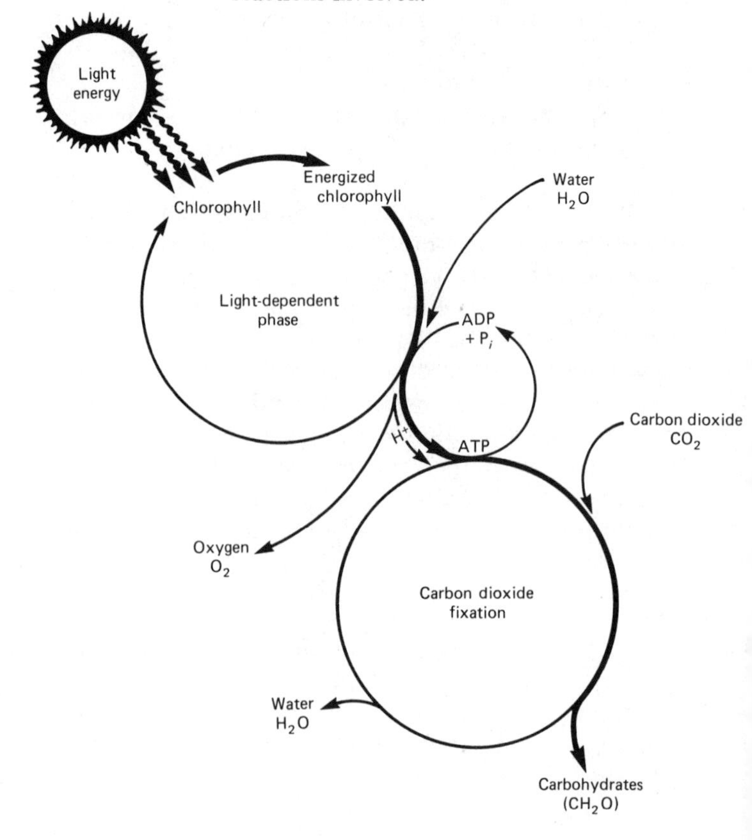

follows the path of energy transfers during photosynthesis.

When information is summarized for simplification and clarification, misconceptions easily arise. At the very least, it should be remembered that neither plants nor animals can utilize light energy *directly* for their own metabolic processes. All depend on the primary product of photosynthesis, sugar, to remain alive.

### Light-Dependent Phase

You may already know that light is composed of radiations varying in wavelength. Although visible light (as detected by the human eye) is interpreted by the brain as white, it can be broken down into a rainbow-like color continuum by passage through a prism. The different colors perceived are the result of different wavelengths reaching the color receptors (cones) within the retinas of our eyes.

Photosynthetic plants can best use wavelengths in the *violet-to-blue* and *red* areas of the spectrum because of the ability of the various photosynthetic pigments to capture these wavelengths. Humans perceive chlorophyll-containing plant parts as green because the pigments do *not* absorb wavelengths in the green portion of the spectrum to any great extent but instead reflect and transmit them.

The function of cellular pigments, particularly chlorophyll, is to absorb light energy. The layered organization of chlorophyll molecules within the chloroplasts enhances light-trapping efficiency. Recent research indicates that individual chlorophyll molecules act as groups of several hundred molecules. Such groups are called "photosynthetic units" or "light traps" and are part of the chloroplast membrane organization (thylakoids, Chapter 3). Light traps absorb and pass along light energy units or **photons**.

During the daytime plants are bombarded by more light energy than they can possibly absorb let alone transform into sugar. The efficiency of energy trapping, however, is enhanced somewhat by an ability to trap and store energy quickly as intermediate compounds, which are used to complete photosynthesis later. ADP is converted to **ATP**; the coenzyme NADP is reduced—accepting both electrons ($e^-$) and hydrogen ions ($H^+$)—to become **$NADPH_2$**. Supplement 4-1 describes the reactions leading to ATP and $NADPH_2$ formation during this **light-dependent phase** of photosynthesis.

### Carbon Dioxide Fixation

After exposure to light the chloroplasts contain ATP and $NADPH_2$ and are ready for the next phase, **carbon dioxide fixation**, which does not depend on the continued presence of light and hence is sometimes called the "dark" phase of photosynthesis.

In the light-dependent phase two raw materials are used—light and water. In carbon dioxide fixation $CO_2$, ATP, and $NADPH_2$ are utilized to produce such simple sugars as six-carbon glucose. In addition to "raw materials," however, factories also need "equipment" to create a product. In chloroplasts this "equipment" consists of many chemical compounds that are constantly reused (enzymes) or regenerated (such as the five-carbon sugar, ribulose).

Carbon dioxide fixation can occur via several pathways. Most plants produce a three-carbon compound, phosphoglyceric acid, as the first apparent product in carbon dioxide fixation. Some, however, produce four-carbon oxaloacetic acid first. A third kind of pathway is exhibited by some succulent plants. The details of these pathways and their distribution among plant groups are included in Supplement 4-2, which also describes a phenomenon called photorespiration.

### Starch Formation

As photosynthesis continues, glucose-phosphates (unstable, or reactive sugar-phosphate compounds) are transformed almost immediately into more complex carbohydrates. Sugars are readily soluble in water and an accumulation of high sugar concentrations during photosynthesis would create osmotic balance problems for cells. Therefore sugar is converted rapidly to insoluble starch within the chloroplasts, at least in higher plants. Once active photosynthesis has ceased, the starch can be broken back down to sugars for transport to other parts of the plant. There it may be resynthesized to starch or some other storage product or be utilized in the making of any of the plant's structural and functional organic compounds.

The net result of photosynthesis is cellular **food**—carbohydrates in whose chemical bonds a portion of solar energy has been trapped. These carbohydrates are the building blocks and energy source for living things, from the plants that manufacture them to the consumer organisms that ingest them. Through elaborate chemical processes, these carbohydrates are also refashioned into thousands of unique secretions, byproducts, vitamins, and other plant components for which humans have devised uses (Chapters 16, 17, and 18).

## FOOD UTILIZATION

Plants utilize the products of photosynthesis in several ways:

1. Cellular respiration, releasing energy for all the activities of the cell.
2. Synthesis of structural and functional organic compounds.
3. Synthesis of such food storage products as starch or oils.
4. Synthesis of such nonfood substances as latex, resins, oils, and alkaloids, many of which are of unknown value to the plants but of economic value to humans.
5. Motility, in a few cases (some species of bacteria, blue-green algae, eukaryotic algae, and some plant gametes).
6. **Bioluminescence**, the emission of light energy, as by certain bacteria, algae, and fungi.

Digestion is a first step in food utilization and breaks down complex carbohydrates, releasing the simple six-carbon sugar, glucose. Cellular respiration involves the disassembly of glucose, releasing energy for cellular activities and liberating carbon dioxide and water.

### Digestion (Hydrolysis)

**Digestion** is the overall process of enzymatically dismantling complex molecules into smaller ones—for example, starches to sugars, proteins to amino acids. Because this process involves addition of water, a more exact term to describe digestive reactions is **hydrolysis** (*hydro* = water, *-lysis* = disintegration). Hydrolysis may occur within the cell (intracellularly) or in the medium outside the cell (extracellularly). The vast majority of plant species rely mainly (if not entirely) on **intracellular digestion**).

Carnivorous plants, bacteria, and fungi utilize **extracellular digestion.** Digestive (hydrolytic) enzymes are released into the substrate outside the cells and digest large food molecules into molecules small enough to permeate or be transported through the cell membrane. Once inside a cell, the products of digestion may be subjected to further hydrolysis (intracellular digestion).

### Respiration

**Cellular respiration** releases energy from sugar. If oxygen is utilized during the process, respiration is said to be **aerobic.** Aerobic respiration is typical of higher plants and animals; its byproducts are carbon dioxide and water. Respiration occurring in the absence of oxygen is called **anaerobic.** The most common type of anaerobic respiration in plants is **fermentation,** which results in formation of various alcohols and organic acids. The distinctive flavor and aroma of aged cheeses, for instance, are created by the byproducts of bacterial metabolism.

It is probable that anaerobic respiration via fermentation evolved earlier than either photosynthesis or aerobic respiration, perhaps in an oxygen-free atmosphere or environment. According to this hypothesis, the earliest organisms were heterotrophic, scavenging on organic compounds in their aqueous medium. When photosynthetic pathways developed, the reactive gas oxygen was released in greater abundance than previously. Its oxidative or burning ability was harnessed in the metabolic machinery of certain organisms. Gradually an increased ability to utilize oxygen for efficient respiration—aerobic respiration—developed. Thus more potential energy could be released from food. Part of this evolution was the development of molecules with an affinity for oxygen—the iron-containing **cytochromes** and their animal descendants, the hemoglobins.

In both fermentation and aerobic respiration the chemical bonds of glucose molecules are broken in *stepwise* fashion by a series of enzymatically controlled reactions (Fig. 4-5). Thus much of the released energy is trapped by phosphorylation of ADP to ATP, although some energy is also dissipated as heat. As in photosynthesis, certain raw materials from the cellular environment are used as well as a number of constantly recycled components of the respiratory apparatus, such as enzymes, coenzymes, and electron acceptors. The initial series of glucose-splitting reactions is called **glycolysis** and is the same for both fermentation and aerobic respiration.

#### Aerobic Respiration

During **glycolysis** a six-carbon molecule of glucose is broken into two molecules of three-carbon pyruvic acid (pyruvate). The next stage, the **Kreb's cycle** or **citric acid cycle**, accomplishes further disassembly of the pyruvic acid molecules

**FIGURE 4-5**
The overall process of **aerobic respiration** is superficially similar to combustion, in that oxygen is used to "burn" (oxidize) organic matter, releasing water, carbon dioxide, and energy. In combustion, however, the energy-releasing reactions involve heat that would be lethal to living cells. Therefore the reactions of cellular respiration are assisted (catalyzed) by **enzymes,** allowing the reactions to occur at safe temperatures and a measured pace. Energy is trapped in **ATP molecules** instead of all being lost as heat and light, as with a fire. For more details see Supplement 4-3.

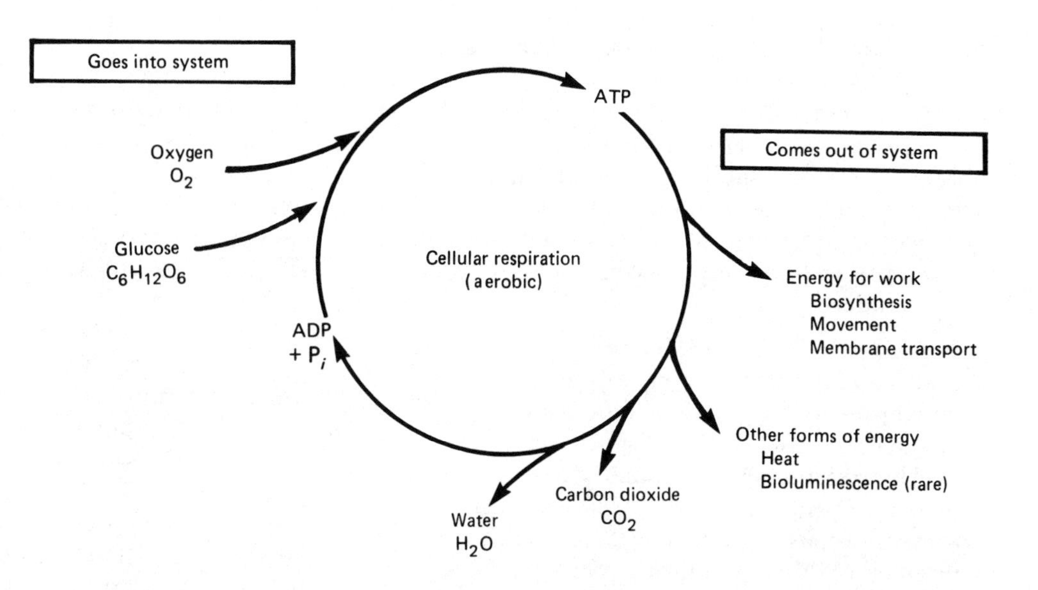


through a cyclic reaction series, releasing carbon dioxide.

The final stage is called **oxidative phosphorylation,** referring to both the combination of oxygen with electrons and hydrogen ions to form water and the phosphorylation of ADP to ATP. Each of these three major phases of typical aerobic respiration is summarized in Supplement 4-3. Table 4-1 summarizes the results of aerobic respiration, using one molecule of glucose.

**TABLE 4-1 The overall balance sheet for aerobic respiration.**

A SIMPLIFIED BALANCED EQUATION DEPICTING THE NET RESULTS OF AEROBIC RESPIRATION

$$\underset{\text{glucose}}{C_6H_{12}O_6} + \underset{\text{oxygen}}{6\,O_2} \xrightarrow{\text{enzymes, etc.}} \underset{\text{carbon dioxide}}{6\,CO_2} + \underset{\text{water}}{6\,H_2O}$$

+ energy (*38 ATP and heat*)

| IRRETRIEVABLY USED | LIBERATED |
|---|---|
| 1 **glucose** molecule | 6 **carbon dioxide** molecules |
| 6 **oxygen** molecules | |

BOTH USED AND GENERATED

**water**
12 molecules formed
− 6 molecules used

6 molecules net gain

**ATP** ($ADP + P_i \rightleftharpoons ATP$)
2 molecules formed during glycolysis
2 molecules formed during the Krebs cycle
36 molecules formed by electron transport system
40
− 2 molecules used during glycolysis

38 molecules net gain

It must be remembered that the myriad individual chemical reactions involve many enzymes, coenzymes, cofactors, and intermediate compounds, which are economically recycled many times. Also, you should think of respiration as a series of reductions and oxidations with oxygen as the final electron (and therefore $H^+$) acceptor. Of the three phases of aerobic respiration, oxidative phosphorylation accounts for the greatest capture of energy as ATP, through a cyclic series of reduction-oxidation reactions, the **electron transport system** (Supplement 4-3).

You will recall that mitochondria are sites of cellular respiration. Biochemical activities within a mitochondrion are segregated by its membrane surfaces and the compartments they create (Chapter 3).

1. The **inner matrix** contains soluble enzymes utilized in the citric acid (Krebs) cycle; enzymes for breaking down fatty acids; ribosomes, DNA, and substances associated with protein synthesis.
2. The **inner membrane** produces enzymes that are needed in the breakdown of pyruvate to enter the citric acid cycle. In addition, the inner membrane is lined with membrane-bound enzymes in the form of tiny units called respiratory assemblies. The membrane-bound enzymes on the inner membrane are associated with oxidative phosphorylation.
3. Both the **outer membrane** and the **intermembrane space** also contain specific enzymes.

#### Alternatives to Aerobic Respiration

Just as there is more than one avenue for photosynthesis, several alternatives to aerobic respiration exist. The three most common alternatives are discussed here: the pentose (pentose phosphate) shunt, alcoholic fermentation, and lactic acid formation. Anaerobic conditions *must* be present for both alcohol and lactic acid formation, for the presence of oxygen will cause pyruvic acid to be oxidized.

*Pentose Shunt.* In one sense, this pathway is actually an alternative to glycolysis in that glucose is converted to a series of intermediate compounds that can enter into aerobic respiration.

The pentose shunt probably has more significance than just being an alternative to glycolysis, however, because the $NADPH_2$ formed is active in a number of biosynthetic activities, such as those involved in lipid metabolism. Thus it is a means of producing $NADPH_2$ by direct utilization of glucose. In addition, this series of reversible reactions involving 2-, 3-, 4-, 5-, 6-, and 7-carbon compounds provides a resource pool for carbohydrate metabolism. See Supplement 4-4 for a diagrammatic summary of the pentose shunt.

*Alcoholic Fermentation.* Glycolysis is the first stage of fermentation. Afterward the way in which pyruvic acid is disassembled differs from aerobic respiration so that the overall results of fermentation differ greatly from those of aerobic respiration. During fermentation some carbon dioxide is liberated, but most of the potential energy remains locked up in chemical bonds of the carbohydrates formed (usually alcohols and organic acids). So fermentation is not as efficient as aerobic respiration and about 80% of the chemical energy remains locked up. A calorie guide for alcoholic beverages (Table 4-2) illustrates this situation. Fermentation by microorganisms has many commercial applications (Chapter 16).

**TABLE 4-2 Calories in one ounce of distilled spirits (whiskies, gin, vodka, rum, and others with no added sugar).**

| | |
|---|---|
| 80 proof[a] | 67 calories |
| 84 proof | 70 calories |
| 86 proof | 72 calories |
| 90 proof | 75 calories |
| 94 proof | 78 calories |
| 97 proof | 81 calories |
| 100 proof | 83 calories |

[a]The "proof" of a distilled beverage is a measure of alcohol content and represents approximately double the alcoholic percentage; so 100 proof spirits have about 50% alcohol content.

The reactions of fermentation are summarized in Fig. 4-6. Notice that two alternatives exist after glycolysis. One sequence for the utilization of pyruvic acid is **alcoholic fermentation,** in which carbon dioxide is liberated and the electrons and $H^+$ ions from $NADH_2$ are transferred to a two-carbon intermediate substance, resulting in the formation of ethanol (ethyl alcohol).

Alcoholic fermentation by living yeast cells has both economic and social importance. The texture of raised-dough baked goods is due to formation of $CO_2$ bubbles, which cause the dough to "rise." When the dough is baked, it hardens around the $CO_2$ bubbles and forms a three-dimensional latticework. The volatile alcohol evaporates during baking. The tantalizing fresh bread odor is that of the yeast cells themselves. Fermented beverages of all types owe their alcoholic content and natural effervescence to these chemical reactions. The carbon dioxide bubbles (carbonation) initially present in wine and distilled beverage mashes are removed during filtration or distillation but are retained in (or restored to) beers and "sparkling" wines. Alcoholic fermentation also occurs to some extent in the tissues of complex plants and animals when they are subjected to anaerobic conditions, as when the respiration rate exceeds oxygen availability.

Interestingly, alcoholic fermentation tends to be a self-limiting process. The conversion of fermenting fruit mash into wine or of apple cider into hard cider (applejack) continues until the alcoholic content reaches a point at which the alcohol-forming yeast can no longer tolerate it. Then other microorganisms may take over, converting the alcohol into organic acids. The result is wine vinegar or cider vinegar, both of which are fine for salads and pickling but can be somewhat of a disappointment to a home winemaker!

*Lactic Acid Formation.* A second type of anaerobic respiration, **lactic acid formation,** occurs in some algae, many higher plants, and the milk-souring bacteria, *Lactobacillus.* Electrons and $H^+$ ions from $NADH_2$ are transferred to the pyruvate molecule, transforming it into lactic acid.

This reaction series is used to make a variety of dairy products, including yogurt and sour cream. It is also used during the initial stages of making certain types of cheeses as well as the fermentation of shredded cabbage into sauerkraut. Lactic acid formation also occurs in animal muscle cells when oxygen demand exceeds the oxygen supply, as during strenuous forms of exercise valued for their "aerobic" aspects, creating an "oxygen debt."

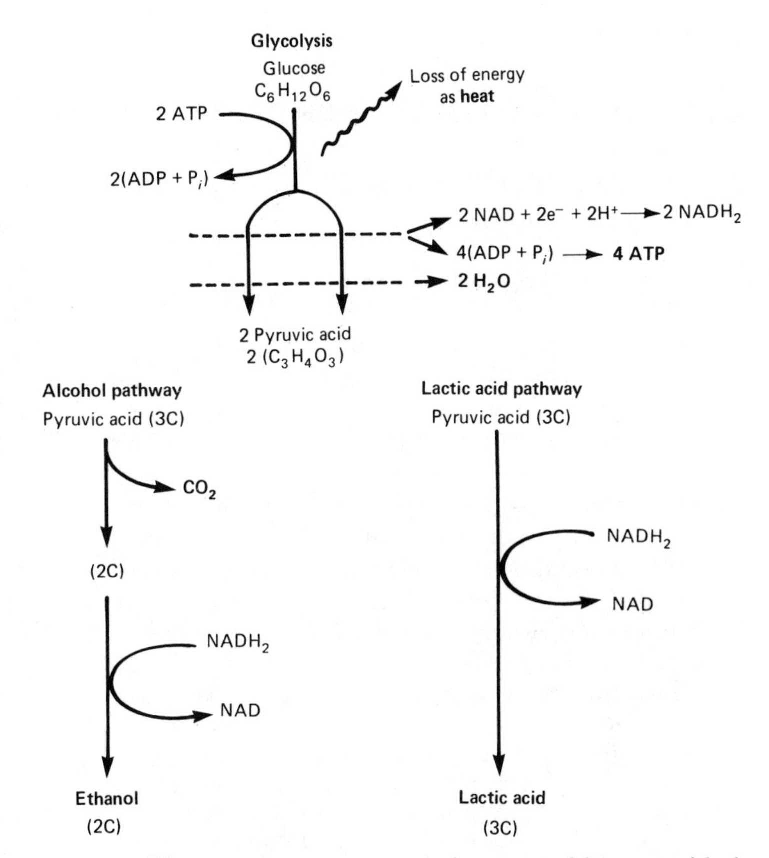


FIGURE 4-6
There are two major pathways of **anaerobic** (oxygenless) **respiration** that can follow glycolysis. In the left pathway pyruvic acid is broken down by the release of carbon dioxide to a two-carbon intermediate substance, which accepts the 2 $e^-$ and 2 $H^+$ carried by $NADH_2$ from glycolysis. The resulting compound is **ethanol (ethyl alcohol)**. In the right pathway pyruvic acid is converted directly into **lactic acid** by the acceptance of 2 $e^-$ and 2 $H^+$ from $NADH_2$. It should be stressed that these reactions are not allowed to occur if sufficient oxygen is present, and that the lactic acid pathway is a reversible reaction. When oxygen becomes available, the lactic acid can be converted back to pyruvic acid, which then enters the citric acid cycle (Krebs cycle) for the completion of respiration via aerobic pathways. Note also that there is a net gain of only two ATP molecules, formed during glycolysis. This represents only about 2.5% of the total energy in the glucose molecule; most of the energy remains locked up in chemical bonds of either ethanol or lactic acid molecules.

An interesting side story to lactic acid formation by *Lactobacillus* arises from the fact that the predominant sugar in milk is lactose and lactose must be digested to glucose (and galactose) prior to glycolysis. Because the first natural food of baby mammals is maternal milk, normal infants are born with the ability to synthesize lactose-digesting enzyme (lactase). In most mammals the young are gradually weaned and lactase production is inhibited so that the ability to digest milk decreases. The human species is often an exception in that many humans continue to drink milk throughout their lifetimes, transferring as children from human milk to that produced by some other mammals.

Milk is high in protein, bone-strengthening minerals, and usually is fortified with added vitamins A and D. When dried cow's milk has been used in hunger-relief programs, it has done literally wonders for malnourished children. Adults in assisted areas, however, have sometimes experienced difficulty because of not drinking milk regularly. Unable to digest milk properly, they developed diarrhea. As a result, powdered milk has occasionally been used to whitewash houses, being obviously "inedible." *Lactobacillus* bacteria *are* able to digest milk, however, and convert it to milk products, such as yogurt and simple cheese, through the lactic acid type of anaerobic respiration. Because these milk products are digestible to humans, perhaps suitable instructions for using powdered milk to make yogurt and cheese could be provided so that the nutrients of milk are available to all age groups receiving such assistance.

## Other Products of Respiration

It would be inadequate to terminate our discussion of respiration without emphasizing that, in addition to energy, many molecular fragments are released for incorporation into essential structural and metabolic compounds. Moreover, some molecular fragments are used to form oils, toxins, and miscellaneous substances, many of which are of no known direct benefit to the plant itself.

Specifically, the breakdown of starch to glucose and glucose to pyruvic acid releases carbon compounds that are used to construct such cell wall components as cellulose and pectins and to construct a variety of compounds, including anthocyanins, lignins, and growth hormones. The conversion of pyruvic acid yields compounds used in the synthesis of fatty acids, cutin, carotenoid and chlorophyll pigments, some aromatic compounds, and other substances. Byproducts of the Krebs cycle are utilized in the synthesis of amino acids and proteins, pigments other than carotenoids (chlorophylls, cytochromes, phytochromes, phycocyanins, phycoerythrins), pyrmidines, and alkaloids.

### The Light Phase: Formation of Intermediate Compounds (ATP and $NADPH_2$)

Figure 4-7 summarizes the flow of electrons and the formation of intermediate energy-holding phosphate compounds that occur while usable light energy is present. This is frequently referred to as the light phase of photosynthesis. The excitation and subsequent reactions caused by absorption of light energy result in formation of high-energy chemical bonds (photophosphorylation), converting ADP molecules to ATP. In addition, the coenzyme NADP (nicotinamide adenine dinucleotide) accepts electrons ($e^-$) and protons (hydrogen ions = $H^+$), becoming transformed into $NADPH_2$.

**FIGURE 4-7** Flowchart of electron transfers and formation of the intermediate compounds ATP and $NADPH_2$ during the light-dependent phase of photosynthesis. The solid N-shaped line indicates the operation of dual photosystems in **noncyclic photophosphorylation** (see text). The broken line designates **cyclic phosphorylation.**

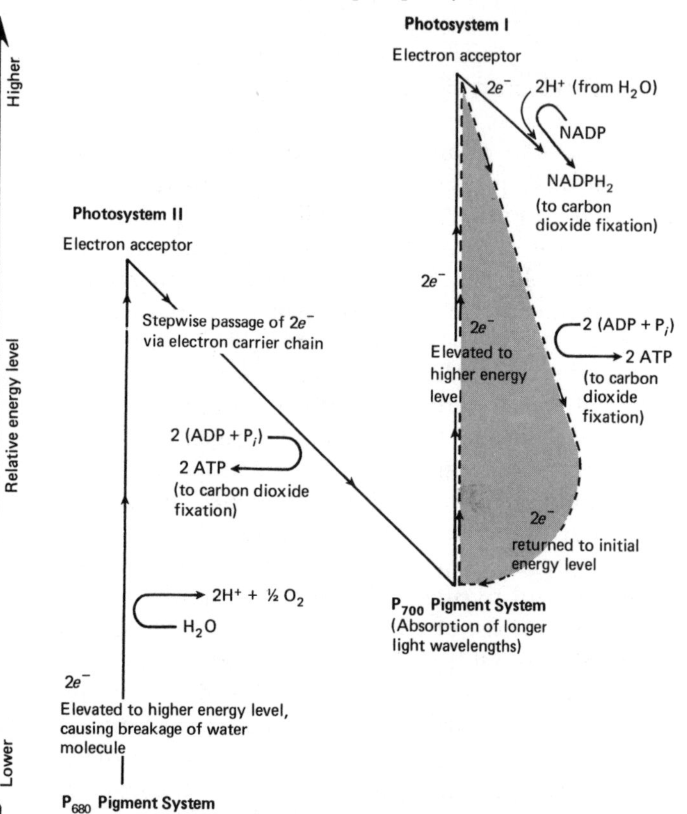


Two types of photophosphorylation are usually involved in the overall photosynthetic process: (a) the kind that occurs by the cycling of electrons from activated pigment molecules and back again and (b) the kind that occurs with electrons originating from a water molecule replacing those liberated from the activated pigment molecules. They are commonly called **cyclic and noncyclic photophosphorylation.**

The terms "Photosystem I" and "Photosystem II" refer to different light absorption systems. In Photosystem I, thought to have evolved first, the pigments absorb longer wavelengths than do those of Photosystem II. Photosystems I and II both utilize complex pigment systems that include chlorophyll *a*, chlorophyll *b*, and other pigments, including carotenoids. Both photosystems are localized on the thylakoid membranes. The product of the two separate but interacting systems is greater effectiveness and efficiency in use of light energy.

The following description is for direct utilization with Fig. 4-7. There are two "starting" positions in Fig. 4-7, indicating the constant and simultaneous bombardment of pigment systems by light energy. In each position units of light energy (photons) are absorbed by pigment molecules, causing two electrons ($2e^-$) to become excited, or elevated to a higher energy state. From this point, the energy held by the electrons could be emitted as heat or light when the electrons return to their stable orbital level. This situation, however, would not result in the synthesis of organic compounds. Instead the electrons become attached to electron acceptor molecules and their energy level is lowered in a stepwise fashion with some of the energy being "tapped off" in the formation of energy-storing ATP bonds. Thus photophosphorylation is an enzymatically controlled, chainlike series of electron-transporting reactions.

Two types of photochemical systems interact during the light-dependent phase of photosynthesis. In Photosystem II of noncyclic phosphorylation elevation of the two electrons to a higher energy state causes a water molecule to split into two hydrogen ions ($H^+$) and an oxygen atom. The two available $H^+$ ions are utilized later during the light phase. The oxygen ultimately unites into a molecular form ($O_2$) and is liberated as a byproduct, not participating directly in photosynthesis. The $2e^-$

are utilized to form two ATP molecules and are accepted at their lowered energy level by the pigment molecules in Photosystem I. Here they presumably replace $2e^-$ transferred to a higher energy level by the absorption of light energy. In a continuation of the noncyclic system, the $2e^-$ will be transferred to a molecule of NADP, giving it a net negative charge or potential that is neutralized by the attachment of $H^+$. Thus the intermediate energy-binding compound $NADPH_2$ is made available for the carbon dioxide fixing phase of photosynthesis. The net result of noncyclic flow from water splitting is formation of ATP, $NADPH_2$ and $\frac{1}{2}O_2$.

A different set of reactions, cyclic photophosphorylation, is indicated by dashed lines. In these reactions, which involve only Photosystem I, the electrons are cycled back to their initial energy level and position within the pigment system. Thus the series of steps involved in this up-down reaction may generate two ATP molecules per $2e^-$. No $NADPH_2$ is formed for subsequent use in carbohydrate synthesis, and no oxygen is liberated.

Although this model is currently acceptable as a proposed scheme of interactions during the light-dependent phase of photosynthesis, many of its details are still unknown. The scheme will undoubtedly be modified as research continues and may be revised later or even rejected for a more accurate format.

## SUPPLEMENT 4-2: Carbon Dioxide Fixation

### Three-Carbon, Calvin-Benson Photosynthetic Cycle

In the "dark reactions" of photosynthesis carbon dioxide is assimilated in the reductive pentose phosphate pathway—reductive because reducing power from $NADPH_2$ is utilized—and the cycle includes several pentose or five-carbon phosphate compounds. The reactions of this pathway, described by Drs. Melvin Calvin, Andrew Benson, and colleagues at the University of California at Berkeley, occur in the majority of species that have been examined. Working in the 1940s, they identified a three-carbon compound called **phosphoglyceric acid (PGA)** as the first apparent product in the fixation of carbon dioxide in the pathway, more recently termed $C_3$ because a three-carbon acid is the first product of $CO_2$ fixation.

Figure 4-8 is a highly simplified diagram of the four major stages of carbon dioxide fixation by the Calvin-Benson pathway. Each stage is composed of a number of separate, enzymatically controlled chemical reactions. In the simplest terms, phosphorylation (transfer of high-energy phosphate bonds) of ribulose by ATP formed during the light phase energizes the sugar, which then reacts with carbon dioxide and water. In the series

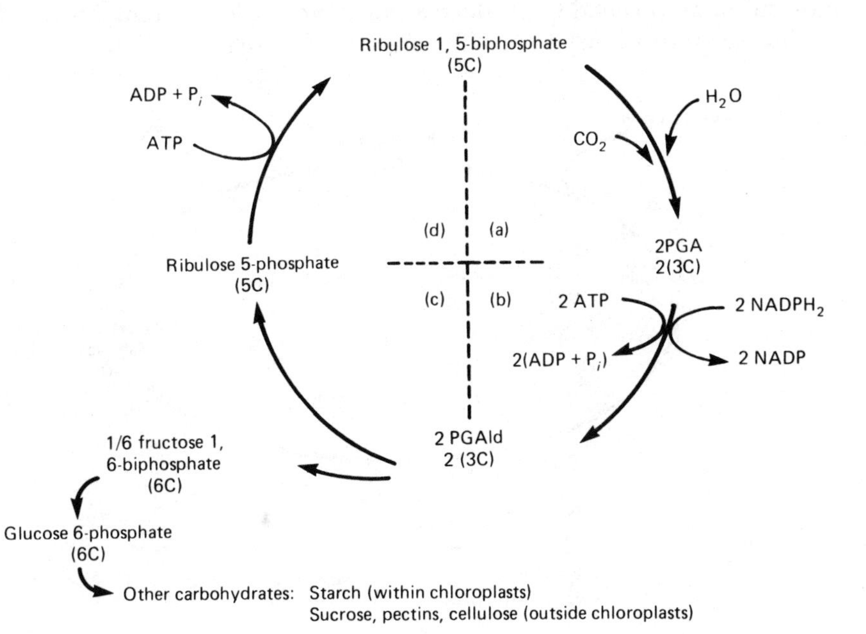


**FIGURE 4-8**
A simplified flow diagram of the four major stages of carbon dioxide fixation found in most plants, the **three-carbon** or **Calvin-Benson** pathway. Subsequent carbohydrate elaboration is also indicated. The stages are (a) Formation of phosphoglyceric acid (PGA). (b) Formation of phosphoglyceraldehyde (PGAld.). (c) Formation of phosphorylated five-carbon and six-carbon sugars. (d) Formation of ribulose 1,5-biphosphate.

of reactions that follows, more energy from ATP is used and $H^+$ ions from $NADPH_2$ are incorporated. The end results are phosphorylated ribulose to replenish that initially used, plus a phosphorylated six-carbon sugar (fructose) that can then be transformed into free sugars, such as sucrose. A more detailed explanation follows.

1. *Ribulose 1,5-biphosphate* (five carbons), sometimes referred to simply as ribulose biphosphate, reacts "spontaneously" (whenever sufficient quantities of materials are present and without additional energy input) with *carbon dioxide* ($CO_2$) and *water* ($H_2O$), in the presence of the enzyme carboxydismutase, to form a six-carbon substance that almost immediately splits into two three-carbon molecules of *3-phosphoglyceric acid (PGA)*.
2. *PGA* combines with *hydrogen from* $NADPH_2$ (formed during the light phase), using *energy from ATP* (formed during the light phase) to form *3-phospho-glyceraldehyde (PGAld.)*.
3. *PGAld.* enters into a series of reactions that results in the formation of several phosphorylated sugars: Illustrated are the five-carbon *ribulose 5-phosphate* and six-carbon *fructose 1,6-biphosphate.*
4. *Ribulose 5-phosphate* is phosphorylated, using *ATP* formed in the light phase, and is thus regenerated for another cycle.

It should be noted that *six revolutions of the cycle,* thus incorporating six carbon dioxide molecules, are needed to produce a *single molecule* of hexose (six-carbon) sugar. Elaboration of starch occurs within the chloroplasts immediately following hexose sugar formation whereas the formation of sucrose and structural carbohydrates (pectins and cellulose) occurs after phosphate compounds (triose, or three-carbon, phosphate) leave the chloroplasts.

In the early stages of photosynthesis, amino acids, organic acids, and other compounds are also produced.

### Four-Carbon Hatch–Slack Pathway

In scientific research there seems to be no happy ending, for each new answer is actually just a new beginning! Thus the processes described by Calvin, Benson, et al., are now known as *one* of the ways that carbon dioxide fixation occurs in plants instead of *the* way. Work with sugarcane in the 1960s indicated that four-carbon compounds (rather than three-carbon PGA) were formed initially during photosynthesis (hence the name $C_4$ plant for such species). M.D. Hatch and C.R. Slack in Australia confirmed this discovery and identified the four-carbon organic acid oxaloacetic acid, formed by the combination of carbon dioxide with a three-carbon compound. Although the beginning is different, the reactions arrive at a stage where carbon dioxide is released from the four-carbon acids (a process of decarboxylation) and then assimilated in the normal reductive pentose phosphate pathway.

Figure 4-9 summarizes the unique portion of the Hatch–Slack pathway (also called the PEP system, for the initial chemical involved, phosphoenol pyruvate).

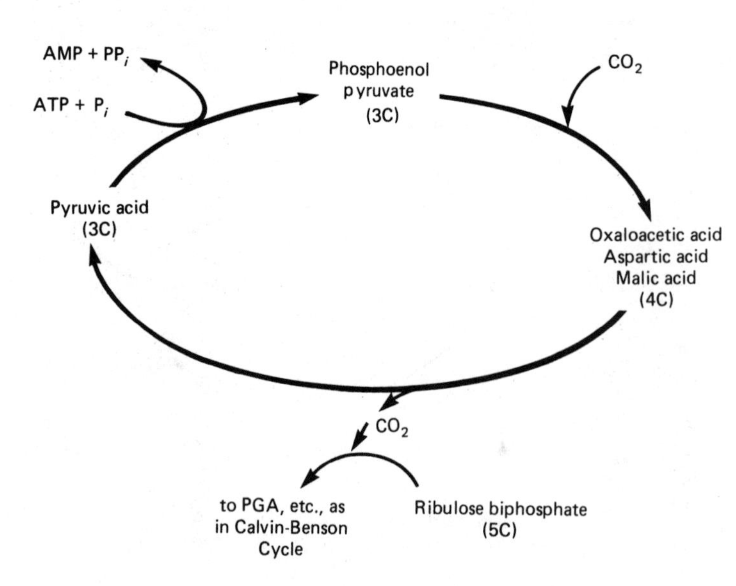


FIGURE 4-9
The **four-carbon Hatch and Slack** photosynthetic pathway, leading to completion of carbon dioxide fixation via the Calvin–Benson pathway. **ATP** is contributed from the light-dependent phase. **Carbon dioxide** is a raw material.

The Hatch–Slack $C_4$ pathway occurs in a number of tropical species in the grass family (Poaceae)—for example, sugarcane (*Saccharum officinale*), corn (*Zea mays*), sorghum (*Sorghum vulgare*), and crabgrass (*Digitaria sanguinalis*). This pathway also occurs in some dicots, such as *Amaranthus* and *Atriplex*. In contrast, such temperate-climate grasses as wheat (*Triticum aestivum*), rye (*Secale cereale*), oats (*Avena sativa*), Kentucky bluegrass (*Poa pratensis*), and creeping bentgrass (*Agrostis palustris*) are $C_3$ plants.

It has been demonstrated that in corn different leaf cells are involved in the $C_4$ system than in the Calvin–Benson ($C_3$) pathway. Carbon dioxide capture via the $C_4$ system occurs in the mesophyll. The four-carbon organic acids (aspartate and malate) derived from oxaloacetic acid are transported to the photosynthetic cells immediately surrounding the vascular bundles for operation of the $C_3$ cycle to complete photosynthesis.

Research shows that, for several reasons, the $C_4$ system is a more efficient pathway for using carbon dioxide under hot, dry environmental conditions. The initial enzyme involved (PEP carboxylase) fixes more efficiently under these atmospheric conditions than its counterparts in the $C_3$ pathway. $C_4$ species have higher temperature optima for photosynthesis and are able to utilize carbon dioxide at lower concentrations than $C_3$ plants. Under hot or dry conditions carbon dioxide is generally the most limiting factor for photosynthesis either because of reduced solubility of the gas at higher temperatures or restricted diffusion of the gas into the leaf as a result of stomatal closure during water stress. The benefit of the $C_4$ pathway is that it concentrates $CO_2$ in the leaf at the site where the four-carbon acids are decarboxylated. Carbon dioxide is then no longer rate-limiting for photosynthesis. Because the concentration of carbon dioxide in our atmosphere is quite low (0.03%), it is rate-limiting for photosynthesis in $C_3$ plants, particularly under the preceding conditions. Dramatic increases in $C_3$ plant growth can be induced by growing them in a carbon dioxide-enriched environment.

#### Crassulacean Acid Metabolism (CAM)

Many succulent plants, such as cacti (Cactaceae) and stonecrops (Crassulaceae), must survive and conduct photosynthesis during extremely dry daytime conditions. To retard water loss the stomates close, limiting atmospheric $CO_2$ availability. To offset this limitation, these species have developed a photochemical system by which stomates are open at night and carbon dioxide is converted to malic and isocitric acids. These organic acids are then broken down during the daytime, when photosynthesis actively occurs, liberating $CO_2$ for carbohydrate formation. This system is known as **crassulacean acid metabolism** (CAM).

#### Photorespiration

A peculiar process called **photorespiration** occurs in $C_3$ plants. Photorespiration is actually counterproductive to the accumulation of carbohydrates during the carbon dioxide fixation phase of photosynthesis. The result is, therefore, reduced efficiency of $C_3$ plants and waste of as much as 50% of photosynthetic carbon dioxide fixation. Photorespiration occurs through metabolism involving the chloroplasts (sites of synthesis of organic compounds) and the peroxisomes and mitochondria (organelles involved in the use of organic compounds).

## SUPPLEMENT 4-3: Aerobic Respiration

This discussion of aerobic respiration corresponds to the diagrammatic summary presented in Fig. 4-10.

*Glycolysis.* The term glycolysis refers to disintegration (lysis) of glucose (glyco-). Glycolysis occurs within the matrix (or ground substance) of the cytoplasm. It involves about ten distinct chemical reactions and at least as many enzymes. Glucose is a stable molecule, meaning that it does not readily disintegrate spontaneously. Thus the first reactions of glycolysis are the transfer of high-energy phosphate bonds from ATP to glucose, changing it into a chemically reactive molecule.

In the succeeding sequence of events, the phosphorylated glucose eventually generates two molecules of a three-carbon substance, pyruvic acid. In these latter stages of glycolysis, some

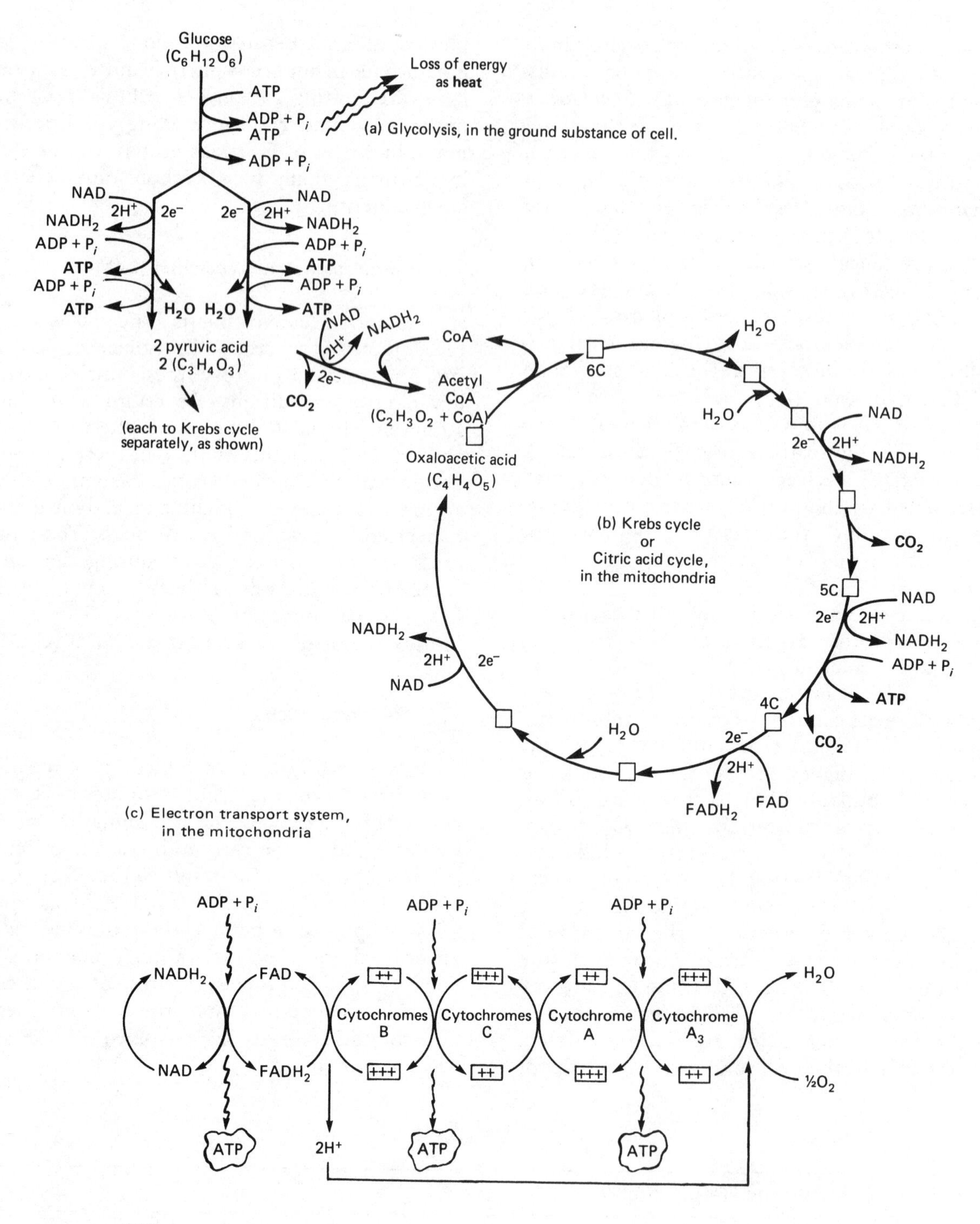


**FIGURE 4-10**

**Aerobic respiration** consists of three stages: **glycolysis, Kreb's cycle (citric acid cycle)**, and **oxidative phosphorylation** via the electron transport system (ETS). Each time NAD or FAD accepts 2 $e^-$ and 2 $H^+$ during glycolysis and the Kreb's cycle, the resulting *reduced* product ($NADH_2$ or $FADH_2$) feeds into the electron transport system. *ATP molecules* are then formed by the stepwise return of the 2 $e^-$ to a lower energy level (a more stable orbital). **Oxygen** (O) is the final electron acceptor, accepting the 2 $e^-$ and the accompanying 2 $H^+$, forming a molecule of **water** ($H_2O$). *The net gain of ATP molecules generated by the utilization of one glucose molecule is 38.* All but two ATP molecules generated during glycolysis and two generated by the Krebs cycle are formed by the ETS. Overall, about *half* the energy in glucose is captured as ATP; the rest is lost as heat or used in the operation of intermediate reactions.

energy is released as heat, some is used in intermediate compound formation, and enough is trapped to convert four ADP molecules to ATP.

Four highly energized electrons ($e^-$) are also emitted. They and their accompanying protons (hydrogen ions, $H^+$) are accepted by the coenzyme NAD (nicotinamide adenine dinucleotide), converting it to $NADH_2$. Whenever $NADH_2$ is formed, it may be "fed into" the interlocking reactions of oxidative phosphorylation, as described subsequently.

*Kreb's Cycle.* A four-carbon organic acid (oxaloacetic acid), a specific coenzyme (Coenzyme A or CoA), and appropriate catalytic enzymes are present in the membrane systems of the mitochondria of respiring cells. They are reusable participants in the Krebs cycle. Pyruvic acid from glycolysis enters the mitochondria and is combined with CoA, forming acetyl CoA, which is the "fuel" of the Krebs cycle. In the process, $CO_2$ is liberated.

With acetyl CoA and appropriate enzymes present, an orderly molecular disassembly occurs. CoA, electrons, hydrogen ions, and carbon dioxide are removed and some energy is trapped to convert ADP to ATP, which occurs during a sequence of transformations from one organic acid to another (indicated by small boxes on the diagram), the first of which is six-carbon citric acid. For this reason, the Krebs cycle is sometimes called the citric acid cycle. The cycle ends with regeneration of oxaloacetic acid, ready to accept the next molecule of acetyl CoA.

The *net results* of the process for *each* pyruvate molecule utilized are removal of three molecules of $CO_2$, formation of one ATP, and reduction of four electron-carrying coenzymes (three $NADH_2$ and one $FADH_2$). Because *two* pyruvate molecules came from the initial glucose, the overall net results equal 6 $CO_2$, 2 ATP, 8 hydrogen-carrying coenzymes. The coenzymes then proceed to the electron transport system phase for generation of ATP by oxidative phosphorylation.

*Oxidative Phosphorylation/Electron Transport System (ETS).* For *each* molecule of the electron and $H^+$ carrying $NADH_2$ or $FADH_2$ produced in the previous stages of aerobic respiration, *molecules of ATP will be generated* by the electron transport system. In the ETS each pair of electrons goes through the system of electron acceptors (respiratory enzymes and coenzymes) by a series of "handoffs," which could be visualized as being analogous to the square dancers' "right-and-left grand." At specific points within the system, sufficient energy is released by the transfer of electrons to *transform ADP to ATP.*

The $H^+$ ions released immediately in the sequence are eventually combined with oxygen to *produce water molecules* ($H_2O$), making oxygen the final electron acceptor. Thus the overall set of reactions in the ETS can be referred to as the oxidation of $NADH_2$ (and $FADH_2$) by oxygen.

Here are the *net results of oxidative phosphorylation* of the products of one glucose molecule: 2 $NADH_2$ from glucose = 6 ATP; 2 $NADH_2$ from breakdown of two pyruvic acid molecules to acetyl CoA = 6 ATP; 6 $NADH_2$ + 2 $FADH_2$ from the Krebs cycle = 26 ATP. In addition, 12 molecules of water are formed.

## SUPPLEMENT 4-4: Respiration—The Pentose Shunt

Figure 4-11 and its caption summarize these reactions. Note that a three-carbon substance is formed that can be altered into pyruvic acid for entrance into the Krebs cycle. Similarly, $NADPH_2$ is formed and can be used to reduce NAD to $NADH_2$ for entrance into the electron transport system and oxidative phosphorylation.

## SUMMARY

When an energized **electron** returns to its stable orbital level, energy is released; this energy can be dissipated as heat or light or can be trapped in chemical bonds. Some of the energy released during photosynthesis and respiration is captured in bonds of **adenosine triphosphate (ATP)**. The terminal phosphate group of ATP may be thought of as an "energy currency unit."

**Enzymes** lower the amount of energy needed to initiate chemical reactions. Each enzyme catalyzes a particular reaction. Enzymes are proteins that are recycled, not destroyed, and that affect the rates but not the products of the reactions. Enzyme functions are directly affected by environmental conditions.

Some enzymes will not function unless a nonprotein **cofactor** is also present. Some in-

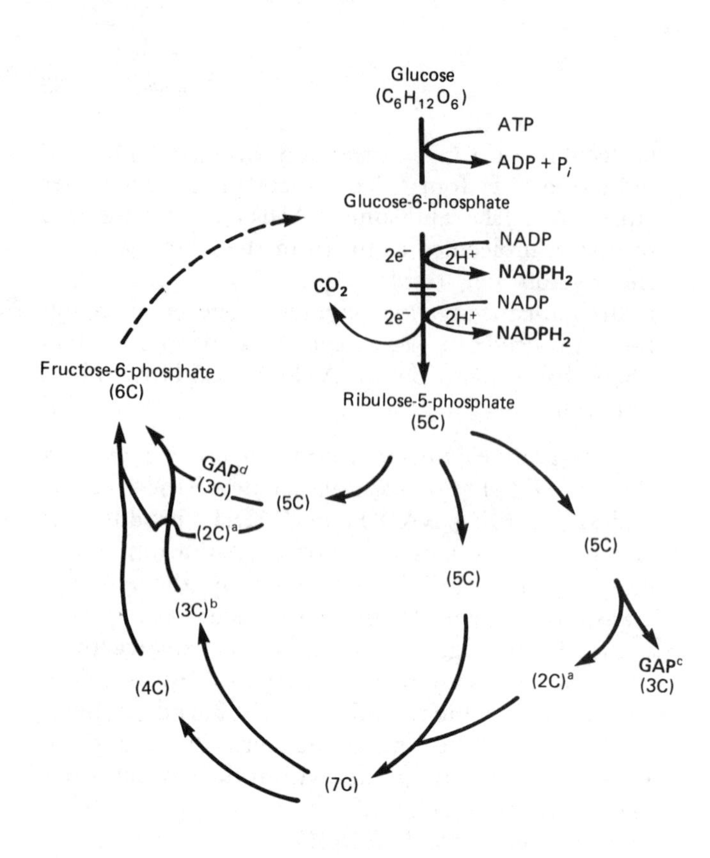


FIGURE 4–11
The **pentose shunt** (pentose phosphate pathway). Instead of being converted directly to three-carbon pyruvic acid, the six-carbon phosphorylated glucose molecule (glucose-6-phosphate) may be altered to a five-carbon molecule (ribulose-5-phosphate), yielding *one molecule of carbon dioxide ($CO_2$) and two molecules of $NADPH_2$*. The $NADPH_2$ can be used to convert NAD to $NADH_2$ for entrance into the ETS; or $NADPH_2$ can be used in such synthetic reactions as fat synthesis. An interlocking series of molecular rearrangements follows, in which molecular *fragments* (indicated here by the number of carbon atoms each transferred group contains) are transferred from molecule to molecule by the enzymes *transketolase*[a] and *transaldolase*[b]. In the process, *one molecule of a phosphorylated three-carbon compound (glyceraldehyde-3-phosphate or GAP) is formed.* The GAP can either be converted to pyruvic acid for entrance into the Krebs cycle[c] *or* be converted to fructose-6-phosphate[d] leading to reentry in the pentose shunt. Both alternatives are shown in this diagram, even though only one may occur at any given time. (Note: The diagram indicates by arrows the flow of chemical reactions in a single direction, as during a "completed" phase. Actually, all the reactions are *reversible* except for the initial phosphorylation of glucose!)

organic cofactors are magnesium, zinc, iron, and copper. Organic cofactors, often vitamins, are called **coenzymes.**

During photosynthesis **chlorophyll**-containing plants combine carbon dioxide, water, and light energy to form carbohydrates.

Plants use light of the violet-to-blue and red wavelengths. **Light traps** within chlorophyll molecules absorb the light energy.

During the **light phase of photosynthesis** energy-holding phosphate compounds are formed. Excitation caused by the absorption of light energy results in the conversion of ADP to high-energy ATP. NADP also accepts electrons ($e^-$) and protons ($H^+$) to form $NADPH_2$.

During **carbon dioxide fixation** carbon dioxide plus ATP and $NADPH_2$ from the light phase are used, forming carbohydrates. There are several pathways of carbon dioxide fixation—the three-carbon, four-carbon, and CAM pathways.

Highly soluble sugars within the chloroplasts are converted almost immediately into insoluble **starch,** which can then be converted back to sugars for transport for storage or use in other parts of the plant.

Plants use the products of photosynthesis for cellular respiration, synthesis of structural and functional compounds, synthesis of food storage products, synthesis of nonfood substances, motility, and, in some species, bioluminescence.

During **digestion** complex molecules are hydrolyzed extracellularly or intracellularly into smaller, usable ones.

**Aerobic respiration** occurs in the presence of oxygen. **Anaerobic respiration,** most commonly **fermentation,** occurs in an oxygen-free environment. During respiration the chemical bonds of glucose are broken in a stepwise fashion, releasing energy in a controlled manner via many enzymatically controlled reactions.

**Glycolysis,** the breaking down of glucose into pyruvic acid, is the first stage of aerobic respiration. In the **Krebs cycle** pyruvic acid is disassembled. **Oxidative phosphorylation** involves the phosphorylation of ADP to ATP and the combining of oxygen and hydrogen to form water. The Krebs cycle and oxidative phosphorylation occur in the mitochondria.

For each molecule of glucose used for aerobic respiration, 38 molecules of ATP are gained.

An alternative to aerobic respiration is the **pentose shunt,** by which glucose is converted to a series of intermediate compounds for use in aerobic pathways and lipid metabolism. It also provides a resource pool of compounds for carbohydrate metabolism.

Glycolysis is also the initial stage of **fermentation.** During fermentation pyruvic acid is disassembled in a less efficient way than in aerobic respiration, for most of the energy remains locked up in organic compounds. Alcoholic fermentation results in the production of ethyl alcohol and carbon dioxide. Lactic acid may also result from anaerobic breakdown of pyruvic acid.

In addition to energy, many molecular fragments are released during respiration for incorporation into essential structural and metabolic compounds.

## SOME SUGGESTED READINGS

Baker, J.J., and G.E. Allen. *Matter, Energy and Life: An Introduction for Biology Students*, 2nd ed. Reading, MA.: Addison-Wesley Publishing Co., 1970. A concise summary of pertinent information for biology students lacking a previous physics or chemistry background.

Bohinski, R.C. *Modern Concepts in Biochemistry*, 3rd ed. Boston: Allyn & Bacon, 1979. A comprehensive college text that focuses on biologically important molecules.

Bonner, J., and J.E. Varner. *Plant Biochemistry*, 3rd ed. New York: Academic Press, 1976. An excellent collection of articles emphasizing the relationship of structure to function.

Dyson, R.D. *Cell Biology: A Molecular Approach*, 2nd ed. Boston: Allyn & Bacon, 1978. A clearly written textbook that emphasizes our emerging knowledge of the dynamic functions and interactions of cell components.

—. *Essentials of Cell Biology*. Boston: Allyn & Bacon, 1975. As the title implies, a less comprehensive treatment of cell biology having a clear summary of the chloroplast and mitochondrion.

Galston, A.W. *The Life of the Green Plant*, 2nd ed. Englewood Cliffs, N.J.: Prentice-Hall, 1964. A brief introduction to plant physiology.

Lehninger, A.L. *Bioenergetics: The Molecular Basis of Biological Energy Transformations*, 2nd ed. Menlo Park, CA.: W.A. Benjamin, 1971. A comprehensive and readable treatment of cellular energetics, including the activities of chloroplasts and mitochondria.

McElroy, W.D. *Cell Physiology and Biochemistry*, 2nd ed. Englewood Cliffs, N.J.: Prentice-Hall, 1964. A brief introduction to the function of cells.

San Pietro, A., F.A. Greer, and T.J. Army. *Harvesting the Sun*. New York: Academic Press, 1967. An interesting and readable treatment of photosynthesis.

White, E.H. *Chemical Background for the Biological Sciences*, 2nd ed. Englewood Cliffs, N.J.: Prentice-Hall, 1970. A summary, available in paperback, to provide a background for the understanding of biological processes.

Zelitch, I. *Photosynthesis, Photorespiration and Plant Productivity*. New York: Academic Press, 1971. Relates photosynthesis within the cell to the leaves and overall productivity of a plant.

# chapter five NEW GENERATIONS OF CELLS AND ORGANISMS

## INTRODUCTION

One characteristic of living things is the ability to reproduce like kind. The basis of reproduction is cellular because all cells are derived from preexisting cells, as stated in Chapter 2. Cells are able to reproduce "like" cells because the genetic heritage of each cell exists as the collection of genes on the DNA of the chromosomes (Chapter 3) and because the chromosomes are transmitted from parent cell to offspring during reproduction. These are the essential facts of reproduction. This chapter deals with the reproduction of both cells and whole organisms.

Each plant species carries in its nuclei a specific set, or **complement,** of different kinds of chromosomes—that is, chromosomes that carry genes for different characteristics. The number ($n$) of chromosomes in a complement varies among species from 1 to 500+, but in most species $n$ equals some number between 3 and 50.

In higher plants chromosomes exist as matched, or **homologous,** pairs; that is, both members of each pair have the *same linear sequence of genes* for specific characteristics, Each pair represents one chromosome from each parent. When a cell contains paired chromosomes, it is said to be **diploid** (double) in chromosomal complement. In contrast, cells that contain only one member of each chromosome pair (chromosomes present in unpaired condition) are said to be **haploid** (single) in chromosomal complement. Diploid and haploid are indicated by the symbols $2n$ ($2 \times n$) and $n$ ($1 \times n$), respectively, showing how many complete chromosomal complements are present.

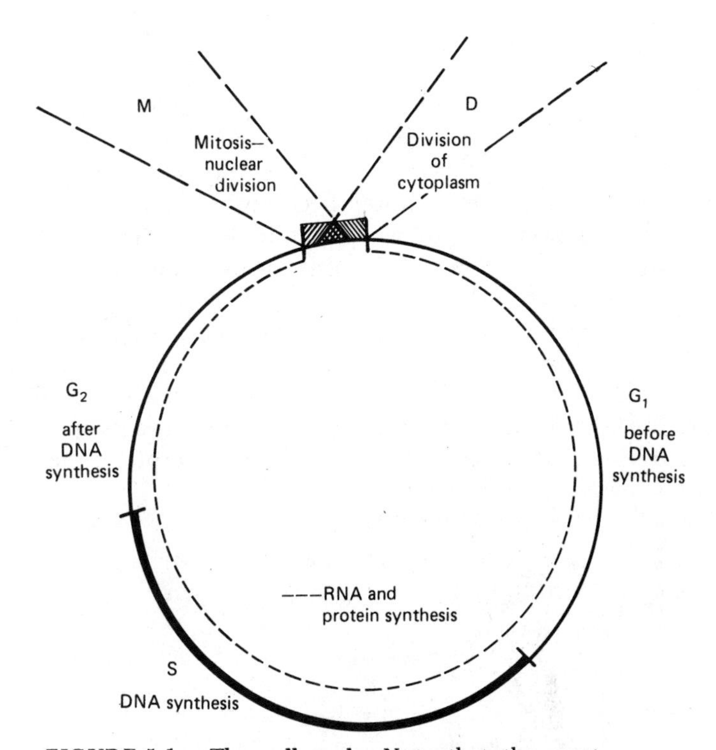


**FIGURE 5-1** **The cell cycle.** Note that the most active period of **metabolic activity** (synthesis) is during the $G_1$, S, and $G_2$ periods, which are collectively called interphase. The events of **nuclear and cytoplasmic division** occur rapidly, during a short period of the cycle (M and D).

## PRODUCTION OF NEW CELLS

Every organism begins small—in most cases, as a single cell! From this point, the development of a mature, multicellular plant, whether lettuce or giant sequoia, occurs by a process of cellular reproduction, which has as its foundation a process called **mitosis.** Mitosis is an orderly nuclear division sequence that results in the formation of two identical nuclei. Mitosis is usually followed by a division of the cytoplasm, **cytokinesis.** The result is two daughter cells of identical hereditary potential as each other and as the parent cell.

Is this a paradox? Or an impossibility? How can a single cell be divided into two that are equal not only to each other but also to the initial cell? The answer, of course, lies in the ability of DNA to reproduce itself (Chapter 3) prior to mitotic division. To visualize the activities of cellular reproduction, it is helpful to consider the total life cycle of a cell.

### The Cell Cycle

The **cell cycle** involves reproduction (replication) of DNA, nuclear division, and cytoplasmic division (cytokinesis) (Fig. 5-1). The cycle is usually divided into five stages, $\mathbf{G_1}$, **S**, $\mathbf{G_2}$, **M**, and **D**, where $\mathbf{G_1}$ and $\mathbf{G_2}$ are the first and second "gaps" separating a period of active DNA synthesis (**S**) from a period of active mitosis (**M**) and cell division (**D**).

The cell actually spends most of its time in the first three stages ($G_1$ + S + $G_2$), which together constitute the period of greatest metabolic activity, involving production of most of the cell's DNA, RNA, and protein. This period ($G_1$ + S + $G_2$) between division phases was originally termed **interphase** and before its true nature was discovered it was erroneously considered to be merely a "resting phase" between mitotic divisions.

The time that a cell spends in each stage of the cell cycle varies among different tissues in different organisms. Some cells remain for a longer time in the $G_1$ stage; however, once they enter the S stage, they usually complete the entire cycle.

Not all cells retain an ability to complete the cell cycle but grow and differentiate to perform specialized functions for the body of the organism. For most cells, differentiation is an irreversible process and the ability to divide is lost.

## Mitosis

When you consider the amount of DNA in a cell, its organization into long and precisely ordered strands, and the fact that the duplicated DNA strands must be separated with a minimum of tangling, breaking, and loss or addition of fragments, the orderly process called mitosis must be regarded as truly astonishing.

Although our understanding of the process has improved considerably since mitosis was described in the 1880s, we still utilize the original terms of that period to separate the process into observable stages or phases. Thus **interphase** describes the nondividing period; **prophase** refers to changes that occur immediately before division; **metaphase** is the between or middle stage; **anaphase** is a stage of returning; **telophase** is the end. Division of the cytoplasm, cytokinesis, usually occurs during telophase. Figure 5-2 (a) through (f) illustrates the events of mitosis, divided into these phases. Although it is helpful to use these phase names to summarize each series of events, you must think "in motion" to understand mitosis as the continuous process it is—a sort of intracellular ballet of moving chromosomes and cytoplasmic components. Figures 5-3 and 5-4 show dividing plant cells.

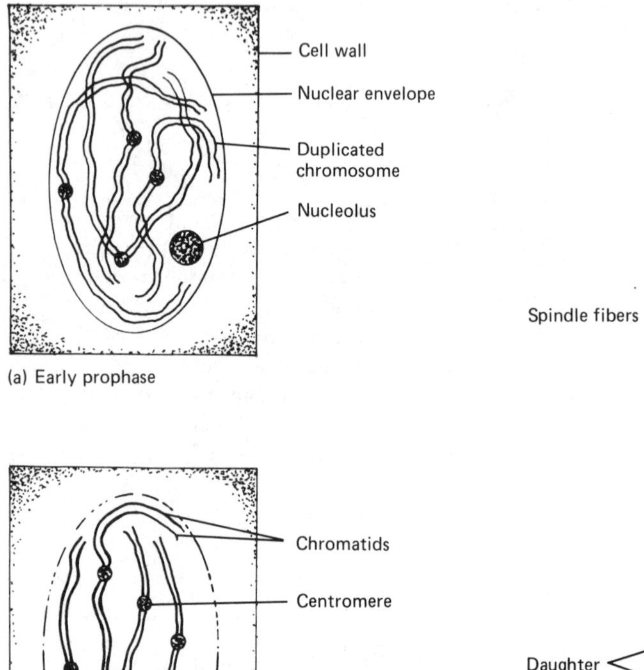


(a) Early prophase

(b) Mid-prophase

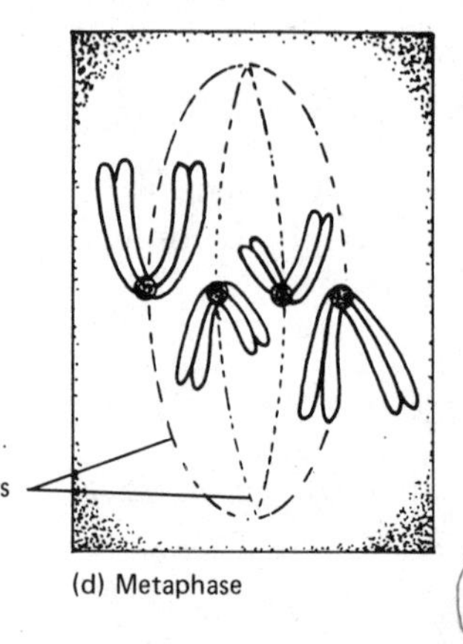


(d) Metaphase

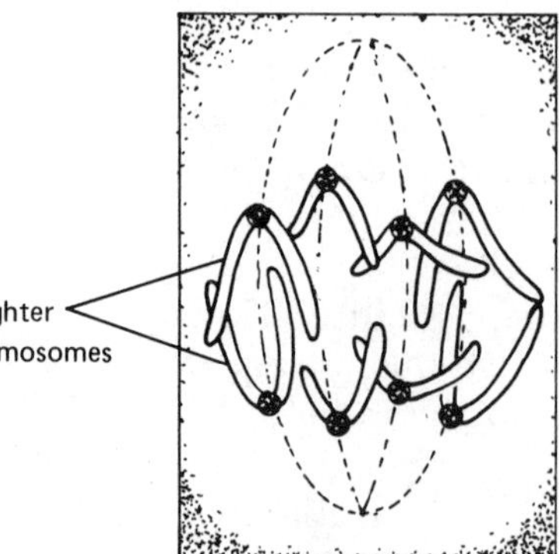


(e) Anaphase

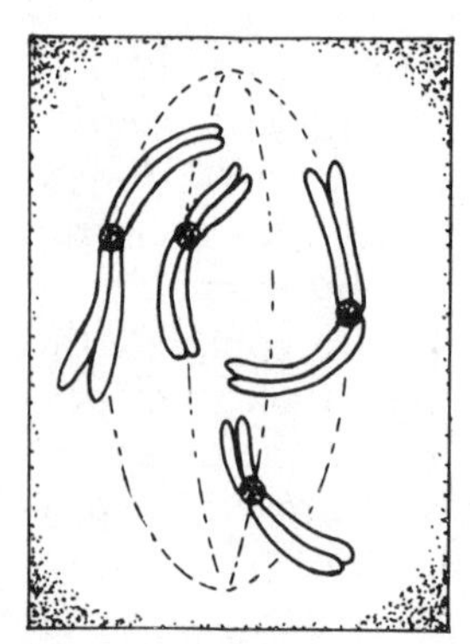

(c) Late prophase

Cell plate

(f) Telophase and cytokinesis

**FIGURE 5-2**
Illustrated sequence of the events of mitosis. (a) and (b) Early and middle **prophase.** DNA strands shorten and thicken, nucleoli begin to disappear, nuclear membrane begins to disappear. (c) Late prophase. Chromosomes clearly visible, nucleoli and nuclear membrane gone, microtubules aligning to form mitotic spindle. (d) **Metaphase.** Chromosomes aligned with centromeres at equatorial plane of spindle; spindle fibers attached to centromeres. (e) **Anaphase.** Centromeres divide, chromosomes separate and move to opposite ends of spindle. (f) **Telophase** and **cytokinesis.** Chromosomes begin to loosen in organization, eventually returning the DNA to chromatin state; spindle disappears, nucleoli and nuclear membranes form, cell plate forms across center of spindle by the coalescence of small vesicles.

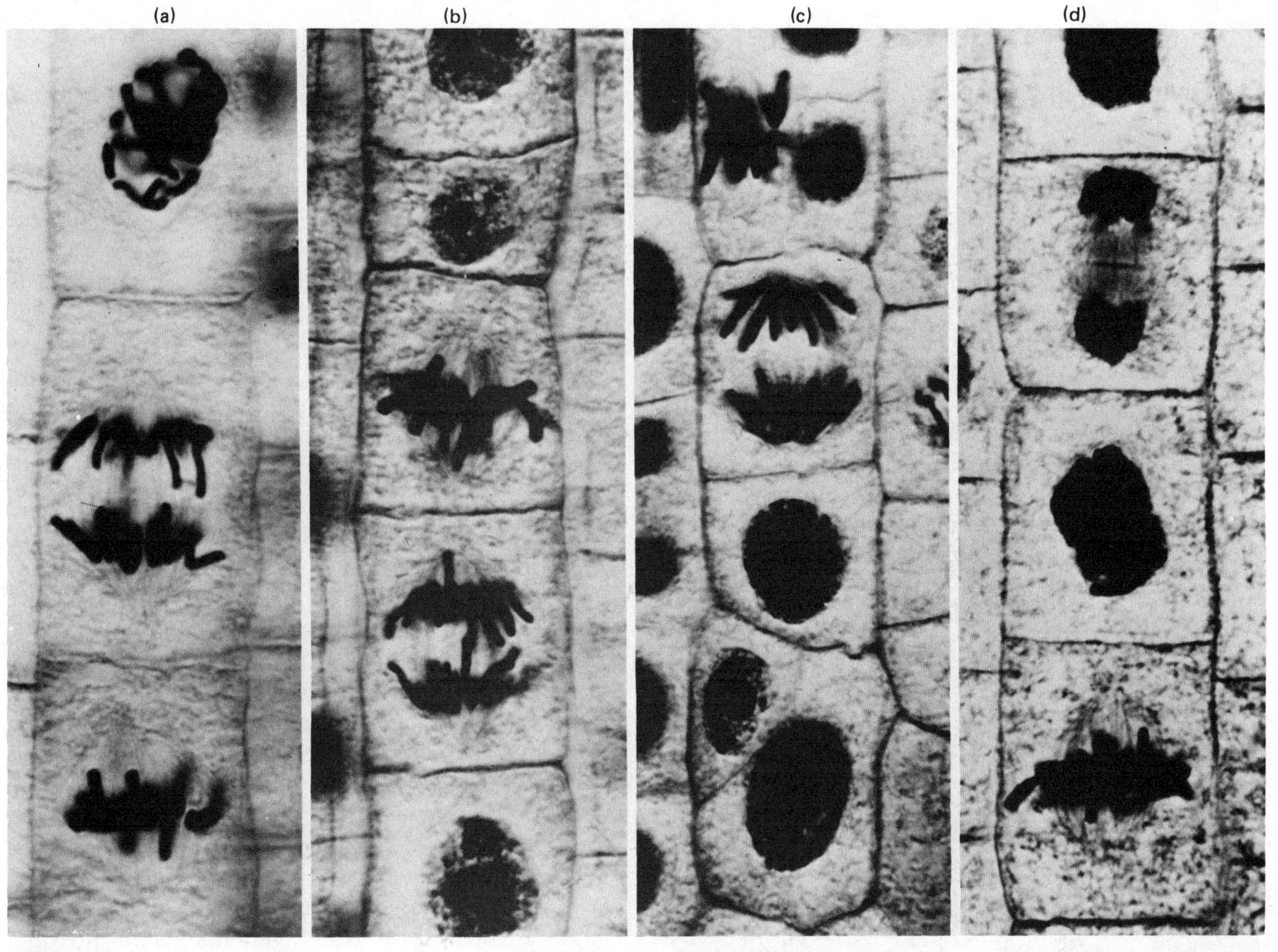


**FIGURE 5-3**
Longitudinal sections (thin slices) of onion (*Allium*) root tip, showing cells in various stages of mitosis. Column (a) shows cells (from top to bottom) in **prophase**, **anaphase**, and **metaphase**. The first whole cell in column (d) is in **telophase**, with a clearly visible **cell plate**, evidence that **cytokinesis** is occurring. (Photos courtesy of Carolina Biological Supply Co.)

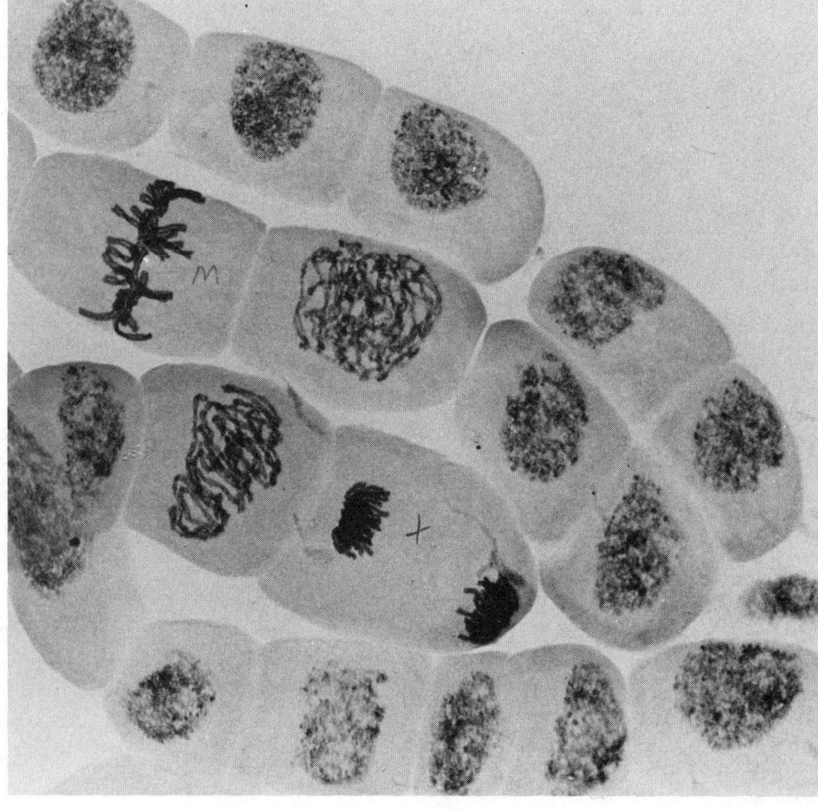

**FIGURE 5-4**
Onion root tip cells in interphase, prophase, metaphase, and telophase. The duplicated condition of the chromosomes is clearly visible in the metaphase cell. (Photo courtesy of Debi Stambaugh.)

### Cell Division: Cytokinesis

New cell wall formation begins with the appearance of a **cell plate,** a row of tiny vesicles containing pectins. The vesicles are oriented along aligned microtubules. The cell plate enlarges from the center, gradually growing outward toward the side walls of the cell as more and more of the pectin droplets coalesce (Fig. 5-5). When the cell plate has reached the side walls, the membranes on each side of the cell plate fuse with the plasma membranes of the new cells, completing the separation.

On both sides of the cell plate the newly separated cells produce microscopic cellulose fibrils, which adhere to and become embedded in the pectins. This delicate wall, composed of cellulose and pectins, is known as the **primary cell wall.** With light microscope observation, the wall resembles a sandwich with cellulose "bread" on each side of the pectin "filling." The middle pectin layer is called the middle lamella.

**FIGURE 5-5** Electron microscope view of the developing cell plate. (Photo courtesy of Robert A. Cecich.)

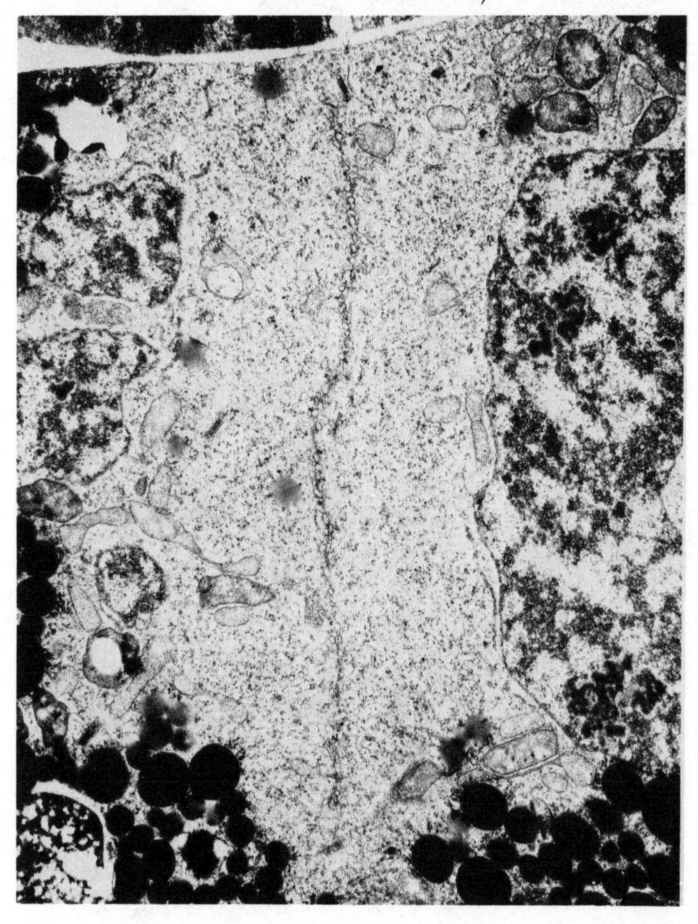

The primary cell wall initially is capable of expansion to accommodate cell growth but gradually reaches a degree of development that prevents further cell enlargement. Secretion of additional cellulose and secondary substances by the cytoplasm completes the cell wall structure.

## PRODUCTION OF NEW INDIVIDUALS

### Sexual and Asexual Reproduction

Most plant species reproduce both sexually and asexually. **Sexual reproduction** involves **syngamy,** the fusion of **gametes** (usually male = sperm, and female = egg) to produce a new individual. Even if the gametes originate from the same parent, each resulting new individual is a distinctive combination of genetic characteristics rather than an exact copy of either parent.

On the other hand, **asexual reproduction** does not involve genetic recombination by fusion of gametes and new individuals of identical genetic potential are produced. All individuals formed asexually from the same parent are called a **clone** and asexual propagation is also called **cloning.** With cloning, a specific variation can be propagated exactly true to type, which generally is not possible in sexual reproduction. Many horticultural varieties (e.g., fruit trees, roses, tulips) have become marketable realities by asexual propagation of some single outstanding parent plant.

There are several means of asexual reproduction. By **fragmentation,** pieces from the parent plant that are broken off can grow into new individuals. This ability is the basis of propagating houseplants by cuttings. Some plants produce special multicellular structures for asexual reproduction, such as miniature plantlets (Fig. 5-6), or bulblets (Fig. 5-7), as well as several other types. A current burgeoning field of applied research is tissue culture or micropropagation (detailed subsequently), in which new individual plants are grown *in vitro* (in glass) from a small fragment or slurry of parental cells. They are true "test tube babies!"

The capacity of plants to reproduce asexually has adaptive significance, for it enables individual plants that are well adapted to their immediate environments to produce many individuals of virtually identical genetic composition. So currently successful genetic combinations are perpetuated.

A life cycle that has both asexual and sexual reproduction combines the advantages of current successful exploitation of the environment with

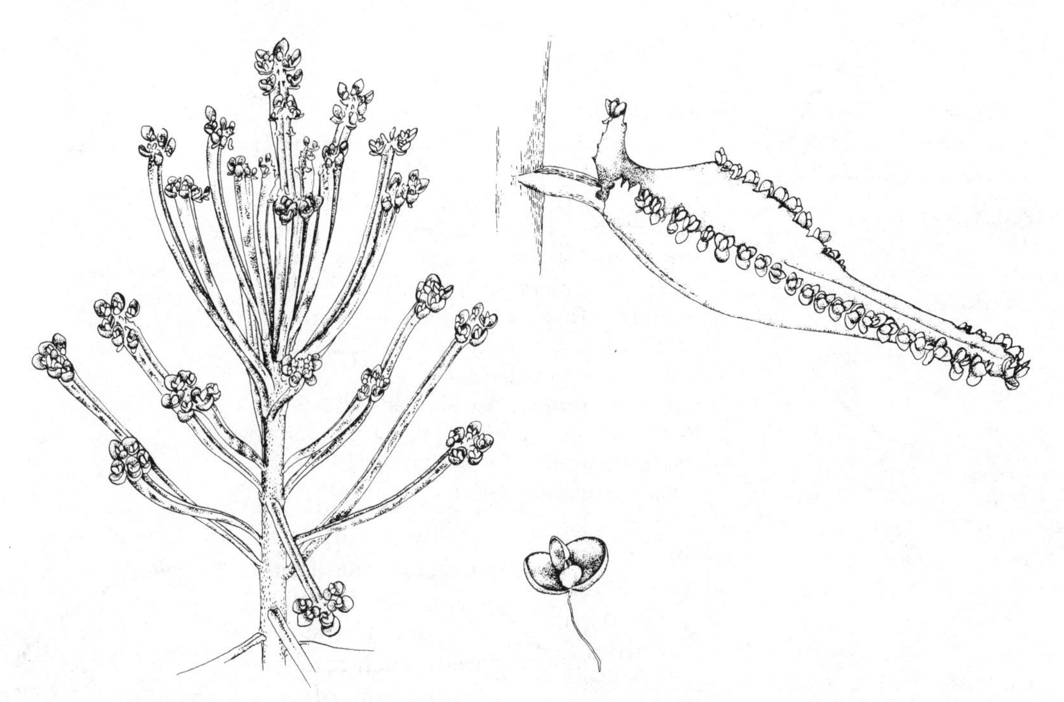

FIGURE 5-6 **Plantlets of *Kalanchöe* and *Bryophyllum* are produced along the leaf margins at vein tips.**

FIGURE 5-7 **Bulblets or offsets of *Ornithogalum* are produced at the base of the parent bulb, a modified stem. (Photo by Mark R. Fay.)**

the potential for variation and subsequent adaptation to changing environmental conditions. This flexibility is possible because of the genetic variation produced by sexual reproduction.

### Alternation of Generations: Meiosis and Syngamy

A typical sexual life cycle in plants involves some degree of alternation between a haploid ($n$) gamete-producing generation and a diploid ($2n$) spore-producing generation (Fig. 5-8). Gamete-producing plants are called **gametophytes** and spore-producing plants **sporophytes.** In the life cycles of higher plants there are separate male (sperm-producing) and female (egg-producing) gametophytes (Fig. 5-9).

As noted, cells reproduce new cells of like chromosomal content by mitosis. In comparison, diploid parent cells produce haploid cells by a special type of *reduction* division, **meiosis.** Figure 5-10 compares the haploid and diploid condition in a theoretical organism, indicating that meiosis is not simply a matter of reducing the total number of chromosomes by half but is the separation of homologous pairs.

Meiosis is an orderly, stepwise process. Like mitosis, it is preceded by DNA duplication. Also like mitosis, its earliest phases are preparation for division, including condensation of chromosomal material into a compact form (the chromosomes) for ease of movement within the cell, dissolution of nuclear membrane, disappearance of nucleoli, formation of a spindle of microtubules (spindle fibers) to assist in orientation and movement of chromosomes, and alignment of the duplicated chromosomes at the center of the spindle.

Once the preliminaries are completed, division begins, as summarized in Fig. 5-11 (a) to (h). In Fig. 5-11 (c) you will notice a difference between mitosis and meiosis—that is, the homologous chromo-

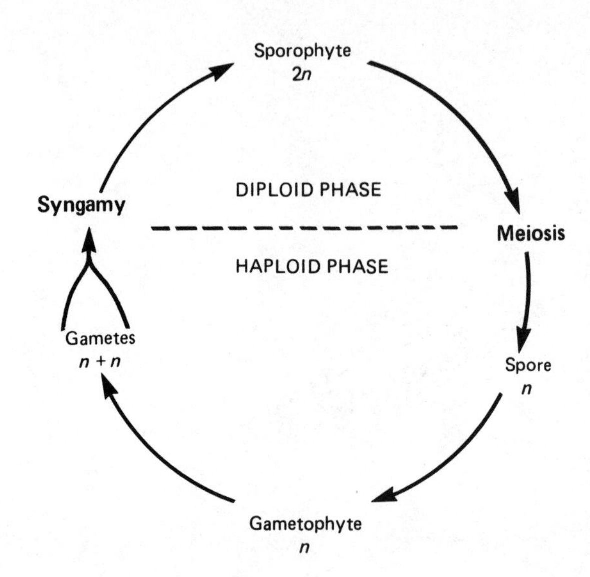


**FIGURE 5-8**
Generalized pattern of **alternation of generations** in a plant life cycle. The diploid ($2n$) organism is called the **sporophyte** because it produces haploid ($n$) **spores** by meiosis. Each spore grows into a haploid **gametophyte**, which produces haploid **gametes** by mitosis. Two gametes, usually male and female, fuse (a process called **syngamy** or fertilization) to form a new diploid organism. Both the sporophyte and gametophyte stages are usually multicellular, with additional body cells produced by mitosis.

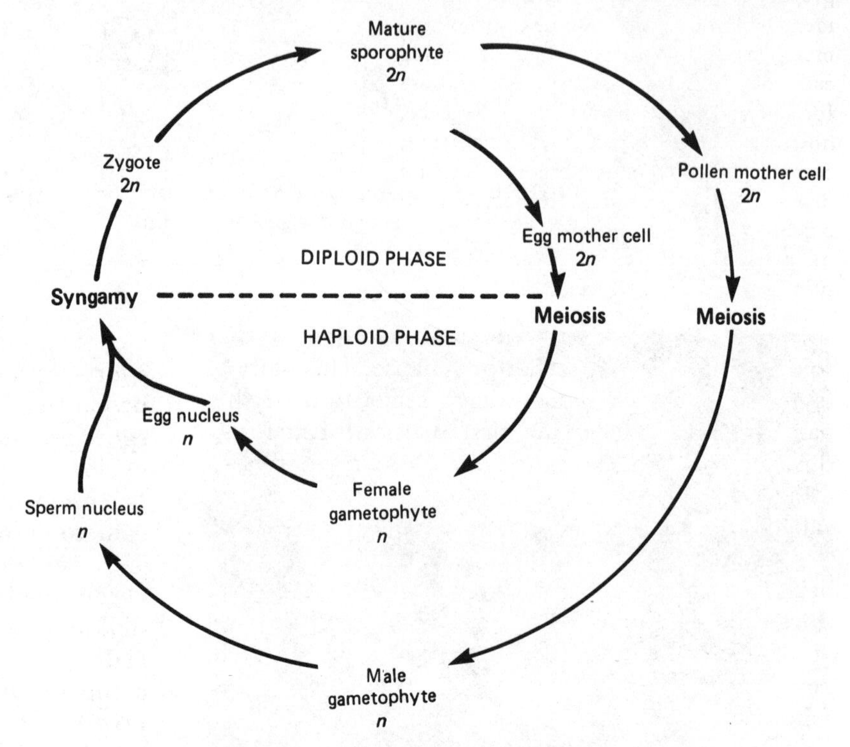


**FIGURE 5-9**
Alternation of generations with separate male and female gametophytes, the typical pattern in higher plants.

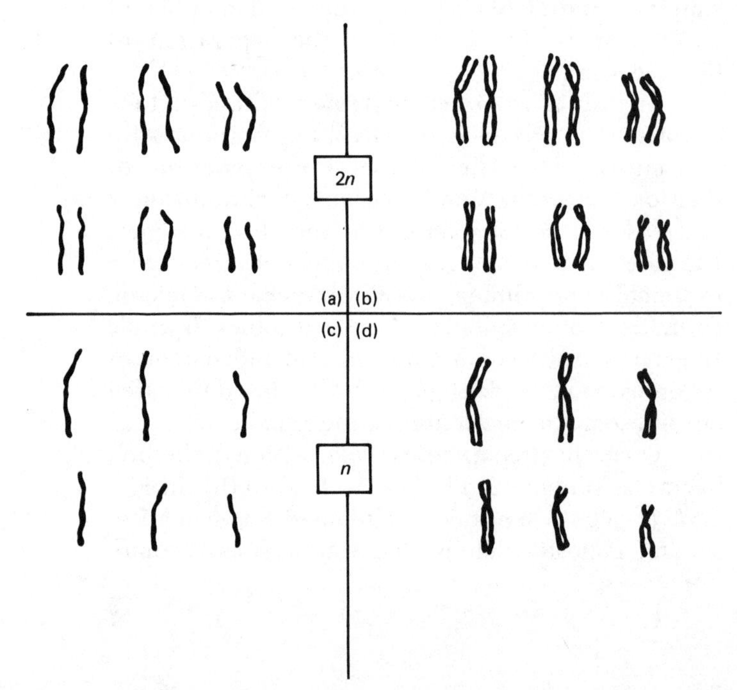


**FIGURE 5-10**
These sets of homologous chromosomes illustrate the concepts of **haploid** ($n$) and **diploid** ($2n$). The theoretical organism in this example has six pairs of homologous chromosomes in each of its body cells, one member of each pair derived from each parental gamete. Because there are six *kinds* of chromosomes, the base number ($n$) equals 6. Thus $n = 6$ and $2n = 12$. Both (a) and (b) are **diploid** cells. In (a) the chromosomes are not duplicated, whereas in (b) each chromosome has duplicated and exists as two identical **chromatids** held together by the one original centromere; thus there are still only 12 chromosomes. Both (c) and (d) are haploid cells, having one member of each of the six chromosome pairs. Each chromosome in (d) is composed of two chromatids.

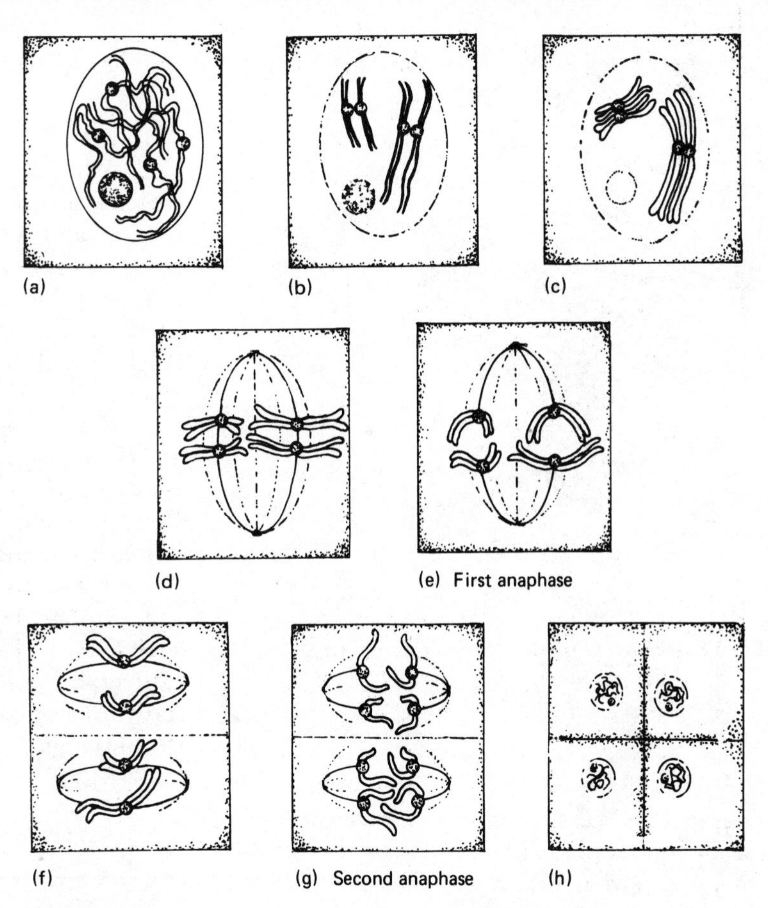


**FIGURE 5-11**
**Meiosis in a plant cell containing two pairs of homologous chromosomes. Meiosis consists of two division stages. The first division is the separation of the homologous chromosomes (chromosome pairs), creating two haploid cells. The second division separates the chromatids of each chromosome, resulting in the ultimate formation of four haploid cells from the original diploid cell. The details of cell preparation for meiosis and the movements of the chromosomes (prophase, metaphase, anaphase, telophase) during each of the two divisions are like those for mitosis. One significant difference, however, is that the pairs of chromosomes come together (synapse) prior to the first division phase, a step that helps to organize the distribution of one member of each pair to opposite ends of the spindle during the first anaphase.**

some pairs line up together or **synapse**. This pairing (synapsis) provides for orderly separation of the pair members during the first division of meiosis. Because each chomosome consists of two strands, attached at the centromere, the synapsed pairs are called **tetrads**, referring to four parts.

During the first division of meiosis one member of each chromosome pair moves to each end of the spindle. Cytokinesis usually follows. Note that at this point a difference is already created. Each new cell is, by definition, haploid! Only one representative of each chromosome pair is present.

During the second division of meiosis the centromeres of the duplicated chromosomes divide and the "daughter" chromosomes separate, migrating to opposite ends of the spindle. Cytokinesis then occurs. Are these cells diploid or haploid?

The overall result of meiosis is four cells, each containing one representative of each chromosome pair and not just any combination of chromosomes. The separation is at least partially random in that either member of a particular pair may end up with either member of another pair. This latter concept is essential to the discussion of genetics in the following chapter.

The opposite of meiosis in a life cycle is syngamy, combining of two gamete nuclei, which restores the diploid condition. (Fertilization is also an acceptable term for syngamy, although syngamy is more descriptive and is free of gender connotations.) The two gametes, each haploid, fuse to form one diploid cell, the **zygote** (Gr., *zygo* = yoked).

We are accustomed to thinking of gametes as they apply to humans, with a nonmotile female egg and a smaller, motile male sperm. This is also true of most plants. Sexual reproduction in some lower plants, however, involves fusion of physically similar gametes.

## THE LIFE CYCLE OF FLOWERING PLANTS

The sexual process in seed-bearing plants involves fusion of male and female gametes (sperm and egg nuclei) *within* protective female reproductive tissues. (See also Chapter 14.) This pattern is analogous to mammalian reproduction as an adaptation for terrestrial reproduction. Following syngamy, the developmental sequence results in a **seed**—an embryonic plant surrounded by stored food for maintenance and initial growth and encased in a water-conserving protective seed coat. (See also Chapter 10.) The seed also contains a combination of growth-regulating substances that either maintain dormancy or stimulate germination (Chapter 9).

Seed-bearing plants are of two major types—those that produce the seeds in cones (conifers and

FIGURE 5-12 Flowers of the Norway maple (*Acer platanoides*).

their relatives) and those that produce seeds in flowers (Fig. 5-12). We shall consider here the details of the flowering-plant life cycle, for the great majority of yard, garden, and house plants are flowering plants.

The stem of a flowering plant is stimulated to produce floral buds by exposure to light for a certain number of hours per day. The critical number of hours of light exposure required to induce flowering in a plant is its **photoperiod** and varies among different species. Floral induction is discussed in Chapter 9.

Some plants are **annuals,** growing from seed, flowering, producing seeds, and then dying in a single growing season. **Biennials** grow and develop vegetatively during their first season and then are stimulated to flower and produce seeds during their second season, after which they die. **Perennials** live indefinitely, generally flowering each year.

## Floral Structure

Most flowers are composed of four basic parts (Fig. 5-13), which represent leaves adapted through evolutionary changes. The reproductive floral parts are the female **pistil** and the male **stamen.** In addition, accessory parts are often present, the **petals** and **sepals.** Floral parts arise from a special shoot, the **receptacle.** The four flower parts are arranged in rings, or whorls, with sepals outermost, then petals, then stamens, then the pistils in the center. The whorl of sepals of a flower is referred to as the **calyx**; the whorl of petals is called the **corolla**; together they are called the **perianth** ("around-anthers").

A pistil has three main regions: the **stigma,** which captures pollen grains; the **ovary,** containing **ovules;** and a connecting portion, the **style.** Each pistil is composed of one or more **carpels** (discussed more completely in Chapter 14). You can generally identify how many carpels are represented in a single pistil by counting lobes of the stigma or chambers of the ovary.

If the base of the ovary is located above the attachment points of other floral parts, the ovary is said to be **superior** in relative position. If the other floral parts are attached above the ovary, the ovary is said to be **inferior** in relative position (Fig. 5-13). An inferior ovary is surrounded by receptacle tissue, which becomes the outer part of the fruit of such plants (Fig. 5-14). Ovary position is an important taxonomic feature, as is the manner in which individual flowers are arranged along a flowering shoot [Figs. 5-15, 5-16, 5-17 (a) and (b), 5-18 (a) and (b), 5-19 (a) and (b)]. The flowers of many

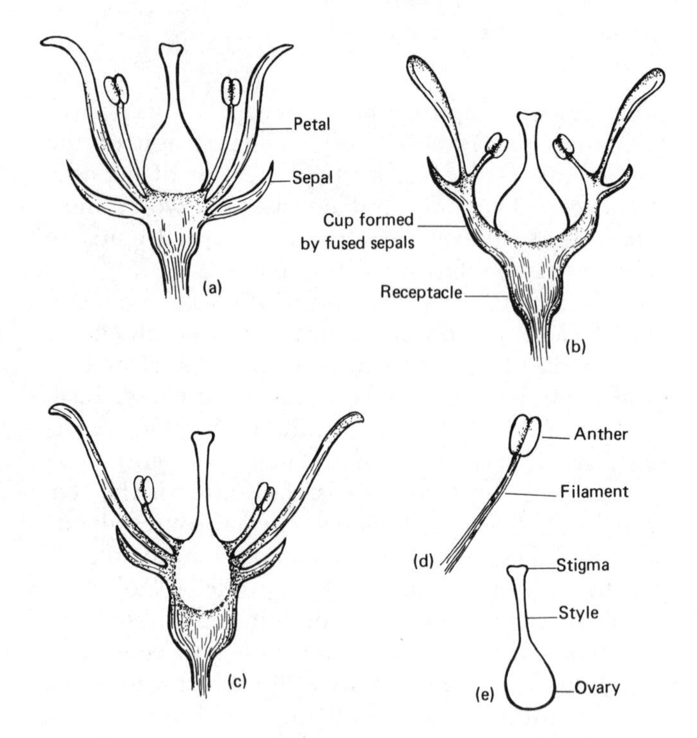


FIGURE 5-13 A flower is composed of four sets of parts, the **pistil** (composed of one or more **carpels**), **stamens, petals,** and **sepals,** in that order from the tip of the flowering axis. They are attached to a **receptable.** Diagrams (a) and (b) illustrate **superior** ovary position; (c) illustrates **inferior** ovary position; (d) and (e) illustrate the parts of a stamen and pistil.

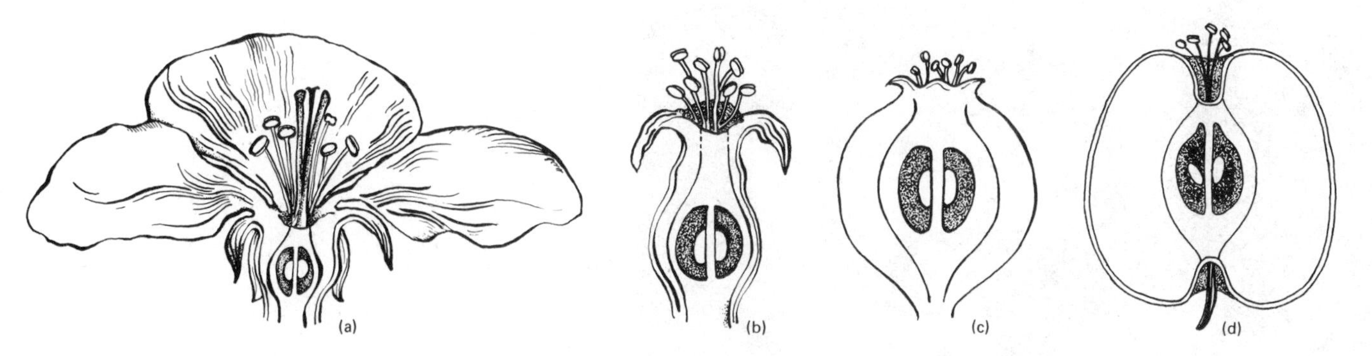


**FIGURE 5-14** An apple develops from the fused tissues of the ovary and the wall of the floral tube, including receptacle tissue. The edible fleshy portion is derived from the floral tube, whereas the "true" fruit (the part that develops from the ovary) is actually the apple core.

**FIGURE 5-15** This arum inflorescence consists of many small flowers packed tightly together on a fingerlike **spathe**, enclosed in a leaflike **bract**. Fruits have developed from the lowest inflorescence.

**FIGURE 5-16** Mullein (*Verbascum*) flowers are arranged in a dense **spike**, the base of each flower attaching directly to the central axis.

(a)

(b)

FIGURE 5-17
In a **raceme** each flower is attached to the central axis by an individual stalk, as in (a) chokecherry (*Prunus*) and (b) Culver's root (*Veronicastrum*).

(a) (b)

**FIGURE 5-18**
(a) Crown vetch (*Coronilla*) flowers are arranged in a **simple umbel,** the flower stalks originating as a whorl. (b) The flat-topped flowering head of dill (*Anethum*) is a **compound umbel** because it is composed of many small umbels.

(a) (b)

**FIGURE 5-19**
(a) The familiar "flowers" of members of the composite or daisy family are actually dense inflorescences called **heads.** (b) In this wild sunflower the outer **ray flowers** are pistillate (female) and the central **disk flowers** have both stamens and a pistil.

species occur in a cluster, or **inflorescence**, rather than singly.

Each stamen is made of an **anther,** which contains pollen-producing chambers attached to the receptacle by a stalk, the **filament.**

Many flowers are bisexual or **perfect,** having both male and female parts (Fig. 5-20); others are unisexual or **imperfect,** having either male or female parts on each flower (Figs. 5-21 and 5-22).

The flowering plants are divided into two major subgroups: the monocotyledonae or **monocots** and the dicotyledonae or **dicots.** Table 5-1 summarizes some major external differences between monocots and dicots. (Internal differences are discussed in Chapter 8.)

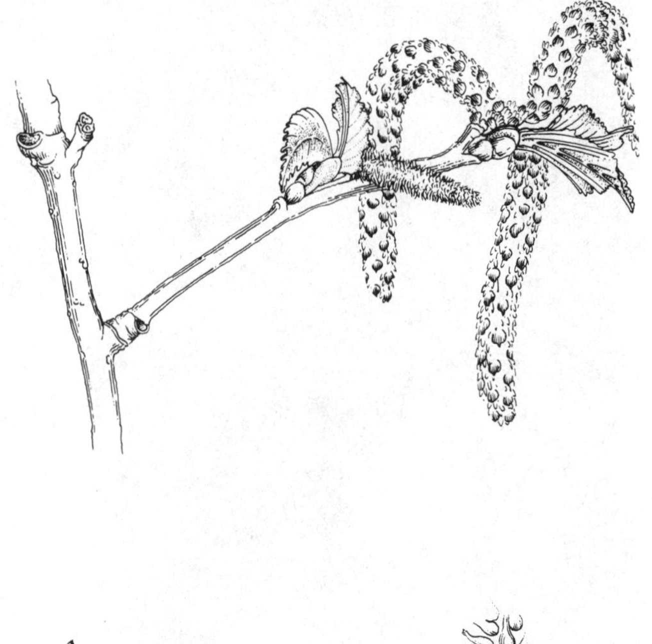

**FIGURE 5-20**
*Fuschsia* flowers are bisexual or "perfect." Note the pistil and stamens that protrude beyond the petals of each flower. (Photo courtesy of David Koranski.)

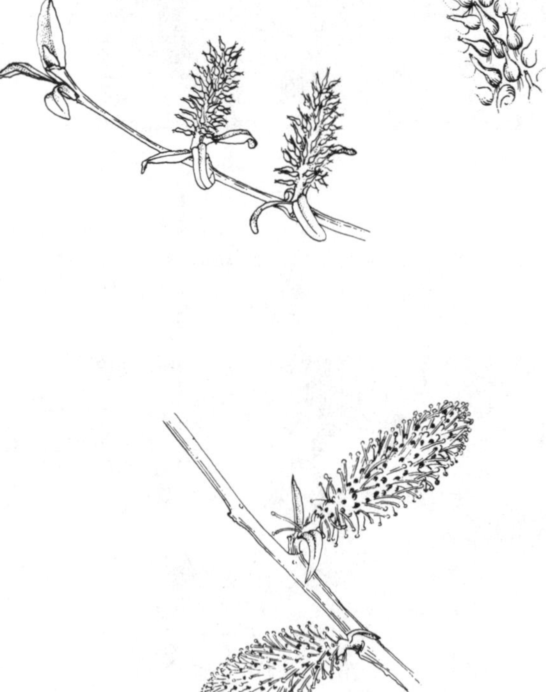

**FIGURE 5-21**
The separate male and female **catkins** of birch (*Betula*) (top) and willow (*Salix*). The pistillate birch catkins are smaller than the staminate catkins. The staminate flowers of willow (bottom) consist entirely of stamens, without accessory parts (petals and sepals); the pistillate flowers (center) consist entirely of individual carpels (simple pistils), shown in closer detail to the right.

FIGURE 5-22 Separate male and female flowers of *Begonia*.

TABLE 5-1 **Some external characteristics used to distinguish monocots from dicots.**

**Monocots**

Floral parts in threes or multiples of threes.
Embryo having a single cotyledon (seed leaf).
Parallel leaf venation.
Fibrous root system.

**Dicots**

Floral parts not usually in threes; in multiples of fours and fives.
Embryo having two cotyledons (seed leaves).
Netted leaf venation.
Most having tap root systems.

Floral structures demonstrate certain consistent characteristics within each plant family and so are an important aid to identification. Table 5-2 summarizes variations in floral structure in six representative plant families.

## Microsporogenesis and Megasporogenesis

As the flower matures much activity occurs, especially in the sexual parts (Fig. 5-23). The devel-

TABLE 5-2 **Some examples of plant family floral characteristics.**

**Monocots**

**Liliaceae** *Lily family:*
3 sepals; 3 petals; 6 stamens; single pistil; superior ovary

**Iridaceae** *Iris family:*
3 sepals (petal-like); 3 petals; 3 stamens; 1 pistil, having three-parted petal-like style; inferior ovary

**Dicots**

**Ranunculaceae** *Buttercup family:*
3 to 20 sepals; 0 to 15 petals; usually many stamens; 1 to many pistils; superior ovary

**Brassicaceae (Cruciferae)** *Mustard family:*
4 sepals; 4 petals (rarely 0); 6 stamens (rarely 2), 4 long and 2 short; single pistil; superior ovary

**Rosaceae** *Rose family:*
5 fused sepals; 5 petals; 10 to many stamens; 1 to many pistils; superior ovary

**Fabaceae (Leguminosae)** *Bean family:*
3 to 5 sepals; 5 petals, modified into a bilaterally symmetrical flower; 10 stamens (rarely 5); single pistil; superior ovary

FIGURE 5-23 Median longitudinal section view of a "perfect" dicot flower—one having both male and female reproductive organs. **Microsporogenesis** occurs within the anthers; **megasporogenesis** occurs within the ovules. These meiosis sites are shaded on the drawing.

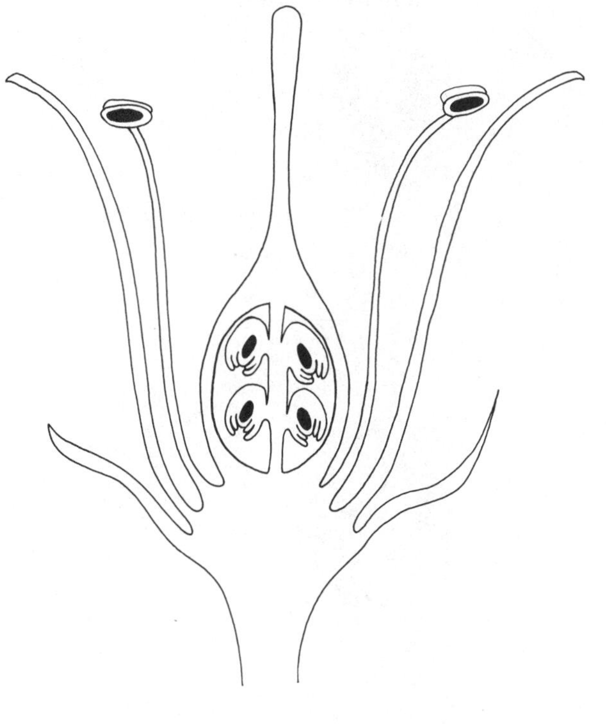

opmental sequence in the anther of the stamen is called **microsporogenesis** (Fig. 5-24) and ends with formation of pollen grains carrying sperm nuclei. Development in the pistil is called **megasporogenesis** (Fig. 5-25) and results in the formation of eggs within ovules. (See also heterospory, Chapters 13 and 14.)

During microsporogenesis special diploid ($2n$) cells called **pollen mother cells** (microspore mother cells) undergo meiosis, producing four haploid ($n$) cells, the **microspores.** The nucleus ($n$) of each microspore divides by mitosis to produce two nuclei—a **tube nucleus** ($n$) and a **generative nucleus** ($n$). Exterior changes also occur as each microspore develops into a mature **pollen grain.** The pollen grain represents the male gametophyte ($n$) of the life cycle because the male gametes ($n$) develop within it.

Egg formation occurs within the ovary in numerous small structures called the **ovules,** sporophytic ($2n$) structures in which the female gametophyte ($n$) develops. In an ovule one diploid cell (**egg mother cell** or megaspore mother cell) divides by meiosis, producing four haploid **megaspores.** The three megaspores nearest the ovule opening (**micropyle**) degenerate and the surviving megaspore increases tremendously in size and volume. During enlargement three mitotic divisions occur, forming eight nuclei ($n$). The eight nuclei rearrange themselves, three moving to each end of the mass of

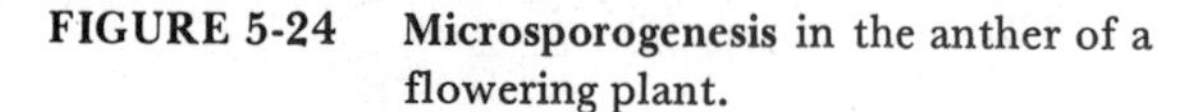

**FIGURE 5-24** **Microsporogenesis** in the anther of a flowering plant.

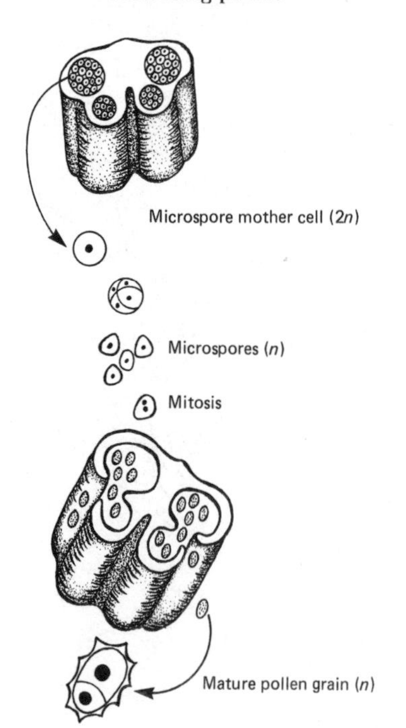


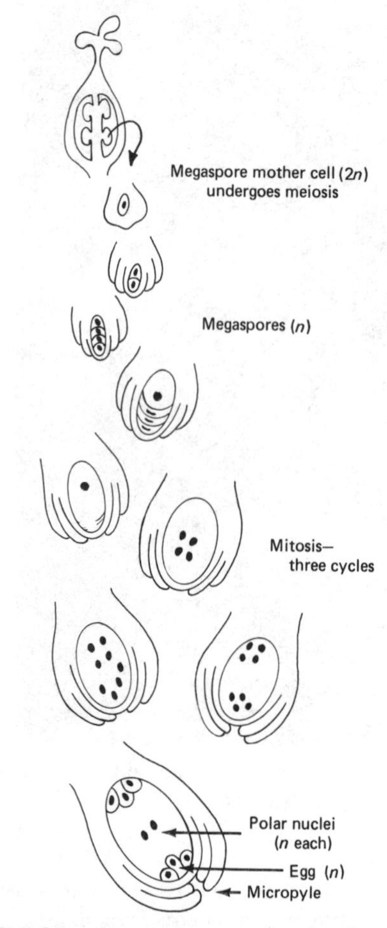


**FIGURE 5-25** **Megasporogenesis,** as it occurs in most flowering plants.

cytoplasm. One of the three nearest the micropyle enlarges to become the **egg nucleus.** Plasma membranes form, separating the nuclei into individual cells. Two nuclei, the **polar nuclei,** remain in the center and become enclosed within the same cell membrane, forming the **endosperm mother cell,** a binucleate cell (one having two separate nuclei).

The seven-celled structure is now called an **embryo sac,** which is the female gametophyte stage of the life cycle because it is the haploid structure that contains the female gamete, the egg. At this stage, the female gametophyte is "mature"—that is, ready for syngamy.

## Pollination and Syngamy

When the pollen grains are mature, the anthers split open, releasing the pollen. Pollen is transferred to the stigma of the pistil, a process called **pollination** (Fig. 5-26). The stigma is often covered with tiny hairlike or fingerlike projections to trap pollen grains. Many pollen grains usually land on a single stigma. Sticky secretions on the stigma help the

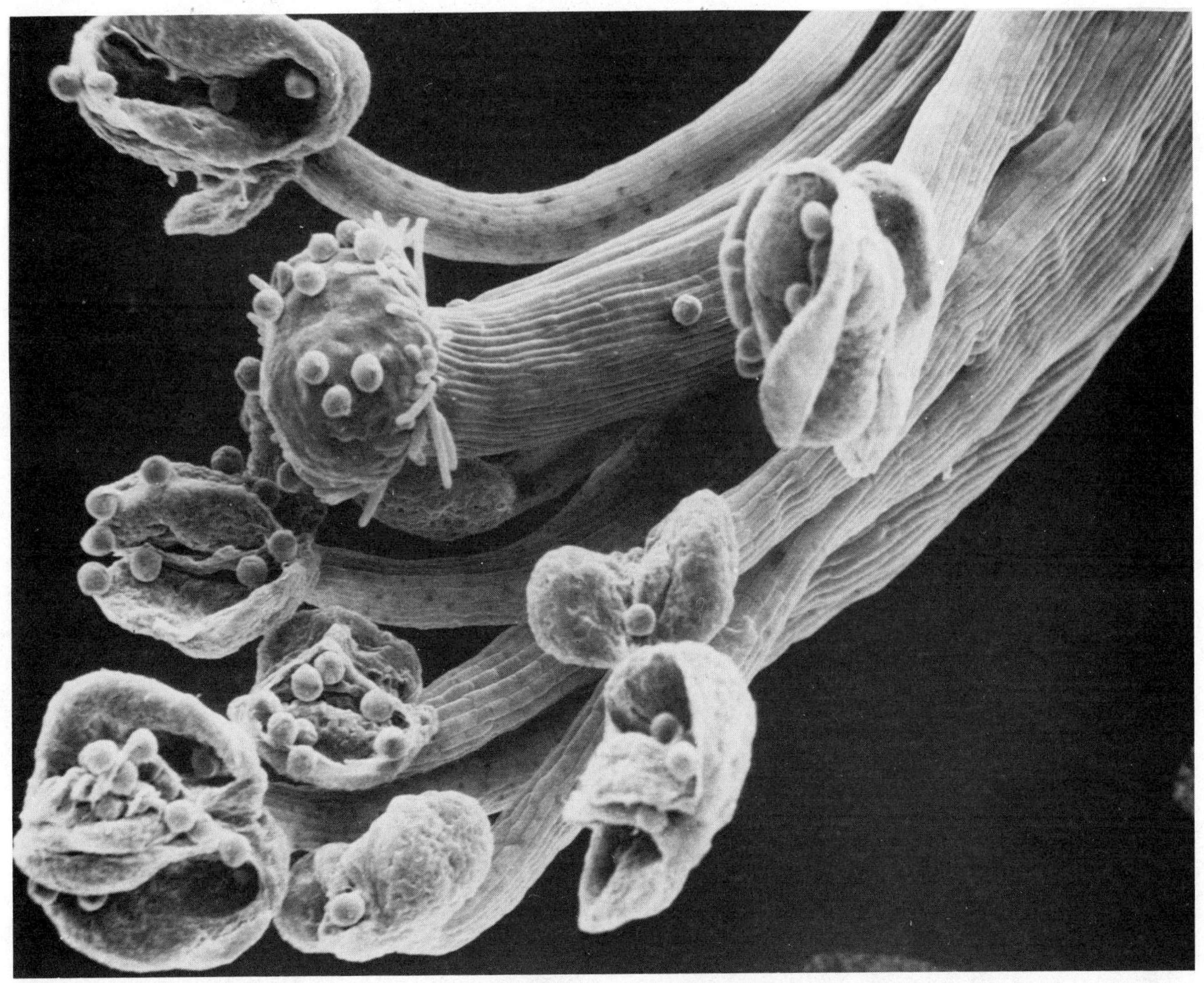

(a)

(b)

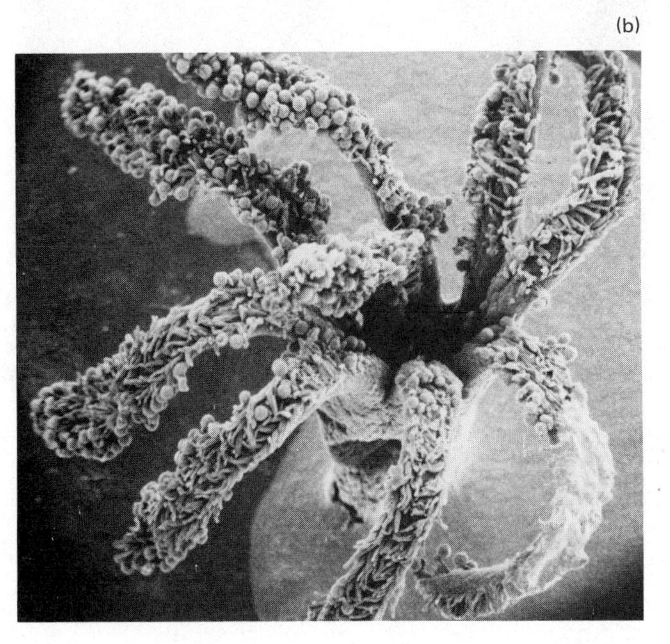

**FIGURE 5-26**
(a) Scanning electron micrograph (SEM) of the stamens and stigma of a soybean (*Glycine max*) flower, showing pollen being released from the anthers and adhering to the stigma. (Photo courtesy of Donald Buntman.) (b) SEM of the branched stigma of moss rose (*Portulaca*). Note the finger-like hairs of the stigma that have caught some spherical pollen grains.

pollen to adhere and stimulate the growth of a tubular extension, the **pollen tube.**

Each pollen tube penetrates the stigma and grows through the style tissues toward the ovary, its metabolism apparently aided by the tube nucleus. During this growth journey the generative nucleus divides once by mitosis to form two **sperm nuclei,** which are the male gametes. Eventually a pollen tube enters the micropyle and penetrates the embryo sac; its end wall dissolves and the sperm nuclei are released into the embryo sac. One sperm nucleus ($n$) fuses with the egg ($n$) to form a diploid zygote ($2n$). The other sperm nucleus ($n$) fuses with both polar nuclei ($n + n$) to form a triploid cell ($3n$) called the primary endosperm cell. The fusion of one sperm nucleus with the egg and the other sperm nucleus with the polar nuclei is referred to as **double fertilization** and is a feature of flowering plants.

## Embryogenesis and Fruit Development

Following fertilization, changes occur that convert an ovule into a seed. The zygote ($2n$) and primary endosperm cell ($3n$) begin a series of mitotic divisions. The result is formation of an embryonic sporophyte, the **embryo** ($2n$), surrounded by multicellular nutritive tissue, the **endosperm** ($3n$). As noted, the cotyledons of many dicot embryos become storage organs (Fig. 5-27) and some or all of the endosperm is used up in the process [Fig. 5-28 (a), (b), and (c)]. In other plants

FIGURE 5-27 An avocado (*Persea americana*) seed has well-developed cotyledons to nourish the young seedling emerging from between them.

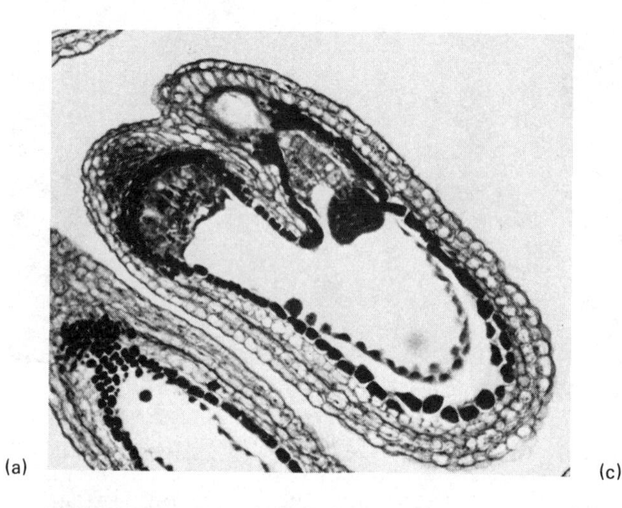

(a)

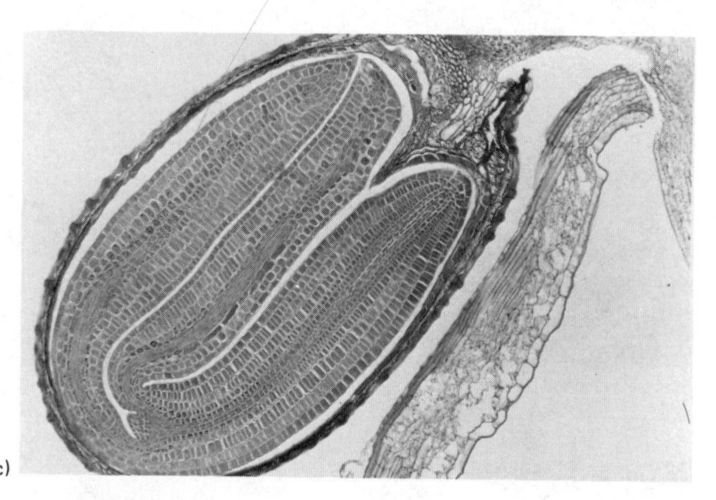

(c)

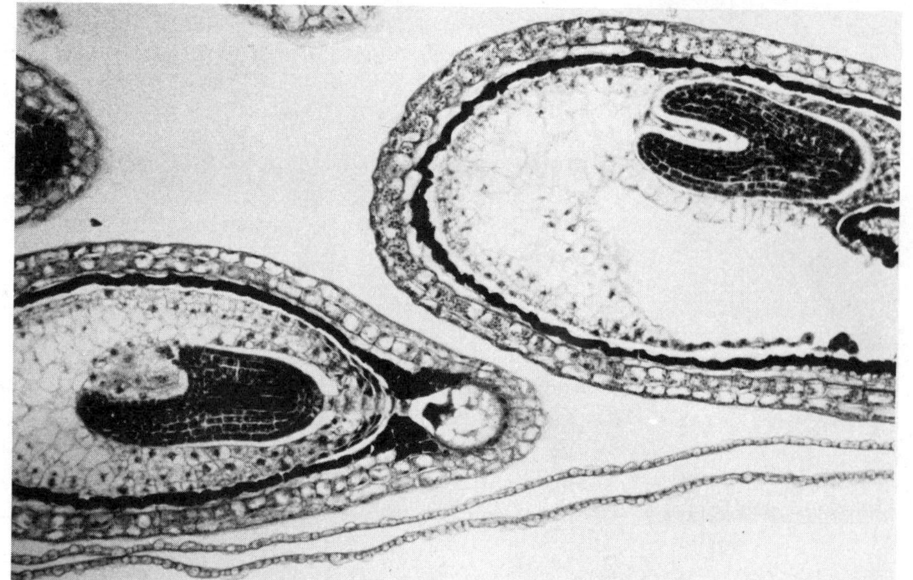

(b)

FIGURE 5-28
(a) and (b) The young embryo in a developing shepherd's purse (*Capsella*) seed digests and absorbs much of the endosperm tissue. (c) The endosperm has been assimilated into the body of the embryo and the food is now stored in the two well-developed **cotyledons.** [Photo (c) courtesy of Carolina Biological Supply Co.]

the endosperm is the main food storage tissue for the embryo. As the embryo matures, the outer ovule wall develops into a protective **seed coat.**

While seeds develop from the ovules, the ovary and associated tissues develop into a **fruit.** The developing seeds and developing fruit produce hormones that influence each other (Chapter 9). Eventually the mature fruit forms a protective structure around the seeds. Many fruits assist in seed dispersal; others fall to the ground and decompose, forming an immediate substrate for subsequent seed germination.

Fruits, like flowers, are variously adapted and can be categorized descriptively. The first basic difference lies in whether the fruit is dry or fleshy and, if dry, if it spontaneously opens or remains closed at maturity. Some fruits are the result of maturation of a single ovary within a single flower (Figs. 5-29 and 5-30). Some, however, develop from one flower that has many ovaries (Figs. 5-31 and 5-32) or from fusion of ovaries from many different flowers within an inflorescence (Fig. 5-33). A few plant groups produce fruits in which the bulk of the fleshy outer portion is actually formed by the proliferation of receptacle tissue around the ovary. The edible part of an apple, for example, is the expanded receptacle whereas the "true" fruit part is the core!

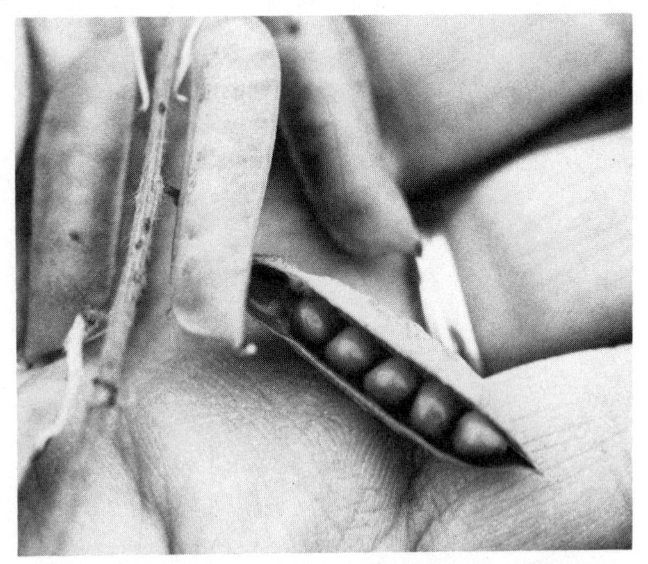

**FIGURE 5-30** Vetch (*Vicia*) is a **legume.** The fruit consists of a pod containing the seeds.

**FIGURE 5-29** Basswood (*Tilia*) flowers and fruits.

## Seed Dispersal

There are many seed-dispersal mechanisms, including tiny lightweight seeds, "wings" [Fig. 5-34 (a-c)], or silky plumes (Fig. 5-35) for air travel; air spaces or corky floats for water dispersal (coconuts); and explosive fruits, such as witch hazel, many legumes, and wild cucumber. Animals act as agents of dispersal in several ways. Spines, barbs, and hooks of fruits become entangled in the hair of mammals (Fig. 5-36). Fleshy fruits are eaten by animals, especially birds, and seeds are scattered with the feces. Nuts and seeds of many types are cached by such rodents as field mice, squirrels, and chipmunks; those that are not recovered germinate in the spring. Seeds are also spread in mud adhering to the feet of aquatic birds. The human species has dispersed seeds of edible and decorative plants all over the world, usually deliberately. Propagating plants from seed is discussed in more detail in Chapters 10 and 11.

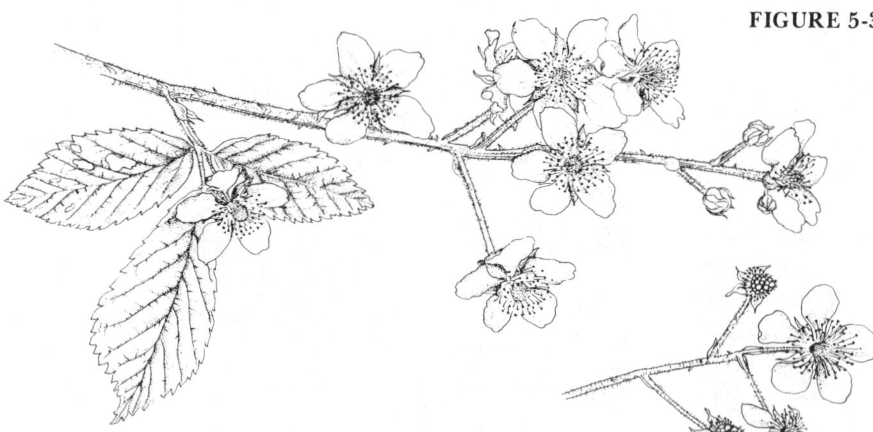

**FIGURE 5-31** Each blackberry (*Rubus*) flower contains many separate carpels, each of which forms a simple fruit. The cluster of simple fruits forms the **aggregate fruit.**

**FIGURE 5-32** A strawberry (*Fragaria*) is also an aggregate fruit, but one in which much of the fleshy tissue is developed from the accessory tissue of the receptacle. (Photo by Mark R. Fay.)

FIGURE 5-33
Pineapple (*Ananas comosus*) has a **multiple fruit,** formed from many separate flowers of an inflorescence; their individual fruits fuse to form the edible tissue. The external remaining flower parts form the diamond pattern on the outside. The central core of a pineapple fruit is part of the floral stalk, along which the flowers were produced, and the top of the pineapple is a cone of stem tissue complete with nodes and apical meristem. (Photo by Mark R. Fay.)

(a)

(c)

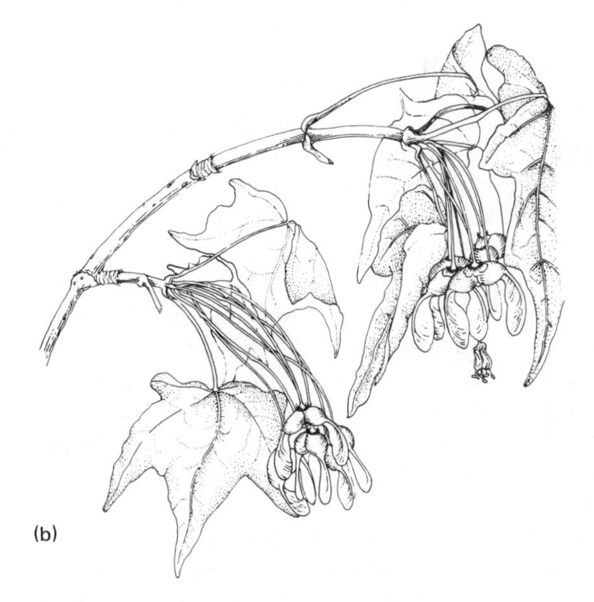

(b)

FIGURE 5-34
(a) The flowers of sugar maple (*Acer saccharum*) are simple wind-pollinated structures that produce the characteristic winged fruit. Drawings (b) and (c) show the maturation of the wings, which are adaptations for wind dispersal of the fruits. (Compare sugar maple flowers to those of a related species in Fig. 5-38.)

FIGURE 5-35
The seed head of salsify (*Tragopogon*) is composed of many individual fruits, each with a soft plume for catching air currents.

FIGURE 5-36
Burdock (*Arctium*) seed heads have many individual fruits, each with a hooked appendage that can catch in animal fur and on clothing.

FIGURE 5-37 The back of an insect entering a *Monarda punctata* flower for nectar is rubbed by both anthers and stigma, cross-pollinating as the insect moves from flower to flower.

## Pollination Ecology

The study of pollination is also called floral ecology. Some adaptations for pollination are fairly simple whereas others are highly specialized, involving at times intimate "coadaptation" with insect pollinators (Fig. 5-37).

Two types of pollination are possible: **self-pollination** and **open** or **cross-pollination.** There are specific floral adaptations to both types. To ensure self-pollination, some flowers remain closed, ensuring pollen transfer from within the flowers. Some species have flowers that are normally cross-pollinated but that also provide for self-pollination as a "backup" method.

On the other hand, many floral adaptations favor cross-pollination. Among the most common are location of the anthers below the stigma so that the stigma is mainly receptive to pollen transferred from flower to flower, such as by an insect; different times of maturation of the stamens and pistils within the same flower; the presence of male and female structures in separate flowers or on separate plants; and **self-incompatibility** whereby chemical incompatibility between pollen and pistil tissues interferes with pollen tube development or syngamy. Many cultivated fruit tree varieties are self-incompatible so that at least two trees must be planted to ensure pollination.

Flowers that are normally cross-pollinated often show striking adaptations to their mode of pollination—for example, by wind (Figs. 5-38 and 5-39). Some aquatic plants are pollinated by pollen floating on the surface of the water. Table 5-3

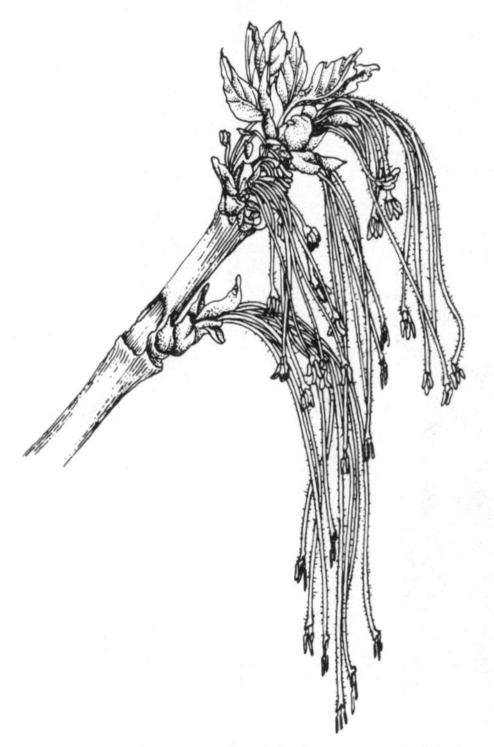

FIGURE 5-38 Wind pollinated flowers of the box elder (*Acer negundo*).

compares some general characteristics of wind pollinated and animal pollinated (usually insect) flowers. In addition to insects, such as bees (Fig. 5-40), flies (Fig. 5-41), wasps, butterflies, and beetles (Fig. 5-42), animal pollinators include birds like hummingbirds and sunbirds. Some flower-feeding bats in the deserts and tropics are pollinators. Several of the Suggested Readings at the end

**TABLE 5-3 A general comparison of wind pollinated and animal pollinated flowers.[a]**

| Wind Pollinated | Animal Pollinated |
|---|---|
| Pollen light, dry, and smooth | Pollen adherent and heavy |
| Stigmas long and feathery | Stigmas rounded, compact, and sturdy |
| Small, inconspicuous flowers | Flowers with showy petals and/or sepals |
| Many male flowers, producing abundant pollen | Flowers scented |
| | Nectar produced in glandular nectaries |
| | Petals marked with "nectar guides", some of which are visible only with ultraviolet light (as seen by bees, for example) |
| | Floral parts sturdier |
| | Flowers having special adaptations for movement of the anthers or style to ensure animal contact with pollen and the stigma |
| Fewer ovules per ovary | Many ovules per ovary |
| One-seeded fruits | Many-seeded fruits |

[a]Compare these features to the photographs of flowers in this chapter and in Chapter 14. Based on these characteristics, would you suppose that such popular flowers as roses, carnations, orchids, mums, and hyacinths are wind pollinated or animal pollinated? The actual structure of a flower is often intimately adapted to the specific animal pollinator.

FIGURE 5-39 The action of wind rocking the anthers of grass flowers sprinkles pollen on the breezes. (Photo courtesy of James R. Estes and Ronald J. Tyrl.)

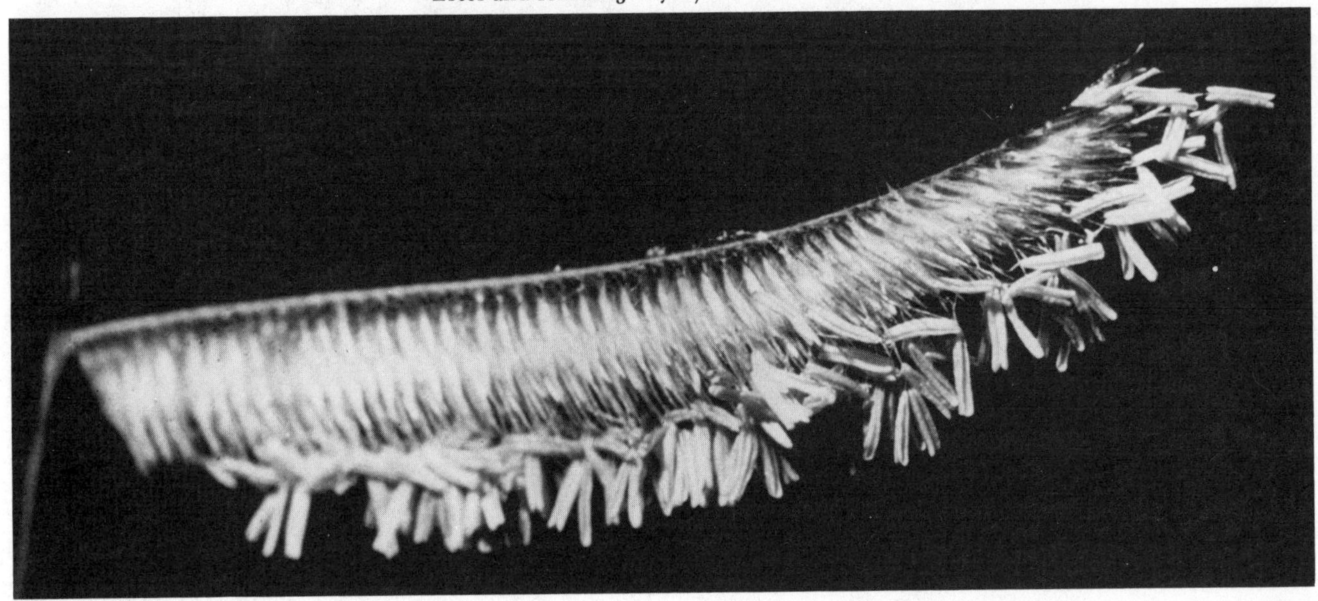

FIGURE 5-40
The dark central markings of the hollyhock (*Althaea*) flower act as a **nectar guide** to attract bees. This bumblebee is covered with pollen from visits to several flowers, serving as an agent of cross-pollination as it seeks food.

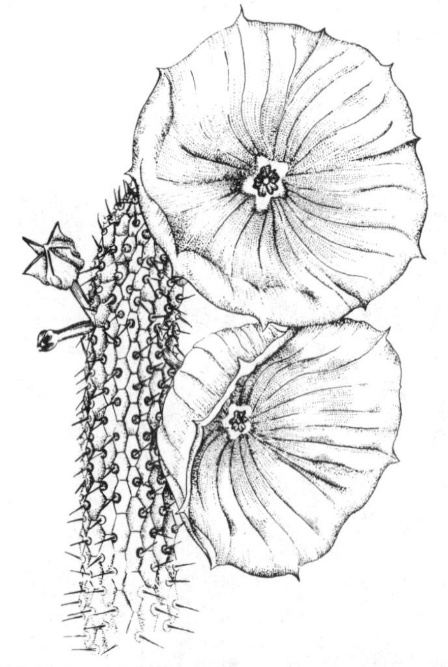

FIGURE 5-41 The blossoms of carrion flower (*Stapelia*) are iridescent purplish green and have the odor of rotting flesh, attracting fly pollinators.

FIGURE 5-42 The movement of the beetles walking among the stamens of this prickly pear cactus (*Opuntia*) flower as they feed on the floral parts stimulates the stamens to thrash vigorously, dusting the beetles and all parts of the flower with pollen.

of this chapter provide fascinating information about the large and interesting subject of coadaptations of flowers and their animal pollinators.

## VEGETATIVE PROPAGATION: CLONING

The natural ability of higher plants to grow new individuals from fragments of the old one is the basis of common vegetative propagation (cloning) practices. Except for budding and grafting, vegetative propagation is based on the formation of **adventitious roots and buds,** those that form where they are not normally expected.

### Tissue Culture: Micropropagation

The aseptic culturing of plant tissues on nutrient media in flasks and test tubes has existed for decades. Today, however, the techniques have become so sophisticated that the potential impact of tissue culture on the propagation of useful plants may be considered seriously.

It is now possible to clone vast numbers of individuals from a single small fragment, or **explant,** of parent tissue. The explants may be single seeds (as with orchids), stem tips (as with orchids, ferns, apples, and carnations), or embryos. One of the most basic methods uses a piece of stem tissue as the explant (tobacco, carrot, endive, asparagus, Dutch iris, citrus, aspen).

During tissue culture the balance of growth regulators (Chapter 9) is carefully controlled so that the explant cells divide repeatedly, producing a mass of undifferentiated **callus** tissue composed mainly of parenchyma cells and a few vascular cells. Pieces of the resulting callus mass can be transferred to separate culture containers and thus be cloned indefinitely!

By altering the balance of growth regulators, technicians stimulate the callus to produce adventitious buds and roots, thereby differentiating into shoot and root systems. The resulting young plantlets are transplanted to petri dishes and eventually to pots with soil.

Tissue culture techniques also permit some direct genetic manipulation. The chemical **colchicine,** for instance, blocks the separation of chromatids during mitosis. Therefore colchicine can be used to produce cells that contain multiple sets of chromosomes. The immature pollen grain (haploid) of some plants can be cultured and then treated with colchicine, causing newly formed cells to be diploid. The young plants that result are normal diploid plants that carry only the genetic characteristics of the chromosomes within the original pollen grain. They are thus known to be homozygous (having the same genes for a characteristic) in every respect (Chapter 6); in contrast, sexual recombination of genes is a random event. The production of individuals of exact known genetic composition has exciting possibilities—for example, in the potential propagation of strains of useful plants that are resistant to specific diseases. There are also experimental methods for fusing diploid (nonreproductive) nuclei from plant body cells.

The nutrient medium used during micropropagation contains an array of essential mineral nutrients, sugar, vitamins, amino acids, growth regulators, and other organic complexes (e.g., coconut milk, banana puree, yeast extract). Obviously the medium, all glassware and equipment, the working environment, and the hands and clothing of technicians must be perfectly sterile during tissue culture procedures. Otherwise undesired bacteria and fungi would be cultured as well.

Herbaceous plants are cultured more than woody plants. Micropropagation of orchids has been commonplace for decades. More and more greenhouse and nursery suppliers are developing tissue culture laboratories for orchids and other herbaceous ornamentals. Extensive research with woody species has been less successful except for the cloning of some rapidly growing strains of lumber-producing conifers.

Overall the field of tissue culture, with its potential for human benefit, is still a true frontier.

### Cuttings

Most houseplants and woody ornamental species are propagated from **cuttings,** segments of the plant that are removed and encouraged to regenerate a new plant body. We will describe the common methods of preparing various leaf and stem cuttings of popular plants. Some plants, such as apples and plums, can be propagated by root cuttings. Generally speaking, cuttings should be rooted at about 20°C (67°F), have uniform moisture and humidity, and be protected from direct sunlight and overheating.

#### Rooting in Soil

The environmental requirements for rooting cuttings are approximately the same as for germinating seeds (see Chapter 10). The rooting medium should be sterile (free of disease-causing organisms), be able to hold water, and yet drain well rather than become soggy. Builder's sand ("sharp" sand) and perlite (a white material of volcanic origin) are excellent rooting media. Newly purchased media are probably sterile; otherwise heat, steam, or fungicide treatment of the medium may be necessary (see Chapter 12).

Adventitious root formation is usually enhanced by treating the cutting with a growth-promoting hormone called **auxin.** Professional horticulturists may use a liquid auxin solution, but **rooting powders** are more convenient for home use. The commercially available rooting powders usually contain crystals of the synthetic auxin, indolebutyric acid (IBA), mixed with talc. Auxins have complex effects on plant growth (see Chapter 9), so dip only the immediate part of the cutting to be rooted in the powder. Otherwise adventitious root development may be promoted, but bud development may be inhibited. *The use of a rooting powder is assumed in all the techniques described subsequently.*

FIGURE 5-43 *Botrytis,* or gray mold, has completely overgrown these geranium cuttings.

Once the cuttings have been placed in the medium, keep the medium uniformly moist by sprinkling or misting. When the root system is well developed, transplant the new plants into their own containers. (Chapter 19 details container plant care.) Some species form extensive roots in a few days, such as Swedish ivy (*Plectranthus australis*). Others, such as *Sansieveria,* require weeks in a rooting bed.

If cuttings develop mushy, brown rotten spots accompanied by gray mold filaments, they are probably infected with *Botrytis* mold (Fig. 5-43). This blight begins on dead tissues and spreads to living ones. Contamination of the rooting medium by fungi is the cause, enhanced by crowding, excess humidity, and water condensation on plant surfaces. It is sometimes helpful to allow fresh cuttings to "heal" for a day or so before beginning to root them. Pretreatment of the rooting medium with a fungicide effectively prevents mold infection.

#### Rooting in Water

Rooting in water is a common practice. Most stem cuttings (and some leaf cuttings) will root in water and some plants can be maintained indefinitely in water if fed and kept rot-free. Rooting in water is not generally as rapid as rooting under optimum conditions in solid media because less oxygen is available and such cuttings are more susceptible to rotting. Also, cuttings rooted in water do not generally transplant to soil as successfully as cuttings rooted in soil, for the roots tend to be more brittle and are easily broken. Light tends to inhibit root formation so that tinted or wrapped containers may be more effective than clear containers.

#### Rooting in "Air"

Some plants will successfully form roots in a self-locking plastic sandwich bag with a piece of dampened paper towel. Examples are the plantlets from such plants as piggyback plant (*Tolmeia*), strawberry begonia (*Saxifraga sarmentosa*), *Kalanchoë* species, and spider plant (*Chlorophytum* spp.) (Fig. 5-44), cuttings that root rapidly, such as those of Swedish ivy, the succulent leaves of jade plant (*Crassula argentea*), and "budded-off" branches of small cacti. Keep bagged cuttings out of direct sunlight to avoid cooking the plants during rooting! As soon as new roots begin to develop, the young plants are ready to be potted.

FIGURE 5-44 Plantlets of the spider plant (*Chlorophytum*) are produced at the tips of stolons. (Photo by Mark R. Fay.)

### Soil and Air Layering

In the layering process adventitious roots form on the stem *before* the cutting is removed from the parent plant. Layering is a successful method for some large plants or those that will not root from cuttings. In **soil layering** a vigorous young shoot is anchored to the soil and kept there until roots have formed. The stem is then removed below the root-bearing portion and transplanted. Brambles (*Rubus* spp.), English ivy varieties (*Hedera helix*), coleus (*Coleus blumei*), Swedish ivy, philodendrons (*Philodendron*), and other vining or climbing plants respond well to soil layering.

Many tropical and semitropical plants are propagated by **air layering** [Fig. 5-45 (a) and (b) and Table 5-4]. Following root formation, cut off and transplant the root-bearing shoot. The remaining stump will usually develop new shoots by activation of existing dormant buds or growth of adventitious shoots from the rootstock. Air layering has two desirable results: new individuals are propagated and original individuals become fuller by development of several main stems.

(a)

(b)

FIGURE 5-45
(a) Air layering of *Ficus*. (b) The closeup shows the roots that have grown throughout the moist sphagnum packing. (Photos courtesy of Robert J. Tomesh.)

**TABLE 5-4 Instructions for air layering.**

1. Notch, slit, or partially girdle the site, preferably just below a node on the previous year's growth.
2. Rub rooting powder into the wound. Secure the wound slightly open with a toothpick if a slit technique is used.
3. Pack slightly moist (wettened, then wrung out) sphagnum around the wound and cover with clear plastic taped snugly to the stem above and below. If the film is adequately sealed, no additional moisture will be required. After several weeks to months white roots will be observed growing within the sphagnum.
4. When roots are well developed, cut the stem below the rooted region and plant the rooted cutting in a container.

### Leaf Cuttings

A typical leaf consists of **blade** (the flat, expanded part) and **petiole** (the leaf stalk). The place on the stem where a leaf is attached is called a **node.** A **bud** is also present at a node. Some leaves are **sessile,** lacking a petiole. There are several methods of preparing a leaf cutting, depending on species, as described in Table 5-5.

Protect leaf cuttings from excessive water loss because, until roots form, they are unable to take in water other than through the exposed tissues. They also require protection from excess heat and brightness, which accelerate metabolism.

**TABLE 5-5 Techniques for preparing leaf cuttings of selected houseplants.**

A. Technique for large *Begonia rex* varieties.
Trim leaf blades or punch disks, making razor crosscuts on the underside of the main vein of the leaf. Adventitious roots and buds will develop at the cut sites.

B. Technique for jade plant, *Kalanchoë*, and similar succulent-leaved plants.
Trim end of leaf and insert into medium. Roots and buds will form at the cut end of the main leaf vein.

C. Technique for *Sansieveria.*
Cut leaves into sections and insert into medium. Roots form first; then a new offshoot.

D. Technique for African violets (*Saintpaulia*), peperomias (*Peperomia*), florists gloxinia (*Sinningia speciosa*), and jade plant.
Use an entire leaf, as in Fig. 5-46.

FIGURE 5-46 African violets (*Saintpaulia*) are propagated from leaf cuttings. New shoots and roots are initiated from the end of the petiole. (Photo by Mark R. Fay.)

### Miniature Plantlets on Leaves

Piggyback plants and some *Kalanchoë* and *Bryophyllum* species (Fig. 5-6) produce miniature plantlets on the leaves as a natural means of self-propagation. When plantlets become obvious on a mature leaf, remove the leaf and secure it against the surface of the medium. Rooting powder is not needed, for the plantlets will rapidly root themselves, especially if "tented" for a week or so.

### Herbaceous Stem Cuttings

The majority of soft-stemmed or **herbaceous** plants are readily propagated by stem cuttings (Table 5-6). The cuttings may be a middle segment of stem, including a node, but the most common type of cutting is a **tip cutting,** using the terminal 5 to 10 cm (2 to 4 in.) of a healthy, actively growing shoot (Fig. 5-47). Because of the vigor of young tissues, tip cuttings are usually the most rapid to form new root systems.

**TABLE 5-6 Techniques for propagating by herbaceous stem cuttings.**

A. Stem segment technique.
For rubber plant, cultivated geraniums (*Pelargonium*), philodendrons, and peperomia, use a stem segment bearing a node and a complete leaf. If the leaf is very large, trim it to reduce transpiration.

B. Tip cutting technique.
Cut a vigorous tip cleanly from the parent plant without tearing or crushing, including at least two nodes. Remove most large leaves, for they are a major source of water loss. To enhance root formation at nodes of such tougher-stemmed plants as *Pelargonium*, make tiny cuts just below each node or into the end of the cut stem before dipping into rooting powder. Insert cuttings into medium so that one or two nodes are buried.

### Woody Stem Cuttings

Most houseplants are herbaceous. Many woody trees, shrubs, and vines, however, are used for landscaping and some are container grown indoors or in window boxes. To propagate woody stems, use stem tips lacking floral buds to direct energy toward vegetative growth.

Successful woody-cutting propagation requires recognition of innate seasonal rhythms. Most vigorous growth occurs in the spring, tapering off to winter dormancy. Newly produced spring and early-summer tissues are still soft, lacking secondary growth (Chapter 8). The new shoots are succulent and easily broken; therefore cuttings taken at this time of rapid growth are called **softwood cuttings** (Table 5-7).

FIGURE 5-47 Both stem cuttings and stem tip cuttings of *Ficus* are in this rooting bed. (Photo by Mark R. Fay.)

**TABLE 5-7 Some genera that are successfully propagated as softwood cuttings, particularly if the rooting beds are misted regularly.**

**Medium: coarse sand**

*Buddleia:* butterfly bush
*Catalpa:* catalpa or Indian bean
*Cercis:* redbud or Judas tree
*Cornus:* dogwood
*Deutzia:* deutzia
*Ginkgo:* maidenhair tree
*Hibiscus:* rose mallow
*Hydrangea:* hydrangea
*Koelreuteria:* goldenrain tree
*Lonicera:* honeysuckle
*Parthenocissus:* Virginia creeper, Boston ivy, American ivy
*Rhus:* sumac, smoke bush
*Sambucus:* elderberry, elder
*Spiraea:* spirea, bridal wreath
*Syringa:* lilac
*Tamarisk:* tamarix
*Viburnum:* viburnum, highbush cranberry and relatives

**Medium: equal parts coarse sand and milled peat moss (to provide acidity)**

*Rhododendron:* azaleas and rhododendrons
*Vaccinium:* blueberries

*Technique:* Cut softwood tips 7 to 12 cm (3 to 5 in.) long and remove most of the leaves except for a cluster at the tip. Use rooting hormones sparingly, if at all, to avoid inhibiting shoot growth. Provide bottom heat, shade, and high humidity. Water by misting.

**TABLE 5-8 Propagating from semi-hardwood cuttings.**

**Conifers**

*Juniperus:* junipers
*Sequoia:* coast redwood, big tree
*Taxus:* yews
*Thuja:* arbor vitae, white cedar
*Tsuga:* hemlock

**Broad-leaves**

*Abelia:* glossy abelia
*Arbutus:* strawberry tree
*Aucuba:* Japanese aucuba
*Buxus:* boxwood
*Calluna:* Scotch heather
*Camellia:* common camellia
*Choisya:* Mexican orange
*Cotoneaster:* cotoneaster
*Erica:* heath
*Escallonia:* escallonia
*Euonymus:* spindle tree
*Hedera:* English ivy and variations
*Iberis:* Evergreen candytuft
*Ilex:* holly
*Mahonia:* Oregon grape
*Osmanthus:* osmanthus
*Pachysandra:* spurge
*Pieris:* Japanese pieris, andromeda
*Pyracantha:* firethorn
*Rhododendron:* rhododendron
*Skimmia:* Japanese skimmia

*Technique:* Use 10 to 15 cm (4 to 6 in.) of the stem tip, removing about $\frac{2}{3}$ of the basal foliage. Rooting hormones can be used effectively. Provide bottom heat, humidity, and shade.

Take **semi-hardwood cuttings** (Table 5-8) when tissues have begun secondary growth and development, in later spring and into early autumn. A distinction between this and the softwood stage is that stems are tougher and bend rather than break. Semi-hardwood cuttings are less susceptible to adverse conditions and disease and their stored food enables them to survive longer during propagation.

**Hardwood cuttings** (Table 5-9) are usually taken off dormant deciduous plants after leaf fall. Their tissues are mature and contain plentiful food reserves.

### Propagation via Specialized Stems

**Runners** or **stolons** are horizontal stems that grow over the ground surface, developing new plantlets at the tips. Strawberries (*Fragaria*), strawberry begonia, spider plant, and *Callisia* are attractive in hanging containers or on pedestals, with the stolons hanging down. To propagate, anchor the stolon tip to the soil surface of a small pot (soil layering). When an independent root system has developed, cut the stoloniferous attachment. Or you may root severed plantlets in a plastic bag, as described earlier.

**Rhizomes** are horizontal stems that grow at or below the ground's surface and produce new shoots and roots at the nodes. Lawn grasses, lily of the valley (*Convallaria majalis*), Iris (*Iris*), ginger (*Zingiber*), bamboos, blueberries (*Vaccinium*), ferns, and others spread by rhizomes. To propagate, section between rooted nodes and transplant the sections.

Many plants grow close to the soil as a **rosette** of leaves that arise from a short, stubby stem region, or **crown**—for example, strawberries, African violets, strawberry begonia, and piggyback plant. The mature crown produces short lateral shoots, called **offsets** or **crown divisions**, which can be re-

**TABLE 5-9 Propagating from hardwood cuttings.**

*Berberis:* barberry
*Celastrus:* shrubby bittersweet, American bittersweet
*Chaenomeles:* flowering quince
*Cytisus:* Scot's broom
*Forsythia:* forsythia, golden bells
*Hydrangea:* hydrangea
*Kerria:* kerria
*Kolkwitzia:* beautybush
*Ligustrum:* privet
*Lonicera:* honeysuckle
*Philadelphus:* mockorange
*Populus:* poplar, aspen, cottonwood
*Ribes:* currants and gooseberries
*Rosa:* roses
*Salix:* willow
*Spiraea:* spirea
*Symphoricarpus:* snowberry, waxberry
*Vaccinium:* blueberry
*Viburnum:* viburnum, highbush cranberry and relatives
*Vitis:* some grapes
*Weigela:* weigela
*Wisteria:* wisteria
also some fruit trees

*Technique:* Take tip cuttings 15 to 30 cm (6 to 12 in.) long and place in rooting medium outdoors (where winter permits), buried except for a single bud. Root hormone tolerance is excellent, but these cuttings may not be as responsive to them. Allow to root under cool field conditions. Transplant in spring before buds begin to swell. Rooted cuttings can be stored in moist sphagnum and plastic wrap until planting. Rooted rose canes may be dipped in paraffin to retard transpiration during storage. Storing rooted hardwood cuttings upside down also will inhibit bud growth.

moved and treated as a leaf cutting until roots form.

Bulbs, corms, and tubers are specialized stems that assist in natural vegetative propagation. Each is distinct in structure. A **bulb** consists of a small mound of stem, bearing overlapping fleshy food-storage leaves (bulb scales) in addition to the above-ground leaves, as in onions and garlic (*Allium*), daffodils (*Narcissus*), hyacinths (*Hyacinthus*), and tulips (*Tulipa*). Bulbs form small new bulbs at their bases. Small bulbs may also develop in the axils of the scalelike leaves. Left in the ground, the original bulb continues to produce small bulbs and older ones are eventually broken away by the younger ones, thus increasing the number of plants in a growing cluster. Some plants, such as the Egyptian onions, produce small aerial bulbs at the tips of aerial shoots. Small bulbs also form in the leaf axils of certain lilies. Whatever the site of origin, young bulbs can be planted directly without rooting.

**Corms** lack fleshy scale leaves and the bulk of the structure is the squatty, swollen stem. Gladiolus (*Gladiolus*) and crocus (*Crocus*) are corm plants. Small corms originate from buds at the nodes and can be removed and planted.

**Tubers** are the expanded tips of rhizomes, as in the Irish potato (*Solanum tuberosum*), Jerusalem artichoke (*Helianthus tuberosus*), tuberous begonias (*Begonia tuberhybrida*), and *Caladium*. Plant whole tubers or cut sections bearing one or more nodes. (Nodes can be recognized by the presence of axillary buds or "eyes.") Cut pieces should be dried at room temperature for several days to allow formation of protective tissue; otherwise rotting will occur.

Some garden and ornamental plants store food in swollen root or stem areas that are referred to as tuberous roots or tuberous stems. **Tuberous root** plants include sweet potato (*Ipomoea*) and dahlias (*Dahlia*). Because they are roots and not modified stems, nodes and internodes are not present so that success of propagation by division depends on whether a bud is present at the stem end of the root. Tuberous begonia, cyclamen (*Cyclamen*), and florist's gloxinia (*Sinningia*) have **tuberous stems**, actually a type of crown.

## Budding and Grafting

If orchardists relied on fruit trees grown from seed, the frustrating result would be constantly unpredictable products. Although some or many of the trees would produce good fruit, the results would be inconsistent because of genetic recombination that occurs during sexual reproduction (Chapter 6). In contrast, vegetative propagation methods ensure exact perpetuation of desirable fruit characteristics. Fruit trees grown from seed are primarily used as hardy and disease-resistant rootstocks and in genetic experimentation.

**Budding** and **grafting** are asexual propagation techniques for woody plants, particularly fruit trees. When an exceptionally good tree is found, buds or **scions** (young shoots) are removed and then grafted onto other related trees (the **rootstocks**). Each scion and bud develops into a branch with the desirable characteristics. An advantage of budding and grafting is that they allow immediate improvement of an existing tree, without having to wait through the juvenile period of a young tree. In addition, the techniques permit development of several different (but related) fruit varieties on a single tree!

There is no magic to budding and grafting—they are a matter of carefully putting the meristematic tissue (vascular cambium) of the bud or scion in contact with that of the rootstock and then allowing natural secondary growth (Chapter 8) to occur, thereby fusing the parts into one.

Grafting is a more drastic method than budding, for it involves cutting off entire branches or making major cuts into branches to complete the union of stock to scion. Occasionally grafting is used to repair a valuable tree damaged by rodents or farm implements (see Supplement 5-5).

Supplements 5-1 to 5-5 describe the methods for T-budding; whip, tongue, and bench grafting; cleft grafting; side grafting; and bridge grafting.

For the average person, **dwarf** varieties of fruit trees have much to offer over full-sized trees. Each tree takes less space, has lower water and nutrient demands, and, with proper management, will produce more than adequately for home use. In addition, dwarf varieties reach fruit-bearing maturity earlier—some as young as 2 or 3 years of age. Dwarf apple, cherry, pear, peach, nectarine, citrus, apricot, and plum varieties are available.

Dwarf variety trees are actually scions of normal-sized trees grafted to **dwarfing rootstocks.** Dwarfing rootstocks apparently exert physiological controls over scions grafted to them, thus limiting vegetative growth (Fig. 5-48). The most notable results in dwarfing have come from use of apple rootstocks developed in England. There are two basic types of dwarf apple rootstocks: stock developed at the East Malling Research Station is designated by "M" and stock developed jointly by East Malling and Merton is designated "MM." Initially the MM series was developed specifically for resistance to woolly aphids.

Possible explanations for the dwarfing effect are still debated in spite of much study. The three basic research approaches have been nutrient uptake and utilization, translocation of nutrients and water, and alterations in internal growth factors. Investigations have shown that all three influence dwarfing, although conflicting reports have caused difficulty in interpreting some experimental results. Contradictions, however, do not necessarily mean that some of the results are inaccurate or invalid, for environmental factors and genetic dominance interactions between rootstock and scion also appear to influence the total picture of dwarfing. Even with the *same* rootstock and scion,

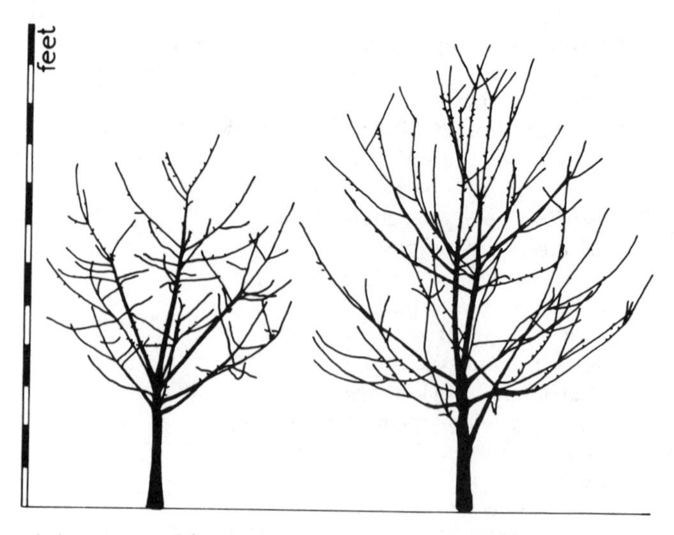


**FIGURE 5-48** Dwarf fruit trees are produced by grafting desirable tops to **dwarfing rootstocks.** The degree of dwarfing influence varies with rootstock variety. (Photo courtesy of East Malling Research Station, Kent, England.)

size and growth rate differ, depending on such environmental conditions as soil fertility, soil depth, soil moisture, exposure, climate, and cultural techniques.

At present, the most acceptable explanation for the dwarfing mechanism appears to be a growth stimulator/inhibitor relationship (Chapter 9). For example, **gibberellic acid** (GA), a growth stimulator, is produced in the roots. The roots of a dwarfing stock produce less GA than those of a full-sized tree. Consequently, less GA is transported upward, providing less stimulation for top growth.

It was also found that leaves of trees with the greatest dwarfing contain substances that stimulate the breakdown of the **auxin** indoleacetic acid (IAA), which is a growth promoter. Studies of dwarfing in apple stock showed a range of relationships (Fig. 5-49), with the most dwarfed stock having lower levels of growth-promoting substances and higher levels of growth-inhibiting substances.

Dwarf trees generally develop a good root system, but the wood produced in the stems is more brittle; consequently, the branches have a greater tendency to break and should be staked for support during fruit bearing, especially for the first five years.

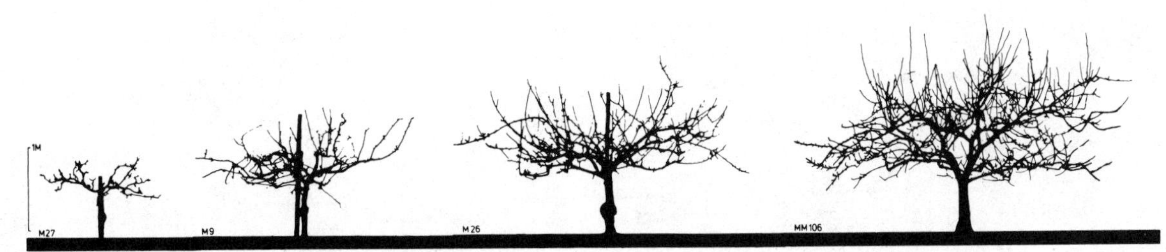


**FIGURE 5-49** The degree of dwarfing caused by different rootstocks is dramatically illustrated by this photographic composite. (Photo courtesy of East Malling Research Station, Kent, England.)

## SUPPLEMENT 5-1: T-Budding

T-budding is the most common method of developing new trees because it is simple, fast, and highly successful [Fig. 5-50 (a) to (e)]. This method can be used in propagating peaches, apricots, prunes, pears, apples, cherries, citrus, nut trees, and ornamentals. Most budding is done during the late summer months of July and August when buds are fully developed. A very sharp pocket knife is adequate equipment.

*Stock.* The stem of the rootstock plant can vary from 6 to 12 mm in diameter. In young saplings (seedlings) the buds are placed 5 to 7 cm above the ground. In more mature trees, such as those 3 to 4 years of age, main branches of approximately 12- to 15-mm diameter can be budded. The most important condition of the stock is that the bark must be **slipping**; that is, the vascular cambium must be actively dividing, producing a zone of soft, differentiating vascular tissue (Chapter 8) so that the bark slips easily from the wood. To ensure optimum condition of the bark to receive buds, water the rootstock heavily 2 to 3 days before budding takes place.

*The Budstick.* The budstick is a twig from the current year's growth. It should be taken from a tree of the variety desired and, if possible, one that has produced fruit with which you are particularly pleased. Cut (not break) the leaves off adjacent to the bud. Keep the budstick wrapped in moist paper towels in a plastic bag. It is best to use the buds as soon as possible, but wrapped budsticks can be stored in a refrigerator for a few weeks.

*Preparing the Stock.* On the outside of the twig (the side away from the center of the tree in older stock trees), cut a longitudinal slit approximately 4 cm long, deep enough to go through the bark and slightly into the wood. At the top of this slit, make a perpendicular cut, to form a T (*ergo*, the name T-budding). To do so, start by cutting into the bark with the middle of the blade at an acute angle approximately 1 to 1.5 cm to the right of the longitudinal line. Using a rolling motion, cut across to 1 to 1.5 cm on the other side. Gently push the bark away from the wood with the angled blade.

*Removing the Bud From the Budstick.* Select only plump, healthy vegetative buds from the middle of the budstick. Hold the budstick with the growing tip toward your body, grasping with the thumb and index finger of your opposite hand and resting the stick against the middle finger. Using a very sharp knife, start a cut approximately 1.5 to 2 cm below the bud. Slice under the bud with a smooth, controlled motion toward your body, cutting the bark and some of the wood and gradually coming out 1 to 1.5 cm above the bud. At this time trap the bud between knife blade and thumb; it is ready for insertion. The bud should have a smooth, flat cut surface with as little bulge as possible to ensure flat contact with the stock for a successful union.

*Inserting the Bud.* Carefully slide the bud (correct side up) into the T-bar portion of the cut. If the bark is slipping easily, the bud will move freely under the bark. If there is a little resistance, place the knife tip approximately 3 to 5 mm below the petiole base and *pull* the bud down into the

(a)

(b)

(c)

(d)

(e)

**FIGURE 5-50** **T-budding.** (a) Cutting out bud from budstick. (b) Three views of the excised bud. (c) T-cut made in the stem to be budded. (d) Bud being inserted into T-cut. (e) Finished appearance of the bud, inserted in place and wrapped with a special rubber strip. (From "Budding and Grafting Fruit Trees in the Home Garden," Leaflet 2990, Division of Agricultural Sciences, University of California.)

slit (being careful not to damage the bud). This method is more effective than pushing the bud, as is sometimes done. Pushing can result in an accordioned bud with a damaged, wrinkled surface. After the bud has been pulled far enough into the cut so that the "tail" is under the bark, the union is ready for wrapping.

*Wrapping the Bud.* Use rubber wrapping strips and begin wrapping at the base of the bud. Start with a self-binding wrap and move up the stem, being careful to wrap tightly and overlap the previous wrap. Continue to the base of the petiole and then move around the bud and continue above it. Complete the wrapping one or two wraps beyond the T-bar cut and stop with a self-binding wrap.

*Caring for the Bud.* Check the buds in a week or two. If plump and full, they are alive and well, but if they have become shriveled, there is still time to do another budding a few inches down the stem. Healthy buds will continue to bind to the stock branch, open, and put out fairly extensive growth. The rubber strips will deteriorate and fall away without restricting or retarding growth.

Once the buds have developed into strong shoots, cut off the rootstock stem approximately 8 to 10 cm above the topmost budded shoot to prevent dieback. After the budded shoot is well established, the remaining excess branch can be trimmed back adjacent to the bud to facilitate healing. Generally the new bud will not need any pruning unless it is excessively vigorous; then it can be headed back in the usual manner.

## SUPPLEMENT 5-2: Whip, Tongue, and Bench Grafting

**Whip** grafting is synonymous with **tongue** grafting, where the stock is attached to the stem portion of the rootstock. In a **bench** graft a scion is attached directly to a piece of root. The procedure is the same for both types of graft (Fig. 5-51). This grafting procedure is used widely in grafting young seedling whips (rootstocks) to a desirable scion for nursery stock to be distributed to retailers. The home gardener can do the same by purchasing strong, disease-resistant, hardy saplings to which favorite scions can be grafted during dormant periods, preferably January through March.

*Preparing the Scion.* Collect scions from 1-year-old wood—measuring approximately 6 to 12 mm in diameter (of varying sizes to match the varying sizes of stock). Select a region on the scion that is approximately the same diameter as the rootstock where the graft is to be made. With a steady, continuous diagonal stroke, cut off the scion. The face of the cut should be about 4 cm long. Then cut straight downward about 1 to 1.5 cm parallel to the long axis of the stem, starting about $\frac{1}{3}$ from the tip of the cut face.

*Preparing the Stock and Making the Graft.* Select an area about 7 cm from the soil line and then make a diagonal cut and a straight cut just as described for the scion. With scion and stock cut with exposed 4-cm face surfaces and matching 1- to 1.5-cm deep cuts, push the two parts together deeply enough to cover all exposed surfaces with the complementary parts.

*Wrapping the Graft.* Wrap the union to maintain contact and lessen dehydration, preferably using rubber wrapping strips, although electrician's or adhesive tape can be used. Waxing is not essential but helps lessen water loss and exposure to pathogens.

*Care of the Graft.* Place the grafted stocks into a bundle of approximately ten plants, wrap with moist paper towels, place in a plastic bag, and store in a cool place or refrigerator until it is safe to plant the rootstock. Because there is a maximum of connecting cambial surface and the connection is tight even before wrapping, the success rate of this graft is high. The cambial regions will continue dividing to produce new cells, thereby binding the plant parts together. As healing continues and tissue area increases, the wrapping material (if not rubber strips) should be cut to allow expansion.

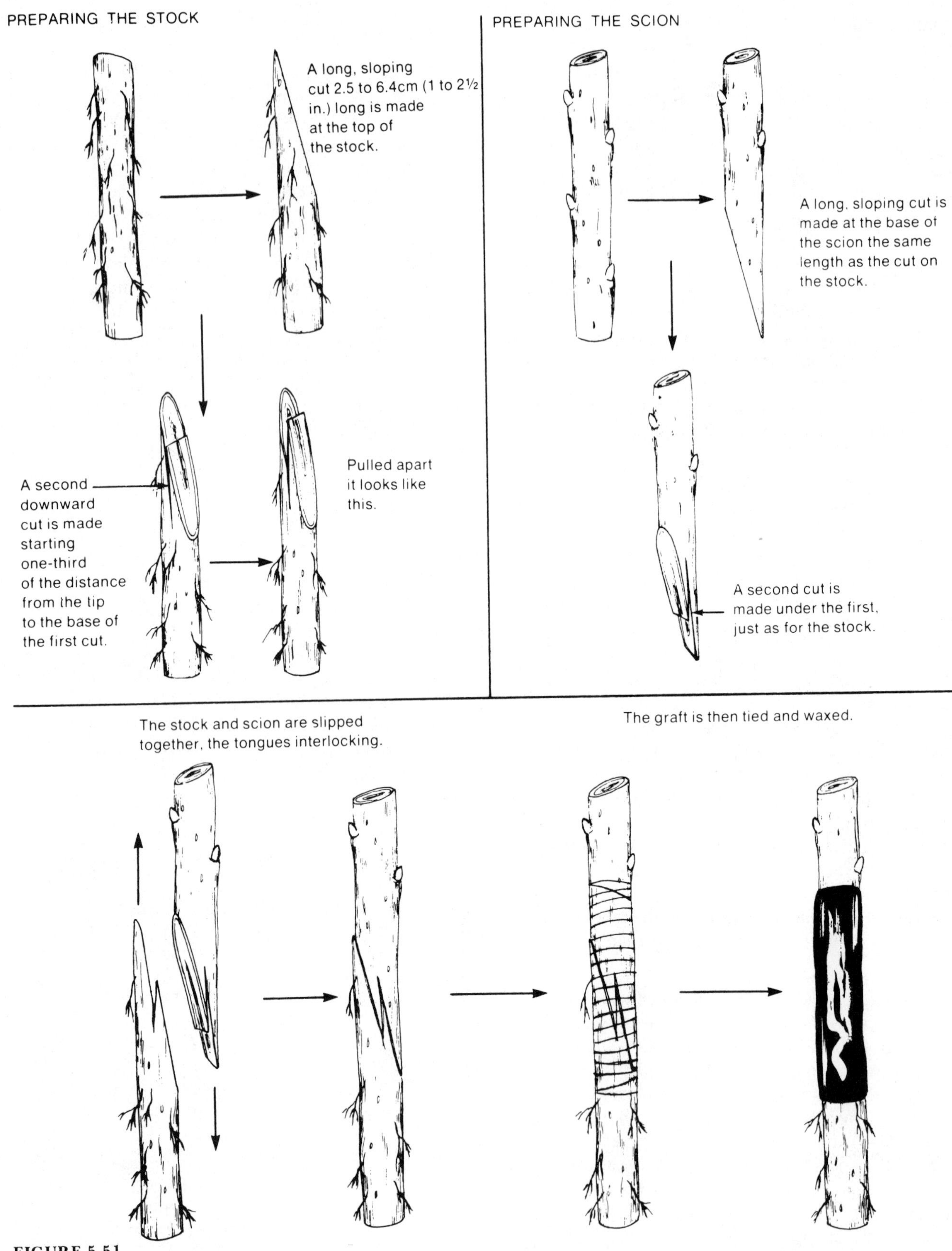


**FIGURE 5-51**
The **whip, tongue,** or **bench graft.** In this series of illustrations, the scion is grafted directly onto the root of the rootstock plant, rather than onto the stem of the rootstock. The method is the same for both sites. (From H.T. Hartmann, W.J. Flocker, and A.M. Kofranek, *Plant Science: Growth, Development, and Utilization of Cultivated Plants*, 1981. Reprinted by permission of Prentice-Hall, Inc., Englewood Cliffs, N.J.)

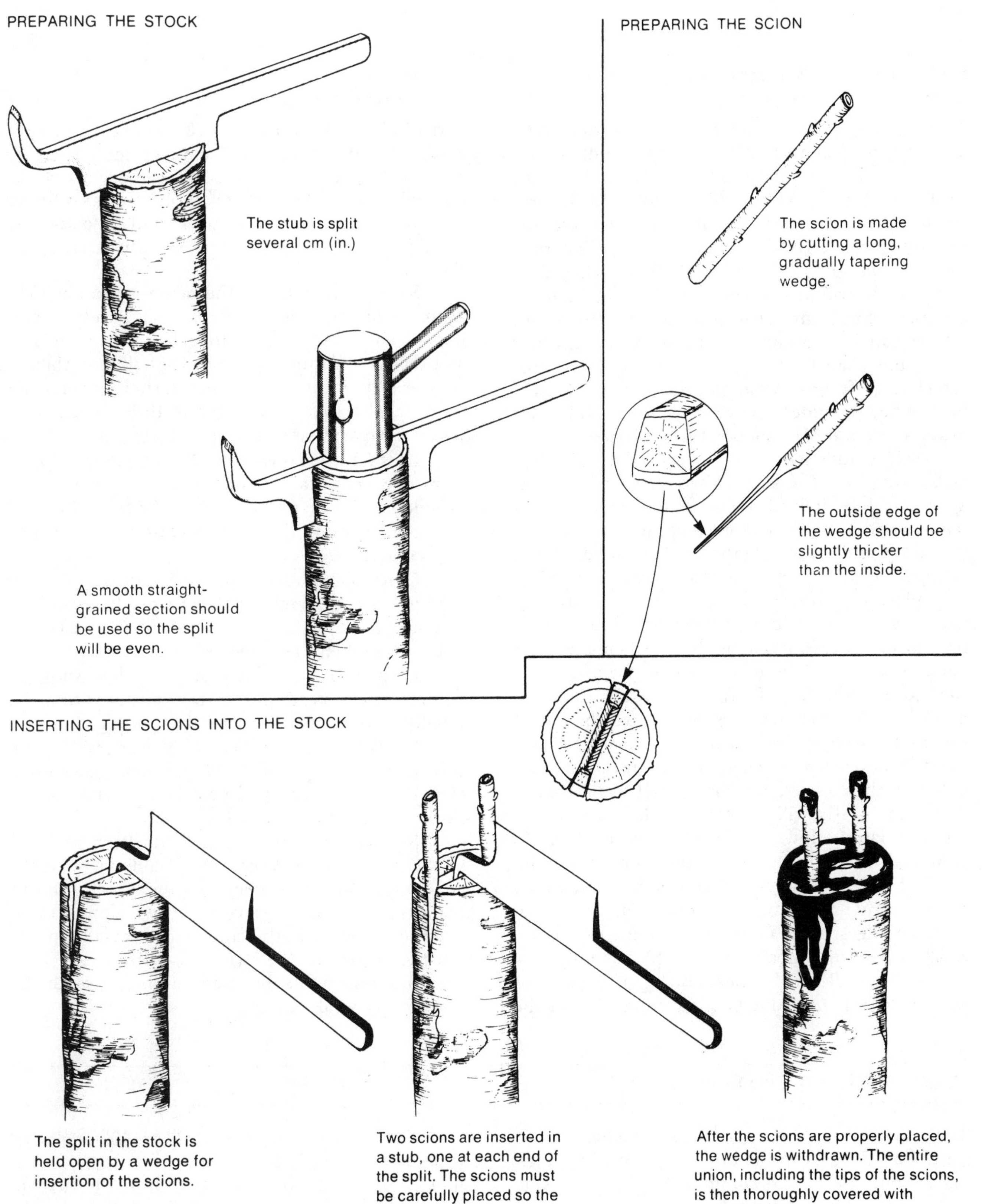


FIGURE 5-52
The **cleft graft.** This method is used often for top-working a tree. For it to succeed, the vascular cambium layers of the scion and stock must be carefully matched, as shown in the small cross-section diagram. (From H.T. Hartmann, W.J. Flocker, and A. M. Kofranek, *Plant Science: Growth, Development and Utilization of Cultivated Plants,* 1981. Reprinted by permission of Prentice-Hall, Inc., Englewood Cliffs, N.J.)

## SUPPLEMENT 5-3: Cleft Grafting

Cleft grafting (Fig. 5-52 on page 103) is utilized on a mature tree when, for instance, fruit production is not sufficient, the variety is not satisfactory, or the gardener simply wants to add varieties to the tree or change branch arrangement. In the last case, the decision is made to "top work" the tree. Generally the tree should be cut back to those branches close to the main trunk. Branches 2 to 5 cm in diameter are preferable, although branches up to about 10 cm can be worked. Smaller branches will heal faster and chance of rot is limited. The best time for cleft grafting is from January through March; in harsher climates, however, the period after heavy snow and cold temperatures is better.

Select budsticks from 1-year-old twigs. Generally they are the size of wooden lead pencils, approximately 0.6 to 1 cm in diameter. Collect scion wood from 1-year-old twigs in the fall after leaf drop and store it until needed, bundled and wrapped in moist paper towels in a plastic bag in a cool place or refrigerator. Appraise the entire tree and make a mental note or sketch of branches to be grafted. Assessing all the 3- to 10-cm-diameter limbs, establish a tentative number of grafts to be worked into the tree. There is no limit to the number of grafts that a tree should have, but they should be well spaced to avoid competition for space and eventual crowding. In this instance, more *is* better than fewer, for there is always a chance that some grafts will not take. Such grafts can be removed later, allowing adjacent ones to fill in. To trim back the branches, use a fine-toothed saw instead of the usual large-toothed pruning saw, which causes tearing of the bark and meristematic tissues. If rough edges occur on the stub, trim with a knife to allow a smooth area for scion replacement.

Use a mallet or hammer and a grafting tool or similar item, like a large butcher knife, to split the end of the stub about 5 to 8 cm deep. The split should be off center to discourage the possibility of too deep a split down the center. Insert the hooked end of the grafting tool or a screwdriver into the crevice to hold open the split. The opening should be wide enough to allow easy insertion of the scion.

Select an area on the scion approximately 1 cm in diameter and make a long, steady, tapering cut about 2.5 to 4 cm long. Turn the stick over and repeat, creating a wedge with a slightly thicker outside. After cutting the wedge, shorten the scion to a length that leaves about three buds. Two scions are usually inserted in each cleft graft.

Insert the scions into the crevice developed from the split, taking special care to match the cambial zones of the scion and the host. Remove the tool to allow the stub to close tightly against the scion(s).

Cover all exposed areas created by the grafting with a commercial grafting wax. Be especially careful of the splits and take time to notice airholes or sunken areas, where wax might be taken into deep areas. Continue to fill in low spots to ensure protection from dehydration and pathogens. Finally, wax the exposed tip of the scion.

Check the graft a day or so after grafting and add more wax if needed. As the scion takes hold, meristematic activity will provide a strong connection between scion and stock.

Because the larger branch has been replaced by two pencil-thin twigs, the juvenile vigor of the new twigs will cause them to grow rapidly. During the first year allow growth of the scion without much pruning. If both scions are successful, prune out the weaker of the two the second year. Third-year pruning should be used to train growth and remove any weak crotches.

## SUPPLEMENT 5-4: Side Grafting

The side graft is used when stock branches are too small for cleft grafting and too large for whip grafting (about 1.5 to 3 cm in diameter). Do side grafting January through March, depending on the climate.

Select a twig that is about 1 cm in diameter. The scion should be about 8 cm long and include two to three buds. Trim the end to a 2.5 cm wedge.

Place a sharp knife at a 20° to 30° angle and strike with a mallet or hammer to make a cut approximately $\frac{1}{3}$ to $\frac{1}{2}$ the way into the branch. Open the cleft by bending the branch opposite the cut and insert the scion. The scion is set at a slight angle to get maximum vascular cambium contact between scion and stock. Release the branch to pressure hold the scion. Taping or wrapping is usually unnecessary. Carefully cut off the stock branch 8 to 15 cm from scion insertion or leave the branch until the scion starts its growth. In either case, thoroughly wax all exposed regions and down the side of the stock about 3 to 5 cm beyond the split.

Watch the graft carefully, especially the day after grafting. Check the wax for airholes and/or sunken areas where the wax has fallen into a deep crack. Remove any lateral buds on the stick that might interfere with growth of the scion.

## SUPPLEMENT 5-5: Bridge Grafting

The bridge graft is a repair graft, used to repair bark damaged by rabbit and rodent feeding, winter cold, disease, and implements. If the trees are small and easily replaceable, it might be wiser to plant another tree. Bridge grafting is usually done to save a mature tree that is already bearing and too valuable to lose.

The damage is often a large patch of bark and vascular cambium that has been removed or, in some cases, girdling (removal of a ring of bark encircling the tree). Several scions are grafted between healthy tissues above and below the wound, reconnecting the vascular tissues. In a few years the scions enlarge and fuse, forming a protective covering over the wood.

## SUMMARY

Chromosomes of higher organisms exist as **homologous pairs.** Cells with pairs of chromosomes are **diploid** ($2n$); those with only one member of each chromosome pair are **haploid** (n).

Each cell has a cycle, including synthesis and division phases. Cells reproduce by **mitosis,** an orderly process of nuclear division. DNA duplicates itself prior to division of the nucleus. Mitosis is followed by **cytokinesis,** separation of the cytoplasm into separate cells. Some cells retain their ability to divide whereas others differentiate into specialized forms and lose their ability to divide.

Most plants reproduce both sexually and asexually. **Sexual reproduction** involves the fusion of gametes (**syngamy**) to form an individual with unique genetic characteristics. **Asexual reproduction** does not involve syngamy and produces **clones** identical to the parent plant.

The life cycle of most species alternates between a haploid gamete-producing **gametophyte** generation and a diploid spore-producing **sporophyte** generation. Diploid parent cells produce haploid reproductive cells by **meiosis,** a process of reduction division with stages similar to those of mitosis. In a life cycle, syngamy restores the diploid condition.

Sexual reproduction in seed plants occurs within protective female tissues. A **seed** is an embryonic plant surrounded by stored food and protected by a water-conserving seed coat. Flowering plants may be **annuals, biennials,** or **perennials.**

Most flowers are made of four basic parts—**pistil (carpels), stamens, petals, sepals,** all attached to a **receptacle.** Some flowers are bisexual, others unisexual.

As a flower matures, **pollen grains** (male gametophytes) are produced in the anthers of the stamen, **eggs** in **ovules** within the carpels.

**Pollination** is the transfer of pollen to the stigma of the pistil. A pollen tube grows down the style to an ovule, where two **sperm nuclei** are released into the embryo sac. One sperm nucleus unites with an **egg,** forming a diploid **zygote.** The other sperm nucleus fuses with the **polar nuclei,** forming a triploid **primary endosperm cell.** The zygote produces the **embryo,** the primary endosperm cell produces the **endosperm** tissue. As the ovule develops into a seed, the ovary forms a protective **fruit.**

Fruits are categorized by whether the fruit is dry or fleshy, remaining open or closed at maturity. There are many seed-dispersal adaptations.

Some flowers are adapted for self-pollination, others for cross-pollination. Flowers have specific adaptations relating to their modes of pollination.

Methods of vegetative propagation are discussed in detail. The phenomenon of **dwarfing** is discussed in connection with budding and grafting.

## SOME SUGGESTED READINGS CHAPTERS 5 AND 6

Darwin, C. *On the Origin of Species by Means of Natural Selection, or the Preservation of Favoured Races in the Struggle for Life.* Garden City, N.Y.: Doubleday and Co., 1960. The best known of Darwin's many works, the argument for his monumental theory of evolution.

. *The Voyage of the Beagle.* Garden City, N.Y.: Doubleday and Co., 1962. Derived from Darwin's journals, the book describes his voyage of discovery. It was a popular seller when published and provided the evidence for his revolutionary theory of evolution.

Faegri, K., and L. Van der Pijl. *The Principles of Pollination Ecology,* 2nd ed. Oxford and New York: Pergamon Press, 1971. A scholarly and thorough treatment of the subject.

Grant, V. *Plant Speciation.* New York: Columbia University Press, 1971. An introduction to the mechanisms of plant evolution, with many examples.

Hartmann, H.T., and D.E. Kester. *Plant Propagation: Principles and Practices,* 3rd ed. Englewood Cliffs, N.J.: Prentice-Hall, 1975. A very complete text on propagating plants, especially by vegetative (asexual) means, incorporating exact methods as well as principles.

Huxley, J.J. (ed.). *A Book that Shook the World.* Pittsburgh: University of Pittsburgh Press, 1958. A series of five essays examining the effect of Darwin's *Origin of Species* on several fields of thought.

Levine, L. *Biology of the Gene,* 2nd ed. St. Louis: The C.V. Mosby Co., 1973. A good, concise genetics text.

Peters, J.A. (ed.). *Classic Papers in Genetics.* Englewood Cliffs, N.J.: Prentice-Hall, 1959. A collection of 28 important genetics papers published through the mid-1950s by the persons involved in many exciting, breakthrough discoveries in genetics.

Proctor, M., and P. Yeo. *The Pollination of Flowers.* St. James Place, London: Collins, 1973. An excellent and beautifully illustrated book on pollination ecology.

Srb, A., K.D. Owen, and R.S. Edgar. *General Genetics,* 2nd ed. San Francisco: W.H. Freeman and Co., 1965. A clear text in classical genetics with many questions and problems to aid the student.

Stebbins, G.L. *Processes of Organic Evolution.* Englewood Cliffs, N.J.: Prentice-Hall, 1966. A brief review of the "how" and "why" of evolution for students with an understanding of basic Mendelian genetics.

Strickberger, M.W. *Genetics,* 3rd ed. New York: The Macmillan Co., 1975. A comprehensive textbook of genetics.

Volpe, E.P. *Understanding Evolution,* 2nd ed. Dubuque, Iowa: William C. Brown Company, 1970. A brief introduction to the mechanisms of evolution.

# chapter six INHERITANCE AND GENETIC CHANGE

photo courtesy of Ingolf Vogeler

## INTRODUCTION

It is appropriate, after a discussion of plant reproduction, to take a closer look at how reproduction is related to inheritance. We have chosen to provide only a brief discussion of genetic principles, including examples of inheritance and evolution by selection as they apply to plant breeding and improvement. Manipulation of the inheritance of plant characteristics is an ancient practice, dating back to the dawn of agriculture. The result has been development of many cultivated varieties that differ in desired ways from their wild ancestors. In the chapter opening art, women in Kumasi, Ghana, display cultivated yam roots of several distinctive types.

We encourage you to explore further the fascinating and complex study of inheritance and the mechanisms of evolution through the Suggested Readings at the end of Chapter 5 as well as through additional materials that may be suggested by your instructor.

## PRINCIPLES OF INHERITANCE: GENETICS

### Gregor Mendel

Nineteenth-century experimentation by an Austrian monk named Gregor Johann Mendel is the foundation of modern **genetics**, the study of inheritance. Mendel was born in 1822 in an area of Austria that is now part of Czechoslovakia. Growing up in a rural farming community, he became interested in the gardening and fruit growing of his region. Following high school, Mendel had difficulty deciding the direction of his life, but in 1843 he decided, finally, to become a priest.

In the next 14 years Mendel unsuccessfully tried to progress in his teaching levels at the monastery and was shuffled between different educational institutions as well as teaching in local community schools. In 1856, at the age of 34, he began what would become monumental experiments in the hybridization of peas. During the next seven years he developed four basic principles: the inherited factor, dominance, segregation, and independent assortment.

Mendel was either very wise or very lucky in choosing the garden pea (*Pisum sativum*) as his experimental organism. Pea flowers are closed during the period of pollination; therefore self-fertilization usually occurs. Because the flowers were naturally protected from cross-fertilization by natural means, he could take pollen from any flower and deliberately place it on the stigma of another flower to make a controlled hybridization or **genetic cross.** He was also very observant to realize the necessity of studying the inheritance of only a few, well-contrasted characteristics through several generations. This choice was a major factor in helping him to succeed; others had included too many characteristics and had failed. Some of the features that Mendel identified and used for his genetic studies were tall or dwarf growth habit, red or white flower color, axial or terminal inflorescences, green or yellow unripe pods, inflated or constricted pods, smooth or wrinkled forms of ripe seed, and yellow or green seed color.

His results were communicated to the Natural History Society of Brüun in 1865 and published in the records of the society. Although this account of Mendel's work was distributed throughout the major libraries of Europe, no understanding or appreciation was forthcoming from his colleagues. It was not until 1900, 16 years after his death, that his work was rediscovered by three researchers working independently, K.E. Correns (Germany), Hugo de Vries (Holland), and E. von Tschermak (Austria).

### The Inherited Factor

During the seven years of experiments Mendel noted the exact perpetuation of very distinct physical characteristics of the peas. Red-flowered plants predictably produced red-flowered plants and tall plants produced tall plants. So he reasoned correctly that there must be innate "factors" within the plants that carry the message of each characteristic from generation to generation. Today we call these inherited units or factors **genes,** which are distributed on the chromosomes. (See also Chapter 3.) Most genes for the same characteristics or traits have two or more alternate forms or **alleles;** so a single characteristic may have two or more **expressions.** More specifically, an allele is an alternate expression for a gene, present at the same position (locus) on a chromosome.

### Dominance

In Mendel's crosses of purebred tall plants with dwarf plants, for instance, he noted that all the **first filial generation** ($F_1$) were tall plants. In reviewing what had taken place, he assumed that the expression of the factor (gene) for tall overshadowed or suppressed the expression of the factor for dwarfness. He then allowed the $F_1$ hybrids (all tall) to self-pollinate, producing the **second filial generation** ($F_2$). Seeds of the $F_2$ generation produced a ratio of approximately three tall plants to one dwarf. He concluded that some factors are **dominant** over others when they occur together. The **recessive** factor can only be expressed when it occurs with another recessive factor for the given characteristic.

In genetic terminology dwarfness in peas occurs when the plant is **homozygous** (having like alleles for a characteristic) for dwarfness. In contrast, tallness occurs when the plant is either homozygous for tallness or **heterozygous** (having genes for alternate expressions of a characteristic).

It is obvious that the visible physical appearance of an organism, the **phenotype,** is not an accurate predictor of the genetic potential of the offspring; however, knowledge of the actual genetic constitution, or **genotype,** usually is. For this reason, pure-breeding homozygous plants are essential to genetic hybridization experiments.

## Segregation

Mendel, in his careful observation of garden peas, realized that the characteristics being studied were inherited as separate entities. They were not simply mixed together but, in fact, remained separate and intact throughout the lifetime of the organism and were passed on to the next generation—and the next—as distinctive units. The evidence was the reappearance of recessive expressions in the $F_2$ generation in predictable mathematical ratios.

Mendel further recognized that the inherited factors occurred in pairs, one member from each parent, and that prior to reproduction the members of each pair separate, or segregate, from one another so that each gamete carries only one member of each pair of factors. We know now that Mendel's principle of **segregation** is based on the mechanism of chromosomal separation called meiosis (Chapter 5).

Table 6-1 shows how to follow the inheritance of genes in a genetic cross, using the device called a **Punnett square** to follow the segregation and recombination of male-originating genes with female-originating genes. [Note: If you do not understand how the haploid ($n$) gametes are derived from diploid ($2n$) parent cells, it is absolutely *essential* that you review meiosis.]

**TABLE 6-1 Dominance.**[a]

**Parent (P) Generation**

| | | | |
|---|---|---|---|
| Female = **TT** (homozygous for tallness) | | | Male = **tt** (homozygous for dwarfness) |
| Potential gametes produced by meiosis: | T T | $2n$ | t t |
| | ↓ | —meiosis— | ↓ |
| | T | $n$ | t |
| | egg | | sperm |

**Use of the Punnett Square to Predict the $F_1$ Generation (TT × tt)**

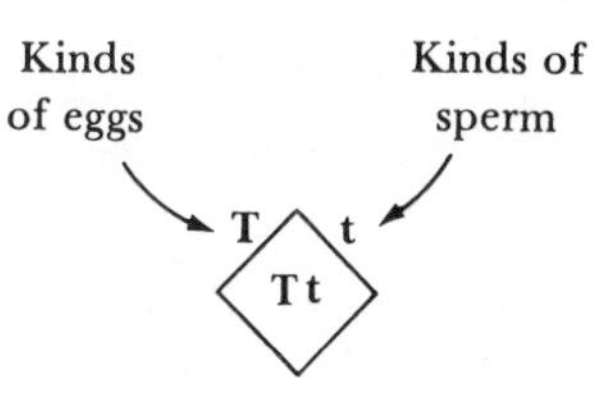


Therefore $F_1$ genotypes = 100% **Tt** (heterozygous)
$F_1$ phenotypes = 100% tall

**Use of the Punnett Square to Predict the $F_2$ Generation (Tt × Tt)**

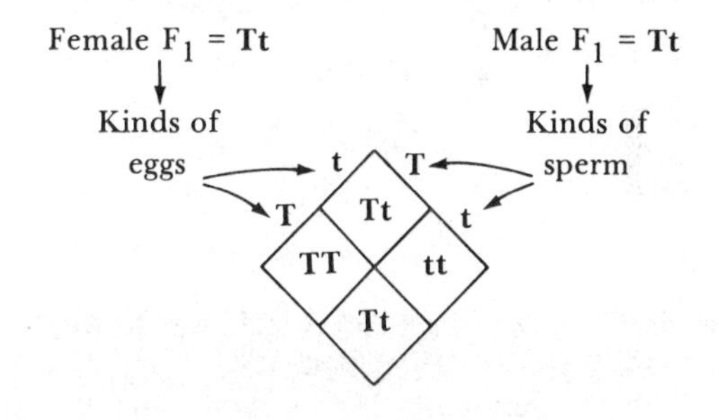


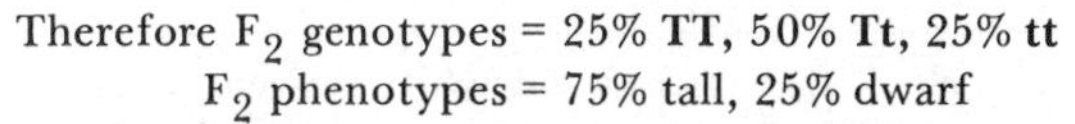

Therefore $F_2$ genotypes = 25% **TT**, 50% **Tt**, 25% **tt**
$F_2$ phenotypes = 75% tall, 25% dwarf

[a] In pea plants tall growth habit is dominant over dwarf growth habit. Therefore the gene for tallness is represented by capital **T** and the corresponding allele for dwarfness is represented by lowercase **t**. When both genes are present, it is customary to write the dominant symbol first, as **Tt**. This checkerboard, called a Punnett square, shows the *potential genes carried by female gametes* on one side and the *potential genes carried by male gametes* on the other side. The squares show *potential genetic combinations in the offspring*. The genetic composition of an individual is its **genotype**; the physical appearance is its **phenotype**. Note that in a simple dominance cross using one characteristic (height) the $F_2$ produces a 1:2:1 genotypic ratio and a 3:1 phenotypic ratio.

## Incomplete Dominance

Mendel's principle of dominance is often demonstrated; however, there are exceptions, such as **incomplete dominance** (blending inheritance) in which the masking of the recessive gene is not total, producing an intermediate effect. If a homozygous red-flowered four-o'clock plant (*Mirabilis*) is crossed with a white-flowered plant (Table 6-2), for example, the $F_1$ plants produce seeds from which grow offspring in the ratio of 1 red:2 pink:1 white in flower color. Because neither allele is dominant over the other, we use uppercase letters distinguished by numerals in working out incomplete dominance crosses.

## Monohybrid Cross

A cross or hybridization in which the inheritance of a single characteristic is being observed is called a **monohybrid cross**. Mendel's experiments discussed so far have been monohybrid crosses. Let us examine some other examples of monohybrid

**TABLE 6-2 Incomplete dominance.[a]**

**Parent (P) Generation**

Male = $R^1R^1$ (homozygous for red flower)
Female = $R^2R^2$ (homozygous for white flower)

**Use of the Punnett Square to Predict the $F_1$ Generation ($R^1R^1 \times R^2R^2$)**

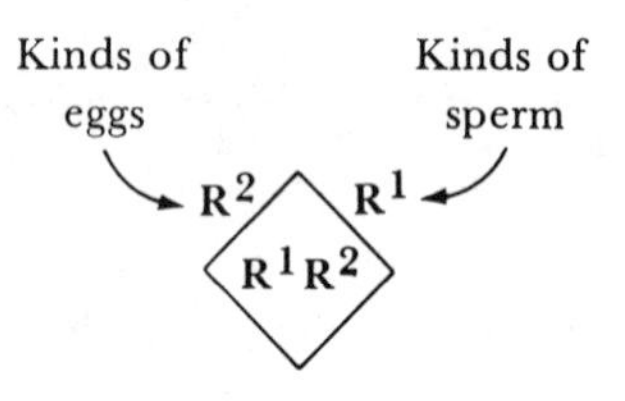


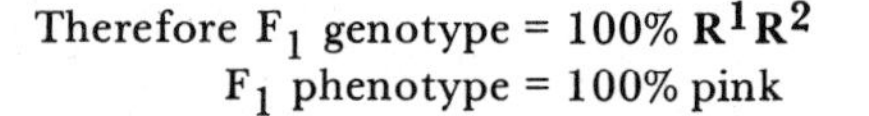

Therefore $F_1$ genotype = 100% $R^1R^2$
$F_1$ phenotype = 100% pink

**Use of the Punnett Square to Predict the $F_2$ Generation ($R^1R^2 \times R^1R^2$)**

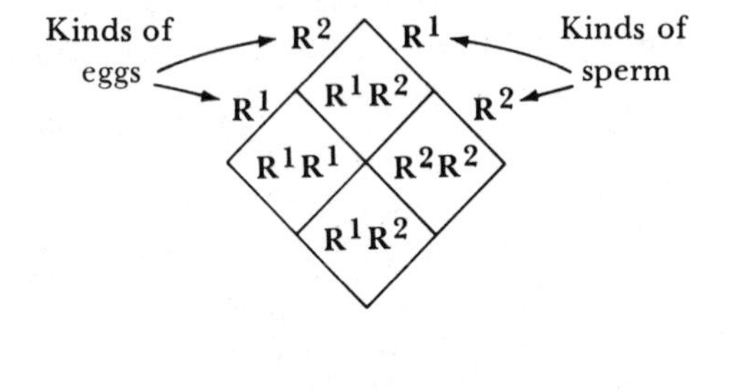


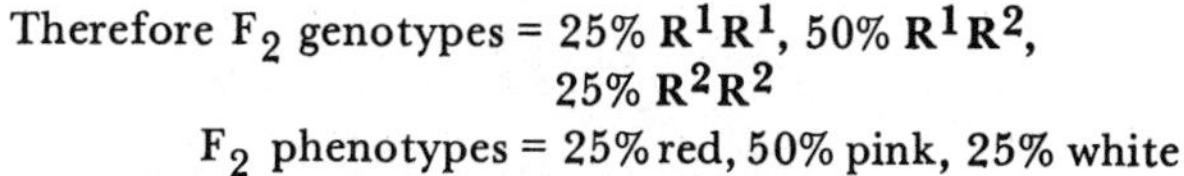

Therefore $F_2$ genotypes = 25% $R^1R^1$, 50% $R^1R^2$, 25% $R^2R^2$
$F_2$ phenotypes = 25% red, 50% pink, 25% white

[a]In four-o'clock flowers red and white flower colors are the result of homozygous conditions. Heterozygotes have pink flowers. Using the Punnett square method, follow and compare the genotypic and phenotypic ratios that are likely to be produced by incomplete dominance. Because neither allele is dominant, their symbols are both capital letters ($R^1$ and $R^2$). Is there a basic difference between complete dominance (Table 6-1) and incomplete dominance in the genotypic ratios or just in the phenotypic ratios?

crosses. The hybridization of a yellow-rooted (dominant) radish (*Raphanus*) with a white-rooted variety leads to a 3 yellow:1 white phenotypic ratio in the $F_2$, as would be expected in a simple dominance cross. In contrast, crossing red- and white-rooted varieties may produce an $F_1$ in which all the roots are purple! The $F_2$ produces a ratio of 1 red:2 purple:1 white. What type of dominance is illustrated?

Although differing in flavor and skin texture, peaches and nectarines are actually varieties of the same species (*Prunus persica*). Their genetic relationship can be demonstrated in simple Mendelian terms (Table 6-3).

## Test Cross

If a plant of unknown genotype is discovered in a garden, a test cross can be performed to determine the genotype. A **test cross** is the hybridization of a plant of unknown genotype to one that is a homozygous recessive for the characteristics being observed. Thus far we have used a simple working definition of homozygous and heterozygous. More completely, a **homozygous** plant is one with identical alleles for a given characteristic. A **heterozygous** plant is one with different alleles for a given characteristic. With a test cross, it is possible to identify the unknown allele in a gene pair, as shown in Table 6-4.

## Independent Assortment

Let us review one of Mendel's hybridization experiments that followed two different traits, a **dihybrid cross**. He crossed a plant that produced only round, yellow seeds with one that produced only wrinkled, green seeds (Table 6-5). All members of the $F_1$ generation produced round, yellow seeds. Self-pollination of $F_1$ plants yielded an $F_2$ generation having individuals that expressed the two original phenotypes *but also* two new combinations (round + green, wrinkled + yellow). Because the genes for seed shape and color appeared in two combinations other than the parental combinations, Mendel concluded that factors for different characteristics are inherited independently of one another. During gamete formation, he reasoned, the factor (gene) pairs behave independently of one another, producing an **independent assortment** of members of each pair of factors in the next generation. Table 6-6 relates independent assortment to meiosis involving two pairs of chromosomes.

**TABLE 6-3 Peach and nectarine inheritance.**[a]

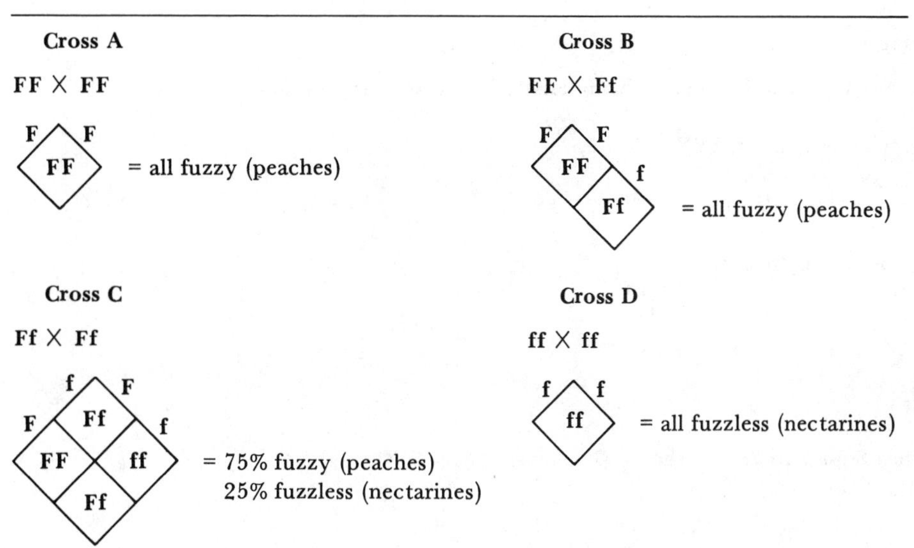


[a]The nectarine is a fuzzless variety of the peach and the fuzzlessness of nectarine skin is due to a recessive allele.

**TABLE 6-4 Test cross.**[a]

**Unknown Plant**

Is tall in phenotype. Possible genotypes = **TT** or **Tt.**

**Test Plant**

Is homozygous recessive, having dwarf phenotype and **tt** genotype.

**Hypothesis A: TT X tt**

If the unknown genotype is **TT**, all offspring of the test cross will be tall.

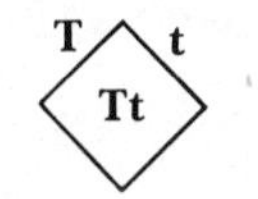


**Hypothesis B: Tt X tt**

If the unknown genotype is **Tt**, both tall and dwarf offspring will be produced by the test cross.

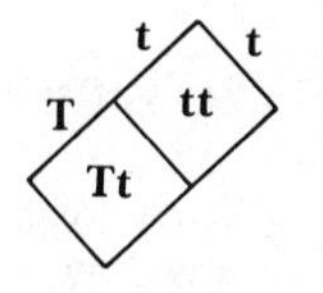


[a]A test cross is used to determine an unknown genotype when the unknown plant shows the dominant expression and thus may be either homozygous or heterozygous for the characteristic. It is crossed with a plant having the recessive phenotype. Why?

**TABLE 6-5 Independent assortment.**[a]

**Parent (P) Generation**

Both parents are homozygous for both characteristics, but one is homozygous dominant:

round, yellow seeds (**RRYY**)

The other is homozygous recessive:

wrinkled, green seeds (**rryy**)

Potential gametes produced by meiosis:

| RRYY | $2n$ | rryy |
|---|---|---|
| ↓ | —meiosis— | ↓ |
| RY | $n$ | ry |

**Use of the Punnett Square to Predict the $F_1$ Generation (RRYY × rryy)**

Therefore $F_1$ genotypes = 100% **RrYy**

$F_1$ phenotypes = 100% round, yellow seeds

**Use of the Punnett Square to Predict the $F_2$ Generation (RrYy × RrYy)**

Potential gametes produced by meiosis:

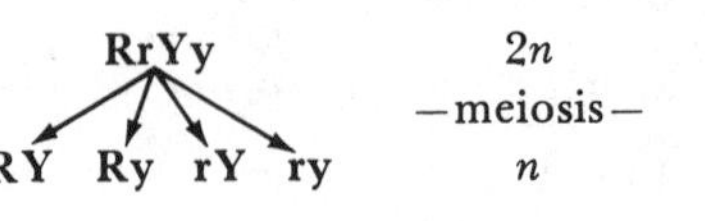


$2n$
—meiosis—
$n$

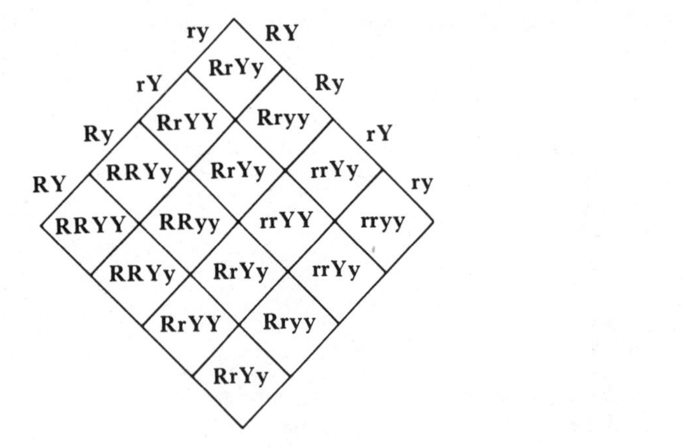


Therefore $F_2$ genotypes = **RRYY, RRYy, RRyy, RrYY, RrYy, Rryy, rrYY, rrYy, rryy**

| $F_2$ phenotypic ratio = | 9 | : | 3 | : | 3 | : | 1 |
|---|---|---|---|---|---|---|---|
| | round, yellow seeds | | round, green seeds[b] | | wrinkled, yellow seeds[b] | | wrinkled, green seeds |

[a]When nonlinked genes are followed through a dihybrid cross, gene combinations in addition to the parental combinations are produced. This situation occurs because the members of each gene pair segregate independently of other pairs during meiosis.

[b]Nonparental combinations.

**TABLE 6-6 Independent assortment as it is related to the movement of chromosomes during meiosis.[a]**

Possible combinations of members of $A^1A^2$ and $B^1B^2$ chromosome pairs in gametes following meiosis:

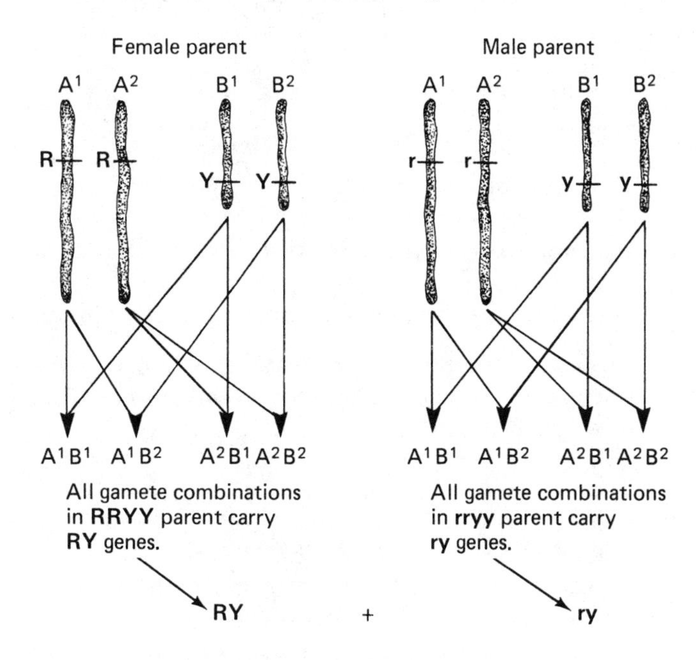


All $F_1$ offspring have the full diploid chromosome. complement—$A^1A^2$ and $B^1B^2$ chromosomes present Genotype is **RrYy** in all $F_1$.

Possible combinations of chromosomes and genes in the gametes of *each* $F_1$ individual following meiosis:

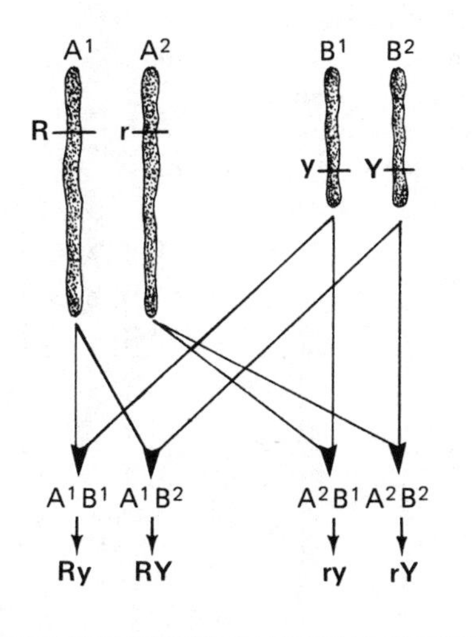


[a] $A^1$ and $A^2$ are homologous members of one pair of chromosomes. $B^1$ and $B^2$ are homologous members of a second pair of chromosomes. Thus, the hypothetical organism is $n = 2$. In this diagram the patterns of gene inheritance outlined in Table 6-5 are identified with actual chromosome segregation during meiosis.

## Linkage and Crossing Over

When Mendel studied the inheritance of more than one characteristic in the same cross, it was fortunate that he chose traits whose genes are located on different chromosomes and so are inherited independently of one another. Doing so enabled him to gather statistical data to develop the principle of independent assortment. Many genetic characteristics, however, are *not* inherited independently of one another because of their occurrence on the same chromosome. All the genes on a single chromosome are physically connected, or linked, so that they are transmitted as a unit. The consistent appearance together of specific characteristics is a demonstration of **linkage** and each chromosome can be referred to as a **linkage group** of genes.

Nevertheless, early twentieth-century genetics investigations revealed some exceptions to linkage, with known linked alleles not always appearing together. These exceptions are caused by physical changes in the chromosomes, such as by **crossing over.** In crossing over, portions of homologous chromosomes are exchanged during meiosis (Fig.

(a)

$A^1$ $A^2$

(b)

$A^1$ $A^2$

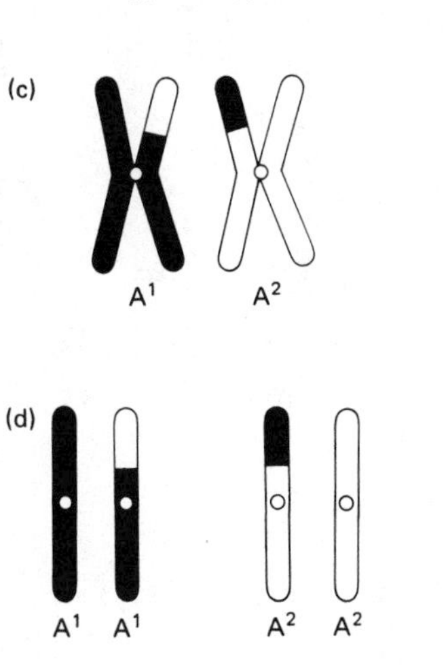


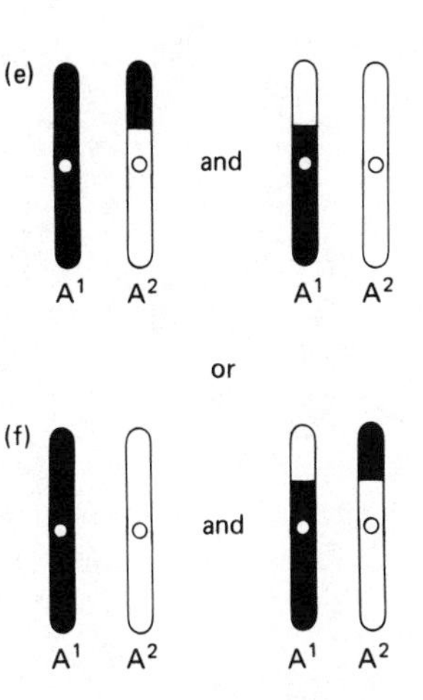


**FIGURE 6-1**
**Crossing over between homologous chromosomes during meiosis. (a) A pair of duplicated chromosomes entering meiosis. (b) Physical entanglement of chromosomes during synapsis. (c) Breakage and fusion, resulting in a reciprocal translocation (exchange) of chromosome segments. (d) Separation of daughter chromosomes. (e) and (f) Different combinations that are possible in the haploid cells formed by meiosis.**

6-1). That such chromosomal exchange occurs is not surprising considering the mass of delicate DNA molecules involved in the dynamics of meiosis. The frequency of crossing over increases with increased distance from the centromere of the duplicated chromosome.

## Mutations and Chromosomal Rearrangements

Occasionally an unpredicted expression of a characteristic appears that cannot be explained as the result of recombination of parental alleles. Such a sudden change in the genetic structure of an individual is called a **mutation** and the individual carrying the mutation is called a **mutant.** Considering the tremendous number of genes present in the genotype of each organism, it is not farfetched to assume that most organisms are mutants for at least one gene. Many mutations remain unnoticed, for they may be ineffectual or have subtle biochemical influences rather than externally visible expressions (science fiction notwithstanding). A mutation may involve a single gene or an entire chromosome and hence a group of contiguous genes.

A **gene mutation** can occur by a change in the nucleotide sequence in the DNA (Chapter 3). Such "small" changes may arise spontaneously or as a reaction to the presence of such disruptive forces as radiation or a chemical substance. Many known carcinogenic (cancer-causing) substances are also **mutagenic** (mutation causing); some studies suggest that this may be a consistent correlation.

A **chromosomal mutation** involves an entire segment of the chromosome [Fig. 6-2 (a) through (e)]. Some types of chromosomal changes are listed here:

1. **Translocation**—the transfer of a chromosome segment from one chromosome to another chromosome.
2. **Inversion**—the reversal of a segment within a chromosome.
3. **Duplication**—the repetition of a segment within a chromosome.
4. **Deletion**—the loss of a chromosomal segment.

Both gene mutations and chromosomal mutations create a nucleus that is slightly different from the original parental genotype. If mutations occur in vegetative (body) cells, they may be passed on to succeeding cell generations by mitosis. A plant, for instance, will occasionally produce a mutation-carrying shoot that is a visible variant or **sport** of the rest of the plant body. The sport can be propagated asexually. If occurring in a reproductive cell, the mutation may be passed on to the next sexually produced generation, thus adding to existing variation within a population (discussed later).

FIGURE 6-2 Drawings (a) through (e) show some types of **chromosomal mutations**—changes in gene sequence—and how they may occur. Letters A–G and numerals 1–7 indicate gene loci on each of two hypothetical chromosomes. Two chromosomes are used here just to show that changes may occur both *internally* and at the *ends.* A **nonreciprocal translocation** between homologous chromosomes may simultaneously cause a duplication on one chromosome and a deletion on another.

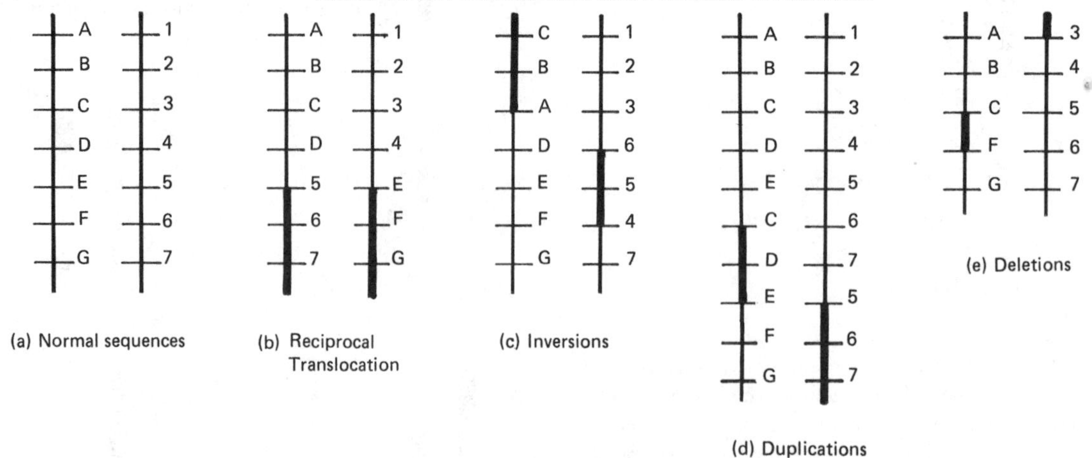

## Pleiotropy and Polygenic Inheritance

Other phenomena create exceptions to Mendel's principles. In **pleiotropy** an individual gene influences several characteristics in an organism. On the other hand, some individual traits are controlled by several genes, a phenomenon referred to as **polygenic (multifactor) inheritance.** In polygenic inheritance a characteristic, such as the color of wheat grains, will show a continuous gradation rather than the clear-cut differences that produce the Mendelian ratios.

## Extranuclear or Cytoplasmic Inheritance

Although the greatest basis for inheritance is nuclear, many examples indicate that certain characteristics are carried from generation to generation in the egg cytoplasm, a phenomenon called **extranuclear** or **cytoplasmic inheritance.** When the egg of a yellow-leaved form of primrose (*Primula sinensis*) is crossed with pollen from a green-leaved form, the progeny are all yellow leaved in the first *and* all subsequent generations. When the reciprocal (opposite) cross is made, however, all progeny have green foliage. Apparently proplastids (immature chloroplasts) of the parent plant are carried over in the egg cytoplasm. The proplastids may be of normal (green) type or pale (yellow) type. Cytoplasmic inheritance occurs fairly regularly in plants and is especially obvious in crosses associated with leaf variegation.

## Polyploidy

You will recall that flowering plant gametes are often haploid ($n$), having a single representative of each chromosome pair. When the egg and sperm are united, the diploid condition ($2n$) is restored. Yet plants commonly have more than two sets (complements) of chromosomes, a condition called **polyploidy.** It is estimated that about half the known 235,000 kinds of flowering plants are of polyploid origin, including many agricultural species! If there are three sets, the organism is triploid ($3n$); four sets, tetraploid ($4n$); five sets, quintaploid ($5n$); six sets, hexaploid ($6n$); and so on. Basically there are two ways for a polyploid condition to develop—autopolyploidy or autoploidy (*auto* = self), and allopolyploidy or alloploidy (*all* = other).

**Autoploidy** occurs when the duplicated chromosome pairs fail to separate during nuclear division, causing doubling of the chromosome number. Autoploidy by a diploid cell during mitosis, for example, causes a tetraploid condition in that cell and its progeny. The result may be a tetraploid portion of the vegetative body. Autoploidy may also occur during meiosis through failure of the duplicated chromosomes (chromatids) to separate. The result is formation of gametes carrying the unreduced chromosomal number. A diploid plant, for instance, could become tetraploid by autoploidy and then produce diploid instead of haploid gametes. If a diploid sperm united with a diploid egg, the resulting zygote would be $(2n + 2n) = 4n$ (tetraploid). If a diploid sperm united with a haploid egg, the resulting zygote would be $(2n + n) = 3n$ (triploid). Tetraploids usually do not interbreed freely with their diploid ancestors, thus enhancing evolutionary divergence between diploid and tetraploid varieties of a species. Because of pairing difficulties during meiosis (at synapsis), triploids have very little success at sexual reproduction.

Tetraploid individuals are often larger and more robust than their diploid relatives. So it is a common plant-breeding practice to block chromosome separation deliberately at meiosis by using the chemical **colchicine**, which interferes with formation of the spindle. Many of the vigorous, large-flowered varieties advertised in nursery catalogs are induced tetraploids. Interestingly, the ability to block cell division also makes colchicine—derived from the autumn crocus (*Colchicum autumnale*)—a potent treatment agent for arresting growth of cancer tumors in humans, although it has drastic side effects because it also interferes with normal cell-replacement processes. (See also Chapter 18.)

In the **alloploidy** process hybridization occurs between two different species and is followed by a doubling of the chromosomes. Because their chromosomes are not homologous (do not carry the same linear gene sequence), hybrids between different species are usually sterile, unable to reproduce sexually. The mechanism of polyploidy, however, produces homologues for all the chromosomes present, creating pairs that *can* match up (synapse) and be distributed during meiosis (Fig. 6-3)! Thus alloploidy (hybridization followed by chromosome doubling) is another important mechanism of creating genetic variation.

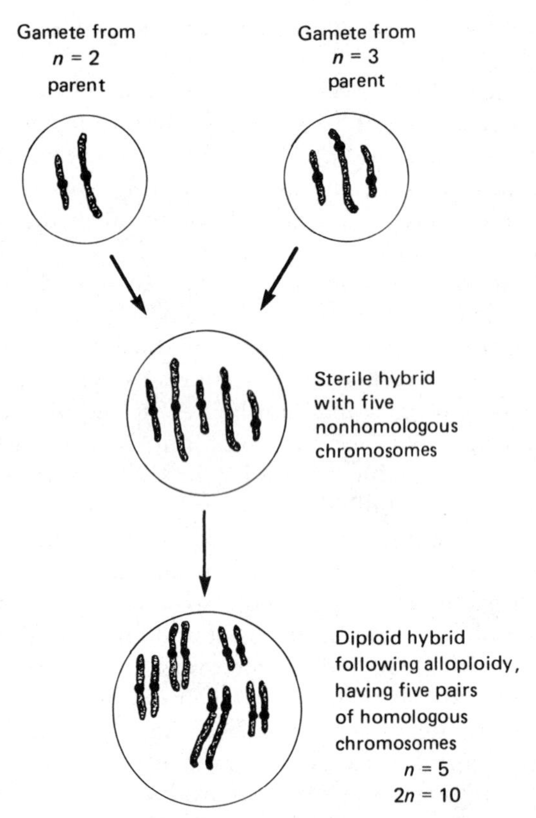


**FIGURE 6-3** **Alloploidy** can be the mechanism for allowing sterile hybrids to become fertile. Here an $n = 2$ gamete combines with an $n = 3$ gamete, producing a hybrid with 5 nonhomologous chromosomes that is sterile because of an inability to synapse during meiosis. Doubling of the chromosomes creates a $2n = 10$ organism that can carry out meiosis, producing $n = 5$ gametes.

## GENETIC VARIATION AND EVOLUTION

### Variation-Producing Factors

Meiosis results in a random (nonbiased) separation or segregation of chromosomes from the various pairs. In an $n = 6$ cell, for example, a total of 64 ($2^6$) meiotic combinations are possible. When meiosis is followed by syngamy, random segregation of chromosomes is followed by random recombination, or reassortment, of genes. The overall result, therefore, is great potential for mixing parental characteristics.

Other sources of variation within a **population** (the collective members of a species living in the same location and thus having an opportunity to interbreed) are gene mutation and changes in chromosome structure or number, as noted. An additional source of genetic variation within a population is the immigration of a spore, pollen grain, seed, or plant fragment from a different population.

### Variation-Reducing Factors

Environmental pressures encountered by a germinating plant at a specific site are complex and include all the physical and biological factors (Chapter 2). Some plants are able to survive; others are not. Those plants that are able to survive and reproduce contribute their genes to the total **genetic pool** of the population whereas those that are unable to reproduce do not. The result is a dynamic creative process of **natural selection** by which adaptively more successful gene combinations become more plentiful than less successful ones. Environmental factors are the selective forces of natural selection.

The evolution of most of our cultivated vegetables, grains, and other crops is the result of **artificial selection** (human selection) acting on the genetic potential of thousands of plant generations. Human decisions regarding the desirable characteristics have been the selecting factors rather than natural conditions. Human selection continues to be the basis of plant improvement.

Selection is not the only process that tends to reduce variation within a population; another prominent one is **inbreeding.** Inbreeding within a population, followed by selection, gradually reduces the degree of variation (heterogeneity) of a population. A high degree of inbreeding occurs in small populations and by self-fertilization. Inbreeding is a common tool for plant improvement.

## PLANT IMPROVEMENT

Early plant improvement was based mainly on the selection of the most vigorous, prolific plants, for families saved the seeds from robust plants for the next year's planting. Although this simple method was certainly responsible for the evolutionary development of our most ancient food plants (Chapter 16), it had limitations because environmental effects can greatly alter growth. Seeds were kept from prospering plants irrespective of their genetic possibilities. In addition, the genetic make-up of the seeds varied widely because of cross-pollination among all adjacent, compatible plants from year to year. Commercial seed companies are now the most reliable source of satisfactory seeds, for they can produce seeds under controlled pollination conditions to yield pure varieties.

## Some Plant Breeding Methods

Controlled plant breeding has led to the development of many cultivated varieties, or **cultivars**, from basic parental species. Today the search for new and "improved" cultivars of food, economic, and ornamental plants continues. Usually the search involves careful research and breeding experiments, using known varieties; serendipity, however, still plays a role as well. It is not unusual to read in a seed catalog, for instance, that a company is offering a variety of pear or apple that originated as a wild seedling. Such a find may be especially important in harsher climates because it demonstrates innate hardiness, thereby revealing a potential for expanding the climatic tolerance range of the species. A horticulturist will watch the promising wildling's manner of growth, time of flowering, quality and quantity of fruit produced, tree and fruit resistance to pathogens, and ability to produce continuously from year to year. If of a high quality, the tree is patented and propagated vegetatively for commercial distribution.

Several basic methods to improve current varieties and develop new varieties of plants propagated from seed exist, including mutation, hybridization, and polyploidy. Naturally occurring or induced mutations are sources of variation.

**Hybridization** is the crossing of plants that differ in one or more feature. Hybridization among *varieties* is usual, but crossing between closely related *species* also occurs frequently among plants (rarely among animals). A phenomenon called **hybrid vigor** or **heterosis** often arises when different varieties or even species are crossed—that is, the hybrid is larger and more vigorous than either parent. The effect is usually lost in the next generation ($F_2$); therefore $F_1$ hybrid seeds must be planted each season to yield predictable results. $F_1$ hybrid seeds are usually advertised as such in vegetable and flower seed catalogs.

Hybridization experiments are followed through at least two generations ($F_1$ and $F_2$) and so require several growing seasons to complete. Controlled outcrossing and inbreeding are important techniques, as is **backcrossing**, the crossing of offspring to a parental type. Backcrossing is used to enhance particularly desirable qualities of that parent in the progeny.

Corn varieties have been improved by controlled inbreeding (selfing) followed by outcrossing (hybridization). The process begins with continued inbreeding and selection for a minimum of 5 years (frequently 7) to develop several purebred, increasingly homozygous strains. During the inbreeding phase vigor, overall size, and yield of the corn plants often decrease, apparently because of the expression of some deleterious recessive genes that come together in a homozygous condition. At the same time, desirable dominant and recessive traits are preserved. After the 5 to 7 years, pure line **A** is crossed with pure line **B**, producing a vigorous $F_1$ **A** × **B** hybrid. Two other pure inbred lines, **C** and **D**, are also hybridized at this time, yielding a vigorous $F_1$ **C** × **D** hybrid. The net result is two "single-hybrid" lines (**A** × **B** and **C** × **D**) from four pure inbred lines (**A**, **B**, **C**, and **D**). To increase the desired heterogeneity and hybrid vigor further, **A** × **B** individuals are crossed with **C** × **D** individuals, thus producing a "double-hybrid" corn variety. Seed catalogs advertise many vigorous, high-nutrition, productive corn varieties described as "single hybrid" or "double hybrid." Hybrid corns obviously will not breed true to type; so hybrid seeds must be purchased for each planting.

The mechanisms of polyploidy have already been discussed, but some examples in cultivated plants may be helpful. Raspberries are of two types, normal summer-fruiting varieties and autumn-fruiting varieties. Summer varieties bear fruit on the previous year's canes (a biennial pattern) whereas autumn-fruiting varieties bear on the current season's growth (an annual pattern). Almost all summer-fruiting varieties are diploid (14 chromosomes, $n = 7$, $2n = 14$) whereas autumn-fruiting varieties are tetraploids (28 chromosomes, $n = 7$, $4n = 28$), apparently autoploid in origin.

An important alloploid example is bread wheat, *Triticum aestivum*, which was developed 7000 to 8000 years ago and has 42 chromosomes. Apparently *T. aestivum* originated from hybridization of a 28-chromosome wheat ($n = 14$) with a 14-chromosome grass ($n = 7$). The sterile hybrid of this cross had 21 chromosomes (14 + 7). Doubling of the chromosomes followed, resulting in a fertile alloploid wheat.

The 38 chromosomes ($n = 19$) of the rutabaga (*Brassica napus*) are also thought to be of alloploid origin. The original hybridization may have been between a member of the cabbage group (*Brassica oleracea*, 18 chromosomes, $n = 9$) and a member of the turnip group (*Brassica campestris*, 20 chromosomes, $n = 10$). The foxglove (*Digitalis mertonensis*) is also considered to be an alloploid (112 chromosomes, $n = 56$) that originated from a natural cross of *D. ambigua* (*grandiflora*) (56 chromosomes) and *D. purpurea* (56 chromosomes).

## The Green Revolution

In the sixties a highly optimistic **Green Revolution** was predicted whereby increasingly improved crop varieties would be produced in increasing abundance so that world agricultural production would keep pace with an "exploding" world population increase. It hasn't happened quite that way. Perhaps the limitlessness of the concept as popularized was simply unrealistic. There is a finite limit to what fertilizers, water, pesticides, and mechanization are able to accomplish even with improved varieties of corn (*Zea*), oats (*Avena*), wheat (*Triticum*), rice (*Oryza*), and other staple crops. Also, a limited portion of Earth's surface is climatically and edaphically suitable for agriculture. Moreover, the fuels to power farm machinery are no longer as cheap and plentiful as in earlier decades. Serious famine still exists in major areas of our globe, especially Asia and Africa. Drought, loss of topsoil, salinization, desert encroachment, suburban growth, and pollution have all taken a massive toll of arable land. The major grain-producing regions of North America and Soviet Asia occasionally experience successive years of unseasonal rains and droughts. Meanwhile the world population continues to increase at a staggering rate.

The overall result of human and environmental circumstances has been a reduction of the world grain surplus and a serious loss of land capable of supporting agriculture. Many forecasters predict disastrous times ahead as the world population overtakes the potential world food supply, perhaps within this decade.

What inspired the Green Revolution concept was the development of "super" varieties that were higher in specific nutrients and that yielded prodigiously under energy-intensive cultivation. For his work in leading the development of corn, bean, and wheat varieties, Norman Borlaug deservedly received the Nobel Prize in 1970. He is often thought of as the "father" of the Green Revolution. Yet most farmers around the world are unable to meet the high cost of fossil fuels that is built into providing energy for the mechanization, fertilizers, pesticides, and irrigation required to tap the maximum potential of the "high-powered" genotypes. They must rely on older varieties that produce lower yields but require less water and fertilizer and have some natural resistance to diseases and animal pests.

Consequently, attention has increasingly focused on the development of high-nutrition and high-yield grains that are also hardier and more disease resistant. Increased effort has also been directed toward identification and improvement of food plant varieties that are tolerant of salinity, drought, cold, and other marginal environmental conditions, varieties that can be produced with less use of fossil fuels, and varieties that can be grown locally, thus reducing the need for expensive transportation of food from a few major production regions to distant markets. Many people believe that for the future we may need to rely more on compact food-producing and delivery systems instead of ever-larger, more centralized, and hence "efficient" production systems that carry a high energy, delivery, (and perhaps political) price tag.

## Germ Plasm Conservation

The development of many fruit and vegetable varieties to meet the needs of large-scale commercial cultivation and mechanical harvesting means that fewer of the older (often tastier) varieties are being grown. As a result, there is a danger that collective species genetic pools, or **germ plasm**, may be seriously reduced as older varieties are lost. There are several good reasons for concern. Speaking somewhat sentimentally, the older, less specialized varieties sometimes have more appealing taste, texture, and appearance. Compare a plump, full-flavored, old-fashioned vine-ripened tomato (*Lycopersicon*), for instance, with the small, hard-walled commercial varieties that have been developed to withstand mechanical picking, storage, and shipment.

A second concern is that by reducing the pool of genetic variations we are reducing the potential for future change—painting ourselves into a corner evolutionarily. When the miracles of irrigation and chemical fertilization opened up desert areas around the world to agriculture, for example, who worried about salinization of the soil? Now, however, strains of older plant varieties that have been preserved yield promise of improved salt tolerance among tomatoes and certain grains. What genetic flexibility must we preserve to meet future conditions? It seems that we must rely increasingly on the productivity of land areas that are less than ideal, having poorer soils, shorter growing seasons, colder (or hotter) climates, and/or less available water. And, of course, the unknown factor in any future formula is change in world climate patterns, both natural and induced by such human activities as deforestation and industrialization.

A final concern about loss of germ plasm is that overreliance on certain genotypes may result in specific plant disease and pest problems of plague-like proportions. Individuals of a single variety have identical strengths and identical weaknesses, including susceptibilities. Under intensive crop monoculture populations of specific animal pests (e.g., insects and nematodes) and disease-causing organisms can increase and spread rapidly. The use of pesticides has been demonstrated to be a form of evolutionary selection for certain insect pests; by eliminating pesticide-susceptible members of a species, we have created populations of pesticide-resistant insects. With many valid health and environmental concerns, continued dependence on chemical controls seems to be limited.

### Looking to the Wild

Today new efforts are being devoted to the identification of wild plants that may be suitable for cultivation. More than a quarter million flowering plant species are known, and yet we rely on only a few for food and other products (Chapter 16).

A few wild plants that may have a potential for future crop use are jojoba (*Simmondsia chinensis*), a northern Mexico shrub with seeds that are 50% liquid wax that is a substitute in every use for whale oil; *Echinochloa,* an Australian grass having potential forage value that can grow from germination to harvest on a single preemergence watering (provided that the watering is adequate to release seed dormancy); and guayule (*Parthenium argentatum*), a desert shrub of Mexico and the southwestern United States that produces rubber and was grown commercially during World War II.

Chapter 17 describes food and other uses that can be made of some of our wild North American plants. Many of the common wild and "weed" species would perform suitably under cultivation, requiring minimal fertilization and pest control.

### Genetic Engineering

The phrase "genetic engineering" describes a wide array of techniques in which the genotype of a cell is physically altered. Once the change has been made, it can be passed on to future generations by normal mechanisms of cellular reproduction. Humans are already benefiting from the laboratory "creation" of a bacterial strain that is able to synthesize insulin to treat victims of diabetes, a disease in which the body's normal insulin production is inhibited. Another "created" bacterial strain produces interferon, a substance that seems to aid the body in combating infections.

During "gene-splicing" bacterial structures called plasmids are used to transfer genes from cell to cell. Such manipulation may make it possible to transfer desirable genes for enhanced growth, disease resistance, protein and vitamin production, and nitrogen utilization. In a recent series of experiments a gene from a French bean (*Phaseolus*) was spliced into the DNA of a bacterial cell that subsequently transferred the gene to the DNA of a sunflower (*Helianthus*) cell it infected. The resulting sunflower cell carrying the bean gene was called "sunbean." The specific gene may have the potential to enhance protein production in sunbean tissue. Researchers hope to culture sunbean plants to analyze what, if any, effect the spliced gene might have.

It has been possible to hybridize body cells of the potato (*Solanum*) and tomato and then culture the hybrid cells to entire plants having some characteristics of each parent. The technique is regarded as a breakthrough with great potential.

In another method potato leaf cells have been cultured, separated, and then cloned and grown into adult plants. Surprisingly, the cells of a single potato leaf (and therefore the clones) are not identical—perhaps due to mutations and chromosomal rearrangements arising during mitosis. The technique, however, permits screening of cloned plants for a variety of characteristics.

## SUMMARY

The work of **Gregor Mendel** is the foundation of modern genetics. He determined that **inherited factors** are present in pairs, one factor from each parent, that there are **dominant** and **recessive** expressions of individual factors, and that factors for different characteristics **segregate** and are **inherited independently** of one another. We now call Mendel's "factors" **genes.**

**Monohybrid, dihybrid,** and **test crosses** are explained. **Incomplete dominance** occurs when neither expression of a gene is dominant, producing an intermediate phenotype. All the genes on a single chromosome are **linked.**

**Mutations** are sudden changes in a gene or chromosome. Translocation, inversion, duplication, and deletion are types of chromosomal rearrangements.

In **pleiotropy** an individual gene influences several characteristics. In **polygenic inheritance** a characteristic is controlled by several genes. Some characteristics are inherited through the egg cytoplasm, a phenomenon known as **extranuclear** or **cytoplasmic inheritance.**

In **polyploidy** chromosomes are present in three or more complete sets (complements) rather than in pairs. Polyploidy is common among seed plants. The mechanisms of autoploidy and alloploidy are discussed.

Gene recombination, gene mutation, chromosomal rearrangement, and immigration encourage **variation** within a population. **Natural selection** and **inbreeding** tend to reduce variation within a population.

**Plant improvement** is the result of evolution shaped by human selection, in contrast to natural selection. Hybridization, inbreeding, and polyploidy are important tools in the development of **cultivars** (cultivated varieties).

The **Green Revolution** was based on the hope that high-yield cultivars would be able to feed the world's increasing population. The high cost of energy and other factors have interfered with its ultimate success.

**Germ plasm conservation** and **genetic engineering** are areas of compelling interest for the future.

# chapter seven PLANT TISSUES

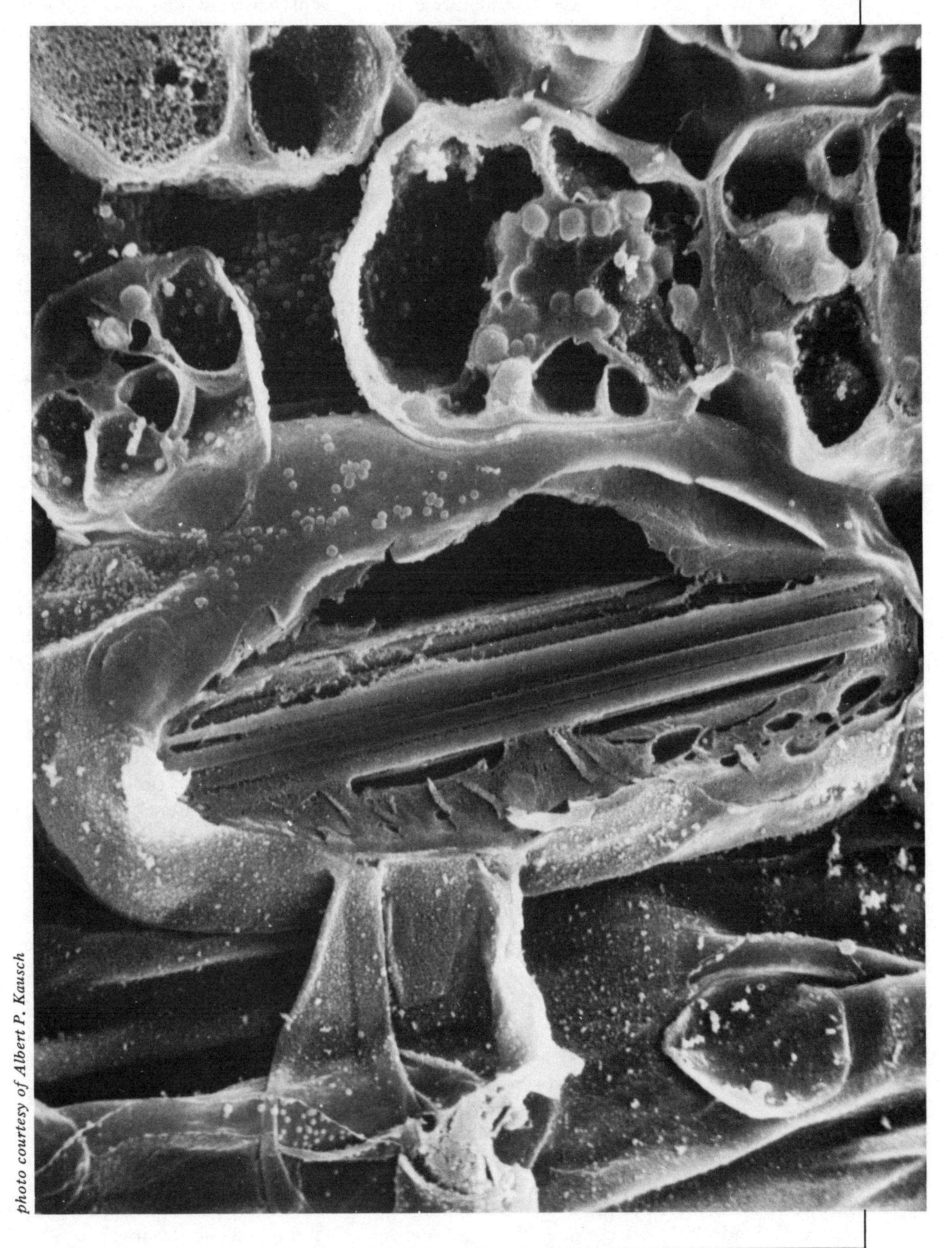

*photo courtesy of Albert P. Kausch*

## INTRODUCTION

As noted in previous chapters, the cell is the basic unit of life. Some organisms are unicellular, but most are multicellular; that is, their bodies are composed of many cells. Multicellularity and greater complexity increase an organism's physiological demands for food and oxygen.

An organism composed of one or only a few cells can fulfill its needs for water, oxygen, nutrients, and waste disposal by direct diffusion between cell surfaces and the surrounding water. The ratio of diffusion surface area to total mass is high. In a larger multicellular organism, however, more cells are internal, not in direct contact with the environment. The surface area to total mass ratio is lower and internal cells must fulfill their needs by diffusion from outer cells that are in direct contact with the environment. This proposition is self-limiting and would not permit attainment of size beyond a certain level.

Therefore an increase in cell number beyond a certain point must be accompanied by an increase in complexity, through specialization. Instead of having basic generalized cells performing all functions throughout the body, multicellular plants usually exhibit specialization of cells into different functional types. Such cells are arranged as **tissues,** groups of cells that perform specialized functions for the organism as a whole.

Plant tissues demonstrate individual and collective models of effective structure/function design and permit further specialization of the plant body into organs. The structure and function of plant tissues are the subject of this chapter. Plant organs are discussed in the following chapter. Based on their embryonic derivation and complexity, plant tissues can be summarized as in Table 7-1. Although the tissue and cell-type names may be unfamiliar, it is *essential* that you learn them well; they are part of your vocabulary for the rest of your study of plants!

## MERISTEMATIC TISSUES

**Meristematic tissues (meristems)** produce new cells by mitosis, thereby giving rise to the plant body. Meristem is derived from the Greek *meristos,* meaning divided. Meristematic cells are usually small and thin walled, have densely staining cytoplasm and a nucleus that is large in relation to cell volume, and lack a central vacuole. When a meristematic cell divides, it produces one cell that remains meristematic and one that may either remain meristematic or differentiate into a specialized cell type. Plant meristems are of two general types: apical and lateral. Although introduced here, their activity is discussed in greater detail in the next chapter.

The meristematic tissue that appears during embryonic development is the **apical meristem.** Apical meristems are found at the shoot and root tips and cause the plant body to grow in length (height). Side branches are formed by apical meristems located in buds that form along the length of existing stems, typically at nodes. Adventitious buds may arise in less typical locations, such as leaves or roots.

The **lateral meristems** produce new cells laterally rather than longitudinally so that they are responsible for increases in the girth or thickness of a stem or root. The **vascular cambium** occurs between the food-conducting tissue and water-conducting tissue and is responsible for production of additional conducting tissues. Vascular cambium cells are longer and narrower, with a greater cell volume in relation to nuclear size than apical meristematic cells. **Cork cambium** develops within various outer tissues, producing protective layers of cork.

## SIMPLE PERMANENT TISSUES

A simple permanent tissue is called "permanent" because most of its cells mature into specific and irreversible forms (rather than remaining meristematic). It is called "simple" because it consists of a single basic tissue type rather than a combination of several tissue types. Several of the tissue types have names ending in "-enchyma," which translates roughly as "in-poured." Thus the first part of the name, combined with "-enchyma," describes the packed-together relationship of the cells to one another within the tissue.

### Parenchyma

**Parenchyma** cells and tissues generally are not highly specialized in shape and often appear as a filler tissue between other more structurally specialized tissues. A literal derivation of the term parenchyma is "poured in along side of." Parenchyma cells are considered the most generalized cell type in a plant body, a factor that contributes to their great functional versatility. Parenchyma cells are often large and **isodiametric,** having somewhat equal dimensions on all sides. The shape is

TABLE 7-1 A summary of plant tissues.

| Tissue Type | Alive or Dead at Maturity | Cell Wall Development at Maturity | Functions |
|---|---|---|---|
| **Meristematic Tissues** | | | |
| Apical meristem | alive | primary | produce new cells at root and shoot tips (apical growth) |
| Vascular cambium | alive | primary | produce new xylem and phloem (lateral growth) |
| Cork cambium | alive | primary | produce cork tissue (lateral growth) |
| **Simple Permanent Tissues** | | | |
| Parenchyma | alive | primary | photosynthesis, storage, support by turgor |
| Collenchyma | alive | primary | support of herbaceous tissues |
| Sclerenchyma | dead | secondary | support, strengthen, form protective coverings |
| Cell types: fibers, sclereids | | | |
| Epidermis | | | |
| Major cell types: epidermal cells, guard cells, trichomes, root hairs | alive | cutin on outer wall (except root hairs) | prevent desiccation of inner tissues, gas exchange (stems and leaves); gas exchange, absorb water and minerals (roots) |
| Cork | dead | secondary | prevent desiccation of inner tissues, mechanical protection, insulation |
| **Complex Permanent Tissues** | | | |
| Xylem | | | |
| Cell types: | | | |
| tracheid | dead | secondary | conduct water solution vertically from roots to rest of plant, support of woody tissues |
| vessel member | dead | secondary | same as for tracheid |
| fiber | dead | secondary | provide strength |
| parenchyma | alive | primary to secondary | storage, radial movement of substances by diffusion |
| Phloem | | | |
| Cell types: | | | |
| sieve cell or sieve tube member | alive, but enucleate | primary | conduct food substances vertically from site of production or storage to rest of plant |
| companion cell or albuminous cell | alive | primary | maintain metabolism of sieve tube member or sieve cell |
| parenchyma | alive | primary | storage |
| fiber | dead | secondary | provide strength |
| sclereid | dead | secondary | provide strength |

sometimes determined by surrounding cells, just as the surface of adjacent soap bubbles in a mass are flattened at each contact point with other bubbles [Fig. 7-1 (a) and (b)]. Some parenchyma cells are boxshaped; others are lobed. The cell wall is thin, mainly primary (composed of cellulose and pectic substances), and encircles a living protoplast that maintains itself as long as the tissue or the organ remains alive. There is usually a large central vacuole. Parenchyma tissue turgidity (Chapter 3) is the primary means of support for herbaceous plants, which lack extensive woody tissue development. This situation is shown by the wilting of plants as water is lost and the cells become flaccid. Parenchyma cells retain their ability to undergo mitosis even after some differentiation and considerable age.

In most parts of the plant parenchyma tissue functions as a storage tissue. Parenchyma tissues that are exposed to light, however, such as in leaves, herbaceous stems, and exposed roots and tubers, develop chloroplasts. Photosynthetic parenchyma is often referred to as **chlorenchyma.** Leaf chlorenchyma tissue is often more specialized in appearance than storage parenchyma (Chapter 8).

Some parenchyma cells in the food-conducting tissue (phloem) perform a special function as **transfer cells** for moving solutes for short distances. Transfer cells are common, especially in seeds, in such glandular structures as nectaries, near small

**FIGURE 7-1** (a) Fresh preparation of **parenchyma** tissue. Calcium oxalate **crystals** are present. (b) Scanning electron micrograph (SEM) of a cattail (*Typha*) leaf. Note the **intercellular spaces** and the bundle of oxalate crystals in the largest cell. [Photo (b) courtesy of Albert P. Kausch.]

(a)

(b)

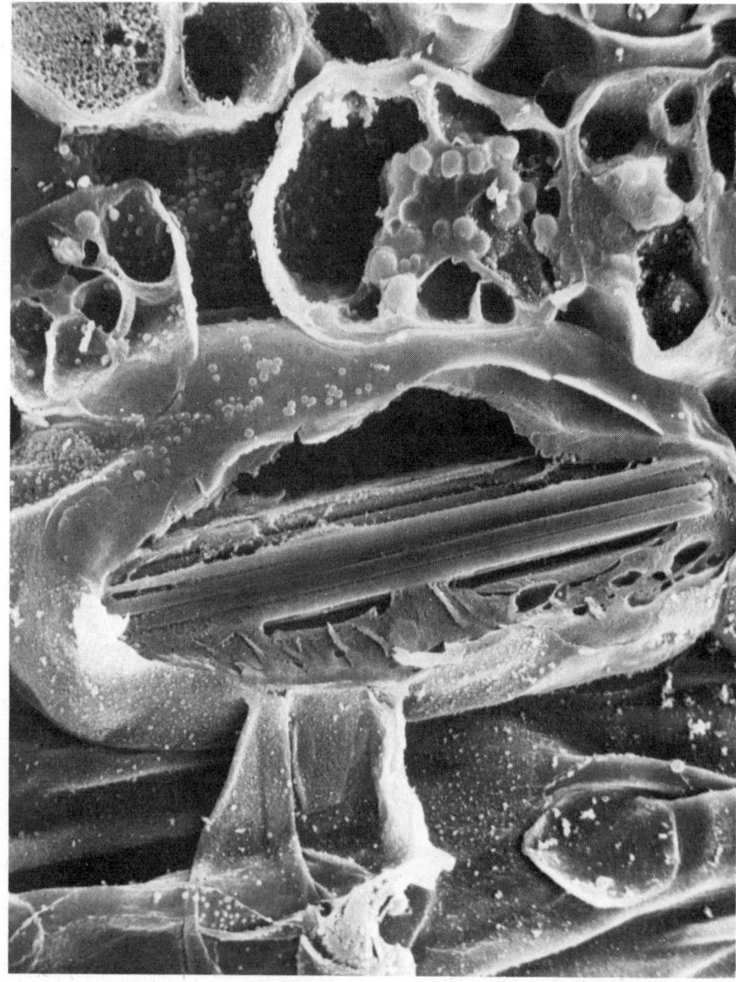

veins embedded in the leaves, and at other potential short-distance transfer sites. They are apparently active in solute transport in dicots.

Under the electron microscope transfer cells are identified by ingrowths of the cell wall and accompanying plasma membrane, giving each transfer cell much greater membrane surface area. Some companion cells (see Phloem section) in the phloem are also transfer cells.

### Collenchyma

**Collenchyma** tissue is adapted for support and strength of upright plant parts, especially herbaceous stems. Collenchyma cells have thicker walls than parenchyma cells, with especially heavy cellulose deposition at certain points, such as "corners" formed where two or more cells meet (Fig. 7-2). Because the adjacent cells are fused together tightly, the Greek-derived *colla-*, meaning glue, is appropriate. The thick walls are made of primary wall materials (cellulose and pectins) rather than secondary material (as lignins); so collenchyma cells retain a potential for growth and occasionally divide by mitosis. This process often occurs as a reaction to injury, enabling even an herbaceous plant to regenerate and reinforce a wounded area.

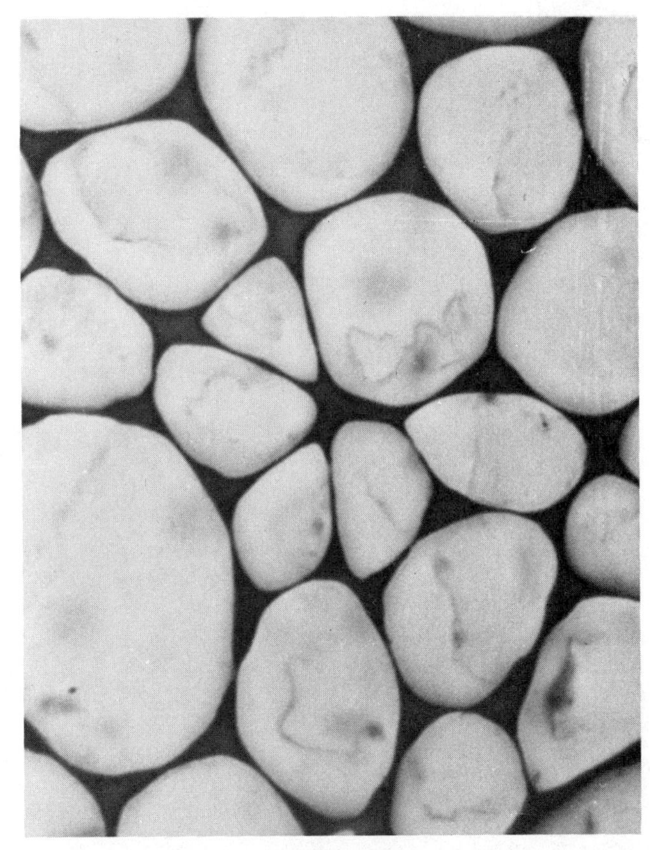

FIGURE 7-2 Collenchyma tissue.

### Sclerenchyma

*Scler-* means hard and **sclerenchyma** tissue is composed of two distinct categories of cell types: fibers and sclereids. Both types have hard secondary cell walls reinforced with lignin. **Fibers** typically differentiate directly from a new cell produced by mitosis and lignified secondary cell wall is deposited as the cell elongates. The mature fiber is a long, thick-walled cell that strengthens and supports the plant body (Fig. 7-3). As the fiber wall develops, the cytoplasm eventually dies so that a functional fiber is actually a cell wall "skeleton" produced by the cytoplasm before its demise. Because of their long, narrow shape and their strength, fibers have many economic uses. This paper, for example, is composed mainly of matted-together wood fibers and conducting cells. Plant fibers are especially easy to recognize in coarse papers, such as paper toweling, facial and toilet tissues, and pulpy tablet paper.

The second type of sclerenchyma cell, the **sclereid**, develops from a parenchyma cell instead of directly from a meristematically produced cell. There are several kinds of sclereids. Sclereids pro-

FIGURE 7-3 Cross section of oak (*Quercus*) wood. Each thick-walled cell having a very small lumen is a sclerenchyma **fiber**.

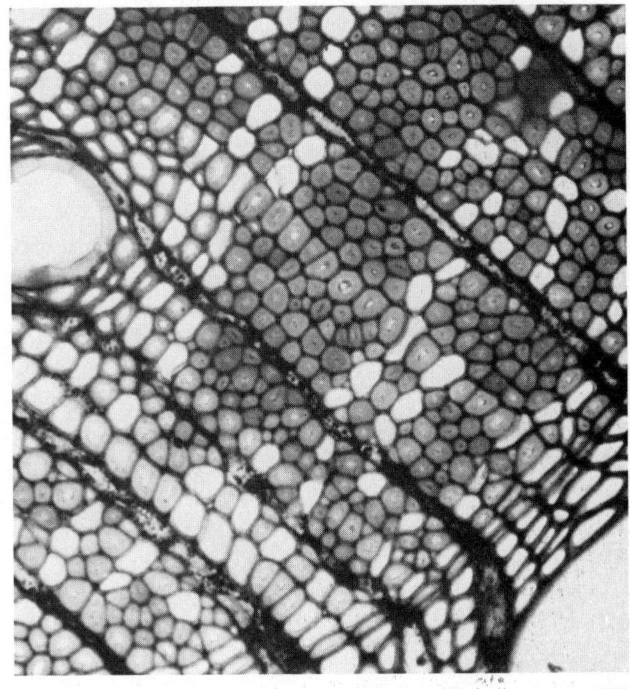

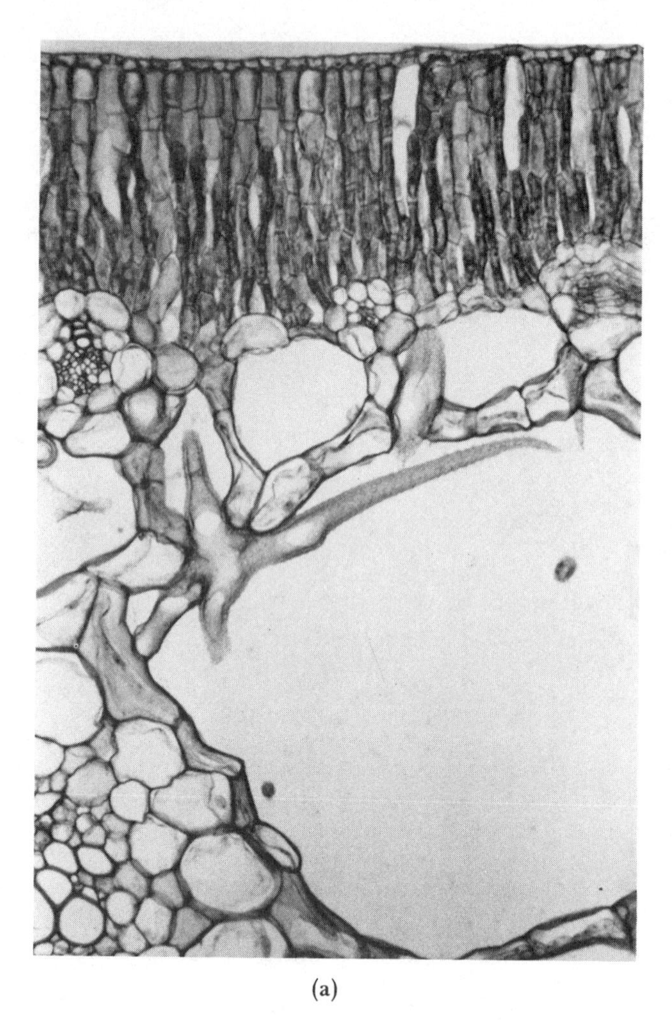

(a)

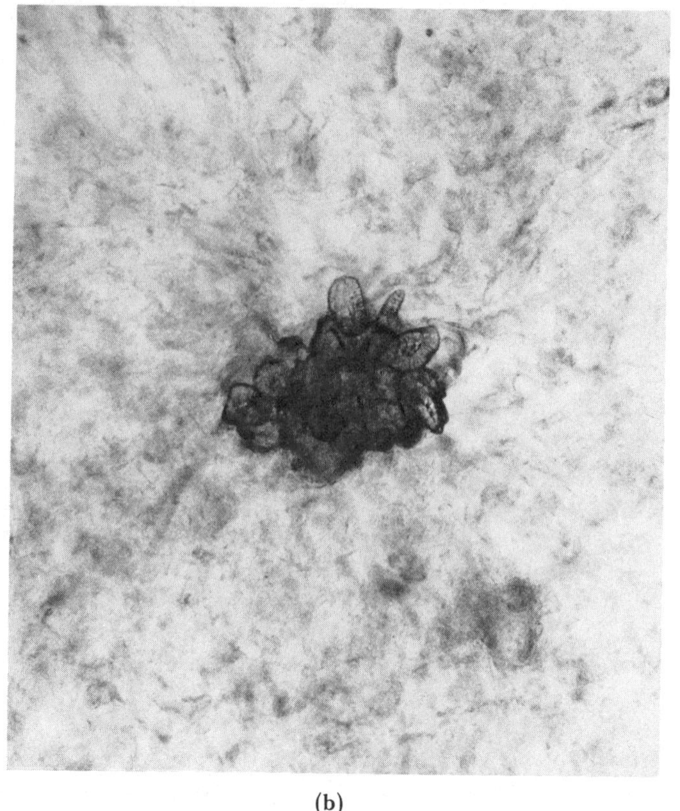

(b)

**FIGURE 7-4**
(a) Starlike **sclereid** in a leaf. (b) Cluster of **stone cells** in the fruit of a pear. [Photo (a) by Jeff P. Carlson.]

vide hardness to plant tissues and are widely distributed in plant bodies, including tree bark, peanut hulls, nutshells, bean seed coats, and the leaves of aquatic plants [Fig. 7-4 (a)]. You are probably familiar with the gritty clusters of stone cells (a type of sclereid) in pear fruits [Fig. 7-4 (b)].

## Epidermis

It is believed that the earliest plants were aquatic. The continuum of increasing complexity among living and known fossil plants suggests definite trends of adaptation to meet the rigors of terrestrial existence. A basic necessity of life surrounded by air instead of water is an outer covering that can substantially reduce water loss and provide mechanical protection; it must, however, exchange oxygen and carbon dioxide on the exposed plant parts. **Epidermis** ("outer skin") tissue serves this purpose.

Usually the epidermis that covers above-the-ground or **aerial** plant surfaces is one cell layer thick and composed of several cell types. Most of the cells are flattened and tilelike and the exposed cell wall is covered with a secreted **cuticle** of a fatty or waxy substance called **cutin.** The cuticle blocks major water loss from the internal tissues and protects them from damage by foreign substances and from invasion by disease-causing bacteria and fungi (Chapter 12).

Because of the cuticle, however, gas exchange between the atmosphere and the inner tissues would be inadequate if there were not some coadaptation to permit effective gas exchange. Thus the epidermis bears special pores, called **stomates** (Gr., *stoma* = mouth). Because of the potential hazard of excessive water loss through the stomates, each stomatal opening is flanked by a pair of specialized cells called **guard cells** (Fig. 7-5, 7-6, and 7-7). The guard cells are usually sausage or kidney shaped, bowed to the outside of the opening. Their cell walls are thickened and resistant to stretching along the inside of the pore and often have small ledges of reinforced wall along the upper and lower edges next to the opening. The cell wall areas away from the pore are thinner and more easily stretched.

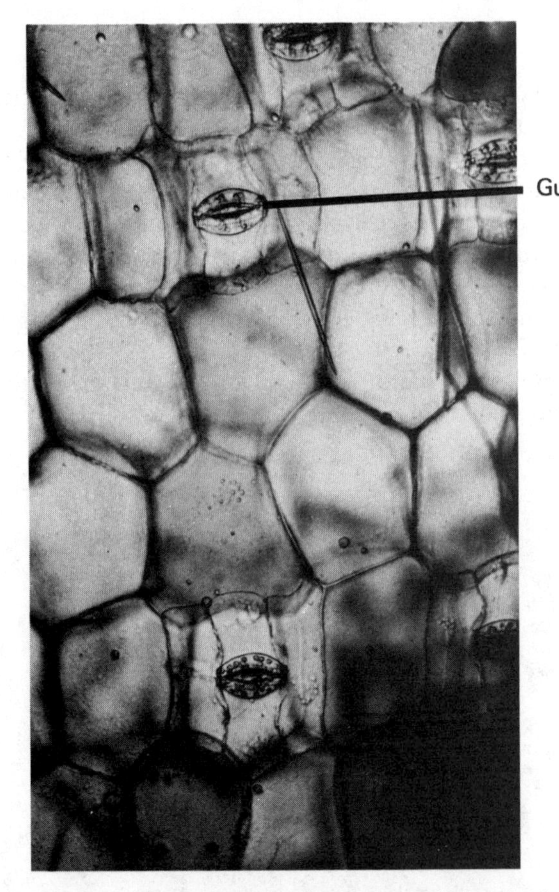


**FIGURE 7-5** Peeled **epidermis** from a living leaf showing chloroplast-containing **guard cells** and regular epidermal cells. (Photo by Jeff P. Carlson.)

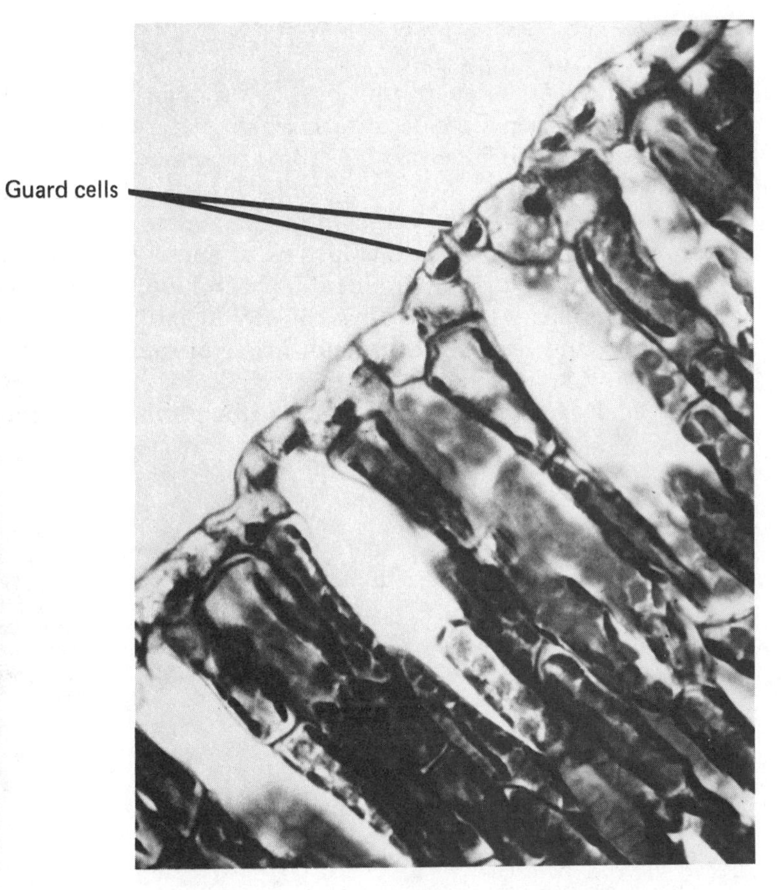

**FIGURE 7-6** Cross section of a water lily (*Nymphaea*) leaf, showing **stomates** and substomatal chambers.

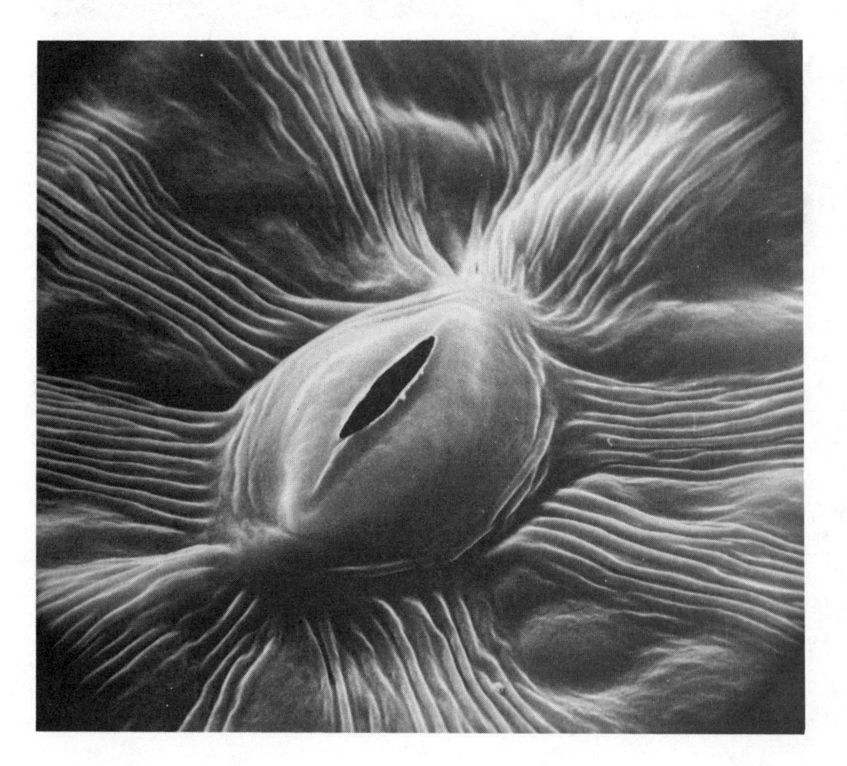

**FIGURE 7-7**
SEM of a stomate. The ridged layer over the epidermal cells and extending over the edges of the guard cells is **cuticle**. (Photo courtesy of Irving B. Sachs, Forest Products Laboratory, U. S. Department of Agriculture.)

When turgid, the guard cells bow apart, opening the stoma. When flaccid, as during high water-loss conditions, their inner margins come together, closing the opening.

Most epidermal cells lack chloroplasts; guard cells, however, contain well-developed chloroplasts. It was thought that the production of soluble carbohydrates during photosynthesis affected guard cell turgidity and hence opening and closing of the stomate. Now, however, it seems that the hormone abscisic acid (Chapter 9) influences influx and outflux of specific electrically charged particles (**ions**), thereby controlling guard cell osmosis.

**Trichomes** are specialized epidermal cells with hairlike or bulbous outer extensions. They are found on such aerial parts as leaves, stems, flowers, and fruits. The **epidermal hair** type of trichome may consist of one or more cells and may be from a millimeter or less to several centimeters in length [Figs. 7-8, 7-9, and 7-10(a) and (b)]. Cotton fibers are epidermal hairs produced on the surface of the cotton seeds.

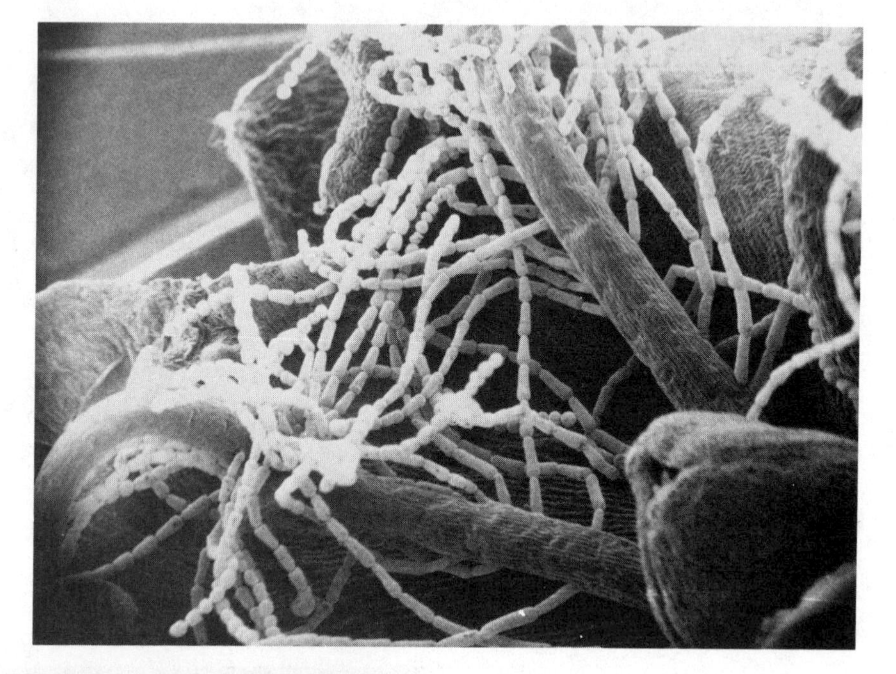

FIGURE 7-8
The light, beadlike strands in this SEM are multicellular **epidermal hairs** found on the stamen filaments of spiderwort (*Tradescantia*). (Photo by Jon M. Holy.)

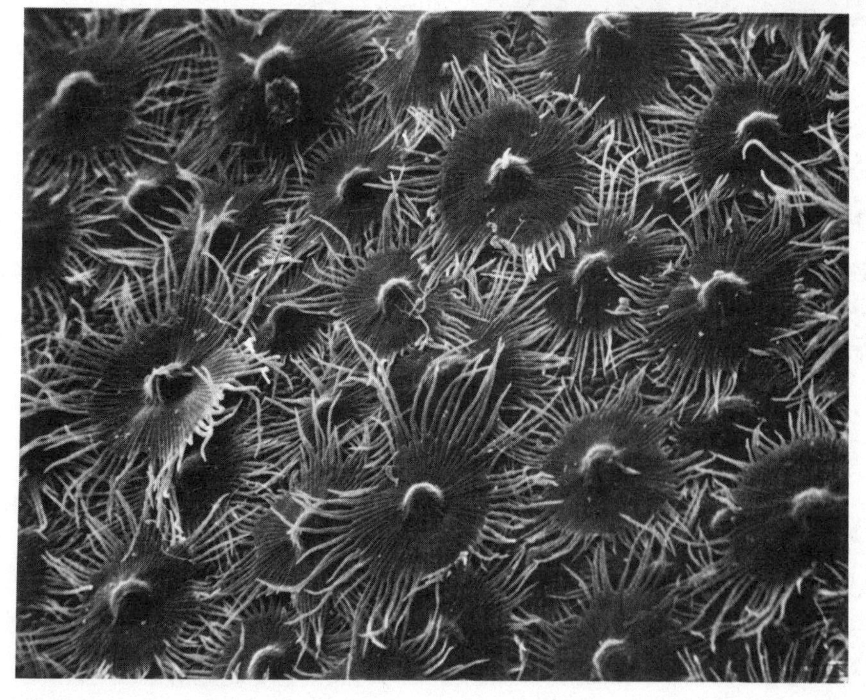

FIGURE 7-9
SEM of the umbrellalike **trichomes** on Russian olive (*Eleagnus*) leaves. (Photo by Jon M. Holy.)

(a)

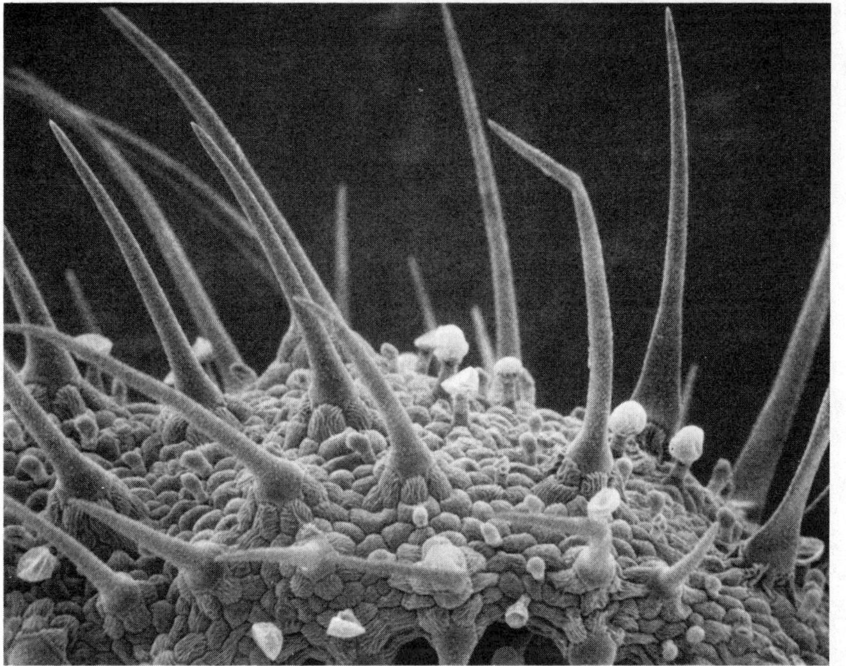
(b)

**FIGURE 7-10**
Mint-scented geranium (*Pelargonium*) leaves are covered with a velvety layer of epidermal hairs (a), some of which are **glandular**, containing the volatile aromatic substances. (b) SEM of the leaf surface. [Photo (b) by Jon M. Holy.]

A second type of trichome is the **glandular hair** (Fig. 7-11). Many glandular hairs contain fragrant volatile oils, such as the various kinds of mints (*Mentha*) and scented geraniums (*Pelargonium*). Some glandular hairs produce sticky substances. Glandular hairs of such carnivorous plants as butterwort (*Pinguicula*) and sundew (*Drosera*) produce sticky secretions and digestive enzymes.

Another example of trichome specialization is the **stinging hair** distributed on leaves and stems of stinging nettle (*Urtica*). The hairs are brittle, sharp tipped, and hollow, containing acetylcholine and formic acid. If you touch a stinging nettle, the ends of the glasslike hairs shatter, piercing your skin. At the same time, the pressure of the contact depresses the shaft of the hair into its bulbous base, thereby injecting the irritating fluid contents. A nettle rash burns and itches intensely.

Other highly specialized epidermal hairs are the touch-sensitive **trigger hairs** found on Venus flytrap (*Dionaea*) leaves. When stimulated by an insect, the trigger hairs cause the leaf halves to fold together, trapping the insect.

Root epidermis tissue differs because its function is to permit water and gases to diffuse through the cell walls. Therefore the outer cell walls are *not*

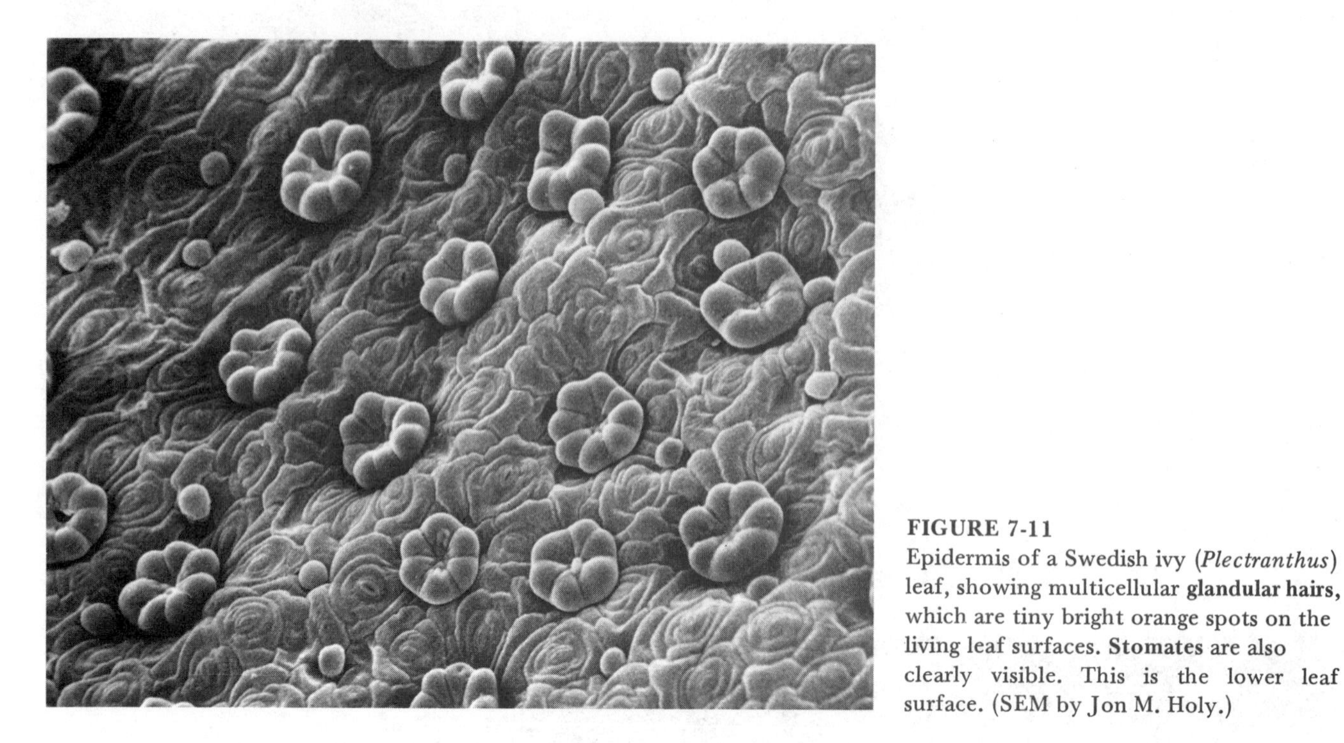

**FIGURE 7-11**
Epidermis of a Swedish ivy (*Plectranthus*) leaf, showing multicellular **glandular hairs,** which are tiny bright orange spots on the living leaf surfaces. **Stomates** are also clearly visible. This is the lower leaf surface. (SEM by Jon M. Holy.)

heavily cutinized and stomates are unnecessary. In addition, the absorptive surface area of root epidermal cells is greatly increased by specialized tubular extensions of the cells, called **root hairs** (Fig. 7-12). (See also Chapter 8.)

**FIGURE 7-12** Section of onion (*Allium*) root tip epidermis showing a single **root hair.**

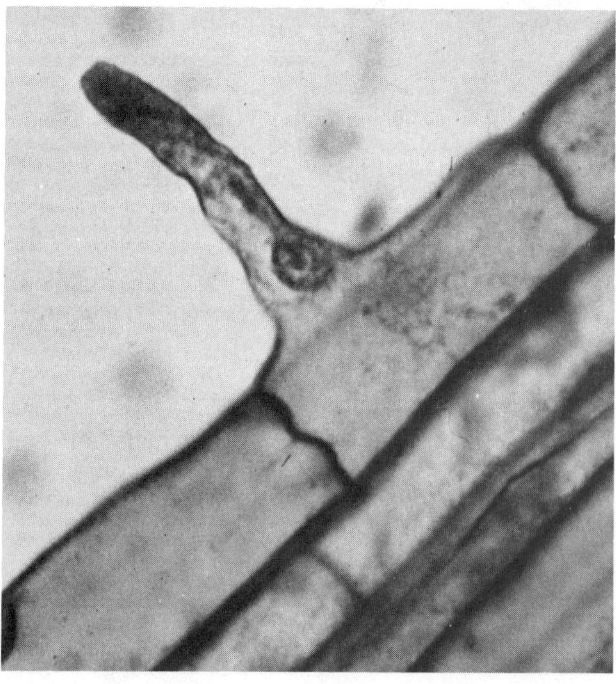

## Cork

In most woody plants **cork** tissue replaces the stem and root epidermis of older parts that outgrow their epidermis by an increase in girth. The epidermis is capable of limited expansion and most epidermises lack the potential to add new cells. So when a plant grows sufficiently in girth, the epidermis stretches until it splits; then it peels and flakes off. Before this loss of epidermis, however, a thin layer of meristematic cells, the **cork cambium,** forms and begins to divide (Fig. 7-13). As cells of the cork cambium divide, the outer cells differentiate into cork cells and the inner cells remain meristematic. Occasionally cells inside the cork cambium differentiate into parenchyma cells called **cork parenchyma.** Both the cork cambium and the cork may become many layered.

Cork tissue lacks intercellular spaces and the thick cell walls become impregnated with a fatty substance called **suberin** that blocks air and water movement. As the cell walls mature, the cytoplasm of each cork cell is sealed off from food, water, and gases and eventually dies. The dead, air-filled cells of mature cork tissue tend to be regular in shape and size, inspiring the early microscopist Robert Hooke to describe them as being similar to rows of prison or monastic cells and giving us the biological term, cell.

Because it is composed of air-filled, suberized cells, cork tissue insulates and cushions in-

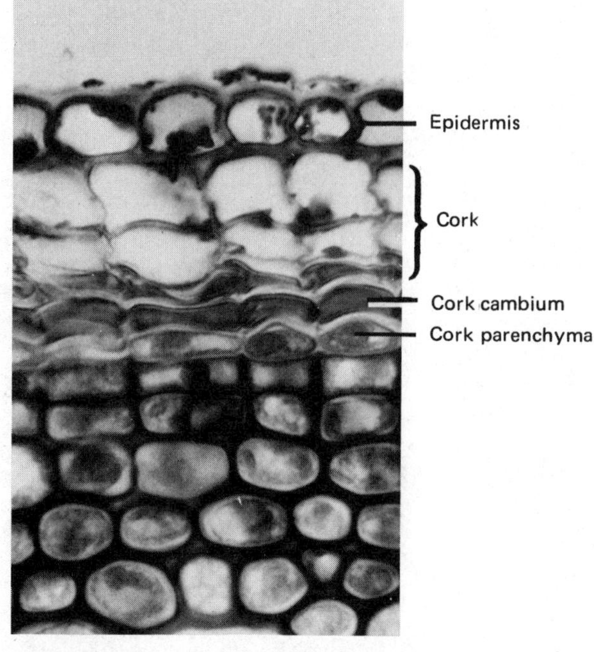


**FIGURE 7-13** Formation of **cork** in the bark. (Photo by Jeff P. Carlson.)

ternal tissues from environmental vicissitudes. It is also buoyant. These features make cork economically valuable, as in wine corks and fishing floats, for example (Chapter 16).

As the stem and root continue to grow, new layers of cork cambium develop in successively deeper tissues. (See discussion of secondary growth, Chapter 8.)

## COMPLEX PERMANENT TISSUES

Complex permanent tissues consist of two or more different cell types and form a permanent part of the plant body. The **xylem** (Gr., *xylos* = wood) conducts water and dissolved nutrients from the root system to other parts of the plant and also provides support and storage. The **phloem** (Gr., *phloios* = bark) conducts substances (e.g., food and hormones) produced in the leaves to sites of use or storage and from sites of storage to areas of use.

Xylem and phloem tissue are found in **veins** or **vascular bundles** in herbaceous plant parts. In woody plant parts xylem is the "wood" portion; the "bark" is composed largely of phloem and cork. Xylem and phloem are separated by the vascular cambium, the meristem that produces new xylem and phloem cells.

### Xylem

Natural buoyancy provided a means of external support for primitive aquatic plants and the tissues were surrounded constantly by dissolved nutrients. As plants that could occupy terrestrial habitats evolved, however, internal support became necessary to attain upright growth for maximum exposure to the sun. Development of a conducting system to provide all plant parts with nutrients and water was also a critical factor in successful exploitation of the land and so specialized water-conducting cells, the **tracheids**, developed.

Tracheids are elongated cells with thick secondary walls that provide internal support (Fig. 7-14). These cells are dead at maturity. Special modified wall areas, the **bordered pits**, occur on side walls and on ends where cells overlap with other tracheids to permit water movement between tracheids (Fig. 7-15). In this way, tracheids form a vertical string of overlapping cells as a conduit for water and dissolved mineral nutrients. Serving the dual function of conduction and sup-

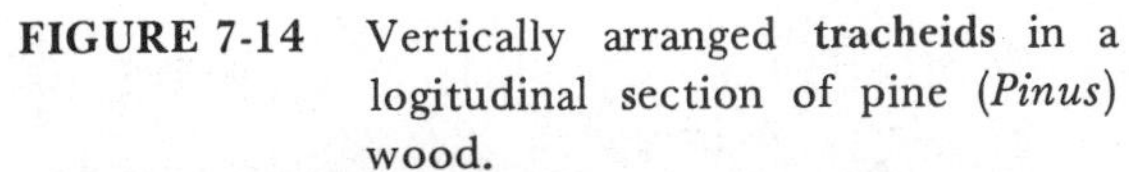

**FIGURE 7-14** Vertically arranged **tracheids** in a logitudinal section of pine (*Pinus*) wood.

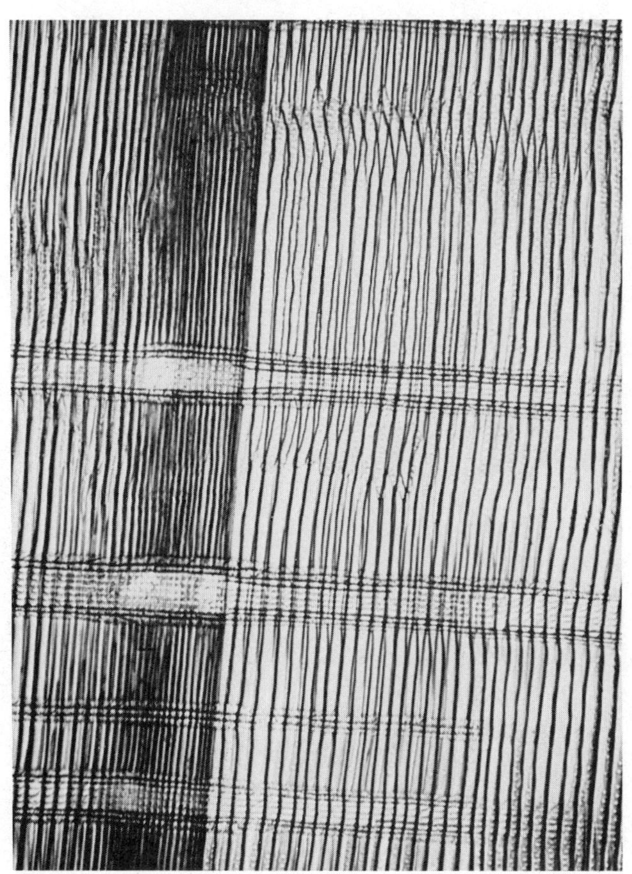

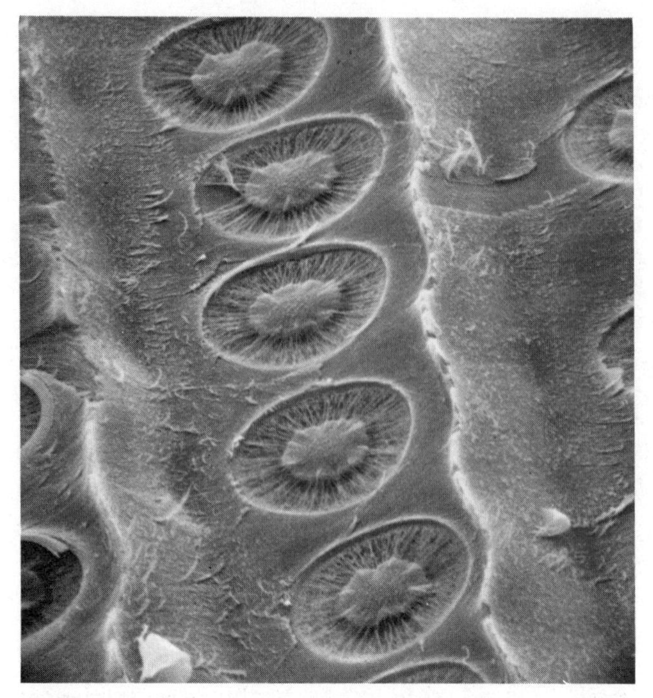

**FIGURE 7-15** Douglas fir (*Pseudotsuga*) tracheids with **bordered pits,** showing the central part or "torus" surrounded by the thin "margo" through which liquids diffuse. (SEM courtesy of Irving B. Sachs, Forest Products Laboratory, U.S. Department of Agriculture.)

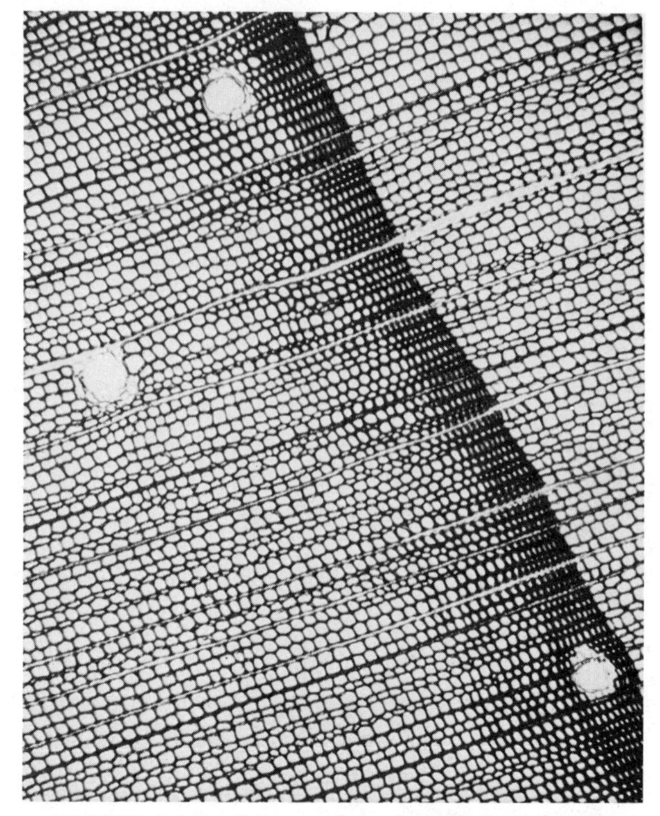

**FIGURE 7-16** Cross section of conifer wood, composed mainly of thick-walled **tracheids.** Three large **resin ducts** are also present.

port, tracheids are considered the most primitive kind of xylem cell and are found extensively in ferns and other more primitive vascular plants (Chapter 14) as well as such conifers as fir (*Abies*) and pine (*Pinus*) (Fig. 7-16).

From this beginning came refinements of both the support and conducting aspects of xylem tissue. In more advanced plants these specializations include evolution of fibers and highly efficient conducting cells, the **vessel members** (also called **vessel elements**) (Fig. 7-17). Vessel member evolution involved a decrease in both cell length and the angle of the end wall, plus more open contact between adjacent conducting cells. Different degrees of vessel member development can be found among existing plant species. The most advanced pattern is seen in vessel members with transverse end walls that are completely

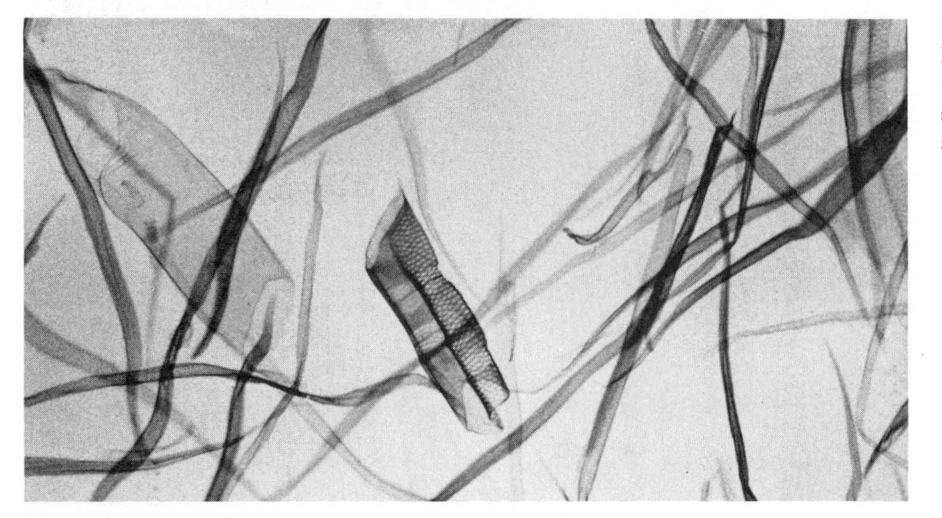

**FIGURE 7-17**
Dicot wood that has been treated (macerated) to separate the cells shows individual vessel members and fibers.

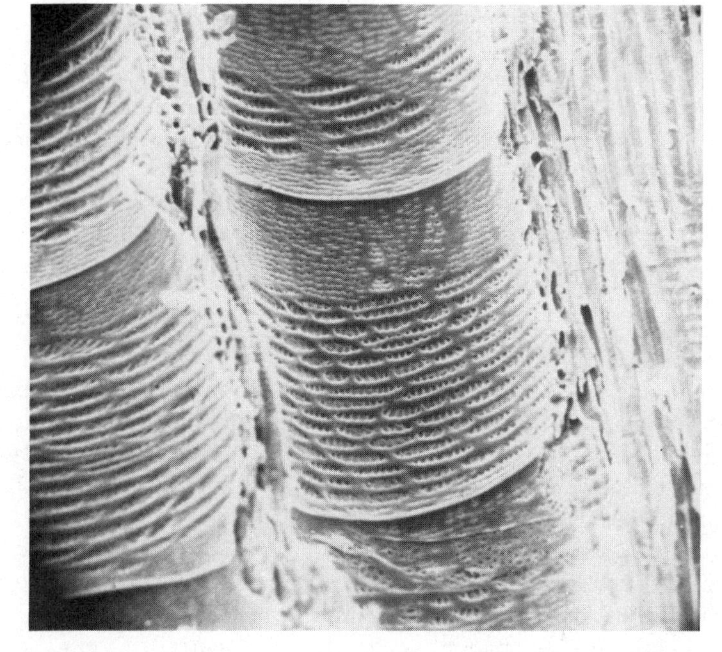

**FIGURE 7-18**
This SEM of the vessel members of red oak (*Quercus*) shows an advanced pattern, with short, barrel-shaped members. The walls are open end to end to permit efficient fluid passage and the side walls have openings that permit movement of fluid between cells. (Photo courtesy of Irving B. Sachs, Forest Products Laboratory, U. S. Department of Agriculture.)

perforated, thus creating very efficient conducting tubes, the **vessels.** Each vessel is formed by individual vessel members whose end walls disintegrate as the cells become functionally mature. The side walls of vessel members are also pitted for lateral water movement (Figs. 7-18 and 7-19). Because vessels facilitate rapid solution movement, they conduct more rapidly than tracheids, which do not have open end walls.

Fibers, vessel members, and tracheids are all dead at maturity. The only other cells found in xylem tissue are parenchyma cells, which are usually alive. Although all four cell types can be found in xylem tissue of many species, they are not all always present. The wood of some conifers, for example, may have only tracheids and parenchyma cells.

## Phloem

Phloem tissue is specialized to transport or **translocate** substances both up and down the plant's axis. As carbohydrates are produced by photosynthesis, for instance, they are distributed to other parts of the plant for use or storage. At a later time the carbohydrates may be moved from storage sites (as in root and stem parenchyma) to actively growing tissues, such as during early spring when the plant comes out of dormancy. The movement of sugars and other substances from storage tissue, for example, is the basis of maple (*Acer*) sap collecting in the spring.

In primitive types of phloem tissue, such as in conifers, conduction is accomplished by individual **sieve cells.** Sieve cells are elongated and have

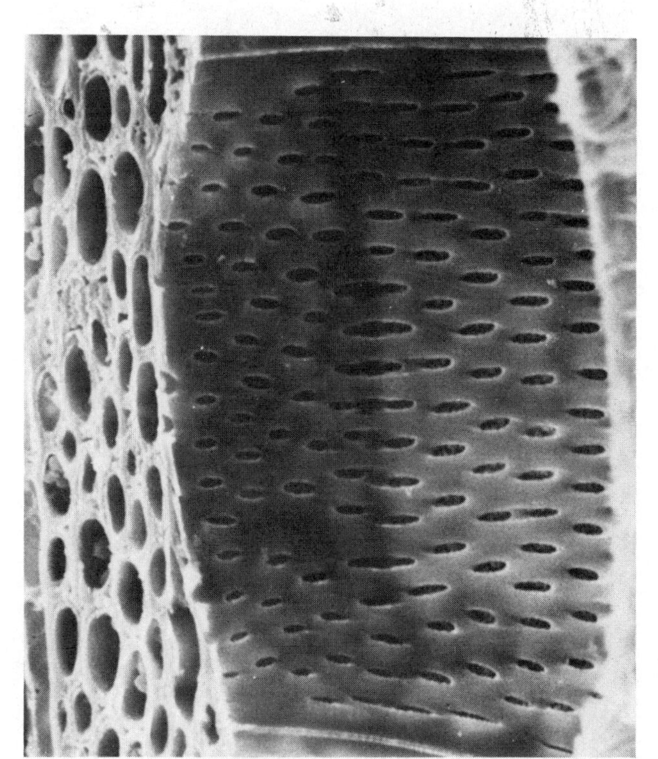

**FIGURE 7-19** Extensive pitting in the lateral vessel member walls permits conduction between adjacent cells in the xylem. (SEM courtesy of Irving B. Sachs, Forest Products Laboratory, U. S. Department of Agriculture.)

tapered ends. The pathway for movement of the food solution from cell to cell is via shared **sieve areas,** specialized regions on the walls of overlapping cell tips where cytoplasmic connections occur. In more highly evolved plants **sieve tube members** (also called **sieve tube elements**) are connected end to end to form **sieve tubes** for increased translocation efficiency. Most flowering plants conduct via sieve tubes.

In sieve tube members the specialized cytoplasmic pores are concentrated in the end walls, forming **sieve plates,** and the degree of end wall angle is almost transverse (Fig. 7-20).

Both the sieve cells of conifers and the sieve tube members of higher vascular plants contain living cytoplasm and both lose their nuclei at maturity (become enucleate). Because nuclear control of metabolism must be maintained somehow, however, specialized "assistance" cells developed. In conifers the pace of metabolic activity for sieve cells is apparently maintained by nucleated **albuminous cells.** For a sieve tube member, the closely related **companion cell** (often a sister cell of the sieve tube member) maintains this function. Other cell types found in the phloem are fibers and parenchyma cells.

**FIGURE 7-20** Conducting cells of the phloem in Virginia creeper (*Parthenocissus*) bark. (Photo courtesy of Jerry D. Davis.)

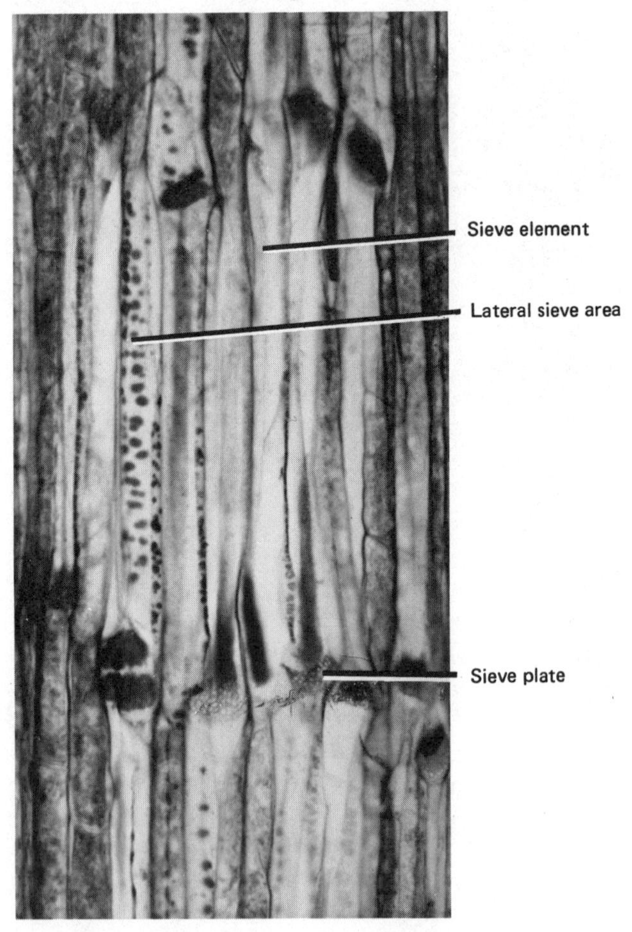


## SECRETORY STRUCTURES

**Secretion,** the separation of substances from the protoplast or their isolation in parts of the protoplast, is an activity that occurs as a normal function in living cells. We already mentioned the secretory nature of the dictyosomes. In addition, many plants have specialized external secretory structures (trichomes, glands, nectaries, and hydathodes) and internal secretory structures (secretory cells, secretory cavities and canals, and laticifers). The value of many secretory structures to the plants that have them is not known although the secretion of bitter, aromatic, toxic, or irritating substances is sometimes useful for discouraging foraging animals.

**Secretory cells** have a wide variety of contents. They may accumulate oils, tannins, resins, mucilages, and crystals. Secretory cells may be found throughout the plant body and in varying sizes and combinations. Many secretory cells have the convoluted cell wall typical of transfer cells.

**Trichomes** and **glands** vary as greatly in structure as in function. They have been discussed with the epidermis.

**Nectaries** are multicellular secretory structures that produce **nectar,** a sugary exudate. Some are especially distributed on flowers. The sugar content of each nectary is directly related to the amount of phloem (higher in sugar) in relation to the amount of xylem (lower in sugar) in the vascular tissue supplying the nectary.

**Hydathodes** express water through stomates at the tips of veins and along leaf margins, a process called **guttation.**

**Secretory cavities** and **canals** are larger than the secretory cells. The contents of most include oily or resinous substances as in citrus plants, *Eucalyptus,* and conifers. Secretory cavities can develop either by separation between cells or by degeneration of a pocket of cells to produce the space.

**Laticifers** are often associated with plants in the spurge family (e.g., *Hevea,* rubber plant), daisy family (e.g., *Taraxacum,* dandelion, and *Lactuca,* lettuce), and fig family (e.g., *Ficus*

**FIGURE 7-21** Milkbush or pencil tree (*Euphorbia*) and its relatives produce an irritating **latex** that exudes freely when the plant is scratched or broken.

*elastica,* Indian rubber plant). Laticifers can contain **latex,** as in the rubber plants and dandelion, or **alkaloids,** as in the opium poppy (*Papaver somniferum*). Laticifer secretions—especially those of members of the spurge family—are sometimes toxic (Fig. 7-21). (See also Chapter 18.)

## SUMMARY

Multicellular organisms require specialized tissues and organs to meet environmental challenges.

**Meristematic tissues** produce new cells by mitosis. **Apical meristems** elongate the plant by adding new cells to root and shoot tips. **Lateral meristems** add to plant girth.

Cells of **simple permanent tissues** mature into specific and generally irreversible forms. They are composed of a single tissue type rather than a combination of tissue types.

**Parenchyma cells** are generally unspecialized in shape and often act as storage tissue, although exposed areas develop chloroplasts. Photosynthetic parenchyma tissue is called **chlorenchyma.**

**Collenchyma** is support tissue found particularly in herbaceous stems.

**Sclerenchyma** is hard, lignified tissue of two general cell types—fibers and sclereids.

**Epidermis** on aerial plant parts reduces water loss while allowing gas exchange via the **stomates. Epidermal hairs** are specialized epidermal cells. **Root hairs** are delicate root epidermis extensions that enhance water uptake.

Protective **cork** tissue forms in most woody plant stems. It is derived from **cork cambium.**

Complex permanent tissues consist of two or more kinds of simple tissues. **Xylem** conducts water and dissolved nutrients from the roots to other plant parts and provides support and storage. **Phloem** conducts carbohydrates produced in the leaves to areas where they are stored or used and from storage sites to areas of need. Xylem and phloem tissues are produced by the apical meristem and the **vascular cambium.**

**Tracheids** and **vessels** are conducting structures in the xylem. **Sieve cells** and **sieve tubes** are phloem conducting structures.

Many plants have specialized external secretory structures—**trichomes, glands, nectaries, hydathodes**—and internal secretory structures—**secretory cells, secretory cavities** and **canals,** and **laticifers.**

## SOME SUGGESTED READINGS (Chapters 7 and 8)

Cutter, E.G. *Plant Anatomy: Experiment and Interpretation. Part I: Cells and Tissues.* Reading, MA.: Addison-Wesley Publishing Co., 1969. A morphogenic approach to plant cells and tissues.

—. *Plant Anatomy: Experiment and Interpretation. Part II: Organs.* New York: Holt, Rinehart and Winston, 1971. As above, an approach that emphasizes the dynamics of development during growth and maturation.

Esau, K. *Anatomy of Seed Plants,* 2nd ed. New York: John Wiley & Sons, 1977. A short anatomy text that is both a condensation and an update of her comprehensive treatise, *Plant Anatomy.*

Lott, J.N.A. *A Scanning Electron Microscope Study of Green Plants.* St. Louis: The C.V. Mosby Co., 1976. An aesthetic and informative atlas of plant ultrastructure.

O'Brien, T.P., and M.E. McCully. *Plant Structure and Development: A Pictorial and Physiological Approach.* New York: The Macmillan Co., 1969. A copiously illustrated short textbook correlating structure and function in the flowering plants.

Raven, P., R.F. Evert, and H. Curtis. *Biology of Plants,* 3rd ed. New York: Worth Publishers, 1981. A recent, well-illustrated botany textbook with an excellent section on plant structure and development.

Ray, P.M. *The Living Plant.* 2nd ed. New York: Holt, Rinehart and Winston, 1972. A brief introduction to the structure and function of plants; in paperback.

Sutcliffe, J. *Plants and Water.* New York: St. Martin's Press, 1968. A concise summary of an important aspect of plant physiology.

Zimmermann, M.H., and C.L. Brown. *Trees: Structure and Function.* New York: Springer-Verlag, 1971. A book that begins to close a gap in our knowledge of whole plants by concentrating primarily on aspects of the structure and function of trees.

# chapter eight PLANT ARCHITECTURE ROOTS, STEMS, AND LEAVES

*photo by Mark R. Fay*

## INTRODUCTION

The most satisfying examples of architecture are those that achieve a harmonious synthesis of aesthetic and functional design. The structure of plant bodies provides many stunning examples of architecture. Everything about the form of a plant body is based on what is necessary for individual and hence species survival.

Whether in water or desert, plants grow in such a way as to make the best use of environmental factors. In plants that grow attached to a substrate the result has been **radial symmetry.** This radiating growth pattern is easily observed by viewing a houseplant from the top or when seeing trees and bushes from the air. The chapter opening art illustrates radial leaf arrangement in the succulent, *Aeonium.*

The aerial part of the plant body produces food by photosynthesis, requiring optimum exposure to the overhead light of the sun. Therefore solar exposure dictates a radial pattern, which can be modified in response to shading by other plants or sun-blocking structures. Even in a relatively small plant, it can be seen that leaves grow to fill in every lighted space, a phenomenon that produces a **leaf mosaic** (Fig. 8-1).

Below the ground, root systems also radiate from the central axis of the plant. Radial symmetry of roots is a response to the demands of two powerful forces: nutrition and gravity. A radiating root system provides support and anchorage. Because most plants are rooted firmly to a substrate, they are at the mercy of the environment, able to utilize only water and nutrients that happen to be in contact with their root systems at any given time. So maximal development is essential within the growth area. The root system radiates in all planes from the ground/stem level, branching and filling the soil with roots laterally and below the plant.

Internally the radial symmetry of plants is strikingly apparent in the arrangement of vascular tissues that carry water and nutrients from the roots to the rest of the plant and that distribute food substances produced by photosynthesis.

The basic body plan in higher plants includes structures that provide for

1. upright growth habit.
2. light gathering and gas exchange.
3. vegetative growth.
4. anchorage.
5. water and mineral absorption.
6. conduction of food, water, and mineral nutrients.
7. food storage.
8. reproduction.

We shall examine the internal development of most structures in this chapter. In the following chapters we will examine control mechanisms and then discuss the whole plant in its growing environment.

## ROOTS

There are two basic types of root systems (Fig. 8-2). In a **taproot system** the **primary root** (first root to emerge from the seed) is much larger and maintains this size differential over **secondary roots** that arise from it. The primary root becomes the **taproot** of the taproot system.

In a **fibrous root system** many roots of approximately the same size provide a rather full complement of thin roots. The primary root usually dies. In monocots, which usually have a fibrous root system, most of the major roots are adventitious, arising from the base of the stem (Fig. 8-3). Plants propagated vegetatively generally have fibrous root systems.

**FIGURE 8-1** Leaf mosaic of *Coleus* plants. Leaf distribution takes advantage of all opportunities for exposure to light.

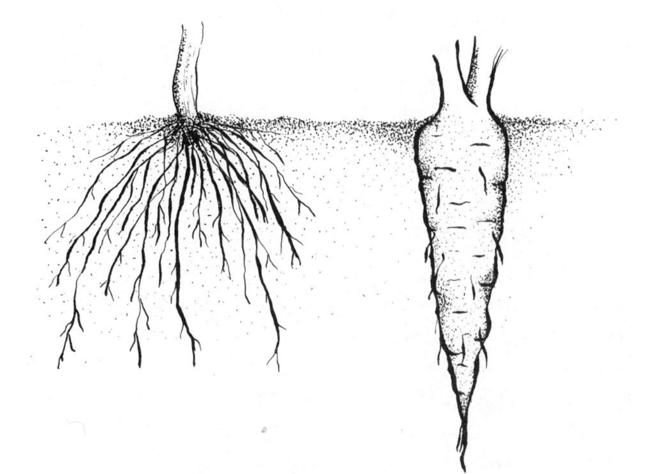

FIGURE 8-2 Generalized **fibrous root system** and **tap root system.**

FIGURE 8-3 **Adventitious roots** arising from the stem of *Agave*, a monocot. The initial root system has been largely replaced by the sturdy adventitious roots.

## The Root Tip

As a root emerges from the seed and enters the soil, its tip consists of

1. a root cap.
2. apical meristem.
3. a region of cell elongation composed of embryonic tissues.
4. a region of differentiation and maturation where cells are beginning to assume their eventual functional form (Fig. 8-4).

One of the most difficult concepts to master in developmental biology is the time/space relationship because it requires thinking "in motion" to

FIGURE 8-4 The apical meristem of an onion (*Allium*) root tip is protected by a thimble-shaped root cap. Just above the dividing cells is a region of cellular enlargement, which merges into a region where the cells are maturing into differentiated tissues.

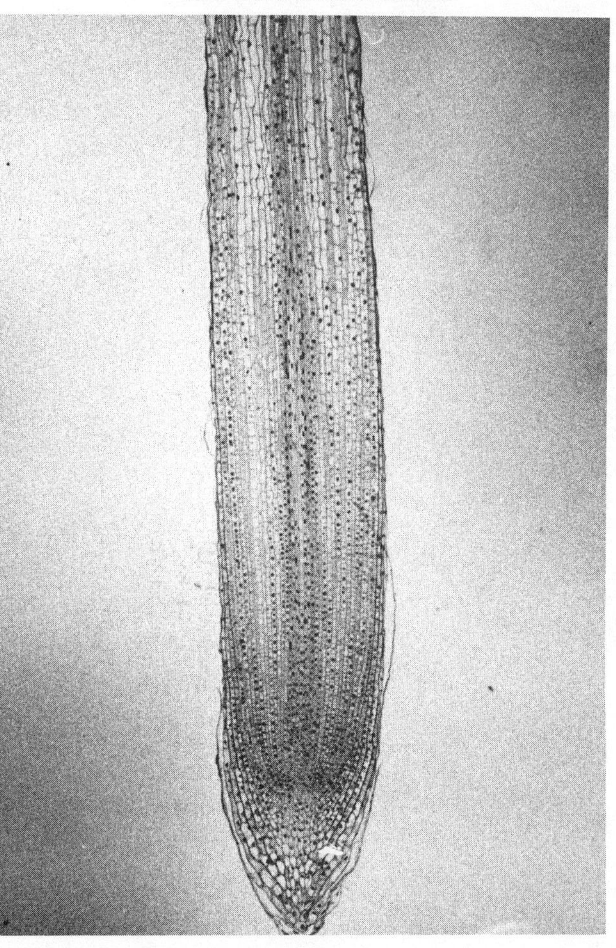

get a true picture. Each of the four regions, for example, is dynamic—a zone composed of changing ranks of cells. Cell "X" is produced by the apical meristem and goes through a sequence of enlargement and maturation in its place. In the meantime, more cells are produced, enlarged, and differentiated, so that the apical meristem is pushed farther away at the front of an advancing wave of mitotic divisions, becoming separated from cell "X" by increasing numbers of cells. Therefore the names of the regions represent human labels to identify stages in a continuing *process* of cellular and tissue development.

#### Root Cap

The **root cap** is produced by the apical meristem. Root cap cells are rather large, thin walled, and provide a continuously replaced shield of cells that protect the apical meristem as it is thrust vigorously through the soil by the forces of cellular expansion in the region of cellular elongation. Outer root cap cells are coated with mucilage, produced by the dictyosomes, which acts as a lubricant to facilitate root penetration between soil particles. The mucilage also protects the root tip from desiccation and provides a moist absorbing surface. There is also some evidence that the root cap is responsible for orienting the root to gravity (Chapter 9).

#### Apical Meristem

The **apical meristem** of the root produces cells in two directions (Fig. 8-5). Cells that mature toward the outside form the root cap; cells produced inwardly form the primary body of the root. In other words, the apical meristem simultaneously adds to and grows through the root cap. Cell division is continuous and with each division one cell typically remains meristematic and the other expands and differentiates. Microscope slide preparations of root tip tissue furnish classical examples of chromosome structure and mitotic activity (Fig. 5-3).

#### Cellular Enlargement and Maturation

Cellular enlargement is a result of water absorption, which creates turgor pressure that ex-

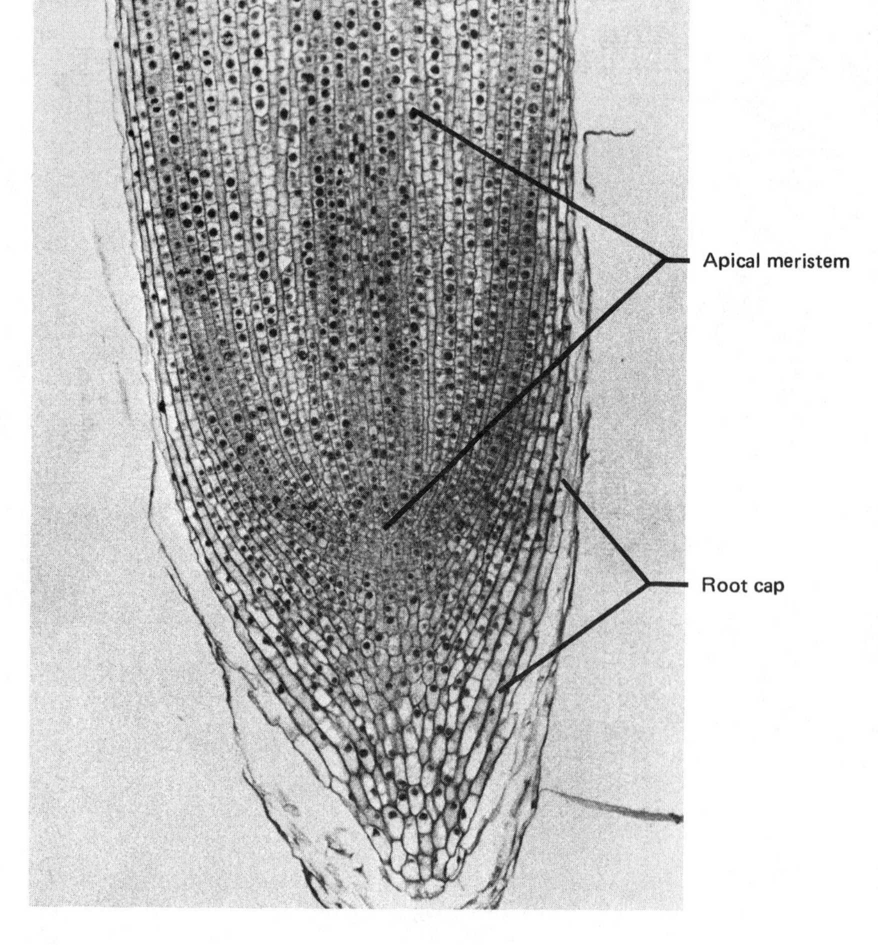


FIGURE 8-5
In an onion root tip the **apical meristem** produces cells in two directions, toward the **root cap** and toward the main body of the plant.

pands the cell. Expansion continues until it is limited by further cell wall development. During enlargement many small vacuoles are formed, eventually coalescing into a central vacuole.

At the region of elongation the immature or **embryonic tissues** (also called primary meristematic tissues) are identified by terms indicating their eventual maturation. The **protoderm** will become epidermis. The **ground meristem** will become cortex, a matrix of parenchyma tissue surrounding other tissues. The **procambium** will give rise to the primary vascular cylinder.

The region of elongation grades into an area where the embryonic cells are differentiating and maturing into primary plant tissues (Fig. 8-6).

## The Primary Root Body

**Primary growth** results directly from the activity of the apical meristem. Therefore the **primary tissues** of the root are those that mature from the embryonic protoderm, ground meristem, and procambium. Table 8-1 summarizes primary and secondary growth in roots. Most higher plants have the same basic developmental organization, although tissue arrangement at the zone of differentiation and maturation differs significantly between the two major subdivisions of flowering plants, the monocots and dicots.

The region of differentiation and maturation is marked by the presence of **root hairs** (Fig. 8-6). Each root hair is the extension of a single epidermal cell. The root hair is logically initiated at this region, for if produced earlier, it would be torn by forward movement caused by cell elongation. Root hairs extend the absorbing surface of the root cells manyfold (Fig. 8-7). A common mistake of beginning students is to believe that a root hair matures into a lateral (secondary) root. This is not so, for the root hairs remain unicellular and are very short lived.

The **cortex** matures from the ground meristem, which is so named because all other tissues are seemingly embedded in it. The cortex is composed mainly of storage parenchyma tissue, with many intercellular spaces and occasional specialized cells containing crystals. The innermost layer of the cortex is the **endodermis**, which plays a significant role in water movement through the root.

The procambium gives rise to several **primary tissues**, including **primary phloem**, **primary xylem**, and part of the **vascular cambium**. A central core

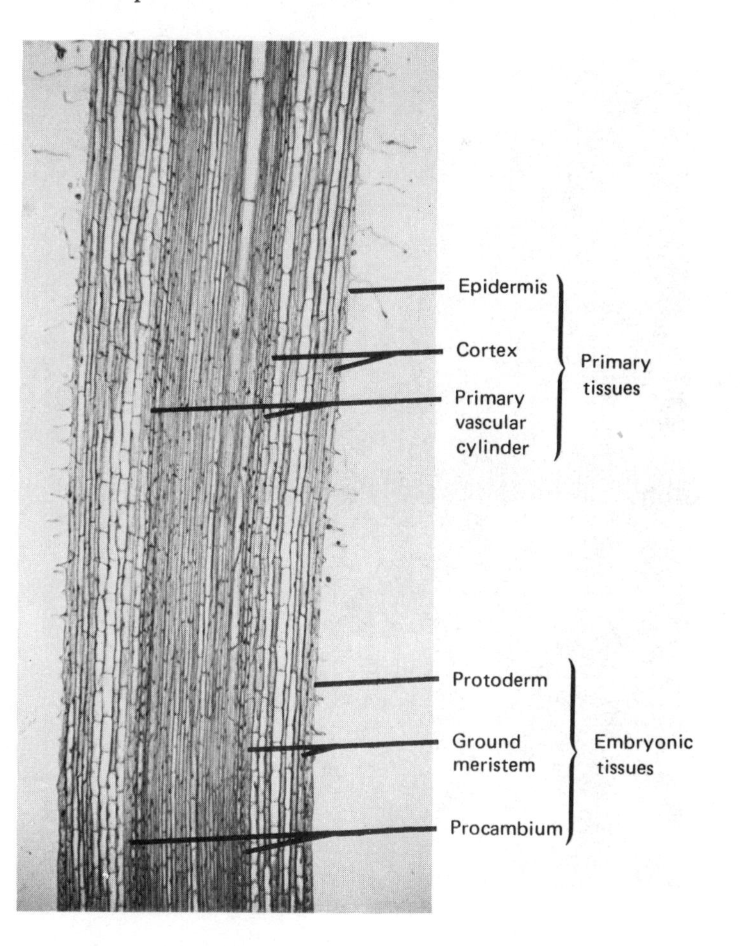


**FIGURE 8-6**
**Embryonic tissues** are found in the region of cellular enlargement. They develop into **primary tissues** in the region of maturation and differentiation, which is identified by the presence of root hairs on some epidermal cells. The primary root of corn (*Zea mays*) is shown here.

**TABLE 8-1 Primary and secondary growth of roots.**

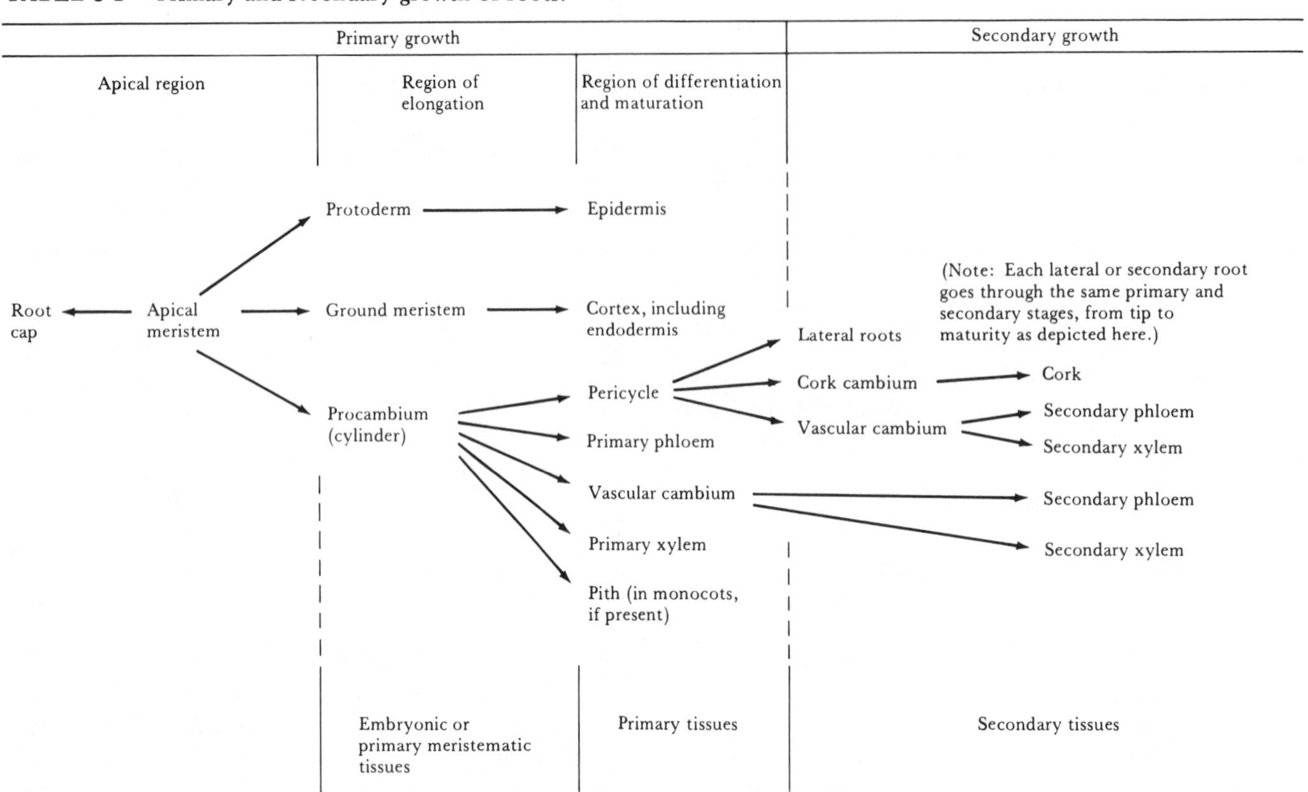

**FIGURE 8-7**
Root hairs dramatically increase the amount of absorptive surface area of epidermal cells. This scanning electron microscope photograph shows root hairs along a primary root. Note how the root hairs penetrate the water-holding spaces between the attached soil particles. (Photo by Jon M. Holy.)

of parenchyma tissue, the **pith**, is present in some monocot roots; it also develops from the procambium. A meristematic tissue called **pericycle** develops between the endodermis and the vascular cylinder.

In monocots and herbaceous annual dicots root-system organization seldom progresses beyond the primary growth stage described (Figs. 8-8, 8-9, and 8-10). Branch roots are produced, but lateral meristem activity does not produce extensive wood and bark tissue. Secondary growth of the root occurs where root reinforcement for anchorage and support is necessary, as with woody perennials.

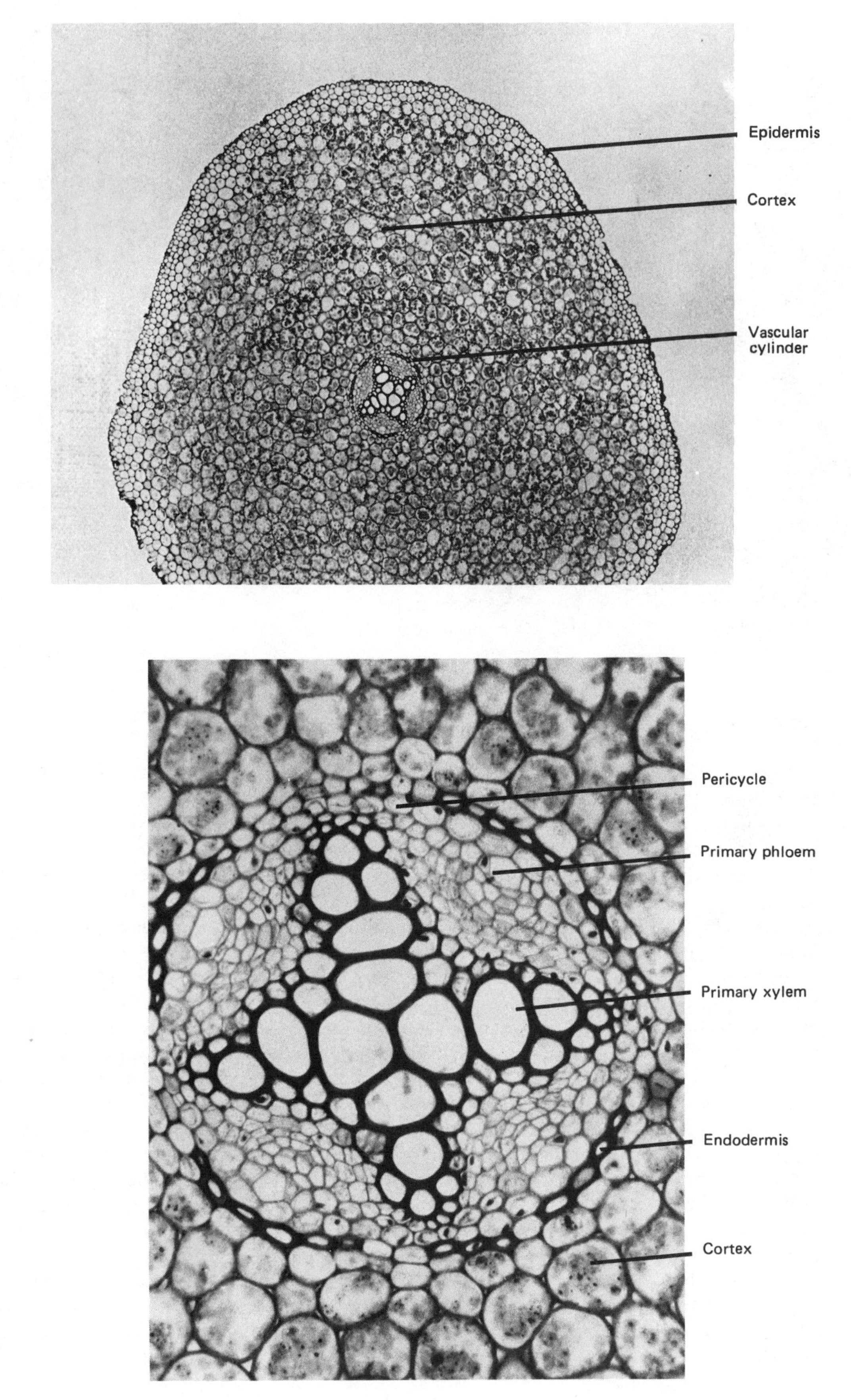


**FIGURE 8-8**
Cross section of a young dicot root.

**FIGURE 8-9**
The vascular cylinder of a young dicot root.

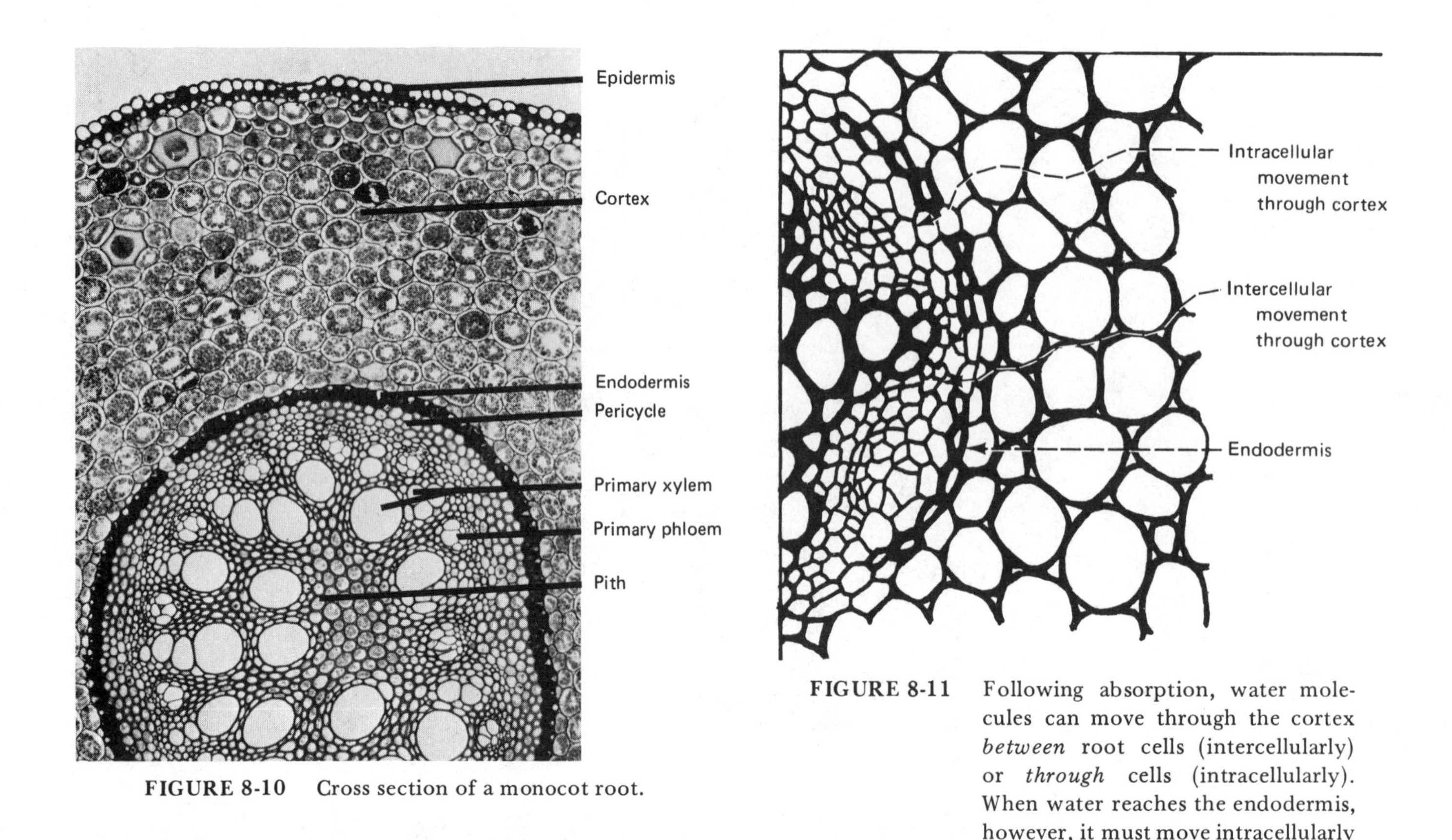


**FIGURE 8-10** Cross section of a monocot root.

**FIGURE 8-11** Following absorption, water molecules can move through the cortex *between* root cells (intercellularly) or *through* cells (intracellularly). When water reaches the endodermis, however, it must move intracellularly to enter the vascular core.

## Absorption

Even in such perennial plants as shrubs and trees, only the young roots (at the primary growth stage) absorb water and dissolved substances, for they bear root hairs.

There are two pathways by which water moves through the cortex and into the vascular system of the root (Fig. 8-11). The first involves **intracellular** movement of water from cell to cell through the cytoplasm of each cell. In the second, water moves **intercellularly** (between cells) along the cell walls until it reaches the endodermis. The **endodermis** is a unique single layer of cells, characterized by reinforced cell wall barriers called **Casparian strips.** Each strip is a thickened band around the side, top, and bottom walls of endodermis cells. The Casparian strip contains suberin and blocks intercellular movement of the root solution into the vascular cylinder. Therefore the only way water and dissolved substances can enter the vascular cylinder is intracellularly, through the front and back walls, plasma membrane, and cytoplasm of the endodermal cells. In this way, the endodermis regulates entrance of water, mineral nutrients, and other substances into the vascular cylinder for conduction to other parts of the plant. As importantly, the endodermis prevents air from entering the xylem, which would disrupt water movement.

## Secondary Growth of the Root

### The Pericycle and Lateral Root Formation

The **pericycle** consists of parenchymatous layer(s) of cells inside the endodermis and may function in as many as three ways. It initiates lateral root formation. If secondary growth occurs, the pericycle initiates the formation of the first cork cambium and part of the vascular cambium.

A **lateral** or **secondary root** is initiated by mitotic divisions in the pericycle (Fig. 8-12). As it grows, it pushes its way through the endodermis, cortex, and epidermis and into the soil. The new lateral root is a typical primary root in its structure, possessing root cap, apical meristem, and zones of elongation and maturation (Fig. 8-13). Lateral roots are connected to the main vascular cylinder and add to the total absorption capacity of the root system.

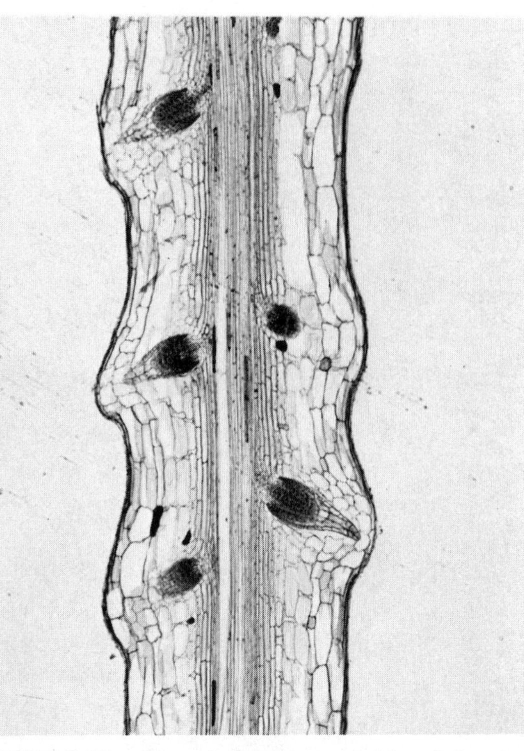

FIGURE 8-12 Lateral roots are initiated in the pericycle and grow outward into the soil. (Photo courtesy of Carolina Biological Supply Co.)

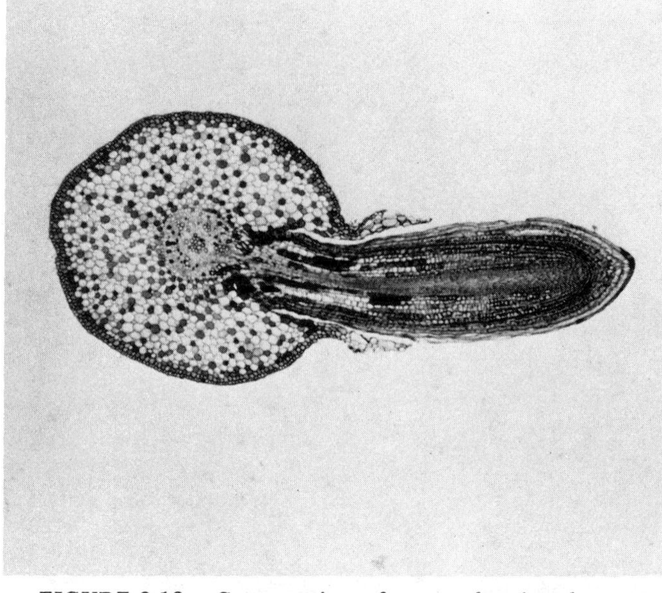

FIGURE 8-13 Cross section of a root showing the emerged lateral root, which has all the characteristics of a typical root tip. (Photo by Jeff P. Carlson.)

### Vascular Cambium and Cork Cambium

In a young dicot root the primary xylem is arranged as a ridged core, with primary phloem in the valleys between the xylem ridges (**poles**). Vascular cambium differentiates from procambium tissue that is present between the primary phloem and primary xylem. Vascular cambium cells also develop from the pericycle over the tops of the xylem poles. The result is a complete sleeve of vascular cambium wrapped around the outline of the primary xylem. The cambium cells divide tangentially,* most rapidly at the sites inside the primary phloem, eventually filling in the valleys with secondary xylem and creating a round core of vascular tissue. The increasing bulk of the new secondary tissues in the vascular cylinder begins to stretch and tear the old primary tissues—endodermis, outer cortex, and epidermis.

As the vascular cambium develops, the outer portion of the pericycle divides both tangentially and radially† to produce a ring of cork cambium cells. Cork cambium cells divide to form a layer of cork, which isolates the outer primary tissues from the vascular cylinder. When the primary tissues eventually slough off, the cork protects the inner tissues.

This part of the root having secondary growth carries on the major functions of conduction, support, and anchorage, but it can no longer absorb water from the soil. The absorbing activity is maintained by the younger, actively growing terminal portions of the root. Therefore during transplanting of seedlings, shrubs, and trees, it is important to maintain as many of the "feeder roots" as possible. If too many of these primary roots are destroyed, an imbalance between transpiration and absorption will cause the top to wilt. Continuation of this imbalance without corrective measures (covering or trimming the top to reduce evaporation; rapid feeder root growth) may cause the plant to die.

Continued enlargement of existing roots coincides with secondary development above ground—

*Tangential cell divisions produce daughter cells to the inside and outside of the vascular cambium.

†Radial cell divisions produce daughter cells that are side by side within the same cell ring.

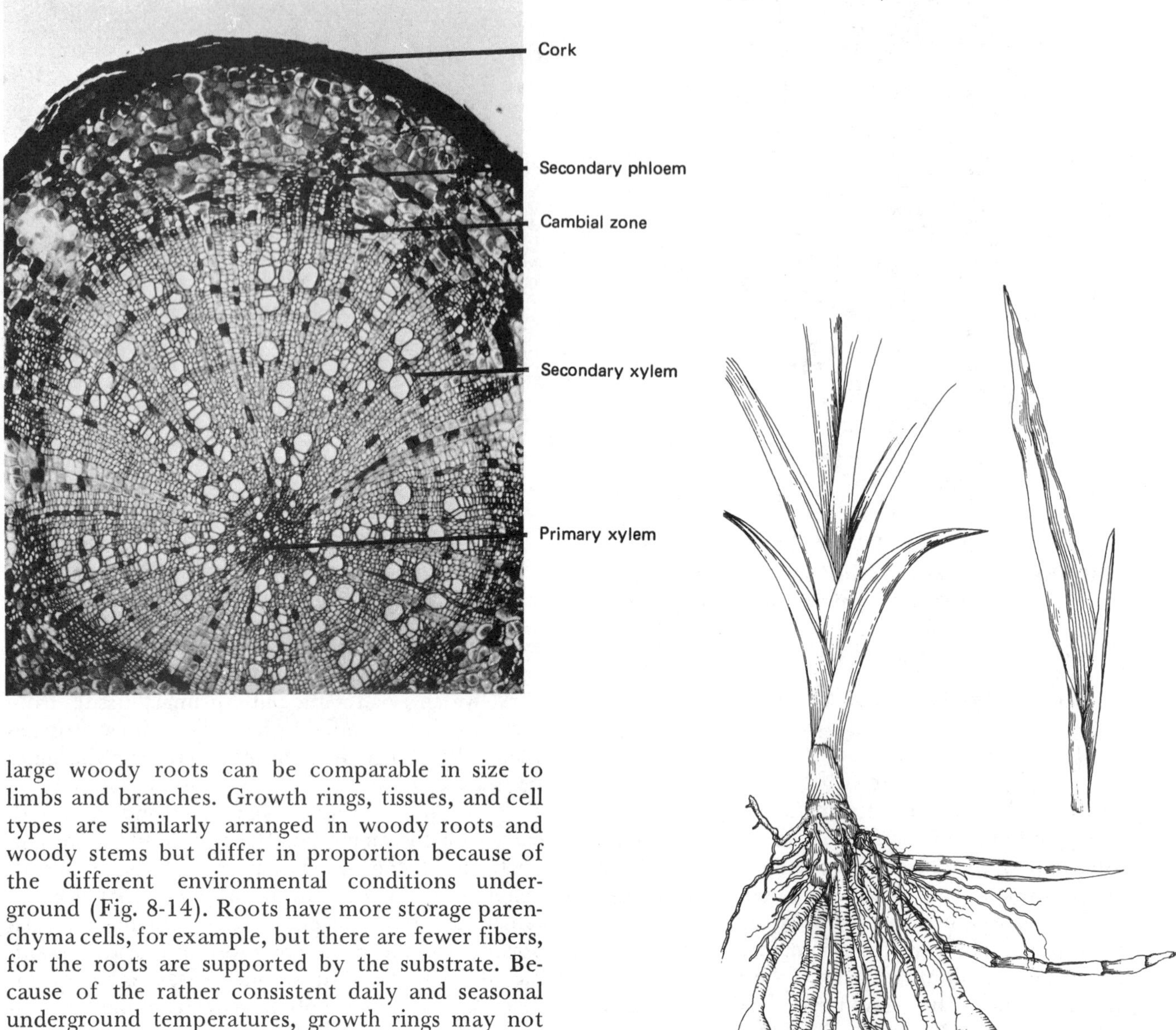


**FIGURE 8-14**
Cross section of a woody root. (Photo by Jeff P. Carlson.)

**FIGURE 8-15**
The daylily (*Hermerocanis*) spreads by horizontal subterranean stems called rhizomes. Two rhizomes extend to the right near the base of the stem in this drawing. The bulbous structures are fleshy storage roots. As with most monocots, the leaves have parallel venation and are sheathlike at their bases.

large woody roots can be comparable in size to limbs and branches. Growth rings, tissues, and cell types are similarly arranged in woody roots and woody stems but differ in proportion because of the different environmental conditions underground (Fig. 8-14). Roots have more storage parenchyma cells, for example, but there are fewer fibers, for the roots are supported by the substrate. Because of the rather consistent daily and seasonal underground temperatures, growth rings may not be as well defined as in the stem.

## Special Root Adaptations

**Prop roots** are adventitious roots that brace the stem, like guy wires supporting a post (Fig. 8-3). **Aerial roots** are adventitious roots that help to anchor climbing plants, such as *Philodendron.* Some biennial and perennial plants have well-developed **storage roots.** The carrot root is a typical storage taproot, having a large amount of cortex. Genetic selection has emphasized sweetness and tenderness and eliminated extensive fiber development.

## STEMS

Plant stems serve several functions. They produce and support leaves, new shoots, and such reproductive structures as cones and flowers. Mitosis by the apical meristem in the shoot tips increases the length of the aerial part of the plant. Stem vascular tissues conduct materials between the root and shoot systems and provide support for the stem and its derivatives. Parenchyma tissues in the stem are storage sites for such diverse substances as sugars, starch, lipids, resins, tannins, and crystals.

**Tubers**, as those of the Irish potato, are specialized storage stems. **Rhizomes** (Fig. 8-15) are subterranean horizontal stems and **stolons**, like those of strawberries, are above-ground horizontal stems. All three types are sources of vegetative (asexual) propagation and are described in Chapter 5.

### External Stem Characteristics

The exterior of a young stem reveals evidence of its functions in producing leaves, flowers, and new shoots (Fig. 8-16). Stems have **nodes**, the points of leaf and bud attachment, and the stem regions between nodes are called **internodes**. The junction of leaf and stem is called a **leaf axil**; so buds at nodes are called **axillary buds.**

In herbaceous stems the buds are unprotected, but the buds of woody stems are protected by hard **bud scales** to prevent drying and injury during winter dormancy. When the bud swells to begin growing into a shoot, the bud scales fall away, leaving scars on the twig. Similarly, a scar remains on the twig after a leaf falls away, revealing the scar-within-a-scar design of the vascular bundles (veins) that connected the leaf and stem conducting tissues.

Herbaceous stems are covered by epidermis

FIGURE 8-16 A young woody twig (maple, *Acer*). Note the bud scale scars on the bark between the new growth and the previous year's growth and on the short lateral branch. Leaf attachment, axillary buds, and lenticels (small white oval structures) are apparent.

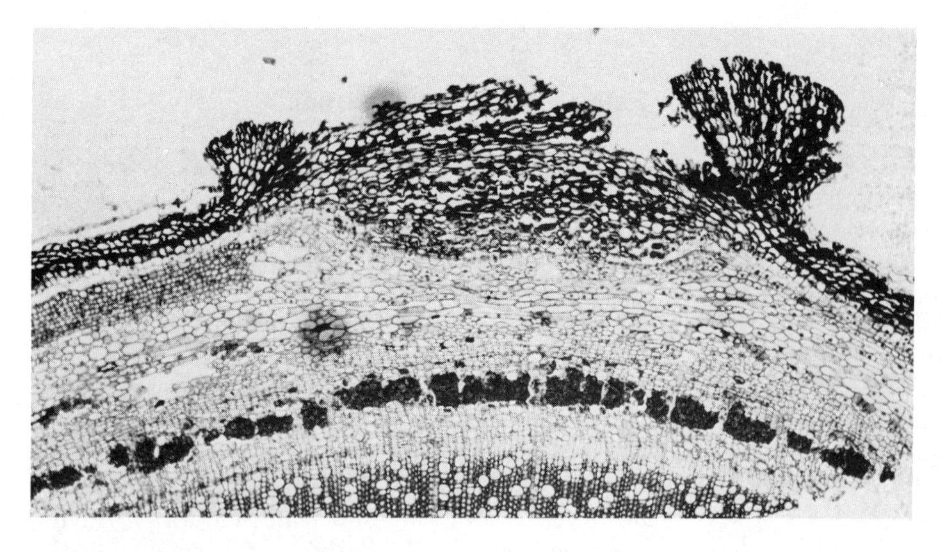

FIGURE 8-17
Section of a woody twig showing a **lenticel** perforating the corky layers of the bark. The loose tissue structure apparently permits gas exchange between the air and stem tissues.

tissue, complete with stomates. In contrast, woody twigs have a waterproof and insulating layer of corky **bark,** perforated in places by lenticels. **Lenticels** are areas of loose tissue that allow gas exchange between the stem and the atmosphere (Fig. 8-17).

## The Stem Tip

Three developmental regions are apparent in the stem tip—the apical meristem, an elongation/expansion zone, and a differentiation and maturation zone where cells begin to mature (differentiate) into functioning primary tissues. These three regions are the same in all stems of higher vascular plants. Table 8-2 summarizes both primary and secondary growth in angiosperm stems.

Of the embryonic tissues produced by the apical meristem, the **protoderm** matures into epidermis, **ground meristem** matures into certain parenchyma tissues, and strands of **procambium** mature into vascular bundles.

**TABLE 8-2 Derivation and differentiation of plant tissues in the angiosperm stem, showing the results of both primary and secondary growth.**

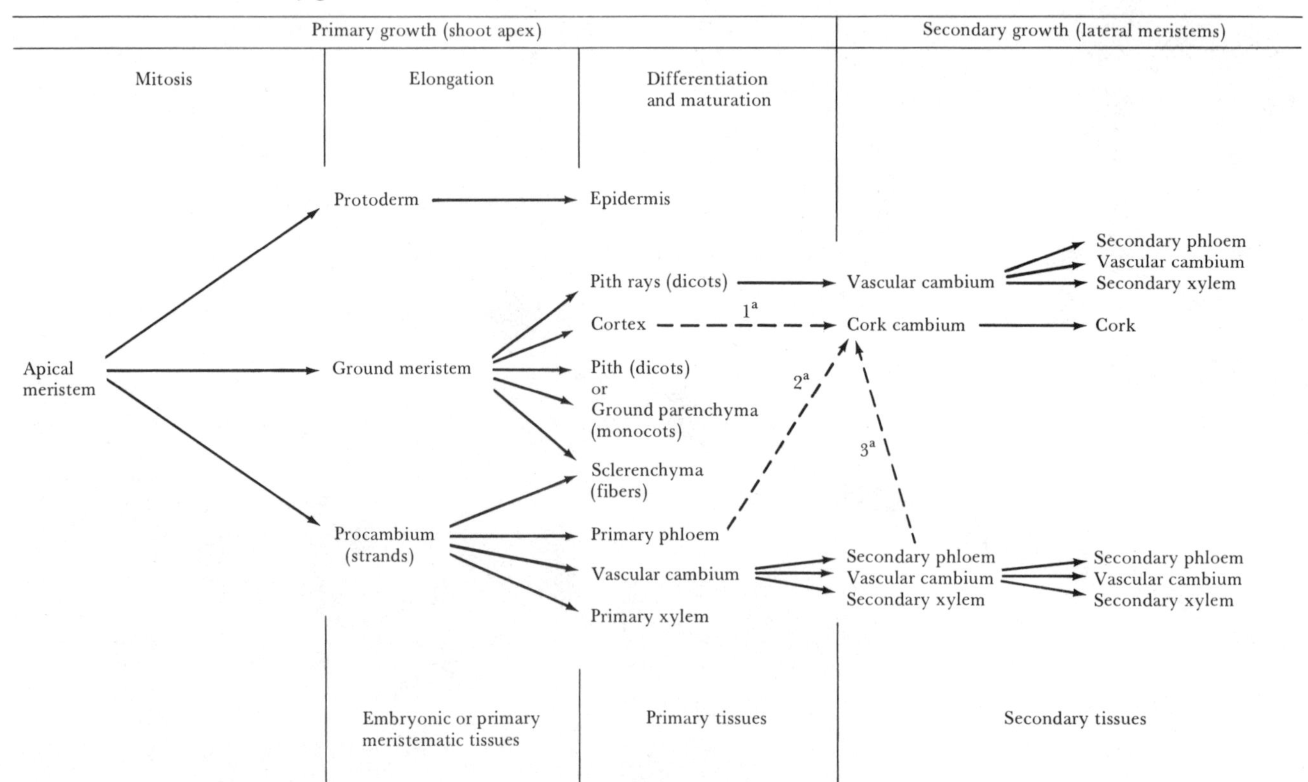


*Cork cambium arises from successively deeper layers of the outer tissues until a true bark is formed, at which time cork cambium continually arises from within secondary phloem tissue.

The stem tip (Fig. 8-18) is very similar to the root tip, although there are significant differences.

1. The stem tip does not have a cap that corresponds to the root cap.
2. Lateral stem appendages develop from bud and leaf primordia in the *outer* part of the meristem in contrast to development of lateral roots from tissue (pericycle) deep within the root.
3. Stem vascular tissues develop as strands, or bundles, embedded in parenchyma tissue rather than as a central core.
4. Stems lack endodermis and pericycle.

## Primary Growth

Internal stem organization at the differentiation and maturation level is similar in all dicots (Fig. 8-19) whether herbaceous or woody. Differences arise during secondary growth if it occurs. Monocot stem organization is different from dicots and yet similar among all monocots (Fig. 8-20). Most monocots do not acquire lateral meristems and hence do not become woody.

In the **primary body** of dicot stems **vascular bundles** (Fig. 8-21) are arranged in a circle, separated and surrounded by parenchyma tissue. The parenchyma tissue that is located between the epidermis and the ring of bundles is called the **cortex** whereas the parenchyma tissue that forms a central core is called the **pith**. The parenchyma regions between the bundles are called **pith rays.** Each bundle is composed of several tissue types, from inside to outside: xylem nearest the pith; a layer of vascular cambium; then phloem; and often a **bundle cap** composed of fibers. The outside of the stem is covered with **epidermis. Collenchyma** tissue may be present just inside the epidermis.

Vascular bundles of monocot stems are scattered within a parenchyma matrix. Because the bundles are not arranged in a ring, as in dicots, the parenchyma is simply called **ground parenchyma.** Each bundle resembles a miniature face, with eyes and nose formed by large xylem vessels (Fig. 8-22); the mouth's a space (lacuna) that used to contain the first vessel. The phloem is on the side of the

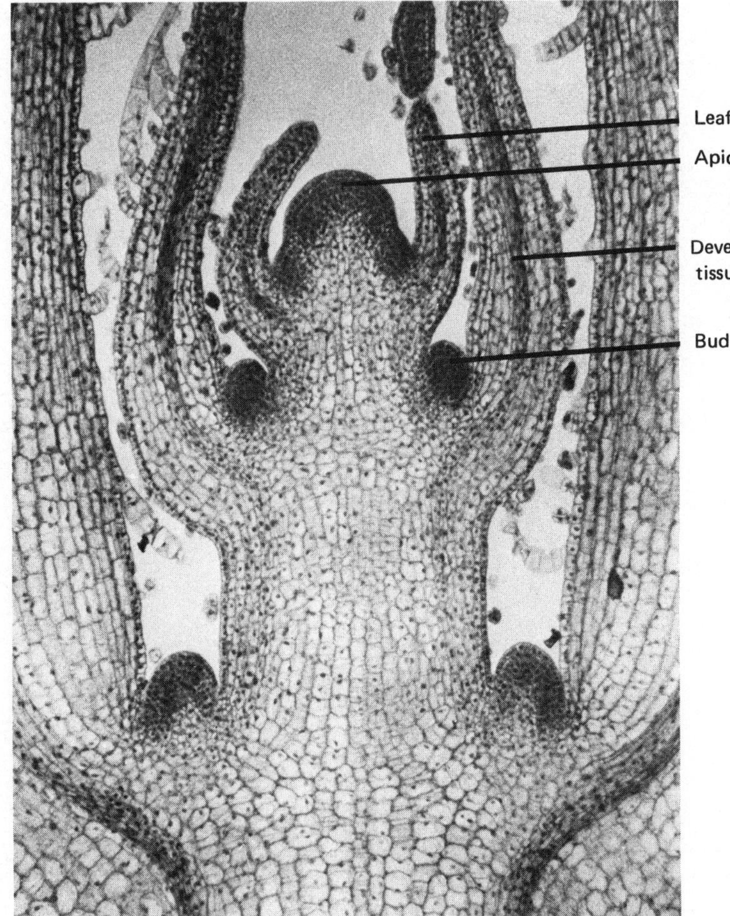


**FIGURE 8-18**
The longitudinal section of a *Coleus* stem tip shows apical meristem in the center, flanked by leaf and bud primordia. Differentiating tissues are apparent in the young leaves and stem.

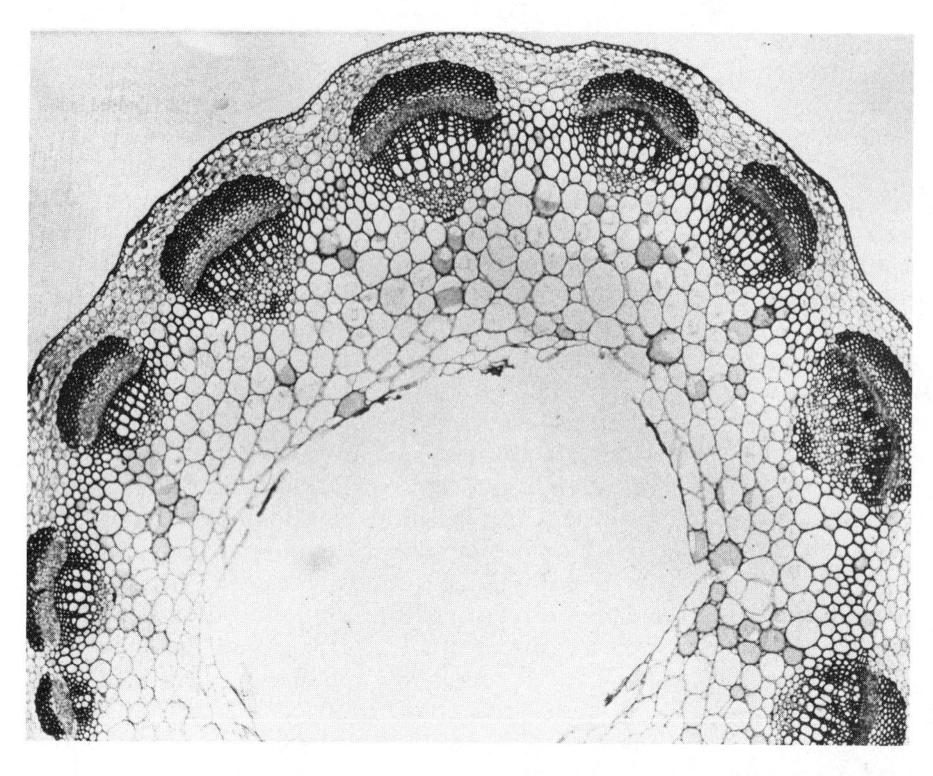

FIGURE 8-19
Cross section of a young dicot stem, showing a ring of **vascular bundles.** The area between the **epidermis** and the outer part of each bundle is **cortex.** The parenchyma tissue between the hollow stem center and each bundle is **pith. Pith rays** of parenchyma tissue fill in between the vascular bundles.

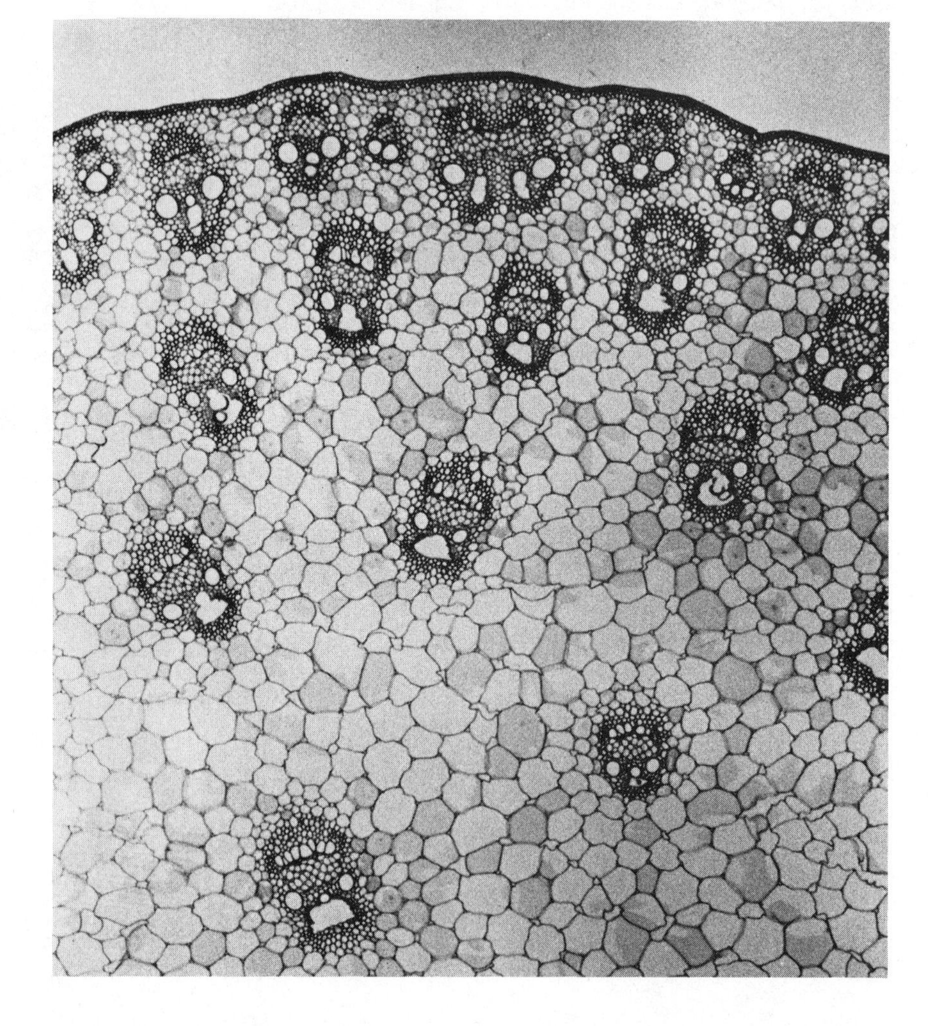

FIGURE 8-20
Cross section of a monocot stem. The stem is covered by **epidermis.** The **vascular bundles** are scattered in a matrix of **ground parenchyma.**

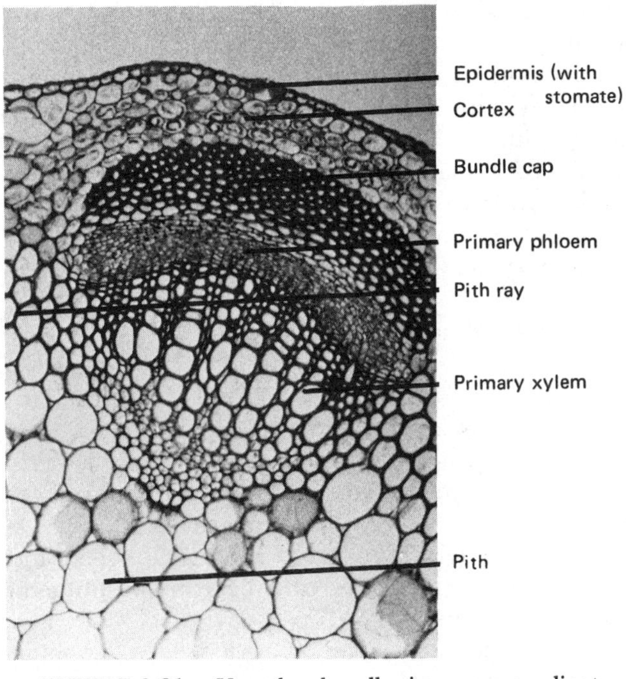


FIGURE 8-21 Vascular bundle in a young dicot stem.

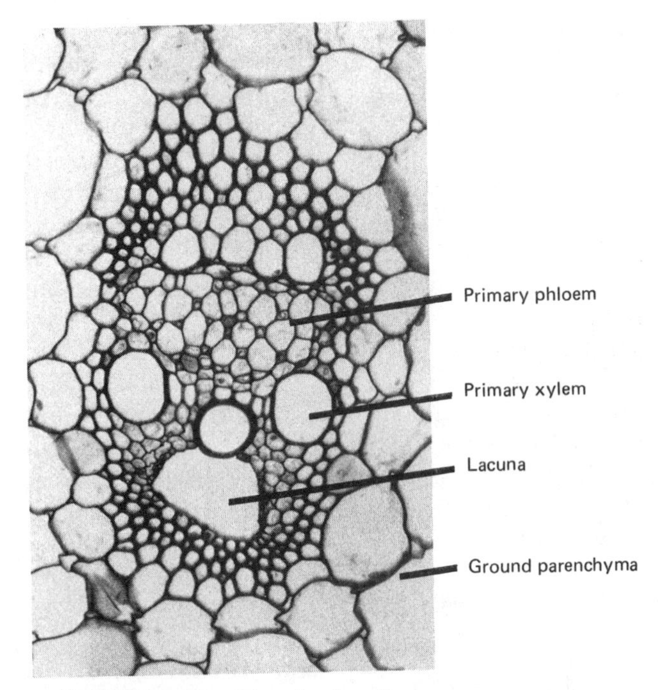


FIGURE 8-22 Vascular bundle in a monocot stem.

bundle away from the center of the stem and opposite the xylem cells. The entire bundle is reinforced and surrounded by fibers. These tissues, plus the epidermis and its underlying fibers, constitute the primary body of a monocot stem.

## Secondary Growth (Dicots)

### The Young Stem

The vascular cambium within dicot bundles continually forms **secondary xylem** and **secondary phloem.** When a cambium cell (**initial**) divides, it produces two cells with different "destinies." If the daughter cell nearer the stem center differentiates, it becomes a xylem cell while the outer cell remains meristematic. On the other hand, if the daughter cell nearer the outside differentiates, it becomes a phloem cell and the inner cell remains meristematic. In this way, the size and complexity of the vascular bundles increase.

Meanwhile certain parenchyma cells between the bundles become meristematic, initiating a cambium layer *between* the vascular bundles. The net result of this activity is formation of a continuous sleeve of vascular cambium, which subsequently produces continuous outer and inner sleeves of secondary phloem and xylem (Fig. 8-23). Thus cambium formation between the bundles is the initial stage toward development of a solid vascular

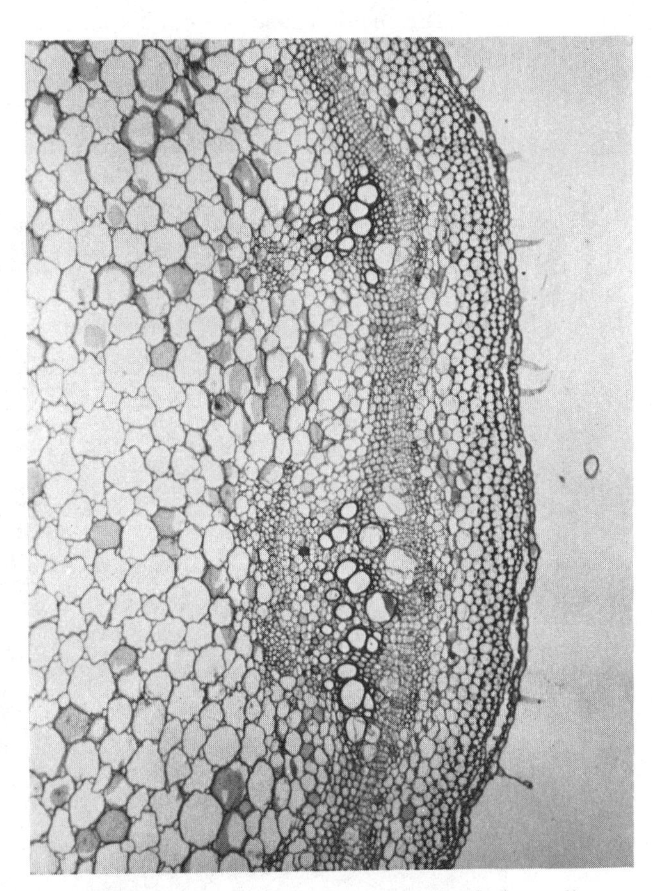

FIGURE 8-23 Early secondary growth in a dicot stem. Notice that a continuous ring of vascular tissue has been formed.

cylinder, composed of **wood** (xylem tissue) surrounded by a thin cambial layer and **bark** (phloem and cork tissue). This condition reaches its maximum development in a tree, the body of which consists mainly of secondary tissues. Even in trees, however, primary tissues exist in the young growing ends of the roots and twigs as well as in leaves and reproductive structures.

So far our understanding of the vascular cambium has emphasized division that occurs on a tangential plane, producing new cells to the inside and outside. If such were the only type of division, however, clearly there soon would be insufficient cambial cells to reach around the increasing girth of the stem! Therefore cambial cells (initials) are also capable of division on a radial plane, producing new cells side by side with existing cambial cells.

#### Loss of Primary Tissues

As a young woody stem grows, some of the primary tissues remain. Tissue proliferation by the vascular cambium, however, compresses and crushes most of the original parenchyma tissue. Growth pressures from the inside cause the cortex and epidermis to stretch and tear. The phloem also tears, forming wedges that become filled in between with parenchyma. Cork cambium layers form in several places within the cortex and down into the primary phloem and the successive layers of cork produced replace the protective function of the outgrown epidermis. In an older stem the successive cork/phloem layers of the mature bark are more apparent (Fig. 8-24).

#### Axial and Radial Transport Systems

It is obvious in Fig. 8-25 that the secondary plant body is actually composed of two systems. The system primarily discussed so far is the vertical or **axial system**, organized to conduct substances vertically. The functional components of this system are tracheids and/or vessels in the xylem and sieve cells or sieve tube members in the phloem. (See also Chapter 7.)

The second system is organized to conduct substances laterally or radially. The **radial system** is composed of longitudinal plates of parenchyma cells that form the **vascular rays** [Figs. 8-26 and 8-27 (a) and (b)]. An example of radial conduction

**FIGURE 8-24** Cross section of a 3-year-old basswood (*Tilia*) stem, showing annual rings in the xylem as well as yearly growth increments in the phloem.

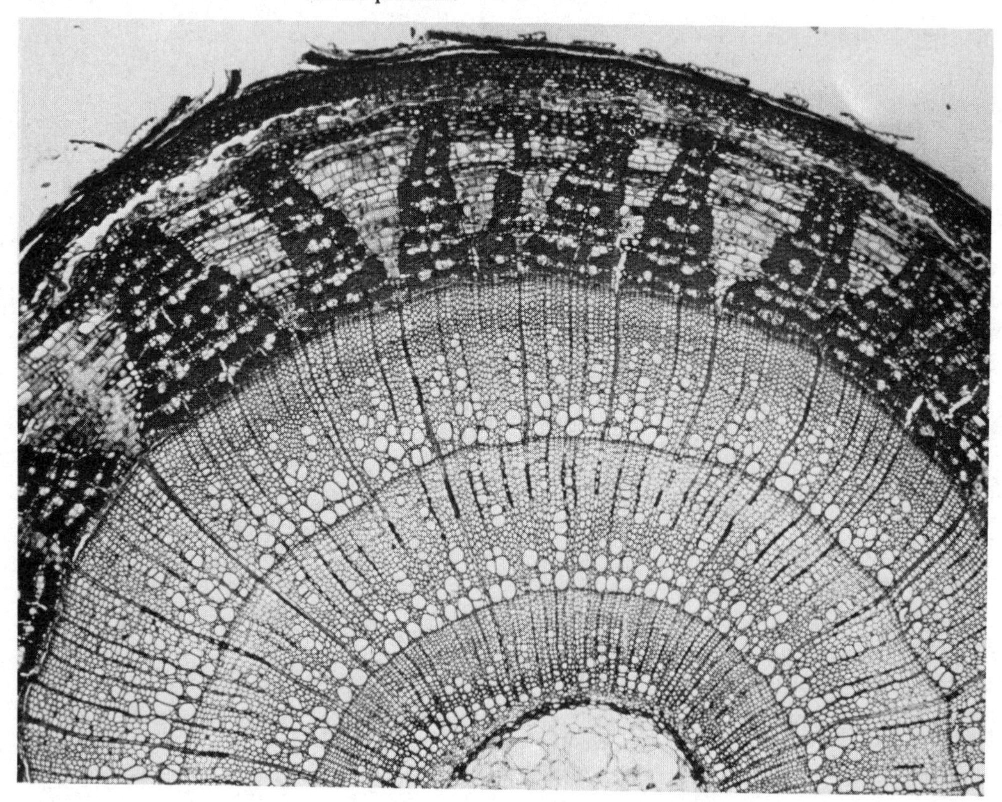

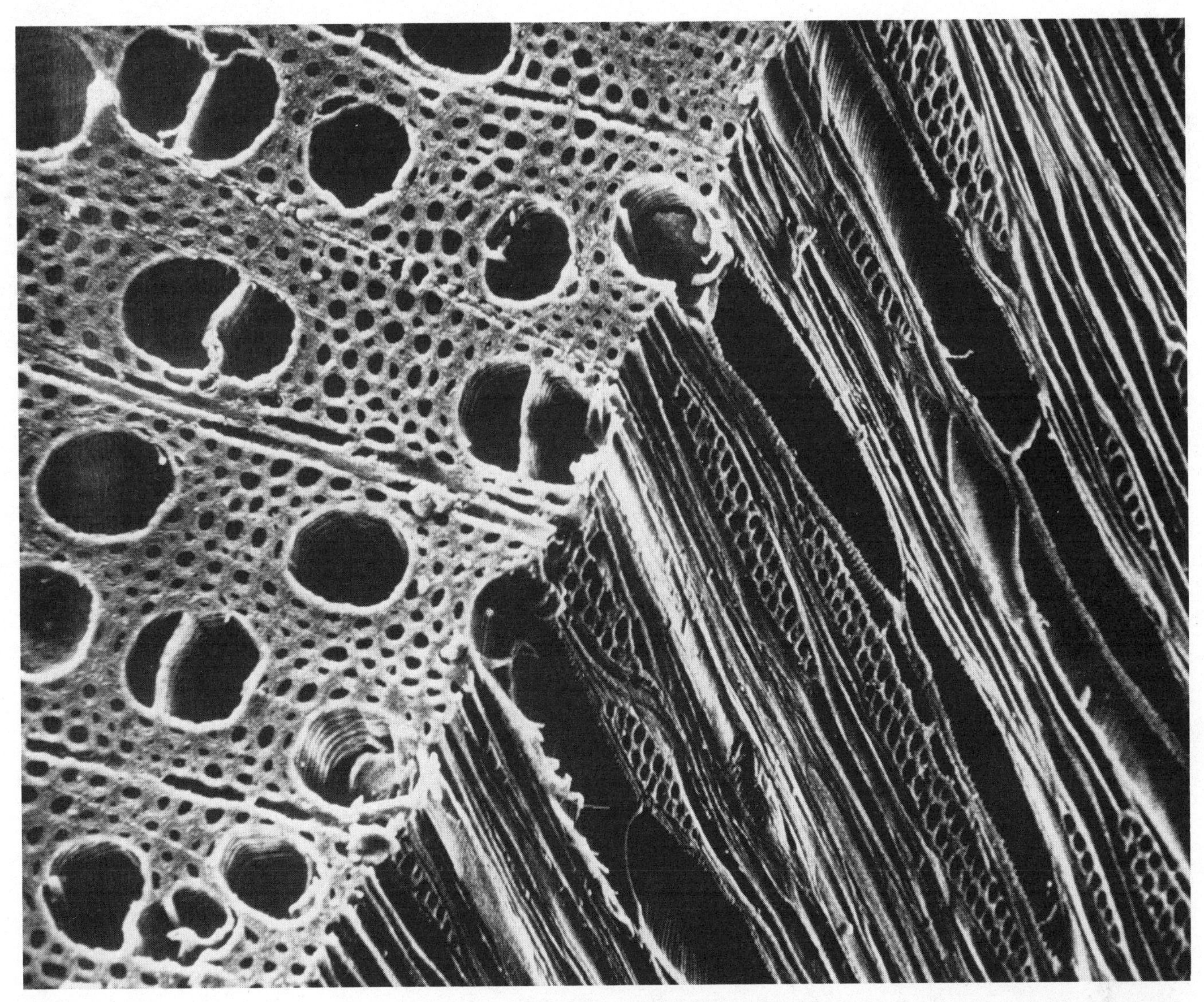

FIGURE 8-25
A scanning electron micrograph of maple (*Acer*) wood, showing the three-dimensional relationship of the vertical conducting system (the vessels) and the radial conducting system (the rays). On the upper (left) surface, vessels and fibers are seen in cross section, and rays appear as narrow, radiating cell rows. On the vertical surface vessels can be seen as long, wide tubes and the rays appear as multicellular plates of parenchyma cells. (Photo courtesy of Irving B. Sachs, Forest Products Laboratory, U.S. Department of Agriculture.)

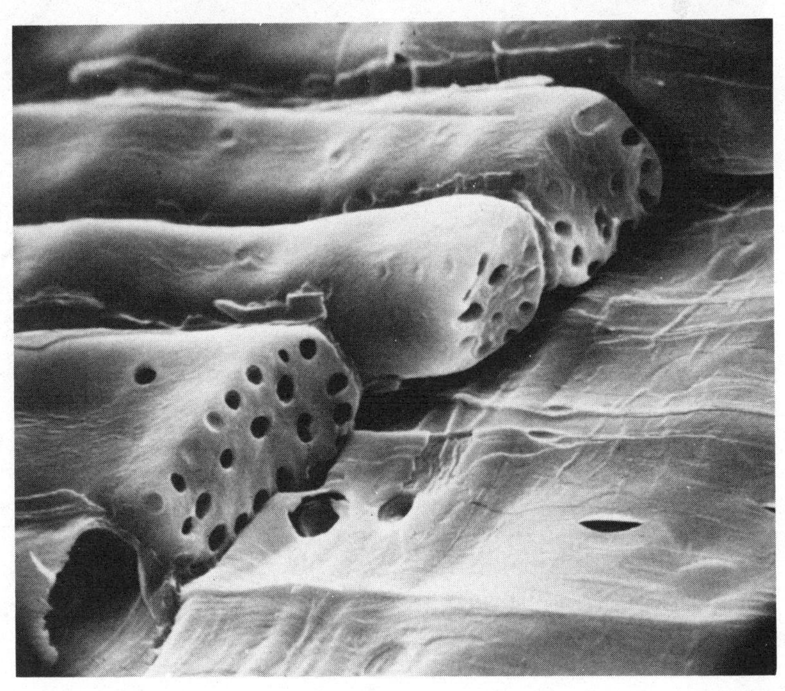

FIGURE 8-26
End-wall perforations in the ray cells of a conifer, Douglas fir (*Pseudotsuga*). (Photo courtesy of Irving B. Sachs, Forest Products Laboratory, U.S. Department of Agriculture.)

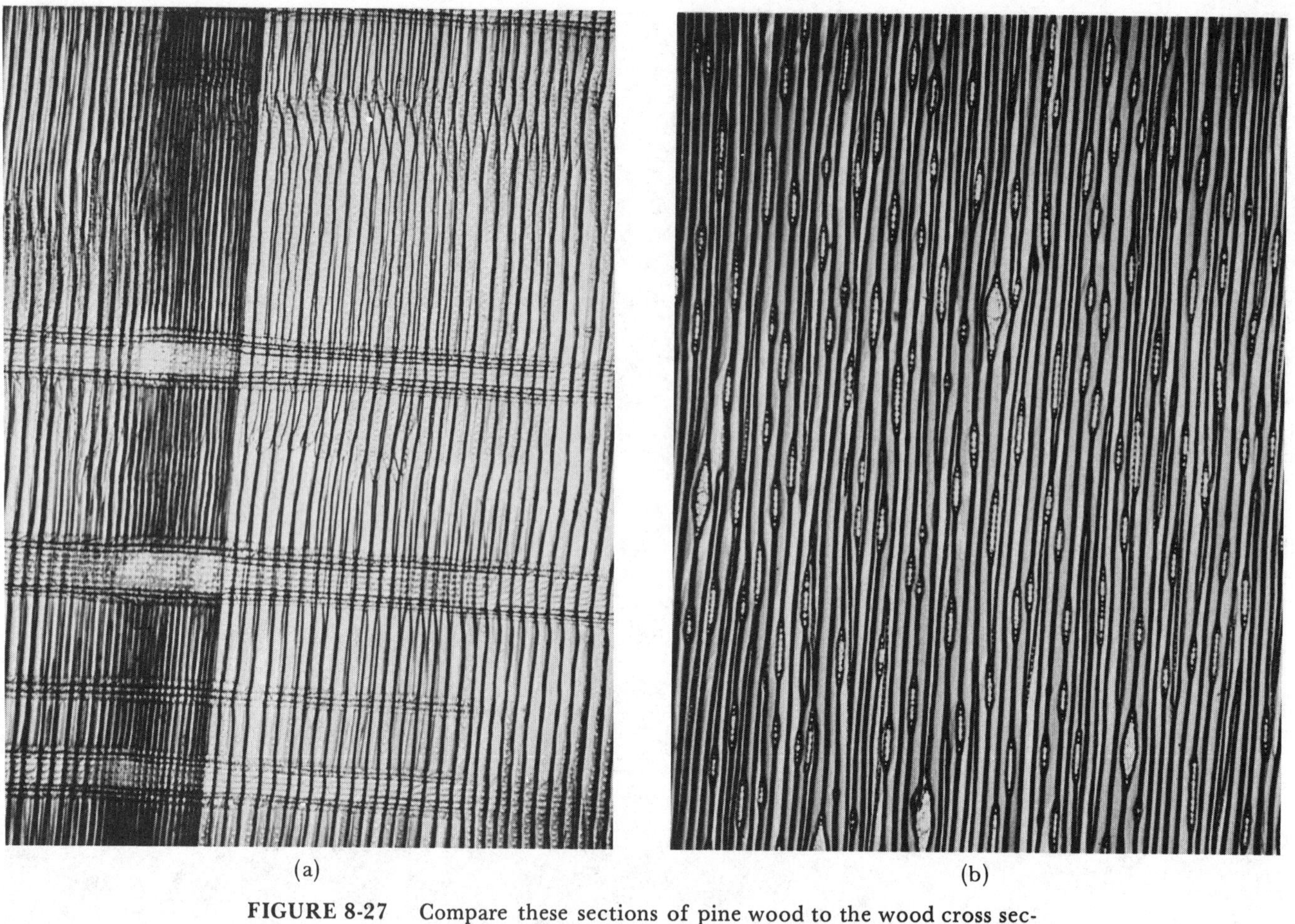

(a) (b)

**FIGURE 8-27** Compare these sections of pine wood to the wood cross section in Figure 8-32. (a) In a **radial** section, the plane of sectioning is parallel to the rays. (b) In a **tangential** section, the plane of sectioning is perpendicular to the rays. Both sections provide a longitudinal view of tracheids.

is movement of substances from the bark and outer portions of the wood into the central portion of the wood for storage.

### Cambial Activity in the Older Stem

The vertical and radial transport systems have different origins. The vascular cambium is composed of two cambium cell types, spindle-shaped **fusiform initials** and box-shaped **ray initials**. Division of the fusiform initials produces the long axial xylem and phloem cells whereas division of the ray initials produces radial xylem and phloem parenchyma cells.

In well-developed woody stems, such as trunks and branches of trees, it is more nearly correct to speak of a **cambial zone**, rather than a cell layer of vascular cambium because at times of rapid cell division—as in the spring—cells are produced by the cambium more rapidly than they are capable of maturing. So at any given time the cambial zone consists of the actual vascular cambium plus cells on either side in various stages of differentiation. Those that are most recent cambial derivatives (and hence show least differentiation) are known as **phloem mother cells** or **xylem mother cells.** In dicots phloem mother cells mature with the possibility of forming sieve tube members, companion cells, parenchyma cells, or fibers; xylem mother cells mature and differentiate into tracheids, vessel members, parenchyma cells, or fibers.

Continued meristematic activity in woody stems permits the tissues of grafted scions and buds to fuse with those of the rootstock, as elaborated in Chapter 5.

### Annual Rings

In temperate climates the activity of the vascular cambium corresponds to the seasons of the year (Chapter 9). When the buds open ("break") in the spring, the cambial initials are stimulated to begin dividing. During spring the tangential planes of division are many, producing numerous xylem and phloem mother cells, in the average ratio of ten xylem to one phloem.

Water is abundant in spring, making nutrients readily available and so vegetative growth is at its most vigorous. Xylem conducting cells produced during this period are larger in radial diameter than cells produced later so that the wood produced during this time (early wood) is lighter in color and density. As the season wears on into drier conditions, xylem cells having a smaller radial dimension but thicker cell walls are produced. This late wood is generally darker and of greater density and hardness.

These differences are macroscopically visible (Fig. 8-24) and each combination of early and late wood constitutes one **annual growth ring**. The number of annual rings is a general indication of the age of a tree or branch. Major fluctuations within a single season can disrupt the annual nature of the rings, however, so that tree-ring experts must be able to discern true annual rings from false rings. Because of the lack of seasonal differences, trees in the tropics may not develop reliable annual rings.

FIGURE 8-28 Tyloses in the lumen of a red oak (*Quercus*) vessel. (Photo courtesy of Irving B. Sachs, Forest Products Laboratory, U.S. Department of Agriculture.)

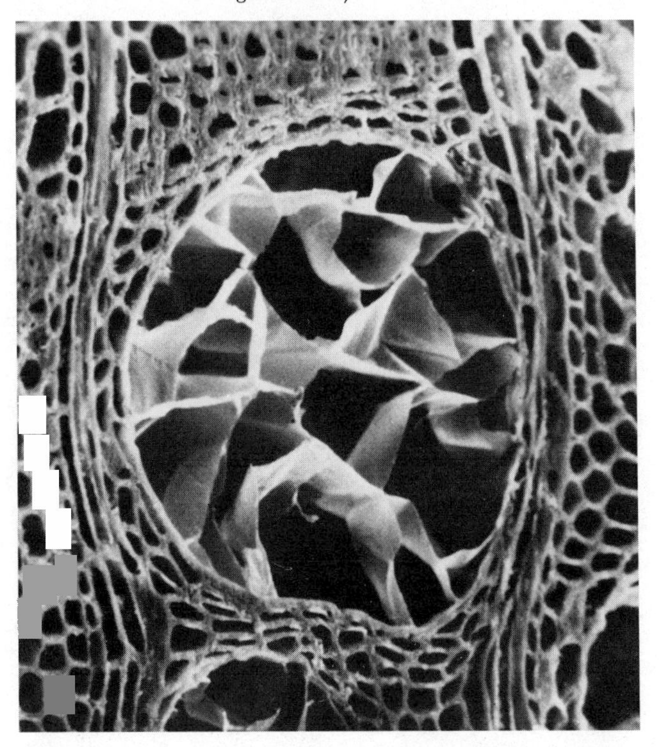

#### Heartwood and Sapwood

In a freshly cut tree a difference between the color of the wood in the center and the wood on the outside is often apparent. The center **heartwood** is much darker than the outer **sapwood.** Sapwood is the actively functioning conducting portion of the wood whereas the heartwood is no longer an active conducting tissue, functioning mainly for support and waste storage.

In the heartwood the parenchyma cells die. The vessels become blocked by the growth of parenchyma cells into vessel lumens, forming bubblelike **tyloses** (Fig. 8-28). The darker color is caused by deposition of minerals and such chemical substances as resins and tannins in the xylem cell walls. It has been suggested that the chemical impregnation of these tissues makes them more resistant to invasion by insects, fungi, and bacteria but does not increase wood strength. The dark color of heartwood makes such woods as black walnut and cherry especially valuable for furniture and paneling.

#### Burls

Occasionally growth abnormalities occur, resulting in formation of bulbous growths called **burls** on the main trunk or branches (Fig. 8-29). Burl formation may be stimulated by injury, disease, or invasion by such parasitic plants as mistletoes. Within a burl, the cambium proliferates adventitious buds and abnormally large amounts of xylem tissue, which becomes rippled and convoluted because it has no room to grow longitudinally. Dense, swirled patterns form in the wood, making burls popular for decorative woodwork, such as bowls, veneer panels, furniture, and pipes.

#### The Bark

Conducting cells of the phloem usually remain functional for only one year; phloem parenchyma cells, however, continue to act as storage tissue until they are cut off by cork cell formation. In most trees the old secondary phloem and cork produce a thick bark. A few notable exceptions are paper birch (*Betula papyrifera*), sycamore (*Platanus occidentalis*), madrone (*Arbutus menziesii*), and *Eucalyptus* trees, which have thin bark that tends to peel off in large patches. The thicker bark of shagbark hickory (*Carya ovata*) comes off in shaggy strips (Fig. 8-30).

On some stems, such as young bur oak

**FIGURE 8-29**
A **burl.** (Photo by Mark R. Fay.)

**FIGURE 8-30** Well-named shagbark hickory (*Carya*).

(*Quercus macrocarpa*) stems and *Euonymus* spp. (Fig. 8-31), cork is proliferated in such a way that it forms corky ridges on the stems.

Conifer barks have **resin cavities** ("blisters") filled with pitch (Fig. 8-32), which is of great commercial value as a source of such organic compounds as turpentine. (See also Chapter 16.)

## Wounds and Girdling

When a patch of bark is stripped off or gouged from a tree, the wound can be healed by the proliferation of new tissues around the perimeter of the wound. Eventually these tissues can fill in the wound (Fig. 8-33), although the tree is vulnerable to insects and diseases at the wound site until healing is completed, perhaps over several years.

**Girdling** occurs when the bark is removed around the complete circumference of a woody plant, such as by rabbit feeding. A girdled plant often dies, perhaps because food is unable to move down to the lower stem and root system—starving the cells—or because growth hormone movement (Chapter 9) is disrupted. If some vascular cambium remains, it will produce new phloem tissue and the plant may survive unless the unprotected cambial tissues perish from dehydration and exposure.

## Shaping and Pruning

Because of the ability of plants to survive loss of parts and to generate new shoots continuously,

FIGURE 8-31
**Corky ridges** on the stem of the ornamental shrub called burning bush (*Euonymus*).

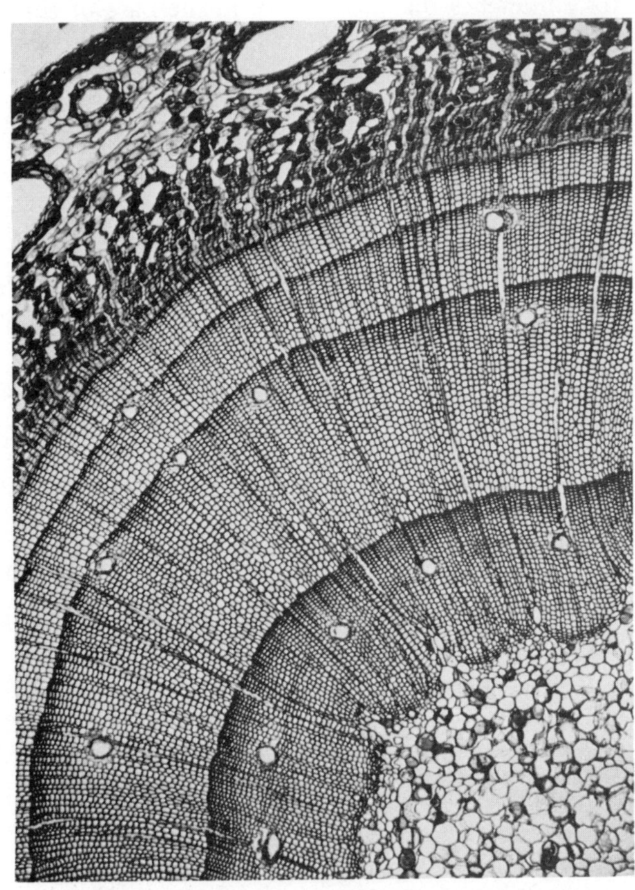

FIGURE 8-32 A 4-year-old pine stem. Note **resin cavities** in the bark and **resin ducts** in the wood.

FIGURE 8-33 The healing of a wound on the trunk of an oak tree. New tissue produced by the vascular cambium is beginning to cover the wound, and will eventually meet and fuse.

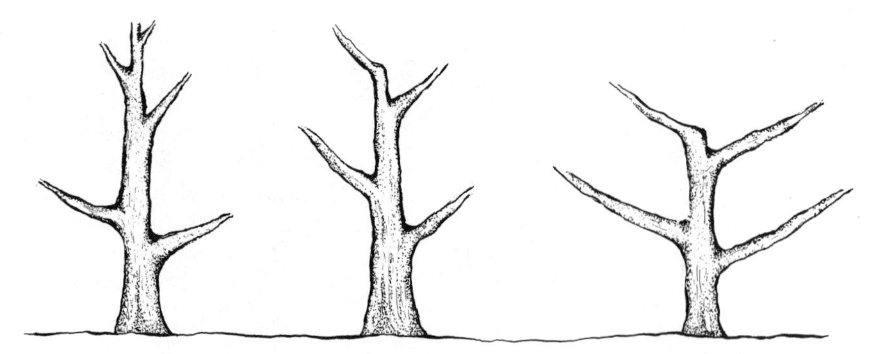

**FIGURE 8-34** Three forms for training fruit trees (left to right): **central leader, modified leader, partly open head.**

**pruning**, the selective removal of branches, can be used to control shape, size, and productivity of cultivated trees and shrubs. The initial pruning of a young fruit tree, for instance, trains it into a form that best fulfills the needs of fruit bearing. Factors that determine desirable tree-growth organization are species or variety, type of branch that produces fruit, fruit color, accessibility for spray penetration, and desired shape and size for landscaping if the tree is a part of the home lot.

Tree or shrub *size* is controlled by limiting top growth, allowing easier harvesting and care, and maximizing sunlight and spray penetration. *Shape* is controlled by limb removal, leaving branches that are well placed and that will not develop weak **crotches** (the junction between the main stem and the limb). Productivity is enhanced by permitting development of the strongest, most advantageously placed major limbs, thus reducing competition for sunlight, space, and nutrients. Careful pruning also encourages the growth of **spurs,** the short side branches that produce flowers and fruit on fruit trees.

When pruning, keep in mind the distance from the ground of the lowest branches that you are developing. Trees elongate only at their tips; therefore the position of low-hanging branches is permanent in regard to climbing, mowing under them, and so on.

The optimum time to prune is in spring, right before bud break. Spring pruning allows removal of limbs that have been winter killed. In milder climates fall pruning is acceptable. In colder climates, however, fall pruning exposes inner tissues to frost and dehydration, thereby increasing the likelihood of winter kill.

Most fruit trees should have only three to five main branches, well distributed around the center. There are three basic plans: central leader, modified leader, and open head (Fig. 8-34). The latter two are the most popular, for they provide more open growth, permitting greater pollen, sunlight, and spray penetration, and maximum fruit production.

To develop a young tree, remove excess branches, a procedure called **thinning.** Each year remove the **suckers** or **water sprouts** (vigorous lateral branches that develop from the base of the tree and from thicker parts of the branches). Remove limbs that cross each other or that grow across the center of the tree.

When pruning, leave a lateral bud to break to the outside of its branch. Encourage development of productive spurs by **heading back**, removing the tips of most branches, thus increasing lateral growth. Most spurs are productive for 2 to 6 years.

Take care not to cut back too much each year, for doing so stimulates vegetative growth at the expense of flowering. In an established pruning program practice a little maintenance each year (Table 8-3).

**TABLE 8-3** **A pruning maintenance program for fruit trees.**

1. Limit top growth by heading back upper leaders.
2. Open up tree.
   - Leave three to five main branches. Remember that branch distance from the ground is permanent.
   - Remove weak crotches.
   - Remove inward growing branches.
   - Remove rubbing branches.
   - Remove water sprouts or suckers.
3. Increase productivity.
   - Trim branch tips to allow development of lateral spurs.
   - Remove old, nonproductive spurs.

Prune mature trees by heading back and by thinning established branches to encourage more vigorous growth of remaining branches. This type of pruning rejuvenates older trees by increasing the amount of productive wood.

## Woods

Woods are often categorized in lay language as "softwoods" and "hardwoods," terms that do not reliably relate to the relative physical hardness of the wood itself but to tree types. Gymnosperms or conifers, such as firs and pines, are referred to as **softwoods**; woody dicot trees are called **hardwoods**.

Other ways to classify woods are by the presence or absence of vessels and the arrangement of the vessels if present. Vessels form large, longitudinal tubes within the wood, visible macroscopically within the wood "grain" so that in cross section vessels appear as "pores" in the wood. If vessels are present, the wood is said to be **porous** and **nonporous** if they are absent. Porous woods can be further classified by vessel arrangement. If larger vessels are produced in the early rather than the late wood, the wood is termed **ring porous** [Fig. 8-35(a)]. If the large vessels are distributed uniformly throughout the annual ring, the wood is termed **diffuse porous** [Fig. 8-35(b)].

Wood is useful to humans in seemingly innumerable ways. The most obvious are as construction materials, varying from dimensional lumber, shakes, and shingles, to thin peeled-wood veneer laminated to plywood to create wood paneling. Other uses include everything from toothpicks to liquor kegs, sporting goods to toilet paper. In addition, many wood extractives are commercially valuable (Chapter 16).

Wood structure varies according to the type

**FIGURE 8-35** Different vessel arrangements in dicot wood: (a) **ring-porous** wood, (b) **diffuse-porous** wood. (Photos by Jeff P. Carlson.)

(a)

(b)

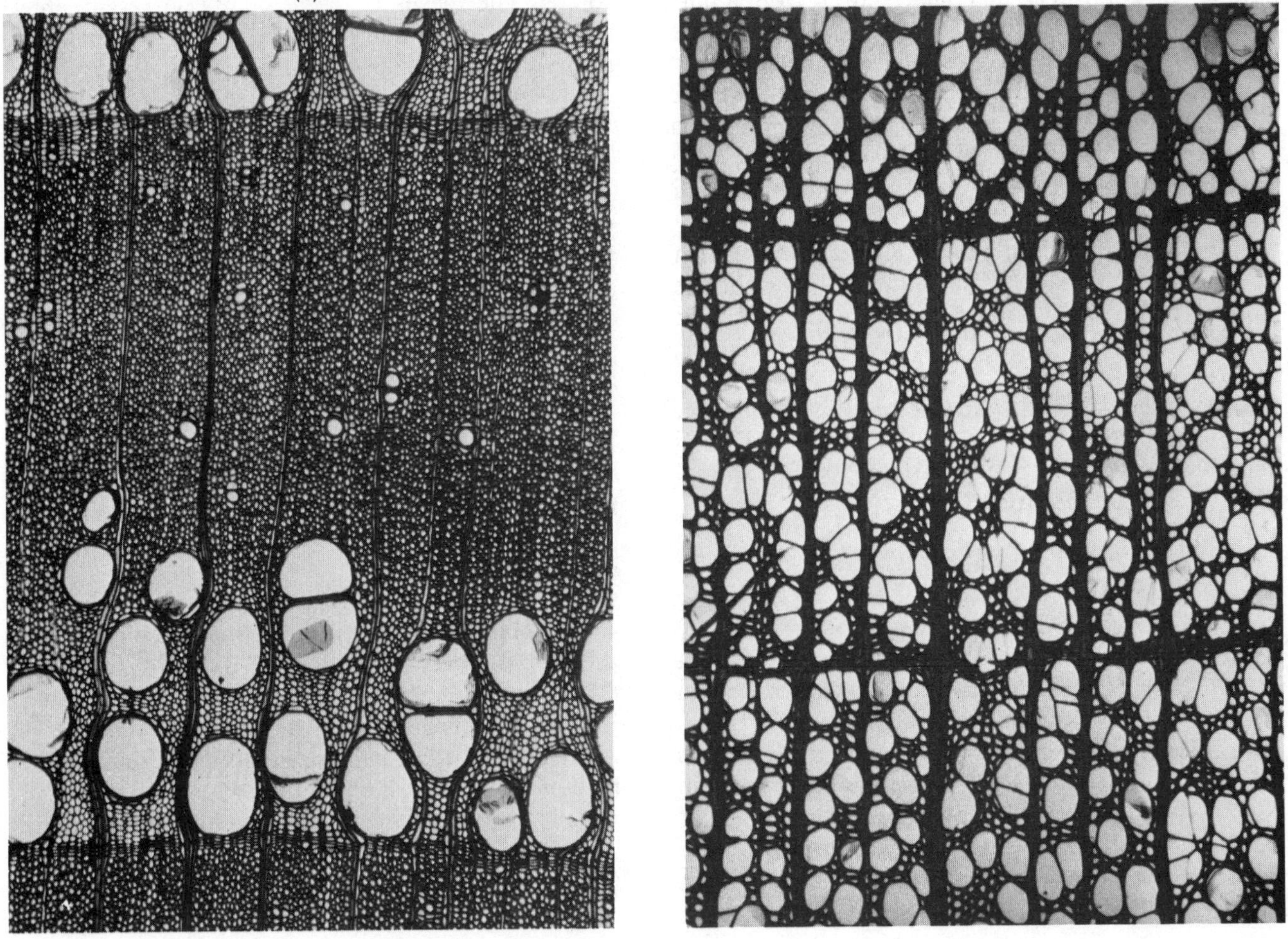

and distribution of cells, reactions to environmental stresses, and the manner in which wood is sawed. Some trees, such as redwoods (*Sequoia*) and bald cypress (*Taxodium*), are especially resistant to decay because of specific chemicals in the heartwood; less resistant woods are poplars (*Populus*) and basswood (*Tilia*). Sapwood of the most resistant species does not, however, compare to heartwood resistance.

Woods noted for extreme hardness, such as hickory (*Carya*), elm (*Ulmus*), maple (*Acer*), and ash (*Fraxinus*), are used as furniture, cabinetry, handles for shovels, hockey sticks, baseball bats, and skiis. Some types of oak (*Quercus*) wood are used for barrels, particularly those for aging liquors, because tylosis formation in the lumens of the conducting cells is so complete that the wood is thoroughly impermeable.

Wood may be cut in various ways, depending on the pattern desired. An "end cut" reveals the cross-sectional appearance of the wood tissue. If a log is cut along its radius, it is said to be "quarter-sawed," which reveals the pattern of rays seen from their sides. You can visualize that a limited number of boards can be quarter-sawed from a particular log. Most lumber is cut tangentially, or "plain-sawed," cutting across the rays in lines that are parallel to a plane marking the diameter of the log. Because of log curvature, most cuts are not perfectly radial or tangential. Baseball players are warned to "keep the bat label up" when batting because bats are labeled on a tangential surface. By keeping the label "up," the point of impact with the ball is perpendicular to the annual rings of the wood. If the ball and bat collide with sufficient force on or opposite the label (tangential surfaces), the bat is likely to break.

The wood "grain" patterns on finished lumber and paneling are due mainly to the differing density and texture of xylem tissue in the annual rings as well as the porosity of the wood.

#### Knots in Wood

You will recall that side shoots (branches) originate as buds on the outside of a stem. Each branch has xylem that is continuous with the main stem, beginning the year that the branch originated. Therefore continued increase in the girth of the main stem causes the secondary tissues of the main stem to overgrow the bases or points of origin of the branches. At the same time, the branches are experiencing secondary growth. When the main stem is sawed for lumber, the overgrown bases of the branches may be visible as knots in the wood, complete with annual rings! These knots are refered to as "tight" knots.

"Loose" knots occur when a branch dies and the adjacent trunk region grows around the dead limb. In this case, the knot in the wood even includes remains of the bark of the dead branch. The knot is loose because there is no continuity of xylem with the main stem, only a sheath of dry bark, so that the knot can fall out of the board.

Knots are especially prominent in wood from conifers, including pines (*Pinus*), firs (*Abies*), spruce (*Picea*), hemlock (*Tsuga*), and Douglas fir (*Pseudotsuga*). Consequently, conifers are valued for decorative paneling, both solid and veneer.

## LEAVES

Leaves manufacture food for the plant and their structure is suited to this function. In addition, they have become modified in many ways through natural selection in response to environmental forces. The leaves of most plants possess **bilateral** (two-sided) **symmetry** rather than the radial symmetry typical of roots and stems.

### Leaf Structure

As you know, a typical dicot leaf is composed of two basic parts, the broader **blade** (lamina) and the **petiole** (leafstalk) that attaches the leaf to the stem. Some leaves, however, lack petioles and are called **sessile** leaves. An axillary bud is usually present on the stem in the **leaf axil.** Most dicot leaves have a netted (branched) vein organization.

Monocot leaf structure differs in several ways. A typical monocot leaf, such as that of a grass, is sheathlike at its base and attached all around the node (Fig. 8-15). In addition, most monocots have parallel leaf venation. Monocot leaves are often reinforced by fibers.

Leaves may be either simple or compound. A **simple leaf** is a single, intact structure even though the margin may be indented in various patterns that range in degree from toothing to deep lobing. In a **compound leaf** incision of the leaf margin extends fully to the major vein(s), creating distinct **leaflets.** Leaflets may be mistaken for individual simple leaves unless care is taken to look for axillary buds. Leaflet arrangement may be **pinnate** (featherlike; see Figure 8-36) or **palmate** (resembling

FIGURE 8-36
Pinnately compound leaves of elderberry (*Sambucus*).

the palm of the hand; see Fig. 8-37). Some leaves are doubly or triply compound; that is, the leaflets are further subdivided.

The flattened shape of a typical leaf is very functional, for it provides a broad exposed surface to collect solar energy and transform it into chemical energy. It is enclosed in cutinized epidermis. In addition to reducing water loss, the smooth surface discourages penetration of the epidermis by pathogenic fungi and bacteria (Chapter 12). As with human skin, infection is more likeiy to enter the plant body through an injury that breaches the protective epidermis. Occasionally **wax** is also deposited to the outer surface of the cuticle – for instance, the waxy "bloom" on cabbage leaves and such fruits as apples, prunes, plums, and grapes.

FIGURE 8-37
Palmately compound leaves of Virginia creeper (*Parthenocissus*).

The main body tissue of a leaf is the **mesophyll** ("middle leaf"), composed of chlorenchyma cells (Fig. 8-38). Mesophyll cells of many dicots are organized in two distinct regions [Fig. 8-39(a) and (b)]. Those nearest the upper (sunward) side of the leaf are columnar in shape and arranged in rows

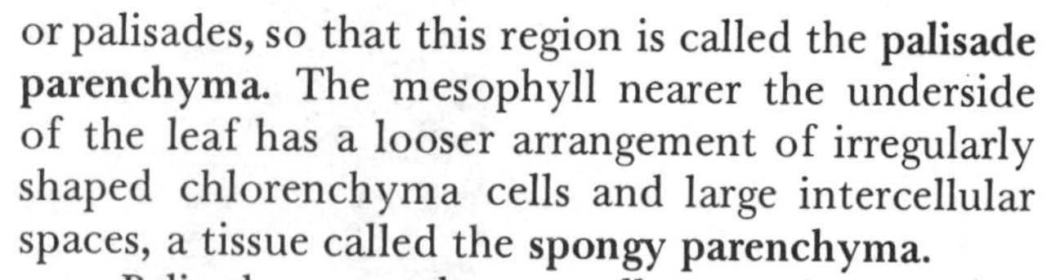

or palisades, so that this region is called the **palisade parenchyma.** The mesophyll nearer the underside of the leaf has a looser arrangement of irregularly shaped chlorenchyma cells and large intercellular spaces, a tissue called the **spongy parenchyma.**

Palisade parenchyma cells contain a higher

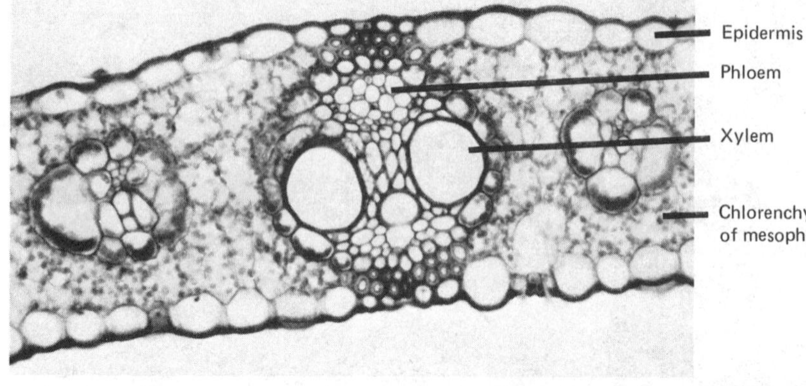


**FIGURE 8-38**
Cross section of a corn leaf, showing a major vein (vascular bundle) and two smaller veins.

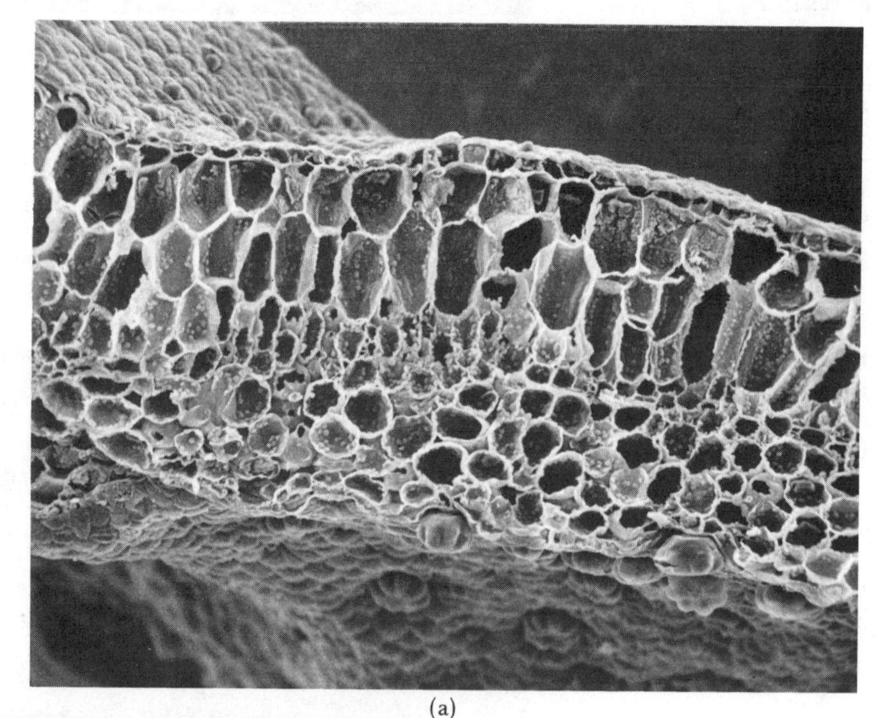


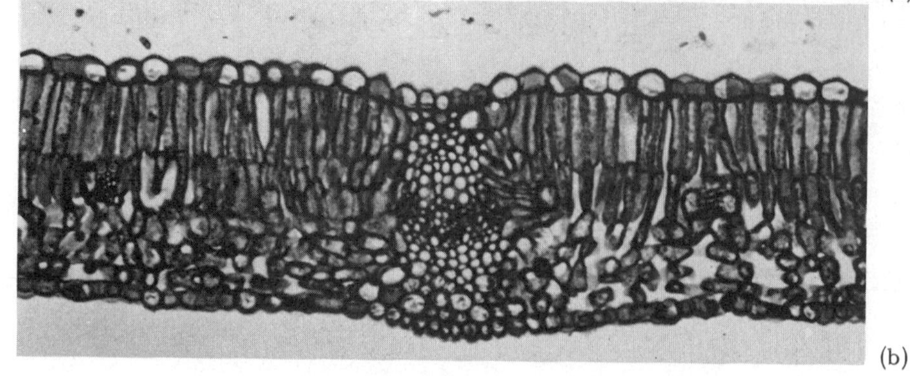


**FIGURE 8-39**
(a) The layers of a dicot leaf are shown in this SEM of a Swedish ivy (*Plectranthus*) leaf. Note the single layer of **upper epidermis,** columnar **palisade parenchyma,** loosely arranged **spongy parenchyma** with air spaces, and the **lower epidermis.** The ovoid objects clinging to the insides of the parenchyma cell walls are chloroplasts. (b) Cross section of a lilac (*Syringa*) leaf as seen with a light microscope. Note the same leaf regions as in (a), plus a major **vein** and two minor veins. [Photo (a) by Jon M. Holy. Photo (b) courtesy of Marshall D. Sundberg.]

density of chloroplasts than spongy parenchyma cells and thus may be identified as the major sites of photosynthesis. In contrast, spongy parenchyma cells provide much surface area for gas exchange between cell surfaces and air spaces. The air spaces, in turn, are open to the external environment via the stomates (Chapter 7). Imagine yourself inside the air spaces of the leaf on a sunny day—a warm, humid, natural greenhouse!

The vascular system of the leaf is embedded in the mesophyll. Each **vein** (vascular bundle) is composed of xylem and phloem tissue and associated cells. In dicots the xylem is commonly on the upper side of each bundle with the phloem below it. A **bundle sheath** of parenchyma cells surrounds each vein. Bundle sheath cells apparently mediate diffusion of food and water between the vascular tissue and the mesophyll. The vascular bundles in monocot leaves are often surrounded by fibers. Leaves may also contain sclereids, crystal-containing cells, and various other specialized cells.

## Leaf Adaptations

So far discussion has centered on a "typical" leaf, one that functions under moderate environmental conditions. Plants adapted to this kind of environment are called **mesophytes**. Earth, however, has extreme environments where plant life exists solely because of special adaptations of not only the leaf but also all plant parts. In addition to environmental adaptations, certain species have leaves that are modified for specific functions, such as storage or reproduction.

### Hydrophytes

Plants adapted to life in the water are called **hydrophytes** (water plants). The evolution of vascular plants involved increasing adaptation to a terrestrial environment. Hydrophytic vascular plants—especially those that live submerged—therefore demonstrate specific adaptations to overcome their own terrestriality! To adapt to loss of essential radiation by the light-filtering effects of water, leaf surfaces may be large and expansive near or above the water surface, like those of the water lily (*Nymphaea*). In contrast, underwater leaves are often finely dissected, such as in the water crowfoot (or buttercup) (*Ranunculus longirostris*). Because water loss is not a problem, leaf area is not limited by this factor. Similarly, a dense cuticle is not critical to survival except for the exposed surfaces of emergent or floating leaves, like those of the water lily. Large intercellular spaces in the leaves enhance gas exchange and provide buoyancy to keep leaves near surface sunlight. Water hyacinth (*Eichhornia crassipes*) is a hydrophyte that floats entirely on the water surface (Fig. 8-40), its feathery root system dangling beneath it.

Hydrophytes obtain oxygen and carbon dioxide by direct diffusion between plant tissues and the water or through stomates on surfaces exposed to the atmosphere. In the water lily, for example, stomates are localized primarily on the upper surface of the leaf. In contrast, terrestrial plant leaves usually have a preponderance of stomates on the lower leaf surface.

FIGURE 8-40
Some hydrophyte species are adapted to living submerged or partially submerged in water. Others, as the water hyacinth (*Eichhornia*) (center) and the small water fern *Salvinia* (left and bottom), are adapted for a floating existence.

Most hydrophytes have reduced water-conducting systems, apparently as an evolutionary response to very restricted (if any) net water loss. In addition, large air channels provide internal gas exchange pathways for stem and root tissues.

### Xerophytes

Plants that demonstrate adaptations to dry environmental conditions are called **xerophytes** (dry plants) and their adaptations center on restricting water loss and enhancing water uptake. Most xerophytic plants have reduced foliage surface area. Leaves, for instance, are less likely to be broad and flat, more likely to be narrowly cylindrical, fibrous, or fleshy if present at all. Xerophyte leaves often have a very thick cuticle and a multilayered epidermis.

Some desert plants produce leaves only during periods of abundant moisture and the rest of the time photosynthesis is carried on in photosynthetic stem tissues. Many xerophytes have very deep taproot systems, such as desert trees and shrubs, or extensive and deep fibrous root systems, such as prairie grasses. (See also Chapter 15.)

The extreme xerophytic adaptation in the New World is represented by members of the Cactaceae (cactus family), which do not produce permanent leaves but carry on photosynthesis entirely in the fleshy, water-storing stem (Fig. 8-41). It may seem a paradox, but cacti generally have shallow root systems, enabling them to absorb surface moisture from precipitation when it is available. Between rain showers or snow melt, cacti rely on the water reservoir of their own tissues instead of the deep soil water table.

Pine needles demonstrate many xerophytic adaptations for reducing transpiration [Fig. 8-42(a) and (b)]. The narrow form reduces surface area, the epidermis is thick and reinforced by an impervious sclerenchyma layer, and stomates are sunken below the external leaf surface.

In some plants stomates are located within **stomatal crypts**, invaginations in the leaf surface. In addition, epidermal hairs may be present within or over the crypts and are thought to reduce transpiration by decreasing air movement in the vicinity of the stomates (Fig. 8-43).

Grasses have several xerophytic adaptations, including a thick, somewhat fibrous epidermis and

**FIGURE 8-41** Desert xerophytes include the creosote bush (*Larrea*) and giant saguaro cactus (*Cereus*). The ribbed stems of many cacti allow great stem expansion for water storage at times of water abundance. (Photo courtesy of John S. Thiede.)

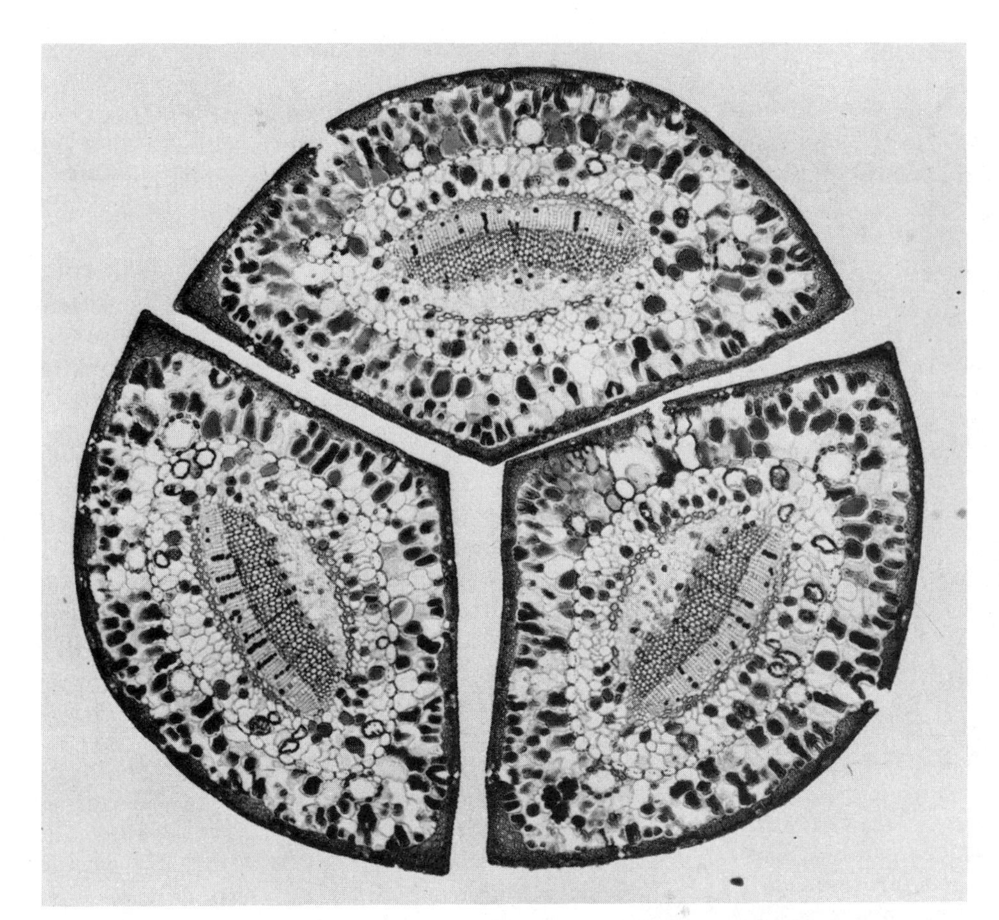

(a)

**FIGURE 8-42**
(a) Cross section of pine needles. The shape of the needle bundle is round so that the shape of the individual needles is a pielike fraction of a circle. This leaf of a three-needle pine shows several xerophytic adaptations (see text). (b) Sunken stomates on a pine needle. [Photo (a) courtesy of Carolina Biological Supply Co. Photo (b) courtesy of Jeff P. Carlson.]

(b)

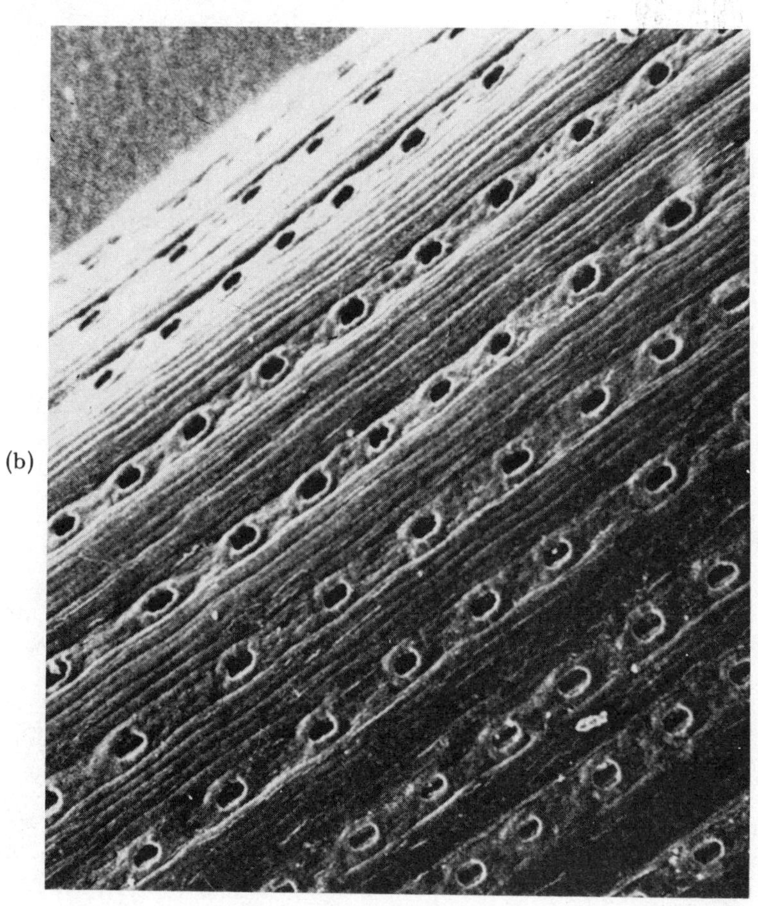

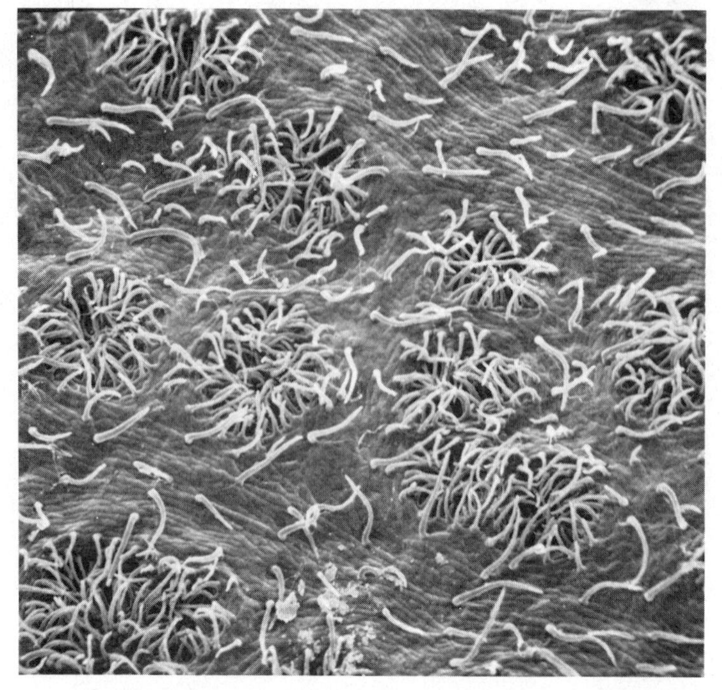

FIGURE 8-43 The lower leaf surface of oleander (*Nerium*) showing hairs in the stomatal crypts. The epidermal hairs are thought to reduce transpiration by reducing air currents across the leaf surface. (Photo by Jon M. Holy.)

a spreading root system that can effectively absorb most of the available moisture within a wide perimeter of the plant. Groups of enlarged epidermal cells called **bulliform** (bubble-shaped) **cells** cause the leaves to fold or roll up when conditions are very dry, thereby further reducing surface area for transpiration.

#### Storage Leaves

The storage leaves that are probably most familiar to you are those of an onion (*Allium*) bulb, whose layered bulb scales are actually modified leaf bases. Onions are biennial, growing vegetatively during the first year, producing food to store through the winter, and then growing, flowering, and setting seed the second year. Their ability to store food (much of it as sugars) has made onions and their relatives a food source to humans.

Another plant that stores food in the modified leaves (bracts) surrounding its flower is the globe artichoke (*Cynara scolymus*).

The succulent leaves of some xerophytes are storage leaves, used for storing moisture. The mesophyll is often gelatinous or mucilaginous and so has great water-attracting and water-holding abilities.

#### Petioles

The edible portion of celery (*Apium graveolens*) and rhubarb (*Rheum rhaponticum*) is the leaf petiole. In both plants the fully developed blade portion is inedible, even poisonous (Chapter 18).

#### Tendrils, Spines, Thorns, and Prickles

**Tendrils** are twining structures found on climbing and creeping species, such as peas (*Pisum*), cucumber (*Cucumis sativa*), watermelon (*Citrullus vulgaris*), and cantaloupe (*Cucumis melo*). Tendrils enable weak-stemmed plants to attain maximum exposure to light and are usually modified leaves, although leaflets, stipules, and even petioles can form tendrils.

In a few species tendrils are formed by modified branches, as in grapes (*Vitis*) (Fig. 8-44), Boston ivy, and Virginia creeper (*Parthenocissus*). The tendrils of both Boston ivy and Virginia creeper develop expanded pads or holdfasts at their tips, which attach the plants firmly to tree bark or masonry.

**Spines** are modified leaves that do not develop a blade portion and become hard at maturity (Fig. 8-45).

In contrast, *neither* thorns nor prickles represent modified leaves. **Thorns** are actually modified branches, developing from axial buds, as can be

FIGURE 8-44 Grape (*Vitis*) **tendrils** are modified leaves.

FIGURE 8-45 **Spines,** modified leaves. Note the presence of axillary buds.

FIGURE 8-46 **Thorns,** modified branches on a *Citrus* tree.

seen in hawthorn (*Crataegus*) and citrus trees (Fig. 8-46). **Prickles** develop as superficial outgrowths of the stem and are common on members of the rose family (Rosaceae), including roses (*Rosa*), blackberries, and raspberries (*Rubus*). Some plants, such as wild currant (*Ribes*), have both spines and prickles (Fig. 8-47).

### Leaf Arrangements

Leaves are initiated in the apical meristem of the buds, first appearing as **leaf primordia.** As the leaves emerge and expand fully, their distinct arrangements become apparent. Although organization of leaves along a stem may at first appear random, with leaves scattered among open spaces to receive maximum exposure to sunlight, a close look reveals a definite and predictable orientation.

Leaves are arranged in three basic patterns: opposite, alternate, and whorled (Figs. 8-48, 8-49, 8-50). In many herbaceous plants (*e.g.,* mints and *Coleus*) and such trees as ash (*Fraxinus*) and maple (*Acer*) there are two axils at each node, located on **opposite** sides of the stem. On the other hand, many plants bear a single axil at each node, thus creating an **alternate** arrangement of leaves [e.g., oaks, elms (*Ulmus*), philodendrons, geraniums (*Pelargonium*), beans, roses]. In a **whorled** arrangement, as in *Catalpa, Auracaria,* and bedstraw (*Galium*), three or more leaves are present at the same node. Many plants have a **basal** leaf arrangement whereby the internodes are so short

FIGURE 8-47 Wild currant (*Ribes*) has both **prickles** (epidermal outgrowths) and spines.

**FIGURE 8-48**
**Opposite** arrangement of maple leaves.

**FIGURE 8-49**
**Alternate** arrangement of cottonwood (*Populus*) leaves.

FIGURE 8-50 **Whorled** arrangement of *Veronicastrum* leaves.

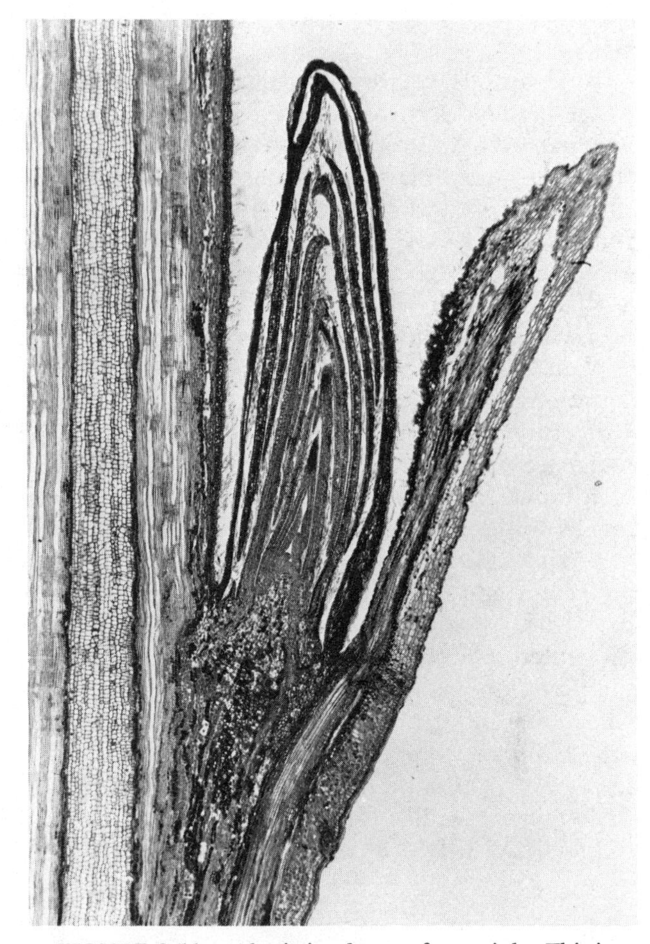

FIGURE 8-51 **Abscission layer** of a petiole. This is a longitudinal section through a node, showing the stem (left), an axillary bud, and a portion of a leaf petiole (right). The abscission layer is the dark band across the base of the petiole. (Photo courtesy of Carolina Biological Supply Co.)

that the leaves form a compact **rosette** near the ground (chapter opening art).

### Leaf Abscission

Leaf drop occurs by an internal cutting away (abscission) of the petiole from the stem. An **abscission layer** of special cells forms at the base of the petiole (Fig. 8-51), cutting off nutrients to the leaf. When the petiole breaks off at the abscission layer, it leaves a cork-covered—and hence protected—scar on the stem. The physiology of leaf abscission is discussed in Chapter 9.

## TRANSPIRATION AND WATER MOVEMENT

The movement of water through the body of a plant is a one-way trip—up! Several physical factors are responsible: osmosis, evaporation (transpiration), cohesion, and adhesion.

Water enters the plant through the root hairs and other epidermal cells of the primary root, mainly in response to an osmotic gradient. The strength of this gradient is maintained by water loss from the aerial portion of the plant. Although some water is used up in photosynthesis, most is lost by **transpiration,** the evaporation of water from plant surfaces. Most transpiration occurs through stomates, although a small amount of water is also lost through the cuticle.

Water molecules have a strong attraction for one another, a force called **cohesion.** In addition, the attraction force between *unlike* molecules, **adhesion,** causes the water molecules to adhere to the walls of the conducting cells of the xylem.

Water molecules escape by diffusion from the humid intercellular spaces within the leaf into the outside air. They are replaced by evaporation of water molecules into the intercellular spaces from wet parenchyma cell walls. This water is replaced from within the cells, then from adjacent cells, and so on, by osmosis. Cells adjacent to the tips of the veins continuously replace the water lost in this way. The process may be likened to the pulling of a chain through the plant, with water molecules representing the links of the chain. Therefore this explanation of water movement is called the **transpiration pull** mechanism.

Water molecules are pulled upward in the xylem cells of the vascular bundles (or wood) in a continuous *cohesive* column, continually renewed by absorption of water into the roots from the soil.

The *adhesive* attraction between the water molecules and xylem conducting cell walls assists the process, for it reduces the gravitational pull on the water column. (Perhaps it is helpful to think of the way climbers brace themselves against the walls while climbing a rock chimney.)

Water in the soil is responsive both to cohesion among the water molecules and to adhesion of the water molecules to the soil particles. When the soil is sufficiently dry that adhesive forces are more powerful than the osmotic gradient, water absorption into plant rootlets falls behind water loss by transpiration. The result is wilting.

As a byproduct of transpirational water movement, minerals are carried into the plant and transported from the root to the stem and leaves. In addition, at least in some plants, amino acids and other substances produced in the roots are transported upward to the rest of the plant body. Evaporation of water from the leaves also cools them, which may be a significant factor on a warm, sunny day.

Transpiration pull accounts for most water movement, especially during the daytime. During the day, for example, temperatures are higher, air currents may be greater, and there is usually lower atmospheric humidity—all factors that increase the rate of transpiration.

Water also moves upward, however, in response to **root pressure**, whereby water is "pushed" up the xylem by the force of water entering the roots in response to an osmotic gradient, with little apparent transpirational influence. This process is demonstrated by the rising of sap in woody plants before their leaves have developed (Fig. 8-52)

FIGURE 8-52 Upward movement of the sap in the spring is demonstrated by this freshly pruned grape stem.

and by guttation. Nevertheless, compared to the forces involved in the transpiration-pull means of water movement, root pressure is a weak force and does not apparently account for significant water movement in the xylem.

## PHLOEM TRANSPORT

The transport of sugars via the phloem is crucial to the survival of plants and hence humans. The more visible results of sugar movement are illustrated by enlargement of fruits and storage organs, such as pumpkins, watermelons, peaches, nuts, and tubers. To divert the sugars produced in the leaves to the developing fruits rather than other plant parts, for example, viticulturists used to girdle the stem (remove a complete ring of bark) below a cluster of grapes. The thinning of tree fruits is a way of directing transported sugars to fewer fruits, thereby producing larger fruits.

At present, the hypothesis of **pressure flow** is the most accepted explanation for phloem transport (Fig. 8-53) from **sources** (sites of photosynthesis or storage) to **sinks** (sites of utilization or storage). Simply stated, chlorenchyma cells of the leaves produce sucrose and other substances during photosynthesis. The products of photosynthesis (photosynthates) are pumped into the sieve tube members of the veins by **phloem loading**, an active transport process (Chapter 3). The high photosynthate concentration in the sieve tubes causes water molecules to move into the sieve tubes by osmosis from adjacent cells. As a result, the pressure of water entering the sieve tubes physically pushes the sugar solution away from the source area and toward a sink. At the sink site the photosynthate molecules are removed or **unloaded** by active transport, maintaining a continuous osmotic gradient to operate the pressure flow. At the sink most of the water reenters the xylem and is recirculated. As you can see, storage sites may be either sinks or sources, depending on whether they are "importing" or "exporting."

During phloem transport the sieve tube apparently plays a passive role as a conduit. The companion cells or parenchyma cells at source and sink sites conduct the active transport of sugars during loading and unloading of the sieve tube. In at least some dicots companion cells and phloem parenchyma cells have the convoluted cell walls and membranes characteristic of transfer cells (Chapter 7).

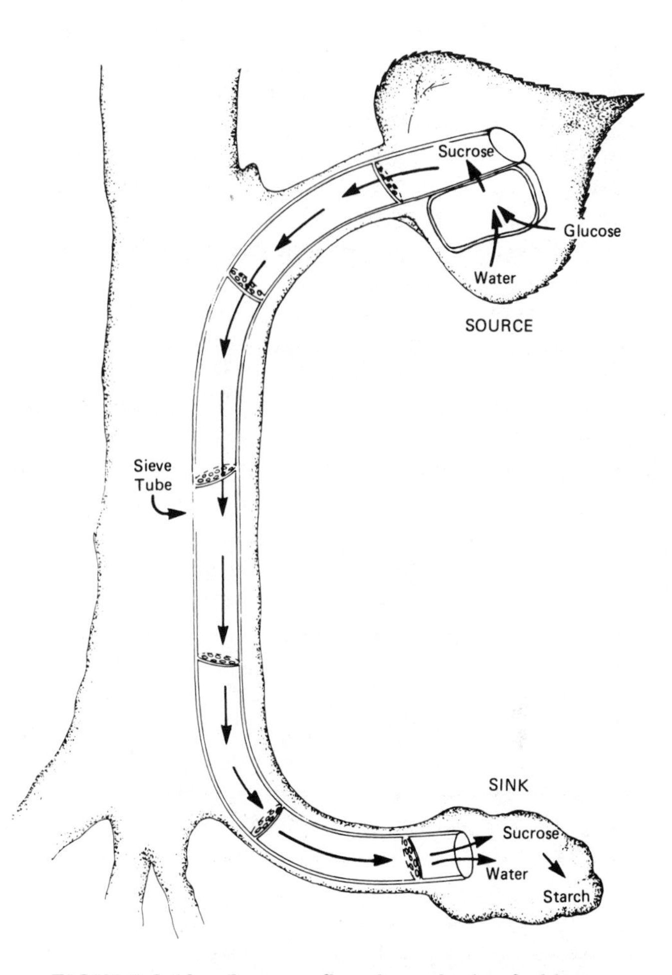


**FIGURE 8-53** **Pressure flow** hypothesis of phloem transport. In this illustration the **source** is a photosynthesizing leaf and the **sink** is a storage organ. See text for an explanation of the forces involved.

## PLANT MOVEMENTS

The fact that certain plant species are rooted in the ground does not preclude mechanical activity. **Tropisms** are bending growth movements in response to environmental stimuli, especially light and gravity. They are discussed in Chapter 9. In addition, vascular plants typically display **nutation.** Time-lapse photography best reveals this motion in which the growing tip executes a spiraling motion as it grows. Nutation is apparently caused by unequal growth near the tip and constant compensation for it.

**Nastic movements** are a response of bilaterally symmetrical plant organs, such as the leaves and flower petals (which are actually modified leaves). Although nastic movements occur as a response to environmental stimuli, they differ from tropisms in that the response is nondirectional and apparently due to turgor changes rather than growth. Examples are the opening and closing of flowers in response to changes in light (photonasty) and temperature (thermonasty). Certain large evening primrose (*Oenothera*) flowers are among the most common to stage a dramatic opening after sunset. Many legumes (beans, peas, and relatives), the prayer plant (*Maranta*), and the shamrocklike *Oxalis* exhibit nyctinasty, in that the leaves fold inward and/or down at night. The sensitive plant (*Mimosa*) displays dramatic response to touch (thigmonasty) by immediately folding its leaflets together and even bending the leafstalks downward (Fig. 2-2). This type of reaction is also utilized by the Venus flytrap.

A few plants (certain bacteria and algae species) are capable of **locomotion,** moving primarily by the use of flagella. The sperm cells of many plants are motile, moving by flagella or cytoplasmic streaming (amoeboid movement). In plant cells that possess locomotory abilities orientation movements known as taxes (singular = **taxis**) occur. As in tropisms, a movement toward the stimulus is a positive taxis. Some major taxes are phototaxis, geotaxis, hydrotaxis (response to water), and chemotaxis. Chemotaxis is important in the movement of lower plants' sperm cells toward the female reproductive organs, which produce chemical attractants (such as sugars).

**Hygroscopic movements** are caused by expansion and contraction of specialized structures in response to changes in atmospheric humidity. They are not due to direct activity on the part of the plant; various plants, however, produce physical adaptations to permit them. Examples are legion. Spore dispersal of some plants (Chapter 13) is assisted by special hygroscopic structures. Certain grass seeds possess a long awn, which coils and uncoils hygroscopically; the result is that the seed is "screwed" into the soil.

## SUMMARY

The architecture of plants enables them to make the most of their environments. Those that grow attached to a substrate tend to have **radial symmetry.** The radial root arrangement is a response to gravity and nutritional needs. The radial shoot pattern maximizes solar exposure. Internal tissues are also radially arranged.

The **primary root** of a **taproot system** is much larger than the **secondary roots** that arise from it. Primary and secondary roots are the same size in a fibrous root system. **Adventitious roots** arise from stems or leaves.

A **root cap,** an **apical meristem,** a **region of elongation,** and a **region of differentiation and maturation** make up the **root tip.** The apical meristem produces cells that form both the root cap and the primary root body.

Cells enlarge as a result of turgor pressure, before enlargement is restricted by the developing cell wall. **Embryonic root tissues,** found at the elongation region, are the **protoderm, ground meristem,** and **procambium.**

Delicate **root hairs** identify the area of differentiation and maturation in which **primary root tissues** mature, including the **cortex, primary xylem, primary phloem, vascular cambium, pericycle,** and **pith.** The **endodermis** of the cortex regulates the movement of water solutions into the vascular cylinder. The pericycle initiates lateral root formation and, in woody plants, gives rise to cork cambium and part of the vascular cambium.

Secondary growth of the root produces a woody interior and corky outer covering. Such root regions provide support and anchorage. The **storage roots** of some species are cultivated for human food.

Besides providing support and storage, stems produce leaves, new shoots, and reproductive structures. Tubers, rhizomes, and stolons are specialized stems.

Shoot tips of higher vascular plants have regions of **apical meristem, elongation,** and **differentiation and maturation.** They differ from roots in that shoot appendages develop from the outer part of the meristem, have vascular strands rather than a central core of primary vascular tissues, and lack endodermis, pericycle, and a protective cap.

The internal arrangement of monocot and dicot stems differs. In young dicot stems the pith is surrounded by a ring of **vascular bundles** whereas in monocots the bundles are scattered throughout the parenchyma matrix. Sclerenchyma fibers strengthen each vascular bundle. Most monocots do not have secondary growth.

During **secondary growth** (in dicots) **vascular cambium** cells produce **secondary xylem** and **secondary phloem.** Ultimately a core of xylem **wood** is surrounded by a **bark** of phloem and cork.

Cambial cells divide both tangentially and radially. **Vascular rays** composed of parenchyma cells move substances radially in the secondary plant body. They are produced by **ray initials. Fusiform initials** give rise to the elongated, vertically transporting xylem and phloem cells.

Xylem cells produced during spring are larger than those produced in the drier season, making spring wood less dense. The combination of spring and summer wood makes up one **annual growth ring.**

Central **heartwood** is darker than outer **sapwood** and no longer conducts materials. The darkness is caused by deposition of resins, tannins, and minerals. The sapwood is the actively conducting xylem.

**Burls** result from abnormal tissue production stimulated by injury, disease, or parasites.

The secondary phloem usually remains functional only for a single season. Conifer barks have **resin cavities.** Injuries to a woody stem are healed by the vascular cambium. If a stem is **girdled,** however, it is likely to die.

**Pruning** practices take advantage of plants' abilities to survive loss of parts and to generate new shoots continually.

Conifers are referred to as **softwoods** and woody dicots as **hardwoods.** Wood tissue may be **porous** (vessels present) or **nonporous** (vessels absent).

Various wood characteristics determine its value to humans.

Leaf structure reflects the leaf's primary role of food manufacturing, with adaptations to trap light and to obtain carbon dioxide and water. The **mesophyll** is composed of chlorenchyma cells, sites of photosynthesis. Leaf structure may be variously modified.

**Hydrophytes** are not limited in size or structure by the threat of water loss. **Xerophytes** may have numerous adaptations to reduce water loss, store water, and increase water uptake.

Onions and globe artichokes have specialized **storage leaves.** Many weak-stemmed plants have leaves (rarely stems) modified as **tendrils. Spines** are also modified leaves whereas **thorns** are modified branches and **prickles** are epidermal outgrowths.

Leaf arrangement can be opposite, alternate, or whorled. Leaf drop occurs by formation of an **abscission layer.**

Water movement into the roots and throughout the body of the plant occurs by a combination of forces, including **osmosis, transpiration, cohesion,** and **adhesion.** Some water moves up the plant body by **root pressure** alone.

The **pressure flow** hypothesis is our best explanation of how phloem transport occurs.

Plants demonstrate several kinds of movement, including **tropisms, nutation, nastic movements, locomotion,** and **hygroscopic movements.**

A **taxis** is a directional locomotory movement in response to an environmental stimulus.

# chapter nine PLANT GROWTH REGULATION

## INTRODUCTION

Imagine yourself as a microscopic visitor within a growing shoot. In the outermost tip of the shoot you observe the rounded, thin-walled meristematic cells, engaged in successive mitotic cycles. Moving deeper, you see enlarging cells recently derived from mitosis, pulsating with cytoplasmic activity. Vacuoles form, move, coalesce. Membranes and organelles form and grow. You are surrounded by movement and synthesis in a viscous liquid environment. Deeper still you see maturing cells differentiating, becoming different from one another even though adjacent.

Your observations are overwhelming, fantastic! These sights would challenge even the most sophisticated movie special-effects artists. What controls this vibrant activity? Given the equivalent genetic potential of each cell, what causes them to become different? Your questions are the subject of scientific research, not science fiction. Yet the mechanisms of plant growth control are but partially understood.

Growth is a fundamental characteristic of living things, involving the reproduction, enlargement, and differentiation of cells. Macroscopically the growth of higher plants is seen as increase in size, followed by physical and reproductive maturation.

The basic anatomical aspects of plant growth and differentiation were described before we had any understanding of *how* growth processes are initiated and controlled. After decades of research we now have many pieces of what has proved an extremely intricate puzzle. We examine some of these pieces in this chapter, with reference to the mysteries yet to be explained.

In addition, we discuss how our knowledge of plant growth regulation has led to such horticultural applications as enhanced food production, marketing, and storage; growth and development of ornamental plants; and control of unwanted plants.

## THE GROWTH CURVE: CHANGES IN PHASE

Complex organisms exhibit a predictable, "typical" growth pattern during their lifetimes, which forms an S-shaped (sigmoid) curve when plotted logarithmically on a graph (Fig. 9-1). Observe that the most rapid growth occurs during the **juvenile phase.**

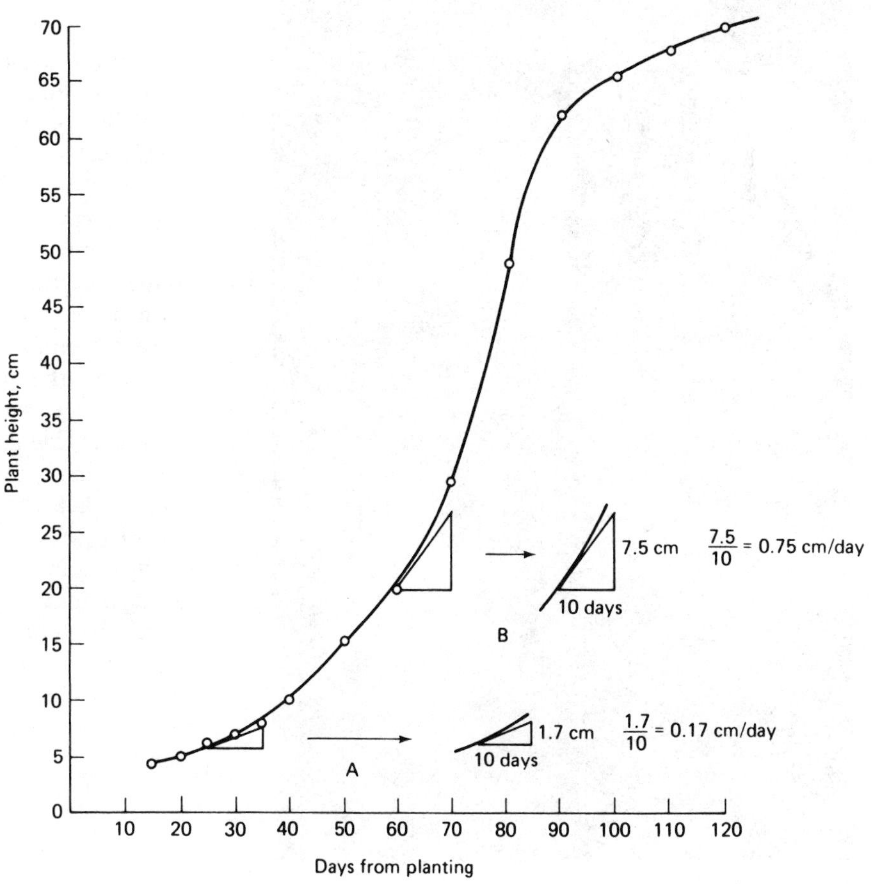


**FIGURE 9-1**
A **typical growth curve** is S-shaped. It includes an initial **lag** phase, a period of rapid **juvenile growth** that tapers off as **maturity** is reached, and ends with a flat growth phase during **senescence.** The annual plant graphed here began flowering about the 85th day after planting. (From G.R. Noggle and G.J. Fritz, *Introductory Plant Physiology,* 1976. Reprinted by permission of Prentice-Hall, Inc., Englewood Cliffs, N.J.)

In plants the juvenile phase occurs during the seedling stage, at which time the size, shape, arrangement, color, pubescence, and other characteristics of the leaves and stem may differ from those of the adult or **mature phase** (Fig. 9-2). Cuttings taken during the juvenile phase tend to root easily whereas cuttings taken after the plant as a whole has reached maturity root less easily.

Sexual reproduction, **flowering** and **seed production,** is an important function of the mature phase of higher plants (chapter opening art). After physical and reproductive maturity are attained, the growth rate tapers off as the organism becomes irreversibly aged (senescent).

**Senescence** is an irreversible period of declining vigor during which specific deteriorative changes occur that lead eventually to cell and tissue death, even death of the entire organism. In annuals, for example, the decline of the vegetative body begins even as the seeds are maturing. Some perennials die back to the ground each year while others lose only their leaves.

FIGURE 9-2 During the juvenile phase leaves and other characteristics of an individual plant differ from those of the adult phase. The **juvenile leaves** of this bean plant are entire; the **adult-form leaves** are divided into three leaflets.

Like other aspects of plant growth, juvenility, functional maturity, reproduction, and senescence are based on tissue responses to chemical substances produced in the plant body in response to both internal and external (environmental) stimuli.

## PLANT GROWTH REGULATORS AND THEIR EFFECTS

In the 1920s Frits Went and his associates conducted experiments with germinating oats in which they found that a substance (or substances) produced in the tip of the **coleoptile** (the sheath that surrounds the shoot) directly affected growth of the entire shoot. Furthermore, they could extract this substance into gelatin blocks and then apply the blocks to the stumps of the coleoptiles of germinating plants from which the coleoptile tips had been removed, causing the plants to behave as though the coleoptiles were still present.

Cautious in his conclusions and yet persuaded by the consistency of his evidence, Went hypothesized that certain aspects of shoot growth are influenced by an innate substance that he called **auxin.**

Although our understanding is still incomplete, we now know that many (perhaps all) aspects of plant growth are regulated by auxin and other plant growth regulators. A **plant growth regulator** is a nonnutrient substance that affects or controls plant growth. Most plant growth regulators are **hormones** that occur naturally within the plant, where they are manufactured, transported, and exert an effect on the function or growth of receptive cells. They are effective in very small amounts. Today we know five general categories of plant hormones: the auxins, gibberellins, cytokinins, ethylene, and abscisic acid.

In addition to the naturally occurring hormones, some have been synthesized and are of great commercial value. Indolebutyric acid (IBA), α-naphthaleneacetic acid (NAA), and the herbicide 2,4-dichlorophenoxyacetic acid (2,4-D) are synthetic auxins.

### Auxin

Although it is also correct to speak of auxins in the plural, for our purposes we will refer to the naturally occurring auxin **indole-3-acetic acid (IAA)** unless otherwise stated. IAA is a product of enzymatic conversion of an amino acid (tryptophan). IAA works at the cellular level and is a major factor in growth and development throughout the plant body, influencing cell enlargement/elongation following mitosis, bending growth movements called tropisms, apical dominance, abscission, floral initiation and development, root initiation, fruit development, bulb and tuber formation, and seed germination.

#### Auxin Production and Movement

Auxin is produced in meristems and enlarging tissues, especially in growing buds and embryos. In a bud it is translocated downward to the elongation region, where it promotes cell enlargement/elongation by affecting cell wall development.

An unusual factor in auxin production is **polar transport**: It is transported away from the shoot tips toward the base of the plant but not upward. The transport of auxin in the subapical region of the root, however, is toward the main plant body.

Experiments have shown that IAA stimulates growth when present in certain concentrations—up to an optimum concentration. At higher than optimum concentrations, auxin seems to inhibit elongation and may even halt it completely. Stem tips generally produce their own optimum amounts of auxin. Because of transport away from the site of production, a higher auxin level occurs below the shoot tips, a factor in apical dominance, discussed later. The cells of the root tips are apparently functionally saturated with auxin and are inhibited by the experimental addition of auxin.

#### Cell Elongation

Research now indicates that auxin stimulates protein synthesis in enlarging cells, apparently by increasing messenger RNA synthesis. To accommodate increased protein synthesis, more ribosomal RNA is also produced. Therefore it seems that long-term, sustained growth responses to auxin probably involve an effect of the hormone on nucleic acid and protein synthesis (Chapter 3).

Auxin seems to promote cellular expansion via its ability to stimulate an acid (low pH) environment in the water that bathes the cell wall, thus increasing cell wall extensibility. The cell wall is composed of long cellulose molecules, which create a rigid wall by the formation of cross-linkages, or bridges, between the molecules. Firmly cross-linked, the cellulose wall resists stretching. Under acid conditions, however, the cell wall can be stretched.

Perhaps our best current explanation is that auxin activates a **proton ($H^+$) pump** in the cell membrane. In other words, it causes the cell membrane to actively transport $H^+$ ions out of the cytoplasm into the cell wall solution. This step lowers the pH in the cell wall solution. It has been hypothesized that the low pH conditions (pH 5) activate a pH-dependent enzyme present in the wall. The enzyme then catalyzes breakage of the cross-linkages. Thus freed, the cellulose molecules are able to slide past one another as the cell expands in response to turgor pressure. The result is an enlarged cell.

As may be imagined, such cellular expansion could cause the existing cell wall to become stretched increasingly thinner. This situation, however, does not occur because cellulose continues to be manufactured during enlargement. Auxin is again the controlling factor, for it stimulates production of cellulose-synthesizing enzymes in both the plasmalemma and the dictyosomes.

There is more to a larger cell than just more water and more cell wall. More cytoplasm is formed, including more endoplasmic reticulum and other organelles. There is a "boom" in protein synthesis, apparently stimulated by auxin.

#### Tropisms

Experiments have shown that when auxin is applied to one side of a growing stem, that side elongates more rapidly than the nontreated side, producing a curvature. This reaction also causes the natural bending growth responses of plants, called **tropisms.** Some tropisms are **phototropism**, bending growth in response to a directional light source (Fig. 9-3); **geotropism**, bending growth in response to gravity; and **hydrotropism**, bending growth in response to a water source.

In phototropism auxin concentrations are higher on the unlighted side of the stem, causing those cells to elongate more rapidly than the cells on the lighted side. This asymmetrical growth produces a curvature in the stem, bending the shoot toward the light source. The result is of obvious functional value to the plant, for it exposes the greatest amount of leaf surface to the light. Because the shoot grows toward the light source, this reaction is called *positive* phototropism.

Plant shoots exhibit *negative* geotropism.

**FIGURE 9-3** Tropisms are bending growth responses to an environmental stimulus, such as light. These shoots are exhibiting **positive phototropism.**

When a plant is grown on its side (even in the dark), the shoot grows upward, away from the source of gravity. Apparently more auxin accumulates on the lower side of the prostrate stem, stimulating greater cell elongation and causing the stem to curve upward. Roots, in contrast, exhibit *positive* geotropism. It is thought that the auxin-sensitive root cells are inhibited by the larger amount of auxin present on the lower side of a horizontal root. Therefore the cells on the upper side elongate at a greater rate, causing a downward curvature. Research suggests that the interaction of auxin with other hormones (ethylene, abscisic acid) may be more important than auxin alone in controlling root tropisms.

Although we can describe the effects, we do not know exactly how plants detect or perceive light and gravitational stimuli. In the stem, the apex, including the youngest leaves, somehow senses the light stimulus. One hypothesis suggests that carotenoid pigments may be involved. Perhaps coincidentally, carotenoid pigments are involved in photoreception in animals, including humans.

In the root the cells of the root cap perceive gravitational stimulus. The sensing mechanism may be related to the presence of very dense starch grains, called **statoliths,** that settle to the bottom of the cells in response to gravity, like so many tiny ball bearings. Conversely, some organelles float upward in the cells. Although it is speculative to draw parallels, the function of small, dense bodies (also called statoliths) in perceiving shifts in balance is well known in the animal kingdom, from the jellyfish medusa to the human ear.

### Apical Dominance

In a vigorously growing shoot the growth of the terminal bud exceeds growth of the lateral buds (Fig. 9-4). This control of lateral bud growth by the tip, called **apical dominance,** is caused by auxin. Apical dominance occurs because auxin produced in the terminal bud is translocated downward in sufficient quantity to inhibit elongation of the cells of nearby lateral buds. As the auxin is translocated farther down the stem, it exerts less influence on the lateral buds produced in previous years, resulting in pyramidal growth form along the main stem, such as in conifers, or along a young branch, such as in many deciduous trees.

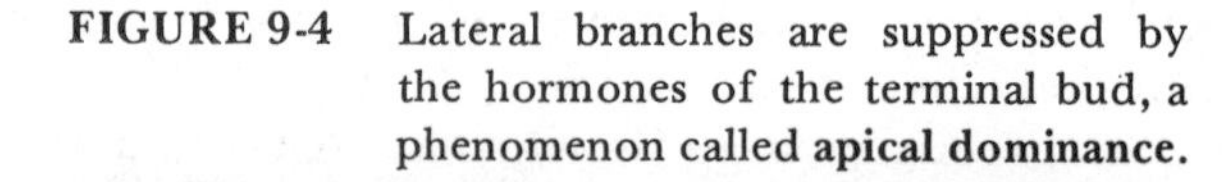

**FIGURE 9-4** Lateral branches are suppressed by the hormones of the terminal bud, a phenomenon called **apical dominance.**

An understanding of apical dominance has many uses. Bushier houseplants and garden plants can be created by pinching out the terminal buds to encourage growth of side branches. Commercial Christmas tree growers shear the branch tips to encourage full, compact growth. Cutting back the ends of young fruit tree branches encourages the growth of lateral shoots, a necessary step in developing the tree. As the lateral shoots grow, they affect each other. Those nearer the tip grow vigorously, developing narrower crotch angles. At the same time, the auxin produced by these upper branches inhibits growth of lower limbs, causing them to grow more slowly and with wider crotch angles. Later the upper limbs that have the physically weaker narrow crotch angles are removed, leaving a tree with strongly developed lower limbs, a more open growth habit, and a desirable height.

#### Other Auxin Effects

Auxin acts with other hormones, especially cytokinins (discussed later), to stimulate cell division. In the spring, for example, the buds of woody plants begin growing, producing auxin. The auxin is transported downward in the phloem to the rest of the plant, initiating cell division in the vascular cambium and the root pericycle (Chapter 8). Auxin also promotes cambial division at the site of a wound or base of a stem cutting, resulting in the formation of an undifferentiated tissue mass called **callus.**

Because auxin also stimulates the formation of adventitious buds and roots from callus tissue, it is used in two types of vegetative propagation/cloning (Chapter 5). Auxin is used to stimulate formation of adventitious buds and roots during tissue culture or micropropagation. The bases of cuttings are dipped into solutions or powders containing synthetic auxin (IBA or NAA) to enhance rooting.

Fruit development is controlled by auxin formed in the floral and fruit tissues. Consider the development of a 1- to 4-mm ovary into a juicy, ripe apple, stimulated by auxins produced by the developing fruit. Auxin produced in response to pollination causes **blossom set,** retention of the flower. Otherwise the flowers would abscise. It also promotes **fruit set,** the retention of developing fruits on the stems. Synthetic auxins are sprayed on some commercial fruit crops to improve both blossom set and fruit set. If pollination is inadequate, the synthetic auxin "fools" floral tissues into reacting as though they have been pollinated. For example, 4-chlorophenoxyacetic acid (4-CPA) is used to improve blossom and fruit set in tomatoes. Another synthetic auxin, β-naphthaoxyacetic acid (BAA), is used to grow seedless tomatoes—because the flowers were not pollinated, seeds do not develop.

Apple, pear, and citrus fruit orchardists have a problem with **preharvest fruit drop**—individual fruits ripen at different times, causing the fruit stalks of many to abscise before harvest time. A diluted spray of the herbicide 2,4-D prevents fruit drop of citrus trees, although it must be used carefully to avoid drifting onto other, sensitive crops. Spraying apple and pear trees with a very diluted naphthaleneacetic acid (NAA) solution keeps most of the fruits on the trees until harvest time.

Paradoxically, NAA is also used for **spray thinning** to reduce excessive fruit set. Spraying apple and olive trees with NAA shortly after blossoming is an effective way to thin out fruits, thereby encouraging maximal development of those remaining. The same auxin has different effects on abscission at different stages in fruit development.

Many ornamental trees produce **stem sprouts** (also called suckers or water sprouts) around their bases. In the past they had to be removed by hand, a time-consuming process. Now a solution of NAA (more concentrated than that used for fruit management) is sprayed on the trunks to inhibit sucker growth.

Auxin may also stimulate differentiation of cambial derivatives into xylem cells.

Some synthetic auxins are important **herbicides** (plant killers). The tremendous productivity of modern agriculture is possible because we are able to grow vast acreages of cereal crops (corn, oats, wheat, barley) with a minimum of competition for water and nutrients from unwanted species. The synthetic auxin 2,4-D kills broad-leaved species (but not grasses, such as the cereals), causing massive disruption of normal growth processes. Discovered in 1946, 2,4-D is used also as a brush killer. Properly handled, it is considered a "safe" herbicide.

Unfortunately, a related herbicide called 2,4,5-trichlorophenoxyacetic acid (2,4,5-T) was also considered "safe." It was recommended for such general purposes as home gardens, lawns, pastures, crops, and brush control. In Oregon it was sprayed aerially over young Douglas fir plantations to kill competing plants. Then reports of abnormally high rates of miscarriage among women near the sprayed areas drew attention to 2,4,5-T and a contaminant, **dioxin,** as potential human health hazards. One dioxin is of particular concern. It is 2,3,7,8-

tetrachlorodibenzo-*para*-dioxin (TCDD), the most toxic synthetic chemical known. Millions of pounds of 2,4-D and 2,4,5-T (the active ingredient of the defoliant called Agent Orange) were sprayed on the tropical forests of Vietnam during the war there. The political, legal, medical, moral, and humanitarian problems created by human exposure to Agent Orange have only begun to emerge. Many Vietnam veterans claim that their severe health problems and birth defects suffered in their children are related to Agent Orange exposure. Unknown numbers of Vietnamese soldiers and civilians were also exposed. The whole story of the effects of both peacetime and wartime uses of 2,4,5-T (the use of which has now been banned in the United States) will only be known as studies and documentation continue.

## Gibberellins

During the 1890s extremely elongated seedlings were observed growing among normal-sized seedlings in Japanese rice paddies. Although they demonstrated vigorous early vegetative growth, the plants died before maturing, leading the condition to be called "foolish seedling" disease. Research in the 1920s by the botanist E. Kurosawa demonstrated that the unusual growth was caused by a fungus infection. The fungus was subsequently identified as *Gibberella fujikuroi,* a sac fungus (Chapter 13), and the growth-promoting substance it produces was named **gibberellin.**

Japanese research during the 1930s led to isolation of several powerful growth-promoting and growth-inhibiting substances as well as initial elucidation of the chemical structure of gibberellin. Because of World War II, western scientists did not learn of the discovery of gibberellin until about 1950, at which time U.S. and English scientists became aware of the Japanese papers and began their own research. Now more than 50 different gibberellins have been isolated, some from fungi and some from healthy tissues of higher plants.

Gibberellin is produced in buds, especially in the leaf primordia, in the embryos of immature seeds, in fruit tissue, and in the roots. The large quantity of gibberellin produced by the root system is made available to the entire plant, for it is translocated both up and down the plant axis. Perhaps the best-known gibberellin is **gibberellic acid** (**GA** or **$GA_3$**).

A wide range of plant responses to gibberellins has been demonstrated. Gibberellins influence cell division, cell elongation, enzyme secretion, flower initiation, sex expression of floral parts, fruit set, fruit growth, maturation and ripening, senescence (especially of leaves), seed and bud dormancy, and germination.

Application of GA to some plants induces extraordinary stem elongation, at times accompanied by decreased leaf size. GA also stimulates elongation of the flower stalk, causing bolting in such long-day plants (see the section on phytochrome) as cabbage and radishes. In high quantities it tends to stimulate flower production in long-day plants along with the bolting, although not always.

Some genetically dwarf plants seem to respond more dramatically to gibberellin application than do related tall plants. It is hypothesized that perhaps the difference between dwarf and normal-sized pea and corn plants, for instance, is related to the genetic inability of the dwarf type to produce critical quantities of gibberellins. This relationship is not consistent, however, among all species having both dwarf and tall varieties.

Gibberellins seem to control seed development, and developing seeds influence the transition of the ovary into fruit. The main commercial use of GA is as a spray to increase the size of table grapes, especially the "Thompson seedless" variety.

Gibberellin is thought to be involved in stimulating fruit development in the absence of pollination, as discussed earlier in connection with auxin. Commercial auxin/gibberellin sprays are now available to stimulate development of unpollinated apple fruit.

Seed germination is also affected by gibberellins. Seeds are held in a dormant condition by the presence of growth inhibitors. The influence of the inhibitors must be removed before growth and germination can occur. (Also see Chapter 10.) Seed dormancy may be broken by the leaching away of the inhibitors, their alteration by chilling or light, and/or a rise in gibberellin content. Gibberellins also appear to assist in the mobilization of food resources in the seed, particularly in grains. In grains, gibberellins apparently stimulate the aleurone layer to produce enzymes that begin rapid digestion of the starch stored in the endosperm of the seed (Chapter 10). There is some commercial use of GA as a presoak for certain seeds to stimulate germination and seedling growth.

Other limited uses of GA are to stimulate development of male flowers in cucumbers, enabling hybridizers to obtain desired pollen, and to reduce or eliminate the chilling requirement for floral bud development in some flowering plants.

The diverse effects gibberellins have on plant

growth and development may arise because there are so many slightly different gibberellins and they may be modified within the plant from one form to another, either more or less active in a particular process, by relatively minor molecular modification. As with auxins, the effects of gibberellins may be related to control of protein synthesis via stimulation and inhibition of specific genes (regions of DNA) to produce specific messenger RNA molecules.

## Cytokinins

**Cytokinins** act mainly to promote cell division, but they are also involved in cell enlargement (especially in young leaves), tissue differentiation, flowering, fruiting, dormancy, and in retarding leaf senescence.

The discovery of cytokinins began with coconut "milk" and herring sperm. A coconut is a very large seed, containing liquid endosperm (a nutritive tissue), the "milk." Early work in culturing plant cells led to the discovery that some substance present in coconut milk apparently promoted cell division. An interesting series of experiments followed, combining coconut milk with auxin, and then combining the nucleotide adenine with auxin. Because adenine is a component of the nucleic acids DNA and RNA, researchers combined auxin with an old sample of DNA extracted from herring sperm. All these combinations promoted cell division. Eventually the substance 6-furfuryladenine was isolated from the old sample of herring sperm DNA and named **kinetin** because of its activity with auxin in inducing cell division (cytokinesis). This initial isolation led to the discovery of an entire group of related compounds called cytokinins. Apparently they are present in all plants in association with the sugar ribose. The first natural cytokinin to be isolated was **zeatin** from kernels of corn (*Zea mays*). It is more active than any of the other naturally occurring cytokinins, although less active than some synthetic ones. Today there are more than a hundred known cytokinins, natural and synthetic.

Cytokinins occur in plant tissues in a "free" state and also as components of transfer RNA. Therefore their activity—like that of other hormones—seems directly related to the control of protein synthesis. It has been found that DNA synthesis occurs in the presence of auxin but that cytokinins must be present in order for cell division to occur. One hypothesis is that cytokinins may be involved in the production of certain proteins necessary to the mitotic process—for instance, those that make up the spindle fibers. This explanation, however, does not account for the full range of cytokinin effects.

Using cytokinins, it is possible to culture plant tissues commercially. (See also Chapter 5.) A sample of isolated plant cells extracted from a desirable parent plant is placed in a flask or test tube containing nutrient medium, cytokinins, and auxin. Cell division proceeds rapidly and a mass of undifferentiated callus tissue is formed. Subsequent differentiation occurs and buds form on the callus tissue. Depending on the balance of nutrients and hormones present, the development of the callus mass may lead to shoot formation, via buds, and/or root formation. If continued callus formation is desired, cytokinins and auxins are maintained at approximately equal levels; a greater proportion of cytokinin to auxin favors bud formation; a greater proportion of auxin to cytokinin favors root formation. By such tissue culture techniques, for example, the western forest industry is producing many generations of "super trees," Douglas firs that have been selected for especially rapid growth to harvestable maturity.

The antagonistic, check-and-balance interaction of cytokinin and auxin now seems related to many aspects of plant growth. Apical dominance, for example, is released by the production of kinetin in lateral buds. "Witches'-broom," in which many buds at the shoot tip grow simultaneously, develops when a pathogen infects the stem tip and produces cytokinins. Cytokinins are sometimes used in greenhouses to stimulate lateral bud growth of such flowering plants as chrysanthemums and roses.

## Ethylene

One of the challenges of fruit storage and transport is to maintain peak freshness, delaying for as long as possible the changes in flavor and texture that occur with overripening. Since 1924 it has been known that ripe fruits produce a gas, **ethylene**, that seems to accelerate the ripening process in adjacent fruits. Therefore a truckload of apples, even when refrigerated, would continue to ripen during shipment.

Other long-known effects are ethylene's ability to release potato tuber buds from dormancy, stimulate leaf abscission, induce flowering in pineapple plants, and induce adventitious root formation in stem cuttings. In a home root cellar, for instance, apples should be stored separately from other harvested foods, such as potatoes, onions, and rutabagas. Otherwise the ethylene produced by the apples will stimulate the development of bud growth, resulting in sprouted vegetables! Ripe apples in a fruit bowl will stimulate continued ripening of other apples, bananas, and so on. A pineapple plant can be induced to flower by ethylene released from a ripe apple or banana (Fig. 9-5).

We now know that *many* aspects of plant growth are affected by ethylene and that, far from being abnormal and economically pesky, the gas is a normally occurring and significant plant hormone.

Ethylene, sometimes called the **senescence** or **aging hormone** and the **ripening hormone,** also influences the formation of abscission layers. In various experiments ethylene has been shown to interfere with such auxin-controlled functions as shoot elongation and tropisms. In some plants it seems to promote cambial division and it acts to maintain the "hook" in the emerging shoot of germinating seedlings, as peas and beans (Chapter 10).

**FIGURE 9-5** Grow a pineapple at home! Cut the top from a healthy pineapple fruit, allow it to cure in the air for a few days, then plant it. After vegetative growth is established bag the plant with a ripe apple or banana for several days. The **ethylene** gas released from the apple or banana will stimulate the pineapple to initiate flowering leading to fruit production, as here.

The interactions of auxin and ethylene seem to be antagonistic. Ethylene, for instance, promotes both ripening and abscission whereas auxin tends to retard both processes. This is another example of **homeostasis,** the maintenance of a controlled dynamic balance in living systems. The effects of ethylene appear related to the *balance* between auxin and ethylene levels in plant tissues. We have discussed the fact that higher than optimum auxin levels inhibit growth. In some cases, the mechanism may be that high auxin levels promote the formation of ethylene, which apparently inhibits growth.

Ethylene ($CH_2{=}CH_2$) is a much smaller molecule than other plant hormones. The fact that it is a gas at physiological temperatures and pressure enables it to diffuse easily through plant tissues. It is effective in minute amounts, but the precise mechanism of its activity is not yet known. One hypothesis is that ethylene affects protein synthesis in some way, either by regulating messenger RNA formation (DNA transcription) or the translation of RNA during protein synthesis in the cytoplasm. Alternatively, it may alter permeability of membranes. It is produced in active meristems, ripening and senescing fruits, senescing flowers, germinating seeds, and in tissues that have been physically stressed by wounding, bruising, or bending.

There are many commercial uses of ethylene—as the gas itself and as an ethylene-releasing compound (2-chloroethyl phosphoric acid or ethephon) that can be mixed into a spray solution. Specific uses are discussed here in connection with ethylene effects.

#### Senescence

The showy but fragile blossoms of irises, orchids, morning glories, and many other plants unfold as a dazzling display of floral structure, color, and scent. Eventually the petals and sepals shrivel, contracting into a limp mass (Figs. 9-6 and 9-7). Some flowers, as the daylily and morning glory, last but a day. An orchid, if left unpollinated, may last for months. Ethylene can be detected when senescence of the petals and sepals

**FIGURE 9-6** An *Iris* in full bloom near several flowers that have already withered.

**FIGURE 9-7** Tissue senescence is related to increasing levels of **ethylene** gas. Degeneration follows, as in these *Iris* blossoms.

begins; and experimental exposure to ethylene can trigger the reactions.

As noted, the aging process leads to senescence, involving both physiological and structural changes. Senescence involves a decline in the metabolic rate as well as in the rates of RNA synthesis and protein synthesis. Senescence may affect the entire plant, leading to its death, or it may involve particular plant parts, such as the flowers, leaves, or fruits. Senescence is an active rather than passive metabolic state. In both plants and animals biologists are tending more and more to the view of cell death as a "preprogrammed" function, carefully built into the life of the tissue or organism as a whole. Some plant cells—as xylem elements, fibers, and cork—are "born to die." The living cytoplasm constructs a functional structure and then self-destructs. Presumably the chemical elements of the cytoplasm are then "recycled" elsewhere in the plant.

Senescence in plants, is controlled by ethylene, interacting with other plant hormones. For example, application of auxin, cytokinin, and gibberellin—in combination or alone—has been shown to delay senescence in various plant parts. The cessation of the influence of these hormones seems a part of the normal aging process.

All cells of a whole plant are related. They are the successive generations derived from specific parent cells and their nuclei contain the same genetic information. With basically identical genetic potential, why do some cells in the plant body become senescent while adjacent cells remain vigorous? This mystery, like that of cellular differentiation, involves the activation and repression of different genes in the various cells. The synthesis of specific messenger RNA by these genes controls the kind and balance of enzymes and other proteins produced in the cell.

**Ripening**

The ovary of a flower enlarges and develops into a fruit by typical growth processes, controlled by such hormones as cytokinins, auxin, and gibberellins. It then ripens. Ripening is a kind of senescence. The aging fruit tissues change, becoming less acid (sour), sweeter, and softer. An unripe apple, for instance, is hard and sour. At the peak of edibility it is softer but still firm, with a nice balance of tartness and sweetness. The crisp tissues are readily crushed by chewing, which releases the flavorful central vacuole contents. An overripe apple is sweet but mealy because substances that cement the wall material of adjacent cells deteriorate. When chewed, the cells of overripe tissues separate from one another rather than rupture; so overripe apples are more suitable for cooking, which breaks down the cells and releases their full flavor.

The developing fruit of some species maintains a high respiratory level during growth (as measured by carbon dioxide release) and then exhibits a lower respiratory rate as it matures (Fig. 9-8). Some fruits (*e.g.*, apples) experience a dramatic peak of respiratory activity (the **climacteric**) at the onset of the ripening process. Ethylene production during the climacteric is about a hundred times the previous level. This information and other research results point to ethylene as the ripening hormone.

It is not known exactly how ethylene stimulates ripening. In some fruits it is present consistently during ripening; in others it only serves to stimulate ripening. Its known metabolic effects include increased membrane permeability—thus permitting freer movement of enzymes, including degradative enzymes—and increased protein synthesis.

There is a good chance that nearly all fruits bought have been treated with ethylene at some time during flowering, fruit development, or after harvesting. Ethylene is of vast economic value.

Ethephon spray is used to stimulate uniform flowering in cultivated bananas. Applied to cucumbers, ethephon increases the ratio of female-to-male flowers and so increases the number of cucumber fruits that will form. It also speeds up fruit maturation so that most fruits can be harvested at one time.

Before harvest, ethephon solution is sprayed on a wide variety of plants, including apples, cherries, tomatoes, cucumbers, blueberries, figs, pineapple, and coffee, to ensure uniform fruit ripeness and coloration at the time of harvest. This step permits very efficient picking, whether by hand or machine. Preharvest ethephon application also stimulates abscission of the fruit stalk, enhancing easy removal from the stems. This factor is important for fruit and nut crops, such as sour cherries, walnuts, and almonds, that are harvested by mechanical tree shakers.

Ethylene gas is released in storage chambers to ripen tomatoes, bananas, and honeydew melons. Green tomatoes are sometimes dipped in an ethephon solution to promote ripening.

Most of us associate a bright orange or yellow color with ripeness and sweetness of oranges, lemons, and grapefruits. The fruit rind, however, sometimes retains relatively large amounts of chlorophyll in spite of fruit ripeness. To meet consumer expectations, therefore, citrus fruits are sometimes "degreened" by exposure to ethylene gas or by dipping in an ethephon solution before

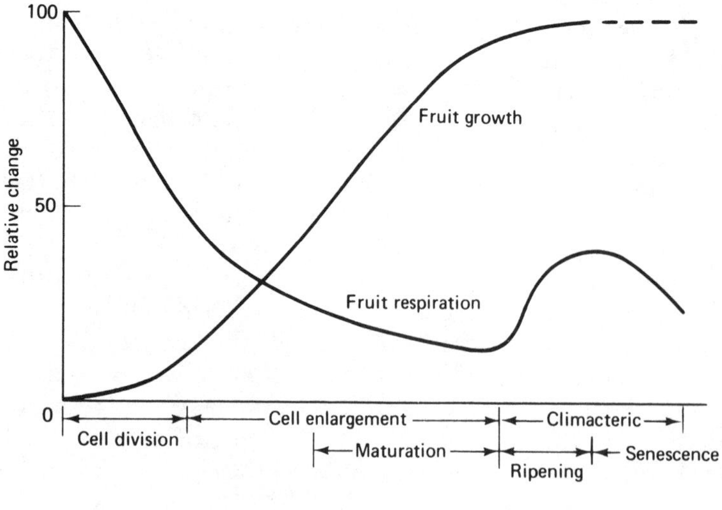

FIGURE 9-8
Developing fruits show changing rates of respiration during growth and ripening, sometimes including a peak of respiratory activity, the **climacteric**. (After D.R. Dilley, "Hormonal control of fruit ripening." Hort Sci. 4: 111–114. Reprinted by permission, American Society for Horticultural Science. Adapted by Noggle and Fritz in *Introductory Plant Physiology*. Reprinted by permission of Prentice-Hall, Inc., Englewood Cliffs, N.J.)

shipping. (Some oranges are even dyed a bright orange.)

Greenhouses also have uses for ethylene. An ethephon solution is used to stimulate floral bud initiation in some ornamental bromeliads. Greenhouse use of ethylene, however, must be carefully controlled because errant ethylene will promote premature senescence of existing blossoms, such as of carnations, roses, and orchids. Ethylene is also a problem during shipment of such bulbs as tulips. If diseased bulbs are present, they will produce enough ethylene to abort flower bud development in the other bulbs.

#### Abscission

As described in Chapter 8, leaf abscission is an orderly part of the life of many plants whether it occurs all at once, as with deciduous trees, or from time to time throughout the life of a plant as the older leaves are shed. You will recall that the abscission layer is a specialized zone of cells across the base of the petiole. It gradually cuts off the dying leaf from its parent stem, allowing the leaf to be shed without damage to the stem and leaving a protective corky scar. Ethylene is the hormone that controls leaf abscission.

In most plants leaf abscission seems to be stimulated by decreasing day length. Such desert plants as the ocotillo, for example, that shed their leaves during the driest parts of the year may have other mechanisms.

Several hormones are involved in leaf abscission, including auxin, cytokinins, and ethylene, but not, apparently, abscisic acid. As a leaf ages, it produces less and less auxin, stimulating changes in the abscission layer. The onset of senescence in the leaf blade results in its producing certain substances called "senescence factors" that trigger senescence in the cells of the petiole. Sensitized by senescence, certain cells of the abscission layer respond to ethylene by producing **cellulase**, an enzyme that specifically attacks the cellulose in the cell walls. As a result, parenchyma tissue in the abscission layer loosens and becomes weak.

A second effect of ethylene is to cause the cells nearest the abscission zone on the stem side to swell, distorting and pushing away the area of contact with the petiole. Held so loosely, the leaf eventually falls away of its own weight, aided by such physical forces as wind, precipitation, and alternate freezing and thawing.

As you know, abscission also occurs with floral parts and fruit, stimulated by ethylene.

### Abscisic Acid

**Abscisic acid** (**ABA**) is a powerful hormone that is categorized as an **inhibitor**, working against the action of auxins and gibberellins. It is formed in fruits and leaves and is readily transported throughout the plant body.

ABA affects winter and summer dormancy of perennial plants, seed dormancy, fruit and leaf abscission, and stomatal closure during stress. Some of ABA's activity seems related to its effect on protein synthesis, but it may also exert some more direct effect.

#### Stress

ABA is sometimes called the **stress hormone**, for it forms when a plant is subjected to unfavorable conditions and initiates specific defensive responses. The best-known role of ABA is stimulation of stomatal closure when a plant is stressed by water shortage. The presence of ABA in the guard cells causes potassium ions to leave the cells, reducing the osmotic potential. As a result, the guard cells become less turgid, closing off the stomatal opening. The stomates remain closed until the amount of ABA present in the guard cells lowers, permitting the return of potassium ions.

#### Summer and Winter Dormancy

Summer and winter are both times of great stress for plants of the northern and mountainous areas of North America. During the winter months plant tissues are subjected to subfreezing temperatures and during the hot and at times windy summer months they must endure limited water availability. These are also unfavorable seasons for seed germination to occur.

To cope, many species have innate periods of **dormancy**, periods of reduced metabolic activity often accompanied by specific structural and chemical changes. The types of dormancy that accommodate the various unfavorable environmental conditions are winter dormancy, summer dormancy, and seed dormancy. The dormant state seems to be controlled in all cases by a balance of growth-promoting and growth-inhibiting hormones.

As noted in Chapter 8, woody plants of temperate regions experience rapid growth in the early spring, putting out new shoots and roots, flowering, and adding new xylem and phloem tissue. As the season progresses, rapid terminal growth subsides as leaves become fully expanded and are at maximum photosynthetic activity. Lateral growth also continues less vigorously—hence the more compressed summer wood of the annual ring. During this **summer dormancy** great metabolic activity is associated not only with food production and storage but also with the initiation of all the buds from which the next season's vegetative and flowering shoots will grow. In other words, the breathtaking cherry blossoms of this spring were actually initiated last summer!

A dramatic improvement in environmental conditions, such as abundant water and nutrients, may release the plant from summer dormancy. The usual pattern, however, is a transition from summer dormancy into **winter dormancy**. Numerous physical and physiological changes are necessary to prepare the plant for winter and the gradual adjustment process is called **acclimation.**

Acclimation is stimulated by decreasing day length, as perceived in some way by the leaves. ABA levels increase and **overwintering buds** are produced. During acclimation structural modifications that occur may include the development of protective bud scales—coated by varnishlike or waxy substances—that protect the buds from water loss and physical damage, and leaf abscission. Physiological modifications include accumulation of sugars and other metabolites in the storage tissues of the bark and wood.

The exact role of ABA in winter dormancy seems related to the relative balance between ABA and growth-promoting hormones also present. Dormancy is maintained until the influence of ABA is overcome, as by increasing amounts of growth promoters and/or transformation of the ABA into an inactive form. These changes seem to take place within the buds themselves. In the North winter dormancy may be broken by exposure of the buds to 1 to 10°C temperatures for 260 to 1000 hours. Even though these requirements may be satisfied before the end of winter, bud growth is usually limited by cold temperatures, which affect the rate of metabolism.

The breaking of dormancy is a critical period for orchardists. Early warm weather, for instance, may stimulate fruit trees to begin active growth prematurely. The tender shoots and flowers are then susceptible to a killing frost. So fruit growers may plant their trees on a north-facing slope where they will be warmed less during the day than on a south-facing slope.

A period of extreme winter cold is one of the most limiting environmental factors, killing plants and hence limiting the northern extension of species ranges. Why is it that some plants survive in the colder northern climates while others are killed? Many kinds of fruit-bearing trees (peaches, apricots, citrus types, almonds, and others), for example, thrive in the southern and western United States. Others, such as some apples, cherries, plums, and pears, can even be grown in the cold upper Midwest and Plains states. The difference is in the innate cold hardiness (or simply "hardiness") of individual species or varieties. **Cold hardiness** refers to the ability of a plant to change its metabolism in order to survive cold temperatures. Figure 9-9 shows regions of annual minimum temperature averages of the United States—necessary information to guide perennial herbaceous plant, shrub, and tree selection.

As you might suspect, cold hardiness and acclimation are important research areas. The possibility of increasing hardiness of economically important plant varieties by even a few degrees would greatly increase their cultivation. This is a legitimate concern as we prepare to provide food and essential plant products for a rapidly increasing world population.

#### Seed Dormancy

As noted in Chapter 5, a seed is the species' link with the future. Growth inhibitors located in the seed or in surrounding fruit tissues prevent germination from occurring until warmth and moisture are favorable for seedling growth. ABA is perhaps the most important inhibitor controlling **seed dormancy.**

Germination occurs only after the influence of the inhibitors is released, such as by leaching (washing away), light, or low temperature. Seeds of desert plants can remain dormant for years until sufficiently heavy spring rains release them. Lettuce seeds and those of some weeds will not germinate unless they are close enough to the soil surface to receive light. Gardeners cannot grow some prairie plants from seed unless they first refrigerate the seeds, a process called **vernalization.** As the inhibitor level is reduced, levels of growth-promoting substances increase and the seed can germinate.

### Other Inhibitors

ABA is a hormone. Some nonhormone substances can also inhibit plant growth, however. Benzoic acid, cinnamic acid, coffeic acid, and coumarin are inhibitors that occur naturally in plants. In addition, synthetic inhibitors having

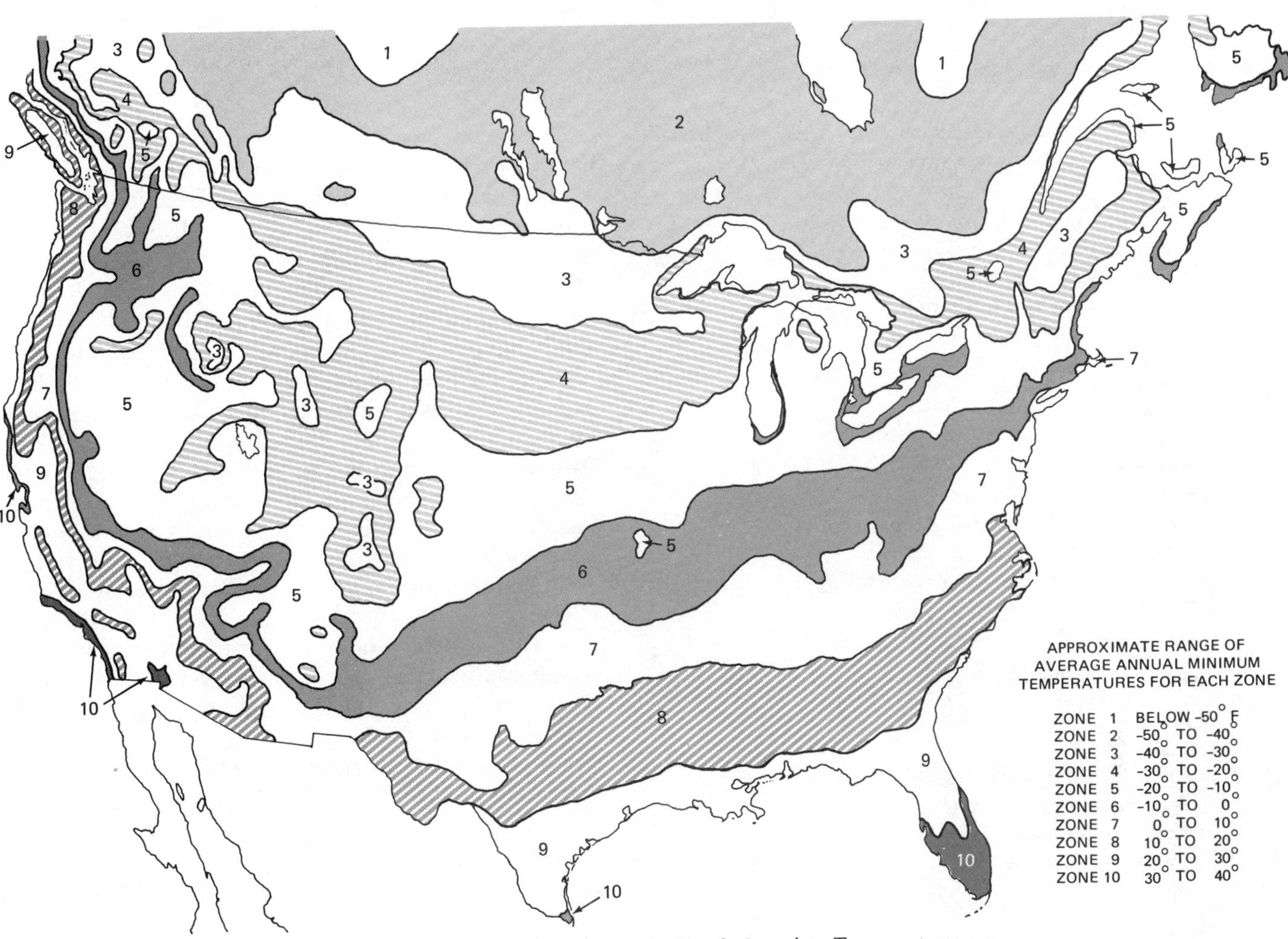


**FIGURE 9-9** Major **temperature zones** in North America. Temperatures are expressed in Fahrenheit rather than Centigrade (Celsius). The limits of the **average annual minimum temperatures** for each zone are indicated. These zones differ somewhat from the "hardiness zones" established by the Arnold Arboretum that are used to rank cultivated plants. Zones 1, 8, 9, and 10 are the same for both systems. Here zones 2, 3, 4, 5, 6, and 7 are arranged in 10 degree intervals, rather than the 15 degree intervals of the Arnold Arboretum zones. (Source: U.S. Dept. of Agriculture.)

commercial uses have been developed. They slow cell division and elongation and are used by floriculturists to produce a more compact growth form. In addition, the treated foliage may be a deeper green. Examples of potted plants treated with synthetic inhibitors are poinsettias, chrysanthemums, azaleas, geraniums, Easter lilies, and tulips.

## PHYTOCHROME

Many plant processes occur in response to a light stimulus. Many of these **photoprocesses** depend on plant reaction to specific wavelengths in the red (660 nm) and far-red (730 nm) ranges. The so-called red and far-red photoreactions are mediated by a molecule called **phytochrome.** Phytochrome is a chromoprotein molecule that consists of two parts: a pigment portion and a protein portion. Table 9-1 lists some of the processes in which phytochrome participates.

Phytochrome is found in many plant groups, including green algae, bryophytes, angiosperms, and gymnosperms, and is probably ubiquitous in green plants. There are two reports of phytochrome in fungi. In the bodies of higher plants phytochrome has been found in the tissues of coleoptiles, hypocotyls, cotyledons, seeds, developing fruits, petioles, leaf blades, stems, vegetative buds, flowers, and roots. The site of phytochrome within cells has not yet been determined.

**TABLE 9-1 Some plant responses to light in which phytochrome participates.[a]**

| **Cells/Tissues in general** |
|---|
| Photoperiodism |
| Elongation of leaf, petiole, stem |
| Plastid morphology and orientation |
| Succulency |
| Formation of tracheary elements |
| Differentiation of stomata |
| Changes in respiration rate |
| Anthocyanin synthesis |
| Increase in RNA synthesis |
| Increase in protein synthesis |
| Changes in rate of fat and protein degradation |
| Auxin breakdown |
| Sucrose incorporation into growing buds |
| Cell membrane permeability |
| **Seeds and seedlings** |
| Seed respiration |
| Germination |
| Straightening of Hypocotyl hook |
| Cotyledon enlargement |
| Epidermal hair formation along cotyledons |
| Differentiation of primary leaves |
| **Stems** |
| Stem elongation |
| Straightening of hypocotyl hook after germination |
| Bud dormancy |
| Flower induction |
| Rhizome formation |
| Bulb formation |
| Epinasty |
| Formation of leaf primordia |
| Sex expression |
| **Leaves** |
| Leaf elongation |
| Abscission |
| Differentiation of primary leaves |
| Differentiation of stomates |
| Unfolding of grass leaf |
| Detection of shading |
| **Roots** |
| Root development |

[a]Some of the responses are sorted out by the major plant structure that is affected; however, it is assumed that the general cellular/tissue responses apply to each of the other categories as well.

The mechanism of phytochrome action is based on the fact that phytochrome exists in two forms that can be transformed from one to the other (Fig. 9-10). In its $\mathbf{P_r}$ (red-absorbing) form phytochrome is rapidly converted in the presence of light to its $\mathbf{P_{fr}}$ (far-red-absorbing) form. In the dark $\mathbf{P_{fr}}$ is slowly converted back to $\mathbf{P_r}$, at least in some dicotyledonous species. Phytochrome responses seem to depend on the ratio of $\mathbf{P_r}$ to $\mathbf{P_{fr}}$ present at a given time. $\mathbf{P_{fr}}$ is the biologically active form of phytochrome. It is hypothesized that $\mathbf{P_{fr}}$ combines with a reaction partner (X), some substance or substances present in the cell, and that this ($\mathbf{P_{fr}} \cdot$ X) complex is the stimulus that induces a physiological response. Therefore the ($\mathbf{P_{fr}} \cdot$ X) complex is sometimes referred to as a **biological trigger.**

One of the major lines of phytochrome research has been on its role in flower initiation. Although the precise role is not known, it seems clear that phytochrome itself is not the flowering stimulus but evidently is involved in triggering or activating the flowering stimulus.

A specific sequence of events is involved in the process of angiosperm reproduction, from floral initiation to fruit-set, involving a complex set of interactions (Fig. 9-11). A plant cannot be stimulated to flower until it has completed a certain vegetative growth requirement (its juvenile phase), which may be a matter of days to years, depending on the species.

Light and temperature seem the most critical environmental factors in inducing flowering. Some grains, for example, will grow, flower, and set fruit earlier in the season if the seeds are subjected to a period of pregermination chilling (vernalization). Plants that germinate from the same kind of seeds that have not been chilled require a longer vegetative growth period before they will flower. Yet temperature alone does not determine flowering; light is a more critical factor for most plants. Specifically, the duration of exposure to light and dark during a 24-hour cycle (**photoperiod**) is a significant factor for floral initiation in most angiosperms. Table 9-2 summarizes the flowering behavior of some species. Those species that flower only if they are exposed to a period of decreasing day length/increasing night length are called **short-day plants.** Those that require increasing day length/decreasing night length are called **long-day plants.** Some plants are called **day-neutral** because they appear unaffected by different photoperiods. Additional categories have now been recognized.

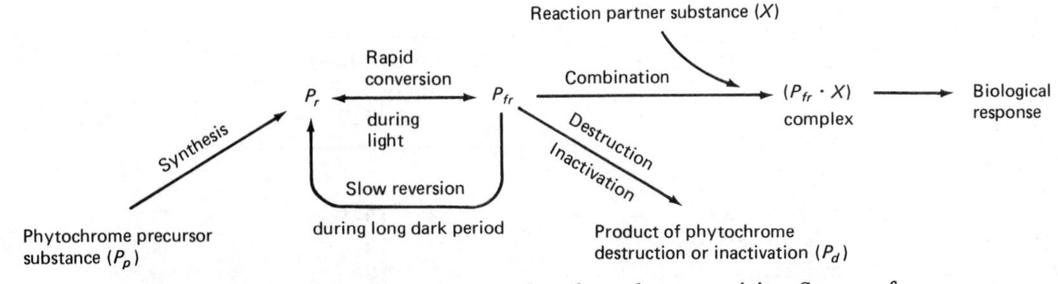


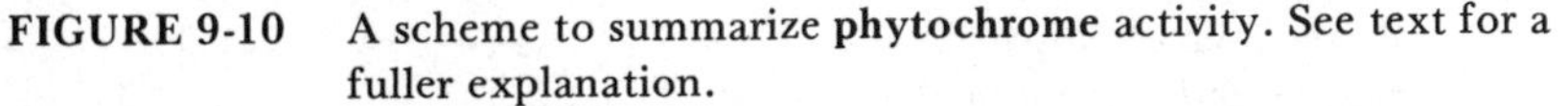

**FIGURE 9-10** A scheme to summarize **phytochrome** activity. See text for a fuller explanation.

Plants that will not flower until exposed to short days followed by long days are called **short-long-day plants**—they generally flower in the late spring or early summer, as candytuft (*Iberis*), and winter rye (*Secale cereale*). Plants that flower after exposure to long days followed by short days are referred to as **long-short-day plants** and generally flower in late summer or early fall, as *Bryophyllum* and night-blooming jessamine (*Cestrum nocturnum*).

Each species with photoperiodic requirements has a certain **critical day length**; that is, the plant will flower only when a specific minimum or maximum number of hours has been reached. Obviously the critical day length (e.g., 11 hours of light followed by 13 hours of darkness) may be the same for some long-day and short-day plants, but the response is different. A long-day plant having a critical day length of 11 hours will flower only when the photoperiod has *exceeded* 11 hours. An

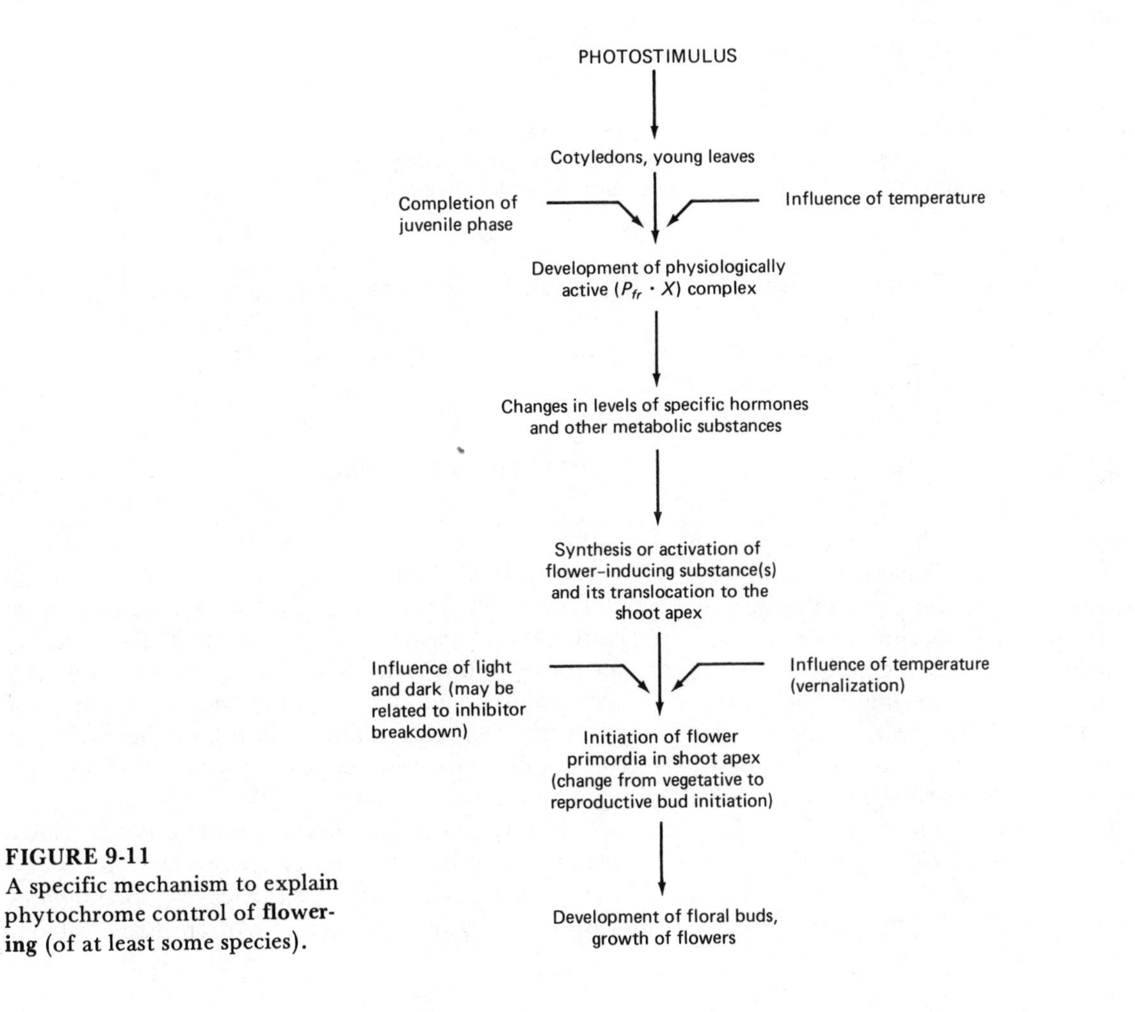


**FIGURE 9-11**
A specific mechanism to explain phytochrome control of **flowering** (of at least some species).

TABLE 9-2 Examples of photoperiod requirements for flowering.[a]

| | | Critical Day Length (in hours) Less Than |
|---|---|---|
| **Short-day plants** | | |
| *Chrysanthemum* | Most species | 15 |
| *Cosmos sulphureus* | Klondyke cosmos | 14 |
| *Euphorbia pulcherrima* | Poinsettia | 12 |
| *Fragaria chiloensis* | Strawberry (most) | 10 |
| *Nicotiana tabacum* | Maryland Mammoth tobacco | 14 |
| **Long-day plants** | | |
| *Anethum graveolens* | Dill | 11 |
| *Rudbeckia bicolor* | Coneflower | 10 |
| *Spinacia oleracea* | Spinach | 13 |
| *Trifolium pratense* | Red clover | 12 |
| **Day-neutral plants** | | |
| *Cucumis sativa* | Cucumber | – |
| *Fagopyrum tataricum* | Buckwheat | – |
| *Fragaria chiloensis* | Everbearing strawberry | – |
| *Impatiens balsamina* | Balsam | – |
| *Lycopersicon esculentum* | Tomato | – |
| *Phaseolus vulgaris* | String bean | – |
| *Zea mays* | Corn | – |

[a]The critical day length may be the same for some long-day and short-day plants, but they may have different responses. For example, red clover will flower after increasing day lengths *exceed* 12 hours whereas poinsettia will flower after decreasing day lengths become *less than* 12 hours.

11-hour short-day plant will flower only when the photoperiod is *less than* 11 hours. The critical day-length requirement limits the latitudinal distribution of some species. The photoperiod stimulus is perceived by young leaves and/or the cotyledons.

It seems that the period of darkness is equally as important as the photoperiod and short-day plants seem to require that the night period be uninterrupted to initiate flowering. Long-day (short-night) plants do not seem to be inhibited by an interruption of the night period and may even flower under continuous light, seeming to require no dark period at all. The flowering of such short-day greenhouse species as *Chrysanthemum* species can be controlled precisely by manipulation of photoperiod and night period to meet the 13-hour critical day length whenever blooming is desired.

Photoperiodic responses are related to the $P_{fr}:P_r$ ratio. The actual series of biological changes that occur during phytochrome-mediated responses, however, seems to be controlled by innate growth regulators, such as cytokinins, auxins, and gibberellic acid, discussed earlier.

## CIRCADIAN RHYTHMS

If you stop to think about it, you must wonder "How does a plant tell time in light and darkness? How does it know that 13 hours of continuous light have elapsed?" In the cocklebur (*Xanthium*) the critical day length seems to be precisely 15 hours and 40 minutes (under one particular temperature regimen). Many plants have been shown to be able to measure time with a 10- to 15-minute degree of accuracy.

Some plant processes occur with a daily (**diurnal**) rhythm or activity cycle, including sleep movements (nyctinasty) by the leaves, opening and closing of flowers, release of carbon dioxide from the leaves, mitosis, enzyme activity, flower petal movement, and nectary secretions. Some rhythms continue even when the plant is continuously illuminated by dim light (at least for awhile). Because most of the innate rhythmic activities vary in periodicity between 22 and 27 hours, they are called **circadian rhythms** (L., *circa* = approximately, *diem* = day). Circadian rhythms must be started or triggered by some cue, such as a critical photoperiod or night-period length. Once initiated, the rhythm becomes "free running," oscillating between phases (such as leaves open, leaves closed) on its own. The pattern of fluctuating internal processes is controlled by an innate timing mechanism called the **biological clock.** The clock remains synchronized with actual 24-hour time by such environmental influences as photoperiod.

One significant function of an internal clock is the response of perennial plants to seasonal changes. As winter approaches, for example, the day and night temperatures drop, day lengths shorten, and leaf abscission is initiated in deciduous plants. If temperature alone stimulated abscission, it is possible that unseasonably cool temperatures could stimulate premature abscission. The interaction of the internal clock and photoperiod, however, may keep abscission on a correct seasonal schedule.

Biological clocks have been demonstrated in many plants and animals. They are probably ubiquitous in all living things except possibly bacteria. In humans, for instance, it is responsible for the "jet lag" phenomenon experienced by travelers who drastically change time zones. Continued research into circadian rhythms and the biological clock mechanism should yield a better understanding of plant and animal life processes.

## SUMMARY

A typical growth curve is sigmoid, with a rapid **juvenile phase.** Sexual reproduction is an important function of the **mature phase** of higher plants. **Senescence** is an irreversible period of declining vigor and deteriorative change, leading to cell and tissue **death,** even to death of the entire organism.

Plant growth is controlled by various **plant growth regulators,** nonnutrient substances that affect or control growth. Most plant growth regulators are **hormones** that are produced in the tissues of the plant, including five general categories: auxins, gibberellins, cytokinins, ethylene, and abscisic acid. Synthetic hormones are of great commercial value and many examples are cited.

**Auxin** was the first plant hormone to be discovered. Produced in meristems and enlarging tissues, it is translocated downward toward the base of the plant. The common naturally occurring auxin is indole-3-acetic acid (IAA). IAA influences cell enlargement, tropisms, apical dominance, abscission, floral initiation and development, root initiation, fruit development, bulb and tuber formation, and seed germination. It acts with other hormones to stimulate cell division.

**Gibberellins** (more than 50 are now known) are produced in buds, embryos of immature seeds, fruit tissue, and roots. The best-known gibberellin is gibberellic acid (GA or $GA_3$). GA produced in the root system is translocated both up and down the transport system; so is available to the entire plant. Gibberellins influence cell division, cell elongation, enzyme secretion, flower initiation, sex expression of floral parts, fruit set, fruit growth, maturation and ripening of fruits, senescence (especially of leaves), seed and bud dormancy, and germination.

**Cytokinins** act mainly to promote cell division but are also involved in cell enlargement (especially in young leaves), tissue differentiation, flowering, fruiting, dormancy, and in retarding leaf senescence. The first natural cytokinin to be isolated was zeatin from corn kernels. There are now more than one hundred known cytokinins. Cytokinin and auxin are essential components of tissue culture techniques.

**Ethylene** is a gaseous hormone that affects a wide variety of plant growth processes. It was first recognized for its role in fruit ripening, for it is produced in the tissues of ripe fruits, accelerating the ripening process of adjacent fruits. Among other things, it influences senescence, ripening, formation of abscission layers, cambial division, and development of a "hook" in germinating seedlings. It seems to act antagonistically to auxin—an important homeostatic feature. It is produced in active meristems, ripening and senescing fruits, senescing flowers, germinating seeds, and in tissues that have been physically stressed by wounding, bruising, or bending. It has wide commercial use.

**Abscisic acid** (ABA) is a powerful hormone that is categorized as an **inhibitor.** It works against the influence of auxins and gibberellins, is formed in fruits and leaves, and is readily transported throughout the plant body. ABA affects winter and summer dormancy of perennial plants, seed dormancy, fruit and leaf abscission, and stomatal closure during stress.

ABA is a hormone that acts as an inhibitor, but other nonhormone substances can also inhibit plant growth.

**Phytochrome** is a chromoprotein molecule composed of two parts: a pigment portion and a protein portion. It participates in many plant **photoprocesses,** processes that occur in response to a light stimulus. Phytochrome action is based on its ability to be transformed back and forth between two different states in the presence of red and far-red light. Its far-red-absorbing form ($\mathbf{P_{fr}}$) is the biologically active form. It is hypothesized that $\mathbf{P_{fr}}$ combines with an intermediate substance or substances present in a cell and that this complex is the **biological trigger** that stimulates a physiological response.

Phytochrome has been studied intensively in relation to floral initiation. Angiosperms bloom in response to day length, or **photoperiod.** Some species are short-day or long-day plants; others seem to be day-neutral.

Innate rhythmic activities, called **circadian rhythms,** have been demonstrated in many organisms. Circadian rhythms demonstrate a fluctuating pattern that runs an approximately 24-hour cycle (22 to 27 hours). This requires the presence of an internal **biological clock** mechanism, which is kept synchronized to actual 24-hour time by such environmental influences as photoperiod.

## SOME SUGGESTED READINGS

Galston, A.W., P.J. Davies, and R.L. Satter. *The Life of the Green Plant,* 3rd ed. Englewood Cliffs, N.J.: Prentice-Hall, Inc., 1980. Available in paperback, this is a well-written and concise explanation of how green plants function.

Leopold, A.C., and P.E. Kriedeman. *Plant Growth and Development,* 2nd ed. New York: McGraw-Hill Book Co., 1975. Includes topics relevant to this chapter as well as those preceding.

Ray, P.M. *The Living Plant,* 2nd ed. New York: Holt, Rinehart & Winston, Inc., 1972. This brief review of plant growth and development has assisted many students in their first encounter with the science of botany; in paperback.

Steeves, T.A., and I.M. Sussex. *Patterns in Plant Development.* Englewood Cliffs, N.J.: Prentice-Hall, Inc., 1972. Replete with illustrations, this well-written text explains plant morphogenesis—the process by which they develop and differentiate.

Thimann, K.V. *Hormone Action in the Whole Life of Plants.* Amherst: University of Massachusetts Press, 1977. Aptly described in its own title, an outstanding and interesting review of this fascinating subject.

# chapter ten FROM SOIL TO SEEDLING

## INTRODUCTION

By now you are well aware of the environmental requirements of green plants for carbon dioxide, sunlight, water, nutrients, and growing space. Soil is the basic substrate of all terrestrial plants, their source of water, nutrients, and a place to carry out their life cycles. In this chapter we discuss how terrestrial plants, including cultivated species, interact with the soil. Then we present seed structure and plant growth from seed to seedling—the vulnerable transitional life phase between "mother plant" and "mother earth" dependency.

## THE SOIL

What is soil? Why, anyone knows! It's the "dirt" that farmers plow and plant seeds in to produce their crops, nothing very complicated. We step on it every day and all the lawn and trees and shrubs are planted in it. In fact, however, soil is a highly complex material. Geologically, **soil** is the weathered surface layer of the earth's crust. It is composed of inorganic substances intermingled with living organisms, the substances they produce, and the products of their decay.

### Soil Composition

When working with soil, wiping up mud, or sweeping away dirt, we tend to consider soil an inert substance rather than a continually changing ecological system. To understand soil dynamics, it is helpful to know the history of a given plot of ground.

Specific soils are combinations of two basic means of derivation. **Residual soil components** develop in place from the parent rock present at the site whereas **transported components** develop from mineral particles brought from their place of origin by glaciers, wind, water, and gravity. Some of our most fertile soils have thick upper layers of transported elements washed down from eroding hills and mountains into river valleys—for example, California's San Joaquin and Sacramento valleys.

The parent material has a distinctive influence on the characteristics of a soil. It may, for instance, be volcanic, sedimentary, or metamorphic in origin. **Physical weathering** of the parent material may begin with periodic freezing of water within small cracks. Because water expands when it freezes, this repeated physical stress causes splitting and eventual crumbling of the rock. The expansion of plant roots growing in tiny cracks and crevices is also a powerful force of rock destruction. Wind has a persistent abrasive action on parent material, especially when it carries sand particles, rain, and sleet. Past and present glaciers are major agents of physical weathering.

**Chemical weathering** occurs when, for example, atmospheric carbon dioxide and water combine to form a weak acid (carbonic acid), which attacks the surface of lime-containing rocks. Plant roots also produce carbonic acid so that the forces of root growth are both physical and chemical. Increased acidity of precipitation by the formation of "acid rain" (Chapter 2) causes rapid chemical weathering.

Development of the soil is continual, as reflected in its structural organization. Most soils have three basic layers that intergrade where they meet. The first layer is referred to as the **topsoil**, which includes a few inches to almost a foot of small inorganic particles with varying amounts of organic material or **humus**. The next layer is the **subsoil**, usually several feet thick with little to no humus and different sizes of weathered parent material or rock. In some soils the subsoil level includes an impermeable layer of hardpan or claypan, which impedes drainage. The deepest layer is the **parent material**, which merges gradually with the subsoil.

Inorganic components of soil come from the parent material and give the soil its texture. "Younger" fragments are larger and rougher surfaced than particles that have been weathered longer. The inorganic particles are classified according to their diameters (Table 10-1) as gravel, sand, silt, and clay. The final product of weathering is clay-sized particles.

**TABLE 10-1 Inorganic soil particle classification.**

| Type of Particle | Diameter | |
|---|---|---|
| Coarse gravel | 5.0 | mm and above |
| Fine gravel | 2.0–5.0 | mm |
| Coarse sand | 0.2–2.0 | mm |
| Fine sand | 0.02–0.2 | mm |
| Silt | 0.002–0.02 | mm |
| Clay | less than 0.002 | mm |

Soils are classified by the percentage component of the various-sized particles. Sandy soils, for example, contain less than 20% silt and clay by weight; clay soils have more than 30% clay particles. **Loam** soils are approximately equal parts of fine (silt and clay) and coarse (sand) particles and are the most suited for agriculture. There are many variations of loam, depending on the relative proportions of the three particle types. Loam composition can be evaluated by shaking two cups of water and a cup of soil in a quart jar. Once the mixture has settled, shake it up again and then set it aside undisturbed for a week to 10 days. Organic material, or humus, will float until it becomes waterlogged. Soil particles will become stratified, with the coarsest sand particles on the bottom and the finest clay particles on the top. A soil sample proportioned 2:2:1 is illustrated in Figure 10-1. The relative proportions of the several types of particles determine the physical and chemical properties of a given soil (Table 10-2).

**Ion exchange capacity** (referred to in Table 10-2) is of particular significance to plant nutrition. Ion exchange capacity refers to the ability of a soil to bind nutrient ions until they can be taken up by plant roots. The surfaces of soil particles tend to have a negative charge that attracts water molecules and positively charged ions (**cations**). This factor is important because many nutrients needed in plant growth are cations, such as potassium ($K^+$), magnesium ($Mg^{++}$), and calcium ($Ca^{++}$). In a given volume of soil, small clay particles offer more overall surface area than do larger sand particles. Nutrients bound to particle surfaces are not easily leached (washed out) from the soil, making clay particles an important factor in soil fertility. One type of ion exchange occurs when hydrogen ions ($H^+$) given off into the soil solution by root hairs displace nutrient cations held on the soil particle surfaces. The nutrient ions can then be absorbed by the root hairs.

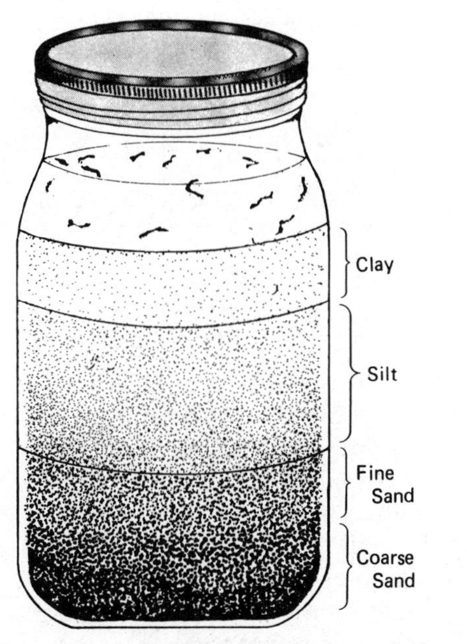


**FIGURE 10-1** A simple soil particle analysis method. A **loam soil** sample was shaken with water in a jar and then allowed to settle for a week or more. Particles settled out by weight, distributing the coarsest particles at the bottom and the finest particles at the top. Humus particles floated for a time, then settled on top. This particular loam sample is 2 parts **sand**, 2 parts **silt**, and 1 part **clay**.

Plant and animal remains (dead bodies and feces) are changed by microorganisms (soil bacteria, fungi, and protozoans) into humus, partially decomposed organic matter that gives "black dirt" its darkness. The reduction of organic matter by saprophytic nutritional processes of soil microorganisms is called **humification.** These organisms are an integral part of the complex soil ecosystem. Eventually all remnants of organic debris (including byproducts and remains of the humification organisms) are broken down even further to liberate carbon dioxide, water, and minerals. This final process is called **mineralization.**

Dark and amorphous humus is a structurally complex but easily identifiable soil component that is extremely important to soil dynamics. Spongy and lightweight, humus improves water infiltration and water-retention capability. (It has been estimated that humus can hold up to nine times its weight in water.) Humus also improves soil aeration and workability.

The colloidal properties of humus are especially notable, dramatically increasing soil ion exchange capacity. Soil **colloids** are tiny particles with a very large surface area per unit of mass. When mixed with water, they remain suspended instead of settling out. These particles are usually large molecules or molecular aggregates, are negatively charged, and attract swarms of cations. Humus colloids in the soil thus provide a vast storehouse of hydrated cation nutrients for plants, keeping them from being washed out of the soil. When cations in solution are used up, those on the colloidal sites are released to the soil solution, thus helping to maintain equilibrium.

Other functions of humus are to provide food for soil organisms and to inactivate toxins produced by certain plants. In addition, humus allows soil particles to aggregate—coating over the individual mineral grains, thereby making them adherent.

## Soil Solution and Plant Nutrition

### Water

As water is applied by rain or irrigation, it first distributes throughout the surface soil mass. Gravity moves water down through porous spaces

**TABLE 10-2** Physical and chemical properties of soil as affected by soil texture.

| Soil Texture | Water Infiltration | Water-holding Capacity | Ion Exchange | Aeration | Workability | Root Penetration |
|---|---|---|---|---|---|---|
| **Sand** | Good | Poor | Poor | Good | Good | Good |
| **Silt** | Medium | Medium | Medium | Medium | Medium | Medium |
| **Clay** | Poor | Good | Good | Poor | Poor | Poor |
| **Loam**[a] | Medium | Medium | Medium | Medium | Medium | Medium |

[a]Loam soil averages out "medium" in all respects because it is of variable composition, depending on the actual proportions of sand, silt, and clay particles that are present.

between soil particles and excess water drains off as **gravitational water.** Water that remains between and around the soil particles is referred to as **capillary water,** which is the main source of water for plant roots. After evaporation or absorption of capillary water, what remains is **hygroscopic water.** This water adheres firmly to soil particles and cannot be absorbed by root hairs.

### pH

Low-pH soil is more acid or "sour" and higher-pH soil is more alkaline. Most houseplants and garden plants prefer soil that is slightly acidic. Table 10-3 explains what a pH number (1 to 14) represents.

*pH Sources and Effects.* Soil pH is initially influenced by the parent rock material. Certain light-colored igneous rocks have silica and sodium and potassium oxides and tend to form acidic soil. Soils that are derived mainly from the weathering of limestone, which contains calcium carbonate, are basic.

Soils become increasingly acidic in several major ways. Two of the most important are continued cropping without renovation of nutrient supply and organic decomposition. The products of decomposition include many organic acids (such as lactic, oxalic, acetic). Also, carbon dioxide released by decomposer organisms readily combines with water to form carbonic acid. Thick layers of surface litter, such as on a forest floor, are particularly acidic. Sulfur dioxide in polluted air combines with precipitation to form sulfuric acid, causing "acid rain," which lowers soil and water pH.

Soil pH is one of the most important factors of plant nutrition because it directly affects the availability of certain plant nutrients. Most plant nutrients are available at around pH 6.5 and any deviation to either extreme reduces availability. On the other hand, some nutrient-containing compounds become more soluble with changes in pH.

Soil pH also has an effect on plant disease control. Certain diseases are more virulent in a basic soil environment—for example, root rot of tobacco and potato scab. Fungi that cause damping off of potatoes and club root of cabbage grow well in acid environments. Therefore alteration of soil pH helps to control these diseases. Finally, plants growing where pH and nutrient conditions are optimal for them tend to be more resistant to diseases.

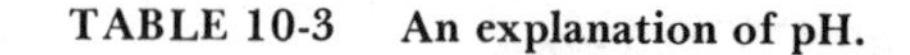

**TABLE 10-3 An explanation of pH.**

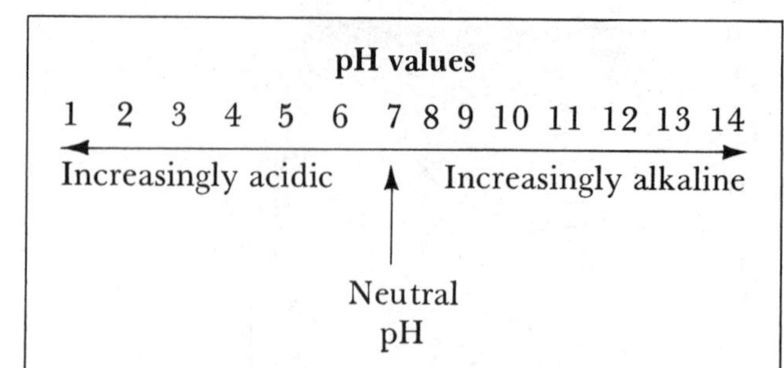


The pH scale is a negative logarithmic scale, expressing the concentration of $H^+$ ions in gram atoms per liter. Each full number differs by tenfold. pH 6 is therefore ten times more acidic (ten times higher $H^+$ concentration) than pH 7; pH 8 is ten times less acidic (more alkaline) than pH 7, and so on.

The concept of pH is frequently confusing. Literally it refers to the potential (p) for hydrogen ions ($H^+$) in a solution. The main source of hydrogen ions is the water molecule, $H_2O$. Water molecules tend to ionize or dissociate slightly, forming positively charged hydrogen ions ($H^+$) and negatively charged hydroxyl ions ($OH^-$). This reaction may be summarized as $H_2O \rightarrow H^+ + OH^-$. When equal numbers of hydrogen and hydroxyl ions are present, a solution is **neutral.** Pure water, for example, has a neutral pH, 7 on a logarithmic scale of 1 to 14. When proportionately more $OH^-$ ions are present, the solution is **alkaline.** When proportionately more $H^+$ ions are present, the solution is **acidic.** The extremes of the scale (near 1 and near 14) are not commonly found in natural environments because such solutions are highly reactive (caustic), being concentrated acids and bases.

A stable pH is essential to normal cytoplasmic activities. In cells fluctuations in proportions of **cations** (positively charged ions) to **anions** (negatively charged ions) are offset by the presence of substances that are able to release cations or anions to maintain a stable pH. These substances are called **buffers.** Cytoplasm is buffered primarily by proteins and amino acids, which can have either a net negative or positive charge. Cell vacuole contents are buffered by organic salts and organic acids (citric, malic, etc.). Buffers are also important factors in maintaining a stable soil solution pH.

*Modifying pH.* Acidic soil (low pH) can be altered by addition of finely crushed limestone (calcium carbonate). The effects of liming soil are complex. In areas of low calcium it acts as a fertilizer. It enhances the availability of some nutrients and decreases the solubility of others. It favors the growth of soil bacteria and limits fungal growth. Although liming a soil has many beneficial attributes, it must be done with care to avoid overliming, which can reduce nutrient availability to the point of causing deficiency symptoms, including chlorosis (loss of chlorophyll).

To adjust basic soils to neutrality or slight acidity, elemental sulfur or sulfates, sphagnum (peat), oak leaf litter, or other acidic compost can be added.

For average garden soils, a desirable pH can be maintained by counteracting acidity through regular restoration of nutrients and periodic liming. Analysis of a soil sample will provide a more precise diagnosis of pH and nutrient levels. Lime (or sulfur) should be applied at recommended levels for a particular area after consultation with a local Agricultural Extension agent.

#### Nutrients

In addition to carbon (C), hydrogen (H), and oxygen (O), other nutrients taken in and utilized by plants in rather large quantities are calcium (Ca), magnesium (Mg), sulfur (S), nitrogen (N), phosphorus (P), and potassium (K), especially N, P, and K. Because all these elements are used in relatively large amounts, they are referred to as **macronutrients.** Although not used in large amounts, iron (Fe) is frequently categorized as a macronutrient because of its early identification as an essential element. A simple saying made up of the symbols of the elements helps students to remember these nutrients: C HOPKNS CaFe Mg—translated, it reads "See Hopkins Cafe, mighty good!"

**Micronutrients** (also called trace elements) were discovered later, using more sophisticated techniques. Micronutrients are elements that are required by plants in small quantities, including boron (B), chlorine (Cl), copper (Cu), manganese (Mn), molybdenum (Mo), and zinc (Zn). Iron (Fe) is also used in trace amounts. We suggest a new mnemonic device for the micronutrients (except iron): CuZn Mo Mn B Cl: "Cousin Molly manages bowling clubs."

Nutrients function as structural components or enzyme activators (cofactors). Carbon, hydrogen, and oxygen, for instance, are found in all organic compounds; nitrogen and sulfur are present in proteins; calcium is a structural component of substances in the middle lamella. **Enzyme activators** temporarily combine with the enzyme structure to enable its participation in a chemical reaction. Calcium, for example, must be present for utilization of ATP. Table 10-4 summarizes common nutrient roles and the consequences of specific deficiencies.

Minerals contribute to the osmotic potential of cells and may directly affect cell membrane permeability. The presence of various cations (positively charged ions) and **anions** (negatively charged ions) also functions as a **buffer system,** maintaining a balanced pH within the cell. (See also Table 10-3.)

Generally plant cells also contain a number of substances that are of no known value to the plant. Some, in fact, may be toxic to the plants or animals that eat them. Examples are such heavy metals as silver, mercury, and lead, plus arsenic, selenium, and aluminum. Most of the micronutrients required for normal growth and development are also toxic in large amounts.

Macro- and micronutrients are present in several forms: in the soil solution, attached to colloidal clay particles, attached to organic colloids, and in certain forms not readily available to plants.

Many plant nutrients in the soil solution are anions, such as nitrate ($NO_3^-$) and sulfate ($SO_4^=$) ions. Phosphate ($HPO_4^=$) ions are easily precipitated by combination with positively charged ions or are attached to certain clays and so are not abundant in the soil solution. In addition, negatively charged ions are more easily lost from the soil during heavy rainfalls and through movement of gravitational water than are cations.

Plants absorb negatively charged ions directly from the soil solution through the root hairs, but absorption of positively charged ions is more complex, involving cation exchange, as described earlier. As mineral cations are removed from the soil, they are replaced by $H^+$ ions. If this process continues without addition of more nutrients, the soil solution continues to increase in $H^+$ ions, becoming acidic.

The use of commercial chemical fertilizers is an integral part of agriculture in the United States today. Fertilizer manufacturing, however, requires energy use and increasing petroleum costs mean greater fertilizer costs. Chemical fertilizers are popular because of compactness of the units, accuracy of ingredient concentrations, and ease of application. Sacks of chemical fertilizers are usu-

TABLE 10-4 **Nutrients essential to plant growth.**

| Essential Elements | Use(s) | Deficiency Symptoms |
|---|---|---|
| **Macronutrients** | | |
| Carbon (C) | structure of all organic compounds | |
| Hydrogen (H) | structure of all organic compounds | |
| Oxygen (O) | structure of all organic compounds; oxidative phosphorylation | |
| Phosphorus (P) | structure of proteins, phospholipids, sugar-phosphates, nucleic acids, ATP, NADP (areas of rapid growth, respiration) | leaves grayish green; dead spots on leaves, fruits; leaves may be malformed; growth stunted; older leaves die; vascular and support tissues weakly developed |
| Potassium (K) | enzyme activator (areas of rapid growth) | mottled leaf chlorosis; leaf tips and margins die; leaves may become curled or crinkled; internodes short; stems weak, maybe with brown streaks |
| Nitrogen (N) | structure of amino acids and proteins (including enzymes), chlorophyll, alkaloids, hormones | older leaves uniformly chlorotic, may die; stunted growth; anthocyanin produced in leaves and stems; stems thin and woody |
| Sulfur (S) | structure of proteins | similar to nitrogen deficiency except that chlorosis first affects younger leaves; chlorosis pale green rather than yellowish or white |
| Calcium (Ca) | structure of middle lamella; maintenance of membranes | leaf tips hook downward; pale green chlorosis; severe stunting as meristematic and rapidly growing areas, such as young leaves, distort, and then die |
| Magnesium (Mg) | structure of chlorophyll; cofactor in energy transfer reactions | chlorosis as for iron, but older leaves first affected; stunted growth; premature leaf abscission; all leaves may become yellowish or white eventually |
| **Micronutrients** | | |
| Iron (Fe) | structure of certain proteins where it is electron carrier in energy transfer reactions | leaves chlorotic except along main veins; younger leaves first affected |
| Boron (B) | normal growth and development | black dead areas involving terminal buds, young leaves, storage roots, tubers, fruits |
| Zinc (Zn) | maintenance of normal auxin levels; enzyme structure or activator in photosynthesis | short internodes; leaves small and distorted; older leaves chlorotic except for along main veins; white dead spots |
| Manganese (Mn) | respiratory enzyme activator | chlorosis between veins, usually on younger leaves, plus many small dead spots |
| Copper (Cu) | present in certain enzyme systems | young leaf tips die, followed by margins |
| Molybdenum (Mo) | enzyme systems re: nitrogen fixation and nitrate reduction | chlorosis between veins of older leaves; leaf margins die; flowers may not form or may drop early; nitrogen deficiency symptoms may also occur |
| Chlorine (Cl) | ionic balance, osmosis; probably oxygen-producing reactions of photosynthesis | leaf tips wilt, become chlorotic, then bronzed, then die back |

ally labeled N-P-K, with numerals like 10-10-10. These symbols indicate the concentration of nitrogen (in several forms), phosphorus (as phosphoric acid, $P_2O_5$), and potassium or potash ($K_2O$). Therefore the numbers 10-10-10 indicate the percentage of each of these three chemicals in the sack and the remainder is filler (Fig. 10-2). Because of the increased costs of packaging, shipping, and storage, fertilizer companies are moving toward higher concentrations in their products. Nitrogen, phosphorus, and potassium encourage different aspects of plant growth (Fig. 10-3).

Organic gardening has experienced a surge in popularity as we have become cautious about possible health effects of herbicides, insecticides, and other agricultural chemicals. Mulching and composting are practices integral to organic gardening; and biological rather than chemical methods of pest control are often adequately effective for the home gardener and plant enthusiast. Many excellent references are available.

It should be noted, however, that such nutrients as nitrates, phosphates, and sulfates are absorbed by plant membranes in inorganic form and membranes do not discriminate between nutrients of inorganic and organic origin. The nutrients are also assimilated within the plant in the same manner so that the assayable food value of "inorganically" and "organically" fertilized plants is the same.

Some primary advantages of organic-style gardening, however, are the holistic emphasis on soil development and replenishment, recycling of organic compounds, freedom from contamination by

**FIGURE 10-2** The numbers on commercial fertilizer bags refer to the relative concentrations of nitrogen (N), phosphorus (P), and potash or potassium (K), the three most important plant nutrients. (Photos by Mark R. Fay.)

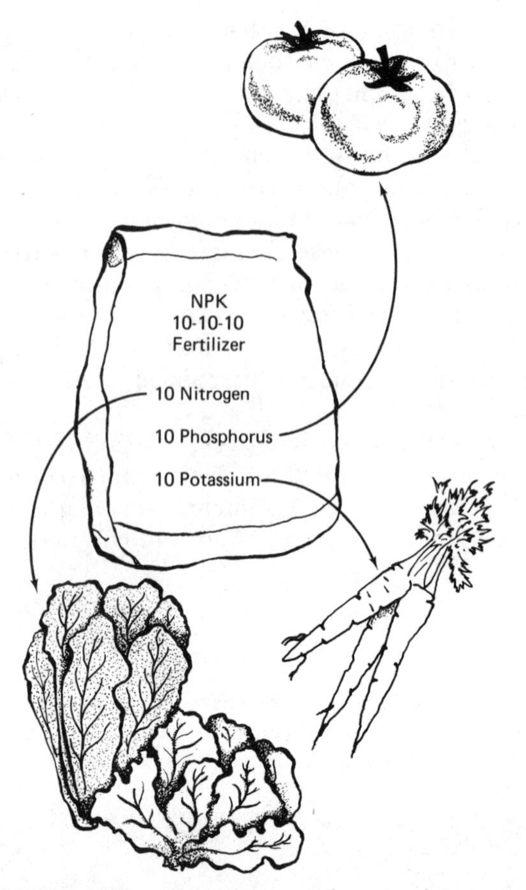


**FIGURE 10-3** The three major nutrients each encourage different aspects of plant growth. **Nitrogen** stimulates foliar growth, **phosphorus** flowering and fruiting, and **potassium** root development.

complex poisons, and freedom from weeding! Organic fertilizers usually contain an array of micronutrients in addition to the macronutrients whereas most chemical fertilizers provide only nitrogen, phosphorus, and potassium.

#### Salinity

Most soils in the United States are sufficiently low in soluble salts (mainly chlorides and sulfates) for successful gardening. Soils of the arid regions of the western United States, however, have a potential salinity problem.

Where rainfall is plentiful, salts accumulated by natural weathering are normally diluted and washed away. Where evaporation exceeds precipitation, however, salts tend to accumulate in the soil surface. (See also Chapter 15.) Irrigation can increase the problem, which has happened in such "fertile desert" areas of southwestern Arizona and California as California's Imperial Valley. Where irrigation water is applied but not allowed to run off, it dissolves mineral salts present in the soil, bringing them to the surface as the water evaporates. The decline of fertility in ancient Mesopotamia—now a desert area of the Middle East—is thought to be due to such salinization.

The same problem can develop with houseplants and home gardens unless there is opportunity for drainage to rinse excess salts out of the soil. Use of chlorinated water and inorganic fertilizers (which contain traces of undesirable minerals as contaminants) increases salinity.

### Soil Air

The soil has many spaces between particles and these pores accommodate air as well as water. The composition of soil air approximates that of the above-ground atmosphere, although concentrations of some components differ, especially humidity. Generally the carbon dioxide concentration is greater in soil air than above ground due to respiration of soil microbes and, to a lesser extent, respiration of plant roots. Associated with this factor is a decrease of oxygen, which is used in the respiration of roots and soil organisms. Most plants absorb oxygen directly from the soil pores by simple diffusion through the epidermis (including root hairs) of the root tips.

Because of the importance of oxygen to soil microbes and respiring roots, farmers and gardeners are careful not to reduce the pore system in the soil by attempting to work the ground too early in the spring. Compressing muddy soil reduces air spaces. In addition, when the compacted soil dries, it is less easily penetrated by roots. As a guideline for average loamy soil, the soil is still too wet if it remains balled after being clumped in your fist. If the soil falls apart after you squeeze it, however, it is dry enough to start working.

Soil compaction is also a problem on building sites, where the surface is leveled with heavy machinery and then further subjected to delivery trucks and heavy stacks of construction materials. If large trees are left on such a site, special care is needed to avoid compacting the soil under the tree (and therefore over the root system). Also, the preexisting soil level must be maintained around the trunk, even if it means building what can be an interesting and attractive "well" around the tree base, to permit adequate soil aeration for the root system.

### Soil Temperature

Because metabolism involves a complex of biochemical reactions that have optimal temperatures for reaction, soil temperature has a direct effect on plant growth. With colder soil temperatures, the growth and absorptive functions of plants decrease until a certain minimum is reached; then these functions come to a near standstill. If the soil temperature falls below freezing, the plant must have adaptations to survive or it freezes and dies. From spring through summer soil temperatures continue to get warmer and the days longer. Beyond an optimum point increasing temperatures may be detrimental to the plant unless it is adapted for growth in high temperatures.

Atmospheric temperature varies considerably during a day whereas soil temperature changes more slowly, remaining fairly level from day to night and during the growing season. Soil temperature can be altered by manipulating the microenvironment, such as by mulching (Chapter 11), thereby influencing the growth rate of cultivated plants.

### Soil Flora and Fauna

During humification and mineralization an imbalance generally exists in the carbon-to-nitrogen ratio because of an overabundance of such plant carbohydrates as cell wall material in relation to nitrogen-containing proteins. (See composting section that follows.) Bacteria, fungi, and green plants all compete for available soil nitrogen. If cultivated plants are deprived of available nitrogen, their growth is limited and it may be necessary to add such high nitrogen-containing materials as manures and chemical fertilizers to get maximum productivity.

Soil fertility depends on the abundant presence of certain bacterial species that participate in the stepwise conversion of atmospheric nitrogen into soil nitrates that are essential for plant growth. (See also nitrogen cycle, Chapter 2.)

Many kinds of small animals are present in the soil, including protozoans, nematodes, crustaceans, mites, adult and larval insects, and the ubiquitous and beneficial earthworm. It has been reported that as much as 25 tons of material are passed through an earthworm and deposited on the surface of one hectare (almost $2\frac{1}{2}$ acres) of soil in a year. Worms, insects, and burrowing mammals mix the soil and provide air spaces for better soil conditions. Even though considered pests, moles and gophers are beneficial to the soil ecosystem because of their aerating and mixing activities.

### Composting

The importance of humus in improving soil texture, water retention, nutrient retention, and various specific physical and chemical soil properties cannot be overemphasized. The humus content of a soil can be increased by cultivating in such basic materials as manure and straw to undergo natural decomposition. Although suitable for field crops, it is not as useful for a flower bed or home garden because it introduces relatively large fragments of material to work around during cultivation, planting, and weeding.

The most refined means of adding humus is as compost. **Compost** is a lightweight, spongy soil and humus mixture high in organic content. The process by which it is produced is called composting, an old and proven method of reducing undesirable organic debris into a useful end product that provides nutrients and ensures soil quality. The variety of raw materials for composting is endless, including lawn clippings, leaves, sawdust, wood chips, bark chips, sludge, garbage, animal manure, hay, straw, ashes from fireplace and grill, shredded newspapers, animal waste, vegetable trimmings, egg shells, and even blood and bone meal. The basic methods of composting are compost pile, pit composting, and sheet composting.

#### Compost Pile

Although you can start with just a "pile," composting is more efficient in an enclosed compost pile. An enclosed compost pile is also less likely to attract dogs, skunks, and other animals. Elaborately designed, efficient prefabricated models are available, but an economical homemade enclosure works well. Cement block, boards, snow fencing, and hardware cloth are commonly used. The size can vary from that of a wire trash burner to large bins approximately 4 by 4 ft.

Important factors in efficient composting are the nature of the organic litter, oxygen, moisture, temperature, microorganisms that will do the deterioration, and protection from excess rain, which will leach out nutrients. Litter fragments should be small, for large units (e.g., branches) require too long a period to deteriorate.

A compost pile begins and continues its existence in layers (Fig. 10-4). Start with approximately a foot of litter, sprinkle on a handful of

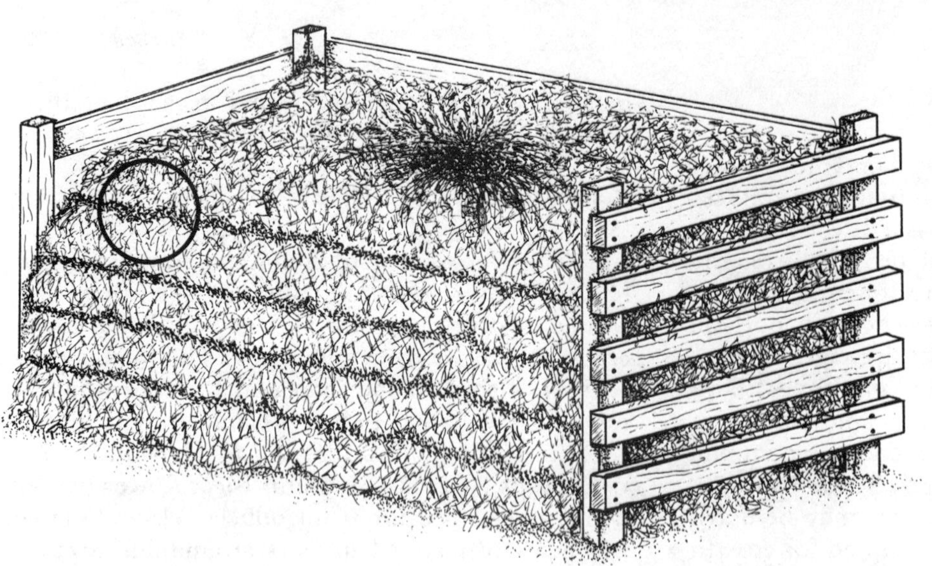

**FIGURE 10-4**
A compost pile requires carbon and nitrogen sources, moisture, and oxygen. Soil is added as a source of decomposer organisms and so that liberated nutrients can be incorporated into soil instead of washing away. The diagram shows an idealized compost pile, having organic litter (high-carbon) alternated with thinner layers of soil and manure or chemical fertilizer (high-nitrogen). The indentation in the center traps precipitation. The approximately 4-by 8-by 4-ft dimensions and slatted framework permit gas exchange to occur.

garden fertilizer or a layer of manure, and then add several inches of soil. Repeat layering sequence, leaving a lower area in the center to catch snow or rain to maintain moisture. Continue layering until the pile is waist high. It is a good idea to start a couple of piles to have compost in different stages of maturation. Compost piles may take from 2 weeks to 3 months to turn litter to usable compost, depending on the type of debris used, temperature, and other environmental variables.

Because of great bacterial activity during decomposition, much energy is released as heat, causing the temperature of the compost pile to increase (Fig. 10-5). The temperature will register 40°C in smaller piles and up to 70°C in larger piles. As the temperature increases, the bacterial populations gradually change from predominantly moderate-temperature species to those that can tolerate and thrive in higher temperatures. Pathogenic bacteria and fungi, weed seeds, and the eggs and larvae of parasitic nematodes (especially important if fecal material is used) are killed by the high temperatures. Enzyme activity is slowed by colder temperatures, which is why composting takes much longer during winter.

As the compost pile ages, there is also a change in pH. Initially the pH is slightly acid from the litter material itself. With bacterial decomposition the pH decreases further. In the latter stages of composting the pH increases to an alkaline range of 7.5 to 8.5, which may be attributed to ion release.

Oxygen is needed for aerobic respiration. Therefore the pile must be turned and mixed at frequent intervals or bacterial action decreases and the speed of composting is retarded.

Nitrogen sources must be added during composting to maintain the **carbon:nitrogen ratio**, an

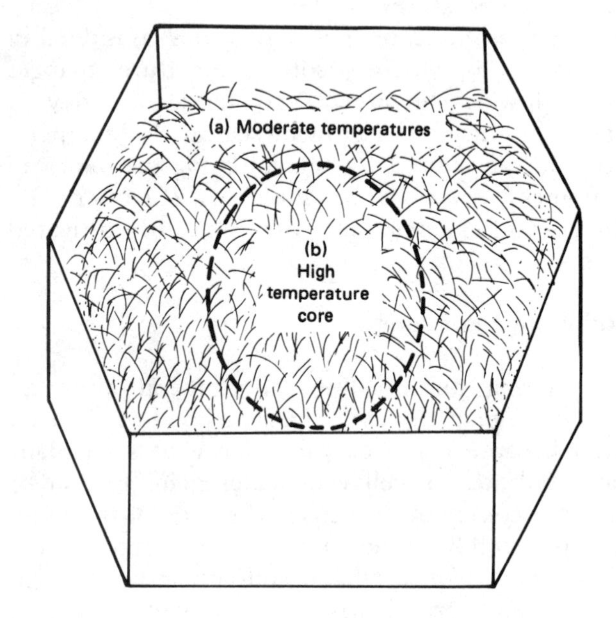


**FIGURE 10-5** Compost pile dynamics. To keep composting proceeding actively, renew oxygen and raw materials throughout the pile by turning and spading once a week—more often in hot weather. This step accommodates different decay organisms that operate at different temperatures. (a) In the outer layers, coarse materials are broken down at moderate temperatures, a relatively slow process accomplished mainly by fungi and mycelial bacteria (actinomycetes). (b) In the center of the pile more easily rotted matter (as that partially decomposed in the outer layers) deteriorates rapidly at high temperatures by action of heat-tolerant bacteria. This phase can be severely limited by lack of oxygen.

integral soil phenomenon in both natural and agricultural soil ecosystems. Most compost litter is plant material, high in carbohydrates (especially cellulose) but low in nitrogen. Bacteria utilize carbohydrates for energy but must have nitrogen to build proteins for new bacterial cells. If too little nitrogen is available, microorganism growth and thus the rate of composting are reduced. Fresh manures (especially of poultry), urine-soaked manure and straw, or such animal wastes as fish carcasses and innards are also sources of nitrogen. Their odors may limit their desirability unless the compost pile is distant from any residences! For many people, chemical fertilizers are the most convenient way to add nitrogen (as well as phosphorus and potassium).

Another advantage of adding fertilizers to compost is that nutrients that become integrated into compost are lost less rapidly by runoff than those applied "bare" to the soil. The compost nutrients are already a part of the colloidal soil system when applied.

#### Pit Composting

This is one of the easiest methods of composting and is a fast way to dispose of such daily kitchen garbage as coffee grounds, peels and rinds, egg shells, and vegetable and animal scraps. Simply dig a small hole in the garden or flower bed, put the material to be composted in the hole, and then cover it over with soil. You may need to add a nitrogen source to prevent the compost-creating microbes from competing with cultivated plants for available nitrogen.

Pit composting requires less effort than a compost pile and the process is completed in a very short time, depending mainly on the size of the raw materials. Because you dig a new pit each time, the compost becomes well distributed throughout the garden or flower bed, without any additional work. Pit composting can be a rapid way to improve small areas of problem soil.

#### Sheet Composting

Sheet composting is done in the fall, after the garden has been harvested, or in the spring prior to cultivation for planting. Organic materials are simply spread out on the garden plot and allowed to decompose for a time; then they are thoroughly tilled into the soil. Rotted hay is often available from farmers not interested in rain-soaked and decaying feed for their cattle. (Nondecomposed hay would introduce weed seeds.) Feedlots have a chronic oversupply of manure. Straw and manure used for livestock bedding, lawn clippings, and raked leaves are other "free" raw materials. Your neighbors may be willing to "donate" garbage bags full of autumn leaves to you rather than pay to have them hauled away.

One note of caution: it is important to know that certain leaves produce a characteristic type of humus. Conifer needles, for example, contain resins that are slow to decay, are very low in calcium, magnesium, and potassium, and produce an acidic compost. You must add fertilizers to supplement these minerals and lime (calcium carbonate) to offset the acidity. Cedar and juniper litter, however, contains relatively high amounts of minerals lacking in most conifers. Most deciduous tree leaves are nonacid forming, contain few resins, and are fairly high in calcium, magnesium, and potassium. There are exceptions. Oaks, although deciduous, have leaf litter that is low in these elements and is more acidic.

## SEEDS TO SEEDLINGS

The soil, as we have discussed its development, is the environment into which a seed falls. In a natural ecosystem the seed must compete with established vegetation for a place to sprout, a struggle that the newly germinated seedling must continue, competing to become established in the soil before the sustaining embryonic nourishment is depleted.

To understand the enormous adaptive significance of the seed in these critical early stages of development, it is necessary to examine more closely how seeds are formed and how the plant makes the transition from embryo to independent seedling.

### Seed Structure

The seed is the end product of the life of a plant. Stated another way, flower production is directed toward continuation and perpetuation of a line of organisms. The seed is well adapted for its responsibility. Most seeds have a tough seed coat for protection, a miniature embryonic plant, a supply of food to allow the embryo to get started, and a set of growth-promoting and growth-inhibiting hormones to control seed dormancy and germination.

The arrangement of food storage material and

the embryo itself differ among various seed plants. In some species the embryo is surrounded by a special nutritive tissue called **endosperm.** Examples of this arrangement are the grasses, such as corn and wheat, conifers, such as pines and firs, and most dicotyledonous plants. In legumes, such as beans and peas, however, most of the endosperm is digested and assimilated by the developing embryo for storage within its own tissue, the seed leaves, or **cotyledons.** The cotyledons are the main edible portion of a peanut, for example.

The bean (*Phaseolus*) (Fig. 10-6) is an excellent example of legumes, a significant food source for most world civilizations, ancient to modern. The seed surface is covered by the **seed coat.** On the seed coat is the **hilum,** a scar where the seed was attached to the inside of the fruit or pod (Fig. 10-7). Above this scar is a pinpoint hole, the **micropyle** (Fr., *micro* = small; Gr., *pyle* = gate), which is the natural opening that received the pollen tube prior to fertilization (Chapter 5). In the open bean seed the largest structures are the cotyledons. Beans have two cotyledons and you will recall that this characteristic is used to separate angiosperms into the dicots (two cotyledons) and the monocots (one cotyledon). The rest of the embryo is cradled between the cotyledons and consists of the **epicotyl, hypocotyl,** and **radicle.**

Corn (*Zea mays*) and wheat (*Triticum aestivum*) are examples of the monocots. The grain is actually a fruit, called a **grain** or **caryopsis,** in which the fruit coat (pericarp) adheres tightly

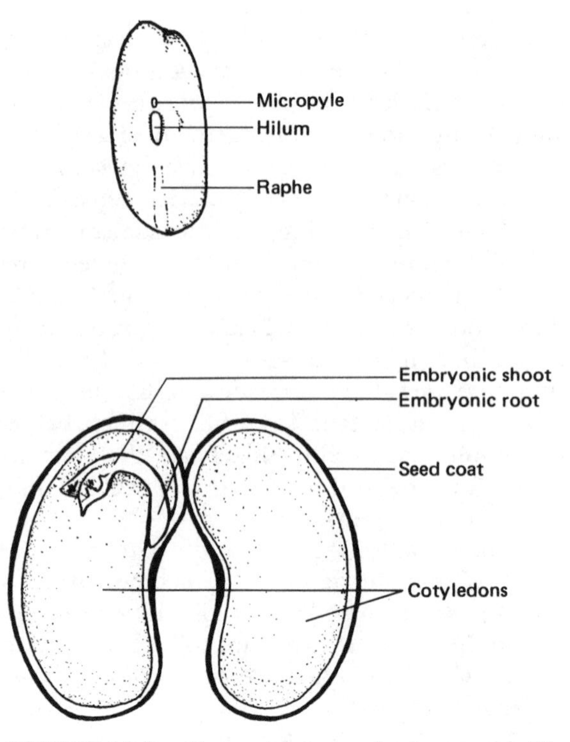


**FIGURE 10-6** Gross structure of a bean seed. The **hilum**—shown in greater detail in Fig. 10-7—marks the former point of attachment to the seed pod. The **micropyle** is located at one end of the hilum, and a ridge called the **raphe** extends along the midline of the seed from the other end of the hilum. (See also Fig. 10-10.)

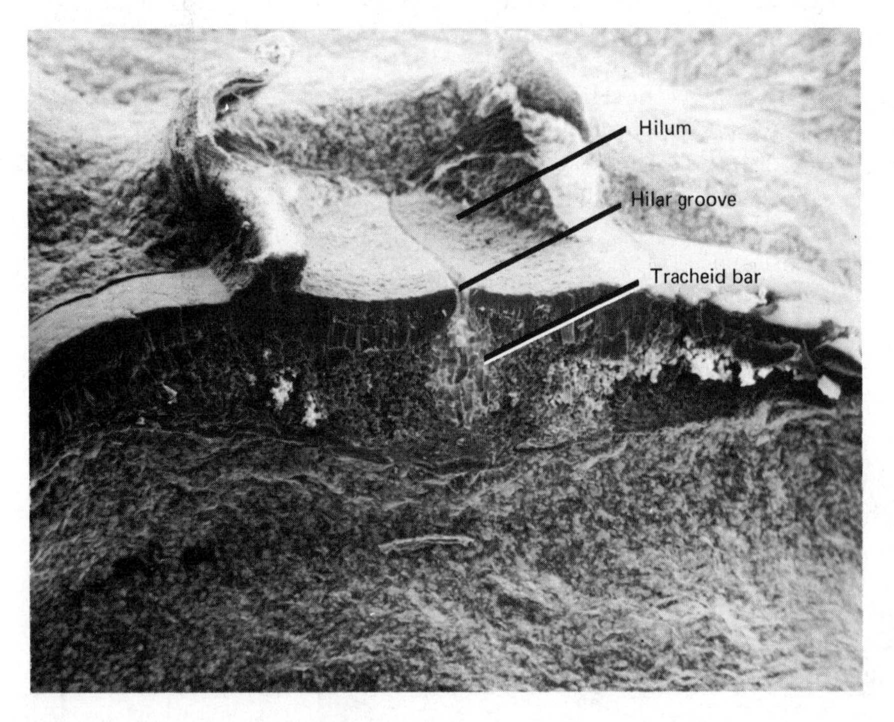


**FIGURE 10-7**
Beans, peas, soybeans and the rest of our most common herbaceous legumes are classified as a major subgroup, the papilionoid legumes, within the family Fabaceae. In members of this group the seed **hilum** has a slit, the **hilar groove,** that is underlain by a **tracheid bar**—a region of tracheids that acts as a hygroscopic "valve" to help the seed dehydrate. When the atmosphere is dry, the tiny groove is open and water evaporates from inner seed tissues through the tracheid bar. When rain or dew is present, the groove swells shut, preventing water from soaking into the seed, which might stimulate premature germination. Note the thickness of the essentially water-impermeable **seed coat.** This is a scanning electron microscope (SEM) view of a pigeon pea (*Cajanus cajan*) seed. (Photo courtesy of Nels R. Lersten.)

to the seed coat. On the surface is a bulge where the style (corn silk) was attached. In longitudinal section (Figs. 10-8 and 10-9) it is easy to see the single cotyledon, or **scutellum,** which is principally an organ for absorbing food from the endosperm for the embryo. The starch-filled endosperm is well developed and a unique layer of nutritive cells, the **aleurone layer,** lines the outer limits of the endosperm. The aleurone is rich in proteins and fats. After stimulation by gibberellic acid (Chapter 9), the aleurone releases enzymes to digest food stored in the endosperm. An interesting feature of grains is the presence of protective sheaths around the embryonic shoot and root, the **coleoptile** and **coleorhiza,** respectively. The coleoptile is an important site of auxin production (Chapter 9).

**FIGURE 10-8** Longitudinal section of a corn grain, or **caryopsis.** Corn is a monocot. As in other cereal plants (grasses), each caryopis is a fruit-and-seed combination having the ovary wall fused to the seed coat. Compare the visible structures with Fig. 10-9. (Photo courtesy of Carolina Biological Supply Co.)

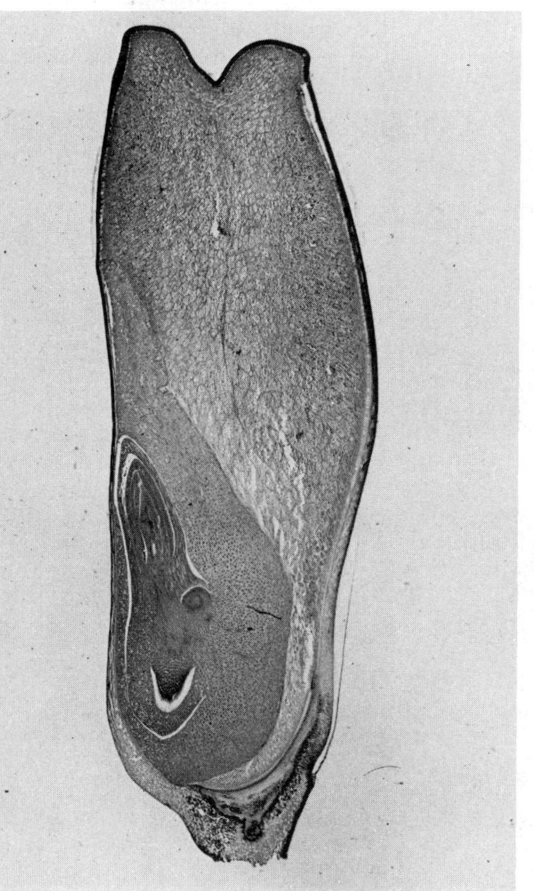

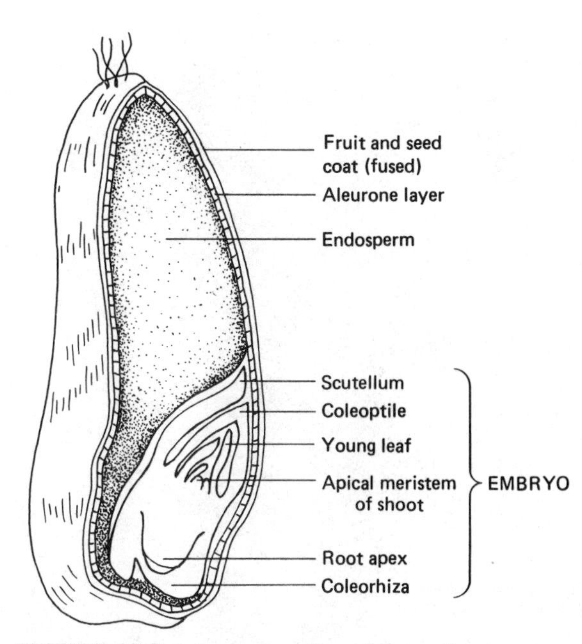


**FIGURE 10-9** Anatomy of a wheat grain, a monocot. The **aleurone** layer is involved in the embryo's utilization of starch stored in the endosperm. The **embryo** part of the grain is the wheat "germ." The covering layers constitute the "bran."

## Germination

### Requirements for Seed Germination

The seed is a nicely packaged unit. It will not germinate until certain environmental factors are present, including favorable moisture, oxygen, temperature, and, in some cases, light. In addition, chilling or leaching may be necessary to remove inhibitory substances and release the seed from dormancy.

Moisture is essential to hydrate the seed (Fig. 10-10). Imbibition of water allows enzymes in the seed to activate, initiating breakdown of stored foods to provide energy and building materials for the embryo. As water is imbibed, the seed swells, breaking the seed coat, which may be hard. As aerobic respiration increases, there is an increased oxygen demand, fulfilled by air in the soil. This is why muddy and compacted soil may retard germination and early seedling growth.

Most plants have an optimum temperature at which germination is most active. Some plants—peas, lettuce, and radishes, for instance—are adapted to early spring germination and will germinate in temperatures around 10°C. Most garden plants germinate later and prefer temperatures

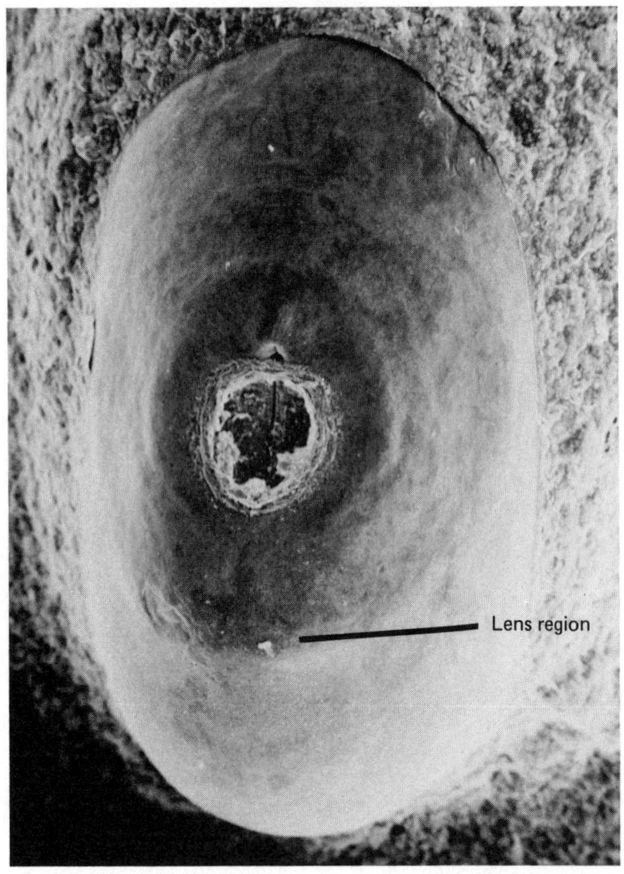


FIGURE 10-10 A specific water imbibition site—the **lens** or strophile—has been identified in the seeds of papilionoid legumes (Fig. 10-7). In this SEM view the lens is visible as a bulge at the end of the darker gray raphe. Epidermal cells in the lens are three to four times longer than elsewhere, hence are a little weaker and tend to swell and separate during imbibition. Note also the micropyle and the hilum with its hilar groove. In the past the hilum was hypothesized as the imbibition site. The seed shown is that of prairie trefoil (*Lotus purshianus*), and is representative of many common legumes. (Photo courtesy of Nels R. Lersten.)

around 20 to 30°C. No matter how early they are planted, seeds of many plants will not germinate until their minimum environmental requirements are met. Certain grasses, weeds, and lettuce also require seed exposure to light for germination.

There are different patterns of germination, which are represented by the following examples.

#### Germination Sequence of the Bean and Pea

With favorable germination conditions, bean and pea cotyledons become hydrated and biochemical processes are activated.

The immediate problem of survival for the young plant is to penetrate the soil and begin to grow and absorb water. The first organ to emerge from the bean seed (Fig. 10-11) is the primary root. It protrudes through the micropyle and grows down, exhibiting positive geotropism (Chapter 9).

The **hypocotyl** (*hypo* = below) is the embryonic stem area below the cotyledon. It begins to grow, pushing apart the cotyledons and bending upward to form a **hook** [Fig. 10-12(a) and (b)]. The hypocotyl hook is extremely important in breaking through the soil, allowing the seedling to emerge into the atmosphere with minimal damage to the meristematic shoot tip. Once it has emerged, the hook straightens, the cotyledons separate, and the **plumule** (seeding shoot) grows, beginning with enlargement of the first foliar leaves.

FIGURE 10-11 Germination of bean seeds, showing the **primary root** emerging from the seed. Note that the root responds to gravity (positive geotropism), regardless of the orientation of the seed in the soil.

(a)

(b)

**FIGURE 10-12**
(a) Germination of bean seeds, showing the formation of a **hook** to break the soil while the delicate stem apex and first true leaves are protected between the cotyledons. (b) As the seedling emerges, the hook straightens out and the **plumule** expands. The cotyledons wither as their stored food is used.

The cotyledons continue to provide energy while the seedling becomes established, gradually withering as stored food is depleted. During this early growth the cotyledons turn green, indicating formation of some chloroplasts for photosynthesis; however, cotyledon photosynthesis is considered to produce only a negligible amount of food for the plant. The true leaves expand and take over food production and the cotyledons eventually fall from the plant.

In peas the process is similar to beans but differs in that the hypocotyl does not elongate. Instead the **epicotyl** (*epi* = above), the region *above* the cotyledon, elongates. As a result of growth above, the cotyledons remain below the soil surface. The epicotyl develops a hook prior to emergence to provide thrust to break the soil After emergence, the epicotyl straightens up and foliar growth begins.

As you can imagine, soil texture has a critical influence on the physical acts of root penetration and shoot emergence. So it is essential to keep the soil as friable (easily crumbled) as possible to facilitate germination.

FIGURE 10-13 Emergence of the **primary root and shoot** from a germinating corn grain. Note that two **adventitious roots** have already grown out from the immature stem region. They and other adventitious roots form the corn plant's fibrous root system.

#### Germination Sequence of Corn

As the grain is hydrated, the embryo synthesizes gibberellin, which diffuses through the endosperm to the aleurone layer. Gibberellin stimulates aleurone cells to produce a starch-digesting enzyme, **amylase,** and other enzymes that break down proteins and lipids stored in the grain. This process has a commercial application in the brewing industry. Amylase is required to break down starch in barley to the sugar maltose, a process called "malting." Maltose is then converted by yeasts into alcohol. Today additional gibberellins are administered to increase amylase production, saving time and money.

The products of digestion are available to the developing embryo, fueling growth. Soon the primary root emerges from the grain, moving into the soil (Fig. 10-13). Next, the plumule emerges from the grain, sheathed within the coleoptile. At its apex the growing coleoptile produces an auxin. The auxin is transported from the tip to the base, inhibiting plumule and adventitious root growth. It becomes dilute enough, however, to permit elongation of the **mesocotyl,** the stem region between the scutellum and coleoptile, pushing the sheathed plumule up and through the soil surface (Fig. 10-14). As the coleoptile reaches the surface, exposure to light decreases auxin production

FIGURE 10-14 Germinated corn, showing the emergence of the plumule through the hormone-producing sheath, the **coleoptile.**

so that growth of the plumule and adventitious roots is stimulated. Because of this interaction, the mesocotyl is very short in shallowly planted grains but longer in more deeply planted grains.

#### Germination Sequence of the Onion

Onions and lilies demonstrate another pattern. First the primary root emerges to penetrate the soil. Next, the tubular cotyledon forms a hook, moving upward to break the soil. The cotyledon starts to straighten up after emergence and often pulls with it the seed coat, clinging to the tip of the cotyledon and containing some endosperm. Eventually the seed coat falls away. The plumule emerges through a small opening at the *base* of the cotyledon.

## STARTING SEEDS INDOORS AND TRANSPLANTING

The best way to study germination and early seedling growth is to plant some seeds! Many plants can be propagated from seed and then be transplanted to a flower bed, garden, or container. For most people, it is easier to buy a few started plants than to start their own. Starting seedlings indoors, however, is fairly easy if a few basic principles are followed and it does have several potential advantages. It allows a greater selection of varieties than are available among commercially available started plants. It may be cheaper if many plants are wanted. Finally, it is personally satisfying, and provides opportunity for close observation of the early growth phenomena described here.

Before beginning, establish some guidelines. Find out the usual dates for setting out transplants in your area and plan to start seeds indoors 6 to 10 weeks prior to the date. List the varieties that you plan to start and the number of days until transplanting. Select containers and a sterile seed-germinating medium. Finally, but critically, be sure you have adequate light—south-facing windows or appropriate plant-growing lights. (See also Chapter 19.)

The germinating medium must drain well and yet hold sufficient moisture, such as milled peat or the inorganic media, sand, perlite, and vermiculite (Chapter 19). Each material has advantages. Vermiculite's moisture-retaining qualities and light texture are good for growing from seed. Also, tender roots are less likely to be torn during transplanting from vermiculite or milled peat than from the heavier media. Individual pressed-peat blocks that expand with moisture are ideal but are more expensive than bulk media (Fig. 10-15).

Absence of pathogens is essential. Commercially purchased media are usually sterile. Nonsterile (such as previously used) media should be heat treated or drenched with a fungicide solution to eliminate bacteria and fungi. Chapter 19 describes a home method for heat sterilization. **Damping off** is a fungal disease that mainly affects young seedlings but can also attack older plants. A thinner region ("wire stem") appears on the stem

**FIGURE 10-15**
The compressed peat pots on the left appear as they are purchased. The center pots have been moistened and a seed has been placed in each one. The pots on the right contain newly germinated seedlings—note the large cotyledons. (Photo courtesy of David Koranski.)

at and just above ground level; then the infected seedling falls over and dies. The disease is caused by the molds *Rhizoctonia* and *Pythium* and there is no cure once the seedlings are afflicted. If damping off occurs, you may be able to save uninfected seedlings by transplanting them into a fungicide-treated medium. Drenching the soil with a fungicide before planting, however, is the best preventive measure.

Any convenient container that provides drainage and aeration can be used (Fig. 10-16). Among the most efficient are wooden seed flats, such as those used in greenhouses. They are correctly proportioned, can be sterilized along with the soil they contain, and can be reused. Peat pots have several advantages, allowing a young plant to germinate and complete its early growth in one container, which is biodegradable and can be placed directly into a new pot or into the ground. The root system need not be disturbed by transplanting.

Place the germinating medium into containers to approximately 3 cm from the top and dampen it moderately. Plant seeds according to size, with medium to large seeds planted three to four times their diameter deep. Small to fine seeds should be pressed lightly into the surface. Be generous with initial seeding, with plans to subdivide to individual containers or to thin mercilessly.

Cover the containers with clear plastic. Fasten the edges loosely or poke small holes to allow excess heat and moisture to escape and to allow gas exchange. There will be some, but not excessive, condensation on the inside of the plastic.

Warm soil temperature enhances germination and most seeds do best at 21 to 24°C (70 to 75°F) as bottom heat. Place the newly seeded container where it will receive gentle bottom heat, such as on top of a refrigerator. Light is not essential during this early phase, except for lettuce.

From the beginning uniform soil moisture and proper drainage are critical. After the seed begins imbibing water and the growth process begins, germination is irreversible. Also, the tender young roots and root hairs have no defense against drying and a single drought can kill the seedlings. On the other hand, overwatering lowers oxygen availability to roots and encourages fungal and bacterial growth.

Depending on species, germination may require 2 days to several weeks. After the seeds have germinated, move the containers to a bright, cooler area, for low light and high temperatures will cause etiolation. Most seedlings do well at 18 to 24°C (65 to 75°F) day temperatures, and 15 to 18°C (60 to 65°F) night temperatures. Before moving the containers into full light, remove the plastic cover (to avoid cooking the seedlings) and apply a mild fertilizer solution.

Seedlings quickly become etiolated if they do

**FIGURE 10-16** A variety of small containers for starting seeds and for initial transplanting. (Photo courtesy of David Koranski.)

not have sufficient light and must be placed in a bright south-facing window or under artificial lighting. Seedlings also exhibit exaggerated phototropism; turn the containers as often as necessary to keep the seedlings straightened out. Periodically adjust artificial fluorescent light to keep it within 10 to 15 cm of seedling tops. Ordinary incandescent light is too hot for this use. Initially, 24-hour lighting enhances growth, but after true leaves develop, adjust the light to 15 to 60 cm above them for 12- to14-hours duration. This lighting schedule will also induce floral bud formation. Etiolation occurs if the light source is too distant whereas burning or yellowing of top leaves means that the light is too near. Continue to use a mild fertilizer solution.

Seedlings will quickly outgrow their germination containers. Thin and transplant seedlings after two sets of true leaves have formed in order to permit vigorous growth with reduced competition. Use good potting soil and start with small pots, moving to larger containers only when each plant has outgrown its previous pot.

To transplant, soak the seedling or rooting flat. Partially fill small pots with premoistened potting mix. Lift a seedling or patch of seedlings from the flat with a fork, keeping a wad of germinating medium attached to the roots. Don't let the root system and medium dangle; otherwise it will break small "feeder" roots. To avoid inadvertently crushing tender seedling stems, you may handle a seedling by a leaf. Gently separate the seedlings (a kitchen fork is helpful) and plant each one with the germinating medium still attached. Add potting soil to the desired level—allowing room for watering—then water thoroughly. Wilting will probably occur because of root damage, so protect the plants with a clear moisture-retaining cover and keep them out of direct light for several days. Once they have recovered, fertilize lightly.

A week or two prior to transplanting flower and vegetable plants in the ground, set the seedlings outside in indirect light during the more pleasant parts of the day. This step permits a process called **hardening** during which carbohydrates are accumulated, toughening the plant to withstand adverse conditions. Reducing temperatures indoors and limiting water prior to setting out of doors will facilitate hardening.

When garden plants have two to six true leaves, they are ready to be transplanted to the soil. Just prior to transplanting, water the plants well. Lay your mulch, if used (Chapter 11), and dig the holes, spacing as recommended. Place a handful of compost, a little soil, and some water in each hole. Gently separate and transfer the seedlings as described earlier. Most plants should be planted slightly deeper than the original soil line of the seedling. Fill in the space around the roots with soil; then soak well.

After the transplants are in, place a collar, approximately 5 cm high, around each plant—pushing it into the ground slightly—to protect each seedling from cutworms. Cutworms are insect larvae that chew seedlings off at ground level. Sections of plastic soda straws slit lengthwise make very effective collars; stiff paper may also be used instead. In areas where evening temperatures can drop to damaging levels, protective "caps" may be needed.

After all the careful observation and tender, loving care, your seedlings are now subject to the natural variations of the out-of-doors environment. Their transition from seedlings to reproductively mature plants is described in the following chapter.

## SUMMARY

**Soil** is the weathered surface layer of the earth's crust, composed of inorganic matter, humus, and living organisms. **Topsoil** consists of humus and inorganic particles. Below it lies the **subsoil**, several feet thick, of weathered parent material with little or no humus. **Parent material**, or rock, is the lowest stratum. The nature of both parent material and **transported components** influences soil characteristics.

Soil **texture** is based on inorganic particle size (silt, clay, sand, gravel) and on the amount of humus present. Fine clay particles participate in **cation exchange**.

Soil fungi and bacteria turn plant and animal remains into **humus**, partially decomposed organic matter having complex water-holding and **colloidal** properties. Being colloidal, humus retains valuable cation nutrients.

Soil water is **gravitational**, **capillary**, and **hygroscopic**. Capillary water beween soil particles is the main source of water for plants.

Nutrients are elements used as structural components or enzyme activators (cofactors). **Macronutrients** are used by plants in large quantities. **Micronutrients** are also essential but in lesser quantity. Most commercial fertilizers contain only nitrogen, phosphorus, and potassium.

Soil **pH** depends on the relative amount of $H^+$ ions in solution—a higher proportion of $H^+$

ions creates an acidic solution whereas a lower proportion creates an alkaline solution. Soil pH affects nutrient availability and soil organisms.

Liming, the addition of crushed limestone, decreases soil acidity. The addition of sulfur, sulfates, and acidic compost increases soil acidity. Most plants prefer neutral to slightly acidic soil.

Excess **salinity** is a problem in some agricultural soils.

**Air** in the soil provides oxygen for microbes and plant roots. Compaction of wet soil can result in oxygen deprivation.

Soil **temperature** directly affects plant growth because it affects the rate of biochemical reactions.

**Soil life** includes the microbes that are essential to soil fertility, many invertebrate animals, and some burrowing vertebrates.

**Compost** is a lightweight, spongy mixture of soil and humus. Composting is a means of converting organic debris to usable form, such as in an enclosed compost pile with successive layers of litter, manure or garden fertilizer, and soil. Heat produced by fungal and bacterial respiration kills weed seeds, parasites, and pathogens in the compost pile. The pile must be aerated and nitrogen must be added to enhance bacterial growth.

A **seed** is an embryonic plant and its food supply enclosed in a protective seed coat. The food supply may be present as **endosperm** tissue or incorporated into **cotyledons**, the seed leaves.

Beans are dicotyledons, having two seed leaves. Corn is a monocotyledon, having a single cotyledon.

Seeds require proper amounts of specific environmental factors to **germinate**. In addition, growth inhibitors that are responsible for seed **dormancy** must be removed.

Imbibition of water activates germination. The embryonic root of beans and peas grows downward in response to **positive geotropism.** A **hook** in the embryonic shoot penetrates the soil while protecting the fragile shoot tip. The cotyledons wither as the food is used up and the true leaves of the seedling begin to photosynthesize.

Within a corn grain, the embryo synthesizes gibberellin, which stimulates the **aleurone layer** to produce amylase. The latter digests starch stored in the endosperm. The shoot is protected by the **coleoptile,** which also exerts **apical dominance** through auxin production. After the coleoptile is exposed to sunlight, it produces less auxin, permitting the growth of the leaves and adventitious roots.

Seeds are easily propagated indoors in a sterile medium that drains well and yet holds moisture. Gentle heat, uniform moisture, humidity, and air are necessary for germination to occur. Once germinated, seedlings need strong light and moderate temperatures. Seedlings exhibit strong positive phototropism. Transplanting proceedures are discussed.

## SOME SUGGESTED READINGS

Brady, N.C. *The Nature and Properties of Soils,* 8th ed. New York: The Macmillan Co., 1974. A standard textbook of soil science.

Foth, H.D., and L.M. Turk. *Fundamentals of Soil Science,* 5th ed. New York: John Wiley & Sons, 1972. A good introduction to the basics of soil science.

Kozlowski, T.T. (ed.) *Seed Biology,* Vols, I, II, and III. New York: Academic Press, 1972. Volume I emphasizes the importance of the seed, its development, and seed germination. Volume II deals with germination control, metabolism, and pathology. Volume III deals with insects, seed collection, storage, testing, and certification.

Rodale, J.I. (ed.) *The Complete Book of Composting.* Emmaus, PA.: Rodale Press, 1960. A compilation of articles from 18 years of *Organic Gardening and Farming Magazine.*

Russell, E.W. *Soil Conditions and Plant Growth,* 9th ed. New York: John Wiley & Sons, 1961. The comprehensive treatise on the effect of soil on plant growth.

Shewell-Cooper, W.E. *Compost Gardening.* New York: The Macmillan Co., 1975. A delightfully written introduction to composting for the amateur gardener.

Thorne, D.W., and M.D. Thorne. *Soil, Water and Crop Production.* Westport, CT.: AVI Publishing Co., 1979. Examines large, uncontrollable environmental factors that affect plant growth, manipulation of environmental factors that can be managed, and present and potential systems to improve crop production in the world's major ecosystems.

# chapter eleven CULTIVATED GROWTH

## FROM SEEDLING TO HARVEST

## INTRODUCTION

The previous chapter described seed structure, germination, and early growth. In this chapter we continue the discussion through the juvenile and mature growth phases, ending in seed and fruit production. The discussion is set in the context of growing cultivated plants, emphasizing controls and alterations of the natural environment that are essential to obtain optimum results. The botanical *reasons* for specific practices are emphasized.

## GARDENING SUCCESS: APPLIED BOTANY

Gardening provides an enormous sense of accomplishment. The idea of maintaining a home yard for lawn only is quickly dissipating in favor of a mixed composition of ornamental flowers and shrubs, lawn, and food plants. Many people have encroached on their leisure lawn and flowers so that their yards are actually vegetable and fruit gardens "dressed" with flowers, lawn areas, aesthetic berry-producing plants, and dwarf fruit trees. Apartment dwellers lacking ground-level sites often have small balcony, patio, or roof-top gardens.

If you do not have your own plot of land, window box, or patio pots, there are many ways to gain access to a site for growing flowers, herbs, and vegetables. Friends may offer to share their property for a "community garden." Many towns now have local park areas set aside for plot rentals and farmers on the perimeters of cities increasingly advertise garden plot rentals. Some cities set aside corner lots for neighborhood gardens; and some colleges and universities provide garden plots for students and staff to use.

The ideal cultivation site is flat and open with at least 6 hours of full sunlight daily. Because most parts of the United States have a rather dry summer growing season, ready availability of water at the garden site is important, especially in late summer when the harvest is peaking.

The preferred soil for most garden and ornamental plants is loam, composed of approximately equal proportions of sand, clay, silt, and humus (Chapter 10). All native soils, however, can produce vegetables, fruits, flowers, and herbs if modified appropriately. The better the soil, light, nutrients, and water, the more productive the garden will be, but you start with what you have and improve from there.

Any gardening effort, whether a meter-wide strip bordering a sidewalk or property line, a 2-by 3-m raised bed, or a hectare garden plot, should receive some planning and forethought. The first step is to match all the growing conditions of the site to the best possible plant varieties for that type of site. This planning stage can be comfortably accomplished indoors during winter.

### Analyzing the Site

Make an outline of the space at hand. If graph paper is available, be accurate by allowing each square to equal 0.5 m of area to be gardened. Note on the paper north, south, east, and west to clarify in your mind the direction of the sun. Be sure to fill in the influence of buildings or trees that surround the garden plot (shade and extension of roots into the garden). Identify the location of available water. Early vegetable crops, such as peas and radishes, will probably not need as much additional water as will longer- and later-growing varieties. Indicate if the plot has a slope to it and plan to plant rows running across the slope to prevent rapid runoff and erosion.

With the garden area mapped out for these factors, decide next on the placement of specific crops or ornamental varieties. For example, it would be poor planning to place an east-west row of tall plants across the south side of a plot and then plant a shorter, sun-loving variety immediately north of it.

A **soil test** will determine if adjustments in nutrients or pH are necessary or desirable. Most garden plants prefer a pH range from 6.5 to 7.0. Because most garden space is rather limited, it is ineffectual to try to provide the exact optimum pH for each plant variety.

### Selecting Appropriate Varieties

An avid gardener initiates seed selection in winter, planning for different varieties, quantities, successions, and unusual types for experimentation. Seed catalogs from various companies allow detailed comparison of the many varieties. In addition, your county Agricultural Extension agent is an excellent source of information and specific local advice through conversation and the many government publications available.

Make a list of the different kinds of plants you want. An herb or flower garden list would include varieties chosen for color, height, texture, scent, time of blooming, and so on. A vegetable list should include varieties that can be enjoyed during the growing season (lettuces, radishes, spinach, summer squash, melons, etc.) as well as those to be preserved for family use during the fall and winter (e.g., corn, beans, carrots, pickling cucumbers, winter squash, beets). In addition to "essentials," it's fun to grow something new and exotic each year.

In selecting varieties, consider the number of days from planting to maturation, especially in colder regions, which have a shorter growing season. Match the **maturation times** (usually given in number of days) to the first and last frost averages for a given location. Take advantage of **genetic resistance** to cold or hot temperatures and diseases by selecting appropriate varieties. Finally, if space is an important factor, select varieties that have compact **growth habits.**

Purchase seeds through seed catalogs, at grocery stores, garden centers, or feed and seed stores. Feed and seed stores often deal in bulk seeds, allowing a purchaser to get seeds more cheaply because there is no expense for elaborate packaging. It is easy to overstock on seeds. By planning on paper, you can visualize the amount of garden space available and set priorities.

### A Planting Schedule

Develop a planting schedule from the list of varieties chosen, based on which kinds can be planted early, middle, and late in the growing season.

Increase the garden's potential for both number and kinds of varieties planted by **succession planting.** First plant varieties that are early maturing and fairly cold hardy; for example—radishes, lettuce, spinach, and garden cress. After harvesting them, rework the soil, add compost, and plant the second set of crops, such as beets, carrots, corn, cabbage, kohlrabi, and chinese cabbage. Your schedule will depend on the length of the growing season in your region. Early vegetables, for instance, can be planted in late January in southern California but rarely before late May in northern Wisconsin.

For slow-maturing plants, such as tomatoes, peppers, and many flowers, gain valuable time on a short growing season by using plants that have been started indoors (Chapter 10). Most vegetable plants and herbs, however, can be sown directly into the ground.

### Starting Seeds in the Garden

There are a number of ways to arrange garden plants. One way is to plant seeds in parallel **rows;** another is to plant them in "hills." A **hill** is not a mound but a group of seeds planted within a localized area approximately 0.5 m in diameter. In fact, it is an advantage in some soils and climates for the "hill" to be slightly *concave* so that it will collect water. Hill planting is often used for vine crops, such as cucumbers, squashes, pumpkins, and melons. The **broadcasting, wide row,** or **bed** method is popular in some parts of the United States for plants like radishes, carrots, beets, and lettuce, as well as for herbs and flowers. The seeds are spread out, or broadcast, over an approximately 0.5-to 1-m wide row, or bed. You may combine two seed types for broadcasting: for example, mix fast-growing radish seeds with slower-growing carrot seeds. The radishes germinate and mature quickly. When the radishes are harvested, it, in effect, spaces the maturing carrots. **Raised bed** gardening is a very efficient way to garden, particularly in a small space, such as a city lot or rooftop, or where it is essential to conserve water as carefully as possible. Raised beds are also easy to work into an overall property landscape plan. The sides of raised beds are frequently framed with 12-in. wide treated boards. The frames are then filled in with soil, the composition of which is easily controlled because it is contained. Drip irrigation and plastic mulch are further refinements to this efficient type of gardening.

Prepare the soil by tilling and raking; then plant the seeds. A general rule of thumb for planting is to plant seeds two to three times the diameter of the seeds; for example, if the seed is 2 mm in diameter, plant it to a depth of 4 mm. In dense (heavy) soils plant seeds more shallowly and in sandy soils more deeply. After covering them, press the soil down on the seeds with the back of a hoe or by walking on a piece of board over them.

A major problem is planting seeds too thickly, resulting in a dense stand of seedlings crowding and competing with one another and thus requiring excessive thinning. To control the sowing of small seeds, mix the seeds with dry sand or dust in a saved spice or seasoned salt jar with a perforated top; then sprinkle the mixture. Some companies produce "seed tapes" that have seeds separated at the ideal distance for good growth; others have "pelleted" seeds.

After seeds are secure in the soil, water the seed beds gently. When they germinate, allow them to develop true leaves; then thin them (ruthlessly if necessary) to space them out according to directions on the seed packets. Leave only the most vigorous plants. If you have room, you may transplant some of the removed seedlings instead of just discarding them.

## SEEDLINGS TO MATURATION

Energy for germination and initial seedling growth is obtained from food sources in the seed. In angiosperms the food is stored in endosperm tissue or the cotyledons. As the embryonic food sources are depleted, the seedling must become capable of producing its own food. It must develop a sufficient root system to draw water and minerals from the soil and a sufficient shoot system to carry on photosynthesis. The water, minerals, and light it needs must be present in adequate quantities. The soil temperatures must be favorable. The seedling's chances for survival and robust development are enhanced if it does not have to compete with nearby plants. Human intervention can provide this optimum environment for the young seedling.

### Mulching

Mulching is a practice that modifies the growing environment of plants. A **mulch** is any material placed on the surface of the soil for several reasons: moisture absorption and retention; temperature distribution and insulation; weed control; protection of the soil surface from rain splashing, runoff, and formation of a crust; fruit protection and cleanliness; and disease protection.

**FIGURE 11-1** **Mulching** with such organic matter as leaves and lawn clippings has many beneficial effects on cultivated plants and the soil.

#### Organic Mulches

Compost, rotted hay, straw, grass clippings, chopped corn cobs, leaves (Fig. 11-1), newspapers, and even paper sacks are all decomposable organic materials used for mulching. Sawdust and bark chips are popular aesthetic mulches for shrubs, trees, and other landscaped areas. Organic mulches have an advantage over inorganic mulches in that they provide nutrients and humus, thus conditioning the soil.

Place mulch between rows and around individual plants after the seedlings are about 10 cm tall. When using a bulky mulch, such as rotted hay or straw, begin with a 15-cm layer, for it will settle and become more compact.

Because mulches are good insulators, mulch around the plants only after the soil has warmed sufficiently. If applied too early, the insulation will keep the soil too cold, thereby retarding growth. This same insulating property maintains good, even growth temperatures during the hot part of summer and protects plants in the cooler fall, thus allowing a slightly longer growth period.

#### Inorganic Mulches

Inorganic mulches include black and clear plastics, tar paper, aluminum foil, and even bricks and paving blocks. Black plastic comes in a variety of sizes and thicknesses. Where spring temperatures are cool, black plastic mulch increases garden yields by providing a warmer soil climate for rapid growth. Wait to lay down black plastic until the frost has gone out of the ground and the soil is dry enough to be worked, so that it won't become compacted (Chapter 10). The soil should still be moist, for it is difficult to deeply irrigate the garden after the plastic is down. Plot the mulched area by cutting holes into the plastic; then transplant through the holes. The holes in the plastic allow water to reach the plants. If planting seeds, lay down the plastic in strips between rows or around beds.

As the growing season passes and hot periods

arise, the black plastic will naturally absorb some solar energy, but most will be reradiated. Air between the soil and the plastic acts as an insulator to prevent transferring extremely high temperatures to the soil and the root system and keeps the soil warmer at night. Thus both plastic and organic mulches help even out soil-temperature extremes. Because black plastic blocks out sunlight, weed seedlings that begin growing become etiolated and soon die.

Clear plastic is especially useful in areas with a generally cooler spring season because it allows sunlight to warm the soil directly. It has been reported that use of clear plastic has speeded maturation of plants by up to 14 days. Although clear plastic admits sunlight and thus allows weeds to grow, temperatures generally get high enough between the soil and the plastic to kill weeds.

FIGURE 11-2 The process of weeding can provide fresh wild vegetables before cultivated varieties are ready to be harvested, including lambsquarters (*Chenopodium*) (on the left) and pigweed (*Amaranthus*) (on the right).

### Care and Feeding

As plants mature, the main gardening effort is to control competing weeds. The usual methods are mulching, hand cultivation (pulling, hoeing), or using a rotary tiller. A tiller is excellent but not useful in tight areas. If using organic mulch, periodically check and replenish it. Some "weeds" are actually "wild vegetables" and can be picked, cooked, and eaten (Figs. 11-2, 11-3, and 11-4).

FIGURE 11-3 A sprinkling of sour dock (*Rumex*) adds lemony flavor to a salad.

FIGURE 11-4
Purslane (*Portulaca*) is considered to be a garden weed by many people but is a time-honored and desirable vegetable to others.

Water the garden adequately to ensure maximum productive growth. The watering time must be long enough to provide deep penetration of water (5 cm or more) rather than just a superficial wetting. Use of drip irrigation systems, soaker hoses, sprinklers, and channel irrigation (watering alongside the rows with a slow rate of flow for deep penetration and minimum soil movement) are acceptable methods. Drip irrigation is the most efficient method in terms of low evaporative water loss. Early morning and late afternoon are the best open watering times in order to minimize water loss by evaporation and prevent sunburning whereby water droplets on leaves and fruit act as lenses, concentrating the sun's rays and causing burn spots.

Provide additional nutrients by side dressing with compost or commercial fertilizer (such as NPK 10-10-10) or by watering with an organic nutrient solution, such as fish fertilizer or "manure tea." Manure tea is made by soaking manure in water and then using the liquid as a fertilizer solution. Potent natural fertilizers (such as poultry manure) and concentrated chemical fertilizers must be applied carefully because direct contact with plant tissues can cause chemical burn. If roots are damaged, the plant will be stunted or killed.

When a plant begins flowering, diminish or discontinue fertilizing or change to a fertilizer lower in nitrogen than in potassium and phosphorus (such as 4-10-10). Nitrogen stimulates continued vegetative growth rather than fruit production (Chapter 10).

## Flowering and Photoperiod

All angiosperms—whether used for foliage, root, stem, or fruit—mature toward a period of flowering, the mechanism of species continuation. Many cultivated flowers and most vegetable garden plants germinate and produce seeds the same year and hence are called **annuals**. Some, however (e.g., parsley, parsnip, and beets), are **biennials**, growing vegetatively the first year, flowering and setting seed the next. Many ornamental and fruit-producing species are **perennial**, living many years and flowering each year. Some plants, such as the century plant (*Agave*), grow vegetatively for many years before they flower; afterward they die. Table 11-1 lists some cultivated annuals, biennials, and perennials.

The process of flowering is influenced by day length, or **photoperiod**, the number of hours of light exposure per day. Photoperiod response is mediated by the pigment **phytochrome** and floral induction is discussed in more detail in Chapter 9. Understanding the influence of day length is important to vegetable gardeners. Some short-day varieties that are planted as summer days are lengthening will rush into flowering and go to seed before the plant body has matured, a process called **bolting**. The result is an unusable vegetable plant.

Other light-related growth processes are influenced by the phytochrome system, including germination of many seeds, stem elongation, anthocyanin formation, and leaf expansion. The impor-

**TABLE 11-1 Selected examples of cultivated annuals, biennials, and perennials.**

| | |
|---|---|
| **Annuals** | |
| Lettuces (*Lactuca*) | Bachelor's button (*Centaurea*) |
| Radishes (*Raphanus*) | Calceolaria (*Calceolaria*) |
| Mustards (*Brassica*) | California poppy (*Eschscholzia*) |
| Spinach (*Spinacia*) | Coleus (*Coleus*) |
| Corn (*Zea*) | Cosmos (*Cosmos*) |
| Peas (*Pisum*) | Dahlia (*Dahlia*) |
| Beans (*Phaseolus, Vicia*) | English daisy (*Bellis*) |
| Squashes (*Cucurbita*) | Morning glory (*Ipomoea*) |
| Cucumbers (*Cucumis*) | Periwinkle (*Vinca*) |
| Broccoli (*Brassica*) | Petunia (*Petunia*) |
| Cauliflower (*Brassica*) | Moss rose (*Portulaca*) |
| Eggplant (*Solanum*) | Sweet pea (*Lathyrus*) |
| Tomatoes (*Lycopersicon*) | Verbena (*Verbena*) |
| Peppers (*Capsicum*) | Wax begonia (*Begonia*) |
| Melons (*Cucumis, Citrullus*) | Zinnia (*Zinnia*) |
| Garden balsam (*Impatiens*) | Marigold (*Tagetes*) |
| **Biennials** | |
| Carrots (*Daucus*) | Money plant or Honesty (*Lunaria*) |
| Beets (*Beta*) | Canterbury bells (*Campanula*) |
| Turnips (*Brassica*) | Forget-me-not, true (*Myosotis*) |
| Rutabagas (*Brassica*) | Evening primrose (*Oenothera*) |
| Parsnips (*Pastinaca*) | Mullein (*Verbascum*) |
| **Perennials** | |
| Asparagus (*Asparagus*) | Bugle (*Ajuga*) |
| Rhubarb (*Rheum*) | American columbine (*Aquilegia*) |
| Jerusalem artichoke (*Helianthus*) | Silver mound (*Artemisia*) |
| | Marsh marigold (*Caltha*) |
| | Painted and Shasta daisies (*Chrysanthemum*) |
| | Lily of the valley (*Convallaria*) |
| | Larkspur (*Delphinium*) |
| | Bleeding hearts (*Dicentra*) |
| | Foxglove (*Digitalis*) |
| | Daylily (*Hermerocallis*) |
| | Iris (*Iris*) |
| | Peony (*Paeonia*) |
| | Garden phlox (*Phlox*) |
| | Black-eyed Susan (*Rudbeckia*) |
| | Blue sage (*Salvia*) |
| | Sweet violet (*Viola*) |

tance of light to germination of lettuce and endive seeds is well known, for example. If light is provided to moist lettuce seeds, they will germinate. Apparently this is an adaptation to ensure germination of these small seeds only when they are close to the surface. Many weed seeds also depend on light for germination—a good reason for mulching!

### Fruit Development

Such garden plants as lettuce, radishes, beets, carrots, spinach, and Brussels sprouts are grown for their vegetative parts. Others, like peas, beans, cucumbers, squash, corn, and tomatoes, are grown for their fruits. Although we tend to think of apples, pears, oranges, melons, and other sweet fruits when the term fruit is used, you will recall that a **fruit** is, by definition, the matured ovary and associated parts. Fruits contain the **seeds**, which develop from ovules (Chapter 5).

Fruits mature in all plants in very similar ways. One factor that causes differences, however, is the location of the ovary in relation to other flower parts (Chapter 5). In plants like tomatoes, peppers, eggplants (Fig. 11-5), beans [Fig. 11-6 (a) and (b)], and peas, the ovary is **superior**; that is, it is located above all the other flower parts (closer to the apex). In cucumbers, zucchinis (Fig. 11-7), and watermelons the ovary is **inferior** in that the attachment of the ovary is below the floral parts (farther from the apex) and the receptacle (floral tube) surrounds and adheres to the ovary. Therefore the receptacle tissue also proliferates to become part of the fleshy edible tissue.

One of the most popular garden fruits is sweet corn. As with other cereal plants, each individual grain or kernel of corn is an entire fruit (caryopsis) in which the outer skin is composed of fused fruit wall and seed coat layers. The juicy, flavorful part of the seed that pops into your mouth with each sweet, buttery bite is endosperm tissue.

## Insect and Disease Control

Cultivated crops are planted as concentrated groupings of single species, but in nature such monoculture is rare. Because of the concentrated food source, insect and pathogen populations are greatly enhanced, thus requiring *preventative* and *control* measures. A number of effective methods favored by "organic" gardeners avoid synthetic pesticides and other poisons.

Prevention involves good sanitation during and after the growing season by immediate removal of diseased plants and/or debris that may harbor insect eggs and larvae and disease-causing organisms. Use of genetically resistant strains also helps to prevent problems.

**FIGURE 11-5**
The developing eggplant fruit is a type of berry.

(a)

(b)

**FIGURE 11-6**
The bean fruit is called a **legume,** and develops from a superior ovary. (a) Note the attachment of the floral parts at the base of the developing fruit. (b) Fruits of various degrees of maturity appear along a single flowering stem (**spur**), indicating the sequence of flowering from the base toward the tip. Note that the shriveled **corolla** (ring of petals) slides off the young fruit and may catch on the stigma end before it falls off. The **calyx** (ring of sepals) remains at the base of the fruit, at least for a while.

**FIGURE 11-7**
Zucchini squash fruit develops from an inferior ovary and most of the fleshy fruit wall is derived from the receptacle.

**Crop rotation**, changing species at specific garden sites each year, limits buildup of diseases and soil insects and is also beneficial to the soil. Some plants are heavier "feeders" (remove more nutrients) than others and rotating allows recovery of the soil. Control of soil pH can also decrease possibilities of certain diseases. (See also Chapter 17.)

In small home gardens direct physical control of pests is sensible and effective. Most larval and adult insects are easily visible and can be picked off and dropped into a container of detergent water, oil, or kerosene. Some organic gardeners spray pulverized insects in a water solution on plants to repel or kill additional insects. A spray solution of crushed garlic and hot chili peppers in water is effective against some insects, such as caterpillars.

**Companion plantings** of such insect-repelling plants as pyrethrum daisies (*Chrysanthemum coccineum*) and marigolds (*Tagetes* spp.) ward off unwanted insects. Onions, garlic, and various herbs may also be useful.

Most gardeners experience at least one season's growth before considering a systematic chemical approach to insect and disease control. If needed, however, the chemical insecticides and fungicides are highly effective.

## Harvest and Storage

During the early growth stages many garden plants must be thinned and some plants removed by thinning can be eaten. Young beets, for example, are good as greens even before the storage roots are well developed.

The major benefit of a home garden is to enjoy the freshness of "just-picked" vegetables and fruits (Fig. 11-8). Asparagus, sweet corn, and peas are especially good fresh because their sugars are converted rapidly to starch after picking, thereby reducing sweetness. In many corn varieties most of the sweetness is gone within 24 hours, which is why "store-bought" corn on the cob is often bland and starchy-tasting.

Garden products in a dormant or semidormant state can be stored for a time without extraordinary methods of preservation, such as in a root cellar or refrigerator. In earlier times root cellars were present in most homes, but increased dependence

**FIGURE 11-8** A fresh-from-the garden harvest.

on supermarkets has almost eliminated such storage sites in modern homes. A cool place (preferably less than 10°C) in the basement or garage will function adequately. Store fruits and vegetables so that they will be protected from light, freezing, drying, and excess moisture—as in plastic bags or covered with wet newspapers or burlap sacks. Root crops can be stored in slightly damp sand but must be tended to avoid dehydration. It is surprising how well and for how long most garden products will remain in excellent edible condition. Store apples separately from other crops because apples release ethylene gas (Chapter 9), which stimulates onions, potatoes, and other vegetables to sprout.

The refrigerator is an effective and efficient storage place and it may be worthwhile to buy an older model for produce storage. Fruits and vegetables in plastic bags will remain refrigerated for months without deterioration.

### Saving Seeds

Most vegetable, flower, and herb seeds can be collected from the garden and will produce true to the parental type, except for hybrid varieties, such as corn and marigolds. Some variation, however, is inevitable among closely related species or varieties within a species as a result of cross-pollination. (See also Chapter 5.) More consistent results are obtained by purchasing "true" variety seeds each season.

To collect seeds, select from the best specimens possible and allow fruits to reach maximum maturity. Remove seeds from the fruits and allow them to air dry. After complete drying, store the seeds in a dark, cool, dry area until the following spring. Labeled jars with airtight lids are ideal. Uniform seed moisture is important, for excess dryness can kill the living cells in the seeds. On the other hand, excess moisture encourages growth of molds and bacteria.

### Putting the Garden "To Bed"

Once crops are harvested or have finished flowering and the plants are no longer usable, cultivate them under to initiate decay (humus production and nutrient release). Jerusalem artichokes, parsnips, and turnips are cold hardy and can survive in the ground under some mulch in most climate zones. In fact, some plants are considered more flavorful after winter chilling—parsnips, for instance.

Clean out flower beds, cutting back the vegetative tops of perennials before mulching for winter (in cold climates). Tubers, bulbs, and roots of nonhardy varieties (e.g., dahlias and canna lilies) may need to be dug out, dried, and stored where they won't freeze.

## SMALL FRUITS

Small fruits are well adapted to home gardens and are easily incorporated into landscape design as well because most types are easy to grow, hardy, and require a minimum of expertise and care. Some strawberry varieties can even be grown in patio pots and window boxes. It is rewarding to grow small fruits, for their products are delicious and often difficult to purchase in peak condition. Winter protection is necessary in some parts of the United States, such as by a straw mulch.

The most commonly grown small fruits are grapes (*Vitis*), strawberries (*Fragaria*), brambles or cane fruits (*Rubus*), and bush fruits (*Vaccinium, Ribes*).

In addition, some wild fruit-bearing plants, such as juneberries (*Amelanchier*), highbush cranberry (*Viburnum opulus* var. *americanum*), Oregon grape (*Mahonia nervosa*), elderberries (*Sambucus*), ground cherries (*Physalis*), salal (*Gaultheria*), buffaloberry (*Shepherdia argentea*), and hackberry (*Celtis occidentalis*), are now becoming popular.

**Grapes** are an increasing part of home gardening plans because of their hardiness to cold climates, ease of culture, and the many uses of the berries. An added interest is use of grape vines for landscaping because of their trainability into interesting forms. The most common landscape uses are for arbors and as visual barriers for separation and privacy.

Grapes have flexible growth requirements but generally do best in well-drained soils and full sunlight. Where there is ample rainfall, summer irrigation is not required. Drier climates do require infrequent but deep irrigation. Fertilization is also important, but excess nitrogen stimulates overly vigorous cane growth, thus reducing fruiting.

Initial pruning at the time of planting is important to encourage the most vigorous buds to grow (Chapter 9, auxins) and to prevent excessive top growth before new roots can be formed to support it. Select the best shoot, cut it back to two or three buds, and remove all other shoots. Afterward prune in autumn (in mild climates) or early spring (before buds open). It is necessary to cut away the abundant shoots of a previous year's growth because all grape flowers (Fig. 11-9)—and hence the berry clusters—arise from 1-year-old wood.

A popular and efficient training method is the **four-cane Kniffin system** by which the mature vine has four major fruit-producing branches that are renewed and replaced each year (Fig. 11-10). By the end of 4 years the vines are trained to a double-wire trellis. Each year after the fourth year select four vigorous 1-year-old shoots, tie them to the wire with soft string or cloth strips, cut them back to about ten buds, select "renewal spurs" at each of the four arms, and then remove all other vegetative growth (Fig. 11-11).

Mention of a strawberry elicits a salivary response in most people. **Strawberries** grow on a wide variety of soils, in many climatic regions, require little time until production, and a small patch takes little care. It is generally the first fruit produced after the long winter of cooler regions.

There are basically four categories of strawberries: regular bearing, ever bearing, climbing, and those that are specialized for growing in pots. Thus strawberry plants can be used not only for fruit but also for decoration and landscaping, as in patio pots, hanging planters, and borders. The strawberry is adaptable, but the ideal soil is a well-drained, moderately fertile, sandy loam with 5.8 to 6.5 pH.

At planting, trim back some of the roots and remove large leaves to reduce transpiration. Place

**FIGURE 11-9**
Buds containing grape floral clusters are produced on one-year-old wood and develop flowers that are simple in structured, lacking showy accessory parts.

**FIGURE 11-10**
A grape plant has just been pruned in accordance with the four-cane Kniffin training method.

**FIGURE 11-11** During pruning **renewal spurs** are left to replace each grape cane the next season.

plants with the soil level even with the crown, the short center stem from which the rosette of leaves arises. Shallowly planted strawberry plants are susceptible to drying and deep planting retards growth of new leaves from the crown. Water the beds thoroughly after planting and during the first few days to prevent drying.

Strawberries form runners (stolons) in the spring, a source of new plantlets. To help the plantlets become established, gently push the tips and nodes of the runners into the soil to assist adventitious root formation. Continued propagation in this way forms solid rows or beds of massed plants. Rooted plantlets can also be transplanted.

In some areas of the United States strawberries can be produced without irrigation because they are an early-season crop, producing flowers and fruit while the soil is still moist. The plants should not be allowed to wilt, especially during the flowering and bearing stages.

Strawberries require abundant nutrients for runner and fruit formation and weed control is important. A 3 cm sawdust or straw mulch between the plants controls weeds and also keeps rain-splashed dirt off the fruits.

**Brambles,** or cane fruits, include black, red, purple, and yellow raspberries, blackberries, dewberries, and their hybrids. Raspberries are the most popular type. Cane fruits are easy to culture, bear at a young age, produce abundantly, and the berries can be used in many ways. Most cane fruits have varieties that are hardy enough to survive even in the coldest regions of the United States.

Cane plants are adaptable but prefer an area with full sunlight and fertile, well-drained, friable soil. Drainage is particularly important, for canes do best on soil that drains to at least 1 m deep. The pH optimum varies among varieties, ranging between 5.5 and 6.8.

Canes require support and can be staked (Fig. 11-12), trellised (Fig. 11-13), or enclosed. The trellis need not be as stout as for grapes.

**FIGURE 11-12** Individually staked raspberries (Photo courtesy of Robert J. Tomesh.)

**FIGURE 11-13**
Raspberries supported by a trellis. These canes have been thinned and the ends trimmed. Dead raspberry canes from the previous year must be removed and the second year canes thinned and trimmed for maximum production, ventilation, and disease prevention.

Unlike grapes, brambles are biennial, meaning that shoots (canes) do not flower and fruit until their second year. While the second-year canes are producing fruit, vigorous new shoots emerge and develop into the canes that will produce the following year. Prune out old canes after the leaves have fallen—burning or otherwise disposing of old leaves and canes to prevent disease. In the spring thin canes from the previous year's growth and cut back those that are particularly rangy. Thinning generally results in larger berries.

Using a deep penetration technique, irrigate cane plants between flowering and fruit maturation as needed to prevent wilting. Because brambles grow so vigorously, regularly replenish soil nutrients. Control weeds by mulching, cultivating between rows or allowing a mowed strip of grass to grow between rows. If using the latter method, fertilize and water to offset the competition of the grass.

**Bush fruits** include blueberries, cranberries, currants, and gooseberries. All are relatively low-growing plants that have native, wild North American relatives. The genus *Vaccinium,* which is especially well represented, includes cranberry, eastern, western, and mountain species of wild blueberry, plus the several species of huckleberries of the western states. (In the eastern United States the term huckleberry refers only to the genus *Gaylussacia.* In the western United States huckleberry is the accepted common name for several *Vaccinium* species. Both genera are in the heath family.) Currants and gooseberries are in the genus *Ribes.*

Blueberries are the most popular cultivated bush fruit but need more care and knowhow than strawberries and brambles. The best site for blueberries provides full sunlight and well-drained soil high in organic matter and with a low (4.2 to 5.5) pH.

Blueberries evolved in moist soils of high acidity and organic material. Their root structure is distinctive, being fibrous and without root hairs. Optimum nutrient absorption is apparently related to symbiotic fungi (mycorrhizae) growing in association with the roots. (See also Chapter 13.) Therefore more care and attention must be given to the soil environment. Organic matter (sawdust, compost, peat moss) and aluminum sulfate can be used to increase soil acidity. Aluminum sulfate is available at garden stores. In some areas of the West, however, the water used for irrigation is so alkaline that soil acidity cannot be effectively maintained.

Very little pruning is necessary during the first few years, but as the plants become older (3 to 4 years), remove some lower branches. Each year continue cutting back tips of the most vigorous branches to develop the plant's shape. After the fifth year the plant should have some well-developed branches. Retain only the five to six most vigorous branches for fruiting. As with other fruits, pruning out wood decreases the total number of flowers—and hence the number of fruits—but results in large fruit.

In harsh climates place additional mulch (such as straw) around the bases of the plants. This organic material, however, may attract rabbits and rodents that will gnaw the bark and stems. With or without mulch it may be necessary to surround the lower part of each bush with some type of wire-fabric collar, as rabbits and mice will feed on the bark both above and below the snowline anyway!

## TREE FRUITS

Tree fruits are classified by the method of fruit development from the flower. Several major categories of tree fruits are cultivated in the United States: pome fruits, stone fruits, nuts, and citrus fruits. In addition, the mild climates of southern California, the Gulf of Mexico region, and the southeastern United States permit cultivation of some exotic tree fruits of several kinds, including figs (*Ficus carica*), dates (*Phoenix dactylifera*), avocado (*Persea americana*), olive (*Olea europea*), guava (*Psidium guajava*), pomegranate (*Punica granatum*), and several others. We will confine our discussion to the most common tree fruits.

### Pome Fruits

In **pome fruits,** such as apples and pears, the ovary is surrounded by and embedded in the floral tube. The mature fruit is composed of seeds and tissues derived from the ovary and the floral tube. The edible portion of a pome is the floral tube tissue; the portion actually derived from the ovary is the core.

Apples grow throughout much of the United States, favoring moderate to cold climates. Ironically, the warm southern climates occasionally limit production because of insufficient winter chilling. The southern climates may also induce excessive vegetative growth.

The cultivated apple (*Pyrus malus*) is thought to have originated in southeastern Europe and the Caspian Sea regions.There are dozens of cultivated varieties of this species. The United States has basically four native species, all of which are collectively referred to as crab apples (*P. coronaria, P. angustifolia, P. rivularis, P. ioensis*).

Pears (*Pyrus communis*) are seldom as hardy as the hardier apple varieties. The pear is believed to have originated in the same region as the cultivated apple. It is also well known in the Orient, where different species have been propagated and cultivated for centuries.

### Stone Fruits

**Stone fruits** are named for the large pit or stone that surrounds the almond-shaped seed. In the flowers the ovary is superior, above and separate from the floral tube. The fleshy portion of the fruit is therefore derived from the ovary wall. The seeds of stone fruits contain varying amounts of the cyanide-releasing substance **amygdalin** (Chapter 18), known today as the substance laetrile, a claimed (but disputed) treatment for cancer. Except for sweet almonds, the seed kernels of stone fruits should be considered too dangerous to eat in spite of their enticing almond fragrance.

Stone fruits include the peach (*Prunus persica*), nectarine (a fuzzless variety of the peach, *P. persica* var. *nectarina*), apricot (*P. armeniaca*), almonds (discussed under nut trees), cherries, and plums. Domesticated cherries are the sour cherry (*P. cerasus*) and sweet or mazzard cherry (*P. avium*); five or six other edible species are wild natives of the United States. Cultivated plums are the European plum (*P. domestica*), Damson plum (*P. insititia*), and Japanese plum (*P. salicinia*), all of which require temperate conditions. In addition, several native species are hardy into the northern United States, where they are cultivated (*P. americana, P. nigra, P. munsoniana,* and *P. hortulana*).

### "Nuts"

In practice, any dry seed of a woody plant is called a "nut," even though the term has a precise botanical definition—a dry, one-seeded *fruit* that does not break open spontaneously (is indehiscent) when it is mature. Table 16-5 (in Chapter 16) provides a classification of common "nuts" by their actual fruit types. This discussion includes both true nuts and other species that are classified commercially as "nuts."

Most of the "nut" trees are native to the United States, including members of the black walnut group (*Juglans hindsii, J. californica, J. nigra, J. major,* and *J. microcarpa*), butternuts (*Juglans cinerea*), hickories (*Carya* spp.), native hazelnuts (*Corylus* spp.), pecan (*Carya pecan*), and American chestnut (*Castanea dentata*).

The most common nonnative species are the English walnut (*J. regia*), cultivated hazelnut or filbert (*C. avellana*), which is a true nut, and almond (*Prunus amygdalus*). English walnuts in the United States are often grafted onto black walnut rootstocks, which are resistant to certain diseases.

Like grasses and other wind-pollinated plants, most nut trees have inconspicuous flowers. Almonds are an exception, producing attractive flowers that are similar to those of its relative, the peach, and that are pollinated by insects. As noted, the fruit of almonds is a stone fruit, not a nut, and the edible portion is actually the seed kernel that is located within the pit.

The wood of nut trees is valuable for furniture and decorative veneers, for it does not swell or warp, is resistant to moisture, and has distinctive color and graining. Because of its beauty and versatility, for example, black walnut lumber has risen in value so rapidly that "tree rustling" from yards, parks, and woodlots is a problem.

### Citrus Fruits

Citrus fruits are oranges (*Citrus sinense,* several varieties), tangerines (*C. reticulata*), grapefruits (*C. paradisi*), lemons (*C. limon*), limes (*C. aurantifolia*), kumquats (*Fortunella japonica*), and unusual hybrids, such as tangelos (tangerine × grapefruit). Most natural citrus species originated in the Mediterranean region and their distribution in North America is restricted to southern climates. Citrus trees typically flower in late summer and the fruits ripen during the fall and winter months, making fresh southern citrus fruits both abundant and welcome during the cold northern winter. This seasonal pattern, however, also makes them extremely susceptible to occasional freezing weather that dips southward. Citrus trees are nondeciduous, losing individual leaves from time to time rather than all at once.

The citrus fruit is classified as a **hesperidium,** a type of berry having a leathery rind. The berry develops from a superior ovary, with the reduced green sepals located on the stem end of the mature fruit. The peel consists of the outer ovary wall (exocarp) and most of the middle ovary wall (mesocarp). The innermost part of the ovary wall (endocarp) produces the many multicellular juice sacs that fill the segments (locules). The individual juice sacs have a slightly waxy surface that keeps them separate. The chloroplasts of the rind change into chromoplasts as the fruit matures.

Citrus seeds are often formed by **apomixis,** a process in which the embryo develops from diploid maternal tissue instead of from a zygote. Because the embryos are derived by asexual means, the apomictic embryos of citrus are therefore clones of the parent plants. Citrus seeds also exhibit **polyembryony,** the presence of more than one embryo per seed. So it is possible for a single citrus seed to contain a sexually produced embryo plus several apomictically produced clone embryos!

Citrus seeds sprout readily and grow into attractive houseplants and yard plants. The only way to be certain of getting fruit of an exact known variety, however, is by budding or grafting onto the seedling. As with apples and pears, most commercial citrus trees are propagated in this way.

### Planting Fruit Trees

Plant fruit trees where they will receive full sunlight and will have as little frost exposure as possible. Side slopes are ideal, but trees should not be planted at the base of a hill because cold air is denser than warm air and will flow like water, collecting in low spots. A general orchard rule of thumb is to plant trees at least 50 ft up from the base of a valley, especially in colder climates.

In our hemisphere a south-facing slope is warmer and drier than a north-facing slope so that buds are likely to open earlier. The colder conditions of north-facing slopes delay bud break, an advantage if late spring frosts are a problem. On flat areas buildings and bodies of water can provide warmth to help prevent frost damage. Orchards, for example, are important in the Great Lakes region. The lakes release heat from the fusion of water molecules prior to freezing. This condition is simulated by the technique of sprinkling or flooding orchards to protect blossoms or young fruits during a freeze.

Other factors to consider when planting a fruit tree are the overall landscape plan, maximum size of the tree, space between trees, distance from fences, buildings, wires, and lot lines, and family yard-use patterns.

Deciduous mail-order trees usually arrive in a dormant condition, bare rooted, in a heavy paper or plastic bag, looking as though they could not possibly be alive! Fine roots and root hairs quickly regenerate after planting, however, provided that the root system is soaked for a day before planting to give the tree a water reserve until the new roots are functional.

Some trees are sold planted in large cans or with an intact root/soil ball ("balled" trees). Find out how long a tree has been in the can, for the root system may be gnarled from confinement.

Planting includes more than putting the root system in the ground, it is also a conscious attempt to do everything necessary to provide an optimum soil environment. Prepare a large deep hole and adjust soil texture if needed. Fruit trees generally do best on sandy-to-silty loam soil (Chapter 10) and require soil depths of about 1.5 to 2 m. If a soil has too much clay or rocks are present, work out the hole with a pick and shovel; then fill it with more acceptable soil during planting. Increase the depth of planting and the distance between trees if water availability is restricted.

Conservatively prune longer roots and those that are broken or damaged. Place the plant in the hole, pour in a bucketful of water, and add soil. Adjust the plant so that the soil level on the trunk is the same as before transplanting, especially with dwarf trees. Dwarf trees have grafted rootstocks. If the dwarf tree is planted too deeply, adventitious roots will grow from the scion and eliminate the dwarfing effect of the rootstock (Chapter 5).

Finish the planting process by pushing soil between the roots with a shovel handle to eliminate air pockets, adding more soil, and so on. Once the soil has all been added, step down around the base of the plant to firm up the soil and provide good root/soil contact.

To complete planting, soak the soil thoroughly and adjust the tree if it shifts from an upright position. Prune away some branches and stem tips to reduce transpiring surfaces, bringing the top into balance with the abbreviated root system. Apply a mild fertilizer, such as liquid fish fertilizer, compost, or well-rotted manure. Concentrated fertilizer at this stage could damage new fine roots.

Wrap the trunk with protective paper and add a wire-fabric guard (Fig. 11-14) to discourage

**FIGURE 11-14** A fabric-wire guard protects shrubs and young trees from rodent and rabbit damage. (Photo courtesy of Robert J. Tomesh.)

rabbits and rodents if there is a problem; they are especially fond of young fruit tree bark and can easily kill a sapling by girdling the stem. Support the tree with a wire looped around the tree and attached at an angle to one stake on the windward side. Use a protective covering, such as a piece of rubber hose, to keep the supporting wire from cutting into the bark.

Dig a shallow trench and dike around the tree about 1 m from the trunk to trap water during the critical stages of initial growth. Control competing weeds. Use an organic mulch around the tree to control weeds, reduce moisture loss, and increase soil fertility.

### Flower and Fruit Production

The ability of fruit trees to self-pollinate varies greatly. Even within a single species different varieties may be both self-fertile and cross-fertile or only cross-fertile. So it is wise to plant more than one tree to ensure pollination and hence fruit production. Nursery catalogs will usually indicate if planting two trees of a specific variety is necessary.

Buds are apparently stimulated to develop flower primordia by some substance produced in the plant's leaves and then translocated via the phloem to the buds (Chapter 9). The flowering of fruit trees does not seem to be as directly determined by the photoperiod/phytochrome system as in herbaceous plants, although day length is a factor.

Flower primordia development in fruit trees seems to be affected by a balance of hormones that are influenced by both environmental and physiological factors. Physical practices, such as bending branches, girdling, root pruning, and branch pruning, will initiate flowering, as will certain chemicals sprayed on the trees. Extremes of hot and dry conditions reduce flower development. Inadequate winter chilling causes flowers to drop in some species.

More fruit is often set on a tree than is considered optimum for health of the tree (branch breakage), fruit size (larger number of smaller fruits), and continuation of yield the following year (hormonal inhibition of flower buds by developing embryos in the seeds). So selective thinning of flowers or maturing fruit is important.

When thinning, remove the smallest, weakest flowers and immature fruits. Next, remove flowers and fruits that are so close together they might push each other off the stems during maturation. Thin early enough in maturation so that cells in the fruits are still actively dividing (Chapter 9). This step will allow remaining fruits to produce more cells, thereby leading to larger fruits.

In addition to thinning, it is often necessary to put wooden props under fruit-laden branches to reduce stress and breakage. As winter approaches, leaves and fruits abscise and buds become dormant. These buds contain the next season's flower and foliage primordia.

## SUMMARY

The chapter describes plant growth from the seedling phase through maturation, flowering, and fruit production. Home gardening is the context, emphasizing the reasons for specific cultural practices.

An ideal site for gardening receives full sunlight daily, is flat and open, accessible to water, and has loamy soil. It may be a yard or a window box.

Matching existing growing conditions to appropriate varieties is the key to success. Important factors are direction of the sun, shade, water location, topography, and soil characteristics. Choose seeds with appropriate growing seasons and temperature and disease tolerances.

Increase yield by **succession planting**, putting early maturing, cold-hardy plants in the ground first, harvesting them, and then planting warmer-season crops.

Starting seeds indoors is helpful with slow-maturing varieties. Plant seeds directly in the garden in rows, hills, or beds. Then **thin** seedlings to reduce competition for nutrients, water, and growing space. Some plants removed during thinning are edible.

**Mulching** with organic or inorganic material contributes to moisture retention and temperature and weed control. Organic mulches contribute to nutrient and humus content.

During the growing season eliminate **competition** from weeds, some of which can be harvested for eating. Apply water in the early morning or late afternoon to avoid excessive evaporation and sunburning. Apply fertilizer carefully to avoid root damage. Discontinue or change fertilizer after flowering begins, for nitrogen stimulates vegetative growth rather than fruit production.

Species may be **annual, biennial**, or **perennial.** Flowering is influenced by day length or **photo-**

**period.** Varieties planted at the wrong time of the season may **bolt** directly to seed without desirable vegetative growth. **Phytochrome** is a key substance in photoperiodism.

**Fruits** vary according to whether the ovary is inferior or superior and the degree to which receptacle tissue is involved.

There are many methods for controlling diseases and pests in the garden.

Some excess crops can be stored in cool places wrapped in plastic bags, newspapers, or burlap.

It is generally better to buy new seeds rather than save them from garden plants because of cross-fertilization.

After final harvesting, turn the soil to initiate decay of plant remains.

Grapes, strawberries, brambles, and bush fruits are popular yard and garden **small-fruit** plants. In addition, many wild species are now sought for home planting.

Grapes are cold hardy, easy to grow, and attractive in the landscape. They require support by stakes, a trellis, or an arbor. **Pruning** is especially important because all flowers—hence fruits—arise from 1-year-old wood. The four-cane Kniffin system is the most common method of training grapes.

Productive, self-sufficient strawberries can be grown for fruit and to dress up a patio or landscape. Plant them up to the **crown**, after trimming large leaves to avoid excess transpiration. Push runners, or **stolons**, into the soil, for new plants will form at the tips and nodes.

Brambles or cane fruits tend to be hardy and fruitful. Support need not be as extensive as for grapes. Fruit is formed on second-year canes.

Bush fruits are low growing and have many native wild relatives. Most often cultivated is the blueberry. Blueberries need an acidic soil, which can be provided by organic matter and aluminum sulfate.

The most commonly used North American **fruit trees** are categorized by the types of fruit that they produce: **pomes, stone fruits, nuts,** and **citrus fruits.** Site selection for fruit trees is important, due to their permanence. Plant carefully after preparing a large hole; then trim, wrap, and stake the tree. Many fruit trees are **self-incompatible,** necessitating the planting of two or more trees of a variety to ensure pollination. Selective removal of some flowers and fruits prevents branch breakage, enhances fruit size, and ensures continuation of yield the next year by reducing inhibitory hormonal effects.

## SOME SUGGESTED READINGS

Anderson, H.W. *Diseases of Fruit Crops.* New York: McGraw-Hill Book Co., 1956. Covers both fruit trees and small fruits.

Atkinson, R.E. *Dwarf Fruit Trees: Indoors and Outdoors.* New York: Van Nostrand Reinhold Company, 1972. A readable introduction to dwarf fruit trees.

Bailey, L.H. *The Pruning Manual.* New York: The Macmillan Co., 1939. The classic book on pruning by one of the most distinguished figures in American horticulture.

Carleton, R.M. *Vegetables for Today's Gardens.* Princeton, N.J.: Van Nostrand Reinhold, 1967. A popularly written guide to vegetable growing with many useful hints.

Clarke, J.H. *Growing Berries and Grapes at Home.* New York: Dover Publications, 1976. A complete guide to small-fruit growing.

Crockett, J.E. *Crockett's Victory Garden.* Boston: Little, Brown and Co., 1977. A popular book by America's best-known gardener.

Davids, Richard C. *Garden Wizardry.* New York: Crown Publishers, 1976. A witty, charming book of gardening basics and sensible ideas, succinctly presented.

Hartmann, H.T., W.J. Flocker, and A.M. Kofranek. *Plant Science: Growth, Development, and Utilization of Cultivated Plants.* Englewood Cliffs, N.J.: Prentice-Hall, 1981. A comprehensive university text that covers everything from indoor plants, gardens, and lawns to agricultural products.

Hartmann, H.T., and D. E. Kester. *Plant Propagation: Principles and Practices,* 3rd ed. Englewood Cliffs, N.J.: Prentice-Hall, 1975. The best of many books on the propagation of plants.

Knott, J.E. *Handbook for Vegetable Growers.* New York: John Wiley & Sons, 1962. A standard textbook for commercial vegetable growing.

Kramer, J. *1000 Beautiful Garden Plants and How to Grow Them.* New York: William Morrow & Co., 1976. Basics of garden designs and skills combined with an illustrated (1000 line drawings) dictionary of ornamental flowers, herbs, vines, shrubs, and trees.

Pirone, P.P. *Tree Maintenance.* Oxford: Oxford University Press, 1972. Eight chapters on general maintenance practices for trees.

Rodale, J.I., and staff (ed.). *How to Grow Vege-*

*tables and Fruits by the Organic Method.* Emmaus, PA.: Rodale Press, 1961. Over 900 pages of data, charts for planting, information, and helpful suggestions from the editors of *Organic Gardening and Farming Magazine.*

Rodale, J.I. (ed.) *Best Ideas for Organic Vegetable Growing.* Emmaus, PA.: Rodale Press, 1969. Incorporates many ideas from test gardens and readers' experiences.

Shoemaker, J.S. *Small Fruit Culture,* 5th ed. Westport, CT.: AVI Publishing Co., 1978. A thorough textbook of small-fruit culture.

Spittstoesser, W.E. *Vegetable Growing Handbook.* Westport, CT.: AVI Publishing Co., 1978. A comprehensive, easily read textbook of organic and nonorganic vegetable growing, from planning to storage.

U.S.D.A. *Gardening for Food and Fun. The Yearbook of Agriculture, 1977.* A collection of articles covering all aspects of home gardening; available from the U.S. Supt. of Documents, Washington, D.C.

Westcott, C. *The Gardener's Bug Book.* 4th ed. Garden City, N.Y.: Doubleday & Company, 1973. An encyclopedia of garden pests, cross-referenced by host plants for easy use; suggestions for control.

Wilson, L. *The Complete Gardener.* New York: Hawthorn Books, 1972. Useful advice for planning, constructing, caring for and enjoying all types of gardens; emphasis on ornamental gardening.

Wyman, D. *Wyman's Gardening Encyclopedia.* New York: The Macmillan Co., 1971. An alphabetic encyclopedia of plants and topics relating to all types of gardening—flowers, vegetables, landscaping, houseplants, and more. An *outstanding* reference book.

# chapter twelve ALTERED GROWTH

## DISEASE MECHANISMS AND RESPONSES

## INTRODUCTION

An enlarging dead spot or an increasingly limp shoot is a personal affront to the amateur horticulturist. To the plant, such symptoms are the final stages of reaction to a pathogen (disease-causing organism), an adverse environmental condition, or the presence of animal pests. The initial reactions of disease are usually on the cellular level, biochemical in nature, and therefore go unnoticed. As the disease progresses, however, tissue changes occur and produce noticeable symptoms. Part of the difficulty in treating plant diseases is that the symptoms often do not become apparent until the disease is well in progress.

The science of **plant pathology** developed directly as a result of the needs of agriculture—including the growing and handling of food, fiber, and ornamental species, plus tobacco—because diseases are a problem at all stages from seed to preharvest to postharvest. Much of the agricultural research conducted in the United States has been financed by federal and state funds at universities and research stations. Noneconomic wild plants have been less studied except as reservoirs or alternate hosts for pathogens of economically important species.

In this chapter some examples of plant diseases are discussed briefly. Our emphasis, however, is on the general principles of plant pathology—especially the effects of disease on the plant victims—instead of the life cycles of pathogens and animal pests. Because the subject is both interesting and of great economic significance, you may wish to explore specific plant diseases in more detail through our Suggested Readings and the many governmental publications available on the subject.

## THE NATURE OF PLANT DISEASES

**Disease** is a composite of two factors—inhibition of normal processes within the plant and initiation of abnormal ones. Thus any condition that results from interference with normal metabolic activities, including hormone balance, can be called a disease. The visible symptoms of the disease are the result of the interaction of the plant with the disease-causing factor(s).

### Infectious Diseases

Infectious diseases are caused by pathogens and can be spread from one host to another. So once diseased plants or plant parts are noticed, they should be treated if feasible or be destroyed to prevent spreading of the disease to other specimens.

Plant pathogens are found in five major natural groups: bacteria (including mycoplasmas), viruses, fungi, nematodes, and parasitic higher plants. (See also Chapters 13 and 14.) As commonly used, there is some overlap of the terminology distinguishing pathogen and parasite. A **parasite** is an organism that exists within or on the body of a host species, deriving its nourishment totally or partially from that host. It may not promote extensive physiological or structural abnormalities and such parasitism is, in a sense, a "refined" form of ingestion. A **pathogen**, however, is a parasite that provokes definite reactions, a disease syndrome, within the host. The distinction is flexible and yet useful in general discussion.

#### Stages in the Life of a Pathogen

*Inoculation.* The progress of an infectious disease consists of several phases. The first is inoculation, the exposure of the host plant to the pathogenic organism. The pathogen source, whether particles, spores, fungal filaments (hyphae), or whatever, is called **inoculum.** Inoculation by viruses, bacteria, and fungi may be by windborne particles or spores, direct contact with another infected plant part, presence in the soil, or by way of animal carrier or **vector.** The most common vectors are herbivorous insects that carry pathogenic inoculum on their mouthparts or within their salivary glands. Viruses, for example, are generally transmitted by sucking insects, such as aphids and leafhoppers.

One example of a fungal disease transmitted by an insect vector is Dutch elm disease (American elm wilt). Two species of bark beetle carry spores of the pathogenic fungus *Ceratocystis ulmi,* which grows and reproduces in the xylem cells of susceptible elms, including the American elm (*Ulmus americana*). The result is blockage of water transport, which causes wilting and death of the shoots of infected branches. Eventually the entire tree dies in spite of chemical treatments that may be attempted (Figs. 12-1, 12-2, and chapter opening art).

*Penetration.* Following inoculation, the pathogen must enter the host. This penetration may be through a wound, such as that produced by feeding insects, blowing dust, bruising, scratching or abrading, broken roots, or ruptures in the epidermis due to growth. Freezing and sun-scald injur-

**FIGURE 12-1**
Healthy and diseased American elms (*Ulmus americana*) lining a drive.

**FIGURE 12-2**
An elm being treated with fungicide in an effort to kill the fungus (*Ceratocystis ulmi*) that has attacked the vascular system.

ies also create wounds that allow penetration. Some pathogens enter by way of natural openings, such as stomates (Fig. 12-3), lenticels, and water stomates. And some pathogens are even capable of direct penetration by mechanical pressure and/or enzymatic digestion of protective surfaces.

*Infection.* The infection stage of disease development is the stage in which the pathogen contacts susceptible tissues and begins its growth, deriving nourishment from the host. At this time, host reactions begin even though they may not become obvious for a matter of days, months, or years. The period of time between infection and manifestation of noticeable disease symptoms is called the **incubation period.** The pathogens may affect the host in several ways:

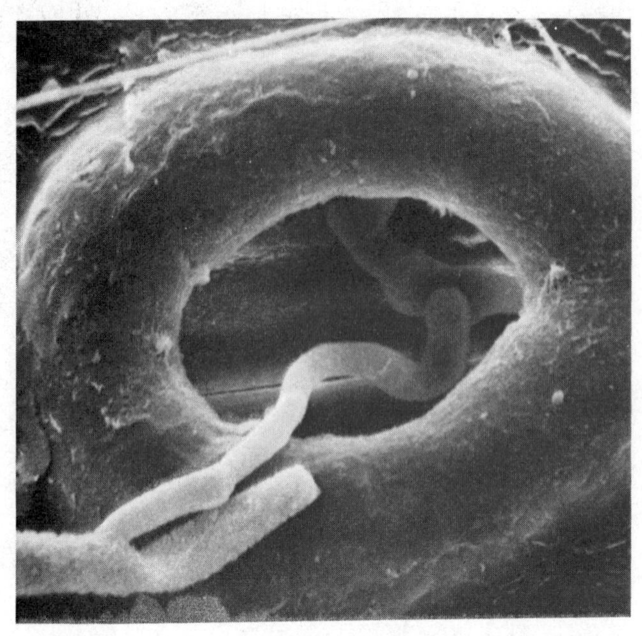

FIGURE 12-3 A scanning electron micrograph (SEM) of a fungal filament entering a stomatal chamber. This fungus (*Scirrhia acicola*) causes brown spot needle blight on several pine (*Pinus*) species. (Photo courtesy of Robert F. Patton.)

1. by consuming host cell contents.
2. by utilizing nutrients produced by the host plant.
3. by producing chemical substances that interfere with normal structural aspects and/or metabolic functions of the host plant, as by interfering with genetic regulation of cellular processes.

Defense mechanisms by the host plant (discussed subsequently) may be expressed passively or actively. In passive resistance innate, genetically controlled structural and metabolic characteristics of the host provide resistance to the full attack of the pathogen. In active resistance the production of inhibitory substances or structural barriers blocks colonization by the pathogen. Successful resistance means survival in a fairly normal state; failure means domination of the plant by the pathogen, resulting in massive disruption of normal function and, frequently, death.

If infection is proceeding successfully (from the standpoint of the pathogen), the pathogen invades many of the host plant's tissues (Fig.12-4 and 12-5). Individual pathogens commonly demonstrate some specificity, invading, for example, only xylem, only phloem, or only cortex. Viruses usually become systemic, infecting the entire plant. The type and degree of symptoms displayed by the host plant depend on the damage and/or alteration of physiological processes that occur during invasion.

*Reproduction.* Again from the standpoint of the pathogen, the successful result of infection is the accomplishment of reproduction. Bacteria, viruses, and fungi all have rapid rates of reproduction. During a season millions of new infective individuals or spores are formed within the body of the host. Reproductive potential is often increased by having a life cycle that has several reproductive phases, involving more than one host (Fig. 12-6).

*Dissemination.* Following reproduction, dissemination of infective individuals or spores occurs. Because they are incapable of significant movement, pathogens depend on passive means of dissemination, such as by wind, water, or animals. Even microscopic nematodes, although motile, are not capable of significant migration. With wind and water, the chance that a single spore will contact a susceptible host is small unless the area is extensively monocultured. So the overproduction of huge numbers of infective individuals or spores is

FIGURE 12-4 Fungal hyphae in butternut (*Juglans cinerea*) vessel members. (Photo courtesy of Irving B. Sachs, Forest Products Laboratory, U.S. Department of Agriculture.)

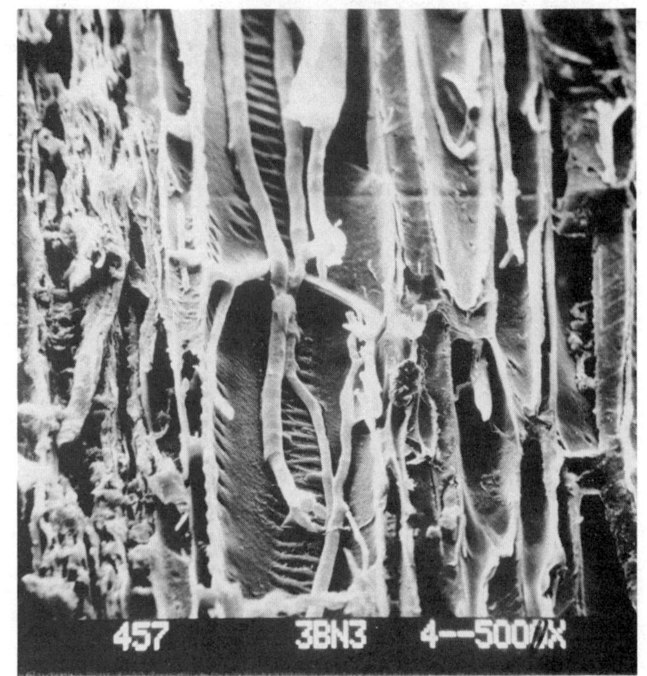

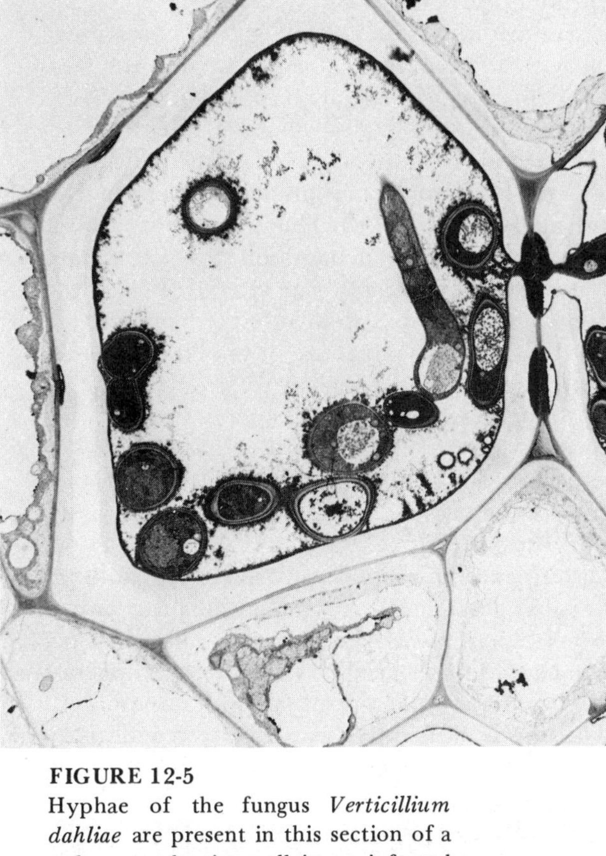

**FIGURE 12-5**
Hyphae of the fungus *Verticillium dahliae* are present in this section of a xylem conducting cell in an infected potato (*Solanum*) root. The dark material coating the inside of the cell wall is produced by the host as a reaction; it may restrict water movement through the xylem, causing **wilting.** Hyphae move into adjacent cells through natural pits in the cell walls. (Photo courtesy of James W. Perry.)

offset. Pathogens which are transmitted by an insect vector have a somewhat higher chance of successfully reaching a susceptible host because insects will certainly attempt to feed and many insect species tend to be specific in what they feed on.

In addition to reproducing within host tissues, some viruses and mycoplasmas further increase their reproductive potential by reproducing within the body of the insect vector or by infecting its offspring. Therefore predation or control measures are often directed at the insect vector.

Organisms that leave their infective stages in the soil rely on the germination of a susceptible host to contact when they themselves germinate (or hatch). Soil contamination is one reason that crop rotation, even within a home garden, is important. In this way, specific organisms cannot accumulate in ever-increasing numbers in the same location on the same host over successive growing seasons.

Some soil fungi accomplish considerable spread by their own growth. Their filamentous bodies are capable of spreading outward several meters and can infect host tissues directly. In addition, some fungi have specialized spore-scattering release mechanisms.

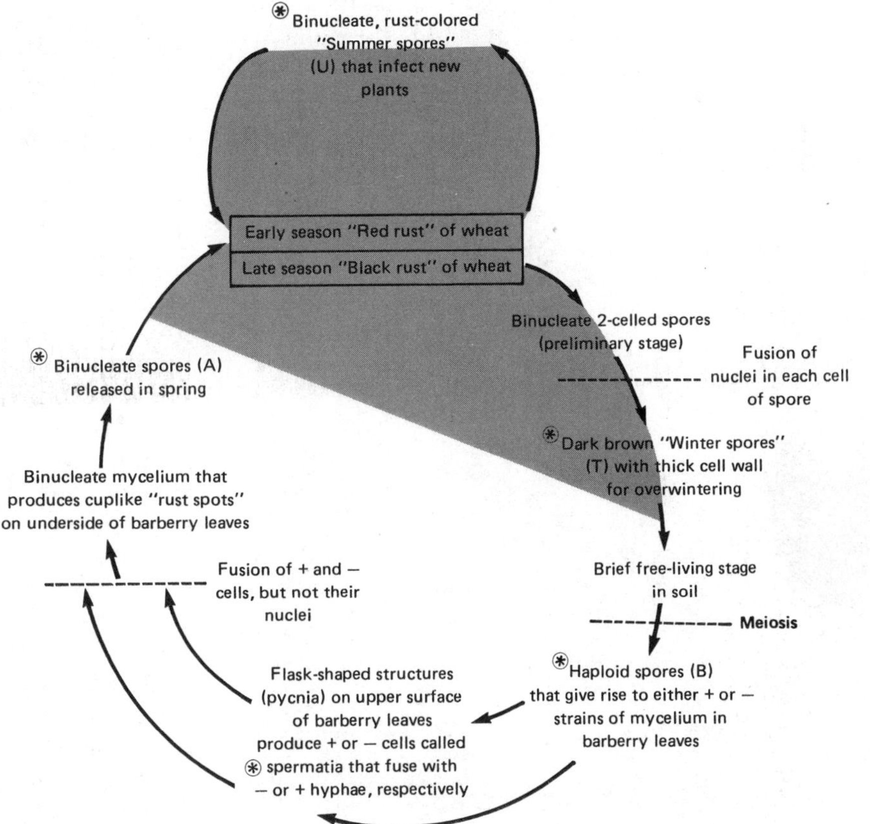


**FIGURE 12-6**
Life cycle of **what rust** (*Puccinia graminis*). Three types of **binucleate** spores are produced, aeciospores (**A**), uredospores (**U**), and teliospores (**T**). In addition, **haploid** basidiospores (**B**) are produced. The complex life cycle, with an **alternate host** (barberry, *Berberis*) and many spore-producing stages, is very efficient in producing large numbers of infective individuals. The shaded region shows the part of the life cycle that takes place in wheat. The several types of reproductive cells are identified by an asterisk * .

*Overwintering.* Once dissemination has been accomplished, pathogens must overwinter or survive the winter season. Some must also survive the summer until the host reaches a susceptible stage. The geographic distribution of many pathogenic species is limited by such factors as cold temperature, excess moisture, or inadequate moisture. Because they can overwinter successfully, soil nematodes are a severe problem in mild climates. In Oregon's mild Willamette Valley, for example, grass crops are grown for seed, and the only practical nematode control has been yearly burning of the fields.

Resistant spores or cysts are common adaptations to surviving unfavorable seasonal conditions. The seeds of parasitic higher plants are especially well adapted to overwintering. Some pathogens overwinter in the bodies of reservoir or alternate host species, either a perennial plant or an animal.

#### The Host's Responses

Disease may be manifested in ways that can be grouped into two major categories: tissue damage and tissue proliferation. Tissue damage interferes with whatever function is performed by the tissue, such as photosynthesis, transport, mitosis, and reproduction.

**FIGURE 12-7** **Tissue proliferation** in response to peach leaf curl disease, caused by *Taphrina deformans*, a fungus. (Photo courtesy of H. Ronald Cameron.)

Tissue proliferation is the result of overstimulated growth in a region. It is manifested in the formation of enlarged, abnormal areas due to increased cell formation and/or abnormal enlargement of cells. The results are distortion of plant tissues and organs (Fig. 12-7) or formation of tumorlike swellings and growths (Fig. 12-8). The deleterious effects of tissue proliferation depend on the extent to which normal plant function is altered. It is common for tissues adjacent to proliferated areas to be crushed, thereby also impairing or eliminating *their* ability to function. In addition to structural difficulties, the excessive combined energy demands of both proliferating cells and pathogenic organisms compete with normal plant activities for food and nutrients. Loss of vigor and general unthriftiness frequently result (Fig. 12-9).

Other external symptoms include cankers, wilts, scabs, leaf spots, rots, and blights. Some symptoms, such as witches'-broom, galls, and lesions are caused by a variety of different pathogens, indicating general types of responses to disease.

The mode and intensity of host reaction to penetration and infection by a particular pathogen vary with a number of factors. The first set of factors is **intrinsic**, related to the host's genetic resistance to providing a suitable environment for the growth and development of the pathogen. One innate factor is the production of inhibitory substances by the host plant; another is the host's ability to wall off the initially infected area from the rest of the plant. This process is normally accomplished by the formation of impervious tissue, such as cork, between the infected and noninfected sites. The walled-off areas may then simply die and new tissue be generated to replace the lost functional region. A third facet of intrinsic resistance is maturity—some plants are susceptible only during germination or early growth. In woody plants, for example, infection of the stem generally occurs in young twigs, unprotected by a substantial bark covering.

Chestnuts provide a genuine example of genetic susceptibility and resistance. The American chestnut (*Castanea dentata*) was almost wiped out as a major forest component by chestnut blight (caused by the fungus *Endothia parasitica*) between 1904 and the mid-1950s. The fungus, native to northern China, was introduced accidentally to the United States, where our native chestnuts had no

**FIGURE 12-8**
**Clubroot** of cabbage (*Brassica*) and its relatives is caused by a slime mold (*Plasmodiophora brassicae*). The slime mold spreads throughout the root, releasing a substance that stimulates excessive cell division and causes some cells to become abnormally enlarged. Some root cells become completely filled with the pathogen. (Photo courtesy of Paul H. Williams.)

**FIGURE 12-9**
A cabbage with clubroot disease is on the left; a normal cabbage is on the right. The disruption of normal function plus the metabolic demands of the pathogen cause **stunted growth**. (Photo courtesy of Paul H. Williams.)

innate resistance to it. Japanese (*C. crenata*) and Chinese (*C. molissima*) chestnuts imported to the United States, however, are genetically resistant to the blight. Now the Chinese chestnut has been crossed with the American chestnut, producing hybrid trees (the "clapper" chestnut) with quality nuts, good tree shape, and blight resistance.

**Extrinsic** factors are the environmental conditions present during initial stages of disease development. Many pathogens require a specific temperature, pH, and humidity. Also, many soil organisms need a specific soil type or nutrient balance. Particular sets of environmental conditions are **predisposing factors** for each disease—that is, the conditions that make a plant most susceptible to maximum disease development. Many host plants, for instance, are more susceptible when at a disadvantage—growing under less than optimum conditions or in a nonvigorous condition. Thus the maximum predisposition for disease development exists when the environmental conditions are optimum for the pathogen and, at the same time, very unfavorable for the host.

#### Symptoms Produced by Viral Diseases

The general effects of viral infection are widespread, usually becoming systemic (affecting the activity of the entire host plant). Dwarfing and stunting of growth are the most common viral disease symptoms (Fig. 12-10). Coupled with them are reduced productivity and shortened life span. Local symptoms of viral infection are distorted growth, proliferation of particular plant parts, development of **leaf mosaic** (leaves develop chlorotic or chlorophyll-lacking areas) [Fig. 12-11 (a) and (b)], or the death of cells (**necrosis**). Necrotic areas appear as spots, streaks, or large patches.

#### Symptoms Produced by Bacterial Diseases

Probably the most familiar type of bacterial disease is bacterial soft rot. It (and fungal rot) can occur in the produce drawers of home refrigerators, for example. General bacterial disease symptoms, which can appear on any part of the plant, involve proliferation of tissues or organs and galls, knots, scabs, or cankers. If vascular tissues are affected, stunting and wilting may occur. Small organisms called mycoplasmas cause "yellows" diseases, which resemble virus diseases.

#### Symptoms Produced by Fungal Diseases

Fungal diseases are many and, like viral and bacterial infections, may cause growth abnormalities, resulting in stunting or excessive growth of parts or whole plants. Localized or general necrosis may result, producing necrotic or chlorotic lesions, galls, and scabs. [Fig. 12-12 (a), (b), and (c)] on the plant. Some of the more common fungal diseases are downy and powdery mildew [Fig.

**FIGURE 12-10** **Prune dwarf virus on Italian prune** (*Prunus domestica*). (Photo courtesy of H. Ronald Cameron.)

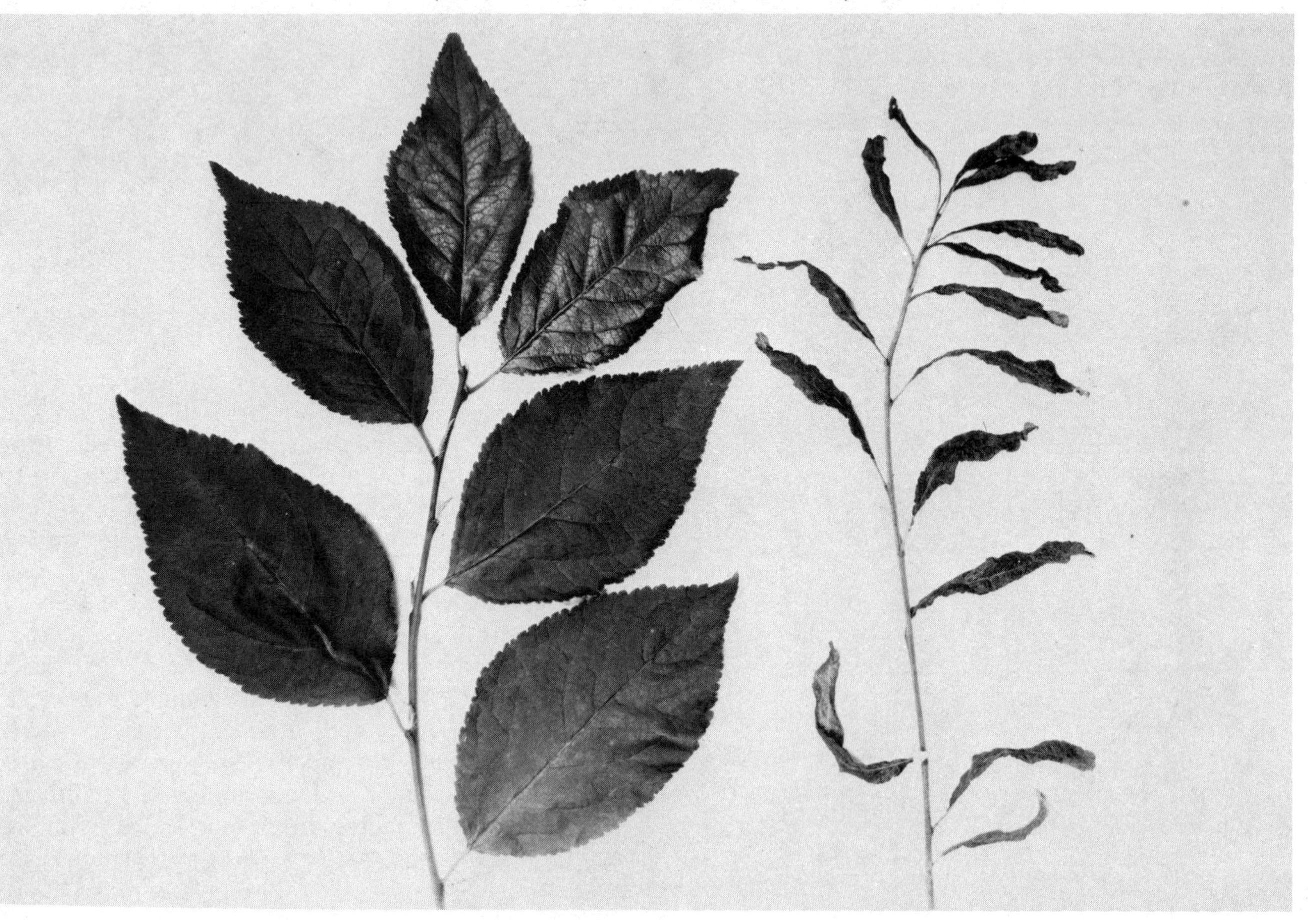

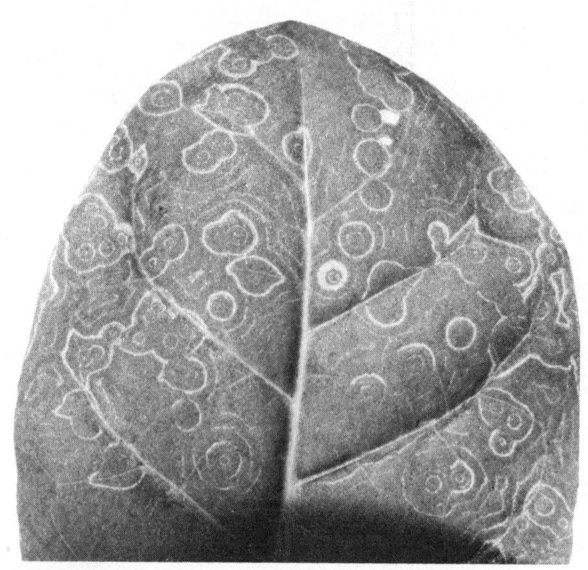

(a) (b)

FIGURE 12-11
(a) **Tobacco mosaic virus** (TMV) is a common, easily transmitted viral disease that causes a mottled or **mosaic** pattern of dark and chlorotic areas on the leaves of tobacco (*Nicotiana*) plants. It is also known to infect about 150 other plant genera, including tomato (*Lycopersicon*). It is a serious economic problem. (b) **Tobacco ring spot virus** (TRSV) causes ring-shaped chlorotic or necrotic lesions. [Photo (a) by Steven A. Vicen. Photo (b) courtesy of Gustaaf A. deZoeten.]

FIGURE 12-12
The barley (*Hordeum*) leaves in (a) show symptoms of **net blotch disease.** The corn (*Zea*) leaves in (b) exhibit symptoms of **southern corn leaf blight.** Both diseases are caused by fungi in the genus *Helminthosporium.* An epidemic of southern corn leaf blight in 1970 destroyed about 15% of the total United States corn crop, affecting all corn varieties containing one specific gene in common. (c) **Scab,** caused by the fungus (*Venturia inaequalis,* is the most important disease of apples, affecting young leaves and developing fruits. [Photos (a) and (b) courtesy of Deane C. Arny. Photo (c) courtesy of Robert J. Tomesh.]

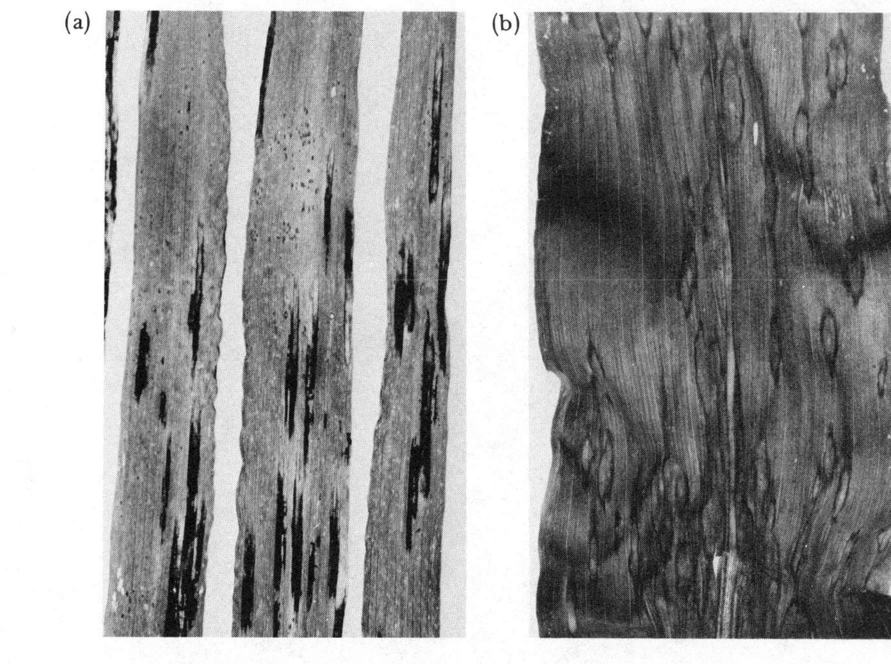

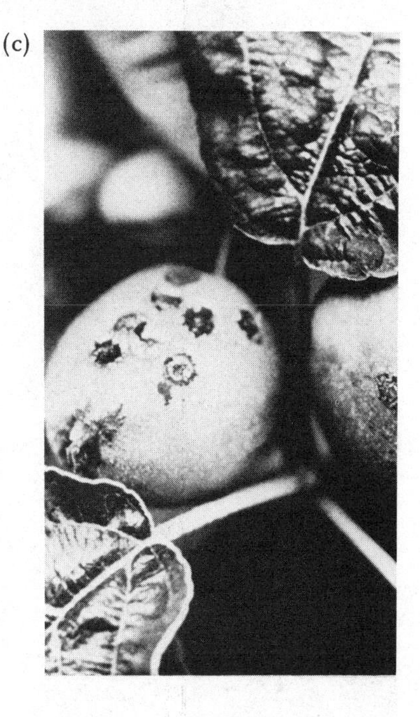

(a) (b) (c)

12-13 (a) and (b)], rusts, fungal rots [Fig. 12-14 (a) and (b)], smuts, and anthracnoses [Fig. 12-15 (a), (b), (c), and (d)].

Tomato, potato, and cucumber wilts are common garden problems (Figs. 12-16 and 12-17). Wheat rust is a significant economic problem and provides a good example of a fungal pathogen life cycle (Fig. 12-6). Table 12-1 describes some common fungal diseases.

(b)

**FIGURE 12-13**
(a) Closeup of **powdery mildew** on the surface of a wild thistle leaf. Powdery mildews are among the most widespread and easily diagnosed fungal diseases. Affected leaves, young shoots, and stems become covered with the grayish, powdery fungus. (b) **Downy mildew** on the fruiting head of oats (*Avena*). Downy mildews mainly infect young, growing tissues, such as leafy and flowering shoots, and may destroy the entire plant. They require a film of water and cooler temperatures, as during a period of cool, rainy weather. [Photo (b) courtesy of Deane C. Arny.]

(a)

(a)

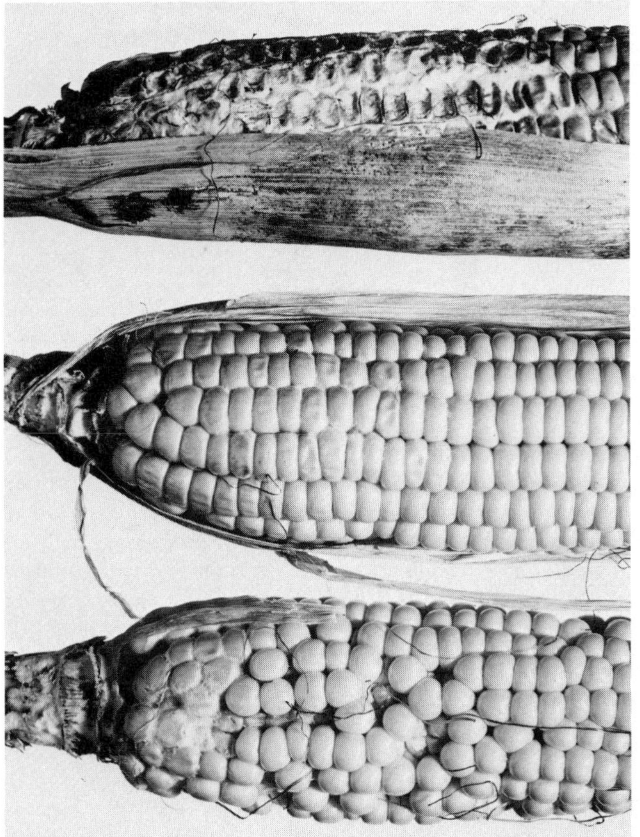

(b)

**FIGURE 12-14**

(a) Root and stem **rot**—as in these carrots (*Daucus*)—is caused by fungi in the genus *Rhizoctonia.* The genus is worldwide in distribution, affecting virtually all wild and cultivated plants and all plant parts. (b) The fungus that causes this **ear rot** of corn grows between the husks and the developing ear, usually beginning at the base of the ear. Eventually the entire ear may shrink and become lightweight. The disease is best controlled by use of resistant corn varieties. [Photo (a) courtesy of Paul H. Williams. Photo (b) courtesy of Deane C. Arny.]

(a)

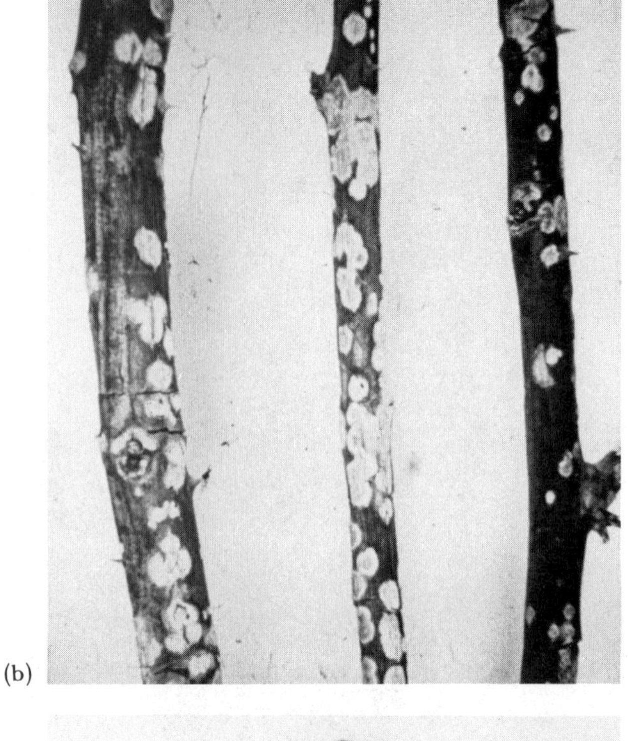

(b)

(c)

(d)

**FIGURE 12-15**

(a) Plant **smuts**, as this one of oats, are serious diseases of grain plants worldwide, attacking the developing fruits. As they grow, the fruits become filled with sooty masses of black, dustlike spores. (b) **Anthracnoses** are diseases that produce ulcerlike lesions on leaves, fruits, and stems, as these on raspberry (*Rubus*) stems. (c) and (d) **Ergot** disease of grains—especially rye (*Secale*)—infect the developing fruits. It can cause an affliction called ergotism or St. Anthony's fire in humans who eat diseased grain (Chapter 18). The fungus is *Claviceps*, especially *C. purpurea*. (c) Infected flowering heads exude a sticky substance, or **"honeydew"**, containing inoculum that spreads to other flowers, infecting them. As the season progresses, each infected grain becomes an ergot **sclerotium**—a hard, purplish-black fungal mass that overwinters. The following spring small, spore-producing stalks develop on the sclerotium (d), disseminating spores to infect blossoming grain plants. [Photos (a), (c), and (d) courtesy of Deane C. Arny. Photo (b) courtesy of Robert J. Tomesh.]

**FIGURE 12-16**
The potato plant in the center is normal. The one on the right was inoculated with infective spore-producing structures of the fungus *Verticillium dahlie*, as was the plant on the left. Both infected plants developed symptoms that include **wilt, curling** of leaflet margins, **necrosis,** and leaf **abscission.** The roots of the left plant were wounded before inoculation, increasing pathogen invasion. (Photo courtesy of James W. Perry.)

**FIGURE 12-17**
Fungal **vascular wilts** are caused by fungi in three genera—*Ceratocystis* (as on elms and oaks), *Verticillium,* and *Fusarium.* The fungal hyphae invade vascular tissues (especially the xylem) where they interfere with water movement. As a result, tissues beyond the blocked vascular tissues wilt and die. Some wilts, as this wilt of a cucurbit, are caused by **bacteria** rather than fungi. (Photo courtesy of Robert J. Tomesh.)

TABLE 12-1 **Some important plant diseases caused by fungi and measures for controlling them.**

| Common Name of Disease | Pathogen | Hosts Attacked | Control Measures[a] |
|---|---|---|---|
| Stem rust of wheat | *Puccinia graminis tritica* | Wheat | Eradicate barberry plants (an alternate host) and use resistant cultivars. |
| Corn smut | *Ustilago maydis* | Corn | Use resistant cultivars. |
| Fusarium wilt | *Fusarium oxysporum* | Tomato, pea, celery, banana, cotton, watermelon | Use resistant cultivars. |
| Powdery mildews | *Erysiphe polygoni* | Many hosts | Foliar fungicides except on cereals, where resistant cultivars should be used. |
| | *Podosphaera leucotricha* | Apples | |
| | *Uncinula necator* | Grape | |
| | *Sphaerotheca pannosa* | Rose | |
| | *Erysiphe cichoraceaurum* | Cucurbits | |
| | *Erysiphe graminis* | Cereals | |
| Rust | *Puccinia striiformis* and *P. graminis* | Turf grasses, especially bluegrass and ryegrass | Keep lawn growing rapidly by fertilization; use fungicides. |
| Brown patch | *Rhizoctonia solani* | Turf grasses, especially bent grass, bluegrass, Bermuda grass, fescues | Avoid nitrogen fertilization. Improve drainage. Use certain fungicides. |
| Brown rot | *Monilinia fructicola* | Stone fruits: peaches, apricots, plums, cherries, almonds, nectarines | Use resistant cultivars where available. Sanitation. Fungicide sprays. Prevent fruit injury during harvest. Refrigerate harvested fruit. |
| Apple scab | *Venturia inaequalis* | Apples | Protectant and eradicant sprays with fungicides. Use resistant cultivars where available. |
| Peach leaf curl | *Taphrina deformans* | Peaches and nectarines | Dormant spray of fungicide after leaf fall and before bud break. |
| Verticillium wilt | *Verticillium albo-atrum* and *V. dahliae* | A wide range of woody fruit and ornamental species; many kinds of herbaceous plants | Plant only resistant species and cultivars. Remove infected limbs or plants and burn. There is no chemical control. |
| Downy mildew of grape | *Plasmopara viticola* | Grape | Foliar sprays with a fungicide. |
| Late blight of potato | *Phytophthora infestans* | Potato and tomato | Use resistant cultivars. Destroy all cull potatoes in field. Spray with fungicides during growing season. Store potatoes at 2 to 4°C (36 to 40°F). |
| Damping off | *Pythium* and *Rhizoctonia* spp. | Seedlings of many species | Fungicidal seed treatments. Soil fumigation or pasteurization. Improve soil drainage. Germinate seedlings at temperatures unfavorable for pathogen growth. |
| Dutch elm disease | *Ceratocystis ulmi* | Elm trees | Remove sources of infection and control European elm bark beetle, which spreads the fungus. |

[a]Detailed directions on control of plant diseases can be obtained from local offices of the Agriculture Extension Service or from commercial garden centers.

(From *Plant Science*, by H.T. Hartmann, W.J. Flocker, and A.M. Kofranek. Englewood Cliffs, N.J.: Prentice-Hall, Inc., 1981.)

FIGURE 12-18
Symptoms of **boron deficiency** in pears. (Photo courtesy of H. Ronald Cameron.)

## Noninfectious Diseases

Noninfectious diseases are not the result of the presence of a pathogenic organism but are due to the presence or absence of specific environmental factors. For this reason, they are sometimes called **environmental diseases.** They involve both the presence of a disease-causing condition and the plant's response to this condition, culminating in symptoms caused by physiological disorders. Although specific symptoms (e.g., wilting or necrosis) may develop on the immediately affected part, more general symptoms, such as lack of vigor, etiolation, chlorosis, and stunted growth, are usually expressed.

Causes may be unsuitable conditions of light, temperature, specific nutrient (Figs. 12-18 and 12-19), oxygen, or pH. (See also Table 10-4 for nutrient deficiency symptoms.) Additional causes are such human-induced problems as air and water pollution and unwise agricultural practices. Information about the prevention of environmental diseases is presented elsewhere in the context of growing plants outdoors and indoors.

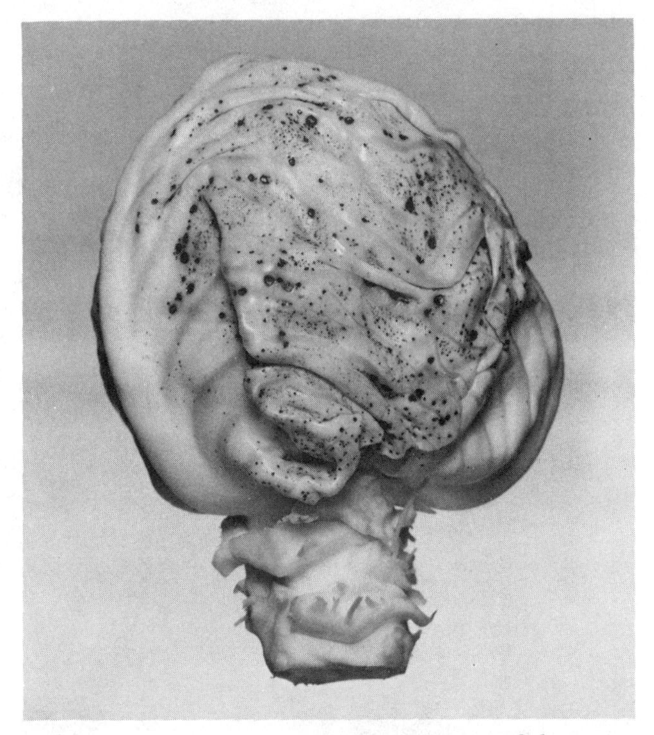

FIGURE 12-19 Some diseases are caused by an *excess* of specific minerals, as this example of **copper toxicity** in cabbage. (Photo courtesy of Paul H. Williams.)

## Parasitic Vascular Plants

Dodders and mistletoes are two groups of parasitic vascular plants that are of minor economic significance when compared to the hundreds of preharvest and postharvest problems caused by viruses, bacteria, and fungi. The characteristics of the parasites, however, provide interesting examples of structural adaptation and specialization; so we will examine them more closely.

**Dodders** (*Cuscuta*) depend on their host plants for all aspects of their nutrition, for they lack chlorophyll and therefore do not photosynthesize. Each dodder plant grows from seed, germinating in the soil and developing a rudimentary root system. The elongating, yellowish, slender vine climbs an adjacent plant. If the plant is not a compatible host species, the dodder may penetrate and live temporarily on it until the growing shoots contact an appropriate host. If the young plant does not establish contact with an appropriate host, it dies.

On a compatible host the dodder plant sends specialized tissue-penetrating branches called **haustoria** into the host's stems and leaves. The haustoria continue to elongate and branch, invading the host's vascular system. Tapped into the host's nutrition, the dodder plant grows and produces seeds (Fig. 12-20). Dodder seeds may be harvested along with the desired crop and be perpetuated as a contaminant among other seeds.

The **mistletoes** include Old World genera and the North American genera *Phoradendron* (about 70 species) and *Arceuthobium* (several species). Infection by *Phoradendron* species (some leafy, some leafless) begins when the sticky seeds are deposited on a young twig or branch. The seeds germinate, forming an attachment organ from which haustoria grow. The haustoria penetrate lenticels or axillary buds; then grow and branch throughout the living bark tissue. Sinkers invade the cambium and actually become incorporated into the wood of the host plant when lateral growth of the host occurs. *Phoradendron* species normally parasitize broad-leaved trees, absorbing water, minerals, and perhaps some carbohydrates from the host. Generally they are dark olive green and require good light for

**FIGURE 12-20**
Dodder (*Cuscuta*) is a parasitic vascular plant, growing here on red clover (*Trifolium*). (Photo courtesy of Deane C. Arny.)

photosynthesis. Leafy species are America's Christmas mistletoe and are usually collected from infested oak trees. They are slow growing and do not kill the host plant as a rule; a single mistletoe plant of this genus may live as long as the host. The presence of the mistletoe may cause degeneration and swelling within the infected region of the branches, however, and the end of the branch commonly dies.

The dwarf mistletoes (*Arceuthobium* spp.) are parasitic on conifers and so are of greater economic significance, especially when the main stem (bole) of young trees is infected. The life cycle is similar to *Phoradendron,* including the nature of host plant infection. Branches swell considerably in response to the parasite and may form burls (Chapter 8). Witches'-brooms may be formed at the ends of branches and there may be copious exudations of resin from the parasitized sites.

Some vascular plants are root parasites. The introduced witchweed (*Striga*) is parasitic on such monocots as corn and sends haustoria from its own root system into the roots of the host plant. Certain species of broomrape (*Orobanche*) are parasitic mainly on the roots of wild plants, including sagebrush (*Artemisia*), saxifrages, and others.

## Animal Pests

Animal pests damage plants by direct tissue destruction during feeding and as vectors of disease. Also, while feeding they create wounds that permit pathogens to enter the plant, thus increasing plant susceptibility to disease.

Nematodes (roundworms) are a serious agricultural problem. They damage plants by their feeding activities, causing tissue destruction, decreased vigor, and often abnormal proliferation of plant tissue. Most nematodes attack the plant roots, causing development of knots, galls, or cysts, but members of two genera live in stems and leaves.

Insects are the dominant life form on Earth today in terms of sheer numbers, diversity, and adaptive specializations. The human species and thousands of herbivorous insect species are in direct competition for food and the success or failure of a crop may hinge on the outcome of the competition. This factor is true in the home garden where careful individual attention can achieve pest removal. In large-scale agribusiness such close individual supervision of the pest population is less feasible. Therefore pests are usually controlled by chemicals that can be applied over large areas quickly. Although effective, chemical pesticides can be dangerous to humans and other animals and their effectiveness is limited by increasingly resistant pest-insect strains.

Direct damage by insect feeding (Fig. 12-21) is seldom considered "disease," although an important aspect of plant pathology. Some insects, such as cotton-boll weevils and corn borers, lay eggs on the developing plants so their larvae may live within and feed on host tissues.

Nevertheless, some insects cause growth abnormalities. Insect larvae may stimulate proliferation of plant tissue around themselves, forming galls (Fig. 12-22). Distorted growth occurs when the balsam woolly aphid feeds on true firs (*Abies*), injecting saliva into the cortex of the young

**FIGURE 12-21** Tent caterpillars can rapidly defoliate an entire small tree or do serious damage to larger trees before they complete the larval stage of their life cycle.

**FIGURE 12-22**
Galls on willow (*Salix*) are caused by an insect larva living in each gall.

branches. Cortical parenchyma tissue reacts to the salivary substance by forming abnormal tumorlike enlargements on the stems. If the apical area of the main stem is attacked, apical buds are killed, resulting in a tree that grows two or more "leaders" if it survives. The vascular cambium is also stimulated, producing abnormal wood and bark. In severe infestations many acres of fir trees die. Forests in northeastern, southern, and northwestern North America, for instance, have suffered extensive losses since the introduction of the balsam woolly aphid into North America from Europe in the 1900s.

Mites, some of which feed on plant tissues, are tiny eight-legged relatives of spiders and ticks. Mite infestations cause direct damage and abnormal growth; for example, mites at a branch tip can cause witches'-broom.

Such animals as nematodes, mites, and particularly insects are of tremendous importance as competitors with humans for food and other plant products. The scope of this text does not, however, permit a more extensive treatment of them. The best and most specific regional information on both pests and diseases is available from local Agricultural Extension services.

## DEFENSES

### Plant Defense Mechanisms

The interaction between a given pathogen and host plant is one of interacting genetic constitutions. The specificity of a pathogen is limited by its own genetic potential. The degree of susceptibility of the host is determined by its genetic potential for structural and biochemical defense mechanisms.

#### Structural Resistance

Genetically determined structural characteristics of a plant provide varying degrees of resistance to penetration by a pathogen. The epidermis resists penetration in several ways. The outer surface is covered with cutin and waxes and the outer cell wall itself is commonly thickened. The waxy coating repels water and spores carried in the water and the thick cell walls provide an impenetrable physical barrier. A dense forest of epidermal hairs may discourage insect feeding. An outer covering of bark protects woody stems and roots.

The natural necessary openings of plants—stomates, water stomates (hydathodes), and lenticels—are weak points in the external defense, as are wounds. Some host species have stomatal structure and distribution that discourage the entrance of water and suspended pathogens. Stomates are present more liberally on lower rather than upper plant surfaces. Some stomates are less susceptible to penetration because they remain closed until the drier part of the day. Small lenticels and those that initiate early cork-layer formation are less susceptible to penetration. It is also believed that dense mats of epidermal hairs on leaves and stems prevent spores from contacting penetration sites. Even following penetration, pathogens may be limited in their spread by such hard tissues as fibers and xylem conducting cells.

Plants are able to react to wounding or pathogen invasion by forming impervious layers that protect as-yet uninjured tissues. Thickening of the cell walls adjacent to the pathogen occurs in some resistant species. Suberin, lignin, or gummy substances may also be deposited in these thickened walls. Damaged regions are commonly sealed off by formation of a layer of cork cells, analogous to the formation of scar tissue in animals. Many plants, including cherries, peaches, and other *Prunus* species, react by depositing gums around wounds or infection sites. The gums rapidly fill intercellular spaces and sometimes cell interiors, blocking further pathogen invasion. Oozing of gums and resins (as in conifers) over wound sites also presents a barrier to inoculation.

Rapid necrosis of an invaded cell following penetration is a defense that quickly deprives the pathogen of nutrients. It is an effective defense against certain fungal diseases, causing fungal hyphae to die along with their host cells. This defense is actually biochemical as well as structural.

Overall the degree to which a plant possesses or can erect structural and biochemical defenses determines its natural resistance to specific diseases.

#### Biochemical Resistance

The most significant innate defense mechanisms seem to be metabolic. Biochemical defenses block disease by preventing penetration or by inhibiting growth and reproduction of the pathogen. Like structural resistance, biochemical defense mechanisms are determined by the genetic constitution of the individual plant rather than the species as a whole. Among cultivated plant species there are many examples of resistant and nonresistant varieties to specific pathogens. Some tomato varieties, for example, are resistant to *Fusarium* and *Verticillium* fungi, which cause wilt diseases.

*Preexisting Defenses.* **Inhibitory exudates**—chemicals produced within the plant and exuded to the plant's external surfaces—inhibit germination and penetration of pathogens. Many inhibitory exudates are known, including some metabolites, such as simple sugars, amino acids, organic acids, enzymes, glycosides, alkaloids, and inorganic acids. In addition, specifically toxic compounds may be exuded. The action of shoot exudates is better understood than that of root exudates.

Other types of preexisting defenses are (a) intracellular inhibitors, which are naturally present in the cells and discourage growth of specific pathogens, and (b) absence of nutrients required by a pathogen.

Research is underway to determine if plants possess some sort of immunological responses similar, at least in effect, to the antigen-antibody reactions of animals. The results are not yet conclusive.

*Induced Defenses.* Although the preceding defenses are innate in resistant plants, most biochemical defenses occur as responses to injury. Like structural defense responses, biochemical defense responses isolate the injury from as-yet healthy tissues and heal it. Whereas a purely physical wound, such as a puncture or scrape, can be healed once and for all, the injury caused by a pathogen tends to be progressive. The success with which a plant resists disease depends on the rate and magnitude of response compared to the growth rate of the pathogen.

After infection begins, host plants commonly enter a phase of higher than normal respiration. The accelerated respiration period may be essential to the activation of biochemical defenses, although the relationship is not yet clearly established.

Some plants produce specific organic inhibitors—most of which are fungitoxic—in response to the presence of a pathogen. The presence of a pathogen may cause alterations in host enzyme systems and protein synthesis, increasing resistance to the pathogen at and near the site of infection, such as by counterbalancing the effects of host versus pathogen enzyme systems. Also, the enhanced enzymatic activity and protein synthesis of the host plant may simply overpower the metabolic activities of the pathogen and thus effectively inhibit its growth.

Some host reactions interfere with pathogen hydrolysis, as by the formation of compounds not readily hydrolyzed by the pathogen's enzymes. Moreover, the host tissue may respond with changes in ion balance that suppress the production of hydrolytic enzymes by the pathogen.

Some pathogens produce toxins that cause disease. Resistant plants are thought to be chemically nonreceptive to the selective toxin activity or able to detoxify them actively.

### Treatment and Control

The most effective cure for plant disease is **prevention** by one of several means. Use genetically resistant varieties—those having effective innate resistance to particularly common and damaging plant diseases. Practice good sanitation to eliminate or substantially reduce pathogenic inoculum. When an individual plant shows disease symptoms, others present probably have also been exposed to the pathogen. Even in the home garden, awareness of possibilities, alertness to symptoms, and advance planning are necessary. Practice crop rotation and avoid large concentrations of a single crop if possible. Treat or destroy infected plants as soon as they show symptoms. Destroy all infected plant parts. Avoid transmitting diseases by contaminated hands, equipment, and work and storage areas.

Various forms of governmental regulation activities have been effectively applied in the prevention and control of plant diseases. Included are the inspection of produce (especially that intended for interstate shipment) and quarantine procedures for plant material imported from other countries.

Because many host species are less susceptible when grown and stored under their own optimum conditions, careful attention to soil moisture, humidity, and temperature can be significant in disease control.

Several lines of **biological control** research have been pursued. The first, already mentioned, is the development of genetically resistant plants. In addition, genetic control of nonpathogenic disorders (environmental diseases) may be improved by developing strains with greater ranges of tolerance for environmental variations. There have also been experiments using microorganisms that are predatory on or pathogenic to the harmful pathogens, such as some fungal species and bacteria-destroying viruses.

There have been notable successes in using insects to control insects—for example, predaceous ladybird beetles (ladybugs) and praying mantids, plus wasps whose larvae parasitize destructive caterpillars. Insect sexual attractants (pheromones) have been used to trap or confuse male insects, effectively preventing them from locating and mating with receptive females. Mating with radiation-sterilized males that are released causes female insects to lay infertile eggs. Further insect research promises to provide answers that may reduce agricultural dependence on synthetic toxic chemicals.

Trap crops have been used with some nematode species. The nematodes infect a special early planting, which is then destroyed before the nematodes are able to reproduce. In the home garden some insect pests can be decoyed from protected crops by planting other species preferred by insects.

Nematodes are also sensitive to a few species of antagonistic plants. Some, but not all, nematode species are sensitive to asparagus and marigolds and are killed by the exudations of these plants.

All these biological control methods may not be economically feasible for large-scale agriculture but have possibilities for adaptation to the home garden.

**Physical control** methods involve killing pathogens by temperature extremes or radiation. Field burning is utilized to control soil nematodes in places. Smaller areas of substrate (such as in greenhouses) are commonly sterilized by steam, hot water, or dry heat. Certain microorganisms and many viruses are susceptible to hot-water or hot-air treatment of their host plant seeds and bulbs. Hot-air curing and drying are common pretreatments for storage of tubers, bulbs, rhizomes, and some hard-walled fruits.

Refrigeration is a common storage method. Cold temperature kills some pathogens; with others it may effectively retard their metabolic processes and hence their ability to initiate diseases. Refrigeration usually begins shortly after harvesting and continues through shipment and storage until consumer use. The most common diseases that would occur without refrigeration are bacterial soft rot and fungal dry and soft rots.

Radiation research has involved ultraviolet, X, gamma, alpha, and beta radiations to kill pathogens. The possibilities for radiation control of diseases during food shipment and storage are enticing, but so far directly applicable practical methods have not been perfected.

**Chemical control** has provided by far the greatest success in controlling both animal pests and infectious diseases of crops. Sprays and dusts that can be applied easily are available for most of the common bacterial and fungal diseases. To be effective, most must be applied to external plant surfaces before infection. Some bactericides and fungicides are systemic—they are absorbed into the host plant. In addition to sprays and dusts applied to the foliage, there are preplanting treatments for roots, seeds, and bulbs. Soil fumigation is another type of control. It must be emphasized that, to obtain maximum effectiveness and safe use of various chemicals, you must *follow the instructions precisely.* Of course, no specific chemical control is effective if it's used against the wrong condition! You should obtain a verified diagnosis from a qualified person, such as an Agricultural Extension agent, or from illustrated leaflets and booklets published by the government and garden chemical companies. Tables 12-1 and 12-2 summarize some diseases and recommended chemical treatments.

TABLE 12-2 Pest control recommendations for selected crops.[a]

| Crop | Critical Time(s) | Pest and What To Do |
|---|---|---|
| **Fruits** | As a general rule, critical times for preventing the most likely fruit pests are from the start of growth in spring through fruit set. | |
| Apples and pears | Early spring when buds have $\frac{1}{4}$-in. green tip | Apple scab. Spray with fungicide every 7 days through bloom. |
| | Early spring when buds have pink tips | Overwintering insects on tree. Spray with dormant oil spray. |
| | Late spring at petal fall | Codling moth, leaf roller, plum curculio, apple scab. Spray with insecticide and fungicide. |
| | Normal or wet summer | Apple maggot fly (railroad worms). Spray every 10 to 14 days with insecticide. |
| Brambles raspberries, blackberries, gooseberries, currants, etc. | Spring when leaf buds are $\frac{1}{2}$-in. long | Anthracnose, spur and cane blight. Apply fungicide. |
| | Spring just before blossoms open | Sawfly, borers, anthracnose, spur and cane blight. Spray with fruit combination spray containing fungicide and insecticide. Repeat 3 days and then in 10 days. |
| Grapes | Spring when new shoots are 6 to 8 in. long | Black rot. Spray with fungicide. |
| | Spring just before blossoms open and after petal fall | Black rot, leafhoppers, rose chafer, moths. Spray with fungicide and insecticide. |
| | Spring just after blossoms fall | Powdery mildew and above insects. Spray with fungicide and insecticide. |
| Stone Fruits cherries, peaches, plums, apricots, etc. | Dormant, before growth starts in spring | Peach leaf curl, black knot. Spray with fungicide. |
| | Petal fall | Brown rot, leaf roller moth, curculio. Spray with fungicide and insecticide. |
| | 7 to 10 days after petal fall | Brown rot, leaf spot, curculio, fruitworm. Spray with fungicide and insecticide. |

TABLE 12-2 (Continued).

| Crop | Critical Time(s) | Pest and What To Do |
|---|---|---|
| Strawberries | Early spring, right after uncovering or just before buds appear | Leaf spot, leaf scorch, aphids, plant bugs, leaf roller. Apply fungicide and insecticide. |
| | While in blossom, every 10 days until 3 days before first picking | Leaf spot, leaf scorch, berry rot, berry mold. Apply fungicide. |
| **Vegetables** | Most vegetables can be put into one of two categories for critical time(s): (a) as seedlings and (b) at blossom and fruit set. Use seeds treated with a fungicide to ensure good seed growth. | |
| Beans | Blossom through fruit set | Leafhoppers, aphids, lygus bugs, beetles. Apply insecticide. |
| Beets and Chard | Before planting | Cutworms. Spray or dust ground with insecticide. |
| Cabbage and related plants, such as broccoli, Brussels sprouts, kohlrabi, cauliflower, collards, kale | As seedlings and mature plants | Aphid, looper, webworm, diamondback, cabbage worm, harlequin bug. Apply insecticide. |
| Carrots | As seedlings | Aster leafhopper. Apply insecticide. |
| Celery | As seedlings | Aster leafhopper, blight. Apply insecticide and fungicide. |
| Cucumber and related vine crops like squash, pumpkins | As seedlings | Striped cucumber beetle. Apply insecticide. |
| Melons | At vining out and/or at blossoming | Striped cucumber beetle, pickleworm, squash vine borer. Apply insecticide. |
| Lettuce | As seedlings | Aster leafhopper. Apply insecticide. |
| Okra | During fruiting | Corn earworm. Apply insecticide. |
| Peas | At blossom and fruit set | Cowpea curculio, pea weevil. Apply insecticide. |
| Peppers | At blossom and fruit set | Pepper weevil, aphid, European corn borer. Apply insecticide. |

TABLE 12-2 (Continued).

| Crop | Critical Time(s) | Pest and What To Do |
|---|---|---|
| Potatoes | Before planting | Grubs, wireworms. Spray or dust ground with insecticide. |
| | Seedlings to blossom | Beetles, aphids, leafhoppers, blight. Apply insecticide and fungicide. |
| Radishes | Before planting | Root maggots. Spray or dust ground with insecticide. |
| Sweet corn | At silking and every 5 days for two more sprays | Earworms. Apply insecticide. |
| Tomatoes | At blossom and possibly at weekly intervals thereafter | Leaf spot, blight, flea beetles, hornworms, leaf miners. Apply fungicide and insecticide. |
| Turnip and mustard | First true leaves | Aphid. Apply insecticide. |

[a]These examples represent a variety of common problems and suggested treatments. Specific chemical controls have not been identified. The most up-to-date recommendation may be obtained from a regional Agricultural Extension agent.

Most chemical controls for pathogens include as active ingredients one or more of the following: sulfur, chlorine, phosphorus, copper, mercury, zinc, chromium, nickel, and cobalt. One of the most versatile and universally used combination bactericide/fungicides is Bordeaux mixture, which contains copper sulfate. Sulfur is especially effective against powdery mildews and other fungi. There are also many organic fungicides.

Use of antibiotics is a form of biological control, for an antibiotic is a substance produced by one organism that is toxic to another, but it is included in this discussion because it is applied as a chemical. Two of the antibiotic groups commonly used in the treatment of animal (including human) diseases are also used in the systemic treatment of certain plants—namely, the streptomycins and tetracyclines among others. Chemical application of another group of biological compounds, the growth regulators, also shows promise in the treatment of some diseases.

It should be recognized that the use of chemical controls is increasingly questioned because of potential long-range damage to beneficial soil microbe populations and therefore soil ecosystems, concern about carryover into foods, and the real hazards of direct human exposure to agricultural chemicals. Finally, chemical controls are immediately effective but do not contribute to the genetic improvement of crop varieties. The development of improved genetic resistance of crop varieties to diseases may be the best eventual solution to at least some of the problems of agriculture. It won't be easy or quick. Breeding experiments for disease resistance require a varied genetic pool, composed of both cultivated species and their wild relatives. The potential but as-yet unexplored value of wild species is a serious argument for their preservation, as emphasized at the end of Chapter 6.

## SUMMARY

Plant **disease** is a reaction to a causative organism or an adverse environmental condition. The science of **plant pathology** developed as a result of the needs of agriculture. Disease is a composite of two factors: the inhibition of normal processes and the initiation of abnormal ones.

A **pathogen** is an organism that causes a disease syndrome within the host. Major groups of organisms pathogenic to plants are bacteria, viruses, fungi, nematodes and parasitic higher plants.

A pathogenic disease follows a known course of events: inoculation, penetration, infection, reproduction, dissemination, and overwintering.

Disease is manifested as tissue damage or tissue proliferation, which interfere with normal tissue functions. Some external symptoms are cankers, wilts, scabs, leaf spots, rots, and blights.

The intensity of host reaction depends on intrinsic factors, such as its genetic resistance to disease. The host may produce inhibitory substances or wall off the infected area from the rest of the plant. Maturity affects resistance capability. Extrinsic factors, such as temperature, pH, and humidity, make a plant more or less susceptible to a given disease.

Viral infection is usually systemic, causing dwarfing and stunted growth as well as such local symptoms as distorted growth, spots, and streaks. Common bacterial and fungal disease symptoms are soft rot, tissue proliferation, galls, cankers, and wilting. Mildews, rusts, and smuts are fungal diseases.

Noninfectious or environmental diseases result from plant response to adverse environmental conditions.

Dodder and mistletoe are parasitic vascular plants. Animals damage plants either by direct tissue destruction or as **vectors** of disease.

Plant resistance to pathogen penetration depends on genetically determined structural and biochemical defenses. Some defenses are preexisting within the plant. Induced defenses are the plant's attempt to isolate and heal infected tissue.

The most effective means of controlling disease is prevention, such as by using genetically resistant varieties of plants, rotating crops, and acting quickly if infection occurs. Many methods of biological, physical, and chemical control are available.

## SOME SUGGESTED READINGS

Agrios, G.N. *Plant Pathology,* 2nd ed. New York: Academic Press, 1978. A very good college textbook on the subject.

Jacobsen, J.S., and C.C. Hill. *Recognition of Air Pollution Injury to Vegetation: A Pictorial Atlas.* Pittsburgh: Air Pollution Control Association, 1970. Symptoms, color photos, and tables of susceptibility for each of the major air pollutants.

Large, E.C. *The Advance of the Fungi.* New York: Dover Publications, 1962. An interesting account of the human battle against pathogenic fungi, from the potato blight to the 1940s.

Pirone, P.P. *Diseases and Pests of Ornamental Plants,* 4th ed. New York: The Ronald Press Company, 1970. Diseases and pests of particular plants and their control.

——. *Tree Maintenance.* New York: Oxford University Press, 1972. General maintenance, diseases, pests, and cures; includes many tree species.

Sprague, H.B. (ed.) *Hunger Signs in Plants,* 3rd ed. New York: David McKay Co., 1964. Effects of nutrient deficiencies, illustrated with numerous photographs.

Stevens, R.B. *Plant Disease.* New York: The Ronald Press Company 1974. An introductory text illustrating principles of plant disease and control.

Streets, R.B. *The Diagnosis of Plant Diseases,* 2nd ed. Tucson: University of Arizona Press, 1972. A practical manual for rapid identification of plant diseases.

Treshow, M. *Environment and Plant Response.* New York: McGraw-Hill Book Co., 1970. A textbook of environmental pathology.

Wellman, F.L. *Plant Diseases.* New York: Natural History Press, 1971. A popular introduction to plant pathology, emphasizing control and the impact of disease on food economy.

# chapter thirteen DIVERSITY PROKARYOTES AND NONVASCULAR PLANTS

photo by Mark R. Fay

## INTRODUCTION

By now you are familiar with certain structural and metabolic details of "the green world" as you most commonly experience it—the gymnosperms and angiosperms. However, the simpler photosynthetic organisms and the mostly "nongreen" world of viruses, bacteria, and fungi are also important to the study of botany. Before beginning to review the great diversity of organisms known as plants, we must pause briefly to consider the relationships among **extinct** (no longer surviving) and **extant** (existing) species.

### On, Off, and From the Record

**Fossils,** preserved remains of ancient organisms, are *facts,* and many extinct organisms are known *directly* by the presence of fossil evidence (Figs. 13-1 and 13-2). Table 13-1 summarizes the geologic history of Earth and its life forms as we know them from fossils that have been discovered and dated.

Studying the ancestry of today's species is complicated because there are enormous gaps in the fossil record, many of which are the result of natural causes that prevented fossilization. To become fossilized, for example, an organism must die in a place where it is protected from weathering and where decomposition occurs slowly, such as in silty riverbeds, mucky bogs, and volcanic ash. Organisms dying on exposed land are unlikely to be preserved because of scavengers, decay, and the ravages of weather. Also, resistant body parts must be present, such as woody tissues, spore or pollen walls, seed coats, bones, teeth, or shells. Many fossils are impressions and casts made in mud that was later transformed into rock. Another major limitation to our knowledge is the lack of adequate financial support to back qualified research and exploration.

**FIGURE 13-1**
Fossil *Sapindopsis* leaf from Indiana. The leaf is shown about 1.3 times its actual size. (Photo courtesy of David L. Dilcher.)

**FIGURE 13-2**
This fossil leaf resembles existing sycamore (*Platanus*) leaves in its lobing, venation, teeth, and petiole, even though it is millions of years old. (Photo courtesy of David L. Dilcher.)

Some of the earliest discovered fossil plants were simple bacterialike or algaelike forms whose soft bodies were not clearly preserved. They are represented by organic layers in ancient sedimentary rocks, dark carboniferous streaks created by the compression and decomposition of the bodies of simple organisms that settled to the bottom of ancient bodies of water.

Studies of chert (a silica matrix rock) deposits in various parts of the world, however, have revealed some well-preserved microscopic fossils (microfossils), especially of bacteria. Analysis of organic residues in these cherts has pushed back the date for which we have direct physical evidence of the existence of living organisms to approximately *3.5 billion years ago.* So it is appropriate to expand the geologic timetable back beyond that of Table 13-1 to include an age of microorganisms (Fig. 13-3).

The relationships among extinct species and their relationships to existing species are also known or hypothesized *indirectly* by extrapolation from structural and biochemical characteristics among extant species.

Comparisons of the structure of living plant organs have provided many clues to plant group relationships. Another promising technique has been the analysis of protein composition, especially of enzymes and other protein-containing substances active in energy-transfer reactions. Visual analysis of plant chromosomes has been helpful in demonstrating relationships among extant species, using techniques that include statistical analysis to verify the probable correctness of hypotheses. Our expanding understanding of DNA and RNA has enabled researchers to compare specific amino acid sequences to determine genetic relationships. DNA and RNA do represent, after all, another kind of historical record!

The work of many investigators has resulted in a number of schemes for identifying and hypothesizing the **phylogeny**, or lines of descent, of plants. Figure 13-4 summarizes plant groups in terms of their possible relationships, with the more primitive lines near the bottom of the diagram. But read on . . . there is more to the story!

## Schemes for Classification of Organisms

The game of "Twenty Questions" has been around for a long time and the first question is traditionally one of classification: "Is it animal, vegetable, or mineral?" This is essentially the

**FIGURE 13-3** An **Age of Microorganisms,** which lasted some 3 billion years, dominates the time span of biological evolution. Microfossils of prokaryotic cells have been found in deposits 3.5 billion years old, and those cells must have been preceded by simpler ones. The earliest eukaryotic fossils are only about 1.3 billion years old. Almost nothing is known about evolution during the age of microorganisms. The macroscopic fossil record goes back only about 600 million years to the time of the earliest metazoans, or multicellular organisms. (Figure and caption adapted from "Archaebacteria," by Carl R. Woese. *Scientific American* 244 (6): 98–122, June, 1981. Copyright 1981 by Scientific American, Inc. All rights reserved.)

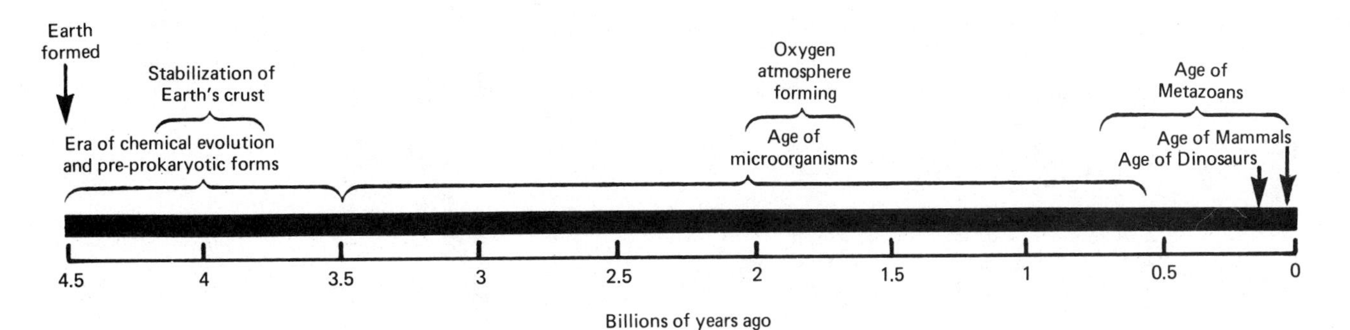

TABLE 13-1 Geologic Eras.

| Eras | Periods | Epochs | North American Climates and Major Physical Events | Biological Events |
|---|---|---|---|---|
| CENOZOIC 65 million years long | Quaternary | Recent Pleistocene | Cold to mild. Four glacial advances and retreats (Ice Age); uplift of Sierra Nevada. | Age of humans. Extinction of many large mammals, including woolly mammoths. Deserts on large scale. |
| | Tertiary | Pliocene | Cooler. Continued uplift and mountain-building, with widespread extinction. Uplift of Panama results in joining of North and South America. | Herbaceous plants abundant as climates diversify. Large carnivores. First known appearance of human-apes. |
| | | Miocene | Moderate. Extensive glaciation begins again in Southern Hemisphere. | Spread of grasslands as forests contract. Grazing animals, apes, whales. |
| | | Oligocene | Rise of Alps and Himalayas. Lands generally low. Volcanoes in Rocky Mountains. | Origin of Asteraceae and many other modern families of flowering plants. Large, browsing mammals. Apes appear. Madro-Tertiary Geoflora expands. |
| | | Eocene | Mild to very tropical. Many lakes in western North America. Australia separates from Antarctica; India collides with Asia. | Grasslands begin to expand. Primitive horses, tiny camels, modern and giant types of birds. |
| | | Paleocene | Mild to cool. Wide, shallow continental seas largely disappeared. | First known primitive primates and carnivores. |
| MESOZOIC 160 million years long | Cretaceous | | Lands low and extensive. Elevation of Rocky Mountains cuts off rain. Africa and South America separate. | Angiosperms appear and become abundant. Age of reptiles. Extinction of dinosaurs at end of period. Marsupials, insectivores. |
| | Jurassic | | Mild. Continents low. Large areas in Europe covered by seas. | Gymnosperms, especially cycads; ferns. Dinosaurs' zenith. Flying reptiles, small mammals. Birds appear. |
| | Triassic | | Continents mountainous. Large areas arid. Eruptions in eastern North America. Continents joined in one mass. | Forests of gymnosperms and ferns. First dinosaurs. Primitive mammals appear. |

**TABLE 13-1 Geologic Eras (Continued).**

| Eras | Periods | Epochs | North American Climates and Major Physical Events | Biological Events |
|---|---|---|---|---|
| PALEOZOIC 400 million years long | Permian | | Extensive glaciation in Southern Hemisphere. Appalachians formed by end of Paleozoic; most of seas drained from North America. | Origin of conifers, cycads, and ginkgos; earlier forest types wane. Reptiles evolve. |
| | Carboniferous Pennsylvanian Mississippian | | Warm. Lands low, covered by shallow seas or great coal swamps. Mountain-building in eastern United States, Texas, Colorado. Moist, equable climate; conditions like those in temperate or subtropical zones—little seasonal variation, water plentiful. | Forests, ferns, lycophytes, sphenophytes, gymnosperms. Major groups of fungi exist. Age of amphibians. First reptiles. Variety of insects. Sharks abundant. |
| | Devonian | | Europe mountainous with arid basins. Mountains and volcanoes in eastern United States and Canada. Rest of North America low and flat. Sea covered most of land. | Rise of land plants. Extinction of primitive vascular plants. Origin of modern subclasses of vascular plants. Age of fishes. Amphibians appear. Mollusks abundant. Lung fishes. |
| | Silurian | | Mild. Continents generally flat. Mountain-building in Europe. Again flooded. | Earliest vascular plants. Modern groups of algae and fungi. Rise of fishes, reef-building corals, and shell-forming sea animals. Invasion of land by arthropods. |
| | Ordovician | | Mild. Shallow seas, continents low; sea covered United States. Limestone deposits; microscopic plant life thriving. | Possible invasion of land by plants. First fungi. Shell-forming sea animals. |
| | Cambrian | | Mild. Extensive seas. Seas spilled over continents. | First primitive fishes. Age of invertebrates. Trilobites, brachipods, other animals. |
| PRECAMBRIAN 4 billion years long | | | Dry and cold to warm and moist. Formation of earth's crust. Extensive mountain-building. Shallow seas. Glaciation in eastern Canada. Planet cooled. Components of different densities separated under influence of gravity. | Origin of life; prokaryotes; eukaryotic cells and multicellularity by close of period. |

Adapted from P. Raven, R. F. Evert, and H. Curtis. *Biology of Plants,* 3rd ed. New York: Worth Publishers, Inc., 1981.

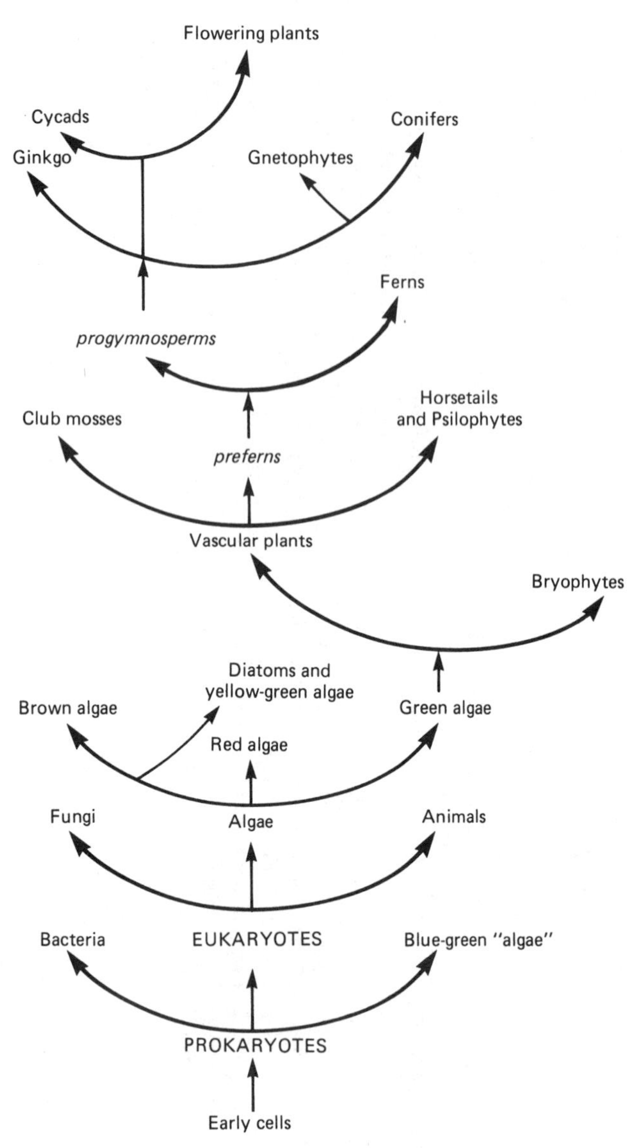


**FIGURE 13-4** A schematic diagram of probable evolutionary relationships among living plant groups. See Fig. 13-6 for a recent modification of the base of this "family tree."

choice of categories utilized by humanity for ages. So it has been natural to confine organisms within two "kingdoms"—Plantae and Animalia. From time to time problems arose, with some nonmotile animals (sponges, corals) temporarily being considered plants. In addition, the invention of the microscope opened a heretofore "invisible" world of animallike microbes. Then bacteria were discovered. Thus with only two choices classification became increasingly arbitrary. The two-kingdom system, after all, was patterned initially on easily recognized, macroscopic, and mainly terrestrial organisms—in other words, higher plants and animals.

To conform all identified organisms to the two-kingdom system has been like the solution attributed to an ancient Greek villain, Procrustes. He placed his victims on an iron bed and created uniformity to the bed's length by stretching or amputating the victims' legs. Thus assignment of many of the single-celled organisms to the "Procrustean bed" of either plant or animal has had adamant and, in many cases, inflexible argumentation.

Closer scrutiny of the system led to the introduction of beds of more appropriate (but not yet perfect) dimensions to fit existing species. To accommodate single-celled organisms, a third kingdom, **Protista**, became popularly accepted. Then a five-kingdom scheme was proposed that listed **Monera** (bacteria and blue-green "algae") and **Fungi** as separate kingdoms (Fig. 13-5). This classification recognizes the significant differences between prokaryotic and eukaryotic organisms.

Cells of existing organisms show these two major patterns of organization. Bacteria and other forms included in the kingdom Monera are **prokaryotic** in organization (*pro* = before; Gr., *karyon* = nut or kernel, for nucleus). Prokaryotic cells are more primitive in structure than eukaryotic cells. They are a thousandfold smaller in volume than typical eukaryotic cells and do not have a discrete nuclear structure bounded by a nuclear envelope. These characteristics (and others; see Prokaryotes section) distinguish them from organisms that are **eukaryotic** (*eu* = true) in organization. The cells of eukaryotes have distinct nuclei in which the chromosomes are located, as well as other advanced characteristics (Chapter 3).

To understand the problems confronting scientists who attempt to discern phylogenetic relationships, let's review a hypothetical scenario for the creation of living cells. It is hypothesized that the first organism ancestors were complex heterotrophic protein molecules that conducted simple metabolic processes utilizing ingredients that surrounded them. In time they developed the ability to reproduce. Eventually a more complex and efficient structural composition evolved, a mass of functionally interdependent and cooperating molecular structures contained within the confines of a membrane—a cell.

Eukaryotic cells may have originated as "composite" structures. There is persuasive evidence that mitochondria and chloroplasts—complex

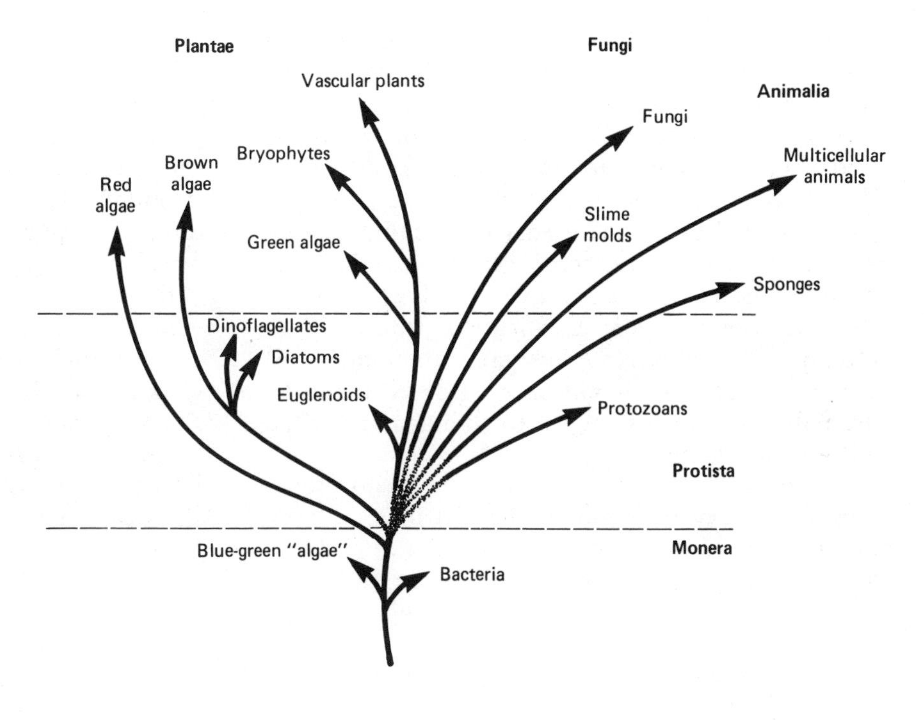


**FIGURE 13-5**
A five-kingdom classification system is based on three levels of organization: **prokaryotic** (Monera); **eukaryotic** and **unicellular** (Protista); **eukaryotic, multicellular,** and **multinucleate** (Plantae, Fungi, Animalia). The three higher kingdoms have widely different body organizations, related to three principal modes of nutrition—**photosynthetic** (plants), **absorptive** (fungi), and **ingestive** (animals). (Adapted from "New Concepts of Kingdoms of Organisms," by R.H. Whittaker, *Science* 163: 150–160, 1969. Copyright 1969 by the American Association for the Advancement of Science.)

cellular organelles that each contain their own DNA, RNA, and special modes of protein translation—are of **endosymbiotic** origin; that is, they descended from separate ancient bacteria that were engulfed by a larger cell and took up permanent residence, evolving along with the host cell. (See also Figure 3-2.) The chloroplast ancestor is thought to have been a photosynthetic relative to the blue-green "algae" (defined subsequently), a hypothesis supported by similarities in RNA nucleotide sequence. Mitochondria are thought to have developed from a respiring bacterium, a hypothesis that seems strengthened by the presence of bacterialike *r*RNA. The engulfing ancestor to the eukaryotic cell is itself thought to have been a primitive bacterial type of cell that lacked a cell wall.

Since the 1960s biologists have classified organisms as either prokaryotes or eukaryotes. Later work, however, suggests strongly that the reason that the prokaryotes are so diverse and have been so difficult to classify further is that the group actually is composed of two major groups of more ancient evolutionary divergence than previously believed. Members of the third group, the **archaebacteria,** are distinct from both eukaryotes and other prokaryotes, or eubacteria (Fig. 13-6).

The archaebacteria include methanogenic (methane-producing) bacteria, extreme halophiles (bacteria that live only in extremely saline environments), and thermoacidophiles (bacteria that live in acidic hot sulfur springs and smoldering coal tailings). The archaebacteria are related to the eukaryotes in at least one way that the eubacteria are not (amino acid sequence of a specific protein). They are distinguished from the eubacteria in several vital ways, including cell wall composition (absence of muramic acid), cell membrane lipids, presence of thymine instead of uracil on their *t*RNA molecules, and certain characteristics of the ribosomes. A comparison of genetic sequencing (amino acid sequences) of *r*RNA further demonstrates that members of the archaebacteria are clearly distinct from the eubacteria.

The consequent hypothesis that arises from all this research is that living cells as we know them were probably not descendants of some prokaryotic ancestor, as suggested in Fig. 13-5, but of some much more ancient common life form or **progenote.** The hypothesis makes further sense when the finely tuned and complex nature of protein synthesis is considered (Chapter 3). In all known life forms it is a consistent and extraordinarily precise mechanism. Logically it is likely that it was preceded by simpler and less accurate mechanisms. These mechanisms made more "mistakes"—produced more variation on which the forces of natural selection could act.

It is generally believed among scientists that Earth's original atmosphere was high in hydrogen, carbon dioxide, and other gases, but not oxygen. Oxygen came later, following the development of photosynthesis, which liberates oxygen from water molecules. Because the environment that was "home" to the earliest progenote organisms was vastly different, it is not unreasonable to assume that the evolutionary process was also different.

If we can accept, on the basis of the latest evidence, that there are three (Fig. 13-6) rather than two basic stems to the "tree of life," perhaps we are a little closer to identifying and understanding the miraculously complex process that has created (and continues to create) Earth's life forms. Because any kind of classification scheme is an attempt of humans to interpret phenomena of nature, no system should be considered final and inflexible; it remains subject to alteration based on additional evidence.

Below the kingdom level, the major taxonomic category for plants is the **division.** Yet even division names often are reevaluated and modified as clearer phylogenetic relationships are discerned. Do not be alarmed or discouraged by seemingly confusing differences of opinion among botanists on how to treat major plant categories. It is an excellent example of scientific process. Table 13-2 presents one acceptable listing of plant categories, including division names that may be used.

Our approach in this text is to concentrate on the *biology* of plants, rather than their classification. We will use the term "plant" in its historic, general sense to refer to all those organisms traditionally included in the study of botany whether classified as Monera, Protista, Fungi, or Plantae in a five-kingdom scheme. In this chapter we discuss single-celled plants and multicellular plants that lack vascular tissues. The following chapter reviews the vascular plants. *This introduction, and the following section on adaptive trends, apply to both chapters.*

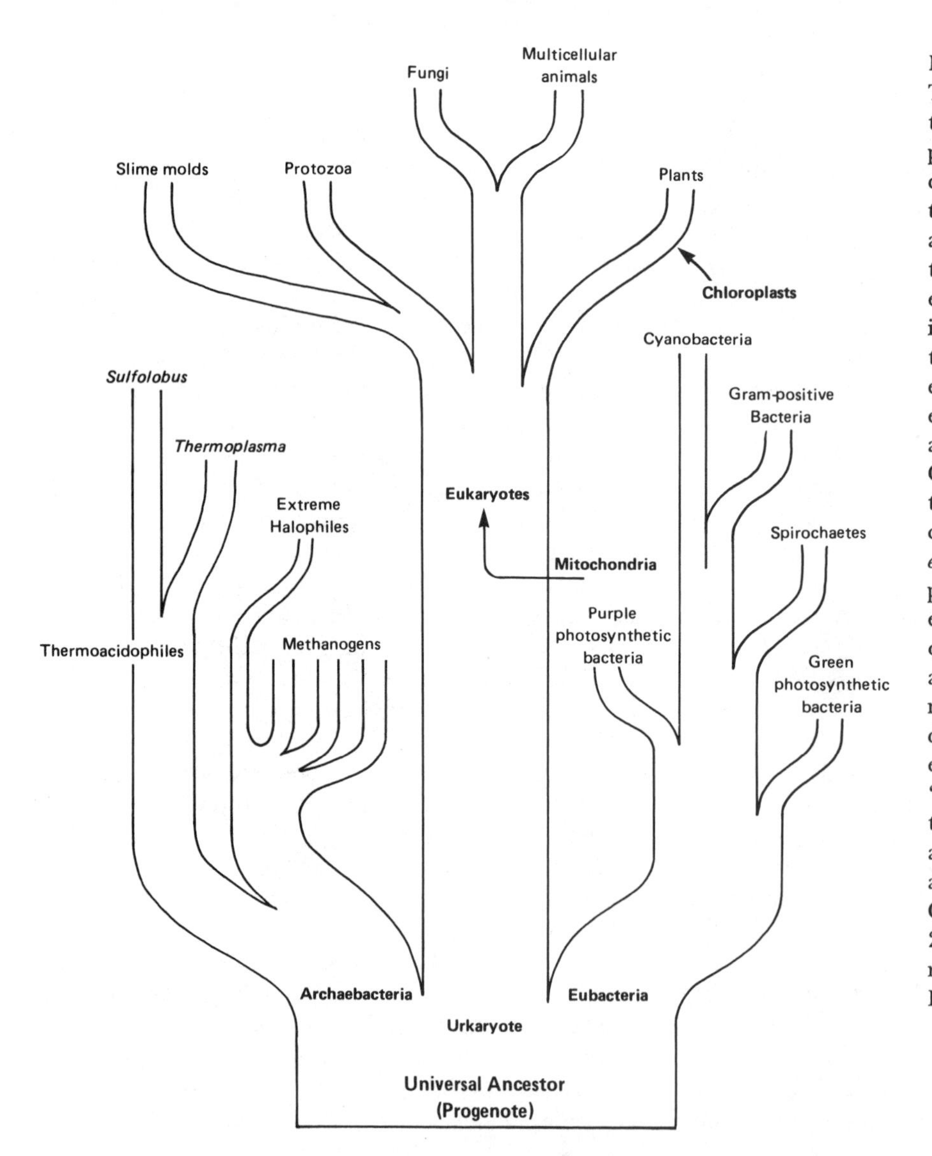


FIGURE 13-6
Three **primary kingdoms** (Archaebacteria, Eukaryotes, Eubacteria) are proposed by C. Woese to accommodate the discovery that the **archaebacteria** are fundamentally different from all other bacteria, which are designated the **eubacteria** (true bacteria). Both eubacteria and archaebacteria are alike in being prokaryotic cells—simple cells that lack a nucleus and are very different in their structural properties from eukaryotic cells, which have a nucleus and several other subcellular organelles. Genealogically, however, archaebacteria and eubacteria are no more closely related to *each other* than *either* group is to eukaryotes. It is proposed that the archaebacteria, the eubacteria, and an **urkaryote**—the original eukaryotic cell—stemmed from a common ancestor (the **progenote**) much simpler than the simplest present-day cells (prokaryotes). **Eukaryotes** evolved after the urkaryote became a "host" for bacterial **endosymbionts** that developed into mitochondrion and chloroplast. (Figure and caption adapted from "Archaebacteria," by Carl R. Woese, *Scientific American* 244(6): 98–122, June, 1981. Copyright 1981 by Scientific American, Inc. All rights reserved.)

TABLE 13-2 One of several acceptable classification schemes for "plants."[a]

| English Name | Taxonomic Division |
|---|---|
| **Non-Embryo-Forming Plants** | |
| **Non-vascular plants** | |
| Bacteria | Schizomycophyta |
| Blue-green "algae" | Cyanophyta |
| Viruses | none |
| Dinoflagellates and cryptomonads | Pyrrophyta |
| Diatoms and golden algae | Chrysophyta |
| Euglenoids (Euglenids) | Euglenophyta |
| Yellow-green algae and chloromonads | Xanthophyta |
| Green algae | Chlorophyta |
| Brown algae | Phaeophyta |
| Red algae | Rhodophyta |
| True fungi and lichens | Eumycota |
| Slime molds | Myxomycota |
| **Embryo-Forming Plants** | |
| Liverworts, hornworts, mosses | Bryophyta |
| **Vascular plants** | |
| *Non-Seed-Bearing Vascular Plants* | |
| Psilophytes (whisk ferns) | Psilophyta |
| Club mosses | Lycophyta |
| Horsetails and scouring rushes | Sphenophyta |
| Ferns | Pterophyta |
| *Seed-bearing Plants* | |
| Gymnosperms | |
| Seed ferns | Pteridospermophyta |
| Cycads | Cycadophyta |
| Ginkgo | Ginkgophyta |
| Gnetophytes | Gnetophyta |
| Conifers | Coniferophyta |
| Angiosperms | |
| Class: Monocots (Monocotyledonae) | |
| Class: Dicots (Dicotyledonae) | |

[a]In a five-kingdom system, the groups are distributed in the kingdoms Monera, Fungi, Protista, and Plantae. (The fifth kingdom is Animalia.)

## ADAPTIVE TRENDS AMONG "PLANTS"

The phylogenetic story of "plants," in the larger sense that we have defined them, is one of increasing complexity and specialization, plus gradual dominance of the terrestrial environment. In this progression several trends can be seen.

### Prokaryotic to Eukaryotic

The differences between prokaryotic and eukaryotic cells have already been noted. The prokaryotic condition is considered more primitive, although it works fine for those organisms that possess it! The greater complexity of the eukaryotic cell was a significant prerequisite to other advances.

### Simple to Complex

When all known related species of plants are compared, there is a trend from simple to complex, generalized to specialized. This trend begins at the cellular level and is evident at all structural levels; it is also expressed in metabolic abilities. Thus more "primitive" organisms tend to be less specialized and more "advanced" organisms more specialized, which is true not only in overall plant life but also within taxonomic groups down to the genus and species level.

### Asexual to Sexual Reproduction

Most plants are able to reproduce asexually. Among prokaryotes and some nonvascular eukaryotes it is the *only* means of reproduction. Genetic recombination occurs in some of these species by transfers of genetic material between vegetative cells, as compared to sexual (gametic) recombination. Most eukaryotic species have true sexual reproduction, involving an alternation of syngamy and meiosis within the life cycle (Chapter 5).

### Haploid to Diploid Dominance of the Life Cycle

In species that possess sexual reproduction and a complete alternation of spore-producing and gamete-producing generations, several types of life cycles are expressed. In many green algae and all bryophytes (mosses and liverworts), the haploid stage is dominant (Fig. 13-7). Thus the gametophyte is the dominant, recognizable vegetative plant and the sporophyte is a less obvious stage. Several algae, such as sea lettuce (*Ulva*), have a life cycle in which haploid and diploid vegetative phases are outwardly identical (Fig. 13-8).

In vascular plants the sporophyte is the dominant, recognizable form of the plant (Fig. 13-9). Ferns and other primitive vascular plants have an inconspicuous free-living gametophyte. The gametophyte of seed plants is not free living (Chapters 5 and 14). A fourth type of life cycle, in which the gametophytic phase is totally absent and the

**FIGURE 13-7** A life cycle with a dominant gametophyte stage in which the diploid generation is represented *only* by a zygote, which undergoes meiosis instead of producing a sporophyte. (Another type of dominant-gametophyte life cycle, one having a distinctive sporophyte, is indicated in Fig. 13-53.)

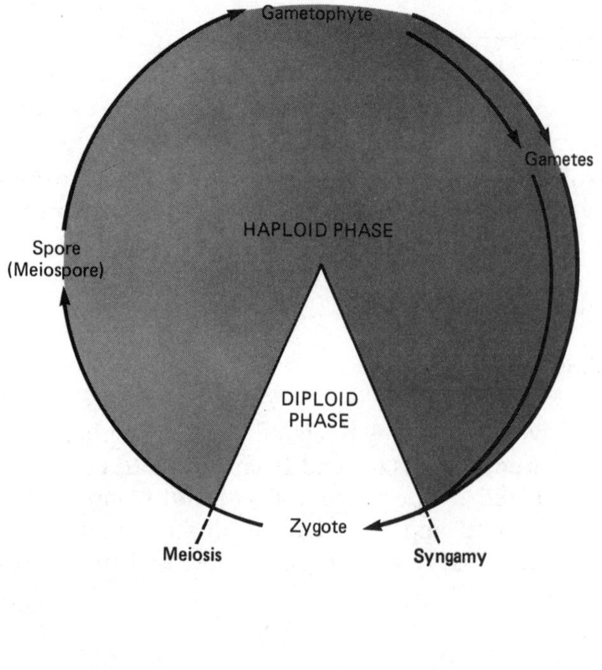

Gametophyte
Gametes
HAPLOID PHASE
Spore (Meiospore)
DIPLOID PHASE
Zygote
Meiosis
Syngamy

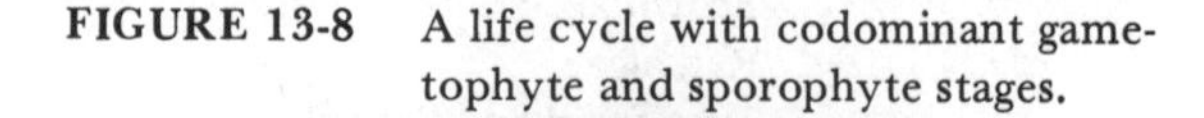

**FIGURE 13-8** A life cycle with codominant gametophyte and sporophyte stages.

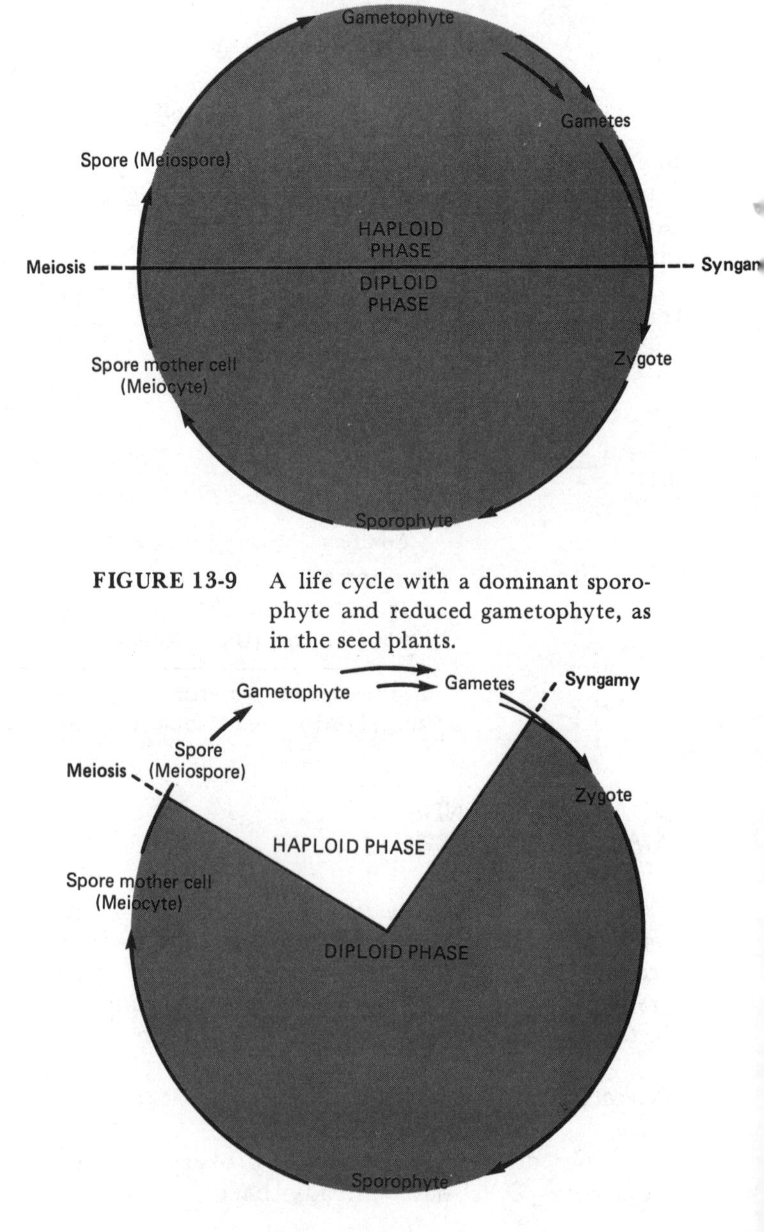


**FIGURE 13-9** A life cycle with a dominant sporophyte and reduced gametophyte, as in the seed plants.

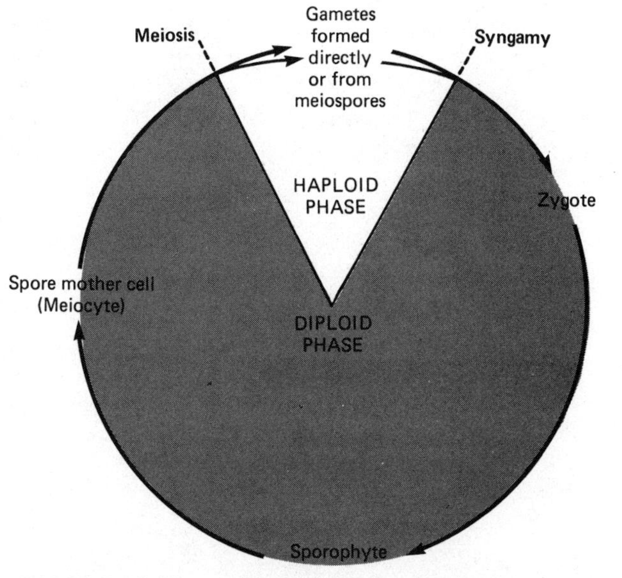


**FIGURE 13-10** A life cycle with a dominant sporophyte stage and *no* gametophyte.

gametes are formed directly by meiosis, is rare in plants but is the normal type of life cycle in higher animals, including humans. The seaweed *Fucus* has this fourth type of life cycle (Fig. 13-10).

### Aquatic to Terrestrial

If there is a major theme or story to the phylogenetic history of plant groups, it is one of increasing adaptation to a terrestrial environment, of populations changing as the physical features of Earth changed. Although there is much debate over whether the first organisms originated in fresh water or saline water, it is assumed that there was first a liquid medium. The transition of plants from total dependence on a liquid medium to a fully terrestrial lifestyle can be traced phylogenetically through the life histories of living as well as fossil groups. The adaptations necessary to this transition are emphasized in this chapter and the following one.

Geological evidence indicates that the surface of our planet underwent many gradual changes over millions of years (Table 13-1). Shifting land masses and volcanic events repeatedly reshaped the surface of the planet and altered the depths and temperatures of Earth's waters. Major climatic changes occurred in every region. Seas came and went and the populations in these ancient waters gradually adapted to the recession of the waters or perished. **Extinction** can be interpreted as a failure of a population to change rapidly enough to survive. In considering the adaptive advances from one plant group to another, therefore, we must try to visualize adaptations as *gradual* shifts in genetic constitution of populations in response to the gradual changes in Earth's environments over these millenia.

The transition from predominantly aquatic to predominantly terrestrial existence involved many adaptations to reduce and prevent desiccation. Lacking a dense aquatic medium to provide buoyancy, land plants also require internal structural means of resisting gravity or else erect growth toward the sun would not be possible.

Plants surrounded by an aqueous solution can absorb water and nutrients by diffusion over the whole body surface. Adaptation to terrestrial life required the development of special absorbing and conducting tissues. The presence of specialized "plumbing" thus enabled plants to attain great size and complexity.

Metabolic processes frequently produce by-products that cannot be further utilized but that may be toxic in sufficient quantities. How plants deal with them is less known than animal mechanisms. Diffusable wastes can be shed directly into an aquatic environment. In higher plants diffusable waste substances may be lost into the soil solution via the root system; some may volatilize from aerial surfaces. Substances that cannot be eliminated by these means, however, are sometimes converted into such chemical forms as inorganic crystals. Leaf fall is an excellent means of waste elimination by terrestrial plants.

In the aquatic environment syngamy occurs between two naked gametes, but the gametes of terrestrial plants require special circumstances to avoid desiccation. Bryophytes (mosses and their relatives) and the lower vascular plants remain dependent on the presence of liquid water from rain or dew to complete fertilization. Fully adapted terrestrial plants (the seed plants) possess an adaptive solution to the problem identical in effect to that reached by fully terrestrial animals—internal fertilization. Fusion of the nuclei occurs completely within the protective confines of maternal tissues, as does initial embryo development. The highest advancement is the development of the seed.

## PROKARYOTES

The kingdom Monera places together all those organisms that are prokaryotic, lacking distinct membrane-bounded organelles, including a nucleus, mitochondria, and plastids. They lack microtubules and the complex 9-plus-2 tubule flagellar organization found in eukaryotes. Nucleic acids, respiratory enzymes, and pigments of prokaryotes are scattered in the cytoplasm or located on the surfaces of membranes within the cytoplasm. Even colonial and filamentous forms lack protoplasmic connections between adjacent cells; thus they are not "truly" multicellular. The major prokaryotic groups are the bacteria and blue-green "algae" (which now may also be classified as bacteria). The viruses are sometimes included in the Monera, but their organization is unique, meriting separate consideration.

## Bacteria

Bacterial cells are of three basic shapes: **rods** (Fig. 13-11), spirals or **spirilli** (Fig. 13-12), and spheres or **cocci** (Fig. 13-13). Bacterial cells may be single, in chains, clusters, or as funguslike filaments. Many bacteria possess simple flagella for locomotion (Fig. 13-14) and some move by gliding.

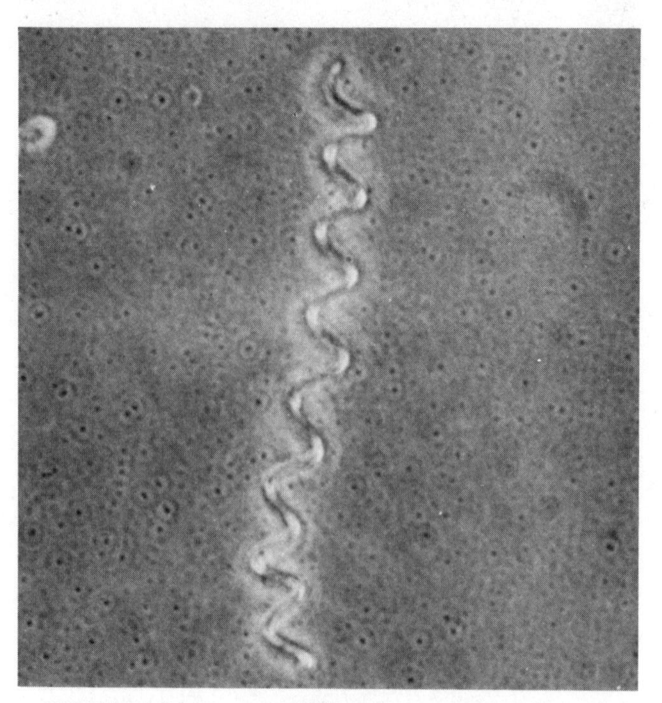

**FIGURE 13-12** A **spirillum** is a spiral form of bacterium. (Photo courtesy of Carolina Biological Supply Co.)

**FIGURE 13-11** Scanning electron micrograph (SEM) of *Escherichia coli*, a **rod** bacterium. The cells are shown about 10,500 times their actual size. (Photo courtesy of Suzanne Moeller.)

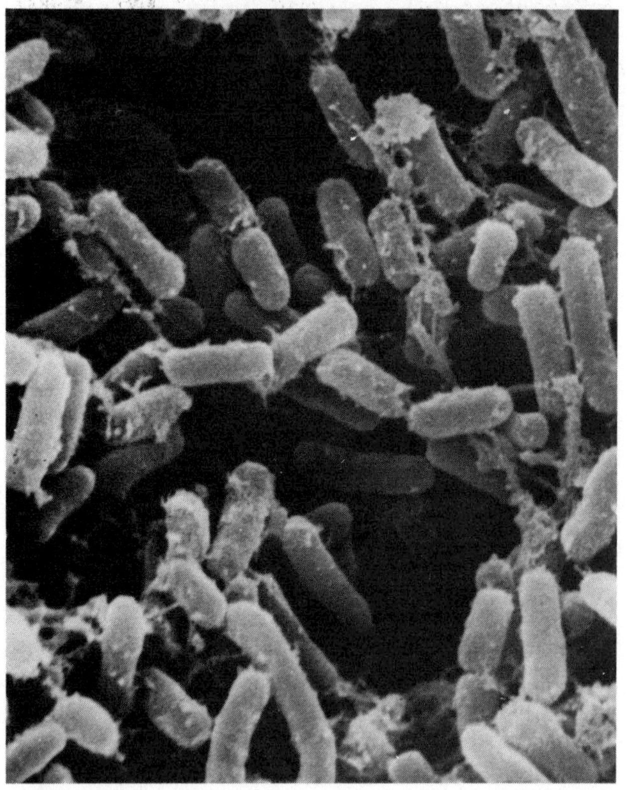

**FIGURE 13-13** SEM of a cluster of **cocci**, spherical bacteria, in an alder (*Alnus*) root nodule cell, magnified about 1,340 times their actual size. (Photo courtesy of Paul A. Schlotfeldt.)

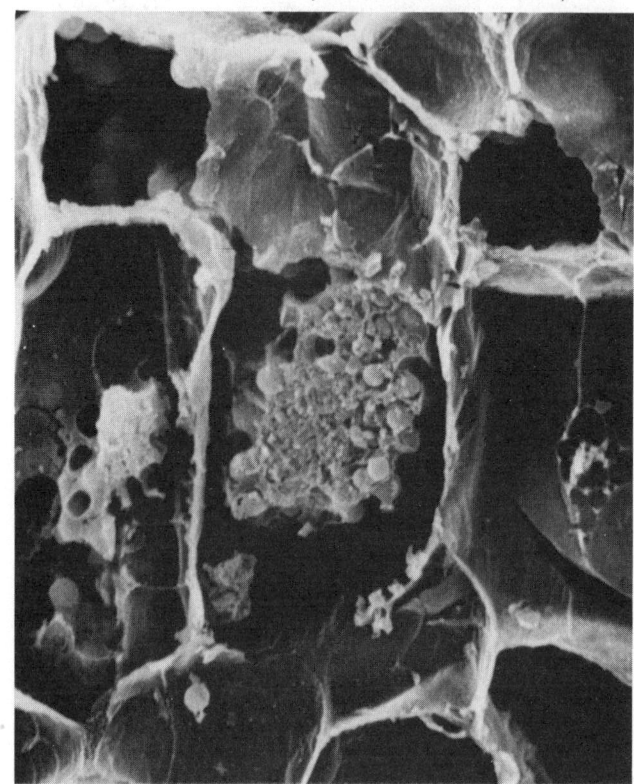

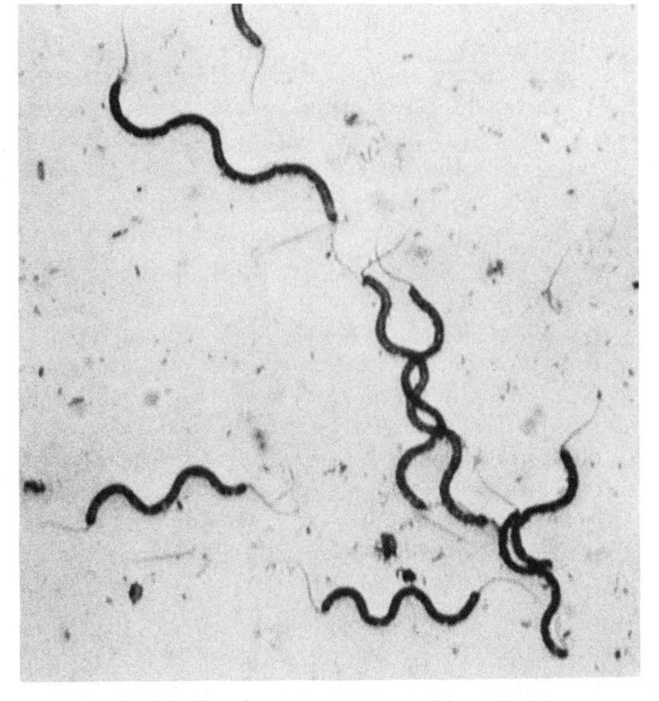

**FIGURE 13-14** Some bacteria, as this one, are flagellated.

Most bacteria are heterotrophic. A few are **chemosynthetic** (chemoautotrophic), oxidizing inorganic molecules (particularly sulfur, nitrogen, iron) to obtain energy. Some bacteria are **photosynthetic**, including sulfur bacteria.

Bacteria differ in their oxygen requirements. Some survive only under **aerobic** conditions; some prefer aerobic conditions but can survive in the absence of oxygen; others survive only under **anaerobic** conditions. (See also composting, Chapter 10.) The bacteria (*Clostridium*) that cause tetanus are anaerobic, which is why puncture wounds, lacking exposure to the air, are so dangerous. If the bacteria thrive, they can produce sufficient quantities of a toxic nerve poison to cause extreme illness or death.

Bacteria reproduce mainly by **fission**, which is preceded by enlargement of the parent cell and DNA replication. Because prokaryotic cells lack spindle fibers to assist in distributing the chromosomes (such as during mitosis and meiosis), the DNA is distributed equally to daughter cells by other means. In many bacteria the DNA molecules replicate, separate, and then attach to involuted areas of the cell membrane (mesosomes) in opposite regions of the parent cell. When the parent cell divides, the duplicated DNA is separated along with the cytoplasm.

Heat and drought-resistant **spores** are formed by many bacteria. (See also the food poisoning discussion that follows.)

Bacterial DNA exists as a ring. In addition, many smaller circles of DNA called **plasmids** may be present. Plasmids are important tools in current "gene-splicing" techniques.

Genetic variation occurs frequently with bacteria, as shown by the relative rapidity with which bacterial strains can become immune to specific antibiotics. The most significant source of variation in bacteria is **mutation**, which can have a rapid effect because bacteria multiply so rapidly. Under optimum conditions the much-studied *Escherichia coli* (Fig. 13-15) may double its population every 12.5 minutes!

Another source of variation is **genetic recombination.** During genetic recombination a DNA fragment is transferred from one cell to another. After introduction to the new cell the DNA fragment may become incorporated into the cell's existing DNA, thereby becoming part of that cell's heritable genetic constitution. Or it may be there as functional (RNA-producing) but not heritable DNA. Genetic recombination occurs in

**FIGURE 13-15** Bacterial cells divide by **fission,** as these *Escherichia coli* cells, magnified about 24,000 times their actual size. (Photo courtesy of Suzanne Moeller.)

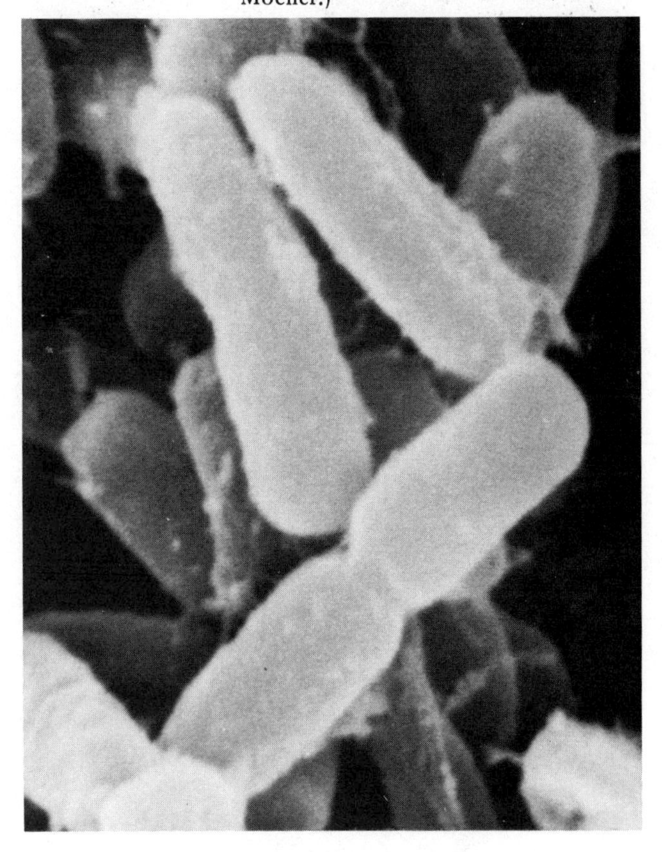

several ways among various bacteria. DNA can be transferred by **conjugation**, the physical exchange of genetic material between cells that form a cytoplasmic bridge through which DNA can move. Plasmids frequently are transmitted in this way. A kind of genetic recombination called **transduction** occurs when bacteriophages (viruses that attach themselves to bacterial cells) take in bacterial DNA, incorporate it with their own DNA, and then transmit the "hybridized" DNA to another bacterial cell, which incorporates the introduced DNA into its own chromosomes. Many bacteria are capable of **transformation** whereby DNA released from bacterial cells that are brekaing up can be assimilated directly into other bacterial cells.

Bacteria surround us, and although we usually think of them as disease causers, the living world depends on them in many ways. Without bacteria (and fungi), elements locked up in complex organic substances of plant and animal bodies would remain there indefinitely. Without decay, the planet would perhaps have run out of molecular building blocks ages ago and its surface would be miles deep in the nondecomposed wastes and bodies of deceased plants and animals. Instead, through bacterial decay and conversions, those elements essential to life are continually recycled through generation after generation of organisms. It is probable that each of us contains carbon atoms once part of a primeval algaelike organism, a tree, a fern, or a dinosaur! Such "chemical reincarnation" is the result of the necessary activities of bacteria.

Bacteria also convert nitrogen and ammonia into nitrates for incorporation into plant proteins and subsequently animal proteins (Fig. 2-7). Many bacterial species participate in human lives by their roles in the production of such foods as yogurt, cheese, vinegar, and sauerkraut. Bacteria are utilized in numerous commercial, industrial ways (Chapter 16). And they have been invaluable in genetic research into the structure and function of DNA molecules.

Bacterial diseases of plants include a variety of wilts and blights, crown gall, and soft rot of vegetables. Serious human diseases caused by bacteria include staphylococcus and streptococcus infections, syphilis, gonorrhea, gangrene, tetanus (lockjaw), bubonic plague, cholera, diphtheria, tuberculosis, leprosy, bacterial pneumonia, rheumatic and scarlet fevers, bacterial dysentery, bacterial meningitis, typhoid and undulant fevers, tularemia, and whooping cough.

When you obtain a small scratch or cut, a staphylococcus infection is possible, even likely. Left untended, the minor wound would most certainly be sore and red within 24 hours because of bacterial infection. Pus would form as white blood cells accumulated to attack the developing bacterial population. Immediate antibacterial treatment, however, allows rapid healing of the wound with no subsequent pain or inflammation. Thorough washing, even with plain soap and water, is thus recommended for minor skin wounds.

One of the most common, most dangerous, and most preventable types of bacterial disease is food poisoning. Food poisoning nearly always results from the unsanitary handling of food during processing or preparation, insufficient cooking, or improper and inadequate chilling. Food processing plants may be the source of the problem when equipment is contaminated by feces of birds, rodents, or the animals being processed. Workers with improperly washed hands are a frequent source of infection.

Restaurant and home kitchens are also responsible for many cases of food poisoning. This situation may be due to use of contaminated equipment (knives, cutting boards) or failure to wash hands adequately after going to the bathroom, touching environmental surfaces, wiping, rubbing, or scratching the face, sneezing, coughing, or blowing the nose. (Observe the hand contacts of waiters, waitresses, and cooks at food establishments as food is prepared and served the next few times you eat out!)

Following contact (inoculation) of food with food poisoning bacteria, the full development of toxins by them is greatly affected by environmental factors, especially temperature and pH. Low temperatures and an acid environment retard bacterial growth; room temperatures and those maintained by insufficient food-warming equipment (such as faulty steam tables) are ideal for the development of bacterial populations and their toxins.

The three most common types of food poisoning are those caused by *Salmonella, Staphylococcus,* and *Clostridium* bacteria. Botulism—caused by the anaerobic *Clostridium botulinum*—is especially dangerous for it usually causes irreversible damage to the central nervous system and is frequently fatal. Botulism is commonly the result of improper canning of such nonacid foods as vegetables and soups. Some tomatoes are low acid and also require heat and pressure canning. Table 13-3 describes certain factors of these three types of food poisoning.

TABLE 13-3 Three common types of bacterial-caused food poisoning.

| | *Staphylococcus aureus* | *Salmonella enteritidus* and *S. choleraesius* | *Clostridium botulinum* |
|---|---|---|---|
| **Description** | nonmotile, aerobic coccus; grapelike clusters | motile, aerobic rod | motile, anaerobic rod |
| **Foods involved** | high-protein foods, such as meats, dairy products; mayonnaise-base salads, such as potato, ham, macaroni; dressing, sauces, gravies, bread pudding, leftovers | mainly meat, poultry, eggs, and their products; also coconut, yeast, smoked fish, dry milk | improperly canned low-acid food; smoked, vacuum-packed fish; fermented foods |
| **Toxin type**[a] | thermostable; exotoxin | thermolabile; endotoxin | thermolabile; endotoxin |
| **Usual incubation period** | 1–7 hours; usually 2–4 hours | 5–72 hours; commonly 12–36 hours | 2 hours–6 days; often 12–36 hours |
| **Usual duration of problem** | 1–2 days | 3–7 days | slow recovery or fatal in 3–10 days |
| **Signs and symptoms** | sudden nausea and vomiting, abdominal cramps, diarrhea, sweating, weakness, prostration, dehydration | abdominal pain, vomiting, diarrhea, chills, *fever*, prostration, headache, loss of appetite, malaise, dehydration | nausea, vomiting, abdominal pain, and diarrhea as early symptoms; then many nervous symptoms; including loss of light accommodation, headache, vertigo or dizziness, dry mouth, difficulty in swallowing, respiratory distress; partial paralysis may persist 6–8 months; irreversible CNS damage |
| **Mortality** | no (rarely) | occasionally | frequently (50–65%) |
| **Infection and source** | *primarily human nasal chambers;* infected wounds; pimples; also mastitic udders; bruised or arthritic poultry tissue | animal feces, including human; fecal traces on egg shells | soil, mud, water, animal feces; spores *highly* heat resistant |
| **Prevention and control** | personal hygiene; sanitize equipment, exclude ill persons from food contact; chill foods rapidly in small quantities; keep picnic foods, sandwiches, etc., chilled | cooking, pasteurization; sanitation in farm, processing and home phases of preparation; chill foods rapidly in small quantities | sufficient heat and pressure during canning; thorough cooking of home-canned foods; sufficient salt in cured or pickled products; proper refrigeration |

[a]**Endotoxins** are toxic substances that remain within the cell. **Exotoxins** are secreted outside the cell. **Thermostable** toxins are not destroyed by heat whereas **thermolabile** toxins are.

Information adapted from *Diseases Transmitted by Foods,* U.S. Department of Health, Education and Welfare, Public Health Service, 1971.

Of the three types, *Staphylococcus* poisoning is the one usually occurring among large groups of people at picnics, restaurants, dormitories, receptions, and so on. Symptoms generally occur within a few hours. *Salmonella* poisoning has a longer incubation and recovery time, is more severe, and normally involves fever in addition to digestive upset. Botulism and *Salmonella* toxins are destroyed by lengthy cooking at boiling temperatures; *Staphylococcus* toxins are not destroyed by heat, even by 30 minutes of boiling. Other food-transmitted diseases are summarized in Table 13-4.

Recognizing the extreme importance of hygiene, adequate cooking, and proper refrigeration in preventing food poisoning, consider the following common practices:

1. leaving food unrefrigerated for several hours between preparation and serving.
2. carrying food to a picnic or potluck in an unrefrigerated container.
3. leaving food out overnight.
4. preparing foods in large quantity; then refrigerating them in a large container.
5. allowing food to cool to room temperature before refrigerating. (The latter practice is probably a carryover from the days of iceboxes, for putting warm food into the icebox would naturally melt the ice!)
6. failing to scrub a cutting board between uses.
7. drinking unpasteurized milk or raw-egg beverages or using raw-egg mixtures in homemade ice creams.
8. stuffing a turkey the evening before roasting.
9. failing to wash hands after touching the face or after visiting the bathroom.

**TABLE 13-4 Other significant bacterial diseases usually transmitted in foods.**

*Salmonella typhi* **(typhoid fever)**

| | |
|---|---|
| Foods: | High-protein foods, raw salads, milk, shellfish; foods that have been handled and then eaten without further treatment |
| Source: | Feces, urine of infected humans; such carriers as infected humans and flies of great importance; also water |
| Control: | Immunization, personal hygiene; protect and treat water; proper sewage disposal; fly control; careful handling and preparation of food; pasteurization; thorough cooking; rapid and adequate refrigeration of foods |

*Salmonella enteritidis* **serotypes (paratyphoid fever)**

| | |
|---|---|
| Foods: | Milk, shellfish, raw salads, eggs |
| Source: | Feces, urine of infected humans; carriers of great importance |
| Control: | Personal hygiene; pasteurization of eggs and milk; protect and treat water; thorough cooking; rapid and adequate refrigeration of foods |

*Clostridium perfringens* **type A**

| | |
|---|---|
| Foods: | Cooked meat or poultry that has been kept at room temperature several hours or cooled slowly; gravy, stew, meat pies |
| Source: | Animal feces, including humans; soil, dust; sewage |
| Control: | Proper cooking and chilling of foods; personal hygiene |

*Clostridium perfringens* **type C (formerly type F)**

| | |
|---|---|
| Foods: | Pork, other meat, fish |
| Source: | Animal feces, plus dietary predisposal |
| Control: | Proper cooking and chilling of foods; eat balanced diet |

*Bacillus cereus*

| | |
|---|---|
| Foods: | Custards, cereal products, puddings, sauces, meat loaf |
| Sources: | Soil and dust |
| Control: | Sanitary preparation and proper chilling of foods; personal hygiene |

*Arizona arizonae*

| | |
|---|---|
| Foods: | Turkey, cream-filled pastry, ice cream |
| Source: | Feces of infected animals, including humans, especially reptiles |
| Control: | Prevent fecal contamination; sanitary equipment and proper chilling |

*Escherichia coli* **(certain groups)**

| | |
|---|---|
| Food: | Coffee substitutes, possibly salmon (?); however, usually transmitted by person-to-person contact |
| Source: | Human feces; infants especially susceptible |
| Control: | Personal hygiene; proper cooking and chilling of foods; protect and treat water; proper sewage disposal |

Information adapted from *Diseases Transmitted by Foods,* U. S. Department of Health, Education and Welfare, Public Health Service, 1971.

**Rickettsias** are structurally similar to bacteria and reproduce by fission; so they are now classified with the bacteria. All are obligate intracellular parasites of animals. Rocky Mountain Spotted Fever is one of several rickettsial fevers transmitted by such "hard" ticks as wood ticks. Typhus, another significant rickettsial disease, is usually transmitted by human lice.

**Mycoplasmas** are small bacteria belonging to a few species that cause plant and animal diseases. The cabbage palms (*Sabal palmetto*) of the southeastern United States are being killed off by a mycoplasmal disease of their crowns.

## Blue-green "Algae"

Although they superficially resemble "true" algae in gross appearance and their presence in aquatic habitats, blue-green "algae" are quite distinct from the true algae. Some scientists now classify the blue-green "algae" (or simply blue-greens) as **cyanobacteria** (Gr., *kyanos* = dark blue), a subgroup of the bacteria. Bacterialike in many respects, they also resemble the eukaryotic algae in some ways. We have given them their own heading because they are still referred to most often as the blue-green algae.

Blue-greens are found in freshwater, soil, rock fissures, a few in the marine environment, and a few in the atmosphere! Some occur in and (along with bacteria) give color to hot springs, such as those in Yellowstone National Park's geyser basins. Some blue-greens are symbiotic with lichens, liverworts, and a variety of vascular plants (in the roots).

Most people know blue-greens as the undesirable green matter that thickens the water of ponds, lakes, and watering troughs during the warm days of summer, particularly in waters that are high in nutrients. They contribute to existing pollution by blocking light to submerged aquatic plants and by producing substances that are toxic to waterfowl, livestock, and humans. Their bodies and those of submergent plants that are killed by lack of light and oxygen contribute to the organic decomposition of bottom sediments, further increasing carbon dioxide and nutrient concentrations.

Not all blue-greens are a human nuisance, however, for some species contribute to soil fertility because of their ability to "fix" nitrogen, converting elemental nitrogen to a form usable by other plants. Blue-greens are deliberately "seeded" into rice paddies so that they will grow and provide nitrogen for the rice plants.

Most blue-greens contain chlorophyll, carotenoids, the blue pigment **phycocyanin,** and often the red pigment **phycoerythrin.** In most cellular respects, however, they resemble bacteria. Most blue-greens exist as filaments, surrounded by a gelatinous sheath or mass in which symbiotic bacteria often live. Some blue-greens are unicellular or colonial. Filamentous blue-greens may move by gliding or by rotation around the longitudinal axis. Figures 13-16, 13-17, and 13-18 illustrate some common blue-greens. The life cycle of *Nostoc,* which usually exists as a mass of filaments embedded in a gelatinous ball, is shown in Fig. 13-19.

**FIGURE 13-16** The gelatinous matrix surrounding cells of the nitrogen-fixing **blue-green** *Gloeocapsa* gives each unit a spherical form. (Photo courtesy of Carolina Biological Supply Co.)

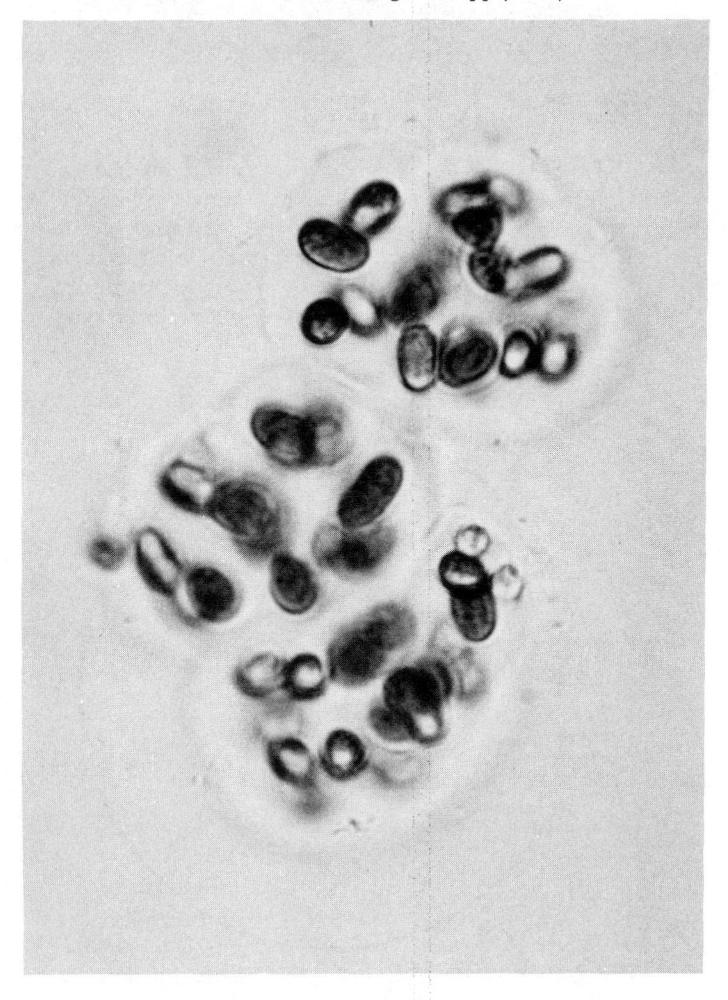

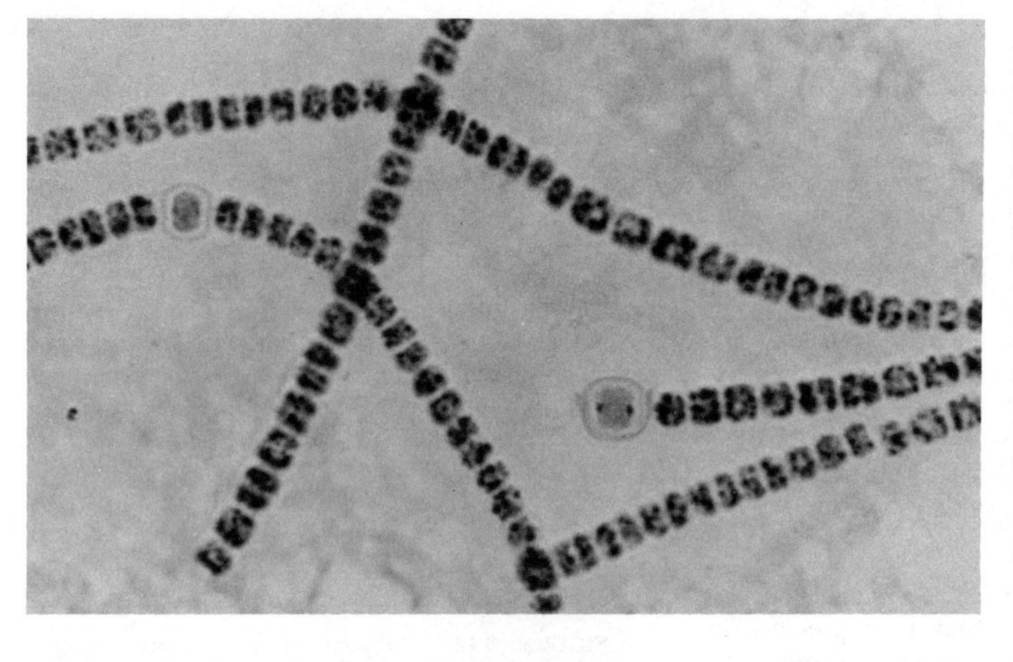

**FIGURE 13-17**
Each *Nostoc* globe consists of filaments embedded within a gelatinous matrix. Filaments are composed of a series of similar cells and occasional special cells called heterocysts and akinetes. An **akinete** is a cell of vegetative origin that develops food reserves and a thickened wall. It can resist a dormant period and then germinate into a new filament. **Heterocysts** are clear walled and function in nitrogen fixation and akinete formation, among other things. This picture shows heterocysts. (Photo courtesy of Carolina Biological Supply Co.)

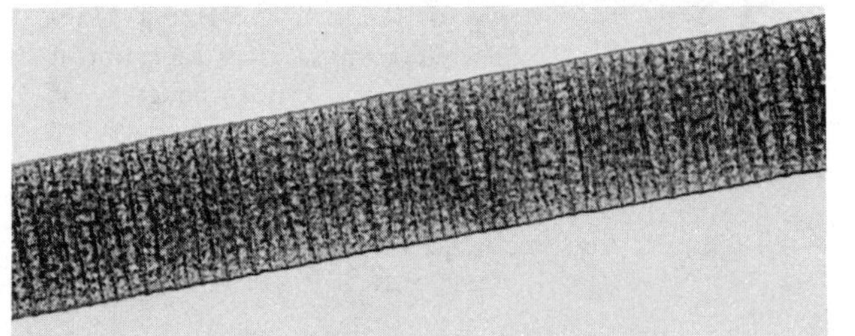

**FIGURE 13-18** *Oscillatoria* is a common filamentous blue-green. Note the very thin gelatinous outer covering, cross-walls of cells that are wider than they are long, and lack of nuclei. (Photo courtesy of Carolina Biological Supply Co.)

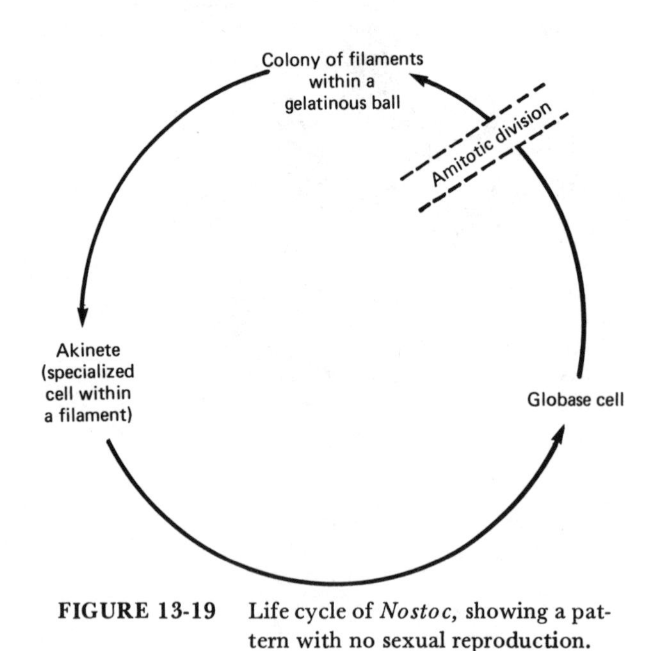


**FIGURE 13-19** Life cycle of *Nostoc*, showing a pattern with no sexual reproduction.

## VIRUSES

Viruses are exceptional entities, highly specialized obligate intracellular parasites of all organisms, which inject their own RNA and/or DNA into host cells and actually take over the metabolic "machinery" of the host cells. Their own reproduction occurs by manipulation of host-cell DNA and RNA synthesis.

The physical structure of viruses is unique and they do not meet the criteria generally used to define living things (Chapter 2). Their structure is noncellular! Each virus particle (virion) consists of a DNA or RNA core enclosed in a protein sheath (Fig. 13-20). Various nonprotein substances may also be associated with the virus particle.

There are *many* significant virus diseases of plants; viruses are therefore of great economic importance. Tobacco mosaic virus was the first virus to be isolated in the laboratory. Because of the ability to cause disease, the name *virus* was given, which is the Latin word meaning poison. Many human diseases are also viral and some have caused fatal epidemics. Examples are smallpox, yellow fever, poliomyelitis, influenza, measles, mumps, and rabies. Immunization vaccines are now used for these serious diseases. Chicken pox is usually such a mild virus disease that preventative immunization is not routinely administered. The common cold remains a problem, for research has not yet pinned down its causative virus(es).

Some dramatic research has demonstrated that viruses cause certain types of cancer in domestic animals and humans. This hypothesis has been around for a long time, awaiting proof. Like

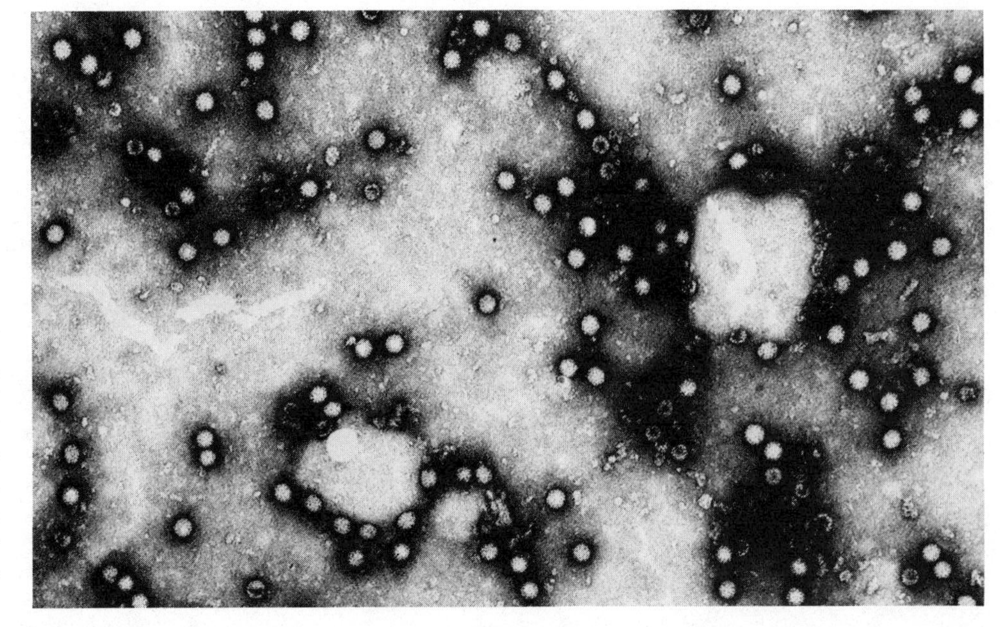

**FIGURE 13-20**
**Virus** particles are tiny, noncellular entities composed of nucleic acids and associated proteins. The round objects here are viruses that cause maize white line mosaic disease, and are shown magnified 60,000 times their actual size. (Photo courtesy of Gary R. Gaard and Gustaaf A. deZoeten.)

the common cold, however, cancer does not appear to be a single disease and it continues to be a highly active subject for biomedical research.

## THE PROTISTA, IN GENERAL

As originally defined in a three-kingdom system, the Protista included all single-celled organisms, even the bacteria. Whittaker's scheme (Fig. 13-5) defined Protista as single-celled and colonial eukaryotes. This definition involved separating the various groups of eukaryotic organisms collectively called "algae" and placing simple forms in the Protista and more complex multicellular forms in the Plantae. Both schemes have justifications as well as practical advantages and disadvantages. As stated earlier, we will discuss all the eukaryotic algae together.

Members of the animal phylum Protozoa are also classified as protists. Some protists possess both plantlike and animallike characteristics. Within two groups (the euglenoids and dinoflagellates) are photosynthetic forms that are classified and studied both as algae and as protozoans. In the same groups are nonphotosynthetic forms that are clearly similar in all other respects to the photosynthetic forms but that are studied as protozoans, not as algae.

Protists are exceptionally common in marine and freshwater environments; some are found in moist soil or other terrestrial environments. Temporary ponds, puddles, and roadside ditches are especially good collection sites. Many protists possess locomotory ability, via **amoeboid movement** (movement by extension and contraction of the cell margin), **cilia**, or **flagella**. Cilia and flagella of protists have the complex 9-plus-2 structure typical of eukaryotes. **Contractile vacuoles**, which assist in expelling excess water, are found in the motile cells of many algae as well as in the protozoa. Many protists exist as symbionts within the tissues of animal species; some protozoan protists are parasitic.

## ALGAE

Photosynthetic protists and multicellular algae are among the most significant plant groups ecologically, for they are the primary photosynthetic producers for life on more than two-thirds of Earth's surface—its waters. As the foundation of aquatic food chains, algae are essential to humans, who consume aquatic animals. As a byproduct of photosynthesis, oxygen is returned to the water and from the water to the atmosphere! More than 70 species of marine algae are used as human food. The green, brown, and red algae demonstrate a wide range of structural and reproductive complexity.

Some of the unicellular algae produce gametes but have no specialized sex organs; some algae have unicellular sex organs; others have multicellular sex organs called **gametangia.** Gametangia (female = **oögonium**, male = **antheridium**) are also characteristic of more advanced groups than the algae but with a significant difference: each cell of algae gametangia forms a gamete whereas the gametangia of more advanced plants consist of an outer layer of sterile vegetative cells surrounding the gametic cells/tissues.

Table 13-5 presents some of the distinguishing characteristics among the various groups of algae.

**TABLE 13-5 Selected characteristics of six eukaryotic algae divisions.**

| Division and Common Name | Approx. Number of Known Species | Pigments | Food Storage Products | Flagella | Major Cell Wall Components | Habitats |
|---|---|---|---|---|---|---|
| Chrysophyta (diatoms and golden-brown algae) | 10,000 | Chlorophyll *a*, chlorophyll *c*, carotenes, xanthophylls, including fucoxanthin | Leucosin, chrysolaminarin, oil | 1 or 2, apical, equal or unequal | Cellulose, pectic compounds, silica, other substances | Marine, freshwater, brackish water; some terrestrial |
| Euglenophyta (euglenoids) | 800 | Chlorophyll *a*, chlorophyll *b*, β-carotene, xanthophylls | Paramylon, oil | 1 to 3, apical or subapical | No cell wall | Mostly freshwater; brackish water, marine; some terrestrial |
| Pyrrophyta (dinoflagellates) | 1000 | Chlorophyll *a*, chlorophyll *c*, β-carotene, xanthophylls | Starch, (oil in some) | 2, lateral, one girdling, one trailing | Cellulose, other materials; or no cell wall | Marine, freshwater, brackish water |
| Chlorophyta (green algae) | 7000 | Chlorophyll *a*, chlorophyll *b*, carotenes, xanthophylls | Starch, (oil in some) | None, 1, or 2-8, apical or lateral, equal | Polysaccharides, sometimes cellulose; sometimes calcified (calcium carbonate) | Mostly freshwater, some marine; few brackish water, terrestrial |
| Phaeophyta (brown algae) | 1500 | Chlorophyll *a*, chlorophyll *c*, β-carotene, xanthophylls, including fucoxanthin | Laminarin, mannitol | 2, lateral, tinsel forward, whiplash behind; in reproductive cells only | Cellulose matrix with alginic acids and other polysaccharides | Almost all marine, especially in cold ocean waters |
| Rhodophyta (red algae) | 4000 | Chlorophyll *a*, carotenes, xanthophylls, phycobilins | Floridean starch (similar to glycogen, an animal product) | None | Cellulose, pectic materials, other polysaccharides; many calcified (calcium carbonate) | Marine, some freshwater; many in tropical seas |

## Diatoms

Within aquatic samples, many of the species that you see will be symmetrical yellowish diatoms (Fig. 13-20) that float freely or grow as epiphytes on submerged plant surfaces. The delicate beauty of diatoms has long fascinated observers [Fig. 13-21 (a), (b), (c), and (d)].

The outer walls of a diatom consist of two overlapping sides, or **valves**, fitted one over the other like a petri dish [Fig. 13-22 (a) and (b)]. During sexual reproduction the valves separate and an exchange of genetic material occurs between adjacent cells. This process is followed by new valve production and asexual division.

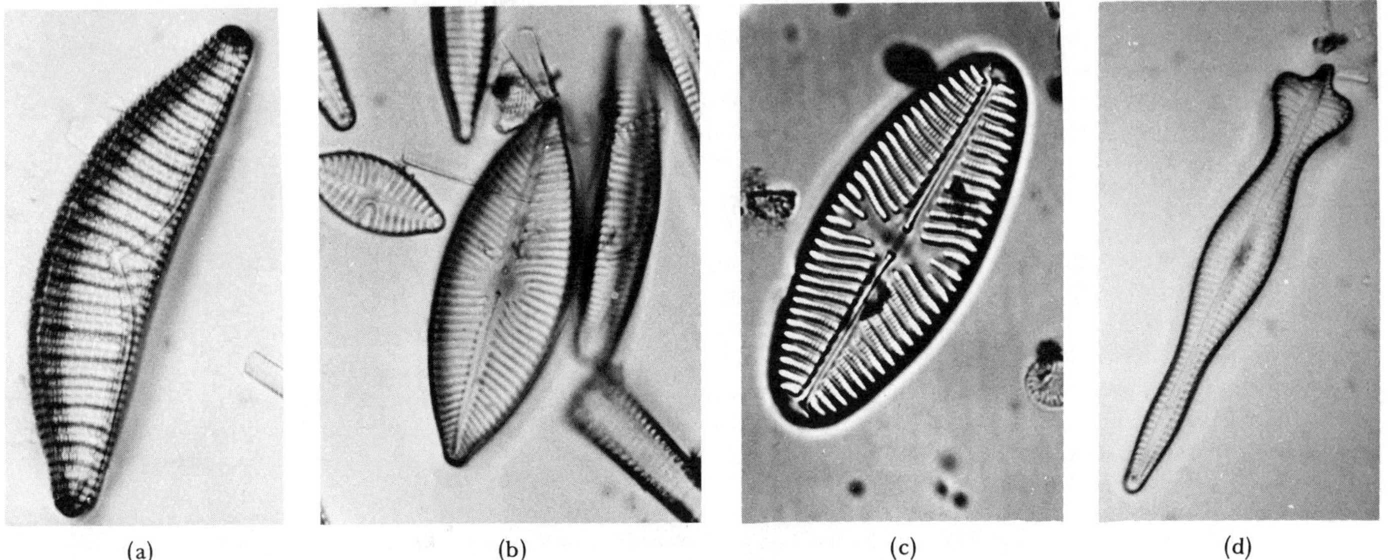

(a) (b) (c) (d)

**FIGURE 13-21** Assorted freshwater **diatoms**, including (a) *Epithemia*, (b) and (c) *Navicula*, and (d) *Gomphonema*. (Photos courtesy of Lloyd E. Ohl.)

**FIGURE 13-22** These SEMs show (a) the outside of an *Anomoeoneis* diatom valve and (b) the inside of a *Cymbella* valve. (Photos courtesy of Lloyd E. Ohl and Christopher A. Ohl.)

(a) (b)

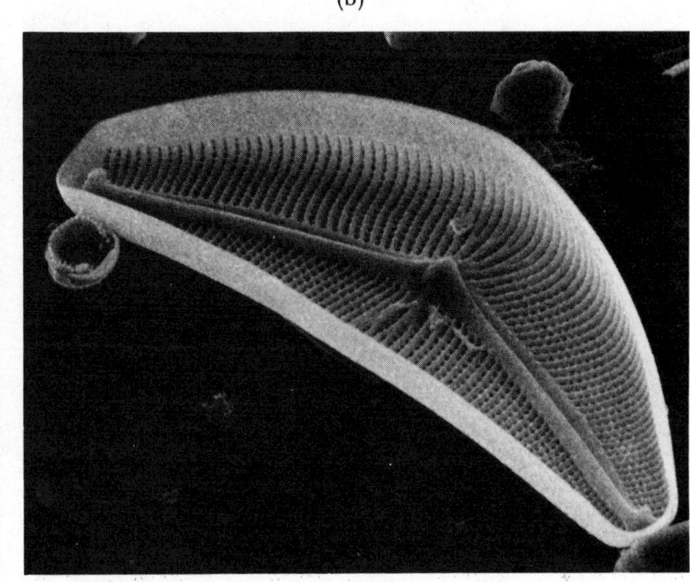

## Euglenoids

In great contrast to the silicified outer boundary of diatoms, that of euglenoids is a flexible **pellicle** that is thicker than a regular cell membrane and semirigid (or semiflexible). Euglenoids characteristically move by anterior flagella (Fig. 13-23), bending to change directions or slither past an obstacle. Some species also possess a limited capacity for amoeboid movement, especially at the nonflagellated (posterior) end. Photosynthetic euglenoids usually possess an "eyespot," or **stigma,** thought to be related to the ability of euglenoids to orient toward a light source, a phenomenon called **phototaxis.**

*Euglena* is the most familiar of these organisms, for it is studied in every introductory biology course from elementary school on up as a "typical" photosynthetic flagellate (Fig. 13-24). Unlike *Euglena,* some euglenoids lack chlorophyll and are completely heterotrophic. Even photosynthetic species are able to survive as heterotrophs when deprived of light, although without light the chloroplasts degenerate.

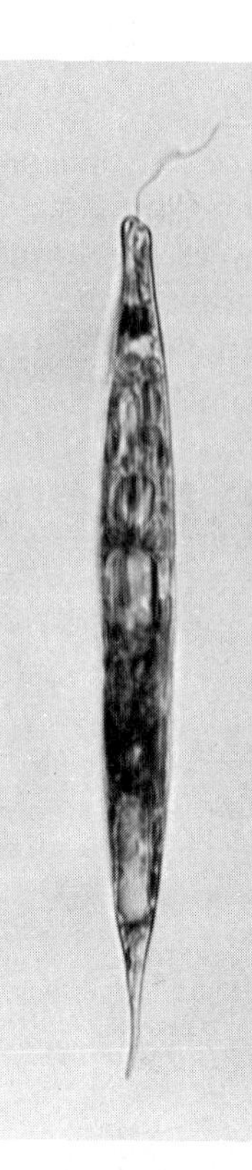

**FIGURE 13-23**
One of several *Euglena* species. (Photo courtesy of Carolina Biological Supply Co.)

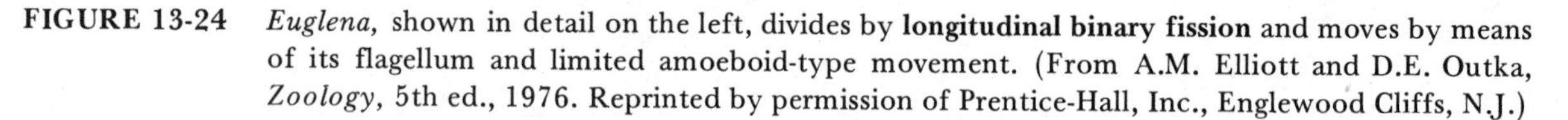

**FIGURE 13-24** *Euglena,* shown in detail on the left, divides by **longitudinal binary fission** and moves by means of its flagellum and limited amoeboid-type movement. (From A.M. Elliott and D.E. Outka, *Zoology,* 5th ed., 1976. Reprinted by permission of Prentice-Hall, Inc., Englewood Cliffs, N.J.)

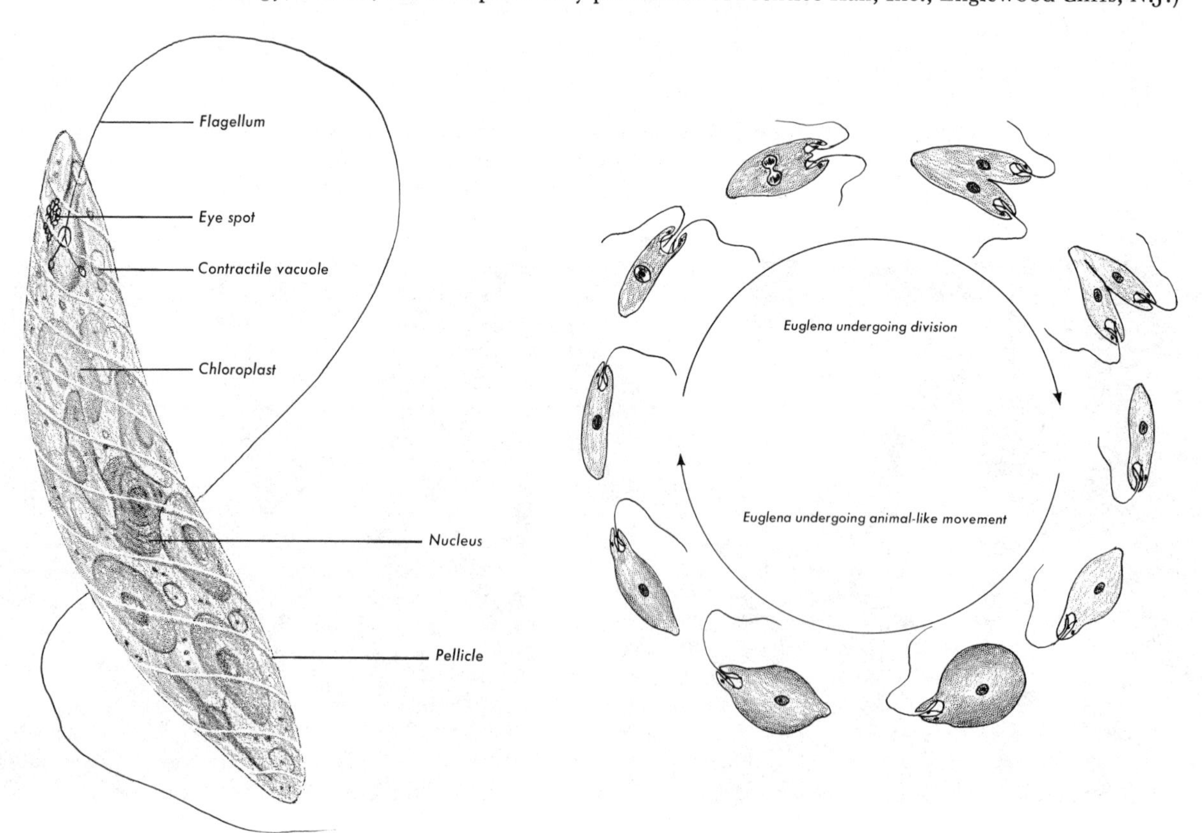

## Dinoflagellates

Dinoflagellates are unicellular forms that, as the name suggests, possess two flagella—one that trails lengthwise along the axis of the body and one that beats within a transverse groove (Fig. 13-25). The body may be naked or armored with cellulose plates. They are common in fresh-, brackish, and marine waters. It is not unusual for photosynthetic and nonphotosynthetic species to exist within the same genus.

Dinoflagellates cause occasional **red tides** in temperate marine waters. During red tides warm environmental conditions permit large dinoflagellate population explosions or **blooms**. The huge populations (especially of *Gonyaulax* and *Gymnodinium*) produce high concentrations of toxins that are poisonous to vertebrates. Thousands of fish may be killed. In extreme cases, a windrow of fish bodies may form where the red tide waters abut another current.

Many humans have died of mussel poisoning caused by red tides when toxins produced by the dinoflagellates accumulate in the tissues of shellfish that ingest them. Humans (or other vertebrates) ingesting the flesh of these mussels, clams, and cockles become victims of **paralytic shellfish poisoning** (mussel poisoning). The active principle is a neurotoxic alkaloid and fatalities are common.

Because red tides occur during the warmer summer months, the old adage about collecting shellfish only during months containing "r's" is helpful, at least in the Northern Hemisphere. Such precautions should be observed, for the toxins are tasteless and not destroyed by normal cooking.

Some photosynthetic dinoflagellates exist as golden symbionts within the tissues of marine protozoans and invertebrates, including sponges, jellyfish, sea anemones, corals, snails and nudibranchs, and flatworms.

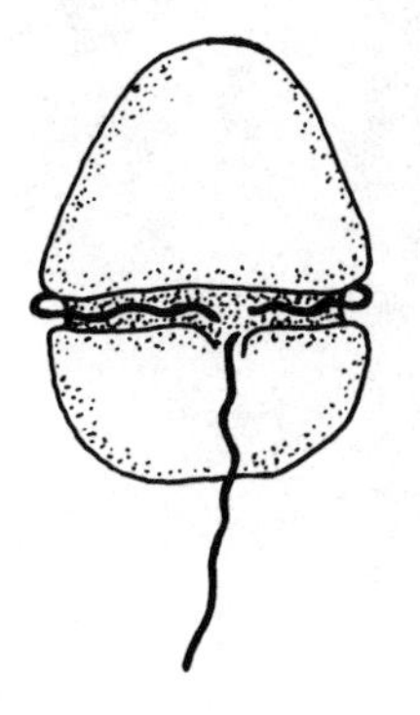

**FIGURE 13-25**
*Gymnodinium* and other **dinoflagellates** are characterized by the presence of a trailing flagellum and one that wriggles within a transverse groove around the cell body.

## Yellow-green Algae

The yellow-green algae have in the past been included in a division with the diatoms. It is a varied group, consisting of biflagellated or nonflagellated unicellular and colonial species and the unusual filamentous *Vaucheria* or water felt. *Vaucheria* is probably the most commonly demonstrated member of this group, consisting of long, multinucleate filaments that apparently lack cross walls (are **coenocytic**).

## Green Algae

The green algae are unicellular, colonial, and filamentous eukaryotes that contain chlorophyll and carotenoid pigments, giving them (usually) a bright green color. Most species live in freshwater habitats and show a beautiful diversity of external form and internal structure when viewed through the microscope. Figure 13-26 illustrates some of the various chloroplast forms that occur among the green algae, including the common filamentous *Oedogonium, Zygnema,* and *Spirogyra.* Some green alga genera are entirely marine. Some green "seaweeds" are sea lettuce (*Ulva lactuca*) and deadman's-fingers (*Codium fragile*). Obviously imagination and the romance of the seas have contributed to the naming of marine algae.

Green algae reproduce sexually by fusion of gametes and asexually by fragmentation. In *Spirogyra* and some others genetic recombination also occurs by conjugation. Reproductive patterns are detailed subsequently.

Three of the adaptive trends introduced at the beginning of the chapter can be seen within the green algae. In complexity of form green algae range from single-celled forms, such as the lovely **desmids** (Fig. 13-27) and flagellated *Chlamydomonas,* to colonial, such as *Platydorina, Volvox* (Fig. 13-28), and the unusual water net, *Hydrodictyon,* to a filamentous or sheetlike multicellular **thallus** (a plant body that is undifferentiated into roots, stems, and leaves). The thallus of marine green algae and some filamentous freshwater species shows some differentiation, for the point of attachment to the substrate consists of a basal cell or cells having gnarled extensions, collectively called a **holdfast.** A gamut of simple to more complex reproductive processes and of gametophyte/sporophyte relationships also exists within the green algae. One of the simplest types of life cycles is that of the unicellular biflagellate *Chlamydomonas* (Fig. 13-29).

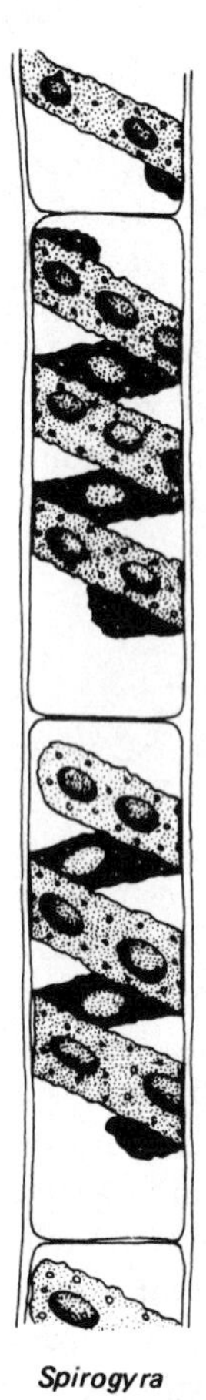


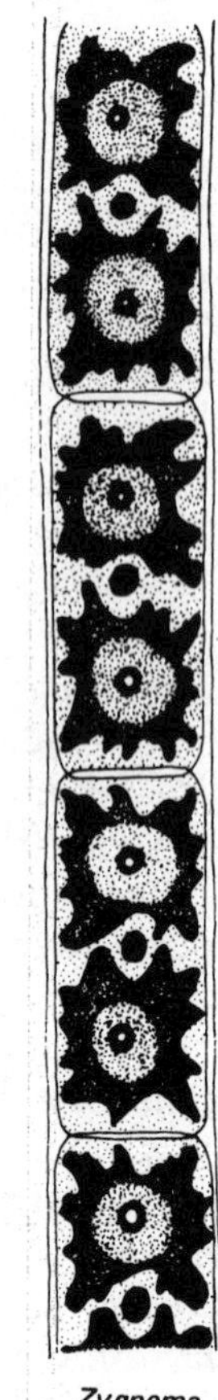


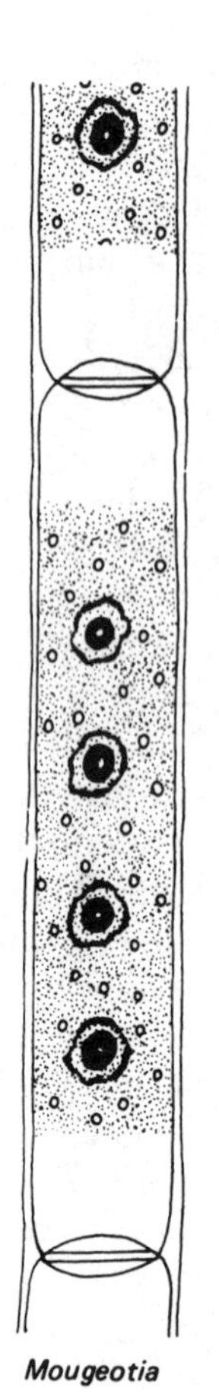


**FIGURE 13-26**
The chloroplasts of **green algae** are variable in size and shape, as demonstrated by *Spirogyra, Zygnema, Mougeotia,* and *Oedogonium.* (From S.R. Rushforth, *The Plant Kingdom—Evolution and Form,* 1976. Reprinted by permission of Prentice-Hall, Inc., Englewood Cliffs, N.J.)

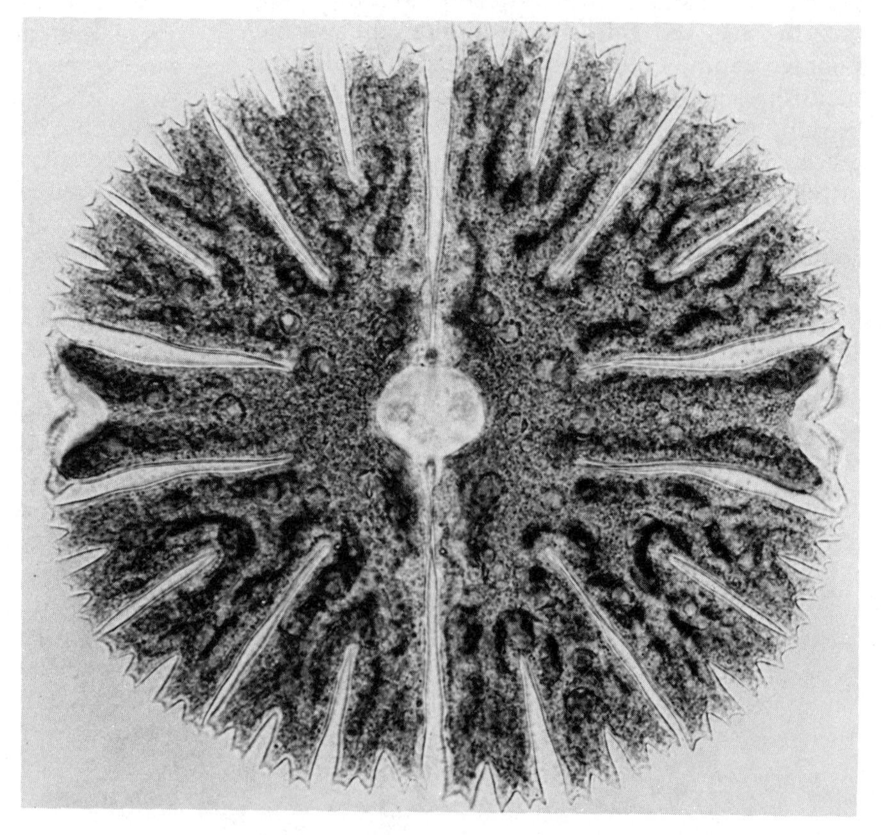

**FIGURE 13-27**
*Micrasterias,* a unicellular green alga (desmid), consists of matching halves. (Photo courtesy of Carolina Biological Supply Co.)

**FIGURE 13-28**
*Volvox* colonies consist of flagellated cells, giving the colonies a tumbling motion in the water. **Daughter colonies** form within mature colonies and are released when the parent colony ruptures. (Photo courtesy of Carolina Biological Supply Co.)

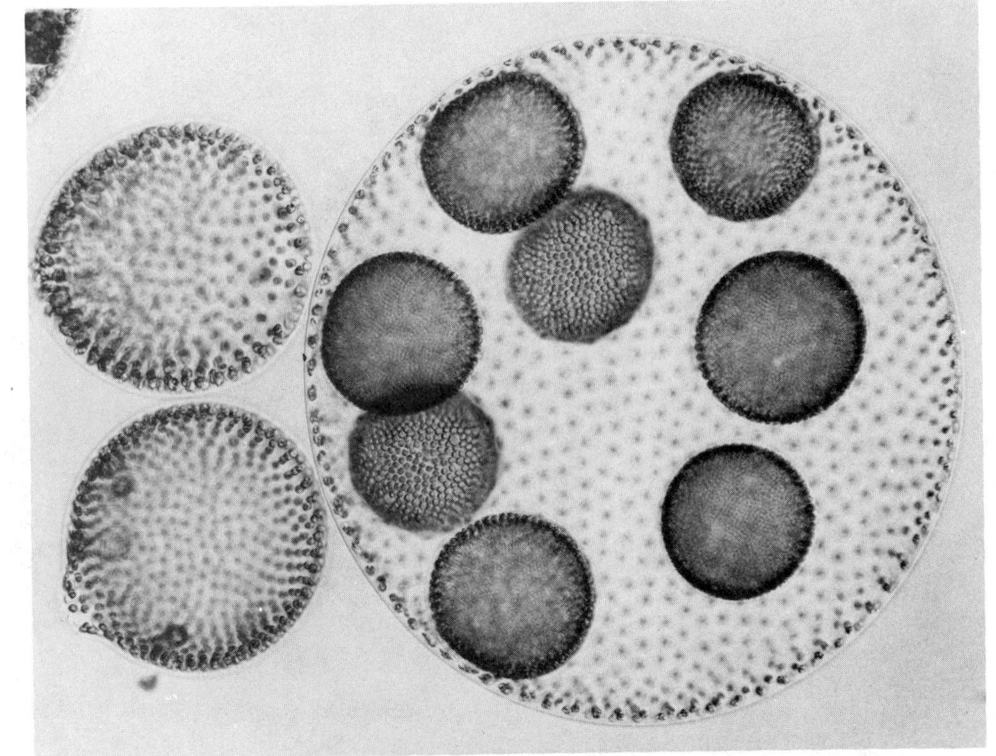

**FIGURE 13-29**
Life cycle of the unicellular green alga *Chlamydomonas,* having similar gametes (**isogametes**) and no distinct gametophyte and sporophyte individuals.

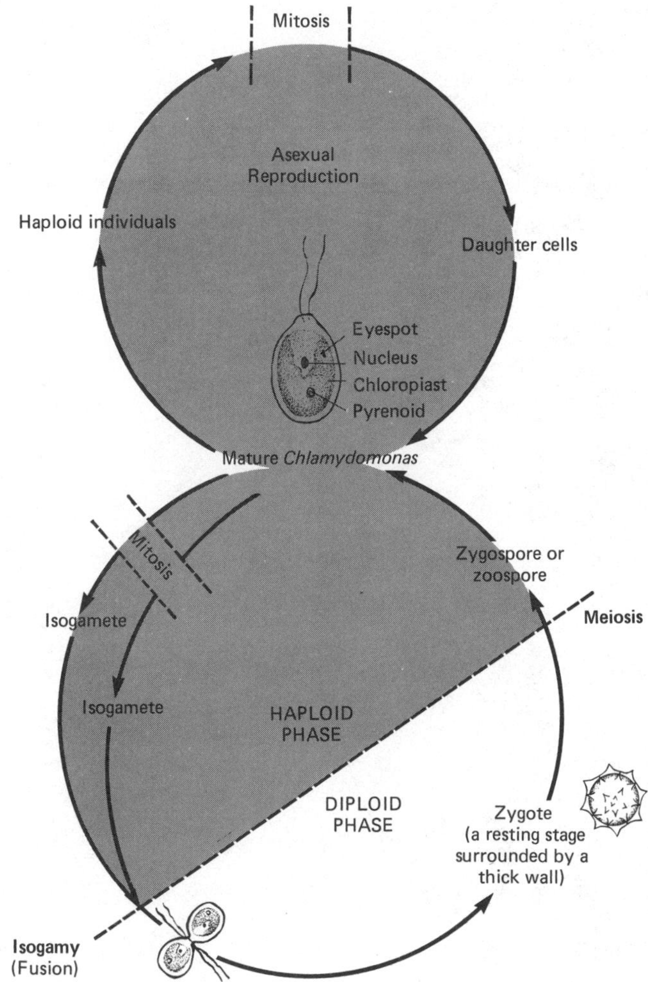


Of colonial green algae, *Gonium* is one of the simplest, consisting of a disk of 4 to 32 flagellated cells (depending on species) within a gelatinous matrix. Colonies of other genera have varying sizes and degrees of complexity, ranging up to the macroscopically visible *Volvox* (L., *volvere* = to roll). Each *Volvox* colony is a hollow sphere formed by 500 to 60,000 individual flagellated cells. The colonies demonstrate strong phototaxis. They are fascinating to observe as they tumble and rotate through the water with the paced grace of objects weightless in outer space! Depending on species, each *Volvox* colony may consist of reproductive cells, plus vegetative cells that are equal in function or have true functional differentiation among the vegetative cells. The life cycle of *Volvox* is reviewed in Fig. 13-30.

The so-called siphonous green algae are distinguished by the fact that they are coenocytic, composed of cells or segments that have a few to many nuclei. *Hydrodictyon* and *Pediastrum* (Fig. 13-31) are freshwater colonies of multinucleate cells. The marine *Codium* and its relatives seem to be completely coenocytic except during a reproductive phase, a particularly amazing fact because *C. magnum* is the largest known green alga, attaining up to a much-branched 8 m in length. *Cladophora,* having both freshwater and marine species, is coenocytic, with septa separating the units.

A few green algae grow in moist terrestrial habitats, on wooden shingles, decking, siding, and

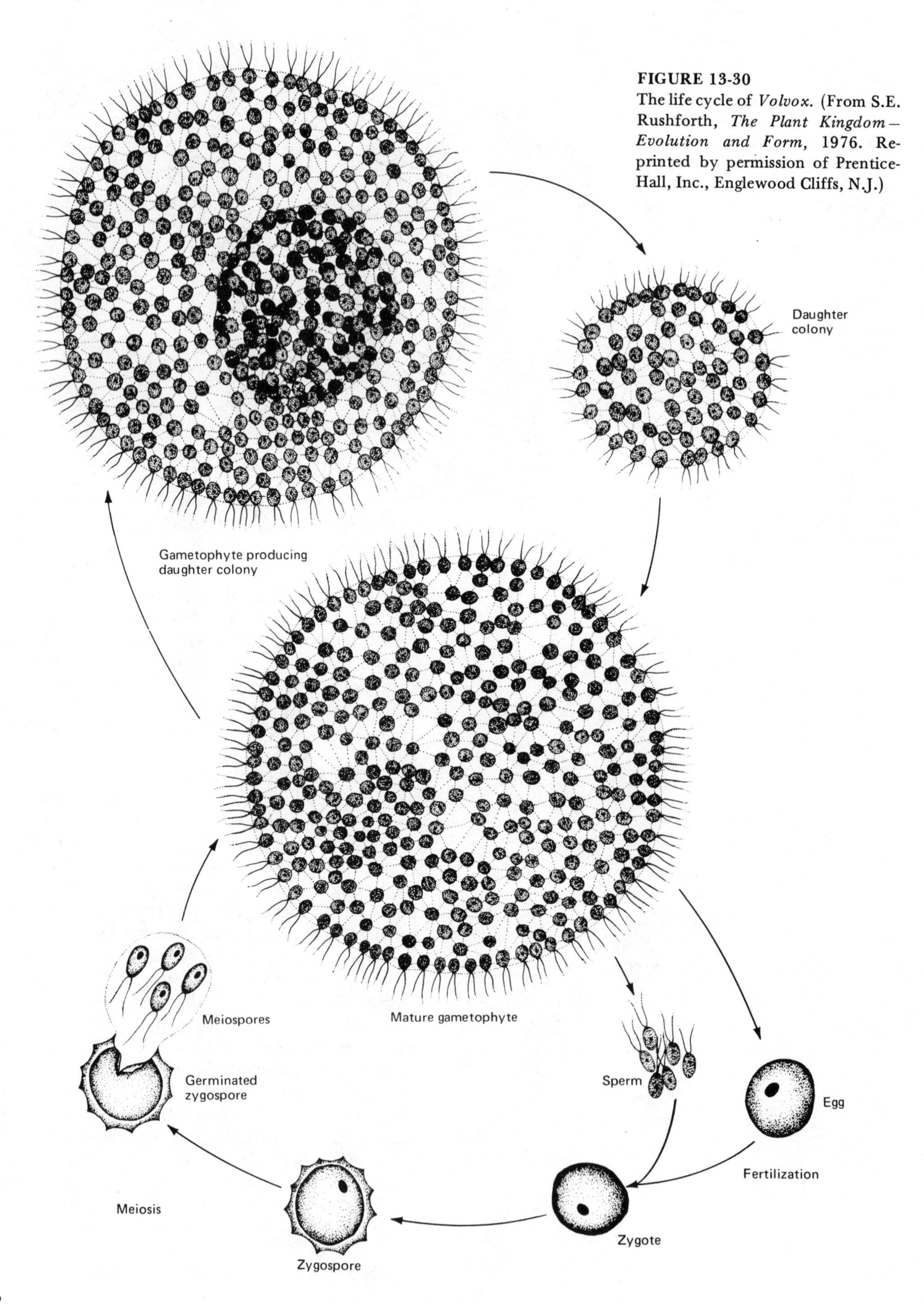


**FIGURE 13-30**
The life cycle of *Volvox*. (From S.E. Rushforth, *The Plant Kingdom—Evolution and Form*, 1976. Reprinted by permission of Prentice-Hall, Inc., Englewood Cliffs, N.J.)

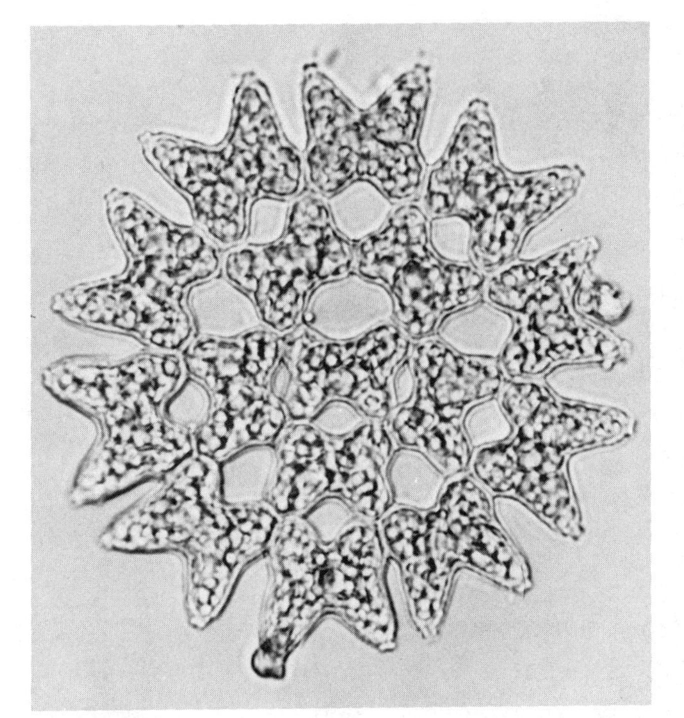

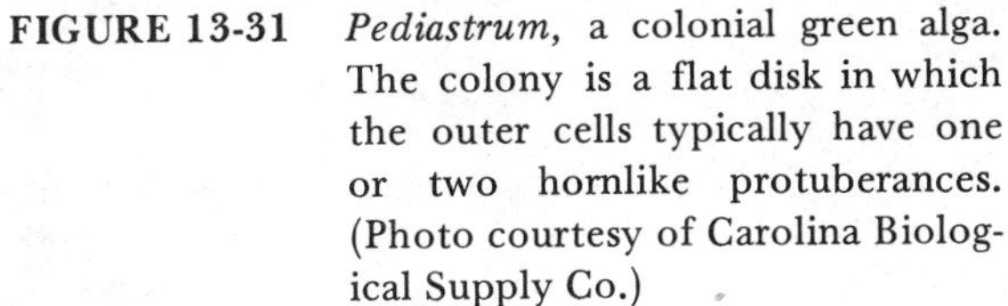

**FIGURE 13-31** *Pediastrum,* a colonial green alga. The colony is a flat disk in which the outer cells typically have one or two hornlike protuberances. (Photo courtesy of Carolina Biological Supply Co.)

outdoor furniture as well as on tree bark and other natural surfaces.

Some green algae exist as symbionts in the tissues of freshwater protozoans and invertebrates; others are symbionts in various marine invertebrates.

Several additional green algae life cycles are worthy of mention, for they show characteristics that may be representative of evolutionary trends in the early evolution of the algae. In freshwater *Ulothrix* species reproduction may be either sexual or asexual. Sexual reproduction involves production of **isogametes,** gametes that are physically similar, although perhaps physiologically different. The resulting zygotes develop thickened cell walls and are an effective dormant phase in which to wait out unfavorable environmental conditions. Note that the life cycle of *Ulothrix* is dominated by the haploid or gametophyte phase and that the zygote is the only representative of the diploid or sporophytic phase.

In sea lettuce (*Ulva*) the life cycle is codominant (Fig. 13-8); that is, the gametophyte and sporophyte generations are equally developed, represented by physically identical thalli. Within the genus, differences in reproduction exist; some species seem to lack sexual reproduction. In those that reproduce sexually, however, the gametophytes produce dissimilar gametes called **anisogametes.** Both are flagellated, but one type is larger than the other. Sporophytes produce motile haploid **zoöspores** that can mature into new gametophyte thalli.

Several hundred species make up the freshwater genus *Oedogonium.* The gametophyte is the dominant generation, but unlike either *Ulothrix* or *Ulva,* sexual reproduction is clearly **oögamous.** Not only are the two gametes morphologically different, but one is a nonmotile **egg** to which the motile **sperm** must swim. Eggs and sperms are each produced in separate specialized gametangia, **oögonia** and **antheridia,** respectively. This pattern is the same as that demonstrated by higher plants (and animals). As in *Ulothrix,* the sporophyte generation is represented only by the zygote, which eventually undergoes meiosis to produce haploid zoöspores (cute little cells with a tuft of about 120 flagella at one end).

Among the many remaining and diverse types of green algae, we conclude by mentioning the common filamentous *Spirogyra,* which does not produce free motile gametes or spores. The filaments are haploid and reproduce asexually by fragmentation. Genetic recombination occurs by **conjugation** (Fig. 13-32), the transfer and subse-

**FIGURE 13-32** Sexual reproduction in the filamentous green alga *Spirogyra* occurs by **conjugation.** Bridges form between the cells of two filaments that are side by side and the cell contents of one filament move across into cells of the other. Fusion of the nuclei produces zygotes. Also, note the spiral chloroplasts for which the genus is named. (Photos courtesy of Carolina Biological Supply Co.)

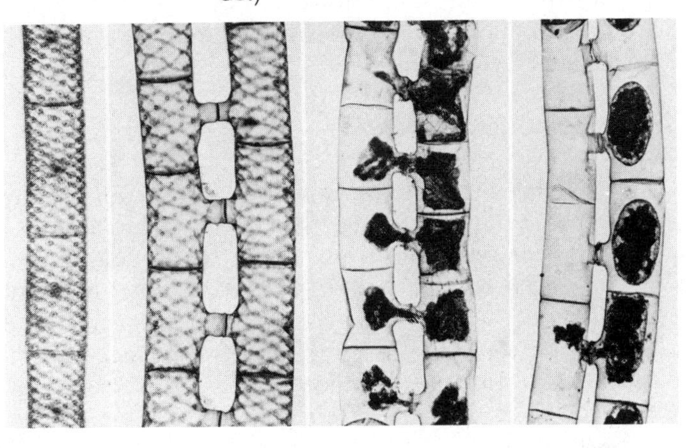

(a)

(b)

**FIGURE 13-33**
(a) An abundance of **marine algae**, mostly browns and reds, in the intertidal zone of a rocky Pacific shoreline. (b) Rockweeds, common brown algae of the intertidal zone. [Photo (a) courtesy of Fred L. Rose.]

quent fusion of cellular contents from one filament with cells of another filament. The resulting zygote develops a resistant cell wall for surviving adverse environmental conditions. As with similar diploid "resting" or "overwintering" structures, this one is called a **zygospore**.

## Brown Algae

Brown algae form a large and economically important (Chapter 16) group of marine algae, or **seaweeds**, ranging from small to moderate-sized intertidal forms [Fig. 13-33 (a) and (b)] to giant offshore forms. The giant kelps, *Nereocystis* and *Macrocystis* (Fig. 15-61), are well named—specimens of *Macrocystis* exceeding 30 m are common. Browns are the most structurally complex algae.

The thallus is typically differentiated into a well-developed holdfast, a central stalk or **stipe**, and a **blade** or blades that have a more or less flattened, sometimes branching upper portion. Many browns have **air bladders** or chambers that buoy them up toward surface light. The kelps also demonstrate cellular differentiation and the stipe of *Macrocystis* even contains sieve-tube-like cells similar to phloem cells of higher plants!

Because of their size, brown algae are not part of marine food chains at the planktonic level. Instead some are grazed on by such animals as snails, limpets, and larger crustaceans. Primarily they provide attachment and shelter for millions of small plants and animals as well as for free-swimming fish, squids, and marine mammals. The largest and most complex kelps are *Laminaria* and related genera (including *Macrocystis*). *Laminaria* species are prominent components of the intertidal and offshore marine communities of the temperate and boreal coastlines of the world. The life cycle of *Laminaria* consists of a large sporophyte and a microscopic gametophyte.

In contrast, some brown algae have a codominant gametophyte/sporophyte life cycle. One of the most unusual life cycles, however, is that of rockweeds, such as *Fucus*. *Fucus* has no independent gametophyte phase and motile gametes are formed directly by the sporophyte (Fig. 13-34). Most brown algae also reproduce by spore production but rarely by fragmentation.

## Red Algae

The red algae, the other main group of seaweeds, actually vary in color from delicate pink to deep iridescent purple. Coralline forms secrete a limy skeletal matrix (calcium carbonate) on which the alga grows so that you might mistake them for coral on first glance. The combination of red tissue over white skeleton produces their pink hues. Some red algae are moderately large and have fairly thick bodies, but most are much more delicate than the brown algae. Branching growth is common. Like browns, they attach to the substrate by well-developed holdfast structures [Fig. 13-35(a), (b), and (c)]. Most red algae are intertidal and many species are of economic importance (Chapter 16).

Red algae reproduce sexually and by spore formation but rarely by fragmentation. The life cycle is dominated by the gametophyte and is often complex, involving up to three distinctive generations (as in *Polysiphonia*) or with no defined alternation (as in some *Porphyra* species). Curiously, red algae do not seem to produce motile cells.

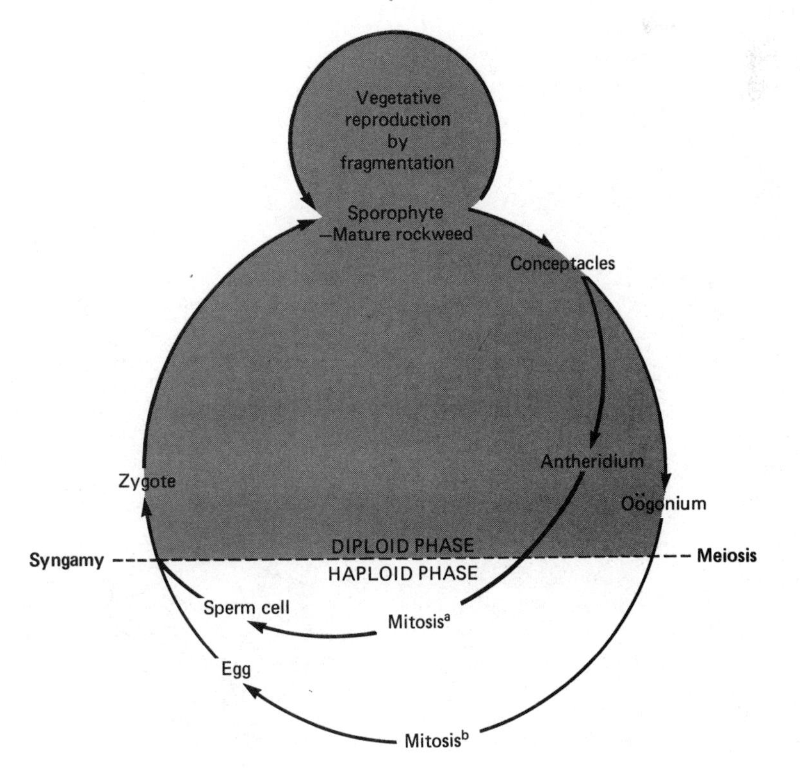


**FIGURE 13-34**
Life cycle of the rockweed *Fucus*. Among plants this is a unique life cycle in that there is no gametophyte. **Gametangia** (gamete-producing structures) develop on the thallus within special pits or cavities called **conceptacles.** The gametangia are one celled and are called **antheridia** and **oögonia.** Unlike the multicellular gametangia of bryophytes and vascular plants, they are derived directly from sporophyte tissue ($2n$). The nucleus of each gametangium divides by meiosis, producing 4 haploid nuclei. A series of mitotic divisions[a] yields 64 nuclei per antheridium, which become incorporated into 64 sperm cells. A mitotic division[b] yields 8 haploid nuclei per oögonium. Of these, 1, 2, 3, or 4 nuclei may develop into egg cells, depending on the species.

(a)

(b)

**FIGURE 13-35**
(a) The **holdfast** of this red alga was cemented so securely to the rock that a piece of rock had to be chipped off to collect the specimen. Herbarium specimens of two red algae, (b) *Plocamium cartilagineum* and (c) *Gelidium nudifrons.* [Specimens (b) and (c) courtesy of James A. Coyer.]

(c)

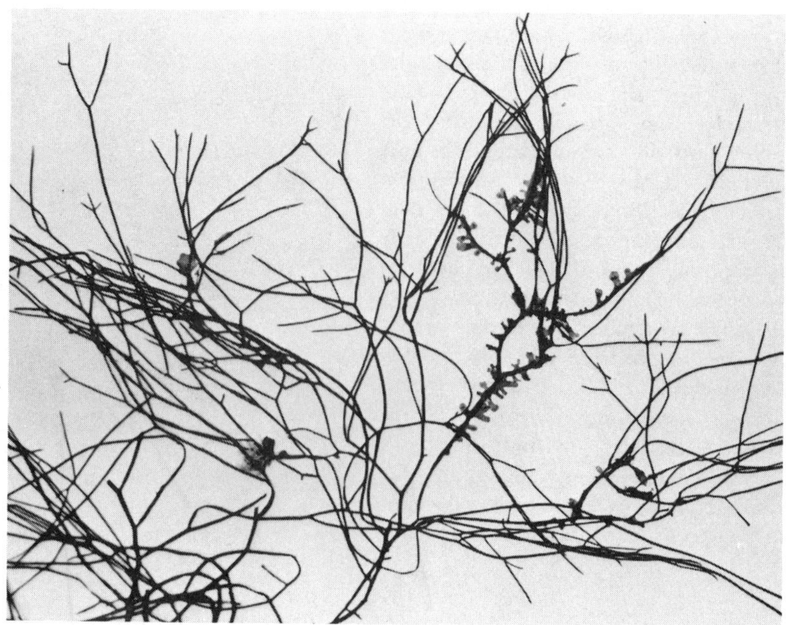

## FUNGI

Fungi are nonphotosynthetic plants that are unicellular, protoplasmic, or consist of a **mycelium**, a mass of filaments called **hyphae**. Fungal cell walls are composed of a complex of polysaccharides, including **chitin**, the primary exoskeleton material of insects and their relatives. An unusual feature of the fungi is that their cells are often **heterokaryotic**, containing two or more genetically different nuclei. Most fungi are saprophytic and feed by decomposing organic matter. Many species, however, are parasites/pathogens, causing plant and animal diseases (Chapter 12). More than 100,000 fungal species have been described, many of which are imperfectly known. The fungi are an extraordinarily heterogenous collection of organisms. They are usually classified into six main groups, based in part on characteristics of the life cycle.

### Chytrids and Oömycetes

Chytridiomycetes are very small fungi, most of which are aquatic. Many are parasitic. They differ from other fungi by having motile spores (zoöspores) as well as motile gametes. Zoöspores and gametes of each species look alike and each possesses a single posterior flagellum. One chytrid, *Coelomyces,* may have potential economic significance for the biological control of mosquitoes. In one stage of the life cycle, *Coelomyces* parasitizes mosquito larvae, which die as a result.

The oömycetes ("egg fungi") are a fairly small group but have had a great impact on humanity. *Phytophthora infestans* is responsible for late blight of white or Irish potato. Its hyphae infect the tuber and then spread up through the stem to the leaves, where spores are produced. The potato plants die. It caused the 1845 to 1851 potato famine in Ireland in which the starvation of approximately a quarter million people was followed by a typhus epidemic and the emigration of thousands of people from Ireland. Between 1845 and 1891 the population of Ireland decreased from about 8.5 million people to 4.7 million. During this period and beyond it the potato famine precipitated major political, social, and economic consequences for Ireland, England, and the United States, to which most of the Irish immigrants came.

Before development of the successful fungicide called Bordeaux mixture, the French winegrape industry was in danger of destruction due to grape downy mildew (*Plasmopara viticola*). Members of the genera *Pythium* and *Rhizoctonia* cause "damping off," a fungal rot of seeds and seedlings.

**FIGURE 13-36** The water mold *Saprolegnia* is shown here growing on a split hemp seed. (Photo courtesy of Carolina Biological Supply Co.)

The common water molds (*Saprolegnia*) (Fig. 13-36) are oömycetes that produce a furry growth on dead plant or animal matter—as in an untended minnow bucket!

During sexual reproduction oömycetes produce differentiated gametes, including a large, nonmotile egg—hence their name. In some classification schemes, the oömycetes are included in a larger category, the phycomycetes ("algalike fungi"), along with members of the next group, the zygomycetes.

### Zygomycetes

Leave bakery products covered at room temperature for a time and the final pieces will probably develop a webby growth and a moldy odor—signs of the gray or black bread mold, *Rhizopus stolonifer* (Fig. 13-37). Therefore most commercial bakery products contain mold-retarding preservatives.

Zygomycetes are characterized by the formation of a zygote ($2n$) surrounded by a thick protective wall; the whole structure is called a **zygospore**. When growth conditions are favorable, the cell contents of the zygospore undergo meiosis, producing a small haploid stalk, at the tip of which is a spore-

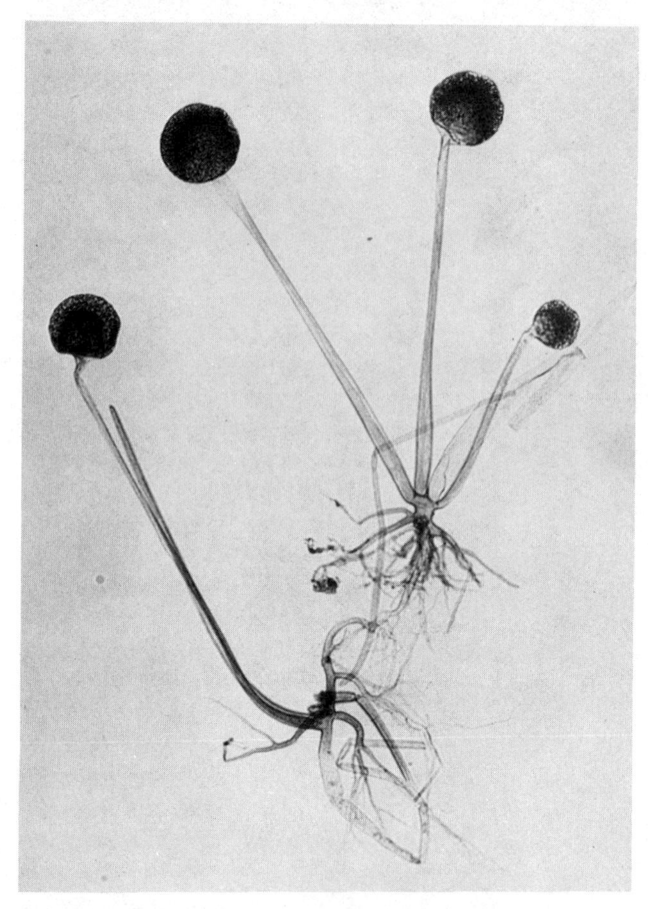

FIGURE 13-37 Sporangia and hyphae of the common bread mold *Rhizopus.*

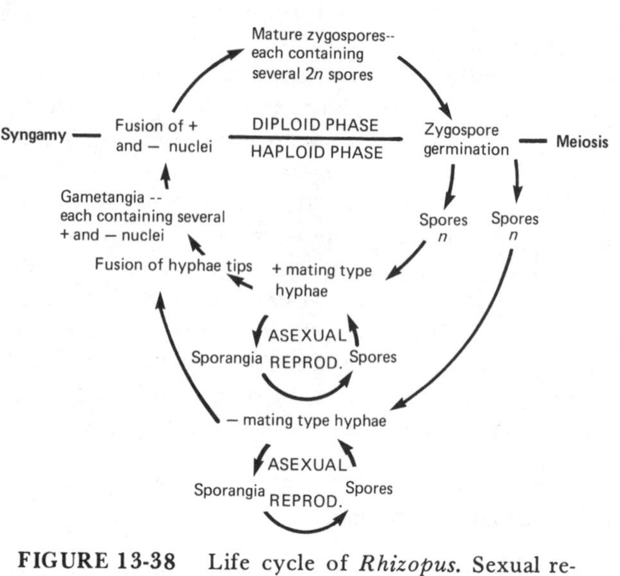


FIGURE 13-38 Life cycle of *Rhizopus.* Sexual reproduction occurs when haploid hyphae of different mating strains (+ and -) meet, fuse, and form a zygote.

FIGURE 13-39 Cup fungi, as these growing on soil, are ascomycetes, which produce spores in tubular saclike structures called **asci.**

producing capsule or **sporangium.** Mitosis in the sporangium produces haploid ($n$) spores that are released for wind dispersal. The life cycle of *Rhizopus* is summarized in Fig. 13-38.

Unlike the oömycetes, zygomycete gametes are not differentiated by size and motile cells are lacking from the life cycle.

### Ascomycetes

The edible morel (*Morchella*) is classified as an ascomycete ("sac fungus") because the spores are produced in sacs, **asci,** that line the morel's invaginated surface. Each ascus produces a definite number of spores, typically eight. Ascomycetes lack motile cells during the life cycle. The hyphae are septate, having cross walls (septa) with a central perforation. Each hyphal segment often contains several to many nuclei. Truffles (some edible), cup fungi (Fig. 13-39), and many wood-rotting fungi are also ascomycetes.

Yeasts (mainly *Saccaromyces*) are economically important single-celled ascomycetes. Yeast fermentation is essential to baking, brewing, winemaking, and the commercial production of certain alcohols and fermentation derivatives. Figure 13-40 shows stages in a yeast life cycle. Yeast cells also reproduce by **budding** off new individuals from the parental cell.

*Claviceps* species are fungi that infest grains with the disease called **ergot.** They produce strong antihemorrhagic alkaloids that are medically useful in controlled amounts but horribly toxic in overdoses. Lysergic acid diethylamide (LSD) is also present in *Claviceps.* A human affliction called ergotism, or St. Anthony's fire, caused by eating bread made from rye grain infected with *C. purpurea,* has been common. Mass cases of ergotism (particularly in the Middle Ages but also as recently as 1951) have been documented. It has even been hypothesized that the seventeenth-century men and women tried for witchcraft in Salem, Massachusetts, actually may have been afflicted with ergotism. Specific incurable and inexplicable symptoms mentioned at the trials include those of ergotism: tingling sensations in the fingers, crawling sensations of the skin, buzzing in the ears, vertigo, hallucinations, and convulsions. These symptoms could account for the distorted and disorderly speech, postures, gestures, and convulsive fits attributed at the time to satanic influence.

The causative agents of Dutch elm disease (American elm wilt), Chestnut blight, and various lung diseases are all ascomycetes.

### Basidiomycetes

Say fungus and most people think "mushroom." Mushrooms, puffballs, bracket or shelf fungi (Chapter opening art), jelly fungi (Fig. 13-41), rusts, and smuts are classified as basidium bearers for the club-shaped, spore-producing **basidia** (sing. basidium) that line the membranes of their exposed structures (basidiocarps) [Fig. 13-42 (a) and (b)]. In fact, the familiar visible mushroom, puffball, or bracket (conch) does not constitute the main fungus body but merely the spore-producing portion. The main part of the mycelium ramifies throughout the substrate from which the basidiocarp arises (Fig. 13-43). Therefore picking a basidiocarp does not eliminate the organism; it merely removes the part that could be responsible for the production of millions of spores.

Mushrooms have long been regarded mystically because of their rapid emergence from the ground as well as their potent toxins. They have been associated with elves, fairies, and "Alice in Wonderland." They have even been used in religious ceremonies and political assassinations. Many spe-

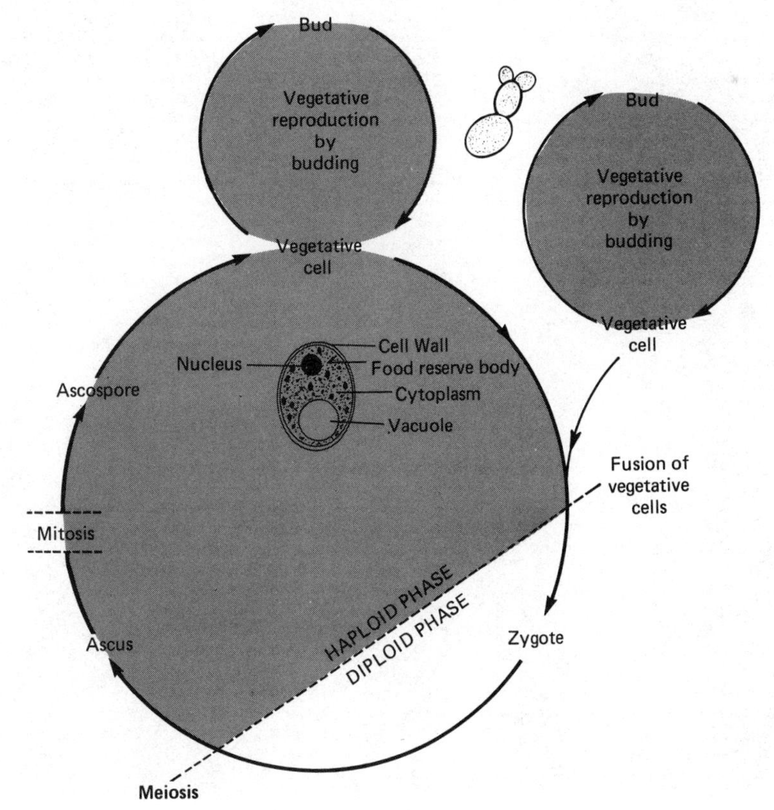


**FIGURE 13-40**
Yeasts, such as *Schizosaccharomycetes,* reproduce asexually by **budding** and sexually by fusion of haploid vegetative cells.

**FIGURE 13-41**
The orange jelly fungus growing on ocean driftwood is also a basidiomycete.

**FIGURE 13-42**
(a) The prized edible shiitake, or forest mushroom, grows naturally on dead wood. These specimens are being cultivated on an artifical "log." The gills on the underside of the mushroom caps are covered with microscopic **basidia.** (b) The bird's-nest fungus grows on branches or soil and consists of a cup in which many small disks form. The disk surfaces are covered with basidia. [Photo (b) courtesy of Carolina Biological Supply Co.]

(a)

(b)

FIGURE 13-43
*Agaricus* mushrooms can be cultured at home from commercially purchased **spawn** (inoculated medium). Various stages can be seen here, from smallest **button** to mature mushroom. The main fungal body, the **mycelium,** ramifies throughout the surface layers of the moist medium. (The half-dollar is in the photo for scale.)

cies are edible; many more, however are classified as unpalatable, unwholesome, or outright poisonous. (See also Chapter 18.) Figure 13-44 outlines the life history of the common commercially grown mushroom, *Agaricus.*

Many fungus life cycles involve two or more stages in more than one host species. The life cycle of wheat stem rust (*Puccinia graminis*) is diagrammed in Fig. 12-6. This fungus is a major problem in the wheat-growing regions of America. Some other important rusts are white pine blister rust (*Cronartium ribicola*) and cedar-apple rust (*Gymnosporangium juniperi-virginianae*). Corn smut (*Ustilago maydis*) produces large, tumorlike growths, especially on developing kernels. The kernels bloat and turn gray as they become filled with

FIGURE 13-44
Life cycle of *Agaricus.* Spores are formed in the many clublike basidia on the gill surfaces of the mushroom cap.

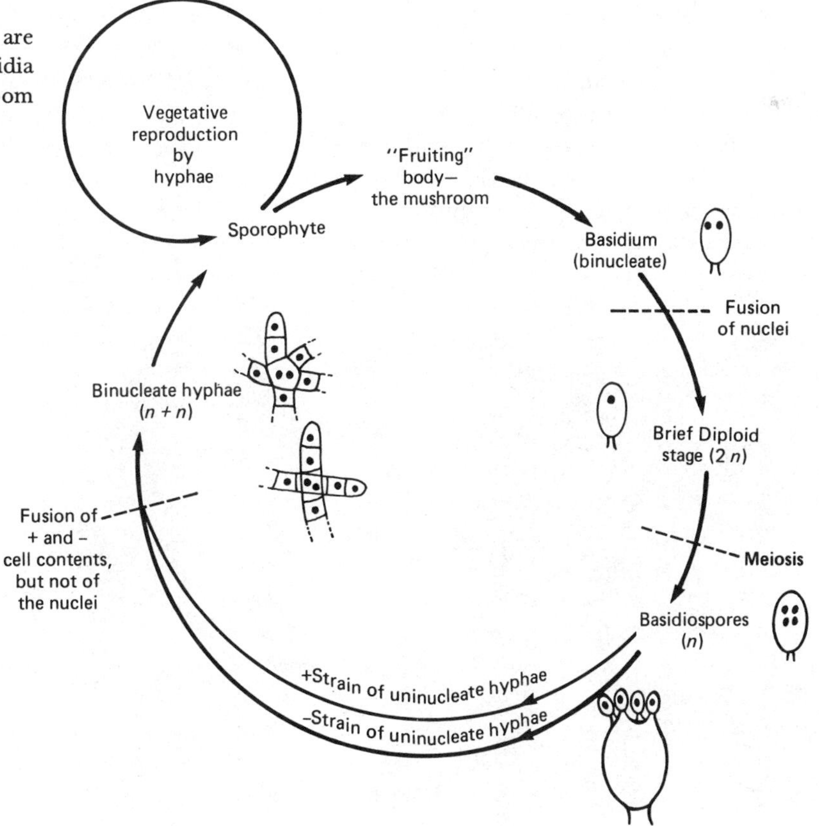

black fungal spores. Loose smut of wheat (*Ustilago tritici*) and covered or stinking smut of wheat (*Tilletia foetida, T. tritici*) are other major economic smut diseases.

## "Imperfect" Fungi

As you are aware, fungal groups are classified by the nature of their sexual reproduction structures. Imperfect fungi is a "wastebasket" category for more than 25,000 species of fungi for which sexual reproduction is unknown; a "perfect" life cycle is defined by the presence of a sexual phase. Perhaps the sexual stages of many species will remain forever unknown due to a complete absence of sexual reproduction from the life cycle as a result of evolutionary loss or because it never developed.

Most of the imperfect fungi appear to be otherwise structurally related to the ascomycetes and a few seem related to the basidiomycetes. Fungi remain in this category until a sexual stage is discovered or induced experimentally.

Imperfect fungi are of major importance as plant pathogens. *Alternaria* causes early blight of tomato and potato and *Fusarium* causes wilts of a wide variety of plants. Other diseases are black spot of roses (*Marssonina*), cucumber scab (*Cladysporium cucumerimum*), and birds-eye rot of grape (*Melanconium fuligenum*).

Molds of the genera *Penicillium* and *Aspergillus* are also economically important. *Penicillium* is the original source of the antibiotic penicillin and some species provide the color and flavor of such blue- and green-veined cheeses as Roquefort and Camembert. Fermentation by *Aspergillus* species produces soy paste (miso) from soybeans and soy sauce (shoyu) from a grain-soybean mixture. *Aspergillus* fermentation occurs during the initial stages of rice wine (sake) production. If the full life cycles of *Penicillium* and *Aspergillus* were clearly delineated, they probably would be classified as ascomycetes.

Important human pathogens within this group are those that cause athlete's foot, ringworm, and jock itch. *Candida albicans* causes the so-called "yeast" infections of the skin and mucous membranes, such as in the mouth ("thrush") and vagina. In addition, the microscopic but multicellular spores of *Alternaria,* which are common outdoors and in house dust, are one of the major causes of hay fever.

## Slime Molds

The slime molds are another group that has been "kicked around" taxonomically. They are very unusual organisms. Some are brightly colored. Some are cellular; others are protoplasmic. Imagine yourself a microscopic human observing the proliferation and advance of a sprawling, restless, amoeboid mass of protoplasm, a **plasmodium** (Fig. 13-45). Suddenly, on communication of some appropriate internal chemical signal, the edges recede and wither as the protoplasm organizes itself from an amorphous branching or spreading mass into heightening heaps or towers. As you watch, the towers take on distinctive and delicate form, with enlarged columnar or bulbous tips supported on slender stalks (Fig. 13-46). Their exteriors solidify before your gaze. Had you entered the scene and observed only the amoeboid phase or only the

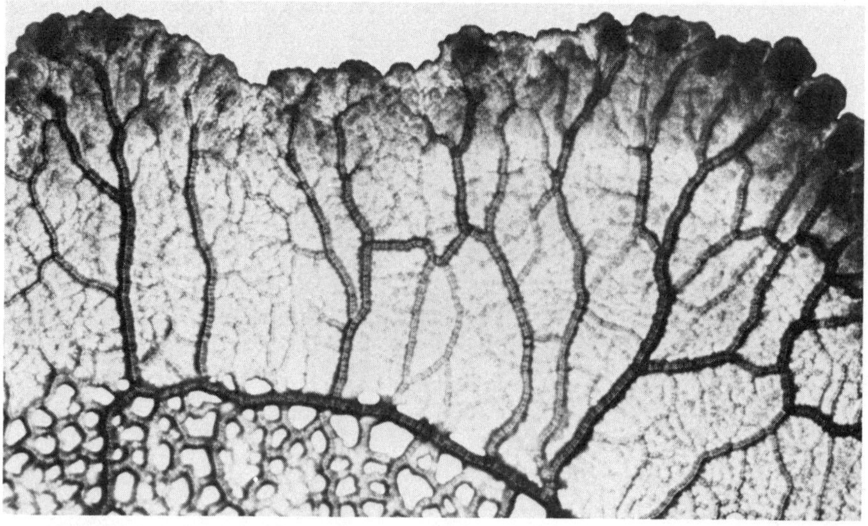

**FIGURE 13-45**
Plasmodium of the **slime mold** *Physarum.* (Photo courtesy of Carolina Biological Supply Co.)

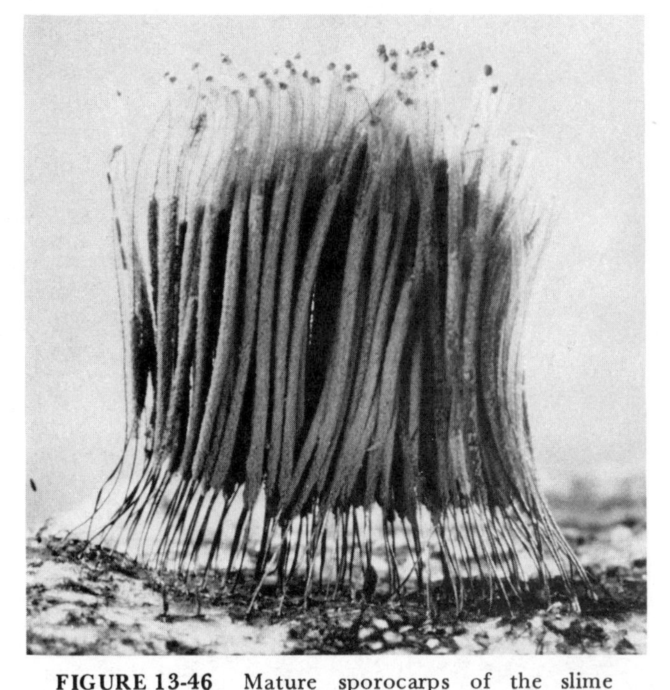

**FIGURE 13-46** Mature sporocarps of the slime mold *Stemonitis.* They are much smaller than shown here but are large enough to be seen as brown brushlike growths on dead wood. (Photo courtesy of Carolina Biological Supply Co.)

stalked phase, you would surely have listed the two as separate life forms—one wet and amorphous, the other hard cased and highly structured.

Actually, it is within the hardened form that meiosis produces spores for a new generation of plasmodia. Spores that happen to fall into water may develop into flagellated "swarm cells" whereas those that germinate in semimoist conditions are commonly nonflagellated amoeboid cells. Both cell types have the ability to encyst if they encounter unfavorable conditions. The swarm cells or amoebae will ingest food particles and each will eventually fuse with another similar cell, producing a zygote. Thus the sexual phase of the life cycle is completed. The zygote proliferates nuclei (by mitosis) and cytoplasm. This multinucleate mass continues to grow and is the new vegetative body or plasmodium.

### Lichens

Slow growing and tolerant of incredibly adverse natural environmental conditions, lichens exist in all of Earth's climatic zones. Their nonvascular bodies, composed of an association of an alga and a fungus, assume one of three forms: tightly encrusting (**crustose**), such as species that grow on rocks or bark (Fig. 13-47); branching, upright, or

**FIGURE 13-47**
**Crustose** lichens have an encrusting type of growth, as those that form small light-gray patches on this old grave marker.

threadlike (**fruticose**), such as species that grow on sandy soil (Fig. 13-48) or attached to tree branches; or as large flattened thalli (**foliose**), such as species attached to tree trunks in moister climates (Fig. 13-49).

The crustose types may be so small as to seem painted onto their substrates and it is impossible to collect them unless portions of the rock bearing them are chipped off.

Because they lack hard-walled vascular and supportive tissues, lichens do not readily fossilize; so their ancestry is impossible to trace by direct observation. Hypothetically the association of a specific fungus with a specific alga may have begun as a loose symbiotic relationship of distinct species. It has now become possible to disassociate and culture separately the algal and fungal components of some lichen species. In nature, however, the vegetative and reproductive stages of the algal and fungal components of a lichen species are totally coordinated. The fungus provides water and a structural framework within which the alga exists; presumably the alga produces food and oxygen that are of benefit to the fungus. If such is indeed the case, the body of a lichen is an example of the most intimate sort of mutualism.

The fungal component is usually an ascomycete, although some lichens contain a basidiomycete instead. Algal components include 26 genera, 17 of which are green algae, 8 blue-green (cyanobacteria), and 1 yellow-green. The green components of about 90% of known lichens, however, are in the green algae genera *Trebouxia* and *Trentepohlia* and the blue-green genus *Nostoc*.

Lichens reproduce asexually by fragmentation of the main body or production of special sand-grain-appearing multicellular body parts (soredia) that contain both alga and fungus and that are broken off by natural causes.

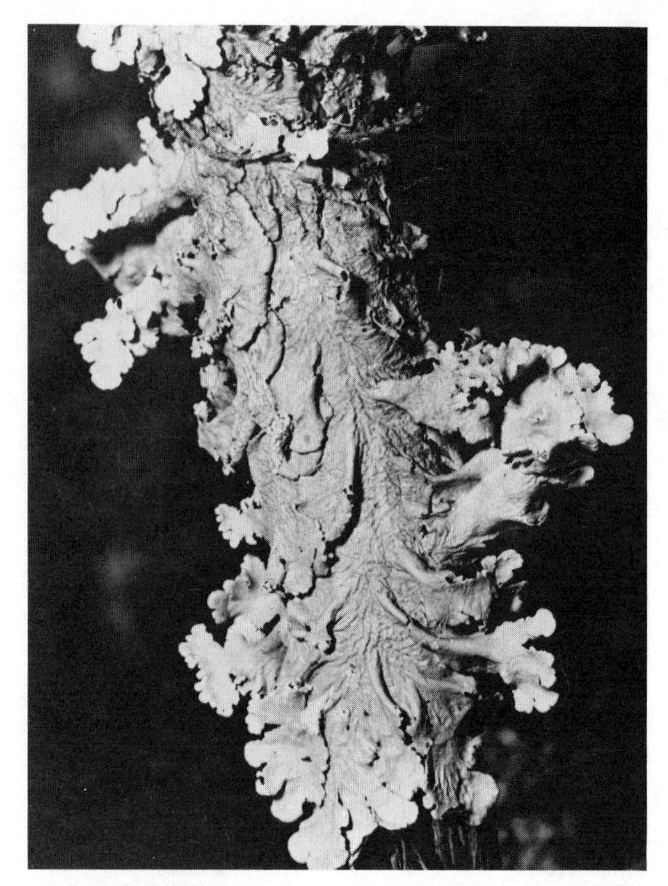

FIGURE 13-49 Flat, leaflike lichens are said to have a **foliose** growth form. (Photo courtesy of Carolina Biological Supply Co.)

FIGURE 13-48
**Fruticose** lichen growth form is branchlike or spiky, as these "British-soldier" lichens growing with moss plants on sandy soil. The cuplike structures are spore-producing ascocarps.

A second mode of reproduction is the indepent reproduction of the individual components. Fungal spore-producing structures are prominent surface features of many lichens (Fig. 13-48). Algal components release abundant motile spores or gametes when free moisture is available. Successful establishment of a new lichen individual, however, depends on contact by a germinating fungal portion with an appropriate alga. Failure to make contact usually results in death of the isolated fungus. The most common algal component (*Trebouxia*) is not known to occur free living, but some other lichen algae do.

Although low in food value, some lichens are collected as food for humans and domestic animals, especially in the Arctic region of North America (Chapter 17). An example is reindeer "moss." Lichens have been used for centuries as the source of certain natural dyes and were the original source of the acid-base indicator dye, litmus.

Lichens also have important research uses—for example, in attempts to determine prehistoric climatic changes in the Arctic and Antarctic and in radioactive fallout studies. Lichens are also indicators of air pollution. They are especially susceptible to such pollutants as sulfur dioxide. Increased amounts of sulfur substances in urban and industrial air cause decreased lichen populations on trees, tombstones, and other rocks. Near a number of larger cities, tombstones dated after the 1950s have no lichen growth and those dated earlier support dead or dying lichens.

Other than such limited uses, lichens have little economic value. Their most significant contribution to the biosphere seems to rest in their role as **pioneer species.** Lichens are among the first organisms to colonize substrates, such as exposed rock.

Growing on these rocky surfaces, lichens absorb moisture directly from the atmosphere and from precipitation. Gradually they etch the surfaces on which they grow by producing acidic solutions and by directly penetrating minute depressions and cracks with their attachment structures. Any such breach in the rock surface is increasingly subject to weathering, such as by the alternate freezing and thawing of water, so that the combination of physical and biological processes initiates breakdown of the rocks to soil. Following formation of suitable nooks and crannies, plus buildup of small amounts of organic material by the body of the lichen, other plants, such as mosses, ferns, and small vascular plants, can become established. The result is a succession of species populations on a given site as the physical conditions continue to be altered. (See also the discussion of succession in Chapter 2.)

## BRYOPHYTES

Bryophytes (Gr., *bryon* = moss) are low-growing green plants in which the gametophyte is the dominant form (Fig. 13-50). In several significant ways, the bryophytes (mosses, liverworts, and hornworts) are analogous within the plant kingdom to the amphibians (frogs and salamanders) within the animal kingdom. Although each group has fully aquatic species as well as species that can survive periodic dryness, they are basically semiterrestrial, requiring free moisture for survival and reproduction. The bodies of both groups are incompletely protected from desiccation so that they must have at least a semimoist environment to survive and

**FIGURE 13-50** Moss plants have modest nutrient and substrate requirements, so are able to colonize marginal sites.

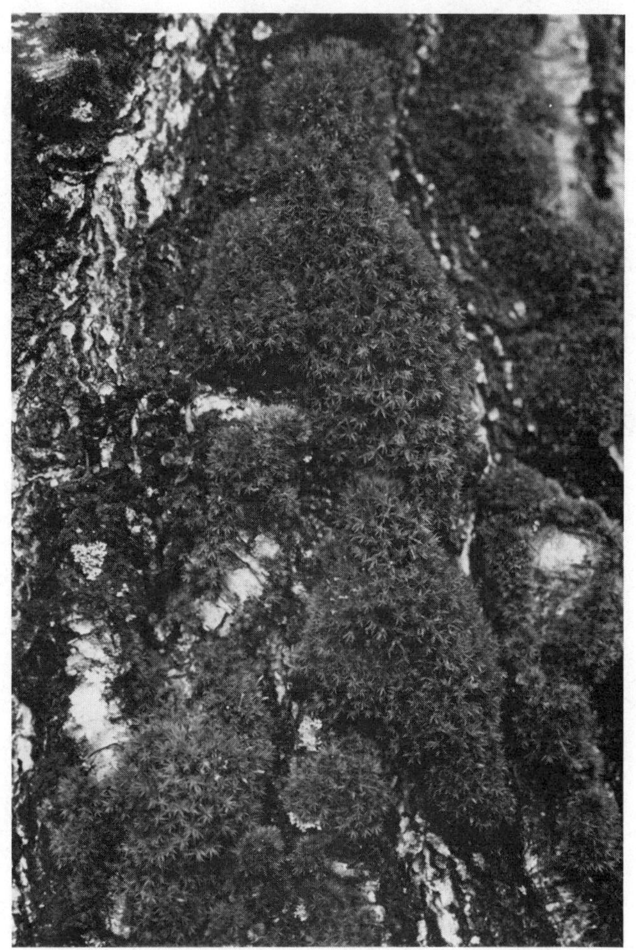

reproduce. When conditions are too dry, they retreat into a state of lessened metabolic activity until conditions are again suitable. Neither bryophytes nor amphibians have fully terrestrial reproduction, for both groups depend on the presence of free moisture for the transfer of gametes.

Lacking well-developed vascular and supportive tissues, bryophytes have low-growing thalluslike or tufted growth. Increased size and erect stature would mean increased bulk to support and increased transpiration, requiring an efficient vascular and support system. The low, compact growth habit also assists the gametophytes in trapping water near the plant body.

Because of their modest requirements, mosses are usually the next stage of terrestrial succession after lichens. Within their own limitations, they are a successful group and can be observed growing in sidewalk cracks, muddy patches, shaded areas, roofs, stone walls, bridges, tombstones, and tree bark. Some mosses are large and spreading in their growth habit and the epitome of moss growth on the North American continent is within the Pacific Coast rain forest of northern California, Oregon, and Washington up to Alaska (Fig. 15-48). (See also Chapter 15.) Some moss gametophytes have fairly complex differentiation of tissues, superficially approximating the true stems, leaves, and rootlets of vascular plants.

Many liverworts have a flattened thallus (hence their name) and are, in general, restricted to moister conditions than mosses. Liverwort antheridia and archegonia are clustered on specialized upright structures (Fig. 13-51). Liverworts also reproduce asexually by producing small multicellular disks called **gemmae** (Fig. 13-52). Because

**FIGURE 13-51** The liverwort *Marchantia*. (a) The **thallus** is the gametophyte, from which the reproductive structures arise. The flat disks on stalks produce **antheridia** on their top surfaces, forming a "splash platform" for rain to scatter the sperm cells. Water landing on the spokelike female structures runs to the undersurface of the spokes, where **archegonia** are located. Two archegonia are visible in (b). Note that they are multicellular. Small **sporophytes** develop at the archegonial sites, maturing and eventually producing spores. [Photo (a) courtesy of Carolina Biological Supply Co.]

(a)

(b)

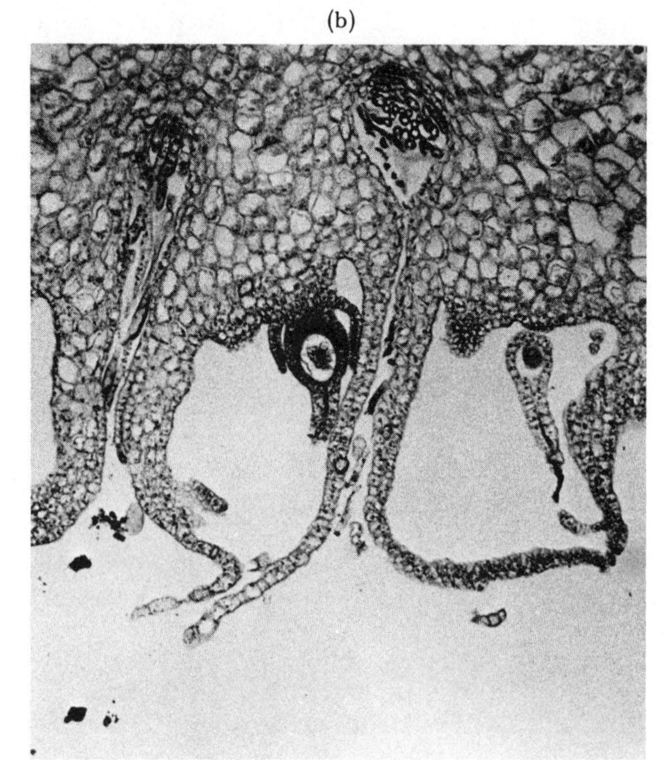

**FIGURE 13-52** Liverworts also reproduce vegetatively, by fragmentation or by the formation of special multicellular structures called **gemmae,** shown here in gemmae cups. Each gemma can grow into a new thallus. (Photo courtesy of Carolina Biological Supply Co.)

liverworts and the other bryophyte group, the hornworts, are simpler in structure and organization than many mosses, some botanists prefer to classify them into a division (Hepatophyta) separate from mosses.

Bryophytes are considered to have shared ancestry with the green algae. The earliest filamentous stage of the moss gametophyte, the **protonema,** is provocatively similar to a green alga and may be misidentified as such in a soil sample. No other plant groups are believed to have originated from the bryophytes.

A bryophyte life cycle is summarized in Fig. 13-53. Gametes are produced in **antheridia** and **archegonia,** specialized gametangia that consist of an outer jacket of sterile cells surrounding the reproductive cells. Flagellated sperm are produced in antheridia, released into a moisture film, and travel along a gradient of chemical attractants released by egg-containing archegonia. Syngamy occurs within each archegonium. Together the presence of complex gametangia and an oögamous life cycle distinguish the bryophytes and vascular plants from the algae and fungi, as does the development of the zygote into a multicellular **embryo** (embryonic sporophyte) within the archegonium.

The embroy develops into a sporophyte that remains attached to the gametophyte, from which it draws water and nutrients [Fig. 13-54 (a) and (b)]. Some sporophytes contribute to their own nourishment by photosynthesis. Sporangia develop with a **capsule** on the sporophyte and meiosis within each sporangium leads to production of many haploid spores, which develop resistant outer coverings. Bryophytes reproduce asexually by spreading growth and fragmentation.

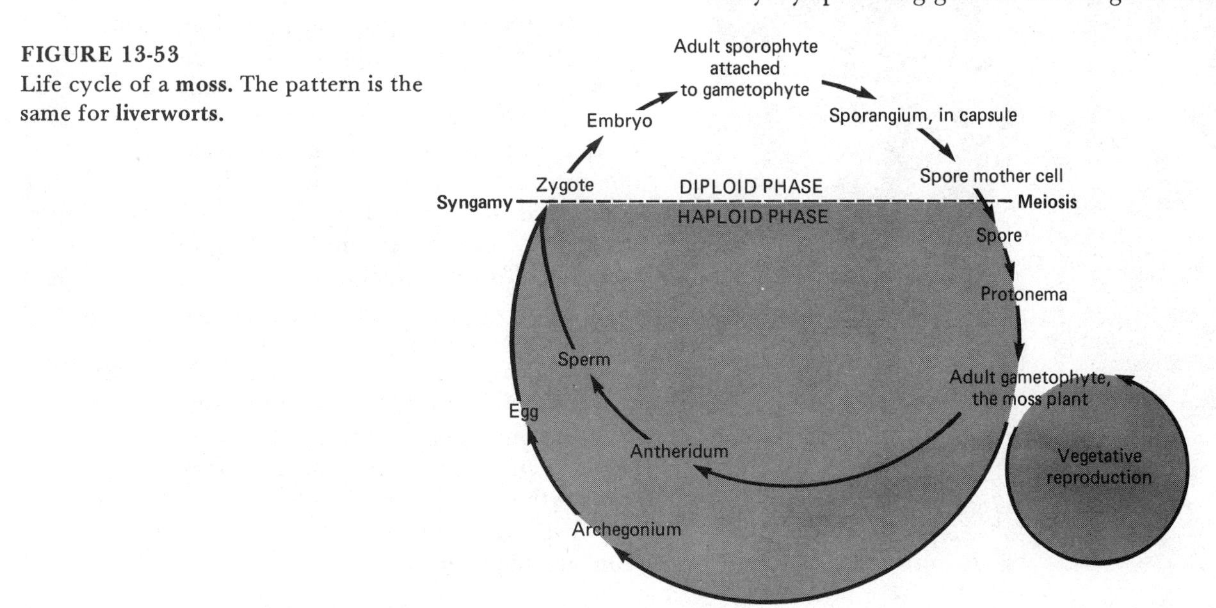


**FIGURE 13-53**
Life cycle of a **moss.** The pattern is the same for **liverworts.**

(a)

(b)

**FIGURE 13-54**
(a) Developing moss sporophytes arise from their gametophytes—the familiar moss plants. (b) Each sporophyte consists of a **foot** (not visible here) that is embedded in the gametophyte, a stalk (**setum**), and a **capsule** that contains spore-producing tissue. The tip of the capsule is a lid or **operculum**. The capsule is usually covered by a sheath called the **calyptra**, as this hairy-cap moss (*Polytrichum*). Eventually, the calyptra slips off and the operculum loosens and falls away, leaving an opening for spore dispersal.

Spores are usually dispersed by wind after they are released from the sporangia by a variety of **hygroscopic mechanisms**—movements powered by the expansion and contraction of specialized structures in response to atmospheric humidity. In the moss *Polytrichum,* for example, the capsule opening is surrounded by flattened wedge-shaped structures, the peristome teeth (*peri* = around, *stoma* = mouth or opening). In response to changes in humidity, the peristome teeth cover or reveal the capsule opening, aiding the dissemination of spores during drier conditions. In some mosses the capsule shrinks as it dries, popping out the spores.

In the capsule of the liverwort *Marchantia* hygroscopic filaments called **elaters** are present among the developing spores. The elaters twist in response to changes in humidity, freeing spores from the capsule.

## SUMMARY

**Extinct** plants are known directly by fossilized remains and indirectly from studying structural and biochemical characteristics of existing (**extant**) species.

Living things were once sorted into two categories—plants and animals—and such a two-kingdom system is still justifiable. Yet a five-kingdom system (Monera, Protista, Fungi, Plantae, Animalia) seems to reflect more closely the natural relationships among organisms. The recent designation of the **archaebacteria** demonstrates significant phylogenetic implications.

Bacteria and blue-green "algae" are **prokaryotes** whereas more complex organisms are **eukaryotes**, possessing a distinct nucleus and other membrane-bounded structures. **Viruses** are unique entities that are noncellular but have DNA and RNA.

Among known plants there are trends from prokaryotic to eukaryotic; from simple to complex; from generalized to specialized. Most plants alternate between spore-producing and gamete-producing stages in the life cycle. The most primitive plants have only asexual reproduction.

In more primitive plants the **gametophyte** is the dominant stage of the life cycle. In vascular plants the **sporophyte** is dominant.

The transfer from aquatic to terrestrial life can be traced through living and fossil groups. The final and most complete break with the aquatic environment was the development of internal reproduction and the **seed**.

**Bacteria** are important members of the kingdom Monera because of their participation in natural cycles. There are many beneficial bacterial species. Bacteria, however, also cause diseases.

**Rickettsias** are intracellular parasites of animals. **Mycoplasmas** are small bacteria or bacteria-like organisms that cause some plant and animal diseases.

Viruses are subcellular in structure and take over the metabolism of host cells, even manipulating host DNA and RNA synthesis.

**Blue-green "algae"** are common in warm, nutrient-rich waters and themselves contribute to water pollution. Soil blue-greens contribute to soil fertility by fixing nitrogen.

Eukaryotic **protists** can be single celled or colonial. Some photosynthesize; others do not. The protists include diatoms, euglenoids, dinoflagellates, yellow-green algae, and protozoans.

Most **fungi** are saprophytic, but some are parasitic. The fungus body may be unicellular—as in yeasts—but is most often a **mycelium** composed of many filaments or **hyphae**. Fungi are classified on the basis of their reproductive structures—basidia, asci, or simple sporangia. Imperfect fungi are those for which sexual reproduction is unknown.

**Slime molds** are characterized by a wandering amoeboid plasmodium stage that metamorphoses into delicate stalked structures bearing the sporangia.

A **lichen** is composed of an alga and a fungus. Because of their low nutritional requirements, lichens can colonize such barren environments as bare rock and so are significant pioneer species in succession.

Nonvascular plants, the algae and bryophytes, lack xylem and phloem. **Algae** (and photosynthetic protists) are the foundations of aquatic food chains.

Green algae are mainly freshwater species whereas brown and red algae are marine "seaweeds." Seaweeds provide food and shelter for millions of microscopic organisms as well as larger ones. Their life cycles are variable and frequently complex. Many species provide human food and economic products.

**Bryophytes** (mosses, liverworts, hornworts) inhibit semimoist environments and the gametophyte is the dominant phase of the life cycle. Specialized gamete-producing structures—**archegonia** and **antheridia**—are present. The bryophyte embryo develops within the archegonium and eventually grows into a sporophyte that remains attached to the gametophyte. Spore release is aided by hygroscopic structures.

## SOME SUGGESTED READINGS FOR CHAPTERS 13 AND 14

Alexander, M. *Microbial Ecology*. New York: John Wiley & Sons, 1971. A readable, interesting text.

Alexopoulous, C.J., and C.W. Mims. *Introductory Mycology,* 3rd ed. New York: John Wiley & Sons, 1979. The standard technical textbook on the fungi, written in a concise, straightforward manner.

Banks, H.P. *Evolution and Plants of the Past.* Belmont, CA.: Wadsworth Publishing Co., 1970. Available in paperback, this introduction to plant evolution emphasizes what is known from the fossil record.

Bold, H.C. *The Plant Kingdom,* 4th ed. Englewood Cliffs, N.J.: Prentice-Hall, 1971. A brief survey of the plant kingdom, which may be supplemented by his more extensive *Morphology of Plants.*

—, and M.J. Wynne. *Introduction to the Algae.* Englewood Cliffs, N.J.: Prentice-Hall, 1978. A comprehensive college text including structural, ecological, and life cycle details; arranged phylogentically.

Brock, T.D., and K.M. Brock. *Basic Microbiology—With Applications.* Englewood Cliffs, N.J.: Prentice-Hall, 1973. A nicely written introduction to microbiology emphasizing practical aspects that affect human affairs, such as health, environmental protection, and food technology. Readable and interesting even for a person without previous science coursework.

Burns, G.W. *The Plant Kingdom.* New York: The Macmillan Co., 1974. A readable, well-illustrated survey of the "plants."

Dawson, E.Y. *Marine Botany: An Introduction.* New York: Holt, Rinehart and Winston, 1966. A textbook surveying the plant types commonly found in the seas surrounding North America.

Delevoryas, T. *Morphology and Evolution of Fossil Plants.* New York: Holt, Rinehart and Winston, 1962. A concise consideration of the major groups of fossil plants.

Emboden, W.A. *Bizarre Plants.* New York: The Macmillan Co., 1974. An entertaining glimpse of the oddities of the plant world and their place in lore and legend.

Foster, A.S., and E.M. Gifford. *Comparative Morphology of Vascular Plants,* 2nd ed. San Francisco: W.H. Freeman and Co., 1974. The standard vascular plant morphology text.

Guberlet, M.L. *Seaweeds at Ebb Tide.* Seattle: University of Washington Press, 1956. Illustrated handbook to assist in identifying seaweeds encountered along the Pacific Coast, especially the northwest coast.

Hale, M.E., Jr. *The Biology of Lichens,* 2nd ed. New York: American Elsevier Publishing Co., 1974. A concise, readable, comprehensive review of our current knowledge of the lichens.

Hudson, H.J. *Fungal Saprophytism, Studies in Biology No. 32.* London: Edward Arnold (Publishers) Ltd., 1972. Paperback that provides an interesting look at the topic, emphasizing fungal ecology.

Pelczar, M.S., and R.D. Reid. *Microbiology,* 3rd ed. New York: McGraw-Hill Book Co., 1972. A standard text that includes prokaryotes, the viruses, and some eukaryotic organisms.

Stebbins, G.L. *Flowering Plants. Evolution Above the Species Level.* Cambridge, MA.: The Bellknap Press (of Harvard University Press), 1974. A fascinating, scholarly treatment of angiosperm origin and diversity, by a renowned authority on plant evolution.

Tiffany, L.H. *Algae: The Grass of Many Waters.* Springfield, Ill.: Charles C Thomas, Publisher, 1958. A popular introduction to the algae, their diversity, and importance.

Watson, E.V. *The Structure and Life of Bryophytes,* 3rd. ed. Atlantic Highlands, N.J.: Humanities Press, Inc., 1971. Well-written coverage of the group; available in paperback.

# chapter fourteen DIVERSITY VASCULAR PLANTS

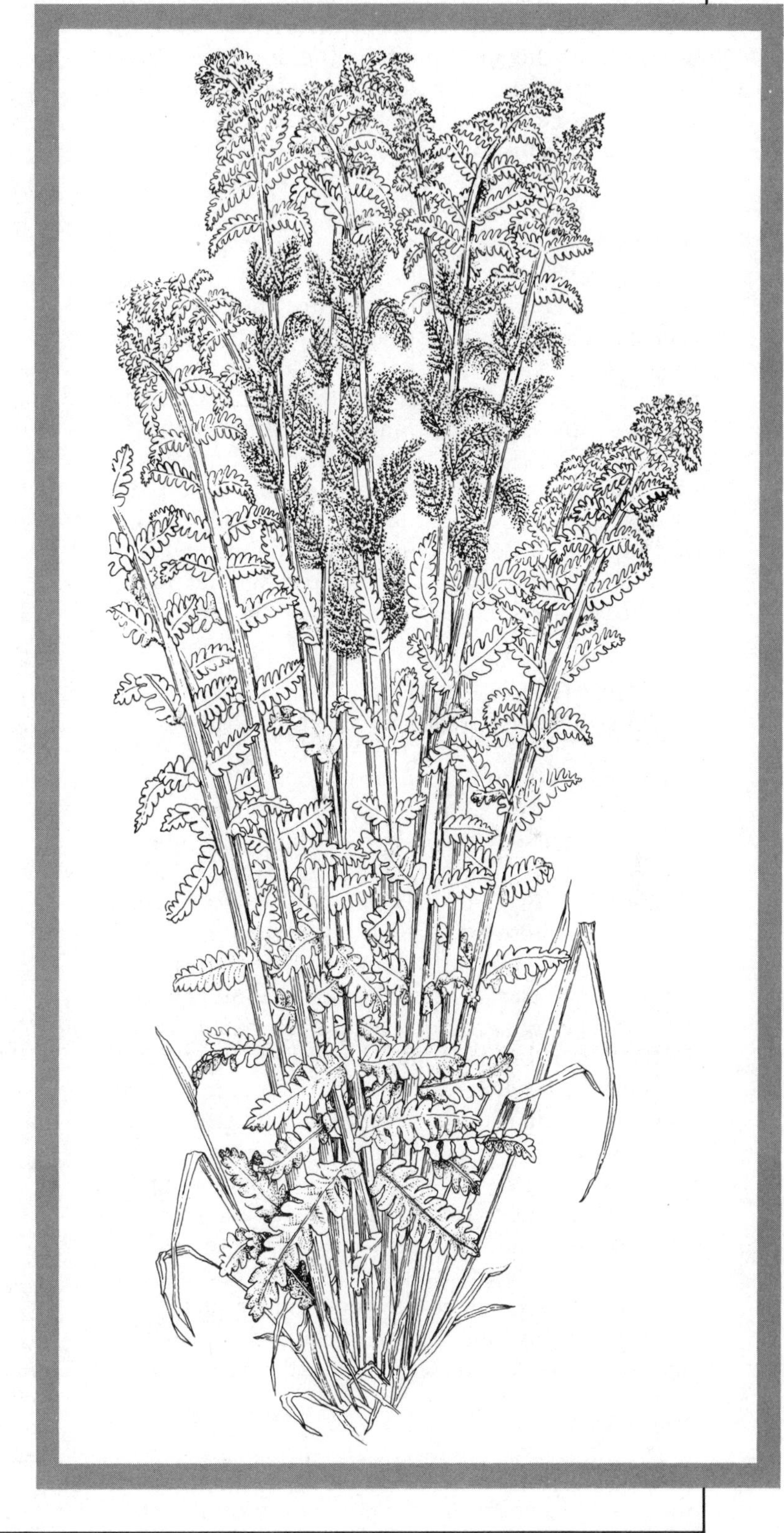

## INTRODUCTION

The plants considered in this chapter are distinguished from those of the previous one by the presence of well-developed conducting or vascular tissues, xylem and phloem (Chapters 7 and 8). In addition, the sporophyte is the dominant phase of the life cycle. Most vascular plants possess a **strobilus,** a cluster of specialized sporangia-bearing leaves called **sporophylls.**

Throughout the first dozen chapters we emphasized the *structure, function, diversity, uses,* and *ecology* of **seed plants,** the gymnosperms and angiosperms. In addition, many examples of specific plants are noted in the chapter on biomes that follows. Therefore this chapter is relatively brief, for we have chosen not to repeat all the preceding types of information. Rather we have decided to relate gymnosperm and angiosperm *life cycles* to the evolutionary and developmental progression that can be traced through other eukaryotic plant groups, as outlined in the adaptive trends section of Chapter 13. Before continuing, you may wish to refer again to Figure 13-5 to review the phylogenetic relationships of vascular plant groups.

## VASCULAR PLANTS THAT DO NOT FORM SEEDS

More than 400 million years ago North America was a warm, low landmass having extensive shallow seas and marshes. Non-seed-forming or "lower" vascular plants dominated the land environment (Table 13-1) (Fig. 14-1). The organic matter of their bodies and those of two primitive seed-producing plant groups—the seed ferns and cordaites—was fossilized and transformed through the millenia into the coal, petroleum, and natural gases that are our primary fuel sources today. Therefore we speak of this era as the Carboniferous Period or Age of Coal. Careful examination of a thin layer of coal reveals fossil evidence of these extinct plants.

Like bryophytes, lower vascular plants require a film of moisture for sperm transfer and development of the inconspicuous gametophyte thallus. Figure 14-2 presents a basic life cycle outline as it applies to most of the lower vascular plants, including the psilophytes, horsetails and scouring rushes, some club mosses, and most ferns.

### Psilophytes

The primitive vascular plants include a large number of extinct species known only from fossils,

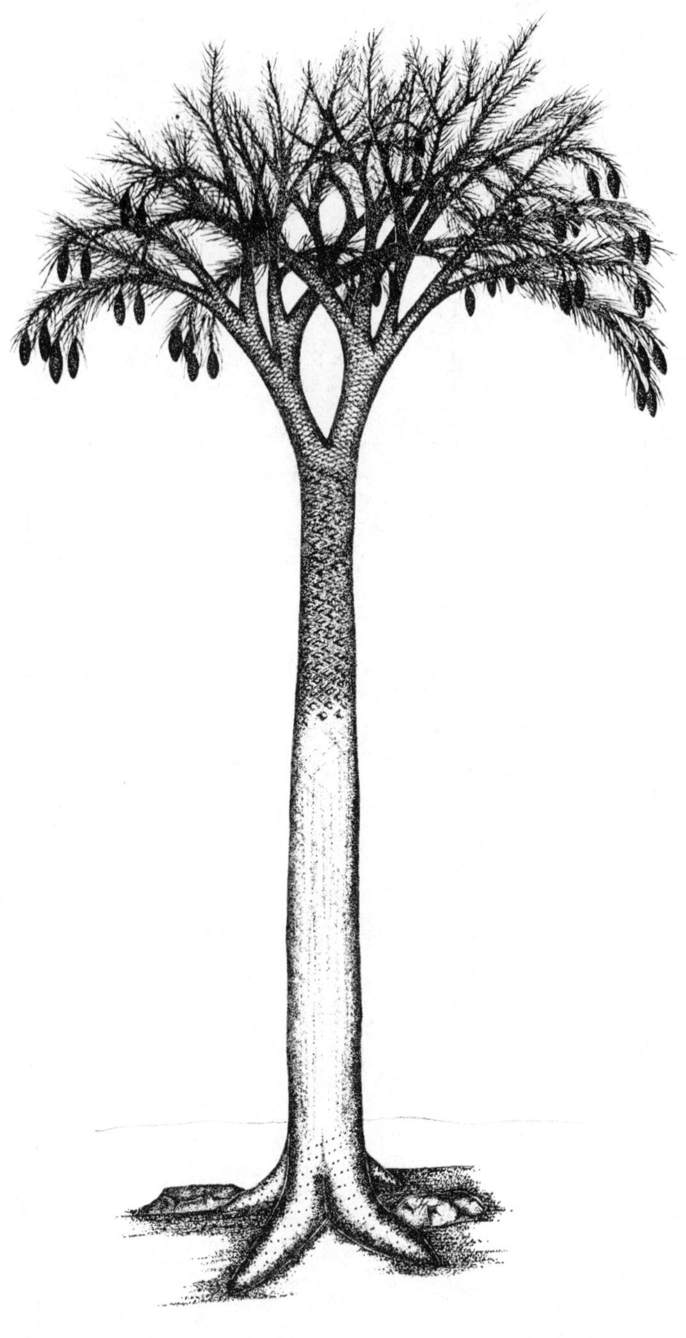

**FIGURE 14-1**
Certain large, extinct plants left abundant and well-preserved fossil remains from which images of the entire plants can be reconstructed. Ancient lycophytes such as *Lepidodendron* were prominent forest components during the warm, humid Carboniferous Period, or Age of Coal. The bodies of these and other plants of the swampy forests of this period were preserved, compressed, and chemically altered over the millenia into coal deposits that are important human fuel resources. Note the prominent **strobili** (cones) on this specimen. (From S.R. Rushforth, *The Plant Kingdom—Evolution and Form,* 1976. Reprinted by permission of Prentice-Hall, Inc., Englewood Cliffs, N.J.)

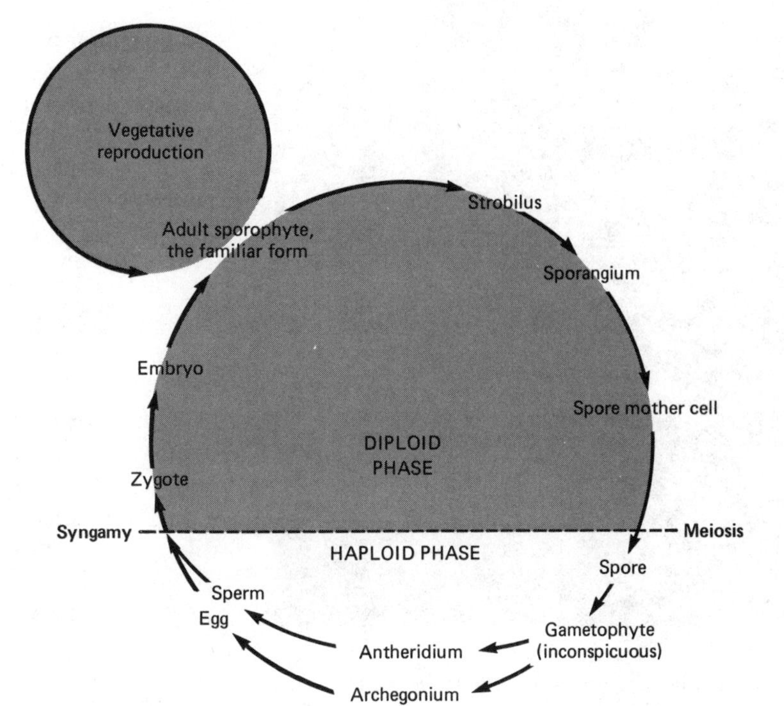


**FIGURE 14-2**
Basic life cycle demonstrated by most non-seed-producing vascular plants, including *Equisetum* and *Lycopodium*. This is a **homosporous** pattern, for only one kind of spore is involved.

with a few additional species that have managed to survive and coexist with the adaptively superior seed plants. Two extant forms, the whisk fern *Psilotum* and a related genus *Tmesipteris,* are usually classified along with some extinct species within the category **Psilophyta.** Some authors, however, contend that it is more appropriate to consider these surviving forms as part of the fern evolutionary line, having structures that regressed to a simpler form rather than following typical fern lines. As evidence, they cite the lack of intermediate fossil forms linking the extinct psilophytes and these two genera, plus the nature of reproductive structures and processes.

Whatever is done with *Psilotum* and *Tmesipteris,* the assemblage of extinct plants, known and unknown, constituting the psilophytes is highly significant to the main line of plant evolution as it has continued to the present. Psilophytes, the simplest and oldest vascular plants, are considered to be of common ancestry with extant vascular plants and possibly even the ancestral group itself.

The *Psilotum* sporophyte (which makes an interesting houseplant) consists of a subterranean rhizome from which the dichotomous (two-branched), nonleafy aerial parts arise (Fig. 14-3). The herbaceous, green aerial parts are usually referred to as stems even though they originate from a rhizome, as do fern fronds. They contain simply organized xylem and phloem tissues. Sporangia form on the aerial stems (Fig. 14-4). The gametophyte is

**FIGURE 14-3** The whisk fern (*Psilotum*) has dichotomous (two-branched), nonleafy shoots arising from a rhizome.

**FIGURE 14-4**
The spore-producing structures of *Psilotum* enlarge and become lighter in color as they mature. Eventually they become transparent, allowing the strikingly bright yellow color of the mature spores to show through.

small but more complex than that of ferns. Some vascular elements have been found in larger gametophytes. Water must be present for the flagellated sperm to swim to the archegonia. *P. nudum* grows in the southeastern United States and *P. complanatum* in Hawaii.

## Horsetails and Scouring Rushes

Today's horsetails and scouring rushes are all in the genus *Equisetum* (*equi* = horse, *setum* = tail). They are remnants of related groups with larger representatives and greater numbers known from fossil sites in North America and elsewhere [Fig. 14-5 (a) and (b)]. They may be the descendants of ancient psilophytes. Silica is present in their cell walls. Because of their gritty texture, scouring rushes were used to scrub pots and pans in pioneer days—still an appropriate camping use!

**FIGURE 14-5**
Shown here are fossils of an extinct tree-form horsetail from Indiana. (a) Whorled arrangement of the shoot components. (b) Strobili of the same species. (Photos courtesy of David L. Dilcher.)

(a)

(b)

The horsetails possess whorled side branches (Fig. 14-6) except for the specialized branches that produce strobili (Fig. 14-7). Scouring rushes are unbranched. The strobilus is composed of small, umbrellalike structures (sporangiophores), each of which bears several sporangia. Each spore has two elaters attached to it; by hygroscopic movement, the elaters assist in unloading the spores from the sporangia (Fig. 14-8). Small green gametophytes are produced, bearing antheridia and archegonia.

## Club Mosses and Quillworts

Club mosses are primitive vascular plants having a sporophyte that superficially resembles moss gametophytes. They are usually distinguished

FIGURE 14-7 *Equisetum* plants with **strobili.** Each strobilus is made up of smaller subunits (sporangiophores) on which the sporangia are produced. (Photo courtesy of Carolina Biological Supply Co.)

FIGURE 14-6 A horsetail (*Equisetum*) with whorled sidebranches. The leaves are represented by a sheath of fused structures at the base of each whorl. The large specimens shown here grow along the cool, humid Pacific coast.

FIGURE 14-8 The spores of *Equisetum* are tiny but active specks, barely visible to the unaided eye. The armlike **elaters** are attached to a spore at one point. When the air is dry they unwrap or extend themselves. A slight increase in atmospheric humidity causes them to recoil around the spore. These movements help spores to fall free of the sporangia when the air is drier and therefore more favorable for dispersal. (From S.R. Rushforth, *The Plant Kingdom—Evolution and Form,* 1976. Reprinted by permission of Prentice-Hall, Inc., Englewood Cliffs, N.J.)

by the presence of a strobilus. Existing lycophytes (club mosses and their relatives) have abundant extinct relatives [Fig. 14-9 (a) and (b) and Fig. 14-1] from the Age of Fishes (Devonian Period) and into the Age of Coal (Table 13-1) but the group consists now of only four living genera. Although superficially mosslike, they have xylem and phloem tissues that extend into true leaves, stems, and roots of the sporophyte. The main stem is a creeping rhizome with upward-growing branches (Fig. 14-10).

The two major living genera, *Lycopodium* and *Selaginella,* have significantly different life cycle variations. *Selaginella* is familiar to us as a decorative ground cover in terraria. The so-called resurrection plant is also a species of *Selaginella.* Where abundant, *Lycopodium* species called "reindeer moss," "caribou moss," or "princess pine" are used as Christmas decorations.

So far the vascular plants discussed are **homosporous**—they produce but a single spore type and the spores grow into gametophytes that produce both male and female gametes (Fig. 14-2). Club mosses of the genus *Lycopodium* are also homosporous.

In the genus *Selaginella,* however, each gametophyte is either male or female, depending on which of two types of spores was its source (Figs. 14-11 and 14-12). Gametophytes developing from the abundantly produced, small **microspores** are male whereas those developing from the fewer, larger **megaspores** are female. Such a life cycle, as seen in this genus and in the seed plants, is termed **heterosporous**, for more than one type of spore is produced.

The quillworts (*Isoetes* and *Stylites*) are distinctive aquatic or semiaquatic plants with greatly reduced shoot structure. They are considered

**FIGURE 14-9** Coal Age **lycophytes.** (a) Typical leaf-scar pattern on the stem. Compare this actual fossil to the drawing in Fig. 14-1. (b) A fossilized strobilus. Compare this to Figs. 14-1 and 14-10. (Photos courtesy of David L. Dilcher.)

(a)

(b)

FIGURE 14-10 The **club moss** *Lycopodium* (with strobili) is part of the forest floor vegetation in a Wisconsin oak woods.

FIGURE 14-11 The life cycle of *Selaginella*. **Megaspores** germinate and develop into female gametophytes. **Microspores** become male gametophytes. After syngamy, the young sporophyte remains attached to the parent plant for a time. The basic features of this **heterosporous** life cycle pattern are carried over into the seed plants.

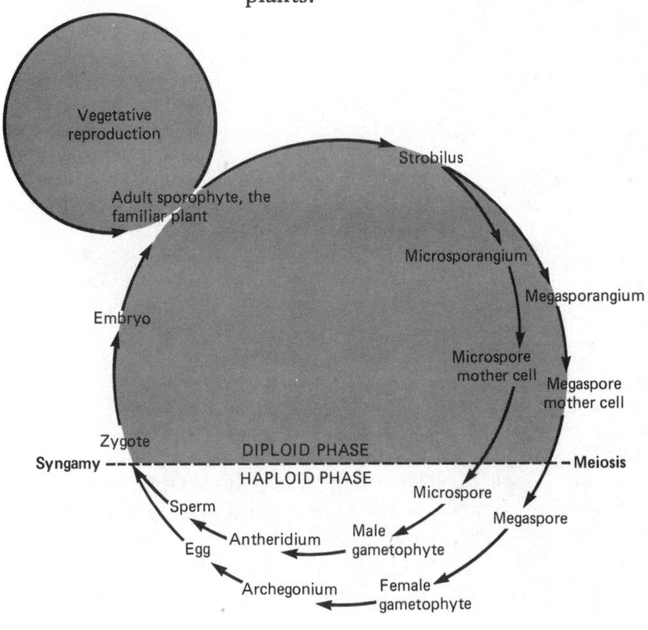


FIGURE 14-12 The *Selaginella* strobilus is made up of many sporophylls, bearing either **microsporangia** or **megasporangia.** Many small microspores are produced within each microsporangium; only four megaspores are produced within each megasporangium.

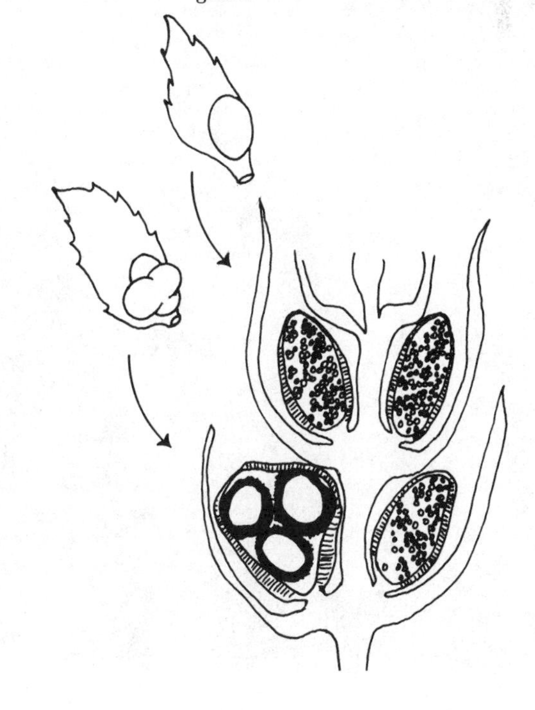

examples of evolutionary reduction from extinct ancestors.

*Lycopodium, Selaginella,* and the quillworts are each considered to have followed separate evolutionary lines from ancient ancestors.

## Ferns

Like other primitive vascular plants, today's ferns have abundant and highly developed fossil relatives (Fig. 14-13). Nearly 10,000 fern species are extant, occupying many habitats and niches (Fig. 14-14). Most ferns are smaller-than-knee-high residents of the forest understory, but some are aquatic [Fig. 14-15 (a) and (b)], some are vinelike, and some—the tree ferns—have treelike growth of up to 20 m in height. Ferns are especially important in the vegetative composition of moist climates, such as the tropics and the Pacific rain forest. Some ferns even manage to exist on highly forbidding landscapes, like the rock brake (*Cryptogramma acrostichoides*), which grows on basalt flows in the Cascade Mountains (Fig. 14-16). Because of their attractive growth forms, many fern species are popular landscaping and container plants.

The main stem of a typical, temperate latitude fern is a horizontal rhizome and the fronds

FIGURE 14-13
*Sphenopteris.* This well-preserved fossil specimen of a Coal Age **fern** resembles its modern relatives. *Sphenopteris* is thought to be ancestral to modern ferns. (Photo courtesy of David L. Dilcher.)

FIGURE 14-14
The maidenhair fern (*Adiantum*) requires humid conditions, such as near stream banks, in grottoes, or on moist north-facing slopes. (Photo courtesy of Fred L. Rose.)

(a)

(b)

**FIGURE 14-15** (a) The water fern *Marsilea* grows rooted to the bottom and sends its fronds to the surface. (b) By contrast, the water fern *Salvinia* floats on the surface, its feathery root system hanging below it. *Marsilea* is also unusual because it is a heterosporous fern [Photo (a) by Mark R. Fay.]

**FIGURE 14-16**
A small splash of green against the black basalt of an Oregon lava flow, the rock brake (*Cryptogramma*) has separate leafy-looking **sterile** fronds and narrower **fertile** (spore-producing) fronds. Compare this photograph to the drawing of interrupted fern (*Osmunda*), the chapter opening art. On *Osmunda* individual fronds have both sterile and fertile regions.

are leaves that grow upward from nodes along the rhizome. Most fern leaves are compound and may be **sterile** or **fertile,** the latter being those on which sporangia are borne. Figures 14-17, 14-18 and 14-19 (a) and (b) illustrate some features of the

FIGURE 14-17 Individual **sporangia** appear as black dots on maidenhair fern leaflets. Each region where sporangia are clustered is called a **sorus.** A sorus may be covered by a flap of leaf tissue called an **indusium,** as shown here. (Photo courtesy of Debi Stambaugh.)

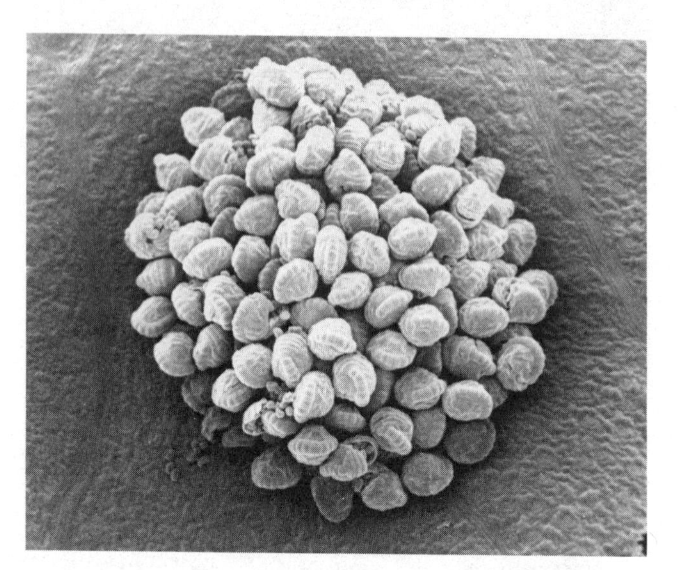

FIGURE 14-18 This SEM shows a sorus from a different fern species. No indusium is present, and the sporangial structure is clearly visible. (Photo by Jon M. Holy.)

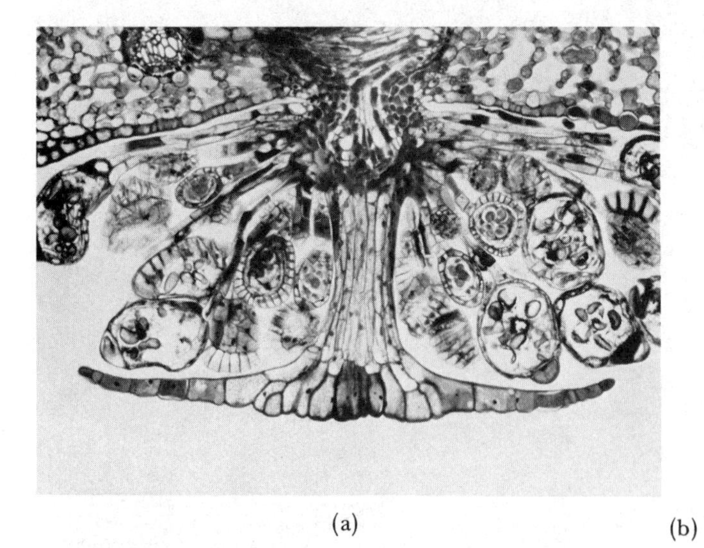

(a)

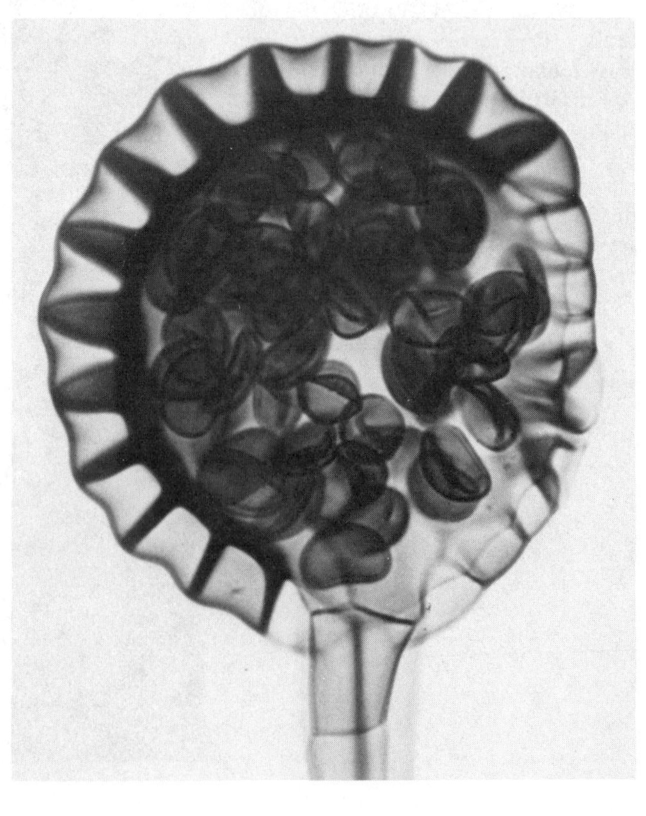

(b)

FIGURE 14-19
(a) Sectioned view of a fern sorus having an indusium. (b) An individual sporangium, showing the thickened inner walls of the **annulus.** As the cells of the annulus dry out, the thinner outer walls contract, ultimately tearing open the sporangium with a sudden snap that flings out the spores. [Photo (b) courtesy of Carolina Biological Supply Co.]

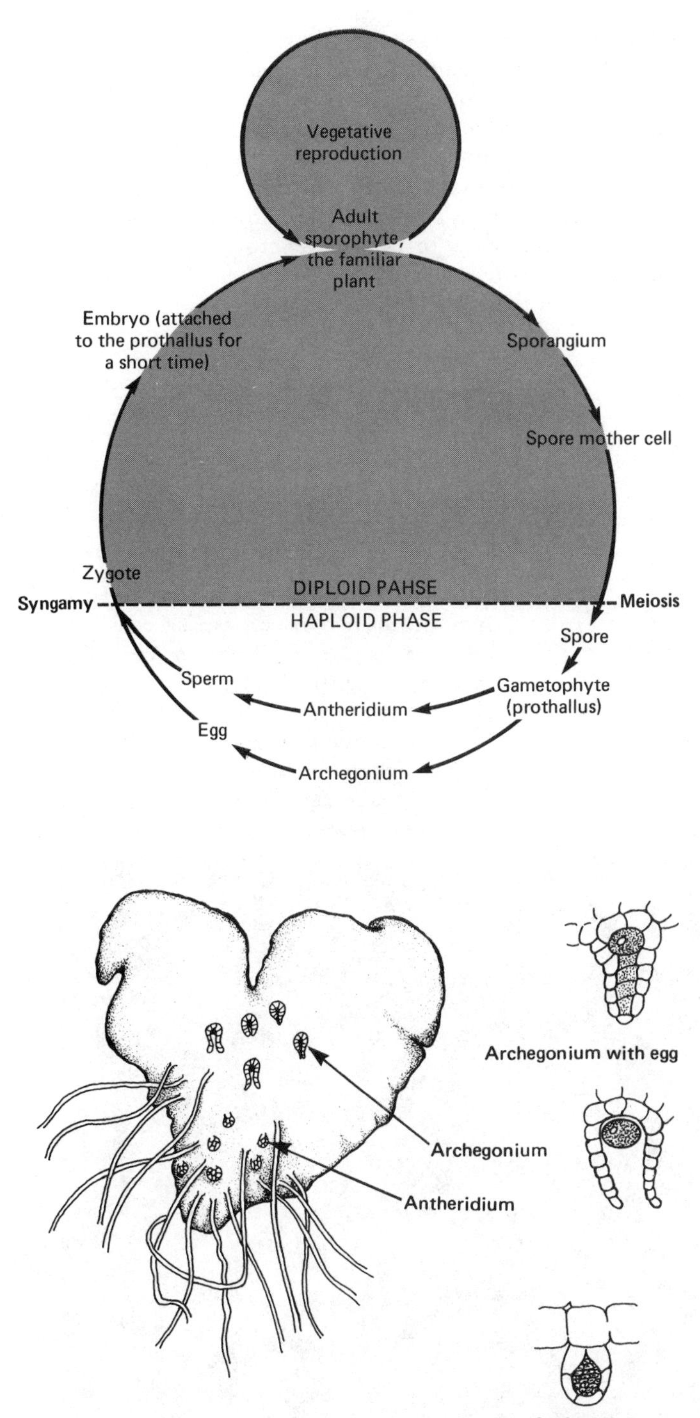


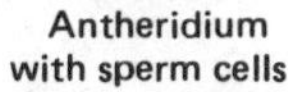


**FIGURE 14-20** The life cycle of a **homosporous fern,** including generalized diagrams of the mature gametophyte or **prothallus, antheridia,** and **archegonia.** Syngamy requires a surface film of water. The sperm swims to the egg, syngamy occurs, and the young sporophyte begins its growth attached to the parent gametophyte.

fern sporophyte. Most ferns have a homosporous life cycle (Fig. 14-20). Like the bryophytes and other primitive vascular plants, the ferns are trapped within a life cycle that requires the presence of a film of water for spore germination (Fig. 14-21), growth of the small, delicate gametophyte or **prothallus,** and sperm locomotion. On a gametophyte, antheridia (Fig. 14-22) generally mature before the archegonia (Fig. 14-23), decreasing the probability of self-fertilization. Gibberellins produced by older prothallia may stimulate development of the antheridia in younger prothallia. Sperm cells seem to follow a chemical gradient of attractant produced by the archegonia.

Initially the young sporophyte remains attached to the gametophyte, but eventually it establishes itself and the gametophyte dies.

**FIGURE 14-21** This SEM shows a fern **gametophyte** as it begins to emerge from the spore case. Many fern species can be identified by the distinctive pattern of sculpturing on the spores. (Photo courtesy of Joan E. Nester.)

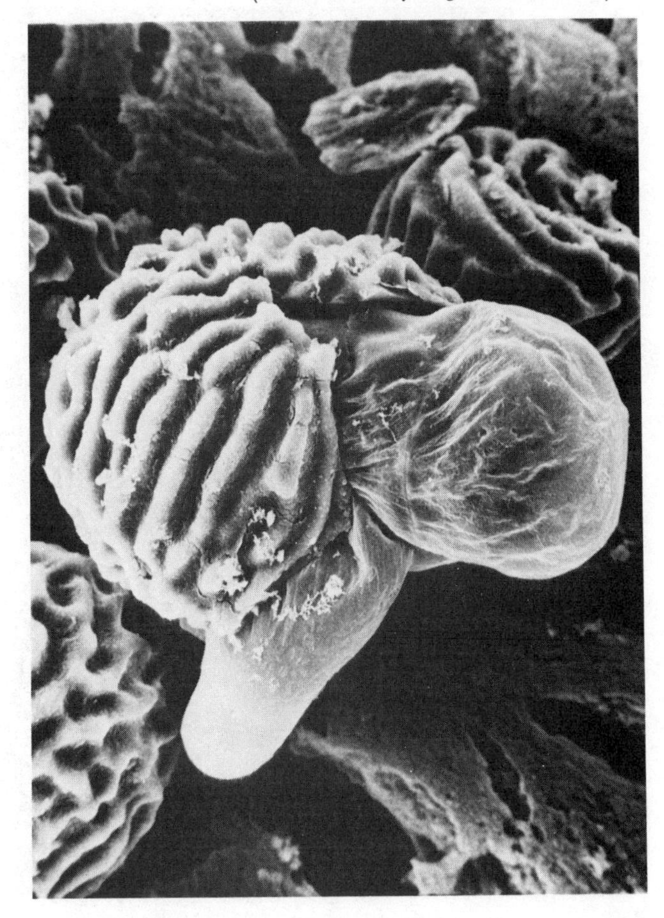

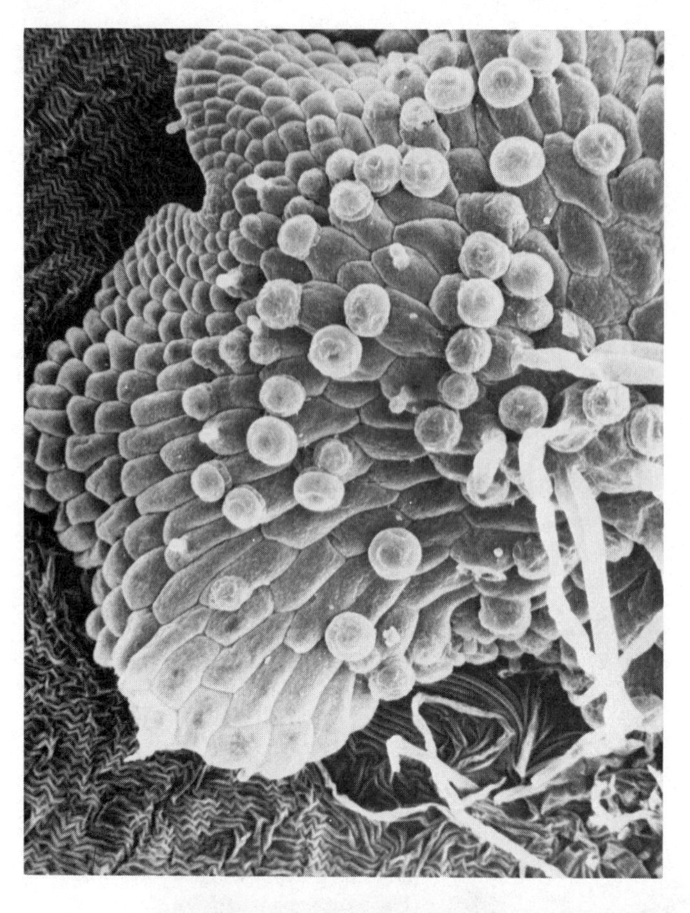

**FIGURE 14-22**
**Antheridia** develop on the fern prothallus near the **rhizoids** (rootlike extensions). Each antheridium has an apical cell that serves as a cap until the sperm cells are ready to be released. The cap cell then loosens and falls or is pushed off, permitting the sperm cells to escape. (Photo courtesy of Jane L. Kotenko.)

**FIGURE 14-23**
**Archegonia** develop near the apex of the fern prothallus. Each vaselike structure contains a single egg. When the egg cell is mature, a canal develops in the neck (venter) of the archegonium (Fig. 14-20). In addition, the archegonium may release a chemical attractant, creating a gradient along which sperm cells may travel. (Photo courtesy of Joan E. Nester.)

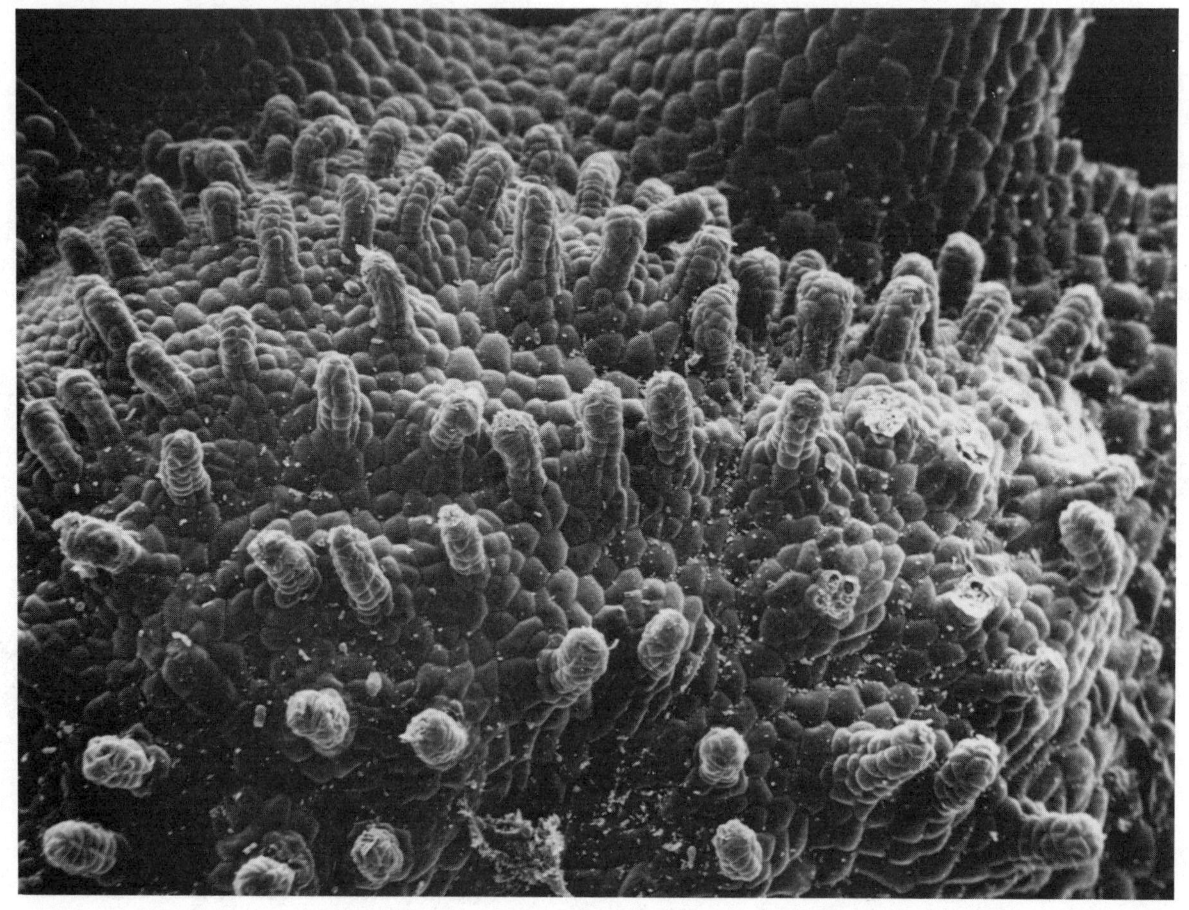

## DEVELOPMENT OF THE "SEED HABIT"

The evolution of *vascular tissues,* followed by *reproductive adaptations* that culminated in *seed formation,* completed the "conquering" by plants of the terrestrial environment by releasing the life cycle from dependence on direct contact with water for both syngamy and embryo development. This reproductive freedom allowed the great **adaptive radiation** of seed-forming plants, their evolution into forms that inhabit virtually every ecological niche, even in the harshest of environments. Table 14-1 summarizes progressive adaptations among terrestrial plants.

The first step toward a "seed habit" was a heterosporous life cycle: megasporangia produce megaspores that grow into female gametophytes; microsporangia produce microspores that grow into male gametophytes.

The second step was the retention of spores within the sporangia so that the gametophyte generation develops within the tissues of the sporophyte. Mitotic division of the megaspore results in the formation of a "captive" female gametophyte that produces its egg(s) while protected within the

**TABLE 14-1 Degrees of adaptation to the terrestrial environment.**

| Terrestrial Blue-greens | Bryophytes | Club Mosses | Ferns | Seed Plants |
|---|---|---|---|---|
| **Desiccation resistance** | | | | |
| Live in moist places | Thinly cutinized epidermis; live in moist places | Cutinized epidermis; live in moist places | Cutinized epidermis; most abundant in moist places | Cutinized epidermis; bark development; leaf and stem modifications as special adaptations to all environments |
| **Support** | | | | |
| Live between or on soil particles | Small or low-growing | Lignified tissues | Lignified tissues | Lignified tissues |
| **Absorption and internal transport** | | | | |
| Diffusion, osmosis | Diffusion, osmosis; some vascular differentiation; absorptive rhizoids | Continuous xylem and phloem connecting roots stems, and leaves | Continuous xylem and phloem connecting roots, stems, and leaves | Continuous xylem and phloem connecting roots, stems, and leaves |
| **Sexual reproduction** | | | | |
| Lacking | Requires water for sperm travel | Requires water for sperm travel | Requires water for sperm travel | Internal sperm transport via pollen tube |
| **Species dispersal** | | | | |
| Fragmentation; spores | Spores; specialized vegetative structures (gemmae); fragmentation | Spores; fragmentation | Spores; fragmentation | Seed containing embryonic sporophyte; fragmentation and specialized vegetative structures |
| **Dominant generation** | | | | |
| Gametophyte | Gametophyte | Sporophyte | Sporophyte | Sporophyte |

megasporangium; the composite structure is known as an **ovule.** Similarly, the microspores were retained within the microsporangia to undergo development into male gametophytes. (See also microsporogenesis and megasporogenesis, Chapter 5.)

A third essential development was a method for sperm nuclei to be transported to the female gametophyte other than by swimming. This problem was surmounted by production of a small, lightweight, desiccation-resistant, and transportable male gametophyte—the **pollen grain.** Each pollen grain carries sperm nuclei.

A fourth improvement was to develop an efficient means of sperm delivery to the egg after pollen transfer. To accomplish sperm transfer, the male gametophyte grows an extension, the **pollen tube,** into the ovule tissues. When the egg is reached, the end of the pollen tube disintegrates and the sperm nuclei travel down it to fuse with the egg, producing a zygote. Developments three and four thus successfully provide for internal rather than external syngamy.

The final development was to protect and nurture the embryo within a **seed** (Chapter 10). Instead of having to take its chances in the environment, the tender embryo of a seed plant develops within protective tissues, surrounded by constant moisture and food, encased within an environmentally resistant seed coat. In addition, the seed is protected by inhibitors in the seed coat that cause the embryo to remain dormant until environmental conditions are suitable for germination. Depending on the species, seeds may remain dormant for a variable length of time—from a few days to several years for most.

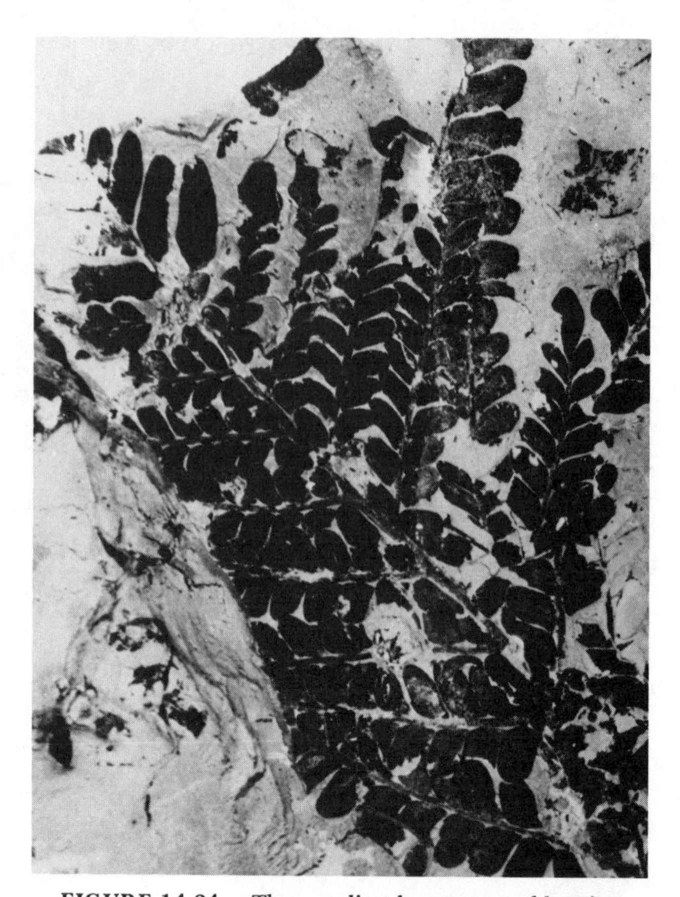

**FIGURE 14-24** The earliest-known seed-bearing plants are the extinct **seed ferns.** This specimen of *Alethopteris* shows several seeds attached along the axis. Collected in Indiana, it is believed to be the only North American specimen that has been found with seeds attached. (Photo courtesy of David L. Dilcher.)

## SEED-BEARING VASCULAR PLANTS

The seed plants are subdivided into two major categories: gymnosperms and angiosperms. To date, the earliest known seed bearers are the extinct seed ferns (Fig. 14-24), which were abundant during the Coal Age and which are usually classified as gymnosperms.

### Gymnosperms

The gymnosperms (*gymn* = thinly clad, *sperm* = seed) are a heterogenous collection of unusual kinds of plants carried over from prehistoric times of greater abundance and diversity within their groups, plus the more "modern" **conifers** typical of our times. Prehistoric gymnosperms are well represented in U.S. coal deposits as well as in the petrified forests of the West. All are perennial and many live for centuries. The oldest known living organism is a bristlecone pine (*Pinus aristata*) in the mountains of California that is, incredibly, nearly 4900 years old. It began its growth in approximately 3000 B.C.!

Most gymnosperms are trees, some of which become very large; others are shrubs (*Ephedra*) or vines (*Gnetum*). They are classified together as gymnosperms because the ovules develop on the *surface* of modified leaves, or sporophylls, within a strobilus. Gymnosperms differ from angiosperms in that the ripening ovule that becomes a seed is not enclosed by the tissue of the sporophyll—no

**FIGURE 14-25**
**Conifer** strobili vary greatly in size and gross structure. All those shown here are megasporangiate cones that have developed to the stage at which seeds are released. Several different *Pinus* species are represented. The smallest structures are hemlock (*Tsuga*) cones and *Juniperus* "berries." The center left cone is of Douglas fir (*Pseudotsuga*); to its right is a cone of giant sequoia (*Sequoia*).

fruit is formed. The strobilus may be fleshy, as that of junipers (*Juniperus*), or dry, as those of pine (*Pinus*), spruce (*Picea*), hemlock (*Tsuga*), fir (*Abies*), and cedar (various genera) (Fig. 14-25).

Pollen-producing strobili are referred to as male or **microsporangiate** cones; ovule-producing strobili are referred to as female or **megasporangiate** cones.

### Cycads

There are nine or ten extant cycad genera. The cycad *Zamia* is native to Florida. Cycads have a palmlike appearance (Fig. 14-26), making both native and imported species popular houseplants. They are used for landscaping in southern climates. *Zamia* reproduction is interesting because each pear-shaped sperm cell develops a spiral band of short flagella around its narrow end even though the sperm are delivered to the egg cell by a pollen tube! This phenomenon is interpreted as a remnant of an ancestral reproductive pattern that required free water for fertilization. Cycads are **dioecious,** having separate male and female plants.

**FIGURE 14-26** **Cycads are primitive gymnosperms.**

### Ginkgo

The dioecious maidenhair tree or ginkgo (*Ginkgo biloba*) has enjoyed renewed popularity as a street and yard plant. Planting of male trees is recommended because the fleshy ovulate strobilus produces foul-smelling butyric acid as it matures. The survival of the ginkgo is credited to its careful culture within Chinese and Japanese monastery gardens over the centuries, as it became extinct (as far as is known) in nature. Ginkgo is often referred to as a "living fossil" because it has widespread and abundant fossil relatives, one of which seems to be identical to this extant species. This gymnosperm is unusual in that it possesses flattened, deciduous leaves that are distinctively fanshaped, having dichotomously branching veins [Fig. 14-27 (a), (b), and (c)].

#### Gnetophytes

The gnetophytes include three remnant genera related by the morphology of the strobilus. Shrubby species of *Ephedra* are xerophytes found on deserts and mountains in various parts of the world. This genus is also found in the deserts of the Southwest, where it is known as Mormon tea or Brigham Young tea, for its early use as an herb tea. An Asiatic species is the source of the drug ephedrine, which is used to constrict blood vessels and to contract nasal membranes. *Ephedra* species are dioecious.

*Welwitschia mirabilis* (the only species of its genus) is also a xerophyte. The strange, slow-growing plant consists of a stubby, woody stem that produces two large, straplike leaves that may be up to 3 m long. The leaves grow by addition of new cells by meristematic tissues at their bases. *Welwitschia* grows only within a restricted, extremely dry region near Africa's southwestern coast.

In contrast to its xerophytic relatives, the genus *Gnetum* consists of broad-leaved shrubs, trees, or liana-type vines of tropical jungles. Vegetatively it resembles dicotyledonous flowering plants. It is dioecious, although the microsporangiate strobili contain nonfunctional ovules.

#### Conifers

Most conifers are xerophytic, having deep root systems, hardened, needlelike leaves with sunken stomates, thick bark, and innate cold hardiness. In the United States, however, at least one conifer of significance—the coast redwood (*Sequoia sempervirens*)—is not cold hardy or xerophytic, requiring mild humid conditions. There are more than 50 conifer genera. Most conifers are evergreen with individual needles usually being retained for 2 or more years. Deciduous conifers include the larches and tamaracks (*Larix*) and bald

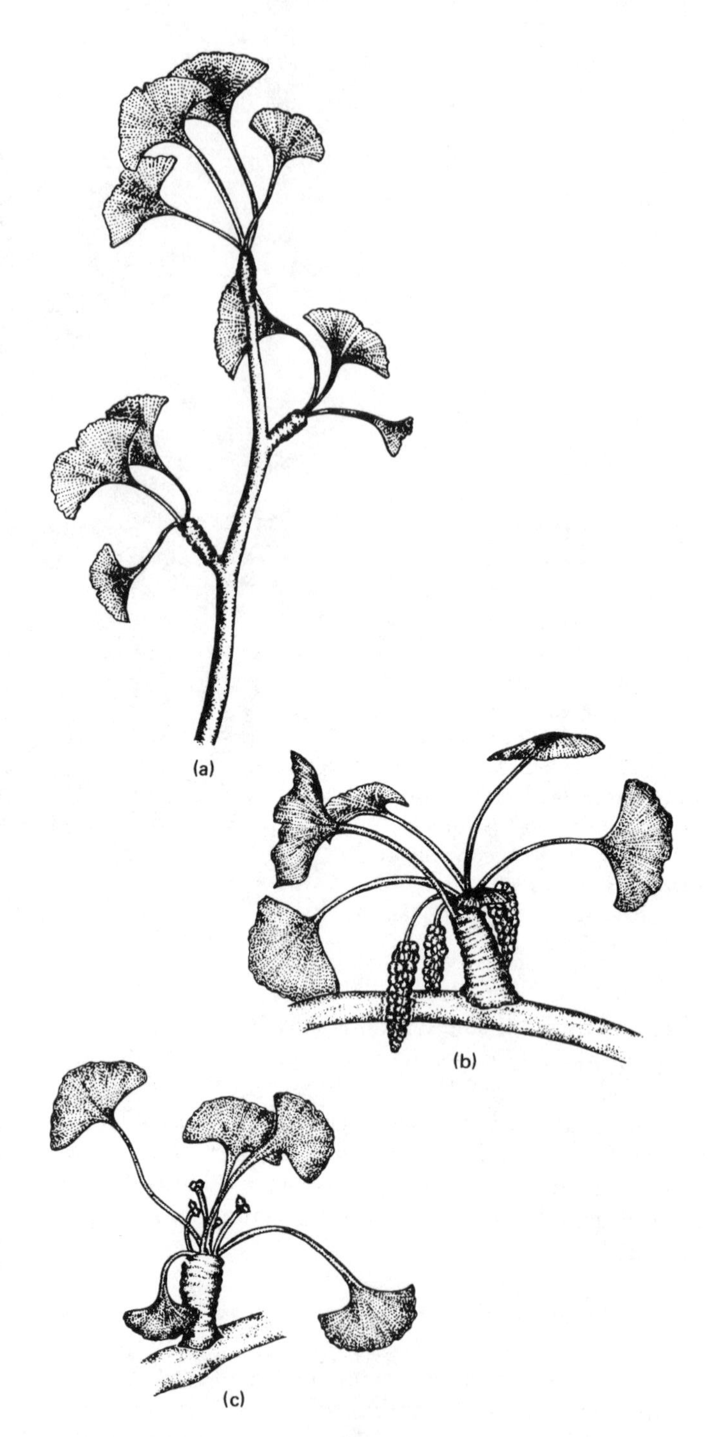


FIGURE 14-27 The **maidenhair tree** (*Ginkgo biloba*) is an unusual gymnosperm that is considered to be a "living fossil." It has deciduous, fan-shaped leaves (a). Male (b) and female (c) structures are borne on different trees. (From S.R. Rushforth, *The Plant Kingdom – Evolution and Form*, 1976. Reprinted by permission of Prentice-Hall, Inc., Englewood Cliffs, N.J.)

**FIGURE 14-28**
Pines are characterized by bearing needles in bundles or **fascicles.** This red pine (*Pinus resinosa*) is a two-needle (per fascicle) pine. The drawing on the left shows a cluster of **microsporangiate** (pollen-producing) cones. The upper right drawing shows the terminal "candle" of new shoot growth in the spring, as well as a young **megasporangiate** (ovule-producing) strobilus. The lower right drawing depicts a 3-year-old megasporangiate cone as it appears when it sheds its seeds—the familiar decorative "pine cone" stage.

cypresses (*Taxodium*). Conifers typically grow as trees that have a single main trunk from which whorls of branches arise.

Conifers, such as the pines, have small microsporangiate cones and larger megasporangiate cones [Figs. 14-28 and 14-29 (a) and (b)]. Figure 14-30 illustrates the life cycle of a typical pine and Fig. 14-31 (a), (b), and (c) shows the development of a pine seed.

The conifers are of vast economic importance as the major source of wood and paper products, plus many organic solvents and other compounds (Chapters 8 and 16). The distinctive structure of conifer wood is illustrated in Chapter 8. Conifers are dominant and subdominant components of several types of ecosystems around the world. The distribution and ecology of North American conifers are discussed in the next chapter.

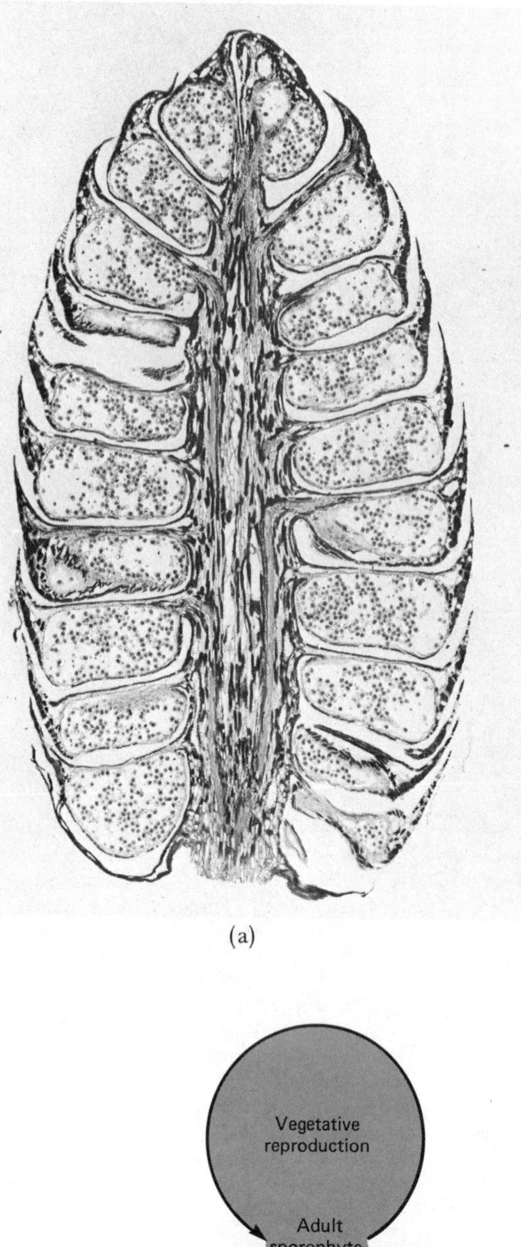

(a)

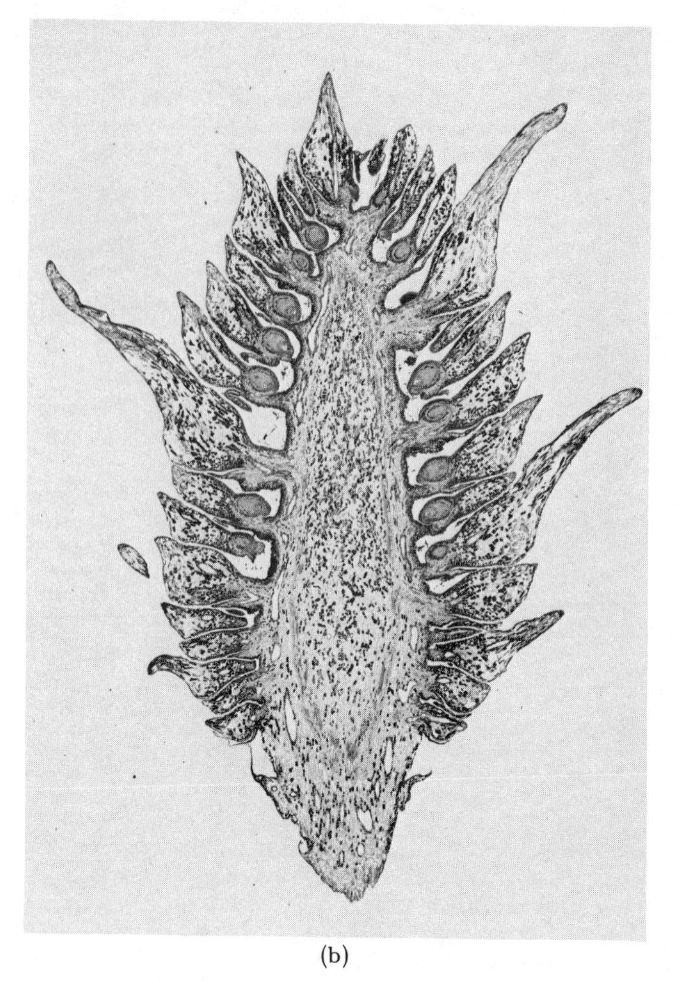

(b)

**FIGURE 14-29**

A strobilus is made up of sporangium-bearing "leaves" called **sporophylls.** In heterosporous plants they may be either microsporophylls or megasporophylls. (a) Microscopic view of a mature microsporangiate cone, sectioned lengthwise. Note the **pollen sacs** (microsporangia) containing pollen grains. (b) Similar view of a young megasporangiate cone at the pollination stage, showing **ovules**—developing female gametophytes within megasporangia—on the surface of **cone scales.** The ovules are seen at different planes in the section because of the spiral arrangement of the scales along the cone axis. (Photos courtesy of Carolina Biological Supply Co.)

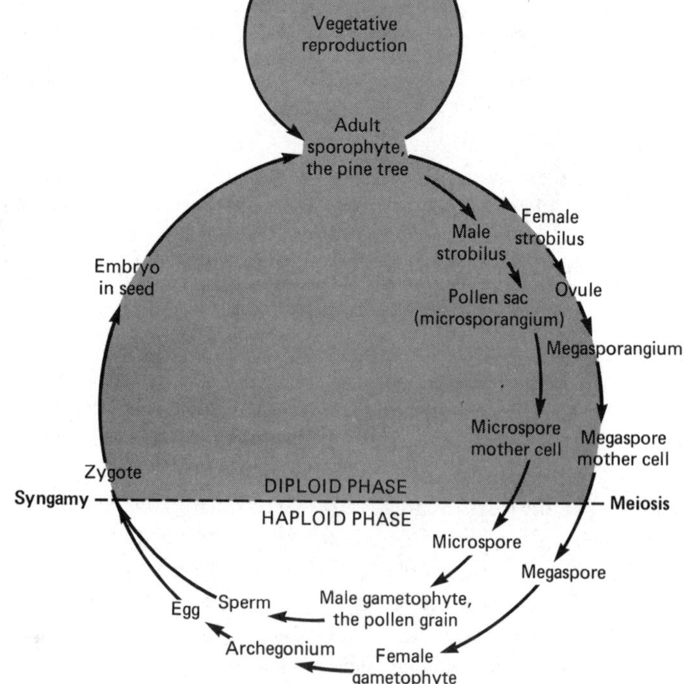


**FIGURE 14-30**

Life cycle of a pine. The production of pine seeds takes three growing seasons, at least in some *Pinus* species. Ovule and pollen formation and pollination all occur during a single season. The microsporangiate cones are then shed. Syngamy occurs during the second year, following the slow growth of pollen tubes to the eggs. The mature seeds are released during the third year.

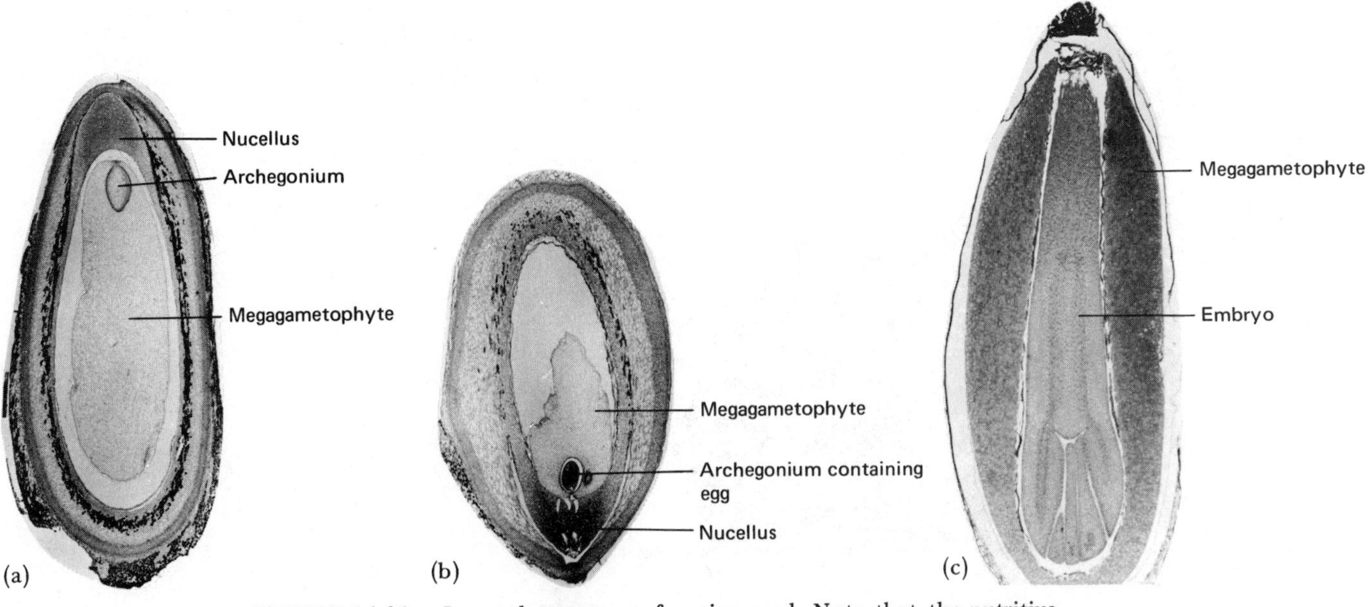


**FIGURE 14-31** Internal structure of a pine seed. Note that the nutritive tissue is derived from the female gametophyte ($n$). Compare this to the angiosperm seeds diagrammed in Chapter 10, having $3n$ nutritive tissue. (From S.R. Rushforth, *The Plant Kingdom—Evolution and Form,* 1976. Reprinted by permission of Prentice-Hall, Inc., Englewood Cliffs, N.J.)

## Angiosperms

The greatest adaptive radiation among plants has been shown by flowering plants, the angiosperms. Angiosperm species range in size from the tiny floating duckweed (*Lemna*) (Fig. 14-32) to enormous and ancient trees. Growth forms and shapes that may seem beautiful, curious, fantastic, or outright bizarre to us reflect hundreds, thousands, even millions of generations of adaptive responses to environmental demands—the result of genetic variation and natural selection (Chapter 6).

Figure 14-33 summarizes a common angiosperm life cycle pattern. Angiosperm megasporogenesis and microsporogenesis are *detailed completely in Chapter 5,* to which you should refer to supplement this discussion.

The angiosperm flower can be regarded as a highly modified strobilus. Although it may seem odd to relate such variable structures as flowers to the relatively simple strobili of conifers, they are both modified shoots, bearing highly modified leaves. In a flower the modified leaves near its base (petals and sepals) are sterile; those near its tip (stamens and carpels) are fertile, being modified sporophylls. **Stamens** contain **microsporangia; carpels** contain **megasporangia** (Fig. 5-23). Flowers may be either bisexual or unisexual, depending on the species.

A stamen usually is composed of two parts: a stalk or filament at the end of which is an anther.

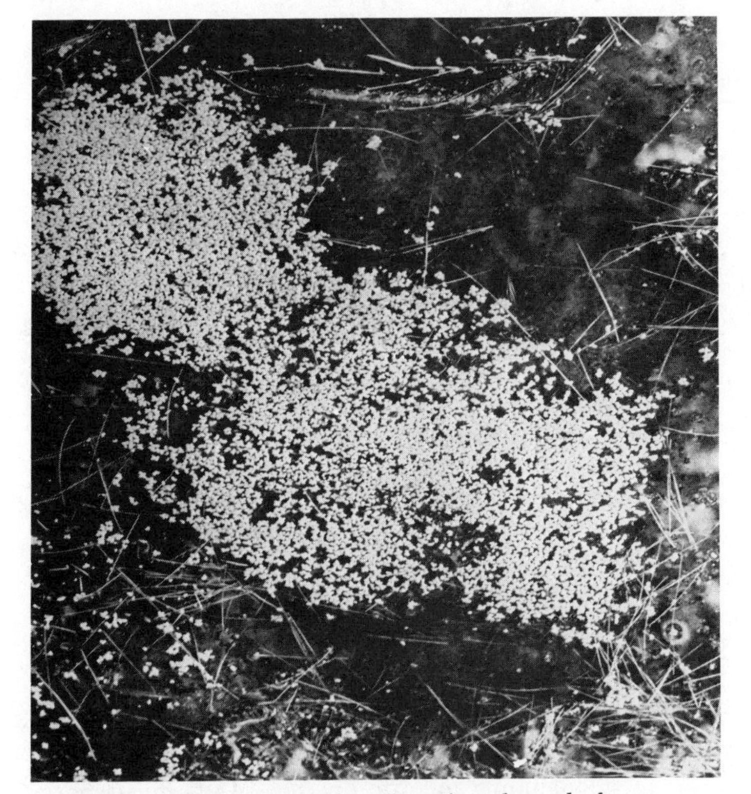

**FIGURE 14-32** The tiniest flowering plant, duckweed (*Lemna*), is familiar as a collective mat of chartreuse-green vegetation on the surface of warm, still water. Even from an airplane you can identify shallow waters by their bright covering of duckweed. (Photo by Mark R. Fay.)

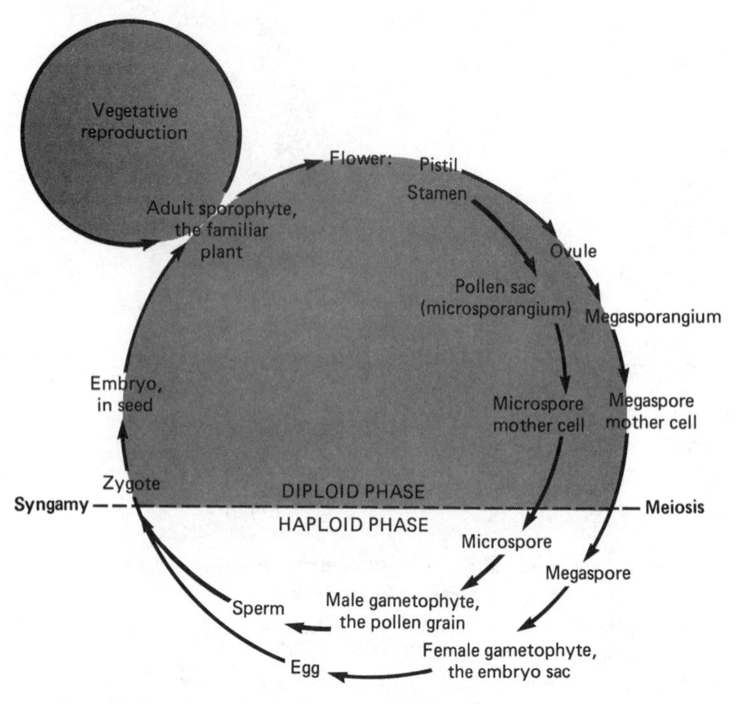


**FIGURE 14-33**
The life cycle pattern of a typical flowering plant, presented in detail in Chapter 5.

An anther encloses one or more microsporangia (pollen chambers), which are sites of microsporogenesis. Microspores are formed by meiosis within the microsporangia and male gametophytes (pollen grains) subsequently develop. The pollen grains are released into the environment by splitting of the anthers or by the activity of an animal pollinator.

Carpels are fused and modified variously to form ovaries. Fossil and developmental evidence demonstrates the carpel to be a modified sporophyll that is folded down the middle and fused around the megasporangial tissues. Carpels are further modified to receive pollen grains, resulting in a differentiated structure called a **pistil.** Each pistil generally has a sticky or feathery tip, the stigma, to which pollen adheres; an elongated stalk or style, which supports the stigma and through which pollen tubes grow; and the enlarged megasporangia-containing chambers of the ovary. Pistils may be composed of one or several carpels. The number of carpels can usually be determined by slicing the ovary (or fruit) crosswise to see how many chambers are present.

As in gymnosperms, ovule formation is complex, including meiotic division of the megaspore mother cell and subsequent tissue development (Chapter 5). Each **ovule** is a female gametophyte enclosed within special layers of sporophytic tissue. The ovule is therefore diploid on the outside, haploid on the inside. Most ovules are large and readily visible within a floral ovary, but microscopic inspection is necessary to distinguish sporophytic and gametophytic tissue layers.

The transition from *ovule to seed* and *ovary to fruit* following syngamy (or appropriate hormonal stimulus) is discussed in Chapters 5 and 11. Angiosperm seed structure is discussed in Chapter 10.

The angiosperms are divided into two general groups, monocots and dicots. You are already familiar with these terms and the differences they represent from previous chapters. As you will recall, the names refer to the presence of one cotyledon or two cotyledons as a part of the embryo. In addition, monocots generally have parallel leaf veins and floral parts in threes or multiples of threes. Dicots normally have netted leaf venation and floral parts in fours or fives or their multiples. (See also Chapters 5 and 8.)

Some flowers are unusual for their complexity, as the orchids, and others for their simplicity, as the bur oak (*Quercus macrocarpa*) (Fig. 14-34). A diversity of floral types is illustrated in Chapter 5.

## SUMMARY

Vascular plants have well-developed **conducting tissues** and the **sporophyte** is the dominant phase. Non-seed-forming vascular plants dominated the land environment more than 400 million years ago.

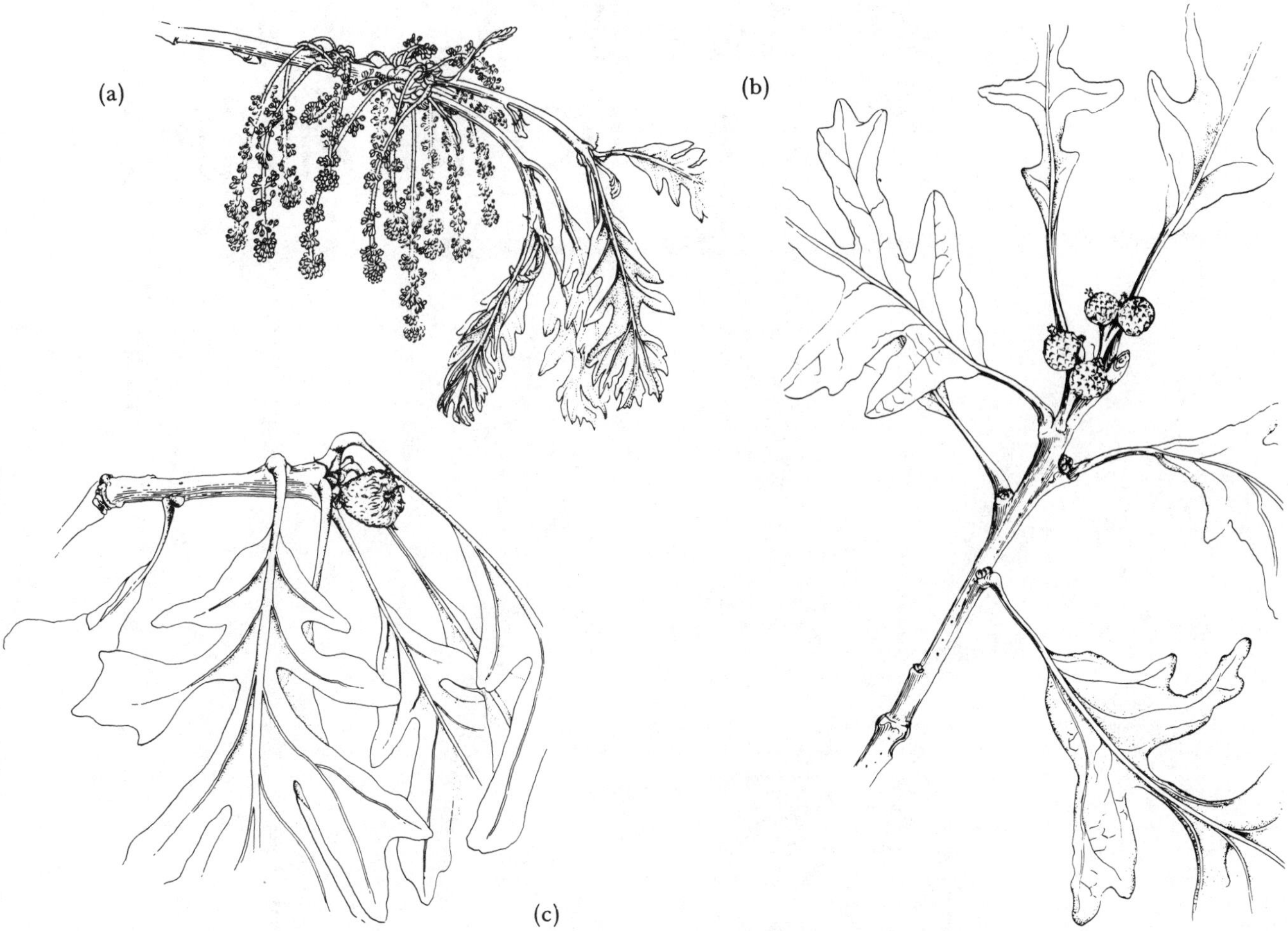

**FIGURE 14-34** Many tree flowers are inconspicuous wind-pollinated blossoms that go unnoticed until the male flowers eventually fall on decks and sidewalks. In this bur oak (*Quercus macracarpa*) the male flowers (first drawing) are a chain of anthers and the female flower is almost completely hidden at the shoot apex. The next two drawings show successive stages in the expansion of the leaves and the development of the acorn, a nut.

In most vascular plants sporangia are borne on **sporophylls,** which are organized into cones or **strobili.** Most **psilophytes** and **horsetails** are extinct. The remaining forms are the simplest and oldest vascular plants and have homosporous life cycles.

Some club mosses are **homosporous;** others are **heterosporous,** a significant pattern that continues into the seed plants.

**Ferns** live primarily in moist climates, producing leaves from rhizomes. Sporangia are borne on some of the compound leaves. With few exceptions, ferns are homosporous.

Among non-seed-producing plants, sperm cells typically are locomotory. Spores of aquatic plants are released into water currents. Land plants take advantage of **hygroscopic movements** to assist in spore release.

**Seed** production completed the adaptation of plants to the terrestrial environment. Several features are part of this adaptation: a "captive" female gametophyte, a transportable male gametophyte, internal sperm transport and fusion of gametes, and, finally, the seed.

Gymnosperm seeds develop on the scales of a strobilus whereas angiosperm seeds develop within ripening ovaries, the fruits.

Some **gymnosperms** are carried over from prehistoric times, as **cycads, ginkgo,** and the **gnetophytes.** Most of our modern gymnosperms are needle-bearing **conifers.**

**Angiosperms** display the greatest adaptive radiation of all plants. The angiosperm strobilus is the **flower**—a stem with evolutionarily modified leaves, the sepals, petals, stamens, and carpels.

# chapter fifteen BIOMES OF NORTH AMERICA

## INTRODUCTION

Most of us, in one way or another, are nomads. Some humans are nomadic in the true sense, wandering from place to place to survive the seasons. Many of us are "vacation nomads." Daily, however, we are nomadic in at least a vicarious sense as television, newspapers, magazines, and books take us to exotic places. Some of us even seek adventure in atlases and road maps, planning for trips we may someday take.

At least once a day we see or hear a national weather report, acquainting us with regional environmental differences. News reports frequently include environmental stories: devastating drought in northern Africa converting arid grasslands and tree-dotted savannahs into desert; wartime defoliation of tropical forests; strip mining in Appalachia; oil and gas "blowouts" and tanker spills; gypsy moth epidemics; fruit fly invasions; channelization of streams and rivers; oil and natural gas pipeline projects; floods and dam disasters; wild coyotes adapting to life in suburban Beverly Hills; hunting and fishing reports.

Based on such encounters, which are a continuing part of our education, we develop mental images of places. The mental image evoked is usually a general impression, an overall sense of the climate and vegetation of an area. In fact, particular climatic and geographic influences do sustain characteristic types of plant and animal life. Hot, arid conditions in both the Old World and New World, for example, sustain plants and animals of the desert. The exact combinations of geography, climate, plant, and animal species differ, but the overall environmental resemblances are great. So it is useful to speak of environments of the world or of a continent in terms of major ecological categories, the biomes. **Biome** is a term that encompasses all the physical and biological factors that constitute a major type of natural area. The face of our planet can be described as an assemblage of biomes.

For our purposes, we limit our consideration of biomes to those associated with the North American continent. More and more Americans (and visitors) are being enriched by travel in North America and so our discussion is directed toward identifying and understanding the biomes that you are likely to explore firsthand some day. Modifying human influences are also mentioned, for we are now a factor in all biomes. Emphasis is placed on *general physical factors, predominant biological factors,* and *geographic distribution* of each biome. (You may find it helpful to review ecological terms and principles introduced in Chapter 2.)

Within a biome in a specific geographical area are a number of smaller environmental units, the ecosystems. An **ecosystem** is an area in nature where the living organisms (the biological community) and nonliving components of the environment interact, producing an exchange of materials and energy.

A **biological community** is composed of populations of different species of organisms. The term **population** refers to all members of a single species present within the community. Every city, for instance, lists a population figure. This figure refers to the population of a single species, *Homo sapiens,* and not to populations of dogs, robins, elms, maples, or woolly bear caterpillars that are also part of the biological community.

So with some ecological principles recalled from Chapter 2, we're prepared to identify and survey North American biomes (Fig. 15-1). Although

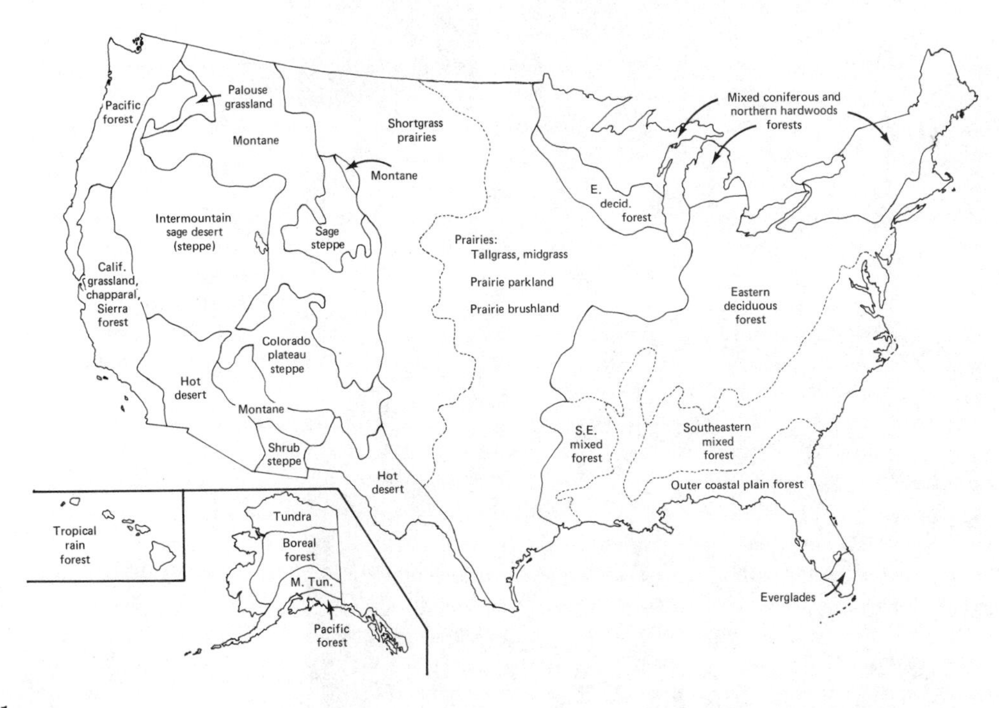


**FIGURE 15-1**
**Biomes** of North America. Such factors as latitude, elevation, soil, local climate, and geologic history have created a variety of ecosystems. The Hawaiian Islands and Alaska are shown in insets. In Alaska the abbreviation "M. Tun." means "mountain tundra." "Steppes" are combined grass-and-shrub areas. Most of North America's steppeland is sagebrush desert. (Adapted from Bailey, R.G. *Ecoregions of the United States.* Published by the U.S. Forest Service, prepared in cooperation with the U.S. Fish and Wildlife Service, 1976.)

our major emphasis is on U.S. regions, much of the following discussion is also pertinent to a Canadian tour.

## ENVIRONMENTS OF NORTH AMERICA

### The Arctic/Alpine Biome

#### Arctic Environments

Much of Alaska and northern Canada is **tundra**, a treeless, flat-to-rolling landscape controlled by the long winters and short summers north of the Arctic Circle. Temperatures of the arctic biome are so cold and so prolonged that the ground is permanently frozen a few inches to more than a thousand feet below the surface and hence is called **permafrost.**

During the short summer season the surface layer of the permafrost thaws, allowing plant and animal life to flourish. This soil tends to be mucky because the water from melting snow and ice cannot drain into the frozen subsoil. It is unstable and slow to recover from disruption (Fig. 15-2). Where there is a slight slope, the melted surface soil may actually flow across the landscape.

Repetitive freezing and thawing cause frost heaving and ridging (Figs. 15-3 and 15-4). Tundra

**FIGURE 15-2**
A road and oil pipeline across the **tundra.** Note the thin soil along the roadside. (Photo courtesy of Fred L. Rose.)

**FIGURE 15-3**
**Frost boils** on the tundra. (Photo courtesy of Fred L. Rose.)

**FIGURE 15-4**
**Frost ridges** on the tundra. (Photo courtesy of Fred L. Rose.)

plants must be able to tolerate root-system disturbances from all this physical activity. Where boggy conditions exist, the soil is acidic; where rivers arising in limestone mountains have filled their valleys with calcareous silts, however, the pH is high, such as in Prudhoe Bay.

Where there is some elevation, soil may be sparse, dry, and rocky, depending on exposure. Abrasion by wind-carried soil and ice adds harshness to the environment. Rivers and streams are common, often shallow, rocky bottomed, and "braided" (Fig. 15-5).

It is not easy to generalize one arctic vegetation type. The very harshness of this ecosystem renders subtle microenvironmental differences important enough to create several distinctly different plant associations. Overall, however, arctic plant life consists primarily of mosses, lichens, small rushes (*Juncus*), sedges, grasses, and dwarfed perennial plants—mainly heath shrubs, willows (*Salix*), and birches (*Betula*) (Figs. 15-6 and 15-7). Peat moss (*Sphagnum*) grows in moist depressions.

Plants have adapted to this harsh life in several ways. Almost all arctic plants are perennial; narrow leaves and small growth habit restrict exposed surface area; food is utilized efficiently and there are ample food storage tissues; a very condensed life cycle allows growth, flowering, and setting seed or

**FIGURE 15-5**
A **braided river** in Alaska. (Photo courtesy of Fred L. Rose.)

FIGURE 15-6
Tundra **tussocks** with dwarf willow (*Salix*). (Photo courtesy of Fred L. Rose.)

FIGURE 15-7 Cotton grass (*Eriophorum*) on the tundra. (Photo courtesy of Fred L. Rose.)

producing bulbils in a few weeks of favorable conditions; vegetative reproduction by production of rhizomes and bulbils is important.

Many animals are adapted to the tundra, including such large herbivores as caribou, moose, and musk ox, and smaller herbivores, such as arctic hare and the many kinds of ground squirrels and mice. Predators and omnivores include grizzly bears and the endangered Kodiak or Alaskan brown bear, wolves, coyotes, arctic fox, and weasels. Migratory birds are abundant in summer, especially waterfowl, but few birds other than the ptarmigan and redpoll are permanent residents. The many shallow wet spots are nurseries for millions of mosquitoes and biting flies.

Many arctic animals, such as some caribou and most of the birds, migrate from the harsh above-ground arctic winter conditions. Others hibernate or remain periodically inactive. Lemmings remain active under the snow. Insects and other invertebrates survive winter as dormant adults or overwintering larvae.

#### Alpine Environments

The conditions of the arctic are created by *latitude* and the resulting changes in solar energy available. Very similar physical conditions, however, in the mountains are created by *altitude.* This arcticlike mountain environment is called **alpine** (Figs. 15-8 and 15-9). The alpine environment differs from the arctic environment mainly in latitude, altitude, precipitation, slope, and daylength.

Because of their lower latitude, alpine areas of temperate zones receive more sunlight than arctic areas. Because of the intensely cold winter temperatures and the heavy snow cover, which is slow in melting, the result, however, is not necessarily a longer growing season for plants. Precipitation is usually greater in alpine than arctic regions. Permafrost is present only at the highest altitudes in alpine areas.

**FIGURE 15-8**
The **alpine zone,** looking toward the Grand Tetons, with interspersed subalpine areas of coniferous trees. The ragged demarcation between alpine and subalpine demonstrates dramatically how topography—rather than latitude—is a primary environmental factor that distinguishes the alpine tundra from the arctic tundra. (Photo courtesy of Fred L. Rose.)

**FIGURE 15-9**
Alpine zone in the Olympic Mountains of Washington. (Photo courtesy of Fred L. Rose.)

High altitude creates some special environmental conditions. The thinner atmosphere filters out less radiation so that organisms must have shielding adaptations to protect them from ultraviolet rays (beware of mountain sunburns!). Greater infrared penetration causes very high daytime ground temperatures in summer. The sparse atmosphere loses heat quickly at night, however, so that diurnal (day-night) temperature variation is great in the summer.

Slope enhances both drainage and air movement. Therefore aridity is a potential problem, depending on atmospheric humidity, degree of drainage, angle of exposure to sun and prevailing winds, and soil humus content. In relatively flat alpine areas moist willow thickets with small rocky-bottomed streams are common. Sloping exposed areas appear as "balds," having vegetation that is low, tundralike, and florally splendid in the spring (Fig. 15-10).

As in the Arctic, mosses and lichens are common, mosses nestling in semiprotected cracks and crannies, lichens growing anywhere, including on smooth rock surfaces.

Flowering plants are primarily miniature species of genera that are represented in other biomes, such as lilies, asters and daisies, larkspurs (*Delphinium*), lupines (*Lupinus*), grasses, and willows. Most alpine species are less than 10 cm tall and many grow as compact "mats" or small mounded "cushions." Adaptations to harsh alpine life center on water conservation, economical vegetative growth, and rapid food and seed production within the short growing season, which is a brief interval of moist favorable conditions between snow melt and thawing of the soil and the hot, dry late summer.

Moss campion (*Silene acaulis*), a flowering plant, possesses many alpine adaptations. It is perennial and has small, thick, darkly pigmented, needlelike leaves. The leaves have a thickened cuticle and are not deciduous. It grows as a "cushion plant," a compact growth form that provides protection from drying, freezing winds and that also traps heat. It has a large, woody taproot that deeply penetrates the substrate for anchorage and moisture and that also contains food storage tissue. The plants are slowgrowing and longlived.

Some alpine plants are annual rather than perennial, completing their entire growth from germination to seed production in the short growing season. Most, however, are perennial.

Several species rely mainly on vegetative propagation, especially by rhizomes. Even though shoot growth is economical, floral structures may be as large and showy as related nonalpine species, seeming extravagant in relation to overall plant size [Fig. 15-11 (a) and (b)]. The large flowers, however, effectively attract insect pollinators.

Some alpine animals, such as mice, pocket gophers, ground squirrels, marmots, and pika, are permanent residents that find winter protection in

**FIGURE 15-10**
An alpine meadow in full bloom in June in the Wind River Range of Wyoming.

(a)

(b)

**FIGURE 15-11** (a) The alpine columbine (*Aquilegia*) has extravagantly showy flowers that attract pollinators. Contrast the large flower size to the tiny, water-conserving leaves, which are about 5-10mm long. (b) Even the alpine willow produces a full-sized flower on a diminutive vegetative body. The male catkin is shown here.

burrows or rocky shelters. Abundant summer insects survive winter as dormant adults or larvae. Most animal life is seasonal, however, including such migratory birds as owls and robins, plus larger animals like moose, elk, deer, bighorn sheep and mountain goats, which use the high country as summer feeding grounds. Also, the alpine area is a feeding ground for many animals that return by evening to their homes in adjacent forested areas.

## The Boreal Forest, Taiga, or Northern Coniferous Forest

Where timberline meets arctic tundra (Fig. 15-12), plant associations expand to include coniferous trees, especially white and black spruce (*Picea glauca* and *P. mariana*), evoking images of the "Northwoods" (Fig. 15-13). Understory vegetation incorporates and intergrades with arctic vegetation,

**FIGURE 15-12**
The **taiga** or **boreal forest** in the foreground extends to its ragged timberline limit in the Brooks Range of Alaska. Note the braided river. (Photo courtesy of Fred L. Rose.)

**FIGURE 15-13**
A typical taiga forest, near Fairbanks, Alaska. (Photo courtesy of Fred L. Rose.)

including alder thickets (*Alnus*) and such berry-producing shrubs as blueberries (*Vaccinium*) and other heaths. Ferns and such small flowering plants as orchids and wintergreen (*Gaultheria procumbens*) abound. Various mice and red and flying squirrels are common. Large mammals include white-tailed deer, wolves, black bear, and, more northerly, moose, wolverine, and Canadian lynx.

Lakes, potholes, and bogs pit the landscape, reflecting Pleistocene glaciation. The open water is ringed by bottom-rooted water lilies (*Nymphaea*), pondweeds (*Potamogeton*), arrowhead (*Sagittaria*), rushes, and wild rice (*Zizania aquatica*). Peat moss thrives as do tamarack (*Larix laricina*), small carnivorous plants, and, of seasonal interest to humans, cranberries (*Vaccinium macrocarpon*).

Waterfowl breed and nest on northern lakes, which also contain such large game fish as walleyes, northern pike, and muskellunge. Water-breeding mosquitoes and biting flies (deer flies, black flies) are abundant.

The land in this biome was sculptured by great prehistoric glaciers beginning a million years ago and ending as recently as 40,000 to 20,000 years ago. Glaciated landforms include gouged depressions, ridges where glaciers dumped loads of sand, gravel, rocks, and boulders along their paths, and large areas of exposed glacially scoured ancient granite. Soils are postglacial, tending to sand and rock, with acidifying humus provided by decomposed plant matter. Much of the soil is classified as bog and half-bog. Farming in cleared boreal forest lands and farther south into the glaciated part of the deciduous forest biome is a yearly process of clearing postglacial rocks and boulders encountered during plowing. Permafrost may be present in places as far south as Fairbanks in Alaska.

The boreal forest is a land of moderate precipitation, most of it snow. The winter season is fairly long (8 to 9 months) and cold (subzero average). The summer growing season is short, with generally mild temperatures.

### Temperate Deciduous Forests

Beginning in Canada from the north and spreading south into the eastern United States and down the Appalachians, the northern coniferous forest is replaced by the several types of communities that make up the temperate deciduous forest biome (Fig. 15-14). South from the Great Lakes, boreal forest gives way to northern hardwood forest. In the southeast it gives way at lower elevations to central hardwood forest. The deciduous forest biome is the setting in which much of the early history of the United States occurred.

Although the regional species composition of the temperate deciduous forest varies, several basic environmental factors determine the biome type. They are the presence of distinct seasonal changes, varying exposures to freezing temperatures, adequate-to-abundant precipitation to sustain a mesophytic flora, and moderate-to-long growing seasons.

**FIGURE 15-14** The **eastern deciduous forest** is a large biome consisting of several major plant associations. (a) Mixed mesophytic forest. (b) Beech-maple forest. (c) Maple-basswood forest plus oak savanna. (d) Appalachian oak forest. (e) Oak-hickory forest. To the south and southeast the eastern deciduous forest intergrades with the outer coastal plain forest (beech-sweetgum-magnolia-pine-oak) and southeastern mixed forest. (See also Fig. 15-1.) (Adapted from Bailey, R.G. *Ecoregions of the United States.* Published by the U.S. Forest Service, prepared in cooperation with the U.S. Fish and Wildlife Service, 1976.)

Precipitation in this eastern North American biome averages 50 to 90 in. per year, distributed evenly throughout the year. Much of the winter precipitation is snow, with temperatures dipping to -30°C and occasionally lower. Summer days occasionally exceed 38°C. Typical midsummer days are warm and humid (even tropical) and typical midwinter days are cold with very low atmospheric humidity.

Such hardwoods (nonconiferous trees) as paper or canoe birch (*B. papyrifera*) and red and sugar maples (*Acer rubrum, A. saccharum*) plus some eastern hemlock (*Tsuga canadensis*) and white pine (*Pinus strobus*), are found where the northern coniferous and northern deciduous forests meet and intergrade. In much of the eastern United States northern hardwood forests are probably seral stages to a white pine climax forest. The total-removal logging practices of the 1900s, however, eliminated the eastern white pine forest as a significant entity, possibly forever (Figs. 15-15 and 15-16).

Beyond the transition zone of northern hardwoods, dominant components of the temperate

**FIGURE 15-15** Lumberjacks in the white pine (*Pinus strobus*) forests of northern Wisconsin harvested a seemingly endless virgin timber resource. (Photo by permission of the Chippewa Valley Historical Society.)

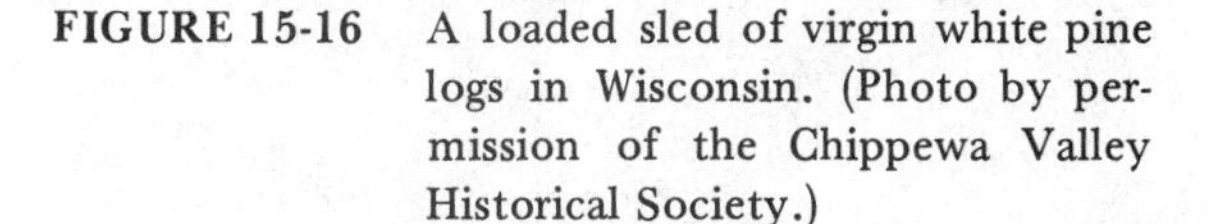

**FIGURE 15-16** A loaded sled of virgin white pine logs in Wisconsin. (Photo by permission of the Chippewa Valley Historical Society.)

deciduous forest are beech (*Fagus grandifolia*), sugar maple, basswood (*Tilia americana*), yellow birch (*B. allegheniensis*), black cherry (*Prunus serotina*), oaks (*Quercus*), and occasional white pines (Fig. 15-17). Each species has its own site preferences and the various species tend to occur in association, with one or two species dominating a stand, depending on the soil, climate, and exposure—for example, sugar maple-basswood, oak-chestnut (*Castanea dentata*), or beech-sugar maple.

Deciduous forest communities exhibit well-developed vertical stratification, with plant and animal species distributed through overstory, understory, ground level (Fig. 15-18), and subsurface levels. Because the biome is tremendously variable, understory and animal life vary according to the dominant overstory associations. Honeysuckles (*Lonicera*), poison ivy (*Rhus radicans*), blackberries (*Rubus*), raspberries (*Rubus*), and sassafras (*Sassafras albidum*) are a few representative understory plant types. And, of course, each individual area has its own beloved and characteristic wildflowers (Figs. 15-19, 15-20 and 15-21).

Some of the more ubiquitous and characteristic animal species are white-tailed deer, gray squirrel, eastern chipmunk, black bear, striped skunk, opossum, raccoon, eastern bluebird, blue jay, cardinal, and ruffed grouse. Many migratory songbirds nest in this biome, and the wild turkey is an indigenous species.

**FIGURE 15-17**
Deciduous woods in the snow. Note the presence of a young white pine among the oaks.

**FIGURE 15-18**
The floor of a deciduous forest is a tapestry of living plants and decomposing plant litter.

**FIGURE 15-19**
Jack-in-the-pulput (*Arisaema*) is a study in subtle browns and greens.

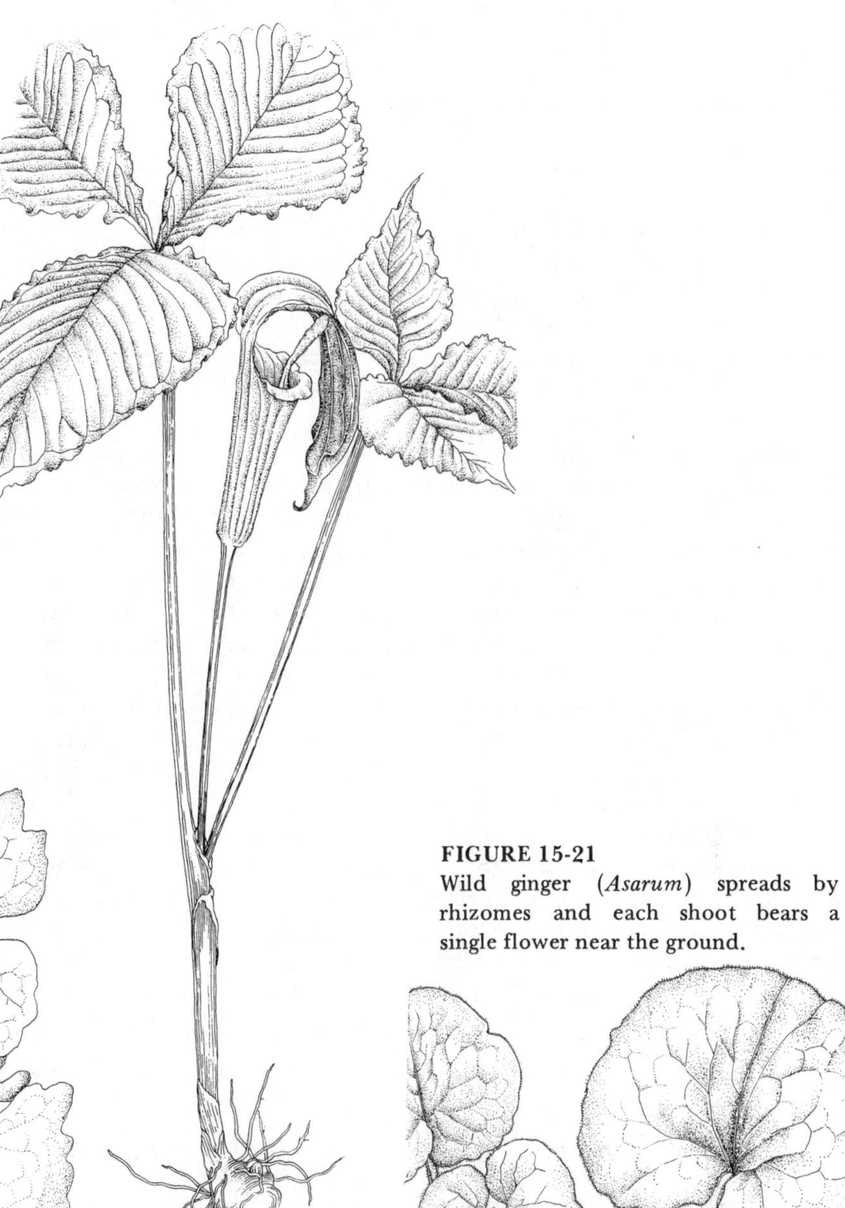

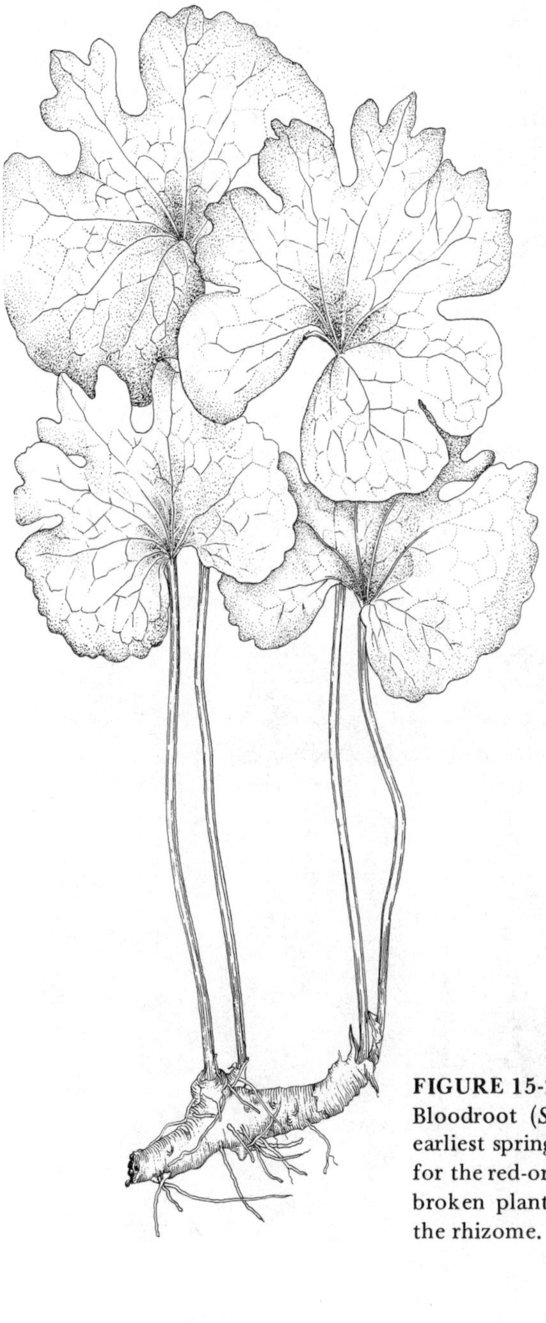

**FIGURE 15-21**
Wild ginger (*Asarum*) spreads by rhizomes and each shoot bears a single flower near the ground.

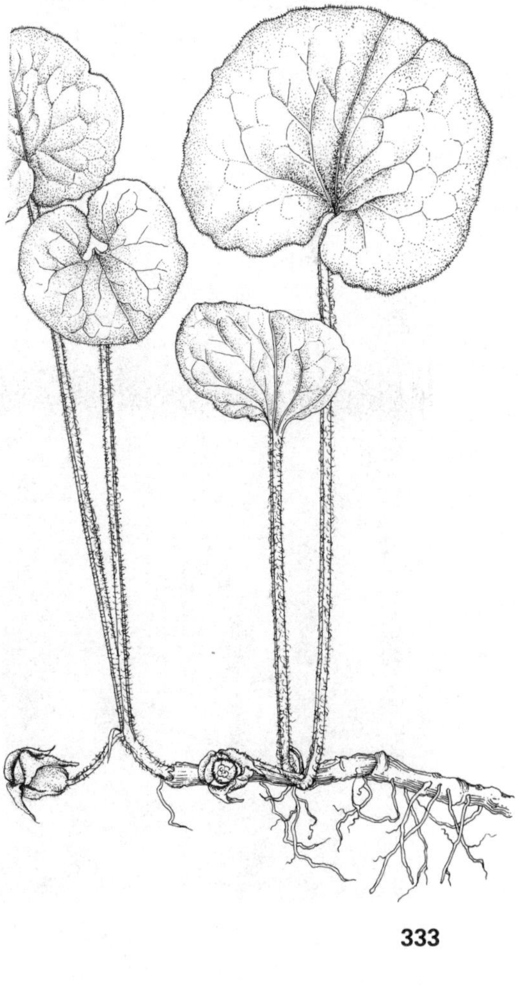

**FIGURE 15-20**
Bloodroot (*Sanguinaria*) is one of the earliest spring wild flowers. It is named for the red-orange fluid that oozes from broken plant parts, as shown here on the rhizome.

## Southern Evergreen Forests

The eastern deciduous forest is replaced by evergreen (nondeciduous) forests along the Atlantic coastal plains from Virginia through Florida, into eastern Texas, and along the Gulf Coast. These forests differ greatly in character from evergreen forests of the western United States because the climate is subtropical and much of the ground is swampy. Here moisture and temperature do not show the marked seasonal differences that characterize the eastern deciduous biome; consequently the vegetation is a combination of stands of several southern pine species (*Pinus*) and areas forested with broad-leaved evergreen trees and shrubs. Many species of the eastern deciduous forests continue as components of the southern forests—for example, the common persimmon (*Diospyros virginiana*) and American holly (*Ilex opaca*).

Some pine forests have a characteristic understory of saw palmetto (*Serenoa repens*). Swamp forests include live oak (*Quercus virginiana*) and other evergreen oaks, magnolias (*Magnolia*), and bald cypress (*Taxodium distichum*). There are many "bays," a common name for several unrelated species, including sweetbay (*Magnolia virginiana*), redbay (*Persea borbonia*), and loblolly bay (*Gordonia lasianthus*). American holly, yaupon (*Ilex vomitoria*) and other hollies are present. Spanish moss (actually a bromeliad, *Tillandsia usneoides*) is an epiphyte that grows in great skeins, especially on the live oaks.

In addition to forested areas, there are many bayous, swamps, and marshes, some primarily of bulrush (*Scirpus*) (Fig. 15-22).

The southern pine forests are considered subclimax communities, developmental stages in succession (Chapter 2), maintained at this stage by human management for lumber and wood extractives (e.g., turpentine and rosin).

Only a small percentage of Florida's land supports the tropical climax "hummock" forests of such broad-leaved evergreen species as strangler fig (*Ficus aurea*), gumbo limbo (*Bursera simaruba*), wild tamarind (*Lysiloma*), and scores of shrubs. Many of the trees and shrubs have toxic sap, exudates, and fruits—for example, the manchineel (*Hippomane mancinella*) and the Florida poisontree (*Metopium toxiferum*). (See also Chapter 18.) Cabbage palm (*Sabal palmetto*) is a striking coastal species. Epiphytic bromeliads, ferns, orchids, trailing vines, and lianas abound.

This unique community is preserved in the Everglades National Park in spite of the diminish-

FIGURE 15-22 A bulrush (*Scirpus*) marsh in Florida. (Photo courtesy of Walter T. Smith.)

ing effects of real estate development, swamp drainage and diversion projects, and in the 1970s the partial completion of a controversial jet airport. The lush vegetation is arranged in numerous strata, creating hundreds of niches for insects, amphibians, reptiles, birds, small mammals, and other animals. Many species are rare or unusual or both. The rare Everglades kite is a hawklike bird that eats snails; it is currently endangered because swamp drainage reduces populations of its preferred food. The small key deer, the rare or extinct ivory-billed woodpecker, the American alligator, the rare American crocodile, and the manatee (an indigenous aquatic mammal) are other unusual life forms for whom the southeastern biome is home.

The tropical wilderness also dominates the Florida Keys, a chain of islands trailing southward from Florida's tip. The keys are inhabited by such magnificent wading birds as American and snowy egrets, great white and great blue herons, wood storks, ibises, and roseate spoonbills. These birds were slaughtered by the thousands to obtain showy plumes for ladies' hats during the 1890s era and several species were threatened with extinction. Protective laws were enacted only after the murder of an Audubon refuge warden by poachers. At the present time, wild habitat for these birds continues to diminish because of human land uses, particularly construction.

## Grasslands: The Prairie

Most of the great American prairie has given way to cornfields, pasture lands, and "amber waves of grain," converted by the plow from sod to agriculture (giving rise to the term "sodbuster"). Yet some ranges are still owned or supervised by the federal government and smaller but excellent tracts are owned or managed by individual states and such conservation groups as The Nature Conservancy. In addition, among the best remaining places to look for native prairie plants are roadsides and railroad rights-of-way (wherever weedkiller allows), long-abandoned fields, and old cemeteries.

Native prairie is the result of abundant spring moisture from melting snow, followed by dry hot summers punctuated by thundershowers. Water loss by evaporation is fairly rapid. A prairie ecosystem produces a deep, well-mulched, fertile topsoil.

The rolling topography of North America's prairies resulted mainly from gradual erosion of the mountain ranges on either side and from the presence of prehistoric inland lakes. Without trees or mountains to stop air movement, strong winds increase evaporation from plant surfaces so that the prairie climate is quite dry during the growing season but still too moist to be a desert. Fire is important in maintaining the prairie, for it eliminates woody species that tend to encroach on the prairie where it meets woodlands.

Adaptations by prairie plants to dry conditions include deeply penetrating root systems and reproduction by growth of underground side branches (rhizomes). The continuous dense layer of grass rhizomes, roots, and the soil is called **sod** (as with popular lawn grasses).

Grass leaves are adapted in various ways to inhibit water loss and many nongrass species have hairy leaves—apparently reducing airflow across leaf surfaces—and swollen subterranean parts for food storage. Most prairie plants are perennial.

Prairies are typically spoken of as tall-grass, mixed-grass, or short-grass prairies, depending on the predominant grass species. Precipitation is the major determining factor in the distribution of these three prairie types.

The **tall-grass prairie** was dominated by sod-forming big bluestem grass (*Andropogon gerardi*), standing head height before flowering and up to about 3.6 m (12 ft) high, including the flowering stems. In balance with shoot growth, roots may penetrate to a depth exceeding 2 m (about 7 ft). With a requirement for moister sites, big bluestem was replaced by certain bunchgrass types as a dominant on drier sites. Tall-grass prairies generally bordered areas of the temperate deciduous forest; and when civilization controlled periodic natural prairie fires, tall-grass prairies were eventually succeeded by oaks as well as by agriculture.

**Mixed-grass prairie** areas are found west of the tall-grass prairie, including the northern Great Plains and Canada's giant "breadbasket." Moisture is at a greater premium in this region and varies greatly from year to year. Grasses of the mixed prairie grow mainly in bunches rather than as sod. Medium-height needlegrass (*Aristida*) and grama (*Bouteloua*) dominate the plant community, with wheatgrass (*Agropyron smithii*) replacing grama in places.

Picture prairie dogs, bison, and Indian tipis and you will probably imagine the **short-grass prairie** or **plains.** Too dry for tall and medium-height grasses, this type of prairie extends south and west of the mixed-grass prairie. Low precipitation, low humidity, drying winds, and hot summer days limit grasses to short-grass species in the plains, which gradually grade into desert.

The soil itself also provides great limitations,

FIGURE 15-23
Ruined prairie, and the human consequences. (Photo courtesy of James R. Estes and Ronald J. Tyrl.)

for there is a permanently dry layer of subsoil that is impervious to plant roots (a limitation similar to that encountered by arctic plants growing above solid permafrost).

Short sod-forming grasses, such as blue grama (*Bouteloua gracilis*), and buffalo grass (*Buchloë dactyloidoe*) are predominant on the short-grass plains, although there are several other important species.

The hot, dry climate of southeastern Texas and adjacent areas supports its own type of grassland, similar to the short-grass plains but consisting chiefly of bunch-forming grasses and occasional woody species, particularly mesquite (*Prosopis juliflora*) or oak. Grassland communities also exist wherever humans have created suitable conditions, as by clearing away woods. Examples are hayfields, pastures, and soil-bank lands.

As with the other kinds of prairie, human activities have monumentally changed the natural face of the plains. The human and environmental disaster of the "Dust Bowl" speaks eloquently of what can happen when the relationship of natural plant communities to the physical environment is unperceived or ignored (Fig. 15-23). Overgrazing by domestic livestock has accomplished similar destruction in other areas of the short-grass plains.

It is not fair to suggest that only grasses grow on a prairie. Nongrass species may actually exceed grasses in numbers, although not in biomass (Figs. 15-24 and 15-25). Such legumes as vetch (*Vicia*), lupine, wild pea (*Lathyrus*), poison vetches and locoweeds (*Astragalus, Oxytropis*), and such composites as asters, daisies, sunflowers (*Helianthus*), and balsamroots (*Balsamorhiza*) are common, as are representatives of many other plant families. Prickly pear cacti (*Opuntia*) occur in many places, especially where the land is overgrazed. Prairie wildflowers are a subject to themselves (Figs. 15-26 and 15-27).

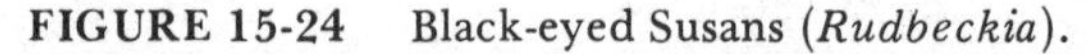

FIGURE 15-24 Black-eyed Susans (*Rudbeckia*).

**FIGURE 15-25** A prairie goldenrod (*Solidago*) and grasses.

**FIGURE 15-26** Bird's-foot violet (*Viola*).

**FIGURE 15-27** Prairie shooting star (*Dodecatheon*).

Unfortunately, humans have tended to change the heterogenous character of the prairie into a monolithic one, with square miles devoted to a single crop or unrelieved grazing. In the past these abuses continued until the land "went bad." Today crops are sustained indefinitely with chemical fertilizers to replace lost macronutrients. The eventual long-range results of this agricultural activity and the erosive consequences of grazing are of constant concern to researchers in ecology, agriculture, and range management. Many scientists fear a second, larger "Dust Bowl" phenomenon because the annual loss of topsoil to wind and water erosion on agricultural land now exceeds by manyfold the loss rate that led up to the 1930s disaster.

Where there is water—whether stream, river, pothole, or manmade stock pond—there is appropriate vegetation, including aquatic and semiaquatic plants, moisture-loving trees, shrubs, and herbaceous species. Willows, cottonwoods (*Populus deltoides*), box elder (*Acer negundo*), and naturalized Russian olive (*Eleagnus angustifolia*) are the dominant trees in these "oases." Where trees follow a watercourse on an otherwise treeless landscape, they constitute a **gallery forest.** Such wet spots in the prairie support many nesting birds, including warblers, magpies, blackbirds, waterfowl, and shorebirds.

Most animals of the prairies are herbivores, including white-tailed deer, bison, pronghorns, domestic horses, cattle, and sheep. Small ground-dwelling herbivorous rodents are legion, the most obvious example being prairie dogs. Via such agricultural crops as wheat (*Triticum sativum*), the human species is a prominent prairie herbivore. Agriculture, however, faces stiff competition from the most numerous herbivores of all—the insects. A monument in Salt Lake City commemorates an occasion when a huge flock of gulls saved Mormon settlers' crops from total devastation by a swarm of voracious locusts.

In the milieu of natural prairie vegetation insects feed many distinctive bird species, such as meadowlarks, lark buntings, and bobolinks. Prairie chickens and sharp-tailed grouse are herbivorous prairie birds. Carnivorous hawks and owls hunt the prairie day and night.

Like the prairie herbivores, most of the predators are burrowers—even such large carnivores as coyotes, foxes, and wolves (formerly), plus badgers, snakes, and a small burrowing owl. Yet the primary predator of large prairie mammals is now the human species.

Humans are not only predators but also affect prairie food chains in other ways. Two of the most serious human impacts on prairie animals are cultivation and introduction of domestic livestock, which compete for forage with native herbivores. A corollary is fencing, which restricts natural animal movements.

## Deserts

West and southwest of the grasslands are the North American deserts, where aridity during the growing season is the critical limiting factor for plants and hence animals.

East of the Cascade, Sierra, and Rocky mountains, the aridity is caused by a **rain shadow effect.** Moisture-laden atmospheric currents moving inland from the Pacific Ocean rise to pass over the mountains. As the air rises, it cools, water vapor condenses, and abundant precipitation falls on the west slopes of the mountain ranges. Thus air passing over to the warmth of the eastern side bears little moisture, being in the rain shadow.

A persistent area of high barometric pressure lingering off coastal California and Mexico diverts moisture-laden air moving from the northwest, contributing to the aridity of the Southwest. This effect is potentiated by the rain shadow of the Sierra Nevada Mountains.

How dry? Deserts are so dry that evaporation exceeds precipitation, which is less than 10 in. per year. Moisture for plant life is available in three main ways: in the spring, as snowmelt from nearby mountains is carried onto the desert in temporarily wet watercourses; from occasional rainstorms that deposit water faster than the baked earth can absorb it, filling dry streambeds (Fig. 15-28) and at times causing flash floods; and from nightly dew. It may seem curious that dew could form in the desert. Desert nights, however, are cool, even cold, allowing condensation of the relatively low amount of water vapor present. This dew is a major water

**FIGURE 15-28** A temporary watercourse in the desert. (Photo courtesy of John S. Thiede.)

**FIGURE 15-29**
White Sands National Monument, New Mexico.

source for nocturnal desert animals (and can cause a soggy sleeping bag!).

Hot days and cool nights are as characteristic of deserts as aridity. Without the filtering effect of atmospheric humidity, the air allows the full onslaught of the sun to heat the earth. Furthermore, the lack of moisture and its heat-holding capacity allows extreme cooling at night as the heat is surrendered back from the earth.

As in the grasslands, winds are a powerful factor in the deserts. The winds increase transpiration, shape dunes, and cause sandstorms that are sufficient to pit and scour automobile windshields and finishes. Summer sand and dust storms are common in such desert cities as Phoenix, Arizona, and Las Vegas, Nevada. The abrasive winds create hard "desert pavement" in places.

Do all deserts have dunes or is this just a romantic Sahara stereotype? Most North American deserts have at least some dune areas. In New Mexico, for example, White Sands National Monument consists of many acres of gypsum dunes (Fig. 15-29). The basic desert surface of the American West, however, is solid soil that originated by erosion. The soil is fertile, as shown when water for irrigation is available for agriculture.

Adaptations of desert plants center on water-conservation measures, such as reducing transpiration, increasing water storage, and restricting growth to the moist, early growing season.

To reduce transpiration, leaves may be greatly reduced in size, altogether absent, or present only during the early part of the growing season (Fig. 15-30). To compensate, stems contain chloren-

**FIGURE 15-30** A plant of the Arizona desert, having photosynthetic stems with some storage tissue. (Photo courtesy of John S. Thiede.)

chyma tissue and perform photosynthesis. Stomates are few and may be shielded from drying air currents by a forest of epidermal hairs. Stomates may remain closed during the heat of the day. Some plants have fibrous or fleshy leaves. Leaves of many species have multicellular epidermal layers or a thick waxy cuticle. Water is stored in the abundant parenchyma tissue of fleshy stems (Fig. 15-31) or taproots. Some xerophytic plants utilize more efficient photosynthetic pathways ($C_4$ and CAM, Chapter 4).

Spines or thorns occur on many desert plants. They are modified leaves or twigs, respectively, that protect the plants from being eaten into by most larger animals, which would expose inner tissues to desiccation (Figs. 15-32 and 15-33). Cactus spines may be dense enough to shield stem tissues partially from the sun and to disperse heat.

Most desert annuals complete their life cycles rapidly during spring moisture. Their seeds have the protection of a thick seed coat in which to endure the dry season, as well as internal growth-inhibiting factors (Chapter 9).

**FIGURE 15-31** Cholla cactus (*Opuntia*) in Arizona. (Photo courtesy of John S. Thiede.)

**FIGURE 15-32** This barrel cactus has abundant water storage tissue, a ribbed stem to allow for expansion as water is absorbed, and spines that protect the soft tissues from feeding animals, dissipate heat, and partially shield the plant body from sun and wind. (Photo courtesy of John S. Thiede.)

Many desert shrubs and trees have tough, deep taproot systems to gather moisture. Roots of the paloverde tree (*Cercidium*), for instance, may extend as far as 24 m (about 80 ft) into the ground, drawing water up from underground sources that originate in the mountains miles away! Early-growing annuals and grasses develop extensive fibrous root systems to trap the moisture that is temporarily available in the upper soil layers in the spring. Cacti also tend to shallow, fibrous root systems, enabling them to capture surface moisture when available.

**FIGURE 15-33** Large *Opuntia* with fruit. (Photo courtesy of John S. Thiede.)

North American deserts are "cool" or "hot," depending on latitude. The cool deserts are the Great Basin area, which is bordered on the east by plains and on the west by the Rocky Mountains, and the Intermountain Basin area between the Rockies and Cascade/Sierras. The interface of the Great Plains and the Rocky Mountains includes "badlands" areas.

The cool desert is blanketed by snow during winter and the dominant vegetation is sagebrush. Various distinct plant communities occur on local sites. Some cool desert plant species are large sagebrush (*Artemisia tridentata*), greasewood (*Sarcobatus vermiculatus*), saltbush or shadscale (*Atriplex*), and hopsage (*Atriplex spinosa*), plus smaller species like green and gray rabbitbrush (*Chrysothamnus viscidiflorus* and *C. nauseosus*), balsamroot, lupine, locoweed, and prickly pear cactus. Desert "tumbleweeds," part of the popular image of the Old West, are plants that die and break off after seed formation and then are rolled by the wind, scattering seeds as they go. The largest western tumbleweed is the nonnative Russian thistle (*Salsola kali*). Distinctive plants are also associated with the numerous volcanic features, such as basalt flows and cinder cones (Fig. 15-34).

The natural sagebrush desert has been utilized extensively as grazing range for beef and sheep. Where irrigation is available, the land may be converted to agriculture for mint (*Mentha*) and such forage crops as alfalfa (*Medicago*). All these uses have a disruptive effect on the native ecosystem. In addition to removal of vegetation by grazing (Fig. 15-35), a controversial practice called "chaining" is used to destroy the sagebrush in an attempt to encourage growth of grasses and herbs (Figs. 15-36 and 15-37).

To the automobile traveler crossing miles of sagebrush desert, the scenery may seem monotonous and devoid of animal life except grazing livestock, an occasional band of antelope, wheeling hawks, golden eagles, and perhaps a flock of turkey-sized sage hens. The same traveler driving at night,

**FIGURE 15-34** Tiny yellow monkeyflowers (*Mimulus*) growing on cinders at Craters of the Moon National Monument, Idaho.

FIGURE 15-35
Grazed sagebrush (*Artemisia*) range on the left, separated by a fence from ungrazed range. Removal of grasses, as by overgrazing, encourages range takeover by sagebrush. (Photo courtesy of Fred L. Rose.)

FIGURE 15-36
The conversion of rangeland to cropland in Idaho is called "Desert Land Entry," and begins with the violent destruction of sagebrush by a process called chaining. This photo shows a sagebrush stand before chaining. (Photo courtesy of Fred L. Rose.)

FIGURE 15-37 The same sagebrush area after chaining. (Photo courtesy of Fred L. Rose.)

however, might see (fences permitting) mule deer, coyotes, bobcats, a rare kit fox, spotted skunks, pocket mice, and kangaroo rats, plus an occasional great horned owl swooping down in search of rodents.

Following rainstorms, water is temporarily available to animals in depressions and ditches; more permanent bodies of water also exist, however. Some are of human creation; some are remnants of larger inland waters that once covered the Great Basin. It is common for such remnant lakes to be saline or alkaline—like those in the old western movies where parched bones surround innocent-looking but deadly desert potholes. Experienced desert hikers know that they must avoid water from potholes that do not support aquatic animals because of a high arsenic content. The Great Salt Lake is an example of a remnant saline lake. Harney Lake in Oregon is an example of an alkaline lake.

The moist freshwater areas in the desert support relatively lush plant life, at least in and for a few inches above water level. Consequently, many nesting birds are attracted, especially if there are trees. A great variety of songbirds may be present, including the nonmusical but raucous magpie. Shorebirds and waterfowl of many species nest along desert lakes and reservoirs, some of which are managed through the National Wildlife Refuge System.

One such refuge, a combination of natural waters and constructed dikes to enlarge and maintain waterfowl habitat, is the Malheur National Wildlife Refuge. Located near Burns, Oregon, Malheur is the oldest of the U.S. refuges. It has been the site of continuing scientific research, a natural laboratory for educational facilities of the state, and a source of pleasure to visiting professional and amateur ornithologists from all over the country. It provides nesting opportunities for a large number of bird species. The trumpeter swan, an endangered species, has been successfully introduced to the refuge.

We might not expect to find amphibians and waterfowl in the desert. The only true cool desert amphibian is the spadefoot toad, which finds shelter by digging under the soil surface. Spadefoots breed in temporary water, such as that which accumulates in cattle tracks and ditches after a rainstorm. Larval development occurs rapidly, the tadpoles feeding on microscopic plant populations that "bloom" in the puddles, and metamorphosed toads emerge in a much shorter time than is typical for most species.

The warmer, lower altitude and latitude desert region of the American Southwest is referred to as a hot desert, extends into Mexico, and can be subdivided into at least three distinct types—the Mojave (as in California), Sonoran (as in Arizona), and Chihuahuan (as in Mexico). These distinctions are based on differences in climate, substrate, and vegetation.

Some plants typical of the hot desert are creosote bush (*Covillea* or *Larrea tridentata*), bur sage (*Franseria*), blackbrush (*Coleogyne ramosissima*), burrobrush (*Hymenoclea*), saguaro cactus (*Cereus giganteus*), paloverde (*Cercidium*), mesquite (*Prosopis*), and acacias (*Acacia*) (Figs. 15-38, 15-39, and 15-40).

The Mojave has fewer plant species than the other two deserts but is home to the unique Joshua tree (*Yucca brevifolia*), found nowhere else in the world. In addition to large plants, many smaller cacti (Fig. 15-42), shrubs, perennials, and annuals enrich the desert.

The Sonoran desert, in particular, is noted for its periodic "desert blooms." In an occasional year winter and spring precipitation is sufficient and timed appropriately to stimulate germination of *millions* of seeds that have been lying dormant in the soil. The abundant moisture washes out growth inhibitors present in the seeds. If moisture conditions remain favorable, germination is followed by growth and flowering, carpeting the desert with the beauty of a "desert bloom."

**FIGURE 15-38** Saguaro cactus habitat in the Arizona desert. (Photo courtesy of John S. Thiede.)

**FIGURE 15-39** An organpipe cactus. (Photo courtesy of John S. Thiede.)

Most animals of the desert have adapted to its hostile daytime environment by being active at dusk, night, or dawn. Even animals that are active during the daytime avoid the hottest portion of the day, resting in burrows, crevices, and whatever shade they can find.

The many species of reptiles, mainly lizards, characteristic of deserts spend early morning warming themselves on rocks—feeding and carrying out their activities—and then retreat to shade or rocky crevices during the afternoon. Some snakes, such as the sidewinder rattlesnake, are active at night.

Birds of the Sonoran desert include the roadrunner and a variety of woodpeckers and owls that live in burrows in large cacti (Fig. 15-42). The sagebrush desert has many seedeaters, such as the sage sparrow. Birds are especially abundant near water.

**FIGURE 15-40** A desert *Acacia* tree. (Photo courtesy of John S. Thiede.)

**FIGURE 15-41** The Joshua tree of the Mojave desert. (Photo courtesy of John S. Thiede.)

**FIGURE 15-42** Saguaro cactus with nest holes. (Photo courtesy of John S. Thiede.)

Mammals show various types of adaptation to desert life, such as being active during cooler times of day. Many species—like the kangaroo rat—burrow. The kangaroo rat and its close relatives share the ability to exist on "metabolic water" derived from the digestion of food. This ability is due to highly efficient reabsorption of water in their kidneys. Some desert animals have enlarged extremities, such as the ears of the jackrabbit and kit fox, to facilitate heat loss.

Invertebrates of the desert include many insects, some of which have life cycles requiring years to complete and which can be completed only if there is sufficient moisture. More familiar invertebrate examples are spiders (e.g., tarantulas) and scorpions.

## Montane Coniferous Forests: Rocky, Cascade, and Sierra Mountains

Two great north-south-running mountain range complexes have been thrust on the western part of the continent by a combination of movements of the earth's crust and volcanic activity. These systems of mountain ranges are the Rocky Mountain complex and the Cascade/Sierra complex. Both of these mighty ranges support the true montane environment, much of which is managed for lumber (Fig. 15-43), mining, and recreation by the U.S. Forest Service, the U.S. Park Service, and the U.S. Bureau of Land Management. Some areas have been selected to remain in a natural condition through congressional designation as wilderness areas.

Various vegetational bands occur, depending on altitude, influenced also by east- or west-facing exposure. At highest elevations timberline plants are subjected to cold, windy conditions only slightly less arduous than those of the adjacent alpine areas. On exposed sites trees and shrubs of this subalpine zone form low, matted islands surrounded by open tundralike areas. This growth form is created by the killing of exposed buds by harsh winter wind and ice, causing lateral growth of the trees. These same tree species assume normal growth shapes below the subalpine zone.

Subalpine-zone species composition differs between the Rockies and the Cascade/Sierras. Predominant Rocky Mountain species are Engelmann spruce (*Picea engelmannii*) and subalphine fir (*Abies lasiocarpa*), plus white-barked and bristlecone pines (*Pinus albicaulis* and *P. aristata*). Remnant stands of Western red cedar (*Thuja plicata*) may be found.

Douglas fir (*Pseudotsuga menziesii*), which is not a true fir, ponderosa pine (*P. ponderosa*), and lodgepole pine (*P. contorta*) predominate in the next altitudinal band, with the pines marching down the slopes to intergrade with and eventually give way to junipers (*Juniperus*) and the sagebrush desert.

Subalpine forests of the Cascades and Sierras are primarily mountain hemlock (*Tsuga mertensiana*), subalpine fir, grand fir (*A. grandis*), noble fir (*A. procera*), Pacific silver fir (*A. amabilis*), and lodgepole pine. On the east slopes the next montane bands incorporate ponderosa pine with lesser numbers of such other conifers as sugar pine (*P. lambertiana*), lodgepole pine, and white fir (*A. concolor*) or grand fir. In the transitional area between montane forest and sagebrush desert, ponderosas are mixed with incense cedar (*Libocedrus decurrens*) and western juniper (*J. occidentalis*).

Fire is an important factor in maintaining montane vegetation. In the ponderosa pine forest of the Deschutes National Forest of central Oregon, for example, studies of annual xylem rings show that prior to fire-controlling activities forest fires occurred approximately every 35 years. These fires

**FIGURE 15-43**
An aerial view of clear-cutting in the Rocky Mountains.

burned rapidly along the ground, fed by abundant dry needle and bark litter, taking woody shrubs, tree seedlings, and saplings. Mature ponderosa pines were virtually untouched by the rapidly burning ground and understory fires, for they lack low-hanging branches and possess more fire-resistant bark than younger trees. The natural mature ponderosa pine forest is therefore a very open stand.

Since fire lookouts and other measures have been instituted, however, periodic burns have not occurred. As a result, there are many thickets of sapling-size ponderosas in logged-over areas that must be manually thinned and that constitute a greater fire danger.

The ponderosa forest in this area is actually considered to be fire climax because it is maintained by such fires. Ponderosa pine seedlings do best in open rather than shaded sites. It does not regenerate as well under an established overstory and would probably be replaced, at least in some areas, by the more shade-tolerant firs.

West-slope forests of the Cascades and northern Sierras are strongly influenced by moisture from atmospheric currents moving in from the Pacific Ocean. Whereas ponderosa pine dominates the drier east slopes, Douglas fir and western hemlock (*Tsuga heterophylla*) may dominate the moister west slopes. In some areas of the California Sierras scattered groves of giant sequoias (*Sequoia gigantea*) exist as remnants of a former wider distribution.

The trembling aspen (*Populus tremuloides*) is found in all mountain communities, especially in moist areas and as an intermediate successional type. Although a typical montane species, it is by no means restricted to this ecosystem and has the widest distribution of any tree species on the continent. In the montane ecosystem it grows on slopes that have been disturbed, as by logging or avalanche.

Aspen is frequently associated with beavers, who are largely responsible for the development of the beautiful mountain–valley meadows (Fig. 15-44). By damming streams and maintaining their canals, beavers cause a lush, fertile bottomland to form, which traps minerals and other nutrients eroding from the slopes and allows the development of a distinctive, willow-thicket habitat for themselves, moose, fish, birds, and other animals. Where beavers are not present, the stream would tend to form a narrow gorge.

Within a forest ecosystem, overstory height and density create strata of greater differences in light, humidity, soil moisture, temperature, and air currents than in lower-growing communities. This variety of habitat types creates a very diverse community.

In the Cascades and northern Sierras striking differences exist in understory on east and west sides because of the rain shadow effect. The drier east slope, mainly open pine forest, has relatively sparse undergrowth that includes such plants as

**FIGURE 15-44**
A Rocky Mountain meadow in full bloom.

rabbitbrush, buckbrush or mountain mahogany (*Ceonothus*), bitterbrush (*Purshia tridentata*), and manzanita (*Arctostaphylos*). Beargrass (*Xerophyllum tenax*) blooms in the mountain passes (Fig. 15-45). On the moister west slopes ferns, berry-bearing plants, such as blueberry, huckleberry (*Vaccinium*), and blackberry, and broad-leaved trees and shrubs like dogwood (*Cornus nuttalli*), vine, and mountain maple (*Acer circinatum* and *A. spicatum*), rhododendron (*Rhododendron macrophyllum*), and western service berry (*Amelanchier florida*) flourish. (Also, there are more gnats and mosquitoes!) Lichens growing on rocks and trees also reflect the moisture differences. On the east slope they are mainly spiky (fruticose) or encrusting (crustose) whereas large foliose lichens occur on the west slope.

FIGURE 15-45 Beargrass (*Xerophyllum*) in bloom in the Santiam Pass, crossing the Oregon Cascades.

In the southern Sierras the west slope is basically arid due to the latitude and the more-or-less stationary high-pressure air cell that affects the southwestern desert. These west slopes are covered not with lush, leafy greenery but with a shrubby vegetation known as **chapparal.** Periodically southern California chapparal makes headlines, for this is also a fire climax community and it burns rapidly and violently. Fanned by offshore breezes, such fires occasionally destroy homes. Ironically, part of the intensity of these fires is a result of fire-control attempts. Without fires, larger than usual buildups of dead leaves and twigs accumulate. They are tinder to the least spark and burn explosively. Some chapparal species are so adapted to fire that burning is actually necessary to weaken the seed coats in order to permit seed germination.

The western montane biome as a whole includes a great variety and abundance of animal life. Some animals are associated with a particular altitudinal zone or a particular layer of vegetation. Birds, in particular, have specific nesting and feeding requirements that distribute species among the available habitats within an ecosystem. Obviously this factor lessens interspecific competition.

Such big-game animals as elk, moose, bighorn sheep, and black and grizzly bears are associated with montane environments and are most readily accessible (in a natural state) to the average individual or family tourist at the oldest of our national parks, Yellowstone National Park in Wyoming. Because the park is enormous and the animals protected, they are less wary and easier to observe than those outside the park. At various sites in Yellowstone it is possible to see many big-game animals from the road. Ospreys and bald eagles fish the Yellowstone River, which is also home for many migratory birds, including the distinctive harlequin duck. Courting and nesting trumpeter swans on Yellowstone Lake can also be observed from the road. Within the park, beaver workings are common and campsites abound with gregarious Uinta ground squirrels, chipmunks, Steller's jays, magpies, Gray jays (also called Canada jays, camp robbers, and whiskey jacks), crows, and ravens.

In many ways, spring (May through June) is the favorite time of year for mountain lovers, who don't seem to mind chilly nights and uncertain weather. The campsites and trails display as-yet untrodden wildflowers in full bloom as well as wildlife that has not yet ascended to summer feeding grounds (Fig. 15-46). Because birds are nesting

**FIGURE 15-46** Elk in Yellowstone National Park are present at lower elevations well into spring, feeding in montane coniferous forest clearings and near thermal areas.

at this time of year, interesting mating and territorial displays are everywhere.

Other montane animals are mule deer, mountain lions (also called cougars and pumas), bobcats, various weasels, mountain cottontail rabbits, golden-mantled ground squirrels, marmots, yellow-pine chipmunks, porcupines, various hawks and owls, Clark's nutcrackers, and western tanagers. A short-eared cousin to rabbits, the pika, lives in rocky talus slopes and piles freshly cut vegetation in the sun to cure as hay for winter feed. Insects and other arthropods are, as usual, the most abundant life forms.

In the western Cascade and Sierra mountain ranges the arid east slopes support populations of mule deer while moister areas of the west slopes are home to a smaller species (or subspecies), the black-tailed deer.

### The Temperate Montane Forest and Pacific Rain Forest

As celebrated in the beautiful Olympic National Park of Washington (Fig. 15-47), the temperate coniferous forest of the northern part of the Coast Range is at many times as wet a place as you can be without being underwater—humidity 100%! The relatively warm, water-vapor-laden atmosphere moves inland from the Pacific Ocean, ascends the western flanks of the Coast Range, condenses as it goes, and drops over 200 to 250 in. of precipitation (most of it as rain) on the Coast Range. When not raining, it is usually foggy for at least part of each 24-hour cycle. This fog may account for as much as 50 in. per year additional precipitation as leaf drip. Because of the great amount of precipitation, the northern Oregon, Washington, and British Columbia region of the coastal coniferous forest is referred to as a temperate rain forest.

These mild, humid conditions create a biome where virtually every stationary surface supports some type of plant life (Fig. 15-48). Lichens form large foliose patches on tree trunks; mosses grow in mattress-thick carpets on the ground and along branches; herbaceous plants (Fig. 15-49), ferns, and berry bushes abound—huckleberries, blackberries, salmonberries (*Rubus spectabilis*), and thimbleberries (*Rubus parviflorus*). Horsetails (*Equisetum*) up to 1.5 m (5 ft) tall are not unusual and a "flatlander" may feel overwhelmed by the vegetation. In the temperate rain forest it is quite possible to be invisible to a person only a meter from you and getting lost is much too easy, as careless hikers and deer hunters regularly discover.

FIGURE 15-47
The Pacific coastal rain forest of the Olympic Peninsula, Washington. (Photo courtesy of Fred L. Rose.)

FIGURE 15-48
Nearly every stationary surface in the coastal rain forest supports some form of life.

**FIGURE 15-49** Western or yellow skunkcabbage (*Lysichitum*), harbinger of spring.

Predominant overstory species are Douglas fir (Fig. 15-50), Sitka spruce (*Picea sitchensis*), western hemlock, and western red cedar in most of the Coast Range, including the rain forest. In northern California the towering coast redwood (*Sequoia sempervirens*) forest represents a remnant of a former range that extended into Oregon (Fig. 15-51).

Logged-over areas are commonly seen in the Coast Range and develop dense stands of red alder (*Alnus rubra*) as an intermediate stage in secondary succession. At lower elevations, as where the Willamette River Valley in Oregon is bracketed by the Cascade and Coast ranges, bigleaf maple (*Acer macrophyllum*) is an important tree species. Several oak species are prominent in the intermountain valleys and on drier sites of the Coast Range, such as the Oregon and California white oaks and the several nondeciduous live oaks in California.

The Coast Range as a whole is extensively managed by the logging industry. Regeneration of the widely used Douglas fir is possible and occurs rapidly in these favorable environmental conditions. One main problem has been in managing steeply sloping sites, for the abundant rain not only promotes growth but also erosion of unprotected ground, making reforestation difficult. Clear-cutting is popular as a means of developing a new and vigorous forest stand but can result in severe erosion. The physical task of removing trees with a maximum of safety and a minimum of downhill wastage is also a problem.

The favorable moisture and temperature of the temperate rain forest also nurture abundant animal life. Of all biomes in North America, this is "heaven" to frogs and salamanders.

Many bird and mammal species are shared with the montane biome, although the Coast Range also supports distinctive associations. The common

**FIGURE 15-50** A 250-foot-tall Douglas fir tree. (Photo courtesy of Fred L. Rose.)

**FIGURE 15-51**
A coast redwood forest.

deer is the black-tailed deer and other coast range animals are Roosevelt elk, fisher (a large weasel), mink, western red squirrel (chickaree), and tree mouse. One endemic species is the mountain beaver—not really a beaver at all—which burrows in stream banks. It suffers a habitat crisis due to competition from released nutria, South American rodents imported for fur farms.

Where the seashore meets the Coast Range there is a transition zone that may be sandy beach, rocky cliffs, shifting dunes, or estuary. Salal (*Gaultheria shallon*) (Fig. 15-52) and imported Scot's broom (*Cytisus scoparius*) (Fig. 15-53) form dense thickets and certain trees also mark this tenuous edge.

Imagine a twisted, flat-topped pine tree silhouetted against an ocean sunset. Just as they occupy drier sites in the mountains, so pines also

**FIGURE 15-52**
Salal (*Gaultheria*), showing the typical bell-like flowers of members of the heath family.

FIGURE 15-53 Scot's broom was imported to help stabilize West Coast sand dunes. It has prospered to the point that it competes with native plants and is gradually moving inland along roadways.

are able to colonize more hostile sites along the coast. Here the soil is sandy, probably salty from surf spray, and persistent winds dry the surface. As in the alpine biome where wind kills buds on the exposed side, buds are killed in this environment by dehydration, sand blasting, salt, and direct physical damage from the wind, producing a contorted tree. Lodgepole pines are the coastal pines of most of the coast, giving way to Monterey pine (*P. radiata*) more southerly. A close look at a hedge-like mat of plant growth on an exposed seaward slope, especially near a cliff, may reveal the victim of these same forces to be a normally tall Sitka spruce.

## The Marine Biome

Oceans of the world provide an environment that, for several reasons, is less variable and less hostile to living organisms than freshwater or terrestrial environments. The natural salinity is relatively constant, approximating that of cytoplasm, so that water balance is less of a problem; temperature, gas, and nutrient levels are fairly constant; body support can be provided by buoyancy; gamete release, syngamy, and embryonic development can occur in a hospitable external liquid medium.

Marine environments are biologically described in terms of their stratification. Note Fig. 15-54 to understand the categories mentioned in the following discussion. Our examples will be of

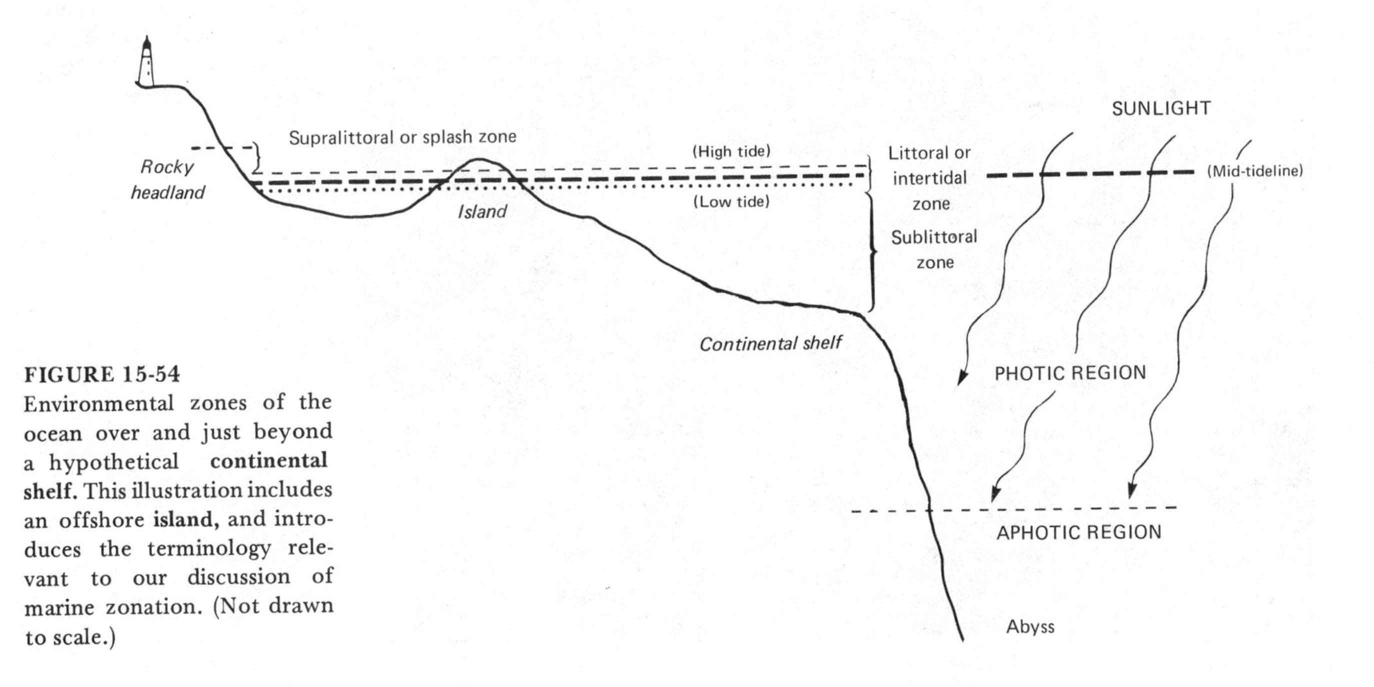


FIGURE 15-54
Environmental zones of the ocean over and just beyond a hypothetical **continental shelf.** This illustration includes an offshore **island,** and introduces the terminology relevant to our discussion of marine zonation. (Not drawn to scale.)

communities above, within, and below the tidal highs and lows, and the open sea.

Several terms describe categories of aquatic organisms by where they live. **Plankton** are small-to-microscopic plants and animals free floating within the surface layers of water. Larger, free-swimming, and chiefly carnivorous animals are classified as **nekton.** The term **pelagic** also refers to this group. Organisms living along the ocean floor make up the **benthos,** the benthic (bottom) community.

#### Shorelines and Littoral Zone

Shorelines are densely populated areas of basically two types—sandy and rocky. Plants and animals living in each type face special problems, especially within the **intertidal** or **littoral** zone that exists between high and low tide lines.

*Sandy Shores.* On the sandy beach organisms are exposed twice daily to desiccation, increased heat and salinity, and constant shifting of sand by waves and wind. Larger plants cannot attach to this unstable substrate and so plant life is mainly microscopic diatoms that live between sand grains. Animals are primarily burrowers or microscopic forms living between sand grains. Even the small free-swimming shrimp and scuds (amphipods) in water that swirls around your feet as you walk along the tideline disappear beneath the surface as the water slides away; then resurface with the next inundation. In this sand-scoured habitat most animals possess an external shell or skeleton. Some remain buried at all times, such as clams.

Higher up the shore, animals burrow, such as sand fleas, or fly, such as the flies and gnats that scavenge among the debris washed ashore. This upper beach line of stranded debris is referred to as the **strand line.**

Inland from the strand line, above the intertidal zone, dunes occur and are constantly sculptured and shifted by the wind. In spite of the sandy substrate, some grasses, herbaceous, and semiwoody or woody plants have adapted to growth conditions in this basically dry environment. Many have fibrous or woody stems for flexible support, extensively spreading or deeply penetrating root systems for maximum anchorage and water absorption, waxy or hairy succulent leaves to retard water loss and damage from salt spray, and reduced leaf surface area. Most plants are low growing. Some are able to excrete excess salt.

Farther from the shore, the dunes are increasingly colonized by plant life (Fig. 15-55). Eventually dunes reach a stage of succession in which pines and then other conifers and a definite understory develop, forming forest vegetation (Fig. 15-56).

*Rocky Shores.* The rocky shoreline is of great interest to amateur beachcombers and professional marine biologists because it supports the greatest density of macroscopic benthic life of any marine habitat. A visit to such a shoreline on either

**FIGURE 15-55**
Dune grasses with extensive fibrous root systems stabilize the sand and are an important stage in dune succession.

FIGURE 15-56
This dune has been colonized for so long that it has become sandstone. Succession has led to a climax community of wind-trained conifers and shrubs. (Cape Perpetua State Park, Oregon)

the Atlantic or Pacific coast is an impressive experience. Not only is nearly every square inch of surface occupied by plants and animals (Fig. 15-57), but the surfaces of larger plants and animals are also colonized by smaller organisms, which are grown on by smaller forms and so on, down to a microscopic level. The density of life is an index of the favorability of the environment and is also partially due to the fact that many intertidal species propagate both sexually and asexually.

Rocky shore organisms face a number of distinct hazards, including problems of remaining in place, enduring wave shock and abrasion (Fig. 15-58), and retaining adequate water. Multicellular algae (seaweeds) have evolved gnarled holdfast structures and have flexible rubbery bodies that

FIGURE 15-57
Rocky shorelines provide a solid substrate for many types of marine algae, the seaweeds, and many associated animal species.

**FIGURE 15-58** Wave action is harsh along a rocky shoreline. Note the **tidepools** on the lower rocks.

give way to the force of the water to avoid being broken or demolished. Furthermore, seaweed bodies contain quantities of gelatinous substances that retain water, protecting inner tissues from desiccation. These algal derivatives are of economic value and are used as thickeners and stabilizers in many food products, including puddings, sauces, dressings, candies, and desserts (Chapter 16).

Animals seek protection in crannies, under rocks, or in the bodies of other organisms or they produce cementing substances or threadlike attachment structures. Motile animals seek shelter during the roughest periods, have soft bodies that are flexible and resilient, or have hard protective shells that not only withstand the pounding surf but are also tapered to distribute wave impact. Some animals have water storage chambers or an ability to close their shells tightly to avoid water loss. Some organisms live only in tidepools, where water remains trapped after the tide goes out (Fig. 15-59).

**FIGURE 15-59**
Sea anemones, sea urchins, encrusting coralline algae, and many other plant and animal species inhabit tidepools that remain after the tide goes out. In this particularly low tide, some organisms that would usually be underwater are exposed until the tide comes in. (Photo courtesy of Fred L. Rose.)

**FIGURE 15-60** The changing face of the shoreline.

Periodic exposure to the atmosphere also places intertidal organisms under stress conditions due to variations in temperature and salinity caused by evaporation. Various physiological mechanisms permit species to tolerate these hazards.

Sexual reproduction via motile gametes (at least the sperm) is necessary for attached organisms. Although swirling intertidal currents help distribute these gametes, they may also wash them away or dilute their concentration, decreasing opportunities for syngamy. To counter this problem, various species, especially animals, spawn synchronously, whereby release of gametes by a single individual chemically stimulates gamete shedding by others, thereby resulting in a sort of intertidal orgy. In addition, some intertidal species release gametes only within the warm, placid confines of a tidepool when the tide is out.

Once reproduction has succeeded, newly formed individuals face a problem of where to settle in this already crowded environment. Many plants, particularly smaller algae, are epiphytic. To aid dispersal, most intertidal animals have motile larvae, generally quite different in form from the adult. They are carried by currents as plankton until they reach an embryonic stage at which attachment to a substrate is possible. Larval forms are a significant component of the plankton and hence of marine food chains.

Life along rocky shorelines exists in clear vertical stratification, from organisms that normally remain inundated (certain large algae, sea urchins, small fish, etc.) to tiny species that populate the steep-faced **splash** and **spray zone (supralittoral zone)** that is never inundated (certain lichens, unicellular algae, bacteria, barnacles, and periwinkles).

Located above the intertidal level on cliffs and islands are nesting sites for such seabirds as gulls, murres, and pigeon guillemots. Dense colonial nesting is an adaptation for species living where space is at a premium but food is abundant. The resulting guano washes into the ocean, returning nutrients for plant growth. Steep embankments overlooking the ocean are also favored homesites for members of the human species (Fig. 15-60).

### Sublittoral Zone

Beyond the intertidal or littoral zone is the **sublittoral zone.** This zone exists approximately between the low tide line and the edge of the continental shelf (refer back to Fig. 15-54). The entire zone is within the range of light penetration. Various marine algae, particularly browns, flourish here with their holdfasts anchored to the bottom and their tops buoyed to the surface light by air-filled bulbs or bladders.

Beds of kelp (large brown algae of several genera) are so dense in places, such as off the California coast, that they are called kelp forests or jungles (Fig. 15-61) and are harvested commercially for a variety of uses. The sea otter, an endangered species, lives above the California kelp beds, diving for abalone and sea urchins. Sea otters sleep while floating, entwined in kelp fronds to prevent being washed ashore.

Sublittoral communities also include such larger animals as octopi, moray eels, and economically important predatory fish—herring, tuna, sharks, barracuda, ling cod, red snapper, and bottom-feeding reef fish. The colorful and varied benthic community includes many examples, such as abalone, lobsters, tubeworms, and sea fans.

Particularly in subtropical and tropical waters, the sublittoral zone may support extensive coral colonies, which form coral reefs. The colonial stony coral animals that are mainly responsible for reef formation form a continuous living mantle of flesh over the surface of the calcareous secreted material in which they live. The soft-bodied animals are frequently brightly colored whereas the limy skeletal material that is built up is usually an ordinary calcareous white. Brightly colored coral-skeleton fragments commonly purchased in souvenir shops usually owe their hues to spray paint. Many coral species exhibit symbiosis with unicellular algae. This seems to be a significant relationship, for coral colonies generally grow only in the zone of light penetration.

### The Open Sea

Beyond the sublittoral zone is the vast open sea. As in the other zones, environments are vertically stratified, with shifts in community composition from level to level. At and near the surface are planktonic forms. The plants (**phytoplankton**) are mainly diatoms of great diversity and abundance. Many animal phyla are represented in the **zoöplankton** and feed mainly on phytoplankton. Most zoöplankton species are small adult forms or larval forms of larger species.

Within the **photic zone** (where light penetration permits photosynthesis), large beds of floating algae like *Sargassum* occur. The photic zone of the open sea, however, is mainly a pelagic community of swimming animals that feed on the plankton and each other. This community of carnivores primarily includes fish, making it the oceanic zone currently most exploited for human food.

**FIGURE 15-61**
A southern California kelp jungle. Like terrestrial forests, this kelp ecosystem demonstrates vertical stratification in regard to physical factors (e.g., light, temperature, nutrients) and the richly diversified biological community. (Photo courtesy of James A. Coyer.)

Other carnivores, such as jellyfish, squid, porpoises, and whales, are also present.

The benthic community within the photic zone is especially rich where the terrain slopes upward along such emergent land masses as islands and the continents. A fairly wide continental shelf is characteristic of both shores of North America and such shelves lie largely if not entirely within the photic zone. Nutrients eroded from the land and carried to the sea by river systems also help to make the continental shelves highly productive.

Below the region of photosynthesis is the dark and mysterious **abyss.** Pelagic species here depend on food that comes to them from the photic zone as prey or, more frequently, as detritus drifting down as a result of predation and death above. This constant "rain" of detritus is food for scavenging species of the dark (aphotic) region. In addition, some spectacularly grotesque (to us) fish species prey on one another in this sparsely populated region. Bioluminescence (the ability to emit light) is characteristic of many of these species, for it is necessary for species recognition during reproduction and is used in luring or pursuing prey. Some species can leave a cloud of glowing discharge to distract a pursuing predator.

The benthos beneath the open seas consists of some organisms that depend totally on detritus from above. Plant life is apparently restricted to saprophytic bacteria, which function in this ecosystem as agents of decay and chemical transformation of nutrients. The nutrients thus transformed are again cycled into the ecosystems of the upper layers by periodic upwelling of deeper layers of water to the surface.

Such benthic animals as sea cucumbers, sea lilies, and various worms collect detritus as it falls or ingest the bottom ooze, digesting what they can (including bacterial cells). Some, such as brittle stars, are carnivorous.

## SPECIAL TYPES OF ECOSYSTEMS

### Estuaries

"Then and now" maps of the great San Francisco Bay and historic Boston Harbor demonstrate how these bodies of water have shrunk. It is a natural phenomenon for silt carried by rivers to build up a delta at river mouths and to fill in bays. In San Francisco, Boston, and many other natural harbors, however, the landfill is chiefly of human origin. Resistant materials, such as trash and demolition byproducts (often concrete), have been used to increase construction sites and wharf areas. Aesthetics aside, why is this of ecological significance?

The ecosystem that forms where the saline waters and sand of the sea meet the fresh water and silt of a river in a semienclosed coastal area is called an **estuary.** Some estuaries are small whereas others are extensive, such as the Chesapeake Bay region.

For centuries American Indian tribes reaped sustenance from estuaries, which continue to be important food resources. In the special semiprotected, nutrient-rich conditions of an estuary live many edible shellfish—oysters, mussels, cockles, and clams—that attach at the surface or burrow into the muddy tidal flats. Crabs, shrimp, and a variety of fish live within estuaries; hundreds of nonhuman-food species are also present.

In the past few decades, however, oceanographic studies have revealed that the importance of estuaries is much greater than simply as sources of immediate harvest. Estuaries are the "nurseries" of the seas. Within their shelter, reproductive and larval stages of many life cycles are carried out, including those of numerous economically important species. Therefore activities that alter the size and/or composition of an estuary, such as land filling, dredging, and pollution, eventually alter its productivity.

Actually, there is no such thing as a "typical" estuary, for climate, individual shore, and contributing river conditions vary so much; however, estuaries are usually classified by the degree to which oceanic and freshwater mix or remain stratified. The stratification of freshwater and saltwater may vary by seasons within a single estuary. In addition, each estuary is constantly filling up with silt from the river. This natural process is offset to various degrees by the flushing action of seasonal storms and floods and, primarily, by dredging.

Two daily tidal flows influence water level, temperature, salinity, and turbidity within estuaries. Tides, however, also affect water levels in the rivers that feed estuaries. Depending on the slope of the riverbed, the river level may be elevated several to many miles up its length by tidal flow and lowered with the ebb. A gradient from predominantly fresh to brackish (mixed fresh and saline) water exists within the lower reaches of such a river.

The biological community of an estuary is varied and stratification into plankton, nekton, and benthic communities is apparent. In addition, there are small semiterrestrial environments of emergent land. Aquatic plant life consists of abundant phytoplankton, particularly diatoms, which are consumed along with zoöplankton by the many filter feeders, such as oysters, clams, and tube-dwelling worms. In addition, larger seaweeds and eelgrass (*Zostera marina*), a vascular plant, grow attached to buoys, pilings, rocks, the mucky bottom, discarded automobile tires, driftwood, and whatever else might be at least partially submerged. Epiphytic diatoms and algae grow on the seaweeds and eelgrass. In the landward regions where open bay gives way to tidal marsh, algae and eelgrass are gradually replaced by saline-adapted grasses, rushes, reeds, sedges, and cattails. The tidal marsh has an overall "tulelike" appearance, with clumps or tussocks of vegetation and meandering channels. (See also Table 15-1.) In tropical regions, including the southwest coast of Florida, mangrove swamps replace tidal marshes.

Birds of many species feast upon the abundance of the estuary. Some, such as cormorants, murres, and pelicans, dive for fish. Egrets and herons spear fish in the shallows. Shorebirds, such as sandpipers and their many relatives, probe exposed moist areas for burrow-dwelling invertebrates. Many ducks fish or forage in the open water or among the vegetation. The romanticized seagull lives as a scavenger, feeding on crushed clams discarded by human diggers, excess bait tossed out by incoming fishing boats, filleted carcasses, garbage dumps and scows, or the remains of casualties to natural causes. Many songbird species occupy the shoreline and marsh vegetation.

**TABLE 15-1 A wetlands classification.**

| **Freshwater Areas** | |
|---|---|
| **Inland fresh areas** | |
| 1. Seasonally flooded basins or flats | Soil is covered with water or is waterlogged during variable periods but is well drained during much of the growing season. In upland depressions and bottomlands. Bottomland hardwoods to herbaceous growth. |
| 2. Fresh meadows | No standing water during growing season but is waterlogged to within a few inches of surface. Grasses, sedges, rushes, broadleaf plants. |
| 3. Shallow fresh marshes | Soil is waterlogged during growing season; often covered with 6 or more inches of water. Grasses, bulrushes, spike rushes, cattails, arrowhead, smartweed, pickerelweed. A major waterfowl area. |
| 4. Deep fresh marshes | Soil is covered with 6 in. to 3 ft of water. Cattails, reeds, bulrushes, spike rushes, wild rice. Principal duck-breeding area. |
| 5. Open freshwater | Water is less than 10 ft deep. Bordered by emergent vegetation. Pondweed, naiads, wild celery, water lily. Breeding, feeding, nesting area for ducks. |
| 6. Shrub swamps | Soil is waterlogged, often covered with 6 or more inches of water. Alder, willow, buttonbush, dogwoods. Ducks nesting and feeding to limited extent. |
| 7. Wooded swamps | Soil is waterlogged, often covered with 1 ft of water. Along sluggish streams, flat uplands, shallow lake basins. Tamarack, arbor vitae, spruce, red maple, silver maple in the North; water oak, overcup oak, tupelo, swamp black gum, cypress in the South. |
| 8. Bogs | Soil is waterlogged. Spongy covering of mosses; heath shrubs, sedges, sphagnum moss. |
| **Coastal fresh areas** | |
| 9. Shallow fresh marshes | Soil is waterlogged during growing season by as much as 6 in. of water at high tides. Deep marshes along tidal rivers, sounds, and deltas on landward side. Grasses, sedges. Important areas for waterfowl. |

TABLE 15-1 (Continued.)

| Freshwater Areas, continued | |
|---|---|
| 10. Deep fresh marshes | Covered with 6 in. to 3 ft of water at high tide. Along tidal rivers and bays. Cattails, wild rice, giant cutgrass. |
| 11. Open freshwater | Shallow areas of open water along fresh tidal rivers and sounds. Little or no vegetation. Important waterfowl areas. |
| **Saline Water Areas** | |
| **Inland saline areas** | |
| 12. Saline flats | Flooded after heavy precipitation. Soil is waterlogged within a few inches of surface during the growing season. Seablite, salt grass, saltbush. Fall waterfowl feeding areas. |
| 13. Saline marshes | Soil is waterlogged during growing season; often covered with 2 to 3 ft of water. Shallow lake basins. Alkali hard-stemmed bulrush, widgeon grass, sago pondweed. Valuable waterfowl areas. |
| 14. Open saline water | Permanent areas of shallow saline water. Depth variable. Sago pondweed, muskgrasses. Important waterfowl feeding areas. |
| **Coastal saline areas** | |
| 15. Salt flats | Soil is waterlogged during growing season. Occasionally to fairly regularly covered by high tide. Landward sides or islands within salt meadows and marshes. Salt grass, seablite, saltwort. |
| 16. Salt meadows | Soil is waterlogged during growing season. Rarely covered with tide water. Landward side of salt marshes. Cord grass, salt grass, black rush. Waterfowl feeding areas. |
| 17. Irregularly flooded salt marshes | Covered by tides at irregular intervals during the growing season. Along shores of nearly enclosed bays, sounds, etc. Needlerush. Waterfowl cover area. |
| 18. Regularly flooded salt marshes | Covered at average high tide with 6 or more inches of water. Along open ocean and along sounds. Salt marsh cord grass along Atlantic. Alkali bulrush, glassworts along Pacific. Feeding area for ducks and geese. |
| 19. Sounds and bays | Portions of saltwater sounds and bays shallow enough to be diked and filled. All water landward from average low tide line. Wintering areas for waterfowl. |
| 20. Mangrove swamps | Soil covered at average high tide with 6 in. to 3 ft of water. Along coast of southern Florida. Red and black mangroves. |

Adapted from S.P. Shaw and C.G. Fredine, *Wetlands of the United States,* U.S. Fish and Wildlife Service Circ. 39, 1956.

## Freshwater Ecosystems

Freshwater ecosystems are highly variable and are studied on the basis of whether the water is flowing, as in streams and rivers (lotic ecosystems), or stationary, as in ponds and lakes (lentic ecosystems).

### Ponds and Lakes

Ponds are generally smaller and shallower than lakes. Plants rooted to the bottom are a predominant vegetational feature of the entire pond. Some ponds persist throughout the summer, but many are temporary, providing a rapid succession of algal and invertebrate life during the spring months. Ponds, whether permanent or temporary, are important breeding sites for frogs and salamanders—so much so that it is possible in the spring to locate a pond by the sound of frog voices!

Aquatic organisms that normally reside in temporary ponds or along the drying edges of a more permanent pond typically pass dry months of late summer and cold winter months encased in a resistant capsule. An interesting project is to collect some dried pond bottom (or edge) soil and place it in a large jar with water. In a moderately warm, well-lit place this jar will soon become a miniature pond ecosystem as plants and animals revive from dormancy.

Lakes and deeper ponds are characterized by temperature stratification. During midsummer a warm upper layer of water usually rests upon a colder deeper layer. In spring and fall differences in temperature—and hence density—between upper and lower layers cause overturn of the strata. Water freezes at 0°C; however, it becomes slightly more dense at 4°C than it is either above or below that temperature. So in spring the warming of surface water (even though there may still be ice on top) to 4°C causes this water to sink and the warmer water to be displaced upward. Convections thus created mix the water layers of different temperatures and bring up bottom nutrients and mix them with those of the top. The same phenomenon may occur in the fall as surface waters cool down to 4°C, although some lakes undergo only a single overturn per year.

Like temperature, oxygen concentrations in a lake may be stratified. Generally surface layers are more oxygenated than lower levels. Lower levels also have higher carbon dioxide levels from benthic decomposition. These differences are of special interest to fishermen desiring to locate species, such as trout, with a high oxygen requirement.

Like rivers, ponds and lakes are influenced by the runoff that feeds them. This runoff picks up particulate matter and dissolved substances from the substrates across which it flows (Fig. 15-62). The particulate matter may contribute to the filling in of the lake or pond; dissolved minerals may affect its pH and "hardness."

**FIGURE 15-62**
This junction of the clear Middle Fork of the Salmon River (draining a wilderness region) and the turbid main Salmon River (draining a multiple use watershed) dramatically demonstrates how water quality downstream is affected by land uses and human activities upstream. (Photo courtesy of James T. Brock.)

Various nutrients, such as phosphates and nitrates, may cause enrichment or **eutrophication**, the addition of nutrients to aquatic environments by natural or artificial means. Due to nutrients and suitable light and temperature, algal population explosions or "blooms" may occur. As a body of water becomes increasingly eutrophic, it is not only unsuitable for human purposes as a water source and for recreation but also continues to feed itself through the decomposition of the now large amounts of organic material. The benthic community changes accordingly, becoming composed of scavengers and detritus feeders who can tolerate higher levels of carbon dioxide generated during decomposition. Animals with high oxygen requirements are inhibited and gradually disappear. Turbidity of the water also increases, thus eliminating intolerant plants.

Once eutrophication begins in a stationary body of water, it does not cease. Eutrophication seems to be a natural successional process in the life of a lake or pond, but human activities accelerate it—drastically at times. Whether it would eventually recover on its own is a moot point, for in most cases it is not feasible to eliminate the causes of the enrichment, particularly without sacrificing agricultural uses of the watershed land. Agricultural soil erosion (including soil, chemical herbicides, pesticides, and fertilizers) is the single greatest cause of water pollution in two-thirds of the river basins in the United States.

In contrast, very deep, cold, nutrient-poor lakes are termed **oligotrophic**. Picture in your mind the brilliant waters of mountain lakes, clear sapphire blue rather than turbid and green tinged. Such lakes support less variety and abundance of organisms, especially algae, and exist in a state of dynamic ecological balance. In this balance, nutrients entering the lake by natural means are assimilated slowly into the ecosystem rather than causing eruptions of growth. Even the relative inaccessibility of such lakes is no guarantee of purity, however. Early stages of eutrophication may be stimulated by recreational activities and high-country grazing. An example of a formerly pristine lake that has been "loved" too much is beautiful Lake Tahoe on the California-Nevada border. Since the development of recreational homes, casinos, and support facilities along its south shore, distinctive eutrophication has occurred, at least in the immediate area of the developments.

The biological community of each lake or pond is stratified. The photosynthetic zone may be divided into two areas: a shallow littoral zone and a lighted but deeper zone, the limnetic zone. In the shallow **littoral zone** live rooted plants whose tops are **emergent**, extending above the water surface (Fig. 15-63), such as rushes, sedges, water buttercup, various species of *Potamogeton* (pondweed), and water lilies (*Nymphaea*); or **submergent**, living below the water surface, such as *Elodea* (water weed), *Myriophyllum* (water milfoil), *Cabomba*

**FIGURE 15-63**
Emergent vegetation of a mountain lake in central Oregon.

(fanwort), and *Valisneria* (tapegrass). Although the names of these submergent plants may seem unfamiliar, you would recognize the plants, for they are sold widely for freshwater aquaria! Bladderwort (*Utricularia*) is an unusual submergent plant in that it is carnivorous (Fig. 15-64).

**Floating plants** are also common in these warmer shallow waters, especially duckweed (*Lemna*). This tiniest of flowering plants compensates for its size by sheer numbers, forming a shiny light-green mat easily recognizable from the highway or the air, particularly in the summer. The underside of the duckweed mat supports its own community of attached and crawling organisms.

In the **limnetic zone**—the region of lighted water beyond the rooted plants—plant life consists mainly of abundant phytoplankton.

FIGURE 15-64 The trapping mechanism of bladderwort (*Utricularia*) is stimulated by physical contact of a prey organism with the hairlike structures. When triggered, a bladder expands abruptly, sucking in the prey. (Photo courtesy of Carolina Biological Supply Co.)

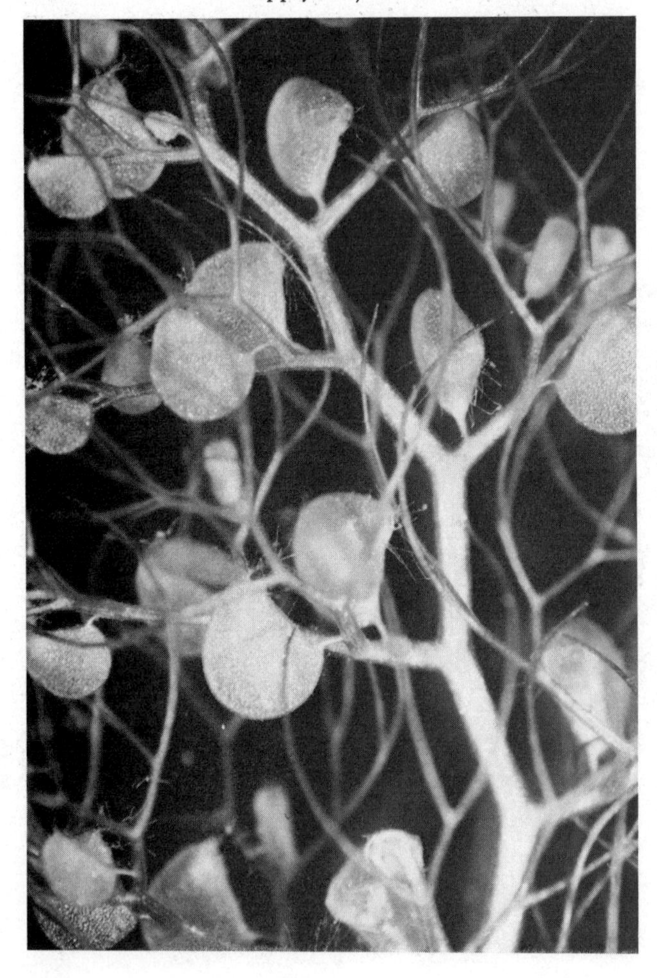

In both zones minute protozoans and tiny multicellular animals abound, feeding on the phytoplankton and organic debris. You can observe the larger of these primary consumer organisms (herbivores) by staring intently into the water using your submerged hand as a backdrop, or by dipping out and observing a sample in a translucent container. Any type of magnifying lens (even a droplet of water on the container) emphasizes the true abundance of organisms present.

Herbivorous frog larvae (tadpoles), small invertebrates, snails, and insect larvae are also primary consumers. The next step in a pond or lake food chain might be small fish, such as minnows, shiners, and juveniles of larger species. Predaceous beetles and aquatic larvae of dragonflies and damselflies eat invertebrates, small fish, and tadpoles. Insects, worms, and even small crayfish are eaten by fish and adult frogs, who may, in turn, be eaten by bullfrogs, fish, snapping turtles, or such birds as herons and egrets.

Beyond the littoral zone and below the world of the plankton, the waters belong to actively swimming (pelagic) species that compose the nekton. Primarily predatory fish, they eat insects alighting on the water, other fish, crayfish, even ducklings and swimming mice. Thus the total energy economy of a lake or pond depends on plant productivity in the shallower regions.

Although they may be associated with the benthos, **periphyton** organisms are those that grow attached to a substrate, whether rocks or debris on the bottom, the underwater surfaces of rooted plants, or dock pilings, for example. Included in this category are various diatoms and other algae, such protozoans as *Vorticella* and *Stentor,* sessile rotifers, such hydroids as *Hydra* and *Obelia,* and sponges. *Euplectella,* a colonial animal, may form colonies the size of a soccer ball and can be observed around docks by underwater swimmers.

Below the depth of light penetration in very deep lakes lies the **profundal zone**, occupied mainly by minute invertebrates and some fish. Due to intense bacterial decomposition on the bottom, the lowest water layers of the profundal zone may be too low in oxygen to support any aerobic (oxygen-requiring) animals.

Even lakes so deep that they are reputed to be "bottomless" have bottoms. The life of the profundal zone consists almost entirely of anaerobic (non-oxygen-requiring) bacteria, which decompose organic detritus that has drifted down from above.

In extremely productive or eutrophic conditions, debris may build up faster than the benthos can use it, resulting in a mucky bottom containing quantities of foul-smelling hydrogen sulfide and methane. Where oxygen is available, however, such animals as protozoans, various small crustaceans, water bears, *Tubifex* worms, midge larvae, and some flatworms exist. These animals filter out small organic particles to feed on, ingest the bottom ooze directly, or release digestive enzymes into their substrate. Then they absorb the products of this extracellular digestion into their bodies.

Some of the same benthic organisms live in the limnetic zone. As the water becomes shallower, however, the benthos becomes more abundant and diverse, reaching its peak in the littoral zone. Clams are the largest filter feeders present. Snails of various types graze along plant and bottom surfaces.

### Flowing Waters

The Continental Divide is a hypothetical north-south line through North America, west of which streams and rivers feed mainly into the Pacific Ocean and east of which they feed mainly into the Atlantic Ocean. Although the line is hypothetical, moving water is not. Moving water has sculpted the mountains and filled intermountain spaces with alluvium. Rivers have subsequently carved the alluvial beds into river valleys, whether wide and fertile or penetrating the earth as narrow gorges.

Flowing-water ecosystems are as varied as the water that forms them. The most influential physical factors are current velocity, type of bottom (bedrock, gravel, mud, sand, etc.), turbidity, pH, oxygen content, temperature, water source, stability of level, and inorganic and organic effluents. Many factors are directly related to the first: velocity. Velocity directly influences not only the nature of the other physical factors but also the nature of the biological community.

*Fast Streams.* Fast, gravely bottomed streams (Fig. 15-65) are more productive than slower streams. They have numerous surfaces for organism attachment and hiding placed for invertebrates and small fish. In addition, periodic riffles provide aeration, increasing oxygen content of the water. Both oxygen and carbon dioxide levels are in approximate equilibrium with the atmosphere. Turbidity in a fast stream may be relatively high during peak runoff times, with silt brought in by tributaries, but is generally low due to the lack of sedimentation permitted by the current. Constant drift of food materials from upstream also contributes to productivity. Fast streams have two basic habitats—riffles and pools, each with its own characteristics (Fig. 15-66).

Primary producers in a fast stream are epiphytic algal cells and colonies (periphyton) rather than

**FIGURE 15-65**
Fast, rocky-bottomed streams are well aerated. (Photo courtesy of Fred L. Rose.)

**FIGURE 15-66** A fast stream consists of **riffles,** as in the center, and **pools,** as near the opposite stream bank. (Photo courtesy of Fred L. Rose.)

planktonic species. Mainstream plants produce cementing secretions for attachment to avoid being swept away. Calcareous or gelatinous coverings may provide further protection. Rooted submergent and emergent species grow along stream banks, especially near pools.

Streamlined body form, powerful locomotion, and/or some means of secure attachment are the main adaptations necessary to animals of fast streams. Aquatic larvae of various insects are abundant, including mayflies, stoneflies, caddisflies, blackflies, and gnats. Fly fishermen along trout streams know how to keep an eye on "the hatch" (adult insects) over the water to guide selection of lures. Insect larvae show many adaptations, including flattened shape, claws and hooks, cemented houses, and burrows. Fish of fast streams, such as trout, have smoothly streamlined bodies and powerful swimming muscles to hold their position in the current. They retreat to quieter pools and eddies between feeding forays.

*Slow Streams.* Slower streams and most rivers (Fig. 15-67) differ from fast streams in several ways. They have sandy, silty, or muddy bottoms that support rooted plants, burrowing benthic organisms, such as freshwater mussels (clams) and bloodworms, plus the wandering crayfish. Oxygen content is lower, even near the surface, and is stratified. Higher carbon dioxide levels are present, particularly along the bottom, as a result of organic decomposition. Water temperature is generally higher and turbidity somewhat greater overall. Vegetation along the shoreline and in sloughs and oxbows is basically similar to that of the littoral zone of lakes and ponds, as described previously. The decreased current even permits a plankton community to develop, which is further nurtured by the warmer temperature and increased nutrient level.

The chemically sensitive insect larvae of clear, highly oxygenated fast streams cannot survive under such conditions. Changes in insect larva populations above and below a source of effluent are even used to monitor river pollution. Strong, torpedo-shaped fish like trout are replaced by such fish as bass and related panfish with tall, compressed bodies for moving among vegetation, as

**FIGURE 15-67** A slow stream has certain characteristics in common with standing water. (Photo courtesy of James T. Brock.)

well as bottom-feeding carp and catfish, perhaps even predaceous muskellunge and pike. The Mississippi River system is also noted for such prehistoric relics as sturgeon, paddlefish, and gar. The "bump on a log" you may see is probably a turtle basking in the sun and Midwest natives are sufficiently wary of snapping turtles that lurk on the bottom, especially the up-to-200-lb alligator snapping turtle.

### Wetlands

Wetlands associated with bodies of water include marshes, swamps, and bogs. Each type of wetland has its own features, as can be seen in Table 15-1, with bogs being most different from other freshwater and related wetland communities.

Bogs support a unique community of plants that live in the acidic conditions associated with *Sphagnum* (peat moss), which is usually the dominant plant. Bog-influenced waters are typically coffee colored due to humic acids from peat. Extensive bogs are a prominent feature of the boreal or northern coniferous forest. The study of bog succession is a "classic" of ecology because it dramatically exemplifies one type of conversion of a former lake or pond to a terrestrial ecosystem. In addition to sphagnum, typical bog vegetation includes cranberry, sweet gale (*Gale*), bog rosemary (*Andromeda*), and the shrubs leatherleaf (*Chamaedaphne*) and Labrador tea (*Ledum*). Carnivorous plants, such as sundew (*Drosera*), pitcher plant (*Sarracenia*), Venus flytrap (*Dionaea*), and cobra lily (*Darlingtonia*), obtain nitrogen by trapping insects and digesting their proteins.

Every year more marshes, swamps, and seasonal floodland disappear as the result of drainage projects to add to agricultural land, for mosquito control, for real estate speculation, or from natural ecological succession. In addition, many more acres are lost to flood control projects that include damming and channelization. Who can blame a farmer who needs the extra 3 acres for his corn or soybeans? It seems like such an insignificant wet place. People who own homes on floodplains certainly desire dams to give them security. Developers can make a profit by converting cheap swampland to recreational property. How about lakeside communities desiring to "clean up" the "weedy" edges, inlets, and outlets of the lakes?

As with loss of topsoil, however, we must become concerned about wetland preservation. America's wetlands have already been diminished *from 120 million acres to less than 75 million,* a habitat destruction of approximately 38%! Why should we leave our remaining wetlands alone and confine our activities to higher ground? Wetlands (small or large) are filters and purifiers of surface waters. When turbid, polluted water enters a swamp or marsh, its movement is slowed and sedimentation occurs, thus removing suspended matter. Ecosystem interactions process nutrients and contaminants. The water is thus purified to enter further bodies of water or the groundwater table. Increasing human populations make increased demands on groundwater sources, which are replenished mainly by surface waters percolating into permeable soils. The loss of natural wetlands and the increase in paved areas both contribute to the decreased availability of groundwater. Wetlands also reduce the degree of erosion that otherwise uncontrolled seasonal flooding would cause.

In addition, wetlands, like estuaries, are natural "nurseries." They are highly productive, creating a large base for numerous food chains. They provide nesting and feeding areas for migratory waterfowl and breeding grounds for fish and amphibians. Dense aquatic vegetation provides protection for small fish as they grow. Extensive wetlands provide sanctuary for hundreds of human-shy animal species, including some that are making a "last stand" in these places.

Because of their basic inaccessibility to human use and development, unaltered wetlands are among the few ecosystems that are still natural. As such, they are precious ecological "laboratories," deserving careful study. Finally, there are aesthetic reasons for preserving natural wetlands (Fig. 15-68).

## SUMMARY

**Biomes** are major types of natural environments. Within each biome there are specific **ecosystems** in which living and nonliving components

**FIGURE 15-68**
Marsh marigolds (*Caltha*) "at home." (See also the chapter opening art.)

interact to exchange matter and energy. All the organisms within a specific ecosystem are referred to as the **biological community**, which is composed of **populations** of many species.

The **arctic** biome is flat, treeless **tundra**, characterized by long, cold winters, short summers, and **permafrost.** Plants inhabiting the arctic biome adapt to the disruption of surface soil caused by alternate thawing and freezing. Plant life consists mostly of mosses, lichens, small rushes, grasses, and dwarf perennial plants. Flowering plants show many adaptations to the rigorous climate.

High-altitude **alpine** conditions resemble arctic conditions, with some important differences. Most alpine flowering plants are dwarf, must conserve water, grow economically, and have rapid seed formation and food production.

The **boreal forest,** where timberline meets arctic tundra, is characterized by coniferous trees intergrading with arctic vegetation as well as many lakes, potholes, and bogs.

In the eastern and central United States variable **temperate deciduous forests** replace the northern coniferous forest, although certain coniferous trees are important forest components in some areas.

The **southern evergreen forests** of the Atlantic and Gulf coasts replace the temperate deciduous forests where the climate becomes more subtropical. The very tip of southern Florida and the Florida Keys support a unique ecosystem, the **tropical hummock forest.**

Our native **grasslands** or **prairies** are the result of abundant spring moisture, followed by hot, dry summers with occasional thundershowers. Prairie ecosystems produce deep, fertile topsoil. Plants of the grassland have many xerophytic adaptations, including deeply penetrating root systems. Precipitation determines whether a given area is tall-grass, mixed-grass, or short-grass prairie. A wide variety of herbaceous and perennial species are characteristic of the prairies.

The **deserts** are west and southwest of the grasslands and are caused in part by a **rain shadow effect.** Desert precipitation averages less than 10 in. per year. Plants have many adaptations to cope with desert conditions. The American deserts include high-altitude "cool" desert in the intermountain West and lower-altitude "hot" desert in the Southwest.

**Montane coniferous forests** characterize the Rocky Mountains and Cascade/Sierra complex. Vegetation varies in these forests according to altitudinal zonation as well as the direction of exposure. The highest mountain zone is the subalpine, beginning at timberline. The species composition of each montane zone in the Cascade/Sierras differs somewhat from comparable zones in the Rockies. The west slopes of the Cascades and northern Sierras are moister than those of the southern Sierras, where shrubby **chapparal** vegetation dominates.

The **temperate montane forest** and **Pacific rain forest** of the Coast Range are dominated by the moist, warm presence of the Pacific Ocean. The rain forest has heavy precipitation and lush vegetation.

The **marine** biome is an assemblage of many kinds of environments. Oceans provide a hospitable environment for living things. Marine environments are distinctly stratified. **Intertidal** or **littoral** organisms on sandy beaches must survive drying, increased heat and salinity, and shifting sand. Above the intertidal zone are constantly shifting dunes on which rooted plants may grow. As succession continues, former dunes become forested.

Rocky shorelines support the greatest density of organisms of any marine habitat, reflecting the favorability of the environment for life. Intertidal organisms of rocky shores are adapted to meet such hazards as being swept away, wave shock, abrasion, and periodic exposure.

The **sublittoral** zone lies approximately between the low tide line and the edge of the continental shelf and is all within the range line of light penetration, the **photic zone.**

Communities of the **open sea** are also stratified. There is a diverse **plankton** community. The **pelagic zone** is mainly a community of swimming animals, or **nekton,** feeding on plankton and each other. Most **benthic** animals and plants live on detritus "raining" from above.

**Estuaries** form where saltwater and sea sand meet river freshwater and silt. Estuaries are fertile "nurseries" for many species. The open bay gives way inland to productive tidal marshes. In the southwest coast of Florida there are mangrove swamps instead of tidal marshes.

Ponds, lakes, and flowing waters are types of **freshwater environments.** Temperatures and oxygen concentrations in lakes and deep ponds are stratified. Nutrients running into a lake or pond have a great effect on its characteristics. Overenrichment causes **eutrophication. Oligotrophic** lakes are deep, cold, and nutrient poor. Ponds and lakes support stratified biological communities, including plankton, emergent and submergent plants, nekton, and benthic species.

Flowing-water ecosystems vary with such

water characteristics as velocity, turbidity, oxygen content, type of bottom, and effluents.

**Wetlands** are unique ecosystems associated with bodies of water, including marshes, swamps, and bogs. Wetlands are constantly being lost to natural and human forces. There are many valid reasons for preserving wetlands.

## SOME SUGGESTED READINGS

Bailey, R.G. *Description of the Ecoregions of the United States.* Ogden, Utah: USDA, Forest Service, Intermountain Region, 1978. With maps, a useful government publication.

Carlquist, S. *Island Biology.* New York: Columbia University Press, 1974. A fascinating study of the ecology of islands in a more technical treatment than his *Island Life,* or *Hawaii, A Natural History.*

Daubenmire, R. *Plant Geography: With Special Reference to North America.* New York: Academic Press, 1978. A brief description of the theory of plant distribution is followed by an extensive survey of the floristic types of North America.

Dawson, E.Y. *Marine Botany.* New York: Holt, Rinehart and Winston, 1966. Structure, ecology, geographical distribution, and uses of marine algae.

Gleason, H.A., and A. Cronquist. *The Natural Geography of Plants.* New York: Columbia University Press, 1964. An enjoyable introduction to the theory of plant geography and the floristic provinces of North America.

Kellman, M.C. *Plant Geography.* New York: St. Martin's Press, 1975. A succinct treatment of principles of plant geography.

MacArthur, R.H. *Geographical Ecology: Patterns in the Distribution of Species.* New York: Harper and Row Publishers, 1972. A good introduction to the theory and practice of geographical ecology, with an emphasis on animals.

Nebel, B.J. *Environmental Science.* Englewood Cliffs, N.J.: Prentice-Hall, 1981. A college text that explains basic precepts of ecology, then goes on to present human interactions with the natural environment. Facts and potential positive and negative consequences of human activities are discussed in topical treatments of soils, agriculture, water, domestic and industrial wastes, pest control, resources, land use patterns and policies, energy, and human population.

Ricketts, E.F., and J. Calvin. *Between Pacific Tides,* 4th ed. Stanford, CA.: Stanford University Press, 1968. A classic account of the intertidal zones of the Pacific Coast by the scientist known affectionately to John Steinbeck as "Doc."

Van der Pijl, L. *Principles of Dispersal in Higher Plants,* 2nd ed. New York: Springer-Verlag, 1972. A comprehensive text illustrating the major trends of seed and fruit dispersal.

Vankat, J.L. *The Natural Vegetation of North America.* New York: John Wiley & Sons, 1979. A compact paperback textbook that provides a basic understanding of vegetation and the human impact on natural vegetation types. Books and papers cited in the excellent reference lists are relevant sources of recent information about North American biomes. Good maps.

# chapter sixteen USEFUL PLANTS AND THEIR PRODUCTS

## INTRODUCTION

The "Wilderness Experience"—modern men and women seek it. A person who experiences it intentionally (or accidentally) immediately realizes the many items it is possible to live without. Existence becomes a goal instead of a way to pass the time. In its most basic terms, survival is usually reduced to fulfilling minimums of three life-sustaining necessities—food, clothing, and shelter—and these necessities can be met by plants. In addition to fulfilling humanity's basic three needs, plants have made vast contributions to the prolongation, enjoyment, and improvement of the quality of life. Plants produce fibers and a wide variety of exudates and extractable substances, such as oils, waxes, resins, gums, and latex. These products are used in many ways—try reading labels of foods, medicines, beverages, even clothing and other commodities.

The science of **economic botany** deals with useful plant species. The history of economic plants is intertwined with the history of ancient and modern civilizations. Consider the historical impact of developing agriculture; of the spice, tea, and opium trades; of natural stimulants, sedatives, laxatives, painkillers, and other medications; of the discovery of antibiotics; of the replacement of parchment from animal skins by paper from plant tissues.

For your convenience, this chapter is organized by use category rather than by taxonomy or geography. Both familiar and not-so-familiar examples have been chosen to emphasize the broad use of plant derivatives worldwide. Because many medicinal plants, such as the narcotics, are discussed in the chapter on hazardous plants, our discussion here is brief and you should refer to Chapter 18 for additional information.

## CODEVELOPMENTS: THE DOMESTICATION OF PLANTS AND THE DEVELOPMENT OF CIVILIZATIONS

It is widely accepted that many prehistoric humans were originally hunters and foragers. Like the animals they hunted, they reacted to seasonal plant growth and were nomadic as they followed their food. No one knows if agriculture was born of a single inventor or if it arose separately in many parts of the world in response to human needs and inventiveness. By accident or intuition, prehistoric humans learned that seeds were the source of new plants similar to the ones from which they were produced and that, by planting seeds, it was possible to grow desirable plants near their camps.

With less need to roam in search of food, it is provocative to hypothesize that groups of primitive people had more leisure time, enabling them to develop creative activities. Some of the earliest creative activities were making clay vessels and woven baskets. Clothing improved as weaving developed. In addition to utilitarian objects, early artisans created sculptures, carvings, and cave paintings. Art developed as a means of recording events and scenes, real or imagined.

The domestication of a few animal species occurred along with the domestication of food plants. Cultures became increasingly sophisticated, for with abundant goods people could live together in larger groups. Cities developed. Specialization arose, with people trading each other for items each had produced. Today most humans have little contact with our food's growth; we trade our work for tokens (currency) and exchange the tokens for foods and goods. In modern countries even the most exotic foods from around the world are available to us in fresh, dried, canned, or frozen form.

Mobility is an important aspect of civilization. Mountain ranges, deserts, and bodies of water are the most serious barriers to movement. Solid surfaces can be walked on, but water requires flotation. Plant materials have provided many means of transportation across rivers, lakes, streams, and oceans, with various cultures utilizing whatever is most at hand—boats of reeds, straw, or bamboo; dugout logs; birchbark canoes; boats of saplings and buffalo hides. These simple types of crafts relied on muscle power, air, and water currents for propulsion, which were also the main sources of power for the great galleys and sailing ships. Eventually mechanical engines powered by burning plant material were invented—machines for travel on land, water, and through the air. Plant fuels are wood and wood products, peat, coal, and, most recently, petroleum products and alcohols.

One of the most interesting aspects of traditional economic botany books and courses is tracing the ancestry of domesticated food species. All domesticated plants had wild ancestors and most domesticated species and varieties will revert to a more primitive or ancestral type if allowed to "go wild" for one or more generations (Fig. 16-1). Studies of many food plants indicate that manipu-

**FIGURE 16-1**
Parsnips that have "gone wild" in southern Wisconsin.

lation of plant genetics is an ancient practice. Sites in Mexico and South America provide a history of maize (corn) from a small-cobbed wild grass to one that was essentially modern in its major characteristics prior to 2000 B.C.! Wheat seems to have been cultivated in the Middle East since at least 4000 B.C. and barley in Europe, Asia, and Africa before that. The nineteenth-century work of Alphonse De Candolle attempted to identify the origins of many cultivated plants (Table 16-1).

**TABLE 16-1** **Cultivated plant origin categories developed by De Candolle, with selected examples.[a]**

| **OLD WORLD SPECIES CULTIVATED FOR OVER 4000 YEARS, according to ancient historians, as known to ancient Egyptians and Chinese, and as can be interpreted from other botanical and word-origin clues** | | | | |
|---|---|---|---|---|
| Almond | Chick pea | Hemp | Peach | Sesame |
| Apple | Cucumber | Lentil | Pear | Sorghum |
| Apricot | Date | Mango | Pomegranate | Soybean |
| Banana | Eggplant | Millet | Purslane | Tea |
| Barley | Fig | Mulberry | Quince | Turnip |
| Broad bean | Flax | Olive | Rice | Watermelon |
| Cabbage | Grape | Onion | Saffron | Wheat |
| **OLD WORLD SPECIES CULTIVATED FOR OVER 2000 YEARS, AND PERHAPS LONGER, known by Greek botanist Theophrastus, found in Lake Dweller sites or having word-origin clues, such as Hebrew or Sanskrit names** | | | | |
| Alfalfa | Celery | Grapefruit | Nutmeg | Rye |
| Asparagus | Cherry | Leek | Oats | Sugarcane |
| Beet | Chestnut | Lemon | Peppers | Taro |
| Bitter orange | Cotton | Lettuce | Plum | Vetch |
| Black mulberry | Garden pea | Lime | Poppy | Walnut |
| Breadfruit | Garlic | Mustards | Radish | Yam |
| Carrot | | | | |
| **OLD WORLD SPECIES CULTIVATED PROBABLY FOR LESS THAN 2000 YEARS, as interpreted by drawings at Pompeii, mentioned by Dioscorides but not Theophrastus, and other clues** | | | | |
| Artichokes | Coffee | Gooseberry | New Zealand spinach | Rhubarb |
| Buckwheat | Common mushroom | Hop | Okra | Salsify |
| Chicory | Currant | Horseradish | Parsley | Spinach |
| Chives | Endive | Luffa | Parsnip | Strawberry |
| Cloves | Gherkin | Muskmelon | Raspberry | Sweet orange |

TABLE 16-1 (Continued).

| NEW WORLD SPECIES OF ANCIENT CULTIVATION, MORE THAN 2000 YEARS, as determined by wide distribution and presence of many known varieties | | | | | |
|---|---|---|---|---|---|
| Cacao (cocoa) | Kidney bean | Maize | Maté | Sweet potato | Tobacco |

| NEW WORLD SPECIES CULTIVATED BEFORE THE TIME OF COLUMBUS, ANTIQUITY NOT KNOWN, as determined by no previous signs of great cultural antiquity | | | | |
|---|---|---|---|---|
| Agave | Jerusalem artichoke | Pineapple | Pumpkin | Squash |
| Avocado | Pawpaw | Potato | Quinoa | Tomato |
| Cotton | Peanut | Prickly pear | Red pepper | Vanilla |
| Guava | | | | |

| NEW WORLD SPECIES CULTIVATED SINCE THE TIME OF COLUMBUS | | | | |
|---|---|---|---|---|
| Allspice | Blueberry | Dewberry | Pecan | Plum |
| Blackberry | Cinchona (quinine) | Gooseberry | American persimmon | Rubber |
| Black walnut | Cranberry | Manioc | | Strawberry |

[a]De Candolle's work, first published in 1882, is a classic in economic botany. As noted in each category, his judgments were based on exhaustive historical investigation, including at times sifting through conflicting and fictitious reports. The task required great scholarship and his work stands as the basis for modern studies of cultivated plant origins.

## CULTURAL USES OF PLANTS

The transition from oral to written language involved development of both language symbols and increasingly effective recording surfaces and instruments. Modern communication owes a great debt to the discovery that plant materials could be pounded in water to produce pulp that could then be spread on screens to dry into flattened sheets—paper. Although the process of papermaking is now highly sophisticated, the basic principle is the same.

Plants have also provided both raw materials and inspiration for the arts, from painting to poetry and dance.

Ancient religions that celebrated natural cycles included mystical recognition of some plant species. The Druids of northern Europe, for example, utilized mistletoe (*Viscum album*) and holly (*Ilex aquifolium*) in their year's-end ceremonies. As Christianity became dominant, these symbols were incorporated into the Christmas celebration. Narcotic and psychoactive plants are important to some religions. In the Americas, for instance, many Indian tribes use or formerly used peyote (a cactus), mescal (a fermented beverage made from the pulp of *Agave*, Fig. 16-2), mescal beans (frijolitos), or certain mushrooms to induce altered states of consciousness, including visions, during religious ceremonies. (See also Chapter 18.) The burning of incense, derived from plants, is an important part of some major religions.

FIGURE 16-2 Different members of the genus *Agave* are the source of sisal fibers and the fermented beverage mescal or tequila. (Photo courtesy of John S. Thiede.)

Besides precious stones and minerals, beautiful woods and aromatic plants are ancient symbols of wealth and esteem. To honor the baby they believed to be the new King of Israel, the Three Kings of biblical history brought gifts of gold and two aromatic plants, frankincense (*Boswellia carteri*) and myrrh (*Commiphora absynnica*).

The ancient Egyptians were sophisticated in the use of cosmetics, pomades, and other preparations derived from plants, animals, and minerals. Their skill at embalming is also legendary, involving the use of spices, oils, and other ingredients.

We are defended from the scent of unwashed humanity by modern soaps, deodorants, and bathing habits. Frequent bathing, however, is a relatively recent development in Western civilization (within the last 50 years). The grand gentlemen and ladies of the French courts became famous for their perfumed waters and plant-derived sachets, which counteracted stale body odors. Both men and women also carried elegant perfume-scented handkerchiefs so that they might pass through filthy streets unoffended by odors of horse and human refuse, fish, meat, and vegetable wares in various states of freshness. In Colonial North America fragrant sweet rush (*Acorus*) leaves deodorized public meeting rooms.

## FOOD AND DRINK

### Fruits—Sweet and Otherwise

Apples, berries, citrus fruits, cherries, avocados, melons, peppers, eggplants, cucumbers, and squashes are all familiar **fleshy fruits**, but some species produce **dry fruits**. (See also Chapter 5.) Tables 16-2, 16-3, 16-4, and 16-5 summarize some economically important fruits and seeds. With many fruits, only the flesh is eaten; the outer rind or peel and the seeds are discarded or prepared in a different manner. Pumpkin and squash seeds are often discarded but may be seasoned, toasted, and eaten.

**TABLE 16-2 Temperate-climate fleshy fruits.**

| Family Name | Common Name | Scientific Name | Geographic Origin |
|---|---|---|---|
| **Pome fruits** | | | |
| Rosaceae | Apple | *Pyrus malus* | Eastern Europe, western Asia |
| Rosaceae | Pear | *Pyrus communis* | Eurasia |
| Rosaceae | Chinese or sand pear | *P. pyrifolia* var. *culta* | China |
| **Stone fruits (drupes)** | | | |
| Lauraceae | Avocado | *Persea americana* | Southern United States, Mexico, Central America |
| Rosaceae | Apricot | *Prunus armeniaca* | Asia |
| Rosaceae | Cherry | *Prunus avium* and *P. cerasus* | Eurasia |
| Rosaceae | Peach | *Prunus persica* | China |
| Rosaceae | Plums, prunes | *Prunus* spp. | United States, China, Japan |
| Ericaceae | Eastern wild huckleberry | *Gaylussacia* spp. | Eastern North America |
| **Gourd fruits** | | | |
| Cucurbitaceae | Melons | *Cucumis melo* | Southern Asia |
| Cucurbitaceae | Watermelon | *Citrullus vulgaris* | Tropical Africa |
| Cucurbitaceae | Cucumbers | *Cucumis sativus*, *C. anguria* | Southern India |
| Cucurbitaceae | Pumpkins, squashes | *Cucurbita* spp. | America or Africa |
| Cucurbitaceae | Chayote or mirliton | *Sechium edule* | Tropical America |

**TABLE 16-2 (Continued).**

| Family Name | Common Name | Scientific Name | Geographic Origin |
|---|---|---|---|
| **True berries** | | | |
| Malvaceae | Okra | *Hibiscus esculentus* | Africa |
| Solanaceae | Peppers | *Capsicum* spp. | Tropical America |
| Solanaceae | Eggplant | *Solanum melongena* | India |
| Solanaceae | Tomato | *Lycopersicon esculentum* | South America |
| Vitaceae | Grapes | *Vitis vinifera* and others | Western Asia, Europe, North America |
| Ericaceae | Blueberries, western huckleberries | *Vaccinium* spp. | North America |
| Ericaceae | Cranberry | *Vaccinium macrocarpon* | North America |
| Saxifragaceae | Currant | *Ribes* spp. | Eurasia, North America |
| Saxifragaceae | Gooseberry | *Ribes grossularia* | Eurasia |
| **Aggregate Fruits** | | | |
| Rosaceae | Strawberry | *Fragaria* spp. | North America |
| Rosaceae | Raspberries, blackberries, boysenberries, loganberries | *Rubus* spp. | North America |
| Moraceae | Mulberry | *Morus* spp. | The Orient |
| **Legumes** | | | |

Some legumes are eaten while the pod is still fleshy, such as green beans and edible-podded peas. The legume group, however, is listed with dry fruits (Table 16-4) because that is how most legumes are eaten.

**TABLE 16-3 Tropical fruits.**

| Family Name | Common Name | Scientific Name | Geographic Origin |
|---|---|---|---|
| Rutaceae | Citrus fruits | *Citrus* spp. | Eastern and southern Asia |
| Musaceae | Banana | *Musa paradisiaca* | India |
| | | *M. p. sapientum* | Malaya |
| Annonaceae | Custard apples sweetsop, etc. | *Annona* spp. | American tropics |
| Arecaceae | Coconut | *Cocus nucifera* | Malay archipelago |
| Arecaceae | Dates | *Phoenix dactylifera* | India, Arabia |
| Moraceae | Figs | *Ficus carica* | Mediterranean, Asia Minor |
| Myrtaceae | Guava | *Psidum guajava* | Tropical America |
| Rhamnaceae | Jujube | *Zizyphus jujuba* | China |
| Sapindaceae | Litchi nuts | *Litchi chinensis* | China, Siam |
| Rosaceae | Loquat | *Eriobotrya japonica* | China |
| Anacardiaceae | Mangos | *Mangifera indica* | Southern India |
| Guttiferae | Mangosteen | *Garcinia mangostena* | Malaya |
| Oleaceae | Olive | *Olea europaea* | Mediterranean |
| Caricaceae | Papaya | *Carica papaya* | West Indies, Mexico |
| Ebenaceae | Japanese persimmon | *Diospyros kaki* | China |
| Ebenaceae | American persimmon | *D. virginiana* | Eastern United States |
| Bromeliaceae | Pineapple | *Ananas comosus* | Northern South America |
| Punicaceae | Pomegranate | *Punica granatum* | Iran |
| Fabaceae | Tamarind | *Tamarindus indica* | Tropical Asia |

**TABLE 16-4** Grains and legumes.

| Family Name | Common Name | Scientific Name | Geographic Origin |
|---|---|---|---|
| **Grains: cultivated species only** | | | |
| Poaceae | Wheat | *Triticum aestivum* | Origin uncertain |
| Poaceae | Maize (corn) | *Zea mays* | South America |
| Poaceae | Rice | *Oryza sativa* | Exact origin uncertain |
| Poaceae | Barley | *Hordeum vulgare* | Asia, North Africa |
| Poaceae | Rye | *Secale cereale* | Southwest Asia |
| Poaceae | Oats | *Avena sativa* | Probably multiple origin |
| Poaceae | Sorghums, broomcorns, milo | *Sorghum vulgare* | Asia, Africa |
| Poaceae | Millets | *Setaria, Echinochloa, Panicum, Pennisetum,* and *Eleusine* species | Europe, Asia |
| Poaceae | Wild rice | *Zizania aquatica* | North America |
| **Pseudograins** | | | |
| Polygonaceae | Buckwheat | *Fagopyrum esculentum* | Central and western China |
| **Legumes** | | | |
| Fabaceae | Peas | *Pisum sativum* | Southern Europe |
| Fabaceae | Garden beans | *Phaseolus vulgaris* | New World |
| Fabaceae | Lima beans | *P. limensis* | Peru, Brazil |
| Fabaceae | Scarlet runner bean | *P. coccineus* | Tropical America |
| Fabaceae | Mung beans | *P. aureus* | Tropical America |
| Fabaceae | Cow peas | *Vigna sinensis* | Central Africa |
| Fabaceae | Soybeans | *Glycine max* | Southeast Asia |
| Fabaceae | Broad beans | *Vicia faba* | Algeria or southwest Asia |
| Fabaceae | Peanuts | *Arachis hypogaea* | South America |
| Fabaceae | Lentils | *Lens culinaris* | Southwest Asia |
| Fabaceae | Alfalfa* | *Medicago sativa* | Southwest Asia |
| Fabaceae | Sweet clover* | *Melilotus alba* | Western Asia |
| Fabaceae | Kudzu* | *Peuraria thunbergiana* | Japan, eastern Asia |
| Fabaceae | Lespedezas* | *Lespedeza* spp. | Japan, China |
| Fabaceae | Vetches* | *Vicia* spp. | Various |
| Fabaceae | Mesquite | *Prosopis glandulosa* | North America |
| Fabaceae | Carob | *Ceratonia siliqua* | Syria |
| Fabaceae | Honey locust | *Gleditsia glandulosa* | North America |

*Those indicated by an asterisk are not grown for human consumption, but as fodder, for erosion control, or for soil improvement.

**TABLE 16-5 Edible "nuts."**

| Family Name | Common Name | Scientific Name | Geographic Origin |
|---|---|---|---|
| **True nuts: hard shell is made up of ovary wall** | | | |
| Anacardiaceae | Cashew | *Anacardium occidentale* | Brazil |
| Fagaceae | Acorn | *Quercus* spp. | Various |
| Fagaceae | Chestnut | *Castanea dentata* | United States |
| | | *C. sativa* | Europe |
| | | *C. crenata* | Japan |
| Fagaceae | Beechnut | *Fagus grandifolia* | North America |
| | | *F. sylvatica* | Europe |
| Betulaceae | Filbert | *Corylus avellana, C. maxima* | Europe |
| Proteaceae | Macadamia nut | *Macadamia ternifolia* | Australia |
| **Drupaceous fruits: shell is made up of inner wall layer of ovary; fleshy or fibrous outer husk formed by outer layers of ovary** | | | |
| Juglandaceae | Hickory nut | *Carya* spp. | Central North America |
| Juglandaceae | Pecan | *Carya illinoensis* | Southern United States, Mexico |
| Juglandaceae | Black walnut | *Juglans nigra* | Central North America |
| Juglandaceae | English walnut | *J. regia* | Iran |
| Juglandaceae | Butternut | *J. cinerea* | North America |
| **True drupes: shell is made up of inside layer of ovary wall; outer layers of fruit are fleshy** | | | |
| Rosaceae | Almond | *Prunus amygdalus* | Eastern Mediterranean |
| Anacardiaceae | Pistachio | *Pistacia vera* | Western Asia |
| **Seeds: shell is actually the seed coat** | | | |
| Lecythidaceae | Brazil nut | *Bertholletia excelsa* | Amazon forest |
| Pinaceae | Pine nut, pignolia | *Pinus* spp. | Western North America |
| **Achene: shell is made up of ovary wall enclosing seed; seed attached to fruit at one point** | | | |
| Asteraceae | Sunflower "seeds" | *Helianthus annuus* | South America |

In cereal plants each individual grain is an entire fruit (**caryopsis**) and there are usually many grains on a fruiting stalk. Cereals were among the first plants to be domesticated and improved by selection, especially corn, wheat, oats, barley, and rice.

Wild peas and beans also show early records of cultivation and improvement. The pod of a **legume** is the fruit. The immature pods of some legumes are edible, such as green beans and peas, whereas only the seeds of soybeans, peanuts, and other beans and peas are eaten (Fig. 16-3).

A **nut** is a hard-shelled fruit and the shell is formed by all layers of the ovary wall. The nut kernel is the seed.

## Seeds

When is a "nut" not a nut? Answer: when it is a seed. Many of the edibles commonly called "nuts" are actually seeds. To prepare a Brazil nut, coconut, or almond for sale and consumption, the outer fleshy or fibrous part of the fruit wall is removed. The shell remaining around the kernel (seed) represents only the inner layer(s) of the ovary wall.

When is a "seed" not a seed? Answer: when it is a fruit! A sunflower "seed" is actually an entire simple fruit (**achene**). The "seeds" of dill, celery, anise, fennel, caraway, and other members of the carrot and parsley family are actually hard, aromatic fruits. The volatile oils of these **savory fruits** are located in structures lining the grooves of each fruit.

## Nonvolatile Vegetable Oils and Fats

Bacon grease, lard, and suet are traditional cooking fats and butterfat from milk is used to produce butter and cheeses. Animal fats are **saturated**, having complex hydrogenated structures. In

FIGURE 16-3
Legumes are a high-protein food. Groundnuts—related to peanuts but in different genera—are sold in this Kumasi, Ghana, marketplace. (Photo courtesy of Ingolf Vogeler.)

general, saturated fats are solid at room temperature. In contrast, most lipids extracted from plants are **unsaturated** and **polyunsaturated** oils. Coconut and cocoa lipids are exceptions, being at least partially saturated. The distinctions are based on structural molecular characteristics but have human health significance.

Saturated fats tend to reconstruct within the body following absorption. Animal fats are also high in cholesterol (a complex lipid known as a steroid). Cholesterol is a normal body substance that is actively synthesized if there is a deficit of it in the body. Excess cholesterol in the circulation, however, accumulates as fatty deposits (placque) on the inner walls of blood vessels, contributing to cardiovascular diseases. So there has been increasing emphasis on reducing dietary amounts of animal fat, substituting such unsaturated and polyunsaturated vegetable lipids as safflower, sunflower, corn, and cottonseed oils. Although the mechanism is not clear, there is also some evidence that a diet incorporating polyunsaturates actually aids in *reducing* cholesterol levels in the body.

The energy potential (calories) of both plant and animal lipids is the same. Therefore substituting margarine for butter eliminates a source of cholesterol but still provides the same number of calories (unless the margarine is a low-calorie formula). In some products unsaturated fats are deliberately hydrogenated, converting them to a saturated form, such as when the naturally unsaturated oils in most peanut butters are hydrogenated to prevent the oil from separating as a layer at the top of the container.

Food and nonfood oils can be extracted from a number of domesticated and wild plants (Table 16-6). Usually the fruit or seed is crushed, heated, and pressed to extract the oils. The residue is often high in food value and can be used as livestock food or for extraction of other components.

For generations born before the 1950s, one of childhood's best-remembered (or worst-remembered) oils is castor oil, extracted from the toxic seeds of the castor bean plant (*Ricinus communis*). Castor oil was administered orally (with predictable results) for ailments suspected to be related to constipation.

## Shoots, Stems, Leaves, and Flowers

The young emerging stems of asparagus and bamboo are examples of edible shoots. Broccoli

TABLE 16-6 Vegetable or nonvolatile oils.

| Family Name | Common Name | Scientific Name | Geographic Origin |
|---|---|---|---|
| **Drying oils** | | | |
| Linaceae | Linseed oil, from flax | *Linum usutatissimum* | N. America, USSR, China |
| Euphorbiaceae | Tung oil | *Aleurites fordii* | China |
| Euphorbiaceae | Candlenut oil | *Aleurites moluccana* | Pacific Islands |
| Fabaceae | Soybean oil | *Glycine max* | China |
| Rosaceae | Oilicica oil | *Licania rigida* | Brazil |
| Lamiaceae | Perilla oil | *Perilla frutescens* | India, China, Japan |
| Juglandaceae | Walnut oil | *Juglans regia* | Iran |
| Papaveraceae | Poppy oil | *Papaver somniferum* | N. Europe, India |
| Asteraceae | Safflower oil | *Carthamnus tinctorius* | Asia Minor, Orient |
| Asteraceae | Niger seed oil | *Guizotia abyssinica* | Tropical Africa |
| Pinaceae | Tall oil | *Pinus* spp. | N. America |
| **Semidrying oils** | | | |
| Malvaceae | Cottonseed oil | *Gossypium* spp. | Various |
| Poaceae | Corn oil | *Zea mays* | S. America |
| Pedaliaceae | Sesame oil | *Sesamum indicum* | India, China |
| Asteraceae | Sunflower oil | *Helianthus annuus* | S. America |
| Brassicaceae | Rape oil, colza oil | *Brassica* spp. | China, Japan, India, Europe |
| **Nondrying oils** | | | |
| Oleaceae | Olive oil | *Olea europaea* | Mediterranean |
| Fabaceae | Peanut oil | *Arachis hypogaea* | S. America |
| Euphorbiaceae | Castor oil | *Ricinus communis* | Africa |
| **Vegetable fats** | | | |
| Arecaceae | Coconut oil | *Cocos nucifera* | Tropical Asia, Central America |
| Arecaceae | Palm oil | *Elaeis guinensis* | W. Africa |
| Arecaceae | Palm-kernel oil | *Elaeis guinensis* | W. Africa |
| Arecaceae | Brazilian palm oils | Various genera | Brazil |
| Sterculiaceae | Cocoa butter | *Theobroma cacao* | Mexico, Central America |

and kohlrabi (Figs. 16-4 and 16-5) are fleshy-stemmed vegetables. With broccoli, eating the stem is only in conjunction with eating the "head," an inflorescence of unopened flower buds. You can demonstrate it by breaking off a small sprig of the broccoli head and setting it aside in water for a few days. Small, four-petaled yellow flowers will develop.

Cauliflower is aptly named because the cauliflower head, like that of broccoli, is actually a developing stem (caul) bearing flowers (Fig. 16-6). Cauliflower is eaten while the flowers are tightly closed, although as it matures the surface becomes "furry" with opening flowers.

Brussels sprouts are fleshy buds. Like its larger relative cabbage, each sprout consists of a very short stem, an apical bud, and developing leaves. The structure of a Brussels sprout is similar to that of a cabbage head, but it is smaller. Many Brussels sprouts are produced spirally along a tall, sturdy stem (Fig. 16-7) whereas cabbage produces one main head per plant.

Cabbage, Brussels sprouts, broccoli, kohlrabi, collards, and kale present an interesting example of plant breeding and variety improvement. All are members of the same species (*Brassica oleracea*), which have been guided in their evolution to distinctively different growth forms. The ancestral type was probably a leafy form similar to kale.

Kale, the many varieties of lettuce (*Lactuca sativa*), cabbage, and such "greens" as spinach (*Spinacia oleracea*), beet (*Beta vulgaris*), chard (*Beta vulgaris* var. *cicla*), radish (*Raphanus sativus*), mustard (*Brassica* spp.), collards (*Brassica oleracea* var. *acephala*), Chinese cabbage or napa (*Brassica pekinensis*), and parsley (*Petroselinum sativum*) are

FIGURE 16-4 Broccoli flowering heads. (Photo courtesy of David Koranski.)

FIGURE 16-5 Kohlrabi ready to harvest. (Photo courtesy of David Koranski.)

FIGURE 16-6 Cauliflower head. (Photo courtesy of David Koranski.)

FIGURE 16-7 Brussels sprouts. (Photo courtesy of David Koranski.)

all vegetables whose leaves are eaten. Only the petioles of celery (*Apium graveolens*) and rhubarb (*Rheum rhaponticum*) leaves are eaten. The blade of the rhubarb leaf is dangerously poisonous and should never be eaten (Chapter 18). The leaves of the tobacco plant (*Nicotiana tabacum*) are also poisonous if ingested but tobacco for smoking, chewing, sniffing, and greenhouse fumigation is a major economic product.

A globe artichoke (*Cynara scolymus*) is an enlarged thistle inflorescence on which the bract bases and receptacle (the "heart") are edible. Many fragrant flowers, such as hibiscus (*Hibiscus*), violet (*Viola*), jasmine (*Jasminum*), and orange (*Citrus*), are used in delicately flavored teas. Hops, used in making beer, are the flowers of the hop plant (*Humulus*), which is related to hemp (*Cannabis*). The resins and other components of the hop flowers give beers and ales their distinctive, slightly bitter flavor. Among the most unusual uses of flowers are as spices. Cloves are the dried, unopened flower buds of *Eugenia caryophyllata.* Saffron is the crushed stigmas of autumn crocus (*Crocus sativus*) flowers.

### Subterranean Stems and Roots

Although it seems logical to label everything that grows underground a root, doing so is not accurate. Rhizomes are underground stems at the tips of which storage organs called tubers may develop. Ginger "root" is the fleshy rhizome of *Zingiber officinale.* The white or Irish potato (*Solanum tuberosum*) is a familiar tuber. Another edible tuber is the Jerusalem artichoke, which is neither an artichoke nor from Jerusalem but a native American sunflower (*Helianthus tuberosus*). A tuber may be distinguished from a root by the presence of nodes, where axillary buds ("eyes") or scalelike papery leaves may develop. Bulbs are specialized stems having fleshy scale leaves, and edible bulbs include garlic, chives, and all the many varieties of onions, shallots, leeks, and so on (*Allium* spp.).

Carrots (*Daucus carota*), parsnips (*Pastinaca sativa*), beets, radishes, rutabaga or Swede (*Brassica napobrassica*), and turnips (*Brassica rapa*) have edible roots. Horseradish (*Armoracia*) roots are grown mainly as a condiment. The sugar beet root (*Beta vulgaris*) is grown for extraction of a food product (sucrose).

### Herbs, Spices, and Condiments

Imagine the blandness of many foods prepared without seasoning. How did the use of seasonings begin? Was it for a change in taste sensations, as an outgrowth of medicinal uses of plants, for preservation of foods, or perhaps to camouflage the ripening flavors of foods (particularly meats) collected in large quantities and consumed over a period of time without the benefit of refrigeration?

Table 16-7 summarizes some of the most commonly used culinary herbs and spices, classified by the part of the plant used.

**TABLE 16-7 Some culinary herbs and spices.**

| Family Name | Common Name | Scientific Name | Geographic Origin |
|---|---|---|---|
| **Subterranean parts** | | | |
| Apiaceae | Angelica | *Angelica archangelica* | Syria |
| Brassicaceae | Horseradish | *Armoracia lapathifolia* | Southeastern Europe |
| Liliaceae | Sarsaparilla | *Smilax* spp. | Tropical |
| Zingiberaceae | Ginger | *Zingiber officinale* | Southeast Asia |
| Zingiberaceae | Turmeric | *Curcuma longa* | China, East Indies |
| **Bulbs** | | | |
| Liliaceae | Garlic | *Allium sativum* | Orient, southern Europe |
| **Barks** | | | |
| Lauraceae | Cassia | *Cinnamomum cassia* | Southeastern China |
| Lauraceae | Cinnamon | *Cinnamomum zeylanicum* | Ceylon |
| Lauraceae | Sassafras, filé gumbo | *Sassafras albidum* | North America |
| **Flowers, flower buds** | | | |
| Capparidaceae | Capers | *Capparis spinosa* | Mediterranean |
| Myrtaceae | Cloves | *Eugenia caryophyllata* | Molucca (Spice) Islands |
| Iridaceae | Saffron | *Crocus sativus* | Asia Minor |

TABLE 16-7 (Continued).

| Family Name | Common Name | Scientific Name | Geographic Origin |
|---|---|---|---|
| **Fruits** | | | |
| Myrtaceae | Allspice, Jamaica pepper | *Pimenta divica* | West Indies, South America |
| Piperaceae | Sweet peppers | *Capsicum* spp. var. *grossum* | American tropics |
| Piperaceae | Paprikas, pimento | *Capsicum* spp. | American tropics |
| Piperaceae | Chilis | *Capsicum* spp. var. *longum* | American tropics |
| Piperaceae | Black pepper, white pepper | *Piper nigrum* | India, Indo-Malayan region region |
| Rutaceae | Citrus peels | *Citrus* spp. | Asia |
| Cupressaceae | Juniper | *Juniperus communis* | Eurasia |
| Magnoliaceae | Star anise | *Illicium verum* | China |
| Orchidaceae | Vanilla | *Vanilla planifolia* | Tropical America |
| **Savory fruits ("seeds") of carrot family** | | | |
| Apiaceae | Anise | *Pumpinella anisum* | Mediterranean |
| Apiaceae | Caraway | *Carum carvi* | Europe, western Asia |
| Apiaceae | Celery | *Apium graveolens* | Various |
| Apiaceae | Coriander | *Coriandrum sativum* | Mediterranean |
| Apiaceae | Cumin | *Cuminum cyminum* | Mediterranean |
| Apiaceae | Dill | *Anethum graveolens* | Eurasia |
| Apiaceae | Fennel | *Foeniculum vulgare* | Mediterranean |
| **Seeds** | | | |
| Zingiberaceae | Cardomom | *Elettaria cardamomum* | India |
| Brassicaceae | Mustards: | | |
| | white | *Brassica hirta* | Eurasia |
| | black | *B. nigra* | Eurasia |
| | Indian | *B. juncea* | Eurasia |
| Myristicaceae | Nutmeg, mace | *Myristica fragrans* | Moluccas or Spice Islands |
| Anacardiaceae | Pistachio | *Pistachia vera* | Western Asia |
| Papaveraceae | Poppy | *Papaver somniferum* | Asia |
| Fabaceae | Tonka beans | *Dipteryx* spp. | South America |
| Fabaceae | Fenugreek | *Trigonella foenum-graecum* | Southern Europe, Asia |
| Pedaliaceae | Sesame | *Sesamum indicum* | Tropical Africa |
| **Leaves** | | | |
| Apiaceae | Chervil | *Anthriscus cerefolium* | Southeastern Europe |
| Apiaceae | Parsley | *Petroselinum crippum* | Mediterranean |
| Liliaceae | Chives | *Allium schoenoprasum* | Europe, Asia |
| Lamiaceae | Balm | *Melissa officinalis* | Southern Europe |
| Lamiaceae | Catnip | *Ocimum basilicum* | India, Africa |
| Lamiaceae | Marjoram | *Marjorana hortensis* | Mediterranean |
| Lamiaceae | Peppermint | *Mentha piperita* | Europe |
| Lamiaceae | Sage | *Salvia officinalis* | Mediterranean |
| Lamiaceae | Summer savory | *Satureja hortensis* | Mediterranean |
| Lamiaceae | Winter savory | *S. montana* | Mediterranean |
| Lamiaceae | Spearmint | *Mentha spicata* | Temperate Europe, Asia |
| Lamiaceae | Thyme | *Thymus vulgaris* | Mediterranean |
| Lauraceae | Bay or laurel | *Laurus nobilis* | Asia Minor |
| Asteraceae | Tarragon | *Artemisia dracunculus* | Mediterranean |
| Ericaceae | Wintergreen | *Gaultheria procumbens* | North America |

## Beverages

Water is essential to the maintenance of life. We do not get enough to survive in the food we eat; so we must drink it. At some time it was discovered that water could be varied by the addition of aromatic leaves, flowers, plant sap, crushed fruit, or honey. The results were teas and what we term soft drinks. It seems a natural extension of this activity that alcoholic beverages were discovered because a liquid solution of organic material containing sugars is nearly certain to ferment if left unrefrigerated for a few days.

From the presumably innocent discovery of alcoholic beverages we have developed many types. They have in common an ability to suppress certain functions of the nervous system, causing temporary lack of inhibitions and insensibility. Alcoholic intoxication is our most popular and socially acceptable form of drug use (with the possible exception of nicotine).

Caffeine-containing plants are the base of popular beverages in many cultures and some are summarized in Table 16-8. Tea, the dried leaves of *Camellia sinensis,* is an anthropologically and politically significant plant product. Originally grown in China, its use in the Orient is a vital part of Chinese and Japanese cultures. After Marco Polo brought tea from the Orient, it became popular in Europe. The early nineteenth-century power of the British Empire rested largely on the commerce of tea with China. The arrangement was complex, with Her Majesty's silver bullion paying for tea (and silks and spices) from sailing China traders,

**FIGURE 16-8** The seeds in cocoa pods are the source of chocolate. As shown, the pods on a plant ripen at different times. This picture was taken recently in Ghana. (Photo courtesy of Ingolf Vogeler.)

**TABLE 16-8 Caffeine-containing beverage plants.**

| Common Name | Scientific Name | Country of Origin | Part of Plant Used |
|---|---|---|---|
| Coffees | | | |
| Arabian | *Coffea arabica* | Abyssinia | Roasted berries |
| Congo | *C. robusta* | Congo region of Africa | Roasted berries |
| Liberian | *C. liberica* | West coast of Africa | Roasted berries |
| Tea | *Camellia sinensis* | India, China | Leaves |
| Cocoa and chocolate | *Theobroma cacao* | Lowlands of tropical America | Seeds |
| Maté,[a] yerba maté, or Paraguay tea | *Ilex paraguariensis* | Brazil, Paraguay, Argentina | Leaves |
| Cola | *Cola nitida* | Africa | Seeds |

[a]We North Americans are probably least familiar with maté, which is the universal drink of millions of South Americans.

who used the profits to buy opium to sell to the Chinese, thus making profits on both sides of the voyage. This China trade was drastically altered by the smuggling of Chinese tea seeds for cultivation in India, a British possession at the time. Coffee brewed from the roasted seeds from *Coffea* spp. berries is now the most popular hot drink in the United States. Many of us drink five or more cups per day. Caffeine, an alkaloid found in coffee, tea, and cocoa (Fig. 16-8) acts as a stimulant on the nervous system and directly or indirectly affects the cardiovascular system.

## CLOTHING AND OTHER USES OF PLANT FIBERS

Literary fiction and motion pictures have popularized the image of prehistoric humans protecting their vulnerable bodies with the skins of animals. Beyond this stage is a large gap in popular notion, at the end of which we visualize advanced civilizations, such as those of Sumer, Mesopotamia, and Egypt, in which humans wore complex garments made from woven and dyed fabrics. Fabrics for the wealthy combined plant fibers and precious metals. The dyes were obtained from plant, animal, and earth sources. Plant waxes and resins and animal oils and fats were used for waterproofing some fabrics.

### Cotton

The most widely used fiber for clothing continues to be cotton. Cotton fibers are the elongated epidermal hairs, or **lint**, on the seeds of the cotton plant (*Gossypium hirsutum*). A machine called a cotton gin mechanically tears lint from the seeds. After the longer fibers have been removed, shorter fibers called fuzz or **linters** remain and are removed by another process. Linters are valuable for their relatively pure cellulose content. Processed cellulose is used in many ways, including **rayon** production. Following ginning, oil is extracted from the cotton seeds and the resulting meal is further processed for use as a high-protein livestock food.

### Flax

Before the mechanization of cotton separation and the textile industry as a whole, flax (*Linum usitatissimum*) was the dominant fiber plant. It was woven into **linen** fabric in ancient Egypt as well as prehistoric Europe. Flax continues to be produced in the United States—particularly in the Northwest—and in Europe. Like cotton, the seed is also valuable. Extracted **linseed oil** is used in paints, varnishes, linoleums, and other products. The remaining pressed seed meal is heat detoxified and then used as livestock food.

Separation of flax fibers from the plant is quite different from the process for cotton. Cotton fibers grow independently of other tissues, for they are epidermal hairs. Flax fibers are actually sclerenchyma fibers of the stem or stalk. The time-honored method of obtaining flax fiber persists, although it is now mechanized. First, mature flax plants are pulled, bundled, and allowed to field dry. After drying, the stems are "retted," either by natural dew or in water tanks. During retting the soft tissues are decomposed by microorganisms, but the lignified tissues resist decomposition much longer. Retting may take several weeks (dew retting) or a few days (tank retting). Overretting produces dark, lower-quality fibers due to overdecomposition.

The retted stem bundles are dried; then bent or rolled to crush the brittle, partially decomposed stem core and epidermis. The bundles are then beaten or "scutched" to knock away crushed tissue fragments and shorter fibers.

Finally, the fiber bundle is combed ("hackled") to align the fibers for spinning. The result of the entire process is a hank of long, silky flax fibers that can be spun into thread to weave linen fabrics.

Some varieties of flax are grown mainly for linseed oil. They produce lower-quality fibers, which are used in cigarette papers because they produce a paper that is free of taste or odor-imparting residues. Flax straw may also be used as livestock feed.

### Other Fibers

Plant fibers are used for mats and hats, such rough fabrics as burlap, and for cordage (twine, cords, ropes). **Hemp** (*Cannabis sativa*) was a major crop in the United States as a source of strong fibers for cordage, which explains the wide distribution and persistence of *Cannabis* plants as weeds, especially around old farmsteads. **Sisal** fiber for cordage is extracted from an *Agave* plant (Fig. 16-9). Selected examples of economic uses of plant fibers are summarized in Table 16-9.

**Wood fibers** are a major economic product, as lumber and for processing into pulp for paper and other commodities (Chapter 8).

FIGURE 16-9
Sisal fibers for twine and other types of cordage are obtained from the leaves of certain *Agave* species. Here workers in Yucatan, Mexico, strip the fibers by hand. (Photo courtesy of Ingolf Vogeler.)

**TABLE 16-9 Some economically important fibrous materials from plants.**

| Uses and Common Names | Plant Sources |
|---|---|
| **Brushes and brooms** | |
| Piassava for brushes | Leafstalks of various palms |
| Broomcorn for brooms | Inflorescence of *Sorghum vulgare* |
| Broomroot for brushes | Root systems of *Muhlenbergia macroura* |
| Coir or coconut fiber for ropes, mats, baskets brushes, and doormats | Husks of the coconut palm, *Cocos nucifera* |
| **Plaiting and basketweaving** | |
| *For straw hats* | |
| Grain stalks | Stems of various grains, such as wheat, rice, barley |
| Toquilla for Panama hats | Leaf veins of *Carludovica palmata* |
| Hat palm for Puerto Rican hats | Leaves of *Sabal causiarum* |
| *For mats and matting* | |
| Stalks and leaves | Various sedges (*Cyperus*), rushes (*Juncus*), and grasses |
| Screw pine | Leaves of *Pandanus* spp. |
| *Baskets* | |
| Sweet grass | Entire plant of *Hierchlöe odorata* |
| Raffia | Strips of lower leaf epidermis of *Raffia pedunculata* |
| *Wickerwork* | |
| Rattan | Stems of *Calamus* spp. |
| Bamboos | Split stems of various bamboo genera |
| Willow | Stems of *Salix* spp. |
| **Fillings and stuffings** | |
| *Soft buoyant fillings, as used in life jackets, stuffed toys* | |
| Kapok | Pod floss of *Ceiba pentandra* |
| Milkweed floss | Pod floss of *Asclepias syriaca* |
| Cattail fluff | Fluff attached to seeds in mature seed heads of *Typha* spp. |

TABLE 16-9 (Continued).

| | |
|---|---|
| *Coarser materials for stuffing, filling, and packing* | |
| Straw, corn husks | Straw of cereals and other grasses, corn husks |
| Spanish moss | Filamentous bodies of this unusual bromeliad, *Tillandsia usneoides* |
| Crin vegetal | Leaves of *Chamaerops humilis* |
| Excelsior | Shredded wood of various trees |
| **Textiles and cordage** | |
| Cotton | Hair on seeds of *Gossypium* spp. |
| Flax for linen | Fibers from stems of *Linum usitatissimum* |
| True hemp for ropes, twines, carpet backing, sacks, bags | Fibers from *Cannabis sativa* |
| Jute for burlap bags, rough weaving | Fiber from secondary phloem of *Corchorus* spp. |
| Abaca or Manila hemp for high-grade cordage, twine | Mainly stalks of *Musa textilus* |
| Mexican sisal or henequen for twine, ropes, cords | Leaf veins of *Agave fourcroydes* |
| Bowstring hemp for cordage | Leaf veins of *Sansevieria* spp. |
| **Miscellaneous** | |
| Bark cloths | Pounded bark of various plants in tropics |
| Luffa or vegetable sponge for washing and scouring | Fibrous skeleton of the fruit (gourd) of *Luffa* spp. |

### Dyes and Sizings

Vegetable and animal dyes and earth substances were predominant prior to the discovery of synthetic indigo and aniline dyes in the 1800s. Nearly all plants will yield dyes and so a complete list would be endless. Some plants, however, produce dyes that are longer lasting or **fast**, withstanding washing and sun bleaching. Dye fastness is enhanced markedly by the use of a **mordant**, usually a metallic salt. The mordant combines with the dye and carries it into the fiber shaft. In addition to causing greater dye impregnation, a mordant can substantially alter the color produced by a particular dye on a particular fiber. See Chapter 17 for a partial list of natural dye plants; also the Suggested Readings list for Chapter 17.

In textile manufacturing, **sizing** is added to cotton and linen fabrics before they are dried and rolled onto bolts. Sizing gives the fabric fullness, or body, by filling in the spaces in the weave. The most common sizing is starch, but fats, wax, paraffin, and such mineral substances as chalk are also used, among other things. Plain linen and cotton fabrics, such as percales, organdy, voiles, and buckram, owe their prewashing crispness and smoothness to the presence of sizing. Once the sizing has been laundered out, the fabric is limper, softer, and wrinkled. The original crisp smoothness must be restored by starching after each laundering. Special "wash and wear" treatments reduce loss of body, as does combining natural fibers with nonwrinkling, nonshrinking synthetic fibers, such as polyesters.

## SHELTER AND OTHER USES OF WOOD

Many forms of human shelter have been provided by plants (Fig. 16-10), from a thornbrush hut with a grassy thatched roof to an elegantly paneled and adorned mansion. Even adobe brick homes have straw worked into the mud to reinforce and strengthen the finished, dried bricks. Under survival conditions in the forest a lean-to of saplings and conifer boughs, together with an insulating bed of boughs, can be lifesaving.

Most homes in North America are constructed primarily of lumber and lumber products even though they may be faced on the outside with masonry. Wood is strong, resilient, and durable (Fig. 16-11). In addition to lumber, plywood, shakes, and shingles that are immediately recognized as timber products, wood contributes to construction in many other ways. Weathersheathing, wood preservatives, floor coverings, adhesives, solvents, exterior and interior finishes all contain wood derivatives.

**FIGURE 16-10** Home construction supplies in Niger consist of support poles and prefabricated walls and roofs. As with most cultures worldwide, the basic raw materials for human shelter are of plant origin. (Photo courtesy of Ingolf Vogeler.)

**FIGURE 16-11**
Wooden logs have long been an important and durable source of shelter, often outlasting by generations the people who harvested them.

"Knock on xylem" for good luck or to prevent a jinx? Wood is xylem tissue in the center of woody plant stems and roots. Different tree species produce woods having individual characteristics, such as hardness, grain, patterns, and color. Wood features are determined by the kinds and density of xylem cells, arrangement of the cells, other organic substances present, and the plane of cutting (Chapter 8).

Woods are commercially classified as "softwoods" and "hardwoods," although these terms do not define actual comparative wood softness or hardness, as noted in Chapter 8. The term softwood refers to coniferous trees, such as pines (*Pinus*), cedars (*Thuja, Libocedrus, Chamaecyparis*), spruces (*Picea*), firs (*Abies*), Douglas fir (*Pseudotsuga*), hemlocks (*Tsuga*), redwood (*Sequoia*), and cypress (*Cupressus*). Hardwoods (some of which actually have quite soft wood!) are such angiosperms as oaks (*Quercus*), maples (*Acer*), ashes (*Fraxinus*), hickories (*Carya*), walnuts (*Juglans*), beeches (*Fagus*), basswood (*Tilia*), and elms (*Ulmus*).

Due to its hardness, durability, and grain, oak is popular for desks, doors, and cabinets, such as those you see in your classrooms and laboratories. Look about you. The large vessels of oak are easily seen in both longitudinal section (the large flat surfaces of desks, doors, and cabinets) and in cross section (end pieces of trim, etc.). The attractive grain pattern of oak is due to the production of much larger vessels early in the growth season in contrast to abundant fibers in the late-season wood.

Many woods are especially prized for their distinctive grain patterns and the value of old homes and antique furniture is enhanced by the type and condition of the woods used to construct them. In addition to the normal wood grain, special parts of a tree may yield unusual grain patterns, called figures. Some examples are ribbon, wavy, curly or fiddleback, blister or quilted, and bird's-eye. In Victorian-era furniture sections of walnut burls were popular as decorative inserts. The crotch wood of a tree branch also yields a distinctive, wavy pattern.

Most lumber is cut from heartwood. Heartwood color frequently varies from black to pinkish or yellowish white. The outer layer of living sapwood is usually light. Furniture constructed of attractive (and more expensive) naturally colored woods (black walnut, cherry, mahogany, and ebony) can be simulated with less expensive woods by staining. The luster—ability of the wood grain to reflect light—also varies, as does the wood's ability to take a natural polish. Modern varnishes and shellacs are now the most common way of obtaining a polished surface.

Knowledgeable antique hunters develop an ability to recognize woods not only by the dealer's or auctioneer's description but also by closely examining the natural color, grain patterns, and hardness of wood.

Lumber species are continually tested for specific structural characteristics as well as appearance, notably at the U.S. Department of Agriculture Forest Products Laboratory in Madison, Wisconsin. Tests include measurements of weathering, decay resistance, thermal conductivity and expansion, electrical properties, effects of chemicals on the woods, weight, working qualities, strength in resisting such forces as tension (stretching), crushing (endwise compression), crossbreaking, shearing, and the shock of blows.

The intrinsic odor of a wood is due to volatile oils, such as oils of sandalwood and cedar. Aromatic cedars are used to line chests and closets because their aroma is fresh, pleasing, and toxic to clothing moth larvae. Not all woods have a pleasant odor, however. Woods usually have odors when first cut, but most odors dissipate as the wood dries or cures. Some odors may be due to decomposition, such as the vinegary (acetic acid) smell of curing oak wood.

Other wood uses are as crates, barrels, posts, and pilings; laminated beams, plywood, and veneer paneling; fuelwood; sports equipment; toothpicks; matchsticks; and excelsior. In addition, wood is converted into products seemingly much removed from the standing tree in the forest, including pulp and paper, rayon and acetate fabrics, and modified woods, such as compressed wood used for airplane propellers.

Wood waste—including sawdust, chips, and wastage from pulp mills—that was formerly burned or otherwise discarded is now used. Some is combined with resins to produce flakeboard and particleboard. Microbes and chemicals are used to extract a wide variety of organic acids and solvents (including ethyl, methyl, and butyl alcohols and acetone), although many substances still remain unrecovered. Waste pulp is sometimes used as a liner for landfill disposal sites. Refined cellulose from wood has been converted into flour for baked goods. Cellulose is indigestible to humans and provides necessary bulk to our diets; so purified wood flour is an effective filler for reducing calorie content of flour products. Breads containing "wood flour" are aptly advertised as "high fiber, low calorie."

Distillation of pine wood produces turpentine and rosin. Charcoal, acetic acid, and methanol (wood alcohol) are derived from hardwoods. Christmas tree growing is a profitable forest industry, and true firs, pines, spruces, and Douglas fir are most commonly cultivated.

Wood remains one of the most important fuels in the world except in the industrialized nations, which rely heavily on the fossil remains of prehistoric plants (peat, coal, petroleum) for fuels. Wood is the only fuel for many peoples and the quest for wood to build fires for cooking and heating has caused disastrous deforestation and ecosystem destruction in the Middle East and North Africa. In response to threatened fossil fuel shortages, many North Americans have returned to wood as at least a secondary fuel for heating (and cooking). Table 16-10(a) and (b) compares the fuel value of some common North American woods.

TABLE 16-10 (a) **Approximate weight, moisture content, and available heat units of selected woods, green and air dry.**[a]

| Species | Weight (Pounds)[b] | | Available Heat Units Per Cord in Million BTUs[d] | | Equivalent Gallons Per Cord of Wood[e] | |
|---|---|---|---|---|---|---|
| | Green[a] | Air Dry[c] | Green | Air Dry | Fuel Oil | L P Gas[f] |
| Ash | 3840 | 3440 | 16.5 | 20.0 | 120 | 167 |
| Aspen | 3440 | 2160 | 10.3 | 12.5 | 75 | 104 |
| Beech | 4320 | 3760 | 17.3 | 21.8 | 130 | 182 |
| Paper birch | 3800 | 3040 | 16.7 | 18.2 | 108 | 152 |
| Yellow birch | 4560 | 3680 | 17.3 | 21.3 | 126 | 177 |
| Elm | 4320 | 2900 | 14.3 | 17.2 | 102 | 143 |
| Hickory | 5040 | 4240 | 20.7 | 24.6 | 146 | 205 |
| Soft maple | 4000 | 3200 | 15.0 | 18.6 | 110 | 155 |
| Hard maple | 4480 | 3680 | 18.4 | 21.3 | 126 | 177 |
| Red oak | 5120 | 3680 | 17.9 | 21.3 | 126 | 177 |
| White oak | 5040 | 3920 | 19.2 | 22.7 | 135 | 188 |
| White pine | 2880 | 2080 | 12.1 | 13.3 | 80 | 112 |

[a]Green wood (wood that has just been cut and split and is burned in the same season) has a moisture content of approximately 60%.

[b]Per standard cord (4 by 4 by 8 ft) containing 80 cu ft of solid wood.

[c]Wood that has air dried for about a year after splitting has an acceptable moisture content of about 20%.

[d]BTU's are British thermal units. One BTU is the amount of heat needed to raise the temperature of 1 lb of water 1°F.

[e]Using air-dry wood at 50% heating unit efficiency, oil 60% and L P gas 65%.

[f]L P gas is liquid petroleum or "bottled" gas.

Adapted from *Wood for Home Heating: Wood as Fuel,* University of Wisconsin-Extension Publication G2874, January 1978, Madison, Wisconsin.

TABLE 16-10 (b) **Rating of firewood heat content, based on a scale in which hickory = 100.**[a]

| Wood | Rating | Wood | Rating | Wood | Rating |
|---|---|---|---|---|---|
| Hickory | 100 | Mulberry | 84–85 | Spruce | 59 |
| Black locust | 101 | Apple | 83–84 | Hemlock | 57 |
| Ironwood | 101 | Ash | 81–82 | Cottonwood | 54–55 |
| Oak | 96–99 | Elm | 71–80 | Balsam fir | 51–54 |
| Beech | 93–96 | Walnut | 74 | Aspen | 53 |
| Hard maple | 83–88 | Soft maple | 67–73 | Basswood | 53 |
| Birch | 79–86 | Tamarack | 70–72 | White pine | 50 |
| | | Cherry | 70–71 | | |

[a]The table assumes that the woods are air dried to a moisture content of approximately 20%.

Table based on U.S. Forest Service Products Laboratory data, Madison, Wisconsin.

**Cork** is a nonwood forest product, being the outer bark of the cork oak (*Quercus suber*). By careful harvesting, cork can be removed from the trees without killing them. Therefore cork production in the Mediterranean region has continued for many centuries and some of the trees are hundreds of years old. Cork is used for stoppers, floats, nonslip treads, and other items, but mainly in insulating board and linoleum.

## MEDICINES

Viruses, bacteria, and fungi cause diseases to which humans, their crops, and domestic animals are susceptible. Consequently, they have been a source of great misery over the centuries. On the other hand, medical research on disease syndromes and life cycles of these organisms has led to constructive revelations about the human body. In addition, discovery of natural antagonisms between specific bacteria and fungi led to the development of antibiotics, which we now take for granted. Even alcohols used to sterilize medical paraphernalia are derived from the metabolic activities of microorganisms.

Since long before the development of modern drugs and medical practices, plants have provided cure and comfort for human ailments (Fig. 16-12). Medicine men, witches, and herbalists of ancient Asia, Europe, and America intermingled medicine and magic. Their repertoires included poultices and extracts for comfort and cure, potions for blessing, cursing, and casting spells of love, fertility, invincibility, success, and wealth.

One of the most ancient medicinal plants is ginseng (*Panax quinquefolius*) [Fig. 16-13 (a) and (b)]. The genus name is derived from the Greek word *panakas,* "a panacea." Ginseng has been used for centuries in oriental cultures to treat stomach, lung, and nerve disorders, aid in recuperation, and soothe the throat. It is also believed to prolong life and sexual potency. The common name is derived from a Chinese word meaning manroot and the most-prized ginseng roots are those that branch to simulate a torso with legs and arms.

Ages ago medicinal plants were assigned **signatures.** By the "doctrine of signatures" some sign or feature of the plant was interpreted as the clue to its relationship to the human body and hence its therapeutic potential. Astrology was also involved. Table 16-11 gives some examples of plant signatures. The emphasis on signatures and herbal medicine was greatest prior to the rapid development of the sciences from the eighteenth century on, but herbal medicine is still practiced in all parts of the world.

**FIGURE 16-12** A market stall in Kumasi, Ghana, displays gourd bowls, as well as a variety of seeds, nuts, roots, rhizomes, and other plant parts to be used for culinary and medicinal purposes. (Photo courtesy of Ingolf Vogeler.)

As the scientific study of human anatomy and physiology superceded religious taboos and superstitions, some of the mysticism associated with medicine disappeared. Diseases were increasingly shown to be the result of such natural causes as nutritional deficiencies, plant toxins, germs, and worms instead of bad vapours, humours, spiritual possession, obsession, and godly retribution.

Yet many medicinal plants have stood the test of time. Even without the magic, they still work and historical remedies have been the source of many modern remedies. Once again, you should browse in a drugstore and read labels to confirm this proposition! Traditional medicinal plant uses are as astringents, demulcents, aphrodisiacs, and

(a)

(b)

FIGURE 16-13
Ginseng roots are now grown as a cash crop in northern Wisconsin, taking several years to grow from seed to maturity. (a) Lath sheds provide needed shade for this forest plant. (b) The prime medicinal plant of the Orient is grown here in a ginseng shed, surrounded by straw mulch. (Photos by Michael W. Hannon.)

**TABLE 16-11 Some examples of plant signatures.**

| Plant Characteristics | "Signature" Uses | Some Examples |
|---|---|---|
| Grows in water and wet lowlands | For "wetness" diseases, such as colds, coughs, rheumatism | Willow, water pepper, mints, verbena, sweet flag, elder, boneset |
| Grows in mucky soil | For mucous excretions | Larch or tamarack |
| Grows in banks of swift-flowing streams and clear ponds | To promote urination and cleanse impurities from urinary system | Horsetails, bedstraw, some mints, smartweed, black elder, water agrimony, hydrangea |
| Grows in gravelly, rocky, sandy places | To remove stone-forming and membrane-irritating deposits from alimentary canal and bronchial system | Bearberry, horsetails, parsley, shepherd's purse, juniper, mayflower, sassafras, saxifrage |
| Convoluted shape | For the brain | Walnut |
| Liverlike shape | For the liver | Liverwort |
| Kidney shape | For the kidneys | Kidney bean |
| Manlike shape | For many uses, including to promote longevity and potency | Ginseng root |
| Yellow color of flowers, fruit, stem, or root decoction | For jaundice, to treat liver and gall bladder disorders | Dandelion |
| Red color of flowers, fruit, stem, or root decoction | For the blood | Rhubarb |

antibiotics; to induce sweating, vomiting, or bowel action; to promote urination; to reduce swelling, agitation, or pain; to stimulate the heart; to heighten sensory sensitivity; to cause visions or delirium; and as poisons. Some plants reduce hemorrhage caused by wounds and childbirth. Literature—old and new—on the subject of folk and herbal medicines is abundant and should be approached with the proverbial "grain of salt." Many specific medicinal plants are discussed in Chapter 18.

The relationship of plant chemicals to human physiology is such that a given substance may have no effect in small amounts, may be medicinal in moderate amounts, and may be poisonous or even fatal in larger amounts. Such a continuum makes it artificial to treat medicinal plants and poisonous plants separately. Most medicinal substances are toxic in sufficient quantities. Old medical books make wonderful reading to anyone fascinated by the history of medicinal plants.

Many popularly recommended and consumed edible wild (and domesticated!) plants are also known to be potentially hazardous in large quantities. Numerous plants now used primarily as seasonings, flavorings, and aromatics have medicinal properties. The spice trade of earlier centuries was based primarily on the procurement of drug plants to provide relief from human disorders and discomforts (Chapter 18).

## OTHER EXUDATES AND EXTRACTIVES

### Sugars and Starches

Sugarcane (*Saccharum officinarum*) stems and sugar beet roots are major commercial sugar sources. Sugar canes are crushed to obtain a sweet juice from which vegetative particles and water are removed. The syrup is further extracted, producing crystallized "raw" sugar that can be partially or fully refined. Brown sugar contains a variety of sugars and other organic substances whereas white (refined) sugar is nearly pure sucrose. Shredded sugar beets are washed and squeezed to dissolve out the sucrose. Many tree saps (e.g., maple and birch) can be made into syrup or fermented into beer or vinegar (Chapter 17). Sap syrups and sugars are manufactured by collecting the sugar-rich spring sap and then evaporating off most of the water by careful cooking.

Starches are usually obtained from such food storage organs as roots, tubers, grains, or seeds. Grains are the most abundant source. Whole grain flour is made by grinding the entire fruit or grain. Refined white flour is made from grains whose outer husk (bran) and embryo (germ) have been removed. Larger cell wall fragments are removed from refined flour and microscopic analysis reveals it to be composed almost entirely of starch grains—and various proteins, vitamins, etc., that can't be seen. Because bran is an important source of bulk in human diets and many proteins and vitamins are found in the wheat embryo, refined white flour has been nutritionally criticized even though it is usually "improved" or "enriched" by secondary addition of vitamins during processing. Some starches (such as wheat) produce a milky sauce or gravy when cooked with liquid whereas others (such as corn) produce clear sauces.

### Volatile or "Essential" Oils

The aroma of a particular flower or spice is due to evaporation (volatilization) of fragrant oil molecules. Aromatic spices and herbs for cooking contain volatile oils that disperse into the air; so spices should be stored in airtight containers in a dark, cool place (not over the stove or refrigerator) to minimize oil loss and chemical alteration. Many cooks prefer to store whole rather than powdered spices; then grind them as needed. Some oils are lost during preparation of spice powders and remaining oils are lost at a faster rate than by whole spices, thus substantially reducing shelf life. Volatile oils also have antiseptic (antibacterial) properties. The earliest recorded use of spices in food was by the Babylonians. It is possible that the use was to retard food spoilage as well as for flavor.

Some oils are commercially extracted to make flavor extracts, perfumes, and medicinal preparations, including the oils of lemon, orange, citronella, cinnamon, sandalwood, and many flowers. Some oils are extracted by cold methods, minimizing oil degeneration. In contrast, mint oil is heat extracted in large stills. Even with good quality peppermint (*Mentha piperata*) many pounds of the freshly cut herb are needed to produce a single ounce of peppermint oil.

Volatile oils are the basic ingredient of the perfume industry. They are blended with other substances to produce interesting combinations; then diluted to different degrees with other oils, water, and/or alcohol. A "perfume" or "parfum" is the most concentrated form of the finished product. "Eau de cologne" or simply "cologne" and

**TABLE 16-12** Some economically important essential, or volatile, oils.

| Name of Oil | Scientific Name | Part of Plant Used and Scent |
|---|---|---|
| **Some volatile oils used in perfumes, sachets, powders, pomades, lotions, soaps, etc.** | | |
| Rose oil, attar of roses | *Rosa damascena, R. centifolia* | Petals; rose |
| Geranium oil | *Pelargonium* | Leaves; geranium |
| Ylang-Ylang or cananga oil | *Cananga odorata* | Petals; exotic floral |
| Cassie or acacia | *Acacia farniesiana* | Petals; violetlike |
| Neroli oil | *Citrus* spp. | Blossoms; orange blossom |
| Oil of orange | *Citrus* spp. | Ripe peel; pungent orange |
| Petit-grain oil | *Citrus* spp. | Leaves, twigs; orange |
| Bergamot | *Citrus aurantium* | Rind; sweet orange |
| Orris | *Iris* spp. | Rhizome; violetlike |
| Calamus | *Acorus calamus* | Distinctive sweet odor |
| Grass oils | | |
| oil of citronella | *Cymbopogon nardus* | Leaves; strong citruslike |
| lemon-grass oil | *C. citrates* | Leaves; lemony |
| palmarosa and gingergrass oils | *C. martimii* | Distinctive fresh odor |
| oil of vetiver | *Vetiver ziganioides* | Roots rhizomes; distinctive odor |
| Oil of bay | *Pimenta racemosa* | Leaves; spicy, masculine odor |
| Lavendar | *Lavendula officinalis* | Dried flowers; lavendar |
| Violet | *Viola odorata* | Flowers; violet |
| Jasmine | *Jasminum officinarum* | Flowers; jasmine |
| Carnation | *Dianthus caryophyllus* | Flowers; carnation |
| Rosemary | *Rosmarinus officinalis* | Leaves, flowering tops; rosemary |
| Hyacinth | *Hyacinthus orientalis* | Flowers; hyacinth |
| Oak moss (lichens) | *Ramalina calicaris, Evernia furfuracea, E. prunastri* | Thallus; element in cosmetics, lavendar perfumes, soaps |
| Linaloe or ois de Rose | *Bursera penicillata, B. glabrifolia, Aniba panurensis* | Wood chips |
| Sandalwood | *Santalum album* | Wood; sandalwood |
| Patchouli | *Pogostemon cablin* | Leaves; patchouli |
| Campaca | *Michelia champaca* | Flowers |
| Heliotrope | *Heliotropium arborescens* | Flowers; heliotrope |
| Lily-of-the-valley | *Convallaria majalis* | Flowers; lily-of-the-valley |
| Jonquil | *Narcissus jonquilla* | Flowers; daffodillike |
| Mignonette | *Reseda odorata* | Flowers; mignonette |
| Clary sage | *Salvia sclarea* | Flowers |
| Tuberose | *Polianthes tuberosa* | Flowers; tuberose |
| Honeysuckle | *Lonicera caprifolium* | Flowers; honeysuckle |
| **Spice extracts also used in perfumes, soaps, etc.** | | |
| Anise | *Pimpinella anisum* | Fruits; anise |
| Caraway | *Carum carvi* | Fruits; caraway |
| Cassia | *Cinnamomum cassia* | Bark; cassia |
| Cinnamon | *Cinnamomum zeylanicum* | Bark; cinnamon |
| Clove | *Eugenium caryophyllata* | Unopened flower buds; cloves |
| Lemon | *Citrus limon* | Rinds; lemon |
| Peppermint | *Mentha piperita* | Leaves; peppermint |
| Thyme | *Thymus vulgaris* | Leaves; thyme |
| Wintergreen | *Gaultheria procumbens* | Leaves; wintergreen |
| Zeodoary | *Curcuma zedoaria* | Rhizomes |
| **Other volatile oils from woods** | | |
| Camphor | *Cinnamomum camphora* | Wood; camphor |
| Cedarwood oil | *Juniperus virginiana* | Wood; cedarlike |
| Eucalyptol | *Eucalyptus dives* | Wood; eucalyptus |
| Oil of turpentine | Various conifers | Wood; used as a solvent |

TABLE 16-13 Some wax-producing plants.

| Family Name | Type of Wax | Scientific Name | Source of Wax and Uses |
|---|---|---|---|
| Arecaceae | Carnauba was | *Copernicia cerifera* | Leaf epidermis; candles, soap, varnish, paints, carbon paper, batteries, sound films, insulation, salves, ointment, car wax |
| Euphorbiaceae | Candelilla wax | *Euphorbia antisyphilitica* | Film on stems; extender wax in other mixtures, cosmetics |
| Myricaceae | Myrtle wax | *Myrica pennsylvanica, M. cerifera* | Berries; candles and soaps |
| Marantaceae | Cauassu wax | *Calathea lutea* | Underside of large leaves; commercial wax |
| Buxaceae | Jojoba wax | *Simmondsia chinensis* | Seeds, which have 50% liquid wax content; waxing products, polishes, candles |

"Eau de toilet" or "toilet water" are diluted versions meant to be splashed on more freely during one's "toilette" (bathing, grooming, and dressing). In addition to making scented waters, volatile oils add fragrance to everything from bath soap, hand lotion, and toothpaste to laundry detergents.

Several volatile oils have medicinal uses. Menthol, eucalyptol, and camphor are primary ingredients in products designed to counteract nasal stuffiness and clove oil is a time-honored treatment for toothache. Table 16-12 lists some volatile oils. Paradoxically, the *irritant* quality of volatile oils is the essence of their medicinal and antiseptic action.

## Waxes

Mechanized carwashes offer an optional "hot wax" misting with carnauba wax. Carnauba wax is a fine quality, hard wax that is also used in floor and furniture polishes, carbon paper, phonograph records, and other products. It is produced on the leaves of a desert palm (*Copernicia cerifera*) to protect leaves from drying in hot desert winds. Many commercial wax preparations, from mascara and lipstick to chocolate candy and hair cream, are blends of plant and animal (particularly beeswax) waxes. A list of selected wax-producing plants appears in Table 16-13.

## Latex, Resins, and Gums

Nearly everyone has observed **latex**. The broken stem of a milkweed or dandelion oozes thick, white juice that hardens into a dark, sticky mass. Other latexes are the raw ingredients from which chewing gum and rubber are made. It is harvested commercially by systematically gashing stems of latex-producing plants; then collecting the latex that is exuded (Fig. 16-14). Because of its source —chicle—chewing "gum" more correctly should be called chewing latex!

**FIGURE 16-14**
A rubber tapper in Liberia makes an incision in a rubber tree. The latex that flows down the incision and spout is collected in the plastic cup attached to the tree. Tapping requires skill so that the successive 1-2 mm bark slices are removed without damage to the underlying cambium. In this way plantations of *Hevea brasiliensis* (a native of the Amazon Basin in South America) remain productive for many years. Although plantation treatment of the collected latex is now very sophisticated, a traditional method used by Amazonians collecting from wild trees is to coagulate the latex over a smoky fire. Small amounts of latex are poured onto the ever-increasing mass that forms around a pole rotated over the fire. Rafts of smoked-ball crude rubber can then be floated downriver to buyers. (Photo courtesy of The Firestone Tire and Rubber Company, Akron, Ohio.)

Coniferous trees are the main commercial source of **resins** or pitch, complex substances that contain oils and that are insoluble in water. Many plants produce **gums,** which are water soluble and mainly carbohydrate in composition.

Tables 16-14 and 16-15 summarize commercially important latex, resin, and gum-producing species.

## Other Extractives

Why is the process of turning raw animal hides into leather called "tanning"? Scraped hides are cured with complex substances called **tannins** obtained from plants (Table 16-16). Tannins precipitate proteins, enhancing leather pliability and toughness and retarding microbial growth. Dif-

**TABLE 16-14** **Some latex and gum products.**

| Family Name | Name of Product | Scientific Name | Country of Origin |
|---|---|---|---|
| **Rubber: derived from caoutchouc in the latex** | | | |
| Euphorbiaceae | Hevea or para rubber | *Hevea brasiliensis* | Amazon region of South America |
| Euphorbiaceae | Ceara or manicoba rubber | *Manihot glaziovii* | Brazil |
| Moraceae | Panama or Castilla rubber | *Castilla elastica* | Southern Mexico to South America |
| Moraceae | Assam or India rubber | *Ficus elastica* | Tropical Asia |
| Asteraceae | Guayule rubber | *Parthenium argentatum* | Southwest United States to Mexico |
| Asteraceae | Dandelion rubber | *Taraxacum koksaghyz* | Turkestan |
| **Nonelastic rubber: used for underwater cable insulation** | | | |
| Sapotaceae | Gutta-percha | *Palaquium gutta* | India to Malaya |
| **Other latex products: chicle, used as the base for chewing gum** | | | |
| Sapotaceae | Chicle | *Achras zapota* | Southern Mexico, Guatemala, Honduras |
| **Gums: breakdown products of internal tissues: water soluble, colloidal, and sugar substances** | | | |
| Fabaceae | Gum arabic | *Acacia senegal* | Northern Africa |
| Fabaceae | Gum tragacanth | *Astragalus* spp. | Iran |
| Sterculiaceae | Karaya gum | *Sterculia urens* | India |

**TABLE 16-15** **Some valuable plant resins.**[a]

| Family Name | Name of Product | Scientific Name | Uses |
|---|---|---|---|
| **Hard resins: contain little if any oil** | | | |
| Fabaceae | Copal | *Trachylobium, Copaifera* | Outdoor varnishes |
| Dipterocarpaceae | Damars | *Balanocarpus, Hopea,* and others | Spirit varnishes, nitrocellulose lacquers |
| Burseraceae | Damars | *Shorea* spp. | Pipe mouthpieces, cigar and cigarette holders, varnishes |
| Pinaceae | Amber (fossil resin) | *Pinus succinifer* | Jewelry, ornaments |
| Anacardiaceae | Mastic | *Pistacia lentiscus* | Varnish, cement in dental work; used in lithography, perfumery, medicine |
| Anacardiaceae | Lacquer | *Rhus verniciflua* | Natural varnish, furniture, decorative wood items |
| Fabaceae | Shellac[b] | *Butea monosperma* | Spirit varnish, phonograph records, inks, glazes |
| Sapindaceae | Shellac | *Schleichera oleosa* | |
| Rhamnaceae | Shellac | *Zizyphus xylopyrus* | |
| Rhamnaceae | Shellac | *Zizyphus jujuba* | |
| Moraceae | Shellac | *Ficus religiosa* | |
| Fabaceae | Shellac | *Acacia nilotica* | |
| Fabaceae | Shellac | *Cajanus cajan* | |
| **Oleoresins: contain oil and are therefore more liquid** | | | |
| Pinaceae | Spirits and rosin | *Pinus* spp. | Paint and varnish thinners, solvents, greases, lubricants |
| Pinaceae | Canada balsam | *Abies balsamea* | Irritant, stimulant, antiseptic; soap, perfume |
| Pinaceae | Spruce gum | *Picea* spp. | Formerly chewed as a gum |

TABLE 16-15 (Continued).

| Family Name | Name of Product | Scientific Name | Uses |
|---|---|---|---|
| **Balsams: contain benzoin or cinnamic acid and are therefore aromatic** | | | |
| Fabaceae | Balsam of Peru | *Myroxylon pereirae* | Cough, syrup, perfumes |
| Fabaceae | Balsam of Polu | *Myroxylon balsamum* | Salves, ointments, cough syrups, perfumes, soap |
| Hamamelidaceae | Storax or styrax | *Liquidambar orientalis, L. styraciflua* | Soap, cosmetics, perfumes, adhesives, lacquers, incense, flavored tobacco, medicines |
| Styraceae | Benzoin | *Styrax* spp. | Similar uses to preceding ones |
| **Gum resins: mixtures of gums and resins** | | | |
| Apiaceae | Asafetida | *Ferula foetida* | Flavorings, drugs, medicines |
| Burseraceae | Myrrh | *Commiphora myrrha, C. erythraea* | Perfumes, tonic, stimulant, antiseptic, incense |
| Burseraceae | Frankincense | *Boswellia carteri, B. frereana* | Incense, perfumes, face powders, pastilles, fumigating powders, as a fixative |

[a]Resins are the oxidation products of essential oils, having complex and variable compositions. They are soluble in such organic solvents as alcohol or ether, but not in water, and have an antiseptic quality.
[b]Shellac is made from the exudate of a scale insect (*Laccifer*) that feeds on these plants.

TABLE 16-16 Some commercial sources of tannins.

| Family Name | Common Name | Scientific Name | Uses in Tanning and Other |
|---|---|---|---|
| **Derived from barks** | | | |
| Pinaceae | Hemlocks* | *Tsuga canadense, T. heterophylla* | Sheepskins, sole leathers, and other heavy leathers |
| Fabaceae | Wattle | *Acacia* spp. | Sole leathers |
| Fagaceae | Oaks* | *Quercus* spp. | Heavy leathers |
| Rhizophoraceae | Mangrove | *Rhizophora mangle* | Various leathers |
| **Derived from woods** | | | |
| Fagaceae | Chestnuts* | *Castanea dentata, C. sativa* | Heavy leathers |
| Anacardiaceae | Quebracho | *Schinopsis* spp. | Various, especially sole leather |
| **Derived from leaves** | | | |
| Anacardiaceae | Sumacs* | *Rhus* spp. | Various |
| Rubiaceae | Gambier or white cutch | *Uncaria gambir* | Leathers; also used as a dyestuff, masticatory, medicine |
| **Derived from fruits** | | | |
| Combretaceae | Myrobalan (from nuts) | *Terminalia chebulia, T. bellerica* | Calf, goat, and sheepskin, sole and harness leather |
| Fabaceae | Divi-divi (from seed pods) | *Caesalpinia coriaria* | Various |
| Fabaceae | Tara (from fruits) | *Caesalpinia spinosa* | High-grade leather; also inks, dyes |
| Fabaceae | Algarobilla (from pods) | *Caesalpinia brevifolia* | Various |
| Fagaceae | Valonia (from acorn cups) | *Quercus macrolepis* | Fine leathers |
| **Derived from roots** | | | |
| Polygonaceae | Canaigre or tanners's dock (wild rhubarb)* | *Rumex hymenosepalus* | Firm, heavy leather; yields a bright orange |
| Arecaceae | Palmetto* | *Sabal palmetto* | Various |

Those indicated by an asterisk are native to North America.

ferent tannins give different shades. In general, they are brownish so that leathers tend to be browned or "tanned" in appearance. (Think of that the next time you are "sun tanned"!) Tannins in coffee and tea also cause brown stains on teeth and cups. Tannins have been used as antidotes in alkaloid poisoning due to their ability to inactivate alkaloids by forming insoluble tannates.

Some plants produce natural toxins (Table 16-17). Unlike environmentally "persistent" pesticides containing chlorinated hydrocarbons as their active ingredients (DDT, chlordane, lindane), biological insecticides are biodegradable (readily broken down in the environment). As a result of congressional action to control persistent pesticides, general insecticides now usually list **pyrethrums** as the major active ingredient. Pyrethrum daisies (*Chrysanthemum* spp.) were the original source of the substance.

**Rotenone** is one of the best known fish poisons. It is used by tribes of the Amazon jungles and wildlife biology specialists to stupefy fish, causing them to float to the surface where they can be captured. It is also used to control a variety of soil insects and external parasites of mammals.

Smoldering **nicotine** pellets are a standard greenhouse fumigant, killing most types of greenhouse pests. The concentrated fumes are also very dangerous to larger animals.

There are several methods of biological control of garden pests on plants. Companion planting involves planting a crop that repels problem animals. Natural sprays of macerated garlic, pepper, or other plants containing irritating oils are effective against some insects and their larvae.

At least two species have been used commercially to extract **vermifuges** to dispel intestinal worms, especially roundworms—the Old World wormwood (*Artemisia absynthium*) and wormseed (*Chenopodium ambrosioides*).

The **herbicide** 2,4-D (2,4-dichlorophenoxyacetic acid and related compounds) was discovered during plant hormone research. It is a synthetic auxin that, when applied to lawns or grain fields, kills broad-leaved weeds without harming the grasses (Chapter 9).

**TABLE 16-17 Some natural plant poisons having economic significance.**

| Name of Substance | Plant Source | Uses |
|---|---|---|
| **Insecticides** | | |
| Pyrethrum,[a] from flowers | *Chrysanthemum cinerariaefolium, C. coccineum, C. marshallii* | Against flies, fleas, garden insects, lice, mosquitoes |
| Quassia, from wood shavings | *Picrasma excelsa, Quassia amara* | Against insects; also internal use for dyspepsia and malaria |
| Citronella, from leaves | *Cymbopogon nardus* | Insect repellant |
| Nicotine, from leaves | *Nicotiana rustica* | Against insects; smoldering pellets used to fumigate greenhouses |
| **Fish poisons** | | |
| Rotenone, from dried roots | *Derris* spp., *Lonchocarpus* spp. | Used as fish poison and against some garden insects |
| Croton oil, from flowers and crushed leaves | *Croton tiglium* | Fish poison |
| Pituri | *Duboisia hopwoodii* | Fish poison |
| **Vermifuges** | | |
| Santonin, from dried unopened flower heads | *Artemisia absynthium* | Expel intestinal roundworms |
| Wormseed oil, distilled from fruits | *Chenopodium ambrosioides* var. *antihelminthicum* | Expel intestinal roundworms |

[a]Synthetic pyrethrums are found in most insecticides available for home use.

## MICROORGANISM-PRODUCED SUBSTANCES

### Antibiotics

Some North American tribes treated wounds by pressing and binding a gray-green mold to the wound. It was not until 1929, however, that **penicillin**, an extract of the gray-green mold *Penicillium,* was "discovered" to be effective in killing infection-causing bacteria. Largely due to the battlefield demands of World War II, the era of antibiotics emerged rapidly. More than 1500 antibiotics were identified within a 25-year period, and the list is now over 2000. Of this number, however, only a few have been demonstrated to be safe and effective for human use. Antibiotics are used to prevent and treat disease in livestock and poultry, in food preservation, as biochemical research tools, and to produce industrial chemicals.

Antibiotics inhibit the growth of other microorganisms by interfering with

1. nucleic acid synthesis.
2. translation of genetic information during protein synthesis (neomycin, streptomycin, tetracyclines).
3. cell wall synthesis and function (bacitracin, penicillins).
4. cell membrane function (nystatin, polymixin, streptomycin).

Some antibiotics, like the penicillins, are narrow in the range of organisms that they will inhibit. Streptomycin is considered of intermediate range and the tetracyclines are truly wide spectrum, effective against both main categories of bacteria (gram negative and gram positive), rickettsia, large viruses, and even some protozoans.

### Steroids

A woman on birth control pills and a person treating poison ivy rash have something in common— the pill hormones and the cortical hormones in the rash cream were probably synthesized by microorganisms. Commercially produced steroids include sterols (e.g., cholesterol), bile acids, androgens (male sex hormones), estrogens and progestins (female sex hormones), and adrenal cortical hormones (cortisone and others).

### Amino Acids, Enzymes, Fermentation

Microbiologically produced essential **amino acids** can be used to fortify breads, cereals, and other foods, especially where inadequate nutrition is prevalent. Currently glutamic acids and lysine are the only amino acids manufactured on a large scale. The food that we eat can be made to taste better by addition of a monosodium salt of an amino acid, called monosodium glutamate (MSG). It is a common ingredient in canned and restaurant foods, although most often associated with oriental cuisine.

Because microorganisms are compact packages of complete metabolic processes, they represent great potential for commercial production of useful **enzymes.** Microbial-produced enzymes are used in bread baking; mash fermentation for alcoholic beverages; cereal, cocoa, chocolate, candy, and animal feed manufacture; coffee bean fermentation; cheese, ice cream, dry and concentrated milk production; enzymatic spot removal in laundry products; flavor restoration and enhancement; production of fruit juices, starches, syrups, vegetables, and wines; bating and dehairing during leather tanning; meat tenderizing; paper, textile, and photographic paper industries; clinical tests, including for diabetes; control of diabetes (insulin production) and indigestion; treatment of abrasions, bruises, inflammation; and as ingredients in drain and septic tank cleaners and wallpaper removers.

With or without an understanding of why it worked, humans have used microorganism enzymatic activities for centuries. Fermentation to produce foods and beverages is known from ancient times. Nondairy foods and beverages produced primarily by **yeast fermentation,** for instance, are beers, wines, hard liquors, bread, and the ethnic foods kumiss (from mare milk, Russia), kefyr (from cow milk, Turkey), taette (from cow milk, Sweden), miso (from soybeans, Japan), and kvass (from rye, Russia). The most commonly used yeast is *Saccaromyces.*

Malt-beverage brewing involves enzymatic processes at several stages: "malting" of the barley; "mashing" of the barley, hops, and water mixture; and fermentation of the resulting strained and sterilized effluent ("wort") into beer. Carbonation is returned to some beers by a second, closed-container fermentation called "Kräusening." Most beers, however, are carbonated by injection into the beer of carbon dioxide collected during fermentation. To avoid the need for refrigeration, most beers are pasteurized before or after packaging. Keg beers and at least one U.S. brand of packaged beer are sold unpasteurized and so must remain refrigerated until consumed. Otherwise, living yeast and bacterial organisms will continue to be active, changing the clarity and taste of the beer.

During beverage fermentation yeast activity ceases when alcoholic content reaches about 18% (by volume). Therefore production of beverages with higher alcoholic content requires **distillation** during which alcohol is evaporated off by heating and then collected and condensed in cooling coils. Volatile aromatic compounds are also present in the distillate, giving each type of liquor a distinctive aroma and flavor. Additional flavor develops by leaching of compounds from aging in wood and charcoal filtering and by direct addition of substances. Table 16-18 summarizes some types of alcoholic beverages.

The molds *Aspergillus* and *Rhizopus* and bacteria *Lactobacillus* and *Streptococcus* are agents of **nonyeast fermentation** for food production, including vinegar (final stages), dairy products (yogurt, sour cream, butter, acidophilous milk), sauerkraut, sausages, pickles, and the ethnic foods tempeh (India), tofu (Japan), and soy sauce (Japan and China). Considering the many known food-spoilage organisms, the list of desirable species is remarkably small.

Cheese is basically curdled milk. The curdling is caused initially by the action of protein-digesting enzymes on milk protein (casein). Rennet, an enzyme-containing extract from calf stomachs, is used to curdle the milk. In addition, *Lactobacillus* bacteria present lower the pH, thus souring the milk. If the milk becomes sufficiently acidic, casein will curdle in response to low pH without rennet. Masses of curdled milk protein (curds) separate from the liquid whey, which contains water, minerals, milk sugar (lactose), and some proteins. The whey is usually discarded but is now being considered as a potential fertilizer, having a mild 1-1-1 nitrogen, potash, and potassium content. Curds are skimmed off and compressed into cakes for aging (ripening). The ultimate flavor and texture of finished cheese is determined by the activity of specific bacteria and mold species present during aging. Other factors are the amount of fat present, how much salt is added, environmental conditions maintained during aging, and duration of aging. Generally the longer a cheese is aged, the stronger or "sharper" in flavor it becomes. Aging is, in some respects, carefully controlled decomposition!

### Single-cell Proteins

Single-cell proteins (SCP) are foodstuffs produced by microbes, mainly yeasts, grown on a variety of substances. The major substrates are petroleum products and a few commercial operations now manufacture animal feeds of SCP. Human food production is still in its initial stages, although the potential is great because bacteria and yeast cells have high protein content, can be grown efficiently and in large quantities, and can utilize waste products of other industrial processes. Some problems exist regarding bacterial cells being too high in nucleic acids and in developing SCP that is comparable in physical food-preparation characteristics (such as texture) to soy proteins. Research is expected to progress in the development of SCP that can provide low-cost/high-nutrient human food protein.

### Negative Economic Aspects of Microbial Action

Many yeasts, bacteria, and fungi are of great economic importance because they cause food spoilage. Spoiled food is recognized by off-colors and flavors, cloudiness or ropiness, sour or putrid odors, mushiness or sliminess, and the physical observation of growing organisms.

Addition of chemical preservatives to bakery products retards mold growth; nitrates and nitrites inhibit bacterial growth on such cured meats as bacon. Home-canning procedures emphasize high temperature and pressure, high osmotic tension (by salt or sugar), and low pH (by vinegar) to prevent growth of spoilage organisms.

From the time that a tree is cut for human uses there is competition from microorganisms, mainly fungi and bacteria. Thus exterior lumber is often treated with wood preservatives, primarily fungicidal. Wood deterioration is an economic and technical problem to the pulp and paper industries at all stages of storage and processing. By the time that a piece of paper is discarded, the wood cells have been altered by many natural and industrial processes and the paper eventually decays into its original organic and inorganic components, plus whatever was added in the manufacturing process.

Even house paints, maintenance paints (ship bottoms and steel tanks), and industrial paints, such as those used on motorized vehicles, are attacked by bacteria and fungi, especially mildews. Paints have been developed whose pigments (such as the white zinc oxide) and synthetic binding in-

TABLE 16-18 **Some alcoholic beverages.**

| Name of Beverage | Plant Sources | Scientific Names |
|---|---|---|
| **Fermented beverages** | | |
| Wines | Grape berries; other fruits | *Vitis* spp. various genera |
| Beers | Grains, such as barley, rye, rice, millet, maize; hop flowers | Various grass genera *Humulus lupulus* |
| Hard cider | Apple fruits | *Pyrus malus* |
| Root beers | Various roots, barks, herbs, such as greenbrier, ginger, wintergreen, burdock, sassafras, sarsparilla | *Smilax, Zingiber, Gaultheria procumbens, Arctium, Sassafras albidum, Aralia* |
| Mead | Honey | |
| Saké | Rice grains | *Oryza sativa* |
| Palm wine | Palm inflorescences | Various species |
| Pulque or mescal | Sap of agave plant | *Agave atrovirens* |
| Chicha | Corn grains | *Zea mays* |
| **Distilled beverages** | | |
| Whiskies and vodka | Various grains, including corn; potato tubers | *Zea mays* and others *Solanum tuberosum* |
| Brandy | Distilled from wine or fermented juice of various fruits, such as apricot, cherry, blackberry | See wines, above |
| Rum | Unrefined products of sugarcane, such as juice and molasses | *Saccharum officinale* |
| Gin | Various grains, especially barley malt and rye; may have juniper strobili for flavoring | Various genera *Juniperus* |
| Tequila | Distilled from fermented roasted stems of agave plant | *Agave tequilina* |
| **Miscellaneous** | | |
| Vermouth | White wine plus herbs | Various |
| Bitters | Herbs and bitter principles in water or alcohol | Various |
| Liqueurs | Distilled spirits, sugar or honey, plant extracts, other flavorings | Various |

gredients are more resistant to attack. Organic microbicides are commonly added.

That microorganisms can exist on petroleum was publicized by the announcement of "oil-eating" bacteria being seeded onto offshore oil spills. The organisms grow where the oil is in contact with water. This natural petroleum-using capacity is used for recovering trapped or spilled oil; however, it can also be a problem because the bacteria also thrive in filter units of furnaces and fuel-storage facilities. Research with microorganisms grown on petroleum extracts has suggested culture methods for obtaining valuable microbial products. The development of single-cell protein (SCP) was an outgrowth of experiments with petroleum-grown microbes.

### Other Uses of Microbes

**Microbial genetics** is a vast research field that continues to provide improved strains of economically important yeasts, bacteria, and fungi. More desirable strains are selected from both naturally occurring and induced variations. Some of the most active research has been with antibiotic-producing microbes by such genetic "engineering" as alteration of DNA/RNA by recombination and physical and biochemical manipulations. (See also Chapter 5.)

Some microorganisms, such as fecal bacteria, are indicators of human pollution from sewage and monitoring of their frequency in surface waters is a standard health procedure. Occasionally when fecal bacterial counts rise during warm summer months, beaches are closed until tests show that the counts are down within safe levels.

## THE FUTURE

The history of economic botany, like that of all human activities, has its roots in the human struggle to survive. Internal rigors include the continual need for food and our own desires for a better life. External demands include adapting our clothing, shelter, and ways of life for protection from the physical and biological elements of nature, including predators, parasites, and pathogens.

We are told that there has always been hunger in the world and that there probably always will be. World agriculture increases yearly, but increases in world population may soon exceed it. Where hunger and poverty exist, there is social unrest, a potentially perilous but at times constructive force in world events. What are the prospects for world agriculture being able to overcome the problem of population growth? Demographers are not sure, but most are not optimistic.

Agricultural and technological research (Fig. 16-15) has produced increasingly productive crop varieties and increasingly efficient ways to grow and harvest them (Figs. 16-16 and 16-17). Basic grains of antiquity, such as maize, oats, barley, wheat, and rice, were domesticated and improved in prehistoric times and this improvement continues. It has resulted in a number of changes, including larger individual grains with smaller accessory parts that must be removed before use and higher amino acid content. Individual vegetable species have also been greatly improved. They are more disease resistant, produce larger fruits that remain on the vines longer and/or mature together, are more tolerant of cold, heat, and drought, are more tender and palatable, have fewer, smaller, or softer seeds, or

**FIGURE 16-15**
Test plots are an important research tool in the development of improved varieties. (Photo courtesy of James R. Estes and Ronald J. Tyrl.)

**FIGURE 16-16** Here the bountiful harvest has exceeded the capacity of storage facilities and wheat is temporarily "stored" on the street. (Photo courtesy of James R. Estes and Ronald J. Tyrl.)

produce no seeds at all. Food plants from all regions of the world are continually tested for cultivation and improvement potential. There is further potential in increased use of wild edible foods. All are active accomplishments of private and governmental research. The average citizen supports these accomplishments through taxes and purchase prices.

Increased productivity is not the only answer, however. Our agricultural resources have both physical and human limitations (Chapter 6). Our dependence on natural resources, including economic plants of other countries, may also become more limited. Changes in individual lifestyle will probably become necessary. The excesses of the past several decades will need to be replaced by more conservative and frugal attitudes of former times. Perhaps individuals will grow more of their own food, a currently popular trend, and will waste less. How many of us, for example, routinely discard the tops of green onions and celery, the outer and less attractive leaves of lettuce and cabbage, the stems of parsley and broccoli? Cook all that together with water, seasonings, broth or bouillon cubes, and perhaps some grain, legumes, or meat and it makes a nutritious soup!

We will need to become more imaginative and sensible. Nonedible organic matter will be composted to increase the fertility of soil to grow more food or be fermented into alcohol for fuel rather than being sent to a landfill site or ground down the garbage disposal unit. Organic sewage residue (sludge) will be recognized more widely for its potential as a natural fertilizer and a source of natural gas (methane) to be burned to generate electrical power and for domestic heat and cooking. In some cities methane runs the sewage plants where it is produced. In other cities methane is simply burned

**FIGURE 16-17** Harvesting wheat. (Photo courtesy of James R. Estes and Ronald J. Tyrl.)

off (wasted) from sewage plants and old landfill sites.

Industries are also involved in more complete resource utilization and a good example is provided by the forest-products industries. In the early development of the nation it was literally impossible to see the forest for the trees. In many cases, the forest was an obstacle to the expansion of the frontier. Not only were the biggest and best trees harvested for lumber, but often trees were also simply felled and burned to make room for agricultural land. Initially neither the technology nor the market existed to utilize much more lumber than was produced. By the turn of the century the emphasis of the logging industry was "cut and get out" and logging practices were wasteful.

The great virgin forestlands of the northeastern and midwestern United States were devastated and loggers moved to the western states, pursuing seemingly endless forest resources. This limitless "frontier ethic" is still deeply entrenched in the minds of many people, who resent what they regard as needless restraints on the full exploitation of forest, mineral, and fossil fuel resources of the western United States—including Alaska—and of Canada.

All aspects of the early lumber industry were wasteful by today's standards. Vast waste occurred during cutting and trimming in the field. At sawmills sawdust, exterior slabs, odd-shaped and odd-sized fragments and bark were discarded and frequently burned. Separate trees were harvested for lumber and pulpmaking and large quantities of pulp fibers were simply washed out as waste in the effluent.

Today wood left over from lumber production is converted to chips. Chips and sawdust are the primary raw materials for production of the most abundantly used grades of paper, from paper bags and boxes to newsprint and notepaper, as well as in chipboard and particleboard. As an alternative to fossil fuels, mill waste that is not suitable for further manufacture into paper, fiberboard, or other useful products can be "hogged" and used as fuel to fire boilers for industries. Forest residuals formerly burned as "slash" on harvest sites are also beginning to assume importance as potential raw material for fuel.

Bark chunks previously discarded are now salvaged, at least from some species, to be milled to various sizes for use as decorative ground covers, mulches, soil conditioners, and in container plant mixes. Wood chips and sawdust are also used as mulches.

Effluents from modern pulp mills contain substantially smaller concentrations of wasted wood fibers and many of the pulp-processing chemicals (such as sulfur compounds) are retrieved. Dimethyl sulfoxide (DMSO) is a byproduct of the paper industry. Chemical plants associated with pulp and paper mills extract such useful substances as lignins from wood "liquor." There are still many valuable sugars, minerals, and other substances that will eventually be salvaged.

Adding to more effective initial utilization and byproduct salvaging is the development of fiber recycling. Although a different-quality paper is produced each time, the same pulp can theoretically be recycled several times. As de-inking processes continue to improve, so do paper-recycling processes. Some mills run almost 100% on recycled fiber. Federal contracts now require that a certain percentage of recycled fiber be used in their papers.

Fibers other than wood fibers can also be used many times. Every second child knows about "hand-me-downs," sometimes as is and sometimes in the form of newly sewn clothes cut from still-good fabrics of older, larger garments. The essence of frugality is captured in a beautiful art form: quilting. Odds and ends of fabric pieces and scraps remaining after the cutting of a pattern or from clothes outgrown or worn out in places are fashioned into decorative blankets to lead another useful life. Woven and braided "rag rugs" have the same raw materials. Patchwork designs for clothing and accessories utilize scraps too small to be used individually but that can be joined to make a piece of fabric large enough for a garment.

If the fabric is not used in one of the preceding ways, it can be recycled in another sense of the word. Melton-type coat fabrics are made of miscellaneous reprocessed fibers. Some high-quality grades of paper (such as rag bond) use natural fibers (cotton, wool, and linen) in their manufacture. Large quantities of old rags are used to make felt, composition roofing materials, and mattress padding.

Wood products and fabrics are only two examples of maximizing utilization of raw materials. Sooner or later human attitudes and activities must be totally converted to a recycling ethic; and it is to be hoped that this conversion will occur before we are forced into it by a plundered planet whose surface has been stripped of its natural ecosystems, its crust churned, honeycombed, and drilled in a search for minerals and fossil fuels.

Evolution of life on Earth is a continual progression. It has taken millenia to attain Earth's life forms and distributions, including those long extinct and those presently surviving. Biologists believe that hundreds, even thousands, of species are yet to be identified. Of plant species that *are* known, only a tiny fraction have been tested for their potential usefulness! Any loss of a natural species is irrevocable. Survival of the human species is closely linked to the continuation of natural environments and their abilities to provide us with food, clothing, shelter, medicines, and all the raw materials that we have learned to use in so many different ways.

## SUMMARY

The history of civilization is interwoven with the history of useful plants. Plant products drastically improve the quality of life, providing not only **food** and **shelter** but also **clothing, medicines, beverages, transportation, industrial products,** and **fuel.**

Plants are also an integral part of human **culture.** The invention of paper was the key to modern communications. Plants have inspired art and poetry. Some ancient religions incorporate plants.

There are many edible fruits and seeds. Fruits range from sweet melons to grains.

Except for coconut and cocoa lipids, which are saturated fats, plant lipids are unsaturated or polyunsaturated and are low in cholesterol.

Shoots, stems, leaves, roots, and flowers of various plant species are eaten directly or used as herbs, spices, and condiments.

Crushed fruits, saps, aromatic leaves, stems, roots, and flowers of many species are the source of flavored beverages. Alcoholic beverages are formed by the fermentation of plant carbohydrates mixed with water.

Cotton **fibers** are elongated epidermal hairs on cotton seeds. Flax fibers are sclerenchyma fibers of the plant stem and are woven into linen. Extracted linseed oil from flax is used in paints and linoleums. Many other plants are valued for their fibers.

Vegetable **dyes** produce subtle colors. Dye fastness is improved by use of a **mordant.**

**Sizing,** such as starch, fats, and waxes, is added to cotton and linen fabrics to make them crisp and smooth.

Most homes in North America are constructed primarily of **wood** and wood products (even such items as weathersheathing and floor coverings).

The types and density of xylem cells, their arrangement, and the presence of specific organic substances determine wood hardness, grain, patterns, color, and aroma.

Wood and wood products are greatly varied. Even sawdust, chips, and waste from pulp mills that were once discarded are used today to make many byproducts and extractives.

Commercial **cork** comes from the outer bark of the cork oak.

Plants have always been important sources of **medicines.**

Other plant **exudates** and **extractives** include sugars, starches, volatile oils, waxes, latex, resins, gums, tannins, and more.

**Microbes** are used to produce many beneficial products, including antibiotics, steroids, amino acids, and food fortifiers. Microbe-produced enzymes have hundreds of applications.

Yeast fermentation is the basis for production of alcoholic beverages, breads, and many foods. Specific bacteria and molds are responsible for the making of sour cream, yogurt, cheeses, soy sauce, sauerkraut, and other foods.

Production by microbes of single-cell proteins (SCP) from industrial waste products is being explored as a means of producing human food.

Unfortunately, many microbes have a negative economic value, for they cause plant and animal diseases, spoil foods, and deteriorate lumber, paper, paints, and other products.

Agricultural research has resulted in major crop improvements. But many people of the world are hungry and resources become fewer, not greater.

With ever-increasing population and diminishing resources, a less wasteful life style must become the norm if resource problems are to improve.

## SOME SUGGESTED READINGS

Anderson, E. *Plants, Man and Life.* Berkeley: University of California Press, 1969. An intriguing little book that presents a possible origin and evolution of crop plants.

Christensen, C.M. *The Molds and Man,* 3rd ed. Minneapolis: University of Minnesota Press. 1965. A delightful account of the impact of fungi on humanity.

Edlin, H.L. *Plants and Man: The Story of our Basic Food.* Garden City, N.Y.: Natural History Press, 1969. A readable text for the general student.

—. *Trees and Man.* New York: Columbia University Press, 1976. A readable account of the growth of trees and their importance on a worldwide scale.

Emboden, W.A. *Narcotic Plants.* New York: The Macmillan Co., 1972. An ethnobotany of the major, and some minor, narcotic plants, arranged according to their effects.

Gray, W.D. *The Relation of Fungi to Human Affairs.* New York: Henry Holt, 1959. A comprehensive text emphasizing the beneficial aspects of the relationship between fungi and humanity.

Heiser, C.B. *Seed to Civilization: The Story of Man's Food.* San Francisco: W.H. Freeman and Co., 1973. The development of agriculture and the domestication of crop plants.

Klein, R. *The Green World: An Introduction to Plants and People.* New York: Harper and Row, Publishers, 1978. A lively text on human uses of plants.

Kreig, M.B. *Green Medicine: The Search for Plants that Heal.* Chicago: Rand McNally and Co., 1964. A popular account of the search for botanical drugs.

Morton, J.F. *Major Medicinal Plants: Botany, Culture and Uses.* Springfield, Ill.: Charles C Thomas, Publisher, 1977. A comprehensive treatment of the important drug-producing plants of the world.

Parry, J.W. *Spices,* Volumes I and II. New York: Chemical Publishing Company, 1969. These companion volumes are a compendium of information about spices—their history, use, structure, and chemistry.

Richardson, W.N., and T. Stubs. *Plants, Agriculture and Human Society.* Reading, MA.: Benjamin-Cummings Publishing Co., 1978. An interesting book covering a wide range of topics.

Schery, R.W. *Plants for Man,* 2nd ed. Englewood Cliffs, N.J.: Prentice-Hall, 1972. The comprehensive textbook of economic botany.

Taylor, N. *Plant Drugs that Changed the World.* New York: Dodd, Mead and Company, 1965. A popular account of the discovery and use of many of our plant-derived drugs.

Uphof, J.C. *Dictionary of Economic Plants,* 2nd ed., rev. Monticello, N.Y.: Lubrecht and Cramer, 1968. Describes the taxonomy, origins, and uses of the world's major and minor economic plants; a useful reference.

# chapter seventeen FOOD AND OTHER USES OF WILD PLANTS

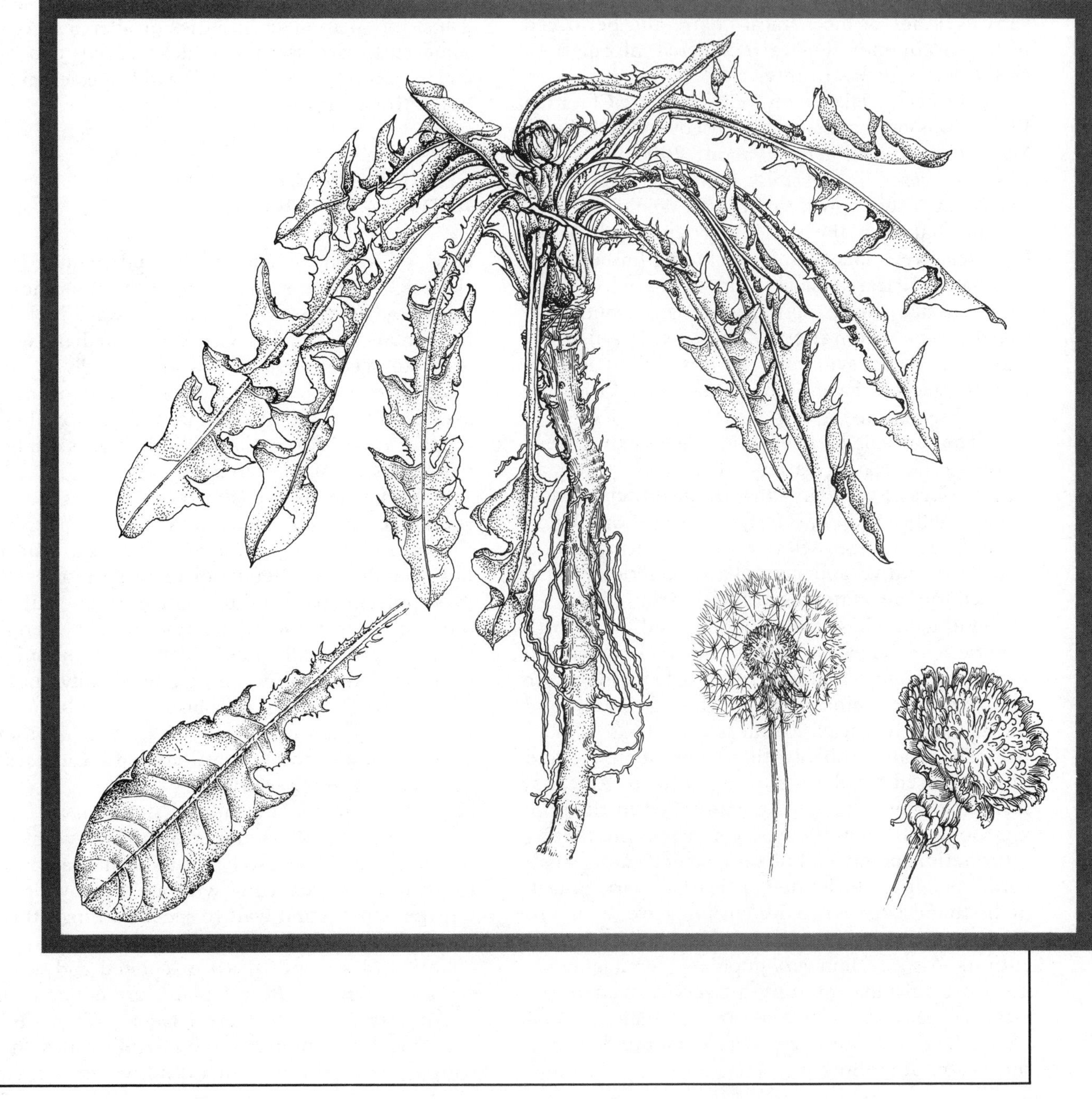

## INTRODUCTION

**Foraging** (seeking food from the wild) is something wild animals and many humans must do to survive. In most of North America, however, foraging has metamorphosed from a necessity to a luxury, a pursuit in which the weekend dilettante can dabble.

In addition to having recreational value, however, awareness of edible and useful wild plants could be a literal lifesaver. People still do become lost while hiking, hunting, or exploring. Motorized vehicles do break down. Even someone who is familiar with a remote area can be isolated by adverse weather or mechanical failure and be forced to rely on common sense, remembered information, and resourcefulness to survive.

The heavy reliance on foraging in pre-Christian Europe was emphasized by the discovery of Tollund Man. This hanged corpse, about 2000 years old, was discovered in a Danish peat bog. The man's body was so well preserved by the acidic bog conditions that even the roughly ground seeds of his last meal were identifiable, including linseed, barley, pale persicaria, black bindweed, gold of pleasure, fat hen, hemp nettle, wild pansy, and corn spurrey. The wide variety of seeds used in his gruel may have been symbolic, for his killing is now hypothesized to have been a ritual part of a Spring Rites observance.

The journals of Lewis and Clark's expeditions record a fascinating variety of uses by American native tribes of the plants that defined their ecosystems, whether prairie, forest, or ocean shore. Much has been done since then to capture information on how both native and naturalized Americans have utilized the raw materials of their surroundings.

Foraging has been popularized in North America by such writers as Euell Gibbons and Bradford Angier. Many illustrated handbooks are available for regions of the United States and Canada (see Suggested Readings).

There are probably thousands of native and introduced edible plants growing in North America, including many not usually mentioned in the popular books. Even in these books there is conflicting information about edibility of some plants. Such conflicts appear to be due, at least in part, to lack of firsthand experience, for authors have tended to pass along information accumulated by previous authors. Also, certain very popular references err in not distinguishing carefully between unrelated species that may have the same or a similar common name. In some cases, errors are compounded by a succession of authors.

In this chapter we have selected only a few examples of food and other uses of wild plants and have grouped them by preparation or use categories. We hope that you will be stimulated to a greater awareness of the "wild vegetables" in your environment.

**Warning:** Because individual sensitivities to specific plants and substances vary, we recommend that *all* the precautions stressed in this chapter be observed carefully. The examples chosen are universally acknowledged as being edible; our discussion, however, does not constitute a recommendation for all individuals, especially those with a history of food sensitivities or allergies. As with some cultivated species (such as wheat, chocolate, and peanuts), some "safe" wild species are *not* "safe" for everyone!

## LEARNING WHAT AND HOW TO COLLECT

How do we recognize edible wild plants? Unlike the cake in *Alice in Wonderland,* they do not bear signs saying "Eat Me." And, anyway, look what happened to Alice! The most immediate way to learn to recognize edible wild plants is to forage with an expert. Such a person is apt to know an assortment of safe foods as well as how best to prepare them. From that point horizons can be expanded by reliable references.

Even the use of illustrated guides does not guarantee correct identification by the novice. Illustrations do not always indicate size and may not include exact details of plant structures. So it is important to combine illustrations with taxonomic and environmental descriptions when identifying a potentially edible plant. A taxonomic key provides additional certainty and is easily mastered with a little practice. It is also a good idea to learn the most common poisonous species in order to avoid them. **See the Precautions and Pitfalls section for more precise precautions.**

In an emergency survival situation it is possible to forage by trial and error (except with mushrooms!). To do so safely, taste a small amount of the raw plant, chewing without swallowing. Spit out the remains and wait to see if burning, stinging, or numbing of the mouth occurs. If not, eat a small quantity (a spoonful) raw or cooked and wait for an hour or more. (Some plants are edible cooked but not raw.) If there are adverse effects, induce vomiting by stimulating the gagging reflex at the back of your throat. Then drink water and vomit again to wash out your stomach and prevent further digestion and absorption.

If things are going well, however, small amounts may be ingested over a period of several hours, after which many authors recommend a long wait (overnight to 24 hours). This process allows a "consensus verdict" from all regions of the digestive tract. When food, water, and energy are at a premium, it is unwise to risk the painful and dehydrating effects of diarrhea!

If all these tests are met, the food may be considered safe to eat in *moderate* quantities. Test only one new food at a time. *Never* experiment with mushrooms. After safe foods have been identified, become accustomed to them gradually by combining them with "civilized" foods that you may have with you. Even when surviving on wild foods, eat balanced and moderate quantitites to avoid digestive problems.

## WILD FOOD PREPARATION CATEGORIES

The following discussion is based on preparation categories, giving basic instructions for preparing foods within each category and more detailed instructions for some specific plants. In general, the examples were selected because they are widely distributed and are good representatives of their categories. Also, space limitation prevents a more complete listing. Several of the Suggested Readings contain good recipes for many wild plants; in addition, wild foods of similar kind can be substituted for garden-variety fruits and vegetables in the preparation of familiar dishes.

Some plants are so versatile that they are listed in several preparation categories. A prime example is the common dandelion (*Taraxacum officinale* agg.) (from French, *dent de lion,* tooth of the lion), illustrated in the chapter opening figure. The early spring crown and young leaves are eaten raw in salads or as a potherb. The flowers and leaves are used for dandelion wine or spring tonic. The root is cooked as a starchy vegetable or is roasted, ground, and used as a coffee substitute.

Most authors also exalt the common cattail (Fig. 17-1) as highly versatile. The spring shoots are prepared as "Cossack asparagus" (which we find less appealing than many descriptions). The immature flowering heads are cooked and eaten like corn on the cob, the pollen is used as flour, and the rhizomes are a source of flour. Some authors recommend using the tender, fresh-tasting new shoots sliced raw in salads. We, however, experienced throat discomfort from raw shoots and found needlelike oxalate crystals in fresh microscope sections.

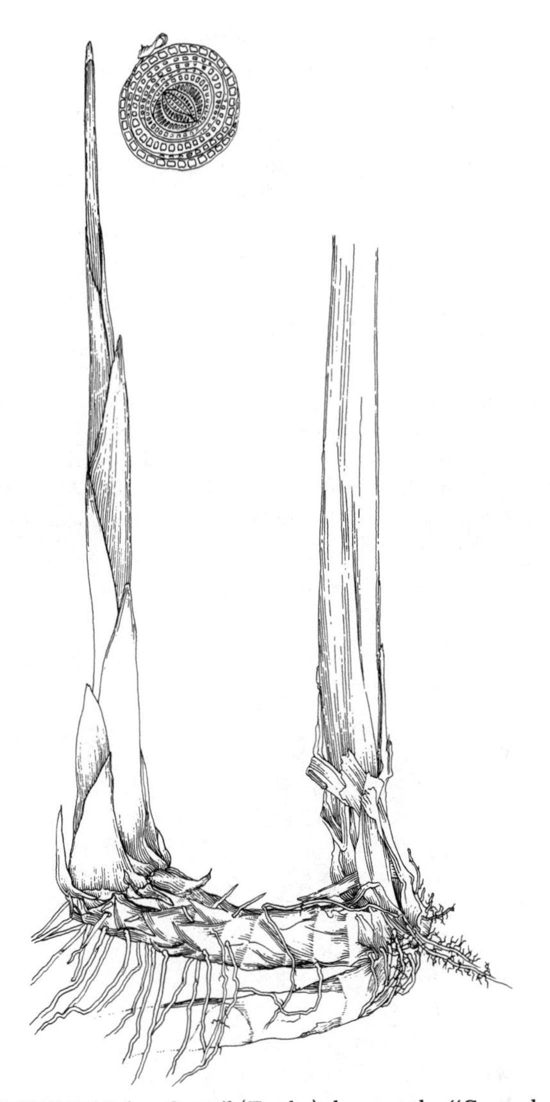

**FIGURE 17-1** Cattail (*Typha*) shoot at the "Cossack asparagus" stage. The cross section shows the rolled leaves.

In our discussion "young stem" or "shoot" refers generally to the main stems or young side branches before flowering; "vegetable" refers to thicker shoots or other parts to be cooked; "leaf" includes the leaf blade and petiole unless otherwise noted.

For some foods, it is necessary to leach away unpalatable substances by repeated soaking and rinsing. Most acorns, for example, are moderately to very bitter and are inedible until the bitter tannins are removed. This process involves boiling the shelled nuts (whole or ground into meal) in several water changes and then soaking and rinsing further until no more brownish color comes into the water. The rinsing may take days.

Certain plants have strong flavors, such as members of the mustard family. Changing water during boiling removes flavor substances, allowing taste adjustment to the eater's palate. Scraping or peeling off the outer covering of subterranean parts, such as rhizomes, roots, and tubers, reduces a strong or "earthy" flavor.

## Salads

Plants usually eaten raw include the pleasantly acid (sour) sorrel (*Oxalis*) leaves and various other plant parts that may need to be scraped or peeled. Flavors vary from the slightly bitter but mostly mild taste of wild lettuces (*Lactuca*) [Fig. 17-2 (a) and (b)] and the soft, fresh, mushroomlike flavor of young plantain leaves (*Plantago*) (Fig. 17-3) to the lively, peppery flavors of such mustards as water-

**FIGURE 17-2**
(a) A wild lettuce (*Lactuca*) with a smooth midrib. (b) Prickly lettuce. Wild lettuces, like garden varieties, have hollow triangular midribs that exude a milky latex when broken, as shown in (b). This is a feature that can be helpful in identifying them.

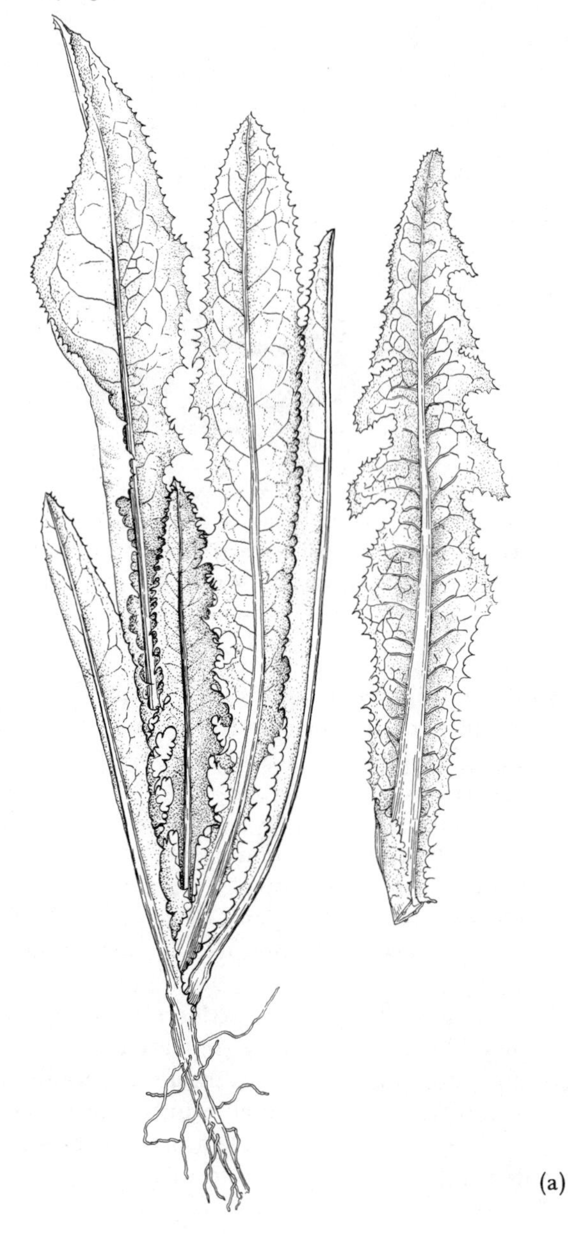


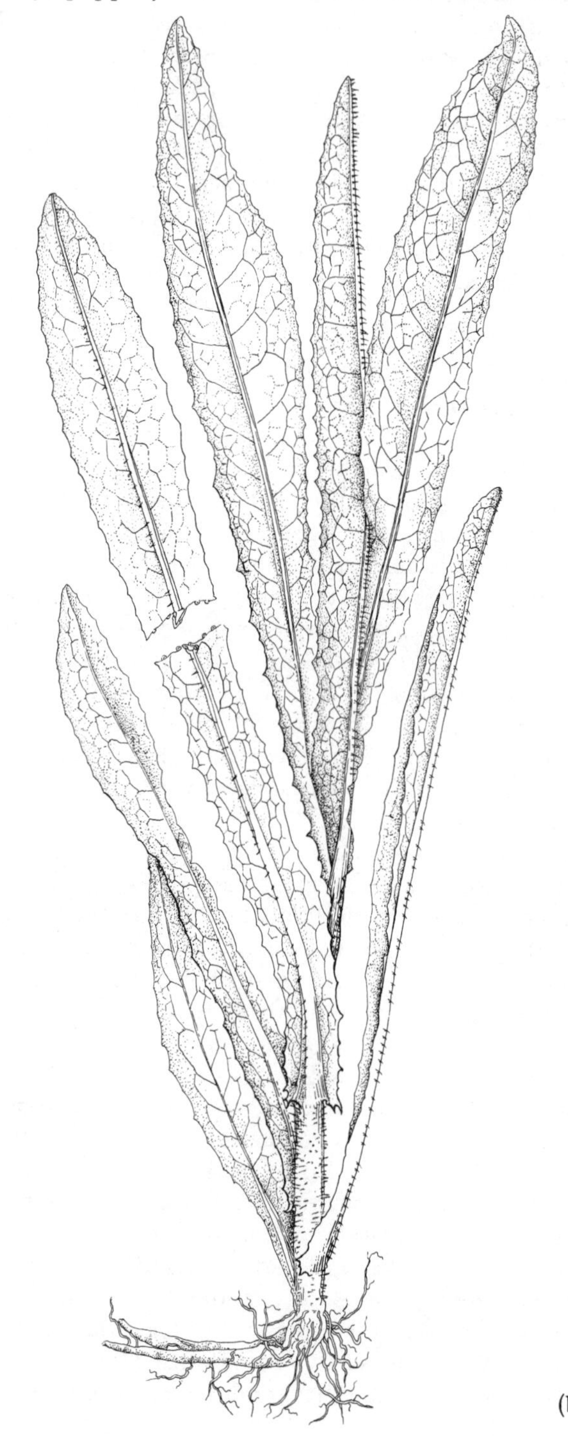

**FIGURE 17-3** Young plantain (*Plantago*) leaves have a sweet flavor in early spring.

**FIGURE 17-4** Sour dock or sheep sorrel (*Rumex*) is tart and lemony.

cress (*Nasturtium officinale*) and wintercress (*Barbarea vulgaris*). For best taste and texture, use younger leaves—either the entire plant in early spring or inner leaves later.

A trip to the yard for young leaves or shoots from such lawn "weeds" as dandelion, plantain, sheep sorrel, or sour dock (*Rumex*) (Fig. 17-4), purslane (*Portulaca oleracea*), and chickweed (*Stellaria media*) (Fig. 17-5) adds interest to a basic lettuce salad. Be careful not to collect plants to which herbicides (as in some lawn conditioners) or insecticides have been applied.

Watercress is an expensive salad delicacy at the grocery store or a luxuriantly rampant weed along small watercourses (Fig. 17-6). In Wisconsin it seems to prefer spring water in limestone areas whether the spring is large and freely flowing or a small seep supporting a muddy soft spot. It usually grows in dense patches. At their mildest during the spring, the growing tips can be collected through fall into winter where freezing does not occur. The small, white, mustardlike flowers and young seed pods accompanying the foliage during summer are also edible.

The plant can be distinguished in several ways from other semiaquatic weeds. Perhaps the most distinctive feature is the presence of white adventitous roots at all submerged nodes (as in Fig. 17-7). When collecting, take only above-water portions of the plant in case the water is contaminated. Several authors recommend the precaution of soaking

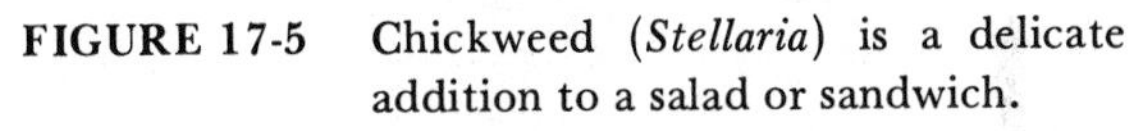

**FIGURE 17-5** Chickweed (*Stellaria*) is a delicate addition to a salad or sandwich.

**FIGURE 17-6** Watercress (*Nasturtium*) growing in a spring-fed watercourse in Wisconsin.

**FIGURE 17-7** Edible shoot of watercress, showing adventitious roots at the nodes.

watercress in water to which purifying tablets (such as Halozone) have been added, especially if the shoots were submerged or in contact with mud.

The crisp-stemmed 5- to 10-cm tips are easily broken or cut off. The entire tips are delicious in salads and sandwiches. Watercress has a spicy taste that blends well with lettuce and tastes especially good when served with a buttermilk/mayonnaise dressing. The plant is also good as a potherb and in soups and meat dishes.

### Potherbs

Potherbs are cooked greens, prepared by boiling until tender. Seasoning may consist of butter and salt, soy sauce, lemon juice, vinegar, cream sauce, or fried fresh or cured meat. Potherbs can be used in casseroles, soups, souffles, or quiches. If a particular green tastes too strong, changing waters during cooking will help.

Experiment with potherb cooking times. For

lambsquarters and other tender greens, 5 to 10 minutes is adequate even though recipes commonly recommend cooking for 20 minutes or more. The longer potherbs are cooked, the mushier and blander they become. With some greens, 2 to 3 minutes is sufficient, such as in stir-fry cooking.

All comments regarding these wild potherbs also apply to such cultivated greens as spinach, New Zealand spinach, collards, beet tops, and chard, even radish tops.

#### Lambsquarters (*Chenopodium album*)

Lambsquarters is a particularly desirable potherb because of mild flavor, ready availability, and ease of collecting (Fig. 17-8). It is one of the earliest and most vigorous weeds in disturbed soil, such as that of a home garden. It is earlier than spinach and delicious when allowed to grow 10 to 15 cm tall and then harvested. Where eradication is not essential, plants can be allowed to produce through spring and early summer, for harvesting the tips encourages equally palatable side shoots.

The goosefoot-shaped leaves are easily recognized. Their undersides have a silvery, water-repelling, granular or mealy coating, as do all surfaces of the newly developing leaves (Fig. 17-9).

**FIGURE 17-9** The silver mealy particles on the leaf and stem surfaces are an aid in identifying lambsquarters.

**FIGURE 17-8** Lambsquarters (*Chenopodium*) compares very favorably with spinach in flavor, cooking qualities, and vitamin content—and it's free!

Lambsquarters is in the same genus as cultivated spinach and the flavors are similar. Lambsquarters leaves, however, are slightly thicker and remain firmer during cooking. Lambsquarters exceeds spinach in some vitamins and minerals (Table 17-2). With lambsquarters around, it is not necessary to plant spinach, which has a tendency to bolt to seed early rather than remaining edible throughout the growing season.

#### Mustards, Wintercress (*Brassica, Barbarea,* and Others)

Mustard greens are early-spring potherbs, at which time the shoots (including flower buds) are mild and tender. A close relative, wintercress (yellow rocket), may be used in the earliest rosette stage [Fig. 17-10 (a)] or later when the flower-bearing shoots [Fig. 17-10 (b)] form. The flowers of these members of the mustard family are identical to those of such cultivated relatives as radishes

(a)

(b)

FIGURE 17-10 (a) The early rosette stage of wintercress or yellow rocket (*Barbarea*), one of the first spring potherbs, with the flowering shoot beginning to emerge. (b) Flowering shoot of wintercress, showing the typical four-petal, six-stamen flower of members of the mustard family.

and broccoli. With mustards and yellow rocket, break off rather than cut the tips so that the break occurs at a naturally tender, nonstringy part.

**Curly Dock (*Rumex crispus*)**

The young leaves of curly dock (Fig. 17-11) may be eaten raw or cooked as a potherb. In addition, the dried fruits (Fig. 17-12) may be ground

FIGURE 17-11 Habit shot of curly dock (*Rumex*) in the early spring when the leaves are tender.

FIGURE 17-12 Detail of a curly dock leaf and seed head. After the papery little "wings" have been rubbed off the seeds, the seeds can be ground for use as a cereal.

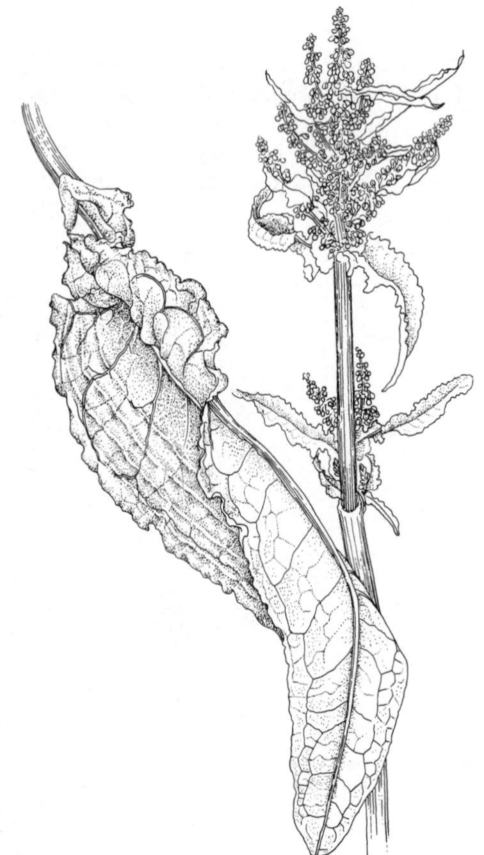

and used as a cereal. Because of the oxalate content, curly dock should be eaten in moderation, as with other foods.

#### Stinging Nettle (*Urtica dioica*)

Bare-skin contact with stinging nettle is painful, but this same plant is a popular potherb and soup green—collected while wearing gloves! Locate young plants in early spring by their proximity to the previous-year's dry stalks (Fig. 17-13). Even in the earliest stages of growth the square-sided stems and stinging hairs are identifying characteristics (Fig. 17-14).

As with lambsquarters, collect only the tender shoot tips, approximately 15 cm long, or wherever they break off easily. Continue to harvest shoot tips throughout the growing season.

On immersion in boiling water the stinging hairs and their irritating acid become harmless. Certain standard cookbooks include recipes for nettle soup, indicating that this is a widely accepted wild edible in both Europe and North America.

**FIGURE 17-13** Dried stalks of the previous year's stinging nettle (*Urtica*) stand are a clue for locating new shoots in the spring.

**FIGURE 17-14** Nettle shoot tip, with newly expanding leaves.

#### Pigweed (*Amaranthus retroflexus*)

Pigweed is an abundant "weed" akin in growth habits to lambsquarters, which is also sometimes called pigweed. Pigweed readily invades home gardens and the young plants (Fig. 17-15) are widely used as a nutritious and mildly flavored potherb.

**FIGURE 17-15** Common pigweed (*Amaranthus*) is rich in vitamins. Cultivated amaranth varieties are attaining garden popularity.

**Pokeweed (*Phytolacca americana*)**

One of the (if not THE) most popular spring potherbs in the southeastern United States is pokeweed. "Poke sallat," as it is colloquially known, is sold in markets and as a canned vegetable. Only the very young shoots are used, cooked with one or more water changes. The mature leaves (often red veined), stems, fruits, and particularly the root are poisonous (Chapter 18). The young shoots contain relatively low concentrations of toxin but still enough to poison if inadequately cooked.

## Cooked Green Vegetables

These vegetables generally differ from potherbs in that stems or leaves are fleshier. The plants are boiled, baked, roasted, or casseroled to be served similarly to domesticated asparagus, broccoli, green beans, and so on.

**Asparagus (*Asparagus officinalis*)**

Asparagus is actually a cultivated species from Europe that has become naturalized. The "wild" asparagus is collected and prepared identically to that found in a garden or at the grocer's. Asparagus is one of the earliest spring vegetables and must be harvested in the spear stage. Mature foliage and summer-maturing red berries are poisonous. Small amounts of toxin are present in the young edible spears, but cooking for even a short time eliminates the problem.

In foraging for wild-growing asparagus along roadsides, railroads, or "weed patches," look for clusters of dried stalks from the previous year. Asparagus is a perennial and shoots (spears) grow from the same root crowns year after year. Consequently, you can develop an asparagus-collecting map.

To harvest, cut or break off the shoot just below soil level, taking care not to damage the shoot-bearing crown. The most mature, vigorous plants produce especially succulent thick spears. As the vascular and support tissues mature, the stem toughens from the base upward. So, before cooking, bend each spear with one hand at the cut end and the other somewhere in the middle. The spear will snap above the region of woody development. Asparagus is sweetest when eaten shortly after harvesting; after picking its sugars are rapidly converted to starch. Cook fresh asparagus only long enough to make it tender and yet still retain its bright green color and firm texture. Asparagus is effectively used in stir-fry cooking.

**Milkweed (*Asclepias syriaca* and Others)**

Common milkweed species having broad leaves (Fig. 17-16) are abundant, growing especially well in disturbed areas, such as along roadsides and neglected lots or fields. Collect the 15- to 20-cm leafy shoot tips, with or without flower buds. The cut or broken hollow stems exude copious amounts of milky-white latex (hence the name "milkweed"), which coagulates into a sticky mess on clothes or seatcovers; so use a bag or newspaper.

**FIGURE 17-16** Broad-leaved milkweeds (*Asclepias*) are edible when carefully cooked. White, sticky latex exudes from broken laticifers, as shown here.

Milkweed must be cooked, for it is poisonous raw. In general, *broad-leaved* milkweeds are safe to eat after cooking, but *narrow-leaved* milkweeds should be avoided! Most references recommend cooking milkweed shoots in one or two changes of water. Their flavor is mild and pleasant and they can be seasoned and served alone or "en casserole."

Sweet-smelling milkweed flowers produce fairly large amounts of very sweet nectar, which was collected and dried for sugar by pioneers. The literature disagrees on the safety of this practice, for the nectar of some species is considered quite toxic.

About midsummer the seed pods develop. They are collected at a young stage and prepared by boiling or batter frying. Eventually the pods toughen and the silk seed "parachutes" develop.

FIGURE 17-17 Jerusalem artichoke (*Helianthus tuberosus*) is a native sunflower that has edible tubers.

## Starchy Vegetables

As the category name implies, these are the "potatoes" of wild fare. Most can be baked, roasted, boiled, and used in casseroles, soups, and stews. Some can be fried if first dipped in batter to enhance tenderness and reduce moisture loss. Most starchy vegetables are food storage organs, especially roots, rhizomes, tubers, bulbs, and corms.

### Jerusalem Artichokes (*Helianthus tuberosus*)

The tubers of this wild sunflower (Fig. 17-17) are so popular that cultivated varieties have been developed. There is a trend toward calling them "sunchokes," for they are not truly artichokes (which is a thistle inflorescence) and are not from Jerusalem. They are native to North America. The name Jerusalem artichoke depicts a language barrier—apparently "Jerusalem" is a phonetic corruption of the Italian word "girasol," which means sunflower. "Artichoke" refers to a similarity between the sweet flavor of the cooked tubers and cooked artichoke hearts.

Jerusalem artichokes store their excess carbohydrates as **inulin** rather than starch. Because inulin does not metabolize into glucose the way starch does, Jerusalem artichokes are recommended for diabetics as a potato substitute.

During late summer the inflorescences appear, with yellow centers instead of the dark centers of other sunflowers. Like other sunflowers, they have rough-surfaced leaf blades and petioles; sunchokes also have rough ("whiskery") stems. By marking the location of sunchoke beds while they flower, you can set the stage for winter and spring collecting. Scrub or peel the sweet, crunchy tubers to eat them as a raw relish or slice them into a salad. In most regions of the United States it is not difficult to establish a bed of this native prairie plant and even a small bed produces prolifically.

### Burdock (*Arctium minus*)

Burdock is a common wild plant with edible roots, which are a component of home-recipe root beers (Fig. 17-18). Burdock is well known for the globes of dry, pesky, hook-bearing burrs that readily stick to clothing and pet fur in the fall. It is a biennial plant, growing a rosette of leaves during the first year, flowering and setting seed the second year, and then dying. The dead fruit stalks frequently stand throughout winter, marking locations of burdock beds for spring and early-summer collecting. This introduced "weed" grows lushly in many habitats, especially where existing vegetation has been disturbed (Fig. 17-19). It is easily recognized by its slightly velvety rhubarblike leaves,

FIGURE 17-18 Burdock or gobo (*Arctium*).

which become enormous by summertime. A cultivated variety, gobo, is a popular Japanese garden vegetable, prepared as tempura. We have found wild burdock to equal the cultivated variety in flavor.

The freshly dug roots are not attractive in appearance. Digging requires a good shovel and a good shoveler because the roots are too long and deep to be pulled out. Roots are commonly 2 to 10 cm in diameter and more than 60 cm long; so they yield abundantly. The roots can be stored in a refrigerator or root cellar.

To use burdock, wash, trim, scrape, and then soak the roots for a few hours. They can be prepared in various ways. One is to cut them into approximately 8-cm segments and then into thin wedges or strips (like very thin carrot sticks). In this way, even somewhat woody parts of the root can be cooked tender. Sections or strips of burdock root can be refrigerated (in water) for several days without loss of flavor. An easy recipe is simply to boil the root pieces and season with butter and salt.

A tasty Japanese-style recipe for cooking the strips of burdock root with meat is as follows: brown thin strips or small cubes of well-trimmed pork in a saucepan; add about 3 tablespoons of soy sauce, a level teaspoon of sugar, and a sprinkling of monosodium glutamate flavor enhancer if desired; add about one cup of burdock root strips and some water; cover pan and simmer gently until the root pieces are tender and have absorbed the soy sauce mixture (about 30 minutes). Take care that the roots do not cook dry and scorch by adding a tablespoon of water from time to time during cooking as necessary.

FIGURE 17-19
Tall flowering stalks from the previous year and the clusters of hooked seeds are aids in locating burdock plants as they begin growing in the spring.

In addition to its sweet, nutty flavor, gobo is reputed in oriental societies to enhance wisdom, virility, and other virtues. It is also noted for its ability to cause flatulence.

Native American tribes utilized the young leaves (less than 10 cm long) as a potherb. They are also popular in some European societies. They have a strong, earthy taste, however, which merits changing waters during cooking and careful seasoning.

#### Salsify (*Tragopogon porrifolius*)

Salsify is known also as goatsbeard or oyster plant. Like sunchokes and gobo, a cultivated variety is available. Salsify is another weed that likes disturbed sites and is most common along roadsides, although it may be difficult to recognize among the grasses until its narrow-leaved stems are joined in early summer by the distinctive yellow flowers. Flowers are succeeded by large, globelike seed heads (Fig. 5-35) that can be collected for dried arrangements (spray with pump-type hair spray to keep seeds from flying away).

Salsify is a biennial and the roots are more tender on the first-year plants. Unlike burdock, wild salsify roots are easily pulled by hand, being smaller and somewhat carrotshaped (Fig. 17-20). They are pale and exude a milky sap from broken rootlets. Whereas a single good-sized burdock root can provide a serving or two, salsify roots are required in quantity. Wild salsify roots are not as tender as burdock roots and, unless collecting is particularly good, much effort is required to make a meal. Salsify roots can also be used as a coffee substitute in the same manner as chicory and dandelion.

FIGURE 17-20 Roots of salsify (*Tragopogon*), also called goatsbeard or oyster plant.

### Cereallike Uses

A wide assortment of plant parts can be dried and ground up for use as meal or flour, including true grains (cereals), seeds, dried flower heads, seed pods, and vegetative parts. In the wild you can prepare unleavened cakes from most foods in this category; and where the comforts and commodities of a home kitchen are available, you can prepare a full range of cakes, breads, muffins, and cookies.

With true cereals (grass fruits), the edible grain must be separated from the dried floral parts, spines, and awns. The grains may also need to be parched, or toasted. In camp both processes can be accomplished at once by carefully shaking the grass seeds in a pan with some small live coals.

Grains, seeds, dried flowers, and nuts are crushed or ground into meal or flour by means ranging from two stones to an electric blender. Extract starch from such fibrous storage tissues as cattail rhizomes as follows: wash thoroughly; crush or pound; shred and manipulate the root or rhizome in a pan of clean water until as much starch as possible has been removed; discard excess fibrous material and set pan aside for the starch to settle out. When the water is clear, gently pour it off, leaving the starch (flour) as a white sediment. With some plants, mucilaginous substances may give the water a slimy feel so that an additional rinse may be desirable. Rinsing or straining will remove unwanted small fibrous matter. The flour may be used wet or be dried for storage.

As with other edibles, avoid diseased or otherwise abnormal parts. This precaution is particularly important with true grains because the toxic fungus ergot (Chapter 18) is common on wild grasses as well as domesticated species. Ergot can be recognized in a seed head by the presence of enlarged, hard, blackened grains among the normal ones.

### Large Seeds and Nuts

Nuts are hard-shelled plant fruits that can be eaten raw, dried, or roasted. Most edible seeds and nuts can also be used like cereals. Wild hazelnuts (*Corylus*) (Fig. 17-21) resemble their relative, the domesticated filbert.

FIGURE 17-21 Hazelnut (*Corylus*) nuts are enclosed in ruffled bracts.

Special care should be taken in identifying large seeds and nuts, for some dangerously poisonous ones are quite attractive. Also, many members of the pea and bean family produce dangerously toxic seeds. Even the seeds of ornamental sweet peas and many species of wild peas are not edible, at least in quantity. (See also Chapter 18.)

## Oils

The seeds of many plants contain oils that can be extracted for cooking use, assuming the same precautions as in the previous paragraph. The edible seed is crushed and then heated in water, causing oil to rise to the surface where it can be skimmed off. Skimming is easier from a narrow-necked vessel. The remaining meal can be used as a cereal.

## Fruits

Many wild fruits can be used in the same way market fruits are used: fresh or dried, as a juice source, and for making syrups, sauces, pies, cobblers, jams, preserves, jellies, and fruit leathers. Such naturally sweet fruits as blueberries (*Vaccinium*) (Fig. 17-22), the western huckleberries (*Vaccinium*), strawberries (*Fragaria*), raspberries (*Rubus*), and blackberries (*Rubus*) require no additional sugar when eaten raw but are usually cooked with additional sweetening. Other fruits, such as cranberries (*Vaccinium*), are too sour for the average person to enjoy raw and unsweetened unless they are mixed with other, sweeter wild or tame fruits. Black elderberries (*Sambucus*) (Fig. 17-23) and chokecherries (*Prunus*) (Fig. 17-24) are inedible until cooked and sweetened, after which they have a delicious, tangy flavor. Wild grapes (*Vitis*) (Fig. 17-25) are often sought for jellymaking.

Some fruits have a more specific use. The fruit clusters of red-fruited sumacs (*Rhus*) (Fig. 17-26) are bruised in water to make a "lemonade," either fresh or after drying. Sumac jelly made from this extract is pink, tart, and lemony! Some fruit pulps may be dried for cereal use.

Native American tribes utilized native fruits to make a nourishing, high-energy, easily stored food called **pemmican.** Fresh or dried berries, such as blueberries or juneberries (*Amelanchier*), were pounded into a mixture with dried, shredded meat and softened tallow—and salt if available. The food provided all major organic categories—carbohydrates, proteins, and lipids. Homemade pemmican is easy to make. To cater to the tastes of modern backpackers, a dried-fruit mixture is marketed in outdoors stores as pemmican.

Some fruits, although attractive, are inedible and even poisonous. Be very cautious in identifying potentially edible kinds.

FIGURE 17-22 Wild blueberry (*Vaccinium*) fruits are delicious raw.

FIGURE 17-23
Black elderberry (*Sambucus*) fruit clusters weigh down the branches until they are harvested by birds or humans. They are unpalatable until cooked and sweetened, as for elderberry syrup.

FIGURE 17-24 Wild chokecherries (*Prunus*) are bitter when eaten raw—hence their name—but have an excellent flavor as jelly.

FIGURE 17-25 Wild grapes (*Vitis*) are smaller and less sweet than their cultivated relatives but are definitely worth collecting for making jelly.

FIGURE 17-26
The red fruit clusters of sumac (*Rhus*) produce a lemony solution.

## Beverages

Liquid refreshment from wild plants is especially abundant during the spring and summer months. Many berries can be crushed and their juice mixed with water. The slightly sweet, freshly drawn sap of maples (*Acer*), birches (*Betula*), and many other deciduous trees can be used "as is" for drinking or as a cooking liquid, especially with fruits needing sweetening. Harvest the sap during the spring by driving hollow tubes (spiles) into the sapwood, with a collecting container suspended below each spile (Fig. 17-27).

Fermented libations—beers and wines—have been concocted by using the various parts of many types of plants. Flowers and fruits are most commonly used. Birch beer, sarsparilla, and root beers utilize twigs, bark, and roots. Recipe books for home winemaking are readily available and include explanations of the temperature and sanitation criteria necessary to produce good wines. Dandelion flowers, wild grapes, plums, cherries, and elderberries are often used for home wines.

Chicory (*Chicorium intybus*) roots and dandelion roots are well-known coffee substitutes. Wash and dry the roots; then roast them slowly in an oven until dark brown. Grind the roasted roots. The grounds can be used alone to brew a hot drink, but they tend to be bitter alone and are usually mixed with ground coffee. At least one brand of Creole-style coffee (a mixture of ground coffee beans and chicory roots) is available commercially.

FIGURE 17-27
Maple syrup collecting in northern Wisconsin. (Photo by Mark R. Fay.)

Chicory was brought to America as a cherished food plant whose crowns and new leaves provide endive for salads. The plant is easily recognized in the summer by its bright-blue ragged flowers (Fig. 17-28) and grows readily in disturbed areas, such as along railroad tracks and roadsides.

Many teas are available from the wild, usually using leaves, stems, flowers, and sometimes fruits (Figs. 17-29 and 17-30). Note also the reference to herb teas (tisanes) in the Precautions section of this chapter.

**FIGURE 17-29** New Jersey tea (*Ceanothus*).

**FIGURE 17-28** The versatile chicory (*Chichorium*) plant.

**FIGURE 17-30** Horsemint or beebalm (*Monarda*), a wild bergamot that has deliciously aromatic leaves.

Mashed young basswood (*Tilia*) fruits, chinquapin (*Castanea*) nuts, and purple avens (*Geum*) roots have been used as cocoa substitutes for making a hot chocolate-type of drink.

### Sugar and Spice and Miscellaneous

Many species are sources of sugars, flavorings, or seasonings. Some tree saps, for example, can be cooked down for syrup and crystallized as sugar. Tree saps can also be fermented to produce vinegars.

Certain plants have distinctive flavors that make them the spices, seasonings, and flavorings of wild edibles—for instance, mustards and peppercress, wild onion and garlic (*Allium*) (Fig. 17-31), and wild ginger (*Asarum canadense*).

Edible species with especially succulent parts, such as cacti, are suitable for making condiments, relishes, and pickles. Some plants, as wild ginger, are simmered in sugar solutions; then partially dried and coated with dry sugar to make candies. Partially dried, coagulated tree sap or gum, as that which occurs at a wound in the bark, is a natural chewing gum. Mucilaginous plants, such as purslane (*Portulaca*) and spiderwort (*Tradescantia*), thicken soups, stews, and other dishes. Probably the most popular thickeners are seaweeds and filé (ground *Sassafras* leaves and twigs), used in gumbo dishes.

A few plants—nettles, cleavers (*Galium*) (Fig. 17-32), and thistle heads—act as rennet to coagulate milk for pudding (junket) and certain soft, unripened cheeses.

FIGURE 17-31 The aromatic potency of wild garlic (*Allium*) is many times that of its milder, domesticated relatives. The plants can be distinguished from wild onions or chives by their triangular rather than round leaves.

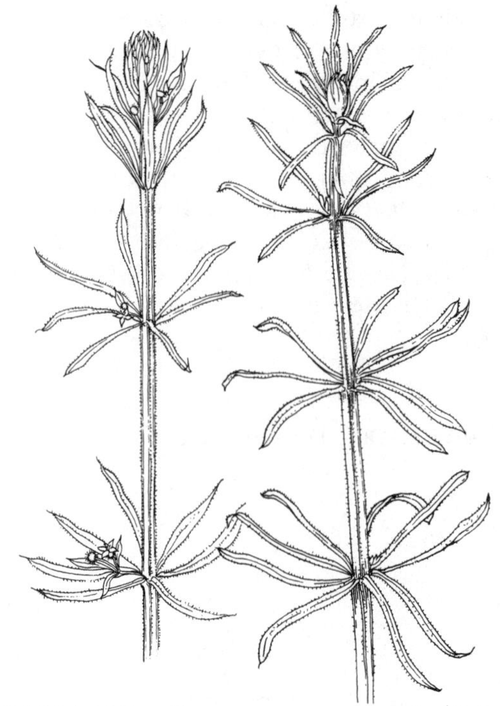

FIGURE 17-32 Cleavers or catchweed (*Galium*) may be used as a potherb or as a kitchen strainer!

## EDIBLE MARINE ALGAE

Aside from ethnic groups that still maintain a strong traditional identity with the sea, the population of North America remains largely ignorant of and prejudiced against the many food uses of marine algae except in a highly refined form. The designation "seaweed" denotes this prejudice just as surely as the term "weed" dismisses the many edible and otherwise appreciable native and naturalized terrestrial plants growing wild. As a group, the large marine algae are free of incidents of human poisoning unless collected from industrially contaminated water.

Marine algae contain gelling substances, **algin**, **agar**, and **carrageenan** (Table 17-1), which are used as thickeners, stabilizers, and fillers in salad dressings, ice cream, chocolate milk, tablets, capsules, hand lotions, and other products. As edibles in their own right, seaweeds have a slight-to-strong marine flavor for which you might have to acquire a taste. Following taste acclimation, many people create special opportunities to eat dishes prepared with marine algae. Most dried seaweeds sold in the United States are from Japan and can be identified by the Japanese names.

The edible marine algae can be used fresh or dried and they are sold dried or pickled in oriental food shops. The thick-bodied kelps (kombu), including several species of brown algae, are used as a vegetable, to make pickles, relishes, or soup stock; they can be fried and seasoned to make a crunchy snack. Some seaweeds are used in salads, typically after being seasoned with sweetened vinegar. Some are especially used in soups. A fragile-bodied red alga called laver or nori (*Porphyra*) is pressed into sheets as it dries, to be used as a wrapper for one type of sushi, a Japanese food of mainly sweet- and sour-flavored rice. Nori is also seasoned and toasted for a snack or condiment. Dried seaweeds, particularly kelp, can be ground into flour.

Edible seaweeds fall within three taxonomic groups—green, red, and brown algae. Most green algae occur in freshwater, but two of the few marine genera are utilized extensively as food. Sea lettuce or aosa (*Ulva*) is a bright-green algae that grows within the intertidal zone along the Atlantic and Pacific shores and is easily recognized and harvested. It is eaten raw or cooked, in salads or soups, and is also dried and powdered as a seasoning. On drying, sea lettuce turns black.

*Enteromorpha intestinalis* is a green alga composed of thin tubes. The Japanese call it ohashinori (hashi is the Japanese word for chopsticks); in Hawaii it is called limu-eleele. One common English name is link confetti. It is frequently used in soups.

**TABLE 17-1 Some marine algae from which commercial algal derivatives are extracted.**

| Product | Algal Examples | Algal Category |
|---|---|---|
| Agar (kanten) | *Gracilaria* spp. | Red |
| | *Gelidium* spp. | Red |
| Algin | *Laminaria* spp. | Brown |
| | *Macrocystis* spp. | Brown |
| Carrageenan | *Chondrus crispus* | Red |

There are several very common red algae genera of the coastal United States with easy-to-collect, abundant representatives. The most common and most versatile is laver or nori (*Porphyra nori*). This purplish-red, thin-sheeted, ruffled-looking red alga is prominent in the intertidal zones of both coasts. It is tender eaten raw either fresh or partially dried, by itself or in salads, and is frequently cooked in soups or meat dishes. It is the one used for sushi wrappers and for a variety of Japanese-style condiments (okazu) that are eaten with rice.

Dulse or dillisk (*Rhodymenia*) is an edible red alga that various authors agree is more tender dried or cooked than fresh. Euell Gibbons pressed partially dried, pliable dulse into layered plugs, which he kept refrigerated until needed, and then shaved off shreds for seasoning salads, slaws, or meat dishes.

Purplish-red, much-branched Irish moss (*Chondrus crispus*) has been used in Europe and the eastern United States as a thickener for blancmange, a pudding. Also an intertidal resident of both seacoasts, Irish moss is marketed in a dried, sun-bleached form. It can be used either fresh or dried, although freshly gathered algae should be rinsed a number of times to remove the marine flavor. Many other red algae contain carrageenan and can be used like Irish moss. The most convenient way to extract the carrageenan is to stuff a quantity of Irish moss into a porous bag and steep it in hot water, stirring occasionally. The longer the soaking, the more carrageenan will be extracted. Remove the bag. The extract gels as it cools. Gentle reheating will reliquify it. Irish moss extract makes gelled fruit desserts and salads, aspic, custards, soups, and blancmange and can be diluted as a beverage when "living off the land." Irish moss can itself be eaten, for it becomes tender when cooked.

Other edible red algae are members of the genera *Gelidium* and *Gracilaria*, both known as limu-maneoneo in Hawaii. They are used as a cooked or raw relish, especially mixed with chopped or mashed tomatoes.

Whereas the common edible red algae are mainly thin bodied, the edible brown algae (kelps) are thick and succulent enough to be utilized as a vegetable and even in making chunky pickles. Commonly eaten genera are *Laminaria, Alaria,* and *Nereocystis.* Dried kelp is known to the Japanese as kombu, a kitchen staple for seasonings, soups, or condiments. In a powdered form it is the basic ingredient for a versatile and frequently used broth (dashi). Kelps can also be nibbled or used raw in salads. Removing the outer membrane and rinsing tone down the marine flavor. One exceptionally common brown alga, the rockweed *Fucus,* is not usually eaten; but it contributes moisture and some flavor to seafood, such as clams, which are steamed in layers of it.

Seaweeds are easily dried for home use. Collect at low tide to select those in the best condition and remove the tough attachment structures (holdfasts) to which sand or rock particles are likely to adhere. Large kelps, such as *Nereocystis* (bull kelp or bullwhip), grow anchored beyond the low tide line. Specimens are frequently washed ashore, however, and may be used if still fresh. Rinse away sand, small animals, and adherent algae with clean seawater. Remove damaged areas or areas where encrusting colonial animals are attached. Most seaweed can be dried after rinsing in seawater, leaving a light powder of flavorful and naturally iodized sea salt. Nevertheless, Irish moss or other reds to be used for carrageenan are better after additional rinsing with nonsaline water. Air dry seaweeds on a rack or screen or by hanging or oven dry at the lowest temperature (drying, not cooking, is the object!) with the oven door partially open. Once completely dried, they should be stored in airtight containers in a cool, dark place for best keeping.

Dried seaweeds are easily reconstituted with water for any seaweed or vegetable recipes. Dried seaweeds are easily shredded and can be powdered by grinding in an electric blender. In addition to use as a seasoning, powdered seaweed can be combined with flour in regular recipes. Algal flour will contribute moistness to the baked product.

## LICHENS

Greatest use of lichens as food occurs in the far North—Canada, Greenland, Iceland, and Alaska. Notable species are rock tripe (*Umbilicaria*) and Iceland moss (*Cetraria*). Wash lichens thoroughly to remove adherent grit and dirt and soak in several water changes for a number of hours before eating. This leaching is necessary to remove bitter-tasting acids, especially **usnic acid**, which are likely to cause indigestion and purgation. Rock tripe is cited as being edible raw if relatively small amounts are ingested. It is best used with meat, however, as a vegetable and thickening ingredient in soups and stews. Roasting by the fire or in a pan before cooking is said to improve the flavor of rock tripe.

Iceland moss is usually soaked, dried, and then crushed into powder to be used in baking cakes and bread, as a thickener in soups and stews, in making a gelatinous gruel, or for blancmange. Reindeer moss (*Cladonia rangiferina*) is another abundant northern lichen that can be used in the same ways.

## FUNGI

This is a large and varied group, ranging from single-celled yeasts and filamentous bread molds to large mushrooms, puffballs, and bracket fungi. Many fungi are edible—in fact, delicious when properly prepared. Most fungi require expert identification. **Morels** (*Morchella*) are such a distinctive form, however, that they are extremely unlikely to be mistaken for harmful species. Even the semilookalike false morels (*Helvella, Gyrometra*) are easily distinguished, their caps being wrinkled and folded rather than pitted like the true morels. True morels are hollow along their full length (Fig. 17-33).

**FIGURE 17-33** Edible morels (*Morchella*), a springtime delicacy.

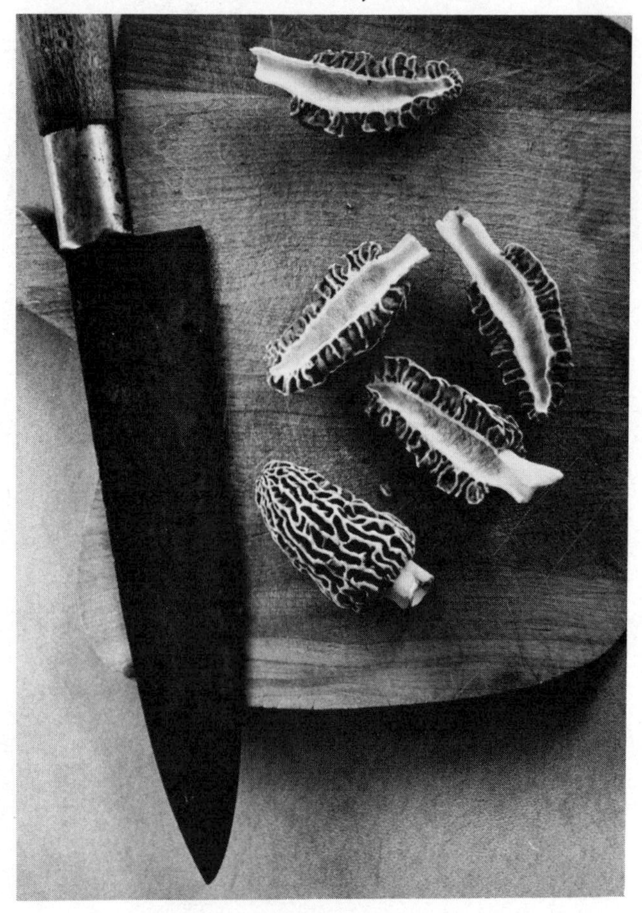

In the upper Midwest morels usually become abundant around mid-May, particularly under dead and dying elm trees (Fig. 17-34). Break or cut morels off at ground level, thus keeping dirt out of the collecting bag and leaving most of the subterranean fungal body intact to generate more morels. Rinse them carefully and soak briefly in a warm, weak salt solution to evict ants or small insect larvae that may be hiding in the hollow interior or pores. Unless the morels are to be stuffed before cooking, it is convenient to cut them lengthwise for rinsing. A simple and delicious mode of preparation is to dredge the morel halves in flour and then fry in butter, salting lightly if desired.

Most **puffballs** are also safe to eat, although some are more palatable than others and certain species have been known to cause indigestion. In general, avoid spiny-skinned puffballs (Fig. 17-35). Precise identification should be made by cutting each puffball lengthwise to be certain that you have not, in fact, collected the button stage of a **mushroom**—such as the white and deadly *Amanita* species (Fig. 17-36).

The giant puffball (*Calvatia gigantea*) cannot be mistaken for anything else, for it attains melon size. Collect it and its smaller relatives before the spores mature so that the interior is still white and not at all darkened. Puffballs can be sliced and prepared in many ways.

The growing body, or mycelium (spawn), of a puffball colony first appears as a white mat of fila-

**FIGURE 17-34** Habitat shot of a morel. Morel collecting requires a sharp eye, as they are tan colored, like dry leaves. A sensitive nose can also be helpful.

**FIGURE 17-35**
Most puffball species are edible, but spiny-skinned species like this one should be *avoided* unless the collector is expert enough to know the exact species.

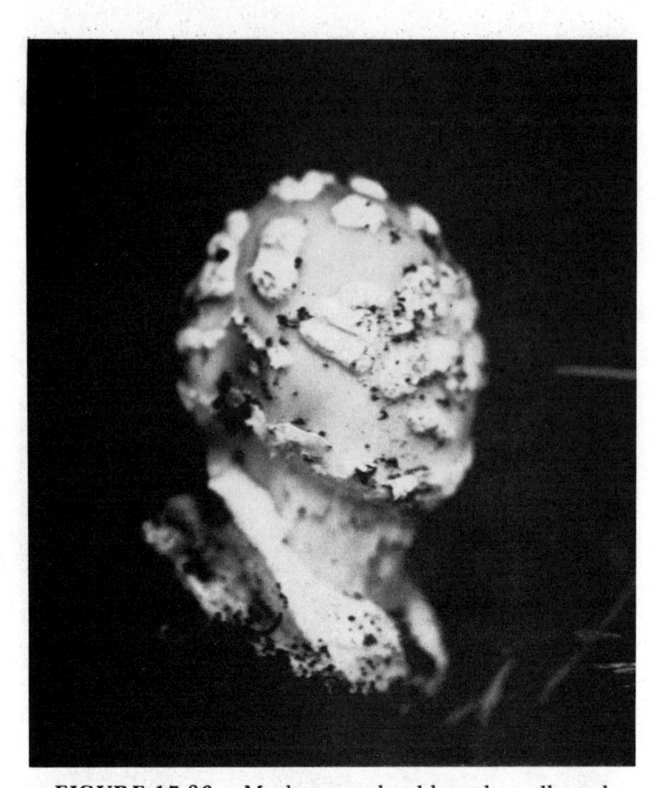

**FIGURE 17-36** Mushrooms should not be collected in the "button" stage, when they may superficially resemble a puffball and when their full identifying characteristics are immature. For example, the identifying characteristics of this poisonous *Amanita* are not all clearly visible until the cap has torn free of the **veil** that attaches its margins to the stalk.

ments (hyphae) at the surface of the soil when conditions are sufficiently moist. Once observed, these mats can be checked periodically to see if the puffballs have formed.

Beyond these two categories of easily recognized edible fungi, expert instruction is essential for the aspiring mycophage (fungus eater). In general, the safest policy is to collect with a knowledgeable and experienced collector who does not tend to be particularly adventurous. In addition, an illustrated handbook and local taxonomic key are essential to provide precise identification and to learn distinctive characteristics and warning signs.

The "taste and wait" precautions that can be used with green plants of uncertain edibility do NOT apply to mushrooms. Even small amounts of poisonous species are certain to induce gastrointestinal agony as well as possible visual and respiratory disruptions and—with some species—paralysis and death or permanent tissue damage (Chapter 18). An additional complication is that some species can be safely eaten by some persons but cause problems for others.

Although many species can be safely collected and eaten, relatively few are typically recommended for the amateur due to their resemblance to dangerous forms. Amateurs should even avoid many excellent genera that are enjoyed by more proficient collectors and that are highly recommended in some foraging books until they have successfully practiced identifying them. Mushrooms should be scrutinized carefully for interpretation of identifying characteristics *at the time of collection* because many species undergo rapid physiological and structural changes after picking.

Certain poisonous species are noted for their similarities to edible species. The common meadow mushroom (*Agaricus campestris*), which is cultivated commercially and solid fresh and canned in nearly all grocery stores, is easily mistaken in the wild for early stages of poisonous *Amanita* species.

A few easily recognized edible species are included in the following brief discussion. An illustrated key should be consulted.

The shaggy mane (*Coprinus comatus*) of lawns, roadways, and gardens is characterized by a conical cap that remains closed and that is covered with "shaggy" whitish-to-brown scales. Common inky cap (*Coprinus atramentarius*) appears near shrubs, trees, and in lawns during later summer and autumn whereas its relative, the glistening inky cap (*C. micaceus*), grows in decaying wood, including old stumps. Inky caps begin to liquify into a black, "inky" mass within a few hours after harvesting, a clue to their identification. They must be used before they self-destruct. Even these easily recognizable and commonly eaten species *should not be eaten with alcohol.* Flushing and shortness of breath are likely to occur when the common inky cap combines with alcohol in the system. This effect can be residual for some time, being triggered by the presence of alcohol even when no mushrooms have been eaten at that time.

Colonial clusters of the oyster mushroom (*Pleurotus ostreatus*) appear on dead tree trunks and occasionally on living trees during summer and fall—reportedly into winter in some localities.

Most people are familiar with the fairy ring mushroom (*Marasmius oreades*), which appears on grassy turf, such as lawns, growing in rings of ever-expanding diameters. This species is a "classic" model of the capped mushrooms that appear in fantasy illustrations replete with elves. In temperate climates this mushroom can be collected year-round. Because it is not the only mushroom that forms a fairy ring, however, other characteristics must be used in identification.

It is to be hoped that this discussion has not discouraged all inclination to collect wild mushrooms but has merely instilled a proper sense of caution. Most handbooks include a list of rules to be applied, especially for the beginning collector. Probably the first rule is to discount any tests for toxic potential that are found in traditional lore: squirrels, mice, deer, and other animals are not safe testers of mushroom species for humans; all poisonous mushrooms do not tarnish silver; certain representatives of both toxic and nontoxic groups peel easily; habitat or substrate preference (e.g., in woods, in meadows, on trees) is not a constant criterion of safety.

In addition, the collector should memorize features of the most poisonous genus, *Amanita,* and avoid any specimens that show *any* evidence of a cup (**volva**) at the base of the stalk (Fig. 17-37). (Look carefully in the ground as well as at the surface.) Do not collect mushrooms before the caps have begun to open, for clear-cut identification is not possible in the button stage. Select firm, newly matured specimens and reject any that appear soggy, have begun to deteriorate, or that harbor tiny maggots (wormlike fly larvae). To check for worminess, cut the stalk and check for pin-sized holes; if one maggot is present, the mushroom is probably riddled with many small tunnels and their slender, pale creators.

Additional precautions include avoiding ill-smelling, bitter, peppery-tasting or tough specimens, mushrooms with milky juice (unless you are certain that they belong to a safe species) and boletes (pore-bearing mushrooms) whose pores are reddish or that turn blue when bruised or cut.

The foregoing are primarily collecting precautions, but there are other commonsense precautions for eating: prepare only one kind of mushroom at a meal so that identification is known in case of difficulty; cook the mushrooms well; eat in moderation; eat only a small amount if trying a new species; allow each diner to be his or her own tester, for individual reactions to certain mushrooms vary; avoid alcoholic beverages when eating mushrooms.

For additional information on mushroom poisoning, see Chapter 18.

## PRECAUTIONS AND PITFALLS

Following the discussion of mushrooms, it seems appropriate to summarize some general precautions applicable to foraging in general.

1. Do not eat anything unless you are confident of its identification. Use an illustrated hand-

**FIGURE 17-37**
White, stately, and potentially deadly—mature *Amanita* mushrooms. Note the **ring** around the stalk. Another identifying characteristic is the cuplike **volva** at or just below the soil line, from which the stalk arises.

book, preferably with a taxonomic key, to be sure.

2. Learn to identify poisonous plants likely to be encountered locally, particularly those that might be mistaken for or are related to edible species. Death camas (*Zygadenus*), for example, has been mistaken for edible camas (*Camassia*) or wild onions (*Allium*) when not in bloom.
3. Learn the specific exceptions to the general rules stated below.
4. Avoid eating bulbs that lack the smell of onions or garlic.
5. Avoid eating plants that have milky or colored juices (members of the poison ivy, spurge, poppy, milkweed families).
6. Avoid all unknown white or red fruits (poison ivy, poison sumac, baneberry).
7. Avoid all fruits that are three angled or three lobed. This step eliminates dangers from the spurge, soapberry, horse chestnut, amaryllis, and lily families, which contain some of the most infamous poisonous plants.
8. Avoid wild seeds, which usually have the greatest concentration of toxin. In general, toxicity is greatest in storage organs like seeds, fruits, tuberous roots, and stems.
9. Young plants and shoot tips are usually less toxic than mature structures. Wilted plants may be more dangerous than turgid ones.
10. Learn which plants require cooking or leaching to make them safe to eat.
11. Beware of plants that may have been treated with herbicides or pesticides. Wilted or distorted-looking vegetation along a country road or railroad right-of-way has probably been sprayed with a herbicide such as 2,4-D. Avoid collecting dandelions and other plants from a lawn treated with lawn-care products. Parks and recreation areas are sometimes routinely sprayed for mosquitoes, flies, ticks, and/or chiggers. Even fruits and vegetables from the grocery store usually retain pesticide or fungicide residues and should be washed thoroughly before use. Exercise the same caution with wild collections.
12. Learn if soil in the area is unusually high in selenium or nitrates, for many plants accumulate toxic amounts of these substances (Chapter 18).
13. Observe the "taste and wait" safety measure recommended at the beginning of the chapter if necessitated by a survival situation—*except never do so with fungi.*
14. Practice moderation with edible wild plants as well as cultivated plants. Many common grocery-variety vegetables contain natural toxic substances that have caused livestock and human poisoning when consumed in more than moderate amounts or when prepared improperly.

It is especially important to use moderation when making teas. Leaves of labrador tea (*Ledum greenlandicum*) of the northern United States and Canada, for instance, contain a resinous carbohydrate that is released to form a fragrant tea when the leaves are steeped in boiling water. Consumption of normal small quantitites of Labrador tea is considered safe. The resinous substance is toxic, however, if consumed immoderately. Teas made from stronger members of the same family have caused severe poisoning and even death.

Tisanes, or herb teas, are increasing in popularity. They are reputed to have certain beneficial qualities—for the heart, digestion, relaxation, and so on. Even Peter Rabbit was comforted with a dose of chamomile tea before bed after his unsettling experiences in Mr. McGregor's garden! Tisanes have been used for specific purposes for centuries and are safe—when consumed in moderate amounts, such as one cup per day. Toxic principles are, in certain amounts, medicinal, having definite chemical effects on an organism's physiological activities. Even beneficial chemicals, however, are poisonous in too large amounts.

## NUTRITIONAL VALUE OF EDIBLE WILD PLANTS

The initial impetus behind experimentation with wild plants is probably curiosity. Later, a person may become more deeply involved in foraging, as a sense of freedom from restrictions to cultivated foods is realized. An impressive case can also be made for the economic and recreational aspects.

Wild food enthusiasts claim that there are nutritional benefits as well. Because of sufficient interest in nutrition, some of the most popular wild plants have been analyzed for nutritional information. Table 17-2 consists of U.S. Department of Agriculture data on selected wild plants as well as some cultivated foods for comparison. Many wild plants compare favorably and sometimes excel in nutritional value. The table also demonstrates how cooking causes certain vitamins and minerals to be washed out of food.

TABLE 17-2 Composition of foods, 100 grams, edible portion.[a]

| Food and Description | Water (Percent) | Food Energy (Calories) | Protein (Grams) | Fat (Grams) | Carbohydrates: Total (Grams) | Carbohydrates: Fiber (Grams) | Ash (Grams) | Calcium (Milligrams) | Phosphorus (Milligrams) | Iron (Milligrams) | Sodium (Milligrams) | Potassium (Milligrams) | Vitamin A Value (International Units) | Thiamine (Milligrams) | Riboflavin (Milligrams) | Niacin (Milligrams) | Ascorbic Acid (Milligrams) |
|---|---|---|---|---|---|---|---|---|---|---|---|---|---|---|---|---|---|
| Amaranth: raw | 86.9 | 36 | 3.5 | .5 | 6.5 | 1.3 | 2.6 | 267 | 67 | 3.9 | — | 411 | 6,100 | .08 | .16 | 1.4 | 80 |
| Asparagus | | | | | | | | | | | | | | | | | |
| raw spears | 91.7 | 26 | 2.5 | .2 | 5.0 | .7 | .6 | 22 | 62 | 1.0 | 2 | 278 | 900 | .18 | .20 | 1.5 | 33 |
| boiled spears | 93.6 | 20 | 2.2 | .2 | 3.6 | .7 | .4 | 21 | 50 | .6 | 1 | 183 | 900 | .16 | .18 | 1.4 | 26 |
| Beet greens: boiled | 93.6 | 18 | 1.7 | .2 | 3.3 | 1.1 | 1.2 | 99 | 25 | 1.9 | 76 | 332 | 5,100 | .07 | .15 | .3 | 15 |
| Blackberries, dewberries, boysenberries, youngberries: raw | 84.5 | 58 | 1.2 | .9 | 12.9 | 4.1 | .5 | 32 | 19 | .9 | 1 | 170 | 200 | .03 | .04 | .4 | 21 |
| Blueberries: raw | 83.2 | 62 | .7 | .5 | 15.3 | 1.5 | .3 | 15 | 13 | 1.0 | 1 | 81 | 100 | (.03) | (.06) | (.5) | 14 |
| Broccoli: boiled | 91.3 | 26 | 3.1 | .3 | 4.5 | 1.5 | .8 | 88 | 62 | .8 | 10 | 267 | 2,500 | .09 | .20 | .8 | 90 |
| Carrots | | | | | | | | | | | | | | | | | |
| raw | 88.2 | 42 | 1.1 | .2 | 9.7 | 1.0 | .8 | 37 | 36 | .7 | 47 | 341 | 11,000 | .06 | .05 | .6 | 8 |
| boiled | 91.2 | 31 | .9 | .2 | 7.1 | 1.0 | .6 | 33 | 31 | .6 | 33 | 222 | 10,500 | .05 | .05 | .5 | 6 |
| Chard, swiss: boiled | 93.7 | 18 | 1.8 | .2 | 3.3 | .7 | 1.0 | 73 | 24 | 1.8 | 86 | 321 | 5,400 | .04 | .11 | .4 | 16 |
| Collards: boiled in small amt. of water | 89.6 | 33 | 3.6 | .7 | 5.1 | 1.0 | 1.0 | 188 | 52 | .8 | — | 262 | 7,800 | .11 | .20 | 1.2 | 76 |
| Corn, sweet: cooked, off cob | 76.5 | 83 | 3.2 | 1.0 | 18.8 | .7 | .5 | 3 | 89 | .6 | trace | 165 | 400 | .11 | .10 | 1.3 | 7 |
| Cucumbers: raw, peeled | 95.7 | 14 | .6 | .1 | 3.2 | .3 | .4 | 17 | 18 | .3 | 6 | 160 | trace | .03 | .04 | .2 | 11 |
| Dandelion greens | | | | | | | | | | | | | | | | | |
| raw | 85.6 | 45 | 2.7 | .7 | 9.2 | 1.6 | 1.8 | 187 | 66 | 3.1 | 76 | 397 | 14,000 | .19 | .26 | — | 35 |
| boiled | 89.8 | 33 | 2.0 | .6 | 6.4 | 1.3 | 1.2 | 140 | 42 | 1.8 | 44 | 232 | 11,700 | .13 | .16 | — | 18 |
| Dock (curly, narrow leaf, broad leaf, sheep sorrel): boiled | 93.6 | 19 | 1.6 | .2 | 3.9 | .7 | .7 | 55 | 26 | .9 | 3 | 198 | 10,800 | .06 | .13 | .4 | 54 |
| Filberts (hazelnuts) | 5.8 | 634 | 12.6 | 62.4 | 16.7 | 3.0 | 2.5 | 209 | 337 | 3.4 | 2 | 704 | — | .46 | — | .9 | trace |
| Grapefruit: raw | 88.4 | 41 | .5 | .1 | 10.6 | .2 | .4 | 16 | 16 | .4 | 1 | 135 | 80 | .04 | .02 | .2 | 38 |
| Jerusalem artichoke: raw | 79.8 | — | 2.3 | .1 | 16.7 | .8 | 1.1 | 14 | 78 | 3.4 | — | — | 20 | .20 | .06 | 1.3 | 4 |
| Kale: boiled | 91.2 | 28 | 3.2 | .7 | 4.0 | 1.1 | .9 | 134 | 46 | 1.2 | 43 | 221 | 7,400 | — | — | — | 62 |

TABLE 17-2 (Continued).

| Food and Description | Water (Percent) | Food Energy (Calories) | Protein (Grams) | Fat (Grams) | Carbohydrates: Total (Grams) | Carbohydrates: Fiber (Grams) | Ash (Grams) | Calcium (Milligrams) | Phosphorus (Milligrams) | Iron (Milligrams) | Sodium (Milligrams) | Potassium (Milligrams) | Vitamin A Value (International Units) | Thiamine (Milligrams) | Riboflavin (Milligrams) | Niacin (Milligrams) | Ascorbic Acid (Milligrams) |
|---|---|---|---|---|---|---|---|---|---|---|---|---|---|---|---|---|---|
| Lambsquarters | | | | | | | | | | | | | | | | | |
| raw | 84.3 | 43 | 4.2 | .8 | 7.3 | 2.1 | 3.4 | 309 | 72 | 1.2 | — | — | 11,600 | .16 | .44 | 1.2 | 80 |
| boiled | 88.9 | 32 | 3.2 | .7 | 5.0 | 1.8 | 2.2 | 258 | 45 | .7 | — | — | 9,700 | .10 | .26 | .9 | 37 |
| Lettuce, iceberg: raw | 95.5 | 13 | .9 | .1 | 2.9 | .5 | .6 | 20 | 22 | .5 | 9 | 175 | 330 | .06 | .06 | .3 | 6 |
| Lettuce, loose-leaf: raw | 94.0 | 18 | 1.3 | .3 | 3.5 | .7 | .9 | 68 | 25 | 1.4 | 9 | 264 | 1,900 | .05 | .08 | .4 | 18 |
| Mustard greens | | | | | | | | | | | | | | | | | |
| raw | 89.5 | 31 | 3.0 | .5 | 5.6 | 1.1 | 1.4 | 183 | 50 | 3.0 | 32 | 377 | 7,000 | .11 | .22 | .8 | 97 |
| boiled | 92.6 | 23 | 2.2 | .4 | 4.0 | .9 | .8 | 138 | 32 | 1.8 | 18 | 220 | 5,800 | .08 | .14 | .6 | 48 |
| Oranges, peeled: raw | 86.0 | 49 | 1.0 | .2 | 12.2 | .5 | .6 | 41 | 20 | .4 | 1 | 200 | 200 | .10 | .04 | .4 | (50) |
| Persimmons, native: raw | 64.4 | 127 | .8 | .4 | 33.5 | 1.5 | .9 | 27 | 26 | 2.5 | 1 | 310 | — | — | — | — | 66 |
| Poke shoots boiled: (Never eat raw!) | 92.9 | 20 | 2.3 | .4 | 3.1 | — | 1.3 | 53 | 33 | 1.2 | — | — | 8,700 | .07 | .25 | 1.1 | 82 |
| Purslane, leaves and stems | | | | | | | | | | | | | | | | | |
| raw | 92.5 | 21 | 1.7 | .4 | 3.8 | .9 | 1.6 | 103 | 39 | 3.5 | — | — | 2,500 | .03 | .10 | .5 | 25 |
| boiled | 94.7 | 15 | 1.2 | .3 | 2.8 | .8 | 1.0 | 86 | 24 | 1.2 | — | — | 2,100 | .02 | .06 | .4 | 12 |
| Salsify: boiled | 81.0 | — | 2.6 | .6 | 15.1 | 1.8 | .7 | 42 | 53 | 1.3 | — | 266 | 10 | .03 | .04 | .2 | 7 |
| Seaweeds: raw | | | | | | | | | | | | | | | | | |
| dulse | 16.6 | — | — | 3.2 | — | 1.2 | 22.4 | 296 | 267 | — | 2,085 | 8,060 | — | — | — | — | — |
| irish moss | 19.2 | — | — | 1.8 | — | 2.1 | 17.6 | 885 | 157 | 8.9 | 2,892 | 2,844 | — | — | — | — | — |
| kelp | 21.7 | — | — | 1.1 | — | 6.8 | 22.8 | 1,093 | 240 | — | 3,007 | 5,273 | — | — | — | — | — |
| laver | 17.0 | — | — | .6 | — | 3.5 | 11.0 | — | — | — | — | — | — | — | — | — | — |
| Spinach | | | | | | | | | | | | | | | | | |
| raw | 90.7 | 26 | 3.2 | .3 | 4.3 | .6 | 1.5 | 93 | 51 | 3.1 | 71 | 470 | 8,100 | .10 | .20 | .6 | 51 |
| boiled | 92.0 | 23 | 3.0 | .3 | 3.6 | .6 | 1.1 | 93 | 38 | 2.2 | 50 | 324 | 8,100 | .07 | .14 | .5 | 28 |
| Strawberries: raw | 89.9 | 37 | .7 | .5 | 8.4 | 1.3 | .5 | 21 | 21 | 1.0 | 1 | 164 | 60 | .03 | .07 | .6 | 59 |
| Sunflower seeds: dry | 4.8 | 560 | 24 | 47.3 | 19.9 | 3.8 | 4.0 | 120 | 837 | 7.1 | 30 | 920 | 50 | 1.96 | .23 | 5.4 | — |
| Walnuts, black | 3.1 | 628 | 20.5 | 59.3 | 14.8 | 1.7 | 2.3 | trace | 570 | 6.0 | 3 | 460 | 300 | .22 | .11 | .7 | — |
| Walnut, English | 3.5 | 651 | 14.8 | 64.0 | 15.8 | 2.1 | 1.9 | 99 | 380 | 3.1 | 2 | 450 | 30 | .33 | .13 | .9 | 2 |
| Watercress: raw | 93.3 | 19 | 2.2 | .3 | 3.0 | .7 | 1.2 | 151 | 54 | 1.7 | 52 | 282 | 4,900 | .08 | .16 | .9 | 79 |

[a]Numbers in parentheses denote values imputed—usually from another form of the food or from a similar food. Dashes denote lack of reliable data for a constituent believed to be present in measurable amount. Calculated values, such as those based on a recipe, are not in parentheses.

Selected from *Composition of Foods*, Agriculture Handbook No. 8, Agricultural Research Service, United States Department of Agriculture.

## NONFOOD USES OF WILD PLANTS

As the saying goes, "Necessity is the Mother of Invention" and a bit of imagination can locate natural materials for manipulation into most of life's essentials.

Many plants, including such tall herbs as nettles, thistles, flax, and certain grasses, are sources of long, strong fibers that can be fashioned into threads for sewing or weaving, cords for fishing lines and nets, and ropes to lash together shelter materials and make snares. A crude method of obtaining the fibers is similar to that used to obtain linen fibers from flax (Chapter 16). Cut, bundle, and dry the mature plants. Soak the dried stems in water for a few days to deteriorate the vegetative tissues. Dry them again; then beat bundles of the dried plants against a post or with sticks to shatter away nonfibrous tissues. The result should be a hank of fibers long enough to be twisted into cordage.

A renewed interest in hand weaving has resulted in the return to many natural fibers and has been accompanied by a surge of interest in experimentation with plant dyes. (See Suggested Readings.) Natural dyes vary from batch to batch and fiber to fiber so that products of natural dyeing are infinitely subtle. In most cases, pretreatment of fibers with a **mordant**, such as alum, will increase their affinity for the dye and hence their colorfastness. Specific mordants result in different colors from the same dyes. Further interest can be pursued in a number of informative, color-illustrated books.

**FIGURE 17-38** Mullein (*Verbascum*), a striking "weed" of grassy areas, is easily identified by its thick, fuzzy leaves. A flowering spike is formed during the second growing season of this biennial.

**FIGURE 17-39** *Yucca,* or Spanish bayonet. (Photo courtesy of John S. Thiede.)

Medicinal uses of wild plants range from first-aid treatment of insect bites with the crushed leaves of mullein (*Verbascum thapsus*) (Fig. 17-38) to using a decoction of laxative cascara bark (*Rhamnus purshiana*).

Cleanliness need not be neglected in the wild, for a number of plants contain saponins, which produce a soapy lather. Saponins should not be taken internally because they are toxic. Roots of soapwort or bouncing bet (*Saponaria officinalis*), various yuccas (*Yucca*) and lambsquarters, and nearly all parts of the agave plant (*Agave*) contain saponin. Indians of the southwestern United States have utilized yucca for centuries (Fig. 17-39) and its use has become commercial in a popular shampoo.

Many plants are used as cosmetics, skin softerners (Table 17-3), or astringents (Table 17-4). Read labels of commercial preparations to see which plant ingredients are included!

**TABLE 17-3 Some emollient plants.**

| Common Name | Scientific Name | Part Used |
|---|---|---|
| Aloe | *Aloe* | Leaves |
| Cattail | *Typha* | Root |
| Fig | *Ficus* | Fruit |
| Groundsel | *Senecio* | Leaves |
| Jimsonweed | *Datura* | Seeds |
| Locust | *Robinia* | Flowers |
| Lotus | *Nelumbo* | Root, leaves, seed |
| Mallow | *Malva* | Foliage |
| Ragweed | *Ambrosia* | Foliage |
| Rose mallow | *Hibiscus* | Seeds (also astringent) |
| Water shield | *Brasenia* | Root |

**TABLE 17-4 Some plants with astringent properties.**

| Common Name | Scientific Name | Part Used |
|---|---|---|
| Ash | *Fraxinus* | Bark |
| Bayberry | *Myrica* | Bark, root |
| Boneset | *Eupatorium* | Root, foliage |
| Chinaberry | *Melia* | Leaves |
| Chinquapin | *Castanea* | Bark |
| Dock | *Rumex* | Root |
| Geranium (wild) | *Geranium* | Bark |
| Goldenrod | *Solidago* | Entire plant |
| Hickory | *Carya* | Leaves |
| Huckleberry | *Vaccinium* | Fruits |
| Knotweed | *Polygonum* | Root |
| Madrone | *Arbutus* | Bark, leaves |
| Oak | *Quercus* | Bark |
| Persimmon | *Diospyros* | Green fruit, bark |
| Pigweed | *Amaranthus* | Leaves |
| Pine | *Pinus* | Needles |
| Plantain | *Plantago* | Leaves, root |
| Rattlesnake fern | *Botrychium* | Leaves |
| Redbud | *Cercis* | Buds |
| Sunflower | *Helianthus* | Leaves |
| Wild azalea | *Rhododendron* | Leaves, flowers, stems |
| Witch hazel | *Hamamelis* | Bark, leaves, twigs |
| Yarrow | *Achillea* | Flowers |

Fragrant oils from plants are used in perfumes and sachets. Among the naturally fragrant wild plants of the United States are wild licorice roots (*Glycyrrhiza*), mint foliage (several genera), rose petals (*Rosa*), sage foliage and wood (*Artemisia*), juniper (*Juniperus*) and cedar wood (several genera), sassafras bark (*Sassifras*), strawberry fruit (*Fragaria*), violet flowers (*Viola*) (Fig. 17-40), wintergreen leaves and berries (*Gaultheria*), and sweet flag rhizomes (*Acorus calamus*). There are many others, as the curious nose is certain to discover.

The sweet flag bears its fragrance throughout the plant. In former times the leaves of this tulelike plant were strewn on the floors of homes, public halls, and churches. Each time the fronds were stepped on, the aroma was released, forming an air freshener to disguise or dispel public odors at a time of casual attention to personal hygiene. When rubbed on the skin, the juice from the rhizome is also reputed to repel insects.

**FIGURE 17-40** These violets (*Viola*) grow from fleshy rhizomes.

Rush chair seats and mats are woven from the mature foliage of the versatile cattail. In addition, tufts of the fully developed seeds with their fuzzy air-dispersal hairs are suitable stuffing for pillows and toys. They have also been used as insulation in quilted clothing, bedding, boots, and shelters. This fluff (and the pollen that precedes it) may also be used as tinder.

Large thorns, such as those of hawthorn (*Crataegus*), can be fashioned into fishhooks, awls, and needles. Catchweed or cleavers (*Galium*) plants (Fig. 17-32) massed together in a bunch are a useful strainer, such as for extracting flour from macerated plant pulp.

Many wild plants have been used as tobacco substitutes, including kinnikinnic or bearberry (*Arctostaphylos uva-ursi*) and sweet coltsfoot (*Petasites palmatus*).

## SUMMARY

Besides providing recreation and dietary variety, knowing how to **forage** is a survival skill. Foraging with an expert is the best way to learn to recognize edible wild plants. Illustrated guides are helpful but should incorporate taxonomic and ecological descriptions. In a true emergency a cautious trial-and-error method will permit identification of safely edible green plants (*not* fungi).

Sorrels, plantain, chickweed, and watercress are examples of plants suitable for **nibbles** and **salads.** They and many other leafy plants make good **potherbs.** One of the best potherbs is lambsquarters, which resembles spinach but exceeds it in some vitamins and minerals. Wild mustards, curly dock, stinging nettle, and pigweed are also good potherbs.

Mature leaves, stems, fruits, and roots of pokeweed are highly poisonous, but the very young shoots ("poke sallatt") are a popular wild potherb in the southeastern United States.

Wild-growing asparagus and broad-leaved milkweeds are examples of cooked **green vegetables.**

The tubers of Jerusalem artichokes (actually a native sunflower) and the roots of burdock and salsify are examples of **potatolike vegetables.**

True grains, as well as seeds, dried flower heads, seed pods, and vegetative parts of specific plants, can be ground into flour for a variety of **cereallike** uses.

Many **nuts** and **seeds** can be eaten directly and are also sources of extractable **oils.**

Many **beverage** sources are available. The freshly drawn sap of certain trees can be drunk or used as a cooking liquid. Wines, beers, coffee and cocoa substitutes, and teas can be made from many plant species.

Edible **marine algae** can be used fresh, dried, or ground into flour and include certain kelps (kombu), laver (nori), Irish moss, and others. Agar, carrageenan, and algin extracted from seaweeds are used as thickeners, stabilizers, and gelling agents in many commercial food products.

Some **lichens** are edible but should be soaked in several water changes before eating to remove acids that cause indigestion.

Neophyte **mushroom** collectors must observe strict precautions to ensure positive identification of safe species. Even "safe" mushrooms may have adverse effects on some individuals.

Specific **precautions** are advised. Foragers, in general, should avoid eating anything that they cannot positively identify. They should know local poisonous plants and not forage in areas treated with herbicides or pesticides or where soils are high in selenium or nitrates. Herb teas, in particular, should be used moderately, for they contain "medicinal" chemicals that may be toxic in quantity.

Many edible wild plants compare favorably to garden species in nutritional value.

There are many **nonfood uses** of wild plants—for instance, as sources of fibers, dyes, "soap," fragrances, and cosmetics.

## SOME SUGGESTED READINGS

Angier, B. *Field Guide to Edible Wild Plants.* Harrisburg, PA.: Stackpole Books, 1974. One of about two dozen titles by Bradford Angier relating to wilderness survival and foraging. This paperback handbook has color illustrations of selected species.

Gibbons, E. *Stalking the Wild Asparagus.* New York: David McKay Co., 1962. An inexpensive paperback field-guide edition of the book that triggered the recent popularity of foraging and that has been available since 1970, as well as Euell Gibbons' other foraging books, *Stalking the Blue-eyed Scallop, Stalking the Healthful Herbs,* and *Beachcombers Handbook.* Emphasis on the eastern half of the United States.

Grae, I. *Nature's Colors: Dyes From Plants.* New York: The Macmillan Co., 1974. Sources of dyes from nature, over 200 recipes, detailed information regarding specific fibers with specific dyes and mordants, and utensils.

Hall, A. *The Wild Food Trail Guide.* New York: Holt, Rinehart and Winston, 1973. Compact, truly pocket sized, and with sound, useful information on food and nonfood uses of wild plants; also helpful lists of food available by seasons.

Harrington, H.D. *Edible Native Plants of the Rocky Mountains.* Albuquerque: University of New Mexico Press, 1967. The identification and preparation of many edible plants from the Rocky Mountain region.

Hedrick, U.P. (ed.). *Sturtevant's Edible Plants of the World.* New York: Dover Publications, 1972. Nearly 3000 species are covered in this edition of Sturtevant's compendium of edible plants; special emphasis on plants of the Americas.

Kirk, D.R. *Wild Edible Plants of the Western United States.* Healdsburg, CA.: Naturegraph Publishers, 1970. One of a few books for the West; many botanically accurate line drawings. Careful warnings about potential hazards distinguish this paperback from most other foraging handbooks.

Krochmal, A., and C. Krochmal. *A Guide to the Medicinal Plants of the United States.* New York: Quadrangle Press (The New York Times Book Co.), 1973. Profusely illustrated, having botanical descriptions, uses, and historical backgrounds of many readily identifiable medicinal plants.

Krochmal, C. *A Guide to Natural Cosmetics.* New York: Quadrangle Press, 1973. Illustrated guide to making cleansers, perfumes, lotions, powders, and so on from plants and other natural ingredients. Includes recipes, historical notes, and an appendix describing raw materials and sources for obtaining them.

——, and A. Krochmal. *A Naturalists Guide to Cooking with Wild Plants.* New York: Quadrangle Press, 1974. Of particular use for the eastern half of the United States, the book emphasizes recipes for common species.

Robertson, S.M. *Dyes from Plants.* New York: Van Nostrand Reinhold Co., 1973. List of dye plants by colors, equipment, materials, explanations of the whats and hows of dying with naural sources; also a chapter on planting a dye garden.

Smith, A.H. *The Mushroom Hunter's Field Guide,* 2nd ed. Ann Arbor: University of Michigan Press, 1963. Well-illustrated, concise, and easy-to-use popular guide to hunting the more common edible fungi.

Stewart, A., and L. Kronoff. *Eating from the Wild.* New York: Ballantine Books, 1975. An inexpensive paperback guide to finding, harvesting, and cooking wild foods; numerous recipes.

# chapter eighteen HAZARDOUS PLANTS

## INJURY, ALLERGY, MEDICINE, POISON

## INTRODUCTION

Among the myriad natural chemicals that plants produce are many affecting animal metabolism. Some give agonizing pain; others relieve pain. Some cause hemorrhage; others staunch the flow of blood. Some help the heart and kidneys; others destroy them. Some forestall death; some kill.

You may be surprised to discover that many common food plants have a toxic potential and that many important medicines are toxins produced by plants. Thus toxin-producing plants are particularly fascinating to study.

Paradoxically, many plant chemicals are medicinal in effect *because* of their toxicity. The line between medicine and poison at times overlaps, such as in the treatment of cancer with colchicine. With most chemicals, however, there is a continuum, ranging from no effect through an effective dosage to a toxic overdose.

Poisonous plants are usually defined as those that have caused serious illness or death. It is more accurate, however, to define **poisonous plants** as the ones that contain specific substances that produce an adverse reaction when ingested in small or moderate amounts.

In addition to truly poisonous plants, our concern here is with hazardous plants in general. Plants that cause contact dermatitis (inflammation of the skin) or allergic reactions are not, strictly speaking, "poisonous" plants, for instance. Yet we discuss them because of their significance to human health and because vernacular application of the term "poisonous" does include them.

There are several fine references on poisonous plants, most of which emphasize livestock poisoning. Our discussion, however, emphasizes human interactions with hazardous plants.

This chapter has several purposes:

1. to acquaint you with some biological basics about hazardous plants.
2. to provide descriptions of some of the most commonly encountered dangerous species.
3. to suggest ways to protect you and your family (now or later) from poisoning incidents.
4. to heighten your awareness of immediate first-aid treatment, species identification, and qualified medical advice.
5. to provide a list of selected references appropriate for field and home use.

## HAZARDS OTHER THAN POISONING

### Direct Physical Injury

Some plant protective parts discourage not only would-be herbivores but humans as well. Generally direct physical injury from brushing against, grabbing, or stepping on a sharp plant part is temporary and produces no systemic effects. This statement is true, for the most part, of the plant kingdom's various thorns, spines (Fig. 18-1), prickles, awns, barbs, and sharp or serrated leaf edges.

**FIGURE 18-1** Prickly pear (*Opuntia*) cactus pads and fruits are edible, but the large spines and many small hooklike spines at the base of each larger spine must be avoided. (Photo courtesy of John S. Thiede.)

## Primary Chemical Irritation

Plants that cause primary chemical irritation affect nearly all humans. Reaction (usually **dermatitis,** inflammation of the skin) occurs shortly after contact. Noncalloused areas of the skin and the eyes are highly susceptible.

The milky sap of many members of the family Euphorbiaceae, including the spurges (*Euphorbia*), crown of thorns (*E. splendens*) (Fig. 18-2), pencil tree or milkbush (*E. tirucalli*) (Fig. 7-12), candelabra "cactus" (*E. candelabrum*), sandbox tree (*Hura crepitans*), and manchineel (*Hippomane mancinella*), causes mild to severe blistering. The sap of the manchineel tree (native to the Florida Everglades) is well known for causing blindness (usually temporary).

Other examples of primary chemical irritants are citrus peel oils and substances found in buttercups (*Ranunculus*), mums and some daisies (*Chrysanthemum*), and maidenhair tree fruit (*Ginkgo biloba*). Direct irritation can also be caused by calcium oxalate crystals. (See Oxalates section.) A protein-digesting enzyme (bromelin) in fresh pineapple juice is mildly irritating.

Some plant species possess stinging hairs and their effects combine mechanical injury and primary chemical irritation. Examples are stinging nettle (*Urtica*) (Fig. 18-3), horse nettle (*Solanum*), wood nettle (*Laportea*), bull nettle (*Cnidoscolus stimulosus*), certain primroses (*Primula*), and *Jatropha* species.

**FIGURE 18-2** Members of the euphorb family, such as this crown of thorns, have an irritating milky sap.

**FIGURE 18-3** Stinging nettle (*Urtica*). A gentle brush with nettle leaves or stems can be temporarily painful because of needle-like epidermal hairs that contain irritating chemicals.

## Photodermatitis

A few plants produce substances that cause **photodermatitis**, a hypersensitivity of the skin to sunlight. The substances contact wet skin, which is then exposed to sunlight. Within a day or two, reddening, increased pigmentation, swelling, and small water blisters appear. Several members of the carrot and parsley family, for example, are known to be phototoxic, including carrot (*Daucus carota*), celery (*Apium graveolens*), parsnip (*Pastinaca sativa*), and wild chervil (*Chaerophyllum bulbosum*).

## Allergies

To survive in the postnatal environment, humans must possess effective innate immunological systems. Allergic reactions are a side effect of immunity, caused by the interaction of allergy-producing factors (**antigens**) with the body's antigen-attacking factors (**antibodies**). During an antigen–antibody reaction cells are stimulated to produce substances that are actually harmful to the body.

The most important of these substances is **histamine**, which has several simultaneous effects. Histamine causes a buildup of pressure in the small vessels, thus producing extensive "leaking" of fluid into tissues from the capillaries. The result is tissue swelling (edema)—in the sinuses, for example. Histamine can also cause spasms of the small breathing tubes (bronchioles) in the lungs, making breathing difficult. We are all familiar with drug preparations that contain antihistamines to counteract the effects of histamines.

In general, an allergic response differs from a primary chemical response in two ways. First, there is a **latent period** between the time of contact and the development of symptoms. During the latent period (more than 24 hours, typically 5 days) immunological changes occur within the body. In addition, an individual must be predisposed or **sensitized** to the substance by inherited tendency, prior exposure, or a combination of both. Sensitivity to allergy-producing factors varies widely among individuals. People usually possess some initial resistance but become sensitized by continued contact with specific allergens. Sensitization is accumulative so that successive exposures increase the potential for a severe reaction.

The development of an allergic response also depends on factors inherent within each plant. Not all parts of the plant may contain equal quantities of the chemical substance. It may be located on plant surfaces, or internally, so that bruising, cutting, or crushing releases it. Hay fever allergens are located on pollen grains or spores. As with primary chemical irritants, stage of maturation, season, and environmental factors also affect concentration and distribution of allergens within a plant body. Allergic responses may progress into secondary complications, including bacterial infections of scratched skin, drug-treatment side effects, generalized edema, and asthma.

There are hundreds of plant species to which humans can develop allergic reactions. Allergic responses to plants usually take two basic forms: contact dermatitis and hay fever.

On a nationwide basis, poison ivy (*Rhus* or *Toxicodendron radicans*) (Fig. 18-4) and poison oak (*R.* or *T. toxicodendron*) (Fig. 18-5) are undoubtedly the leaders in causing human **allergic contact dermatitis** because of an oleoresin (phenol) known chemically as a 3-*n*-pentadecylcatechol, or urushiol. The reaction is of a delayed response type. Several hours to days after contact the skin of a sensitized individual becomes reddened and itchy, with water blisters at sites of contact. Contrary to a popular misconception, the blister fluids do not spread the inflammation, for they are formed inside the body. The inflammation is spread by traces of urushiol on the hands, clothing, or other objects (even pets). Droplets of the substance can also be spread in smoke when the plants are burned. As is typical of allergies, individual sensitivity to urushiol varies greatly.

People who work intensively with particular plants are quite likely to develop sensitivities to them—for example, horticulturists and florists working with English ivy (*Hedera helix*), tulips (*Tulipa*), *Narcissus, Philodendron* species and certain lilies; agricultural employees working with hops (*Humulus lupulus*), onions, carrots, tomatoes, potatoes, and garlic; lumbermen and woodworkers with various types of wood, especially poplar (*Populus*), and certain lichens and leafy liverworts that may be present on tree bark.

Many pollens cause contact dermatitis even though pollen is usually thought of only in terms of hay fever. The two different allergic responses are stimulated by totally different pollen components. Water-soluble substances stimulate hay fever whereas oleoresins cause allergic contact dermatitis. Pollen dermatitis affects exposed skin, such as the face, neck, and arms, forming a sheetlike dermatitis. Ragweed pollen is the most common culprit in the United States. Some other problem pollens are box elder (*Acer negundo*), poplar, maple (*Acer*), ash

FIGURE 18-4
Poison ivy foliage in Wisconsin.

FIGURE 18-5
Poison oak foliage in Oregon.

(*Fraxinus*), cocklebur (*Xanthium*), mare's tail (*Hippurus tetraphylla*), and marsh elders—but *not* poison ivy!

Although several hundred plant species cause allergic contact dermatitis, most of the offenders are found in nine plant families (Table 18-1).

Urticaria, hay fever, asthma, eczema, and allergic gastrointestinal disturbances are examples of **atopic allergies**, experienced by persons who have abnormal immunity reactions to everyday antigens, such as pollen, spores, animal "dander," and organic material in dust.

**Urticaria** ("hives") is characterized by development of large welts all over the body and small edematous areas throughout internal organs. Antihistaminic drugs are very effective in treating urticaria.

**Hay fever** is one of the most common and annoying chronic medical problems. Antigen–antibody reactions occur in the nasal tissues and histamine is released, causing edema. Antihistaminic drugs are effective in treating the edema but do not prevent nasal irritation; so it is probable that other substances in addition to histamine are released by the tissues.

The pollen grains of many plant species cause hay fever, as do the spores of some fungi, especially *Alternaria* (causes blight of potato, tomato, carrot, and other species). Pollens that are lightweight and wind borne cause hay fever—not the heavier insect-borne pollens, such as goldenrod, which has unfairly taken the blame for ragweeds that bloom at the same time [Fig. 18-6 (a) and (b)]. Hay fever causing pollens vary seasonally (Table 18-2).

**Asthma** occurs when antigen–antibody reactions are localized in the lungs, causing constriction of bronchioles and consequent breathing difficulty. Severe asthmatic reactions ("attacks") can quickly reduce the lungs' ability to provide adequate oxygen and emergency medical treatment is necessary. Antihistaminic drugs do not help asthma; so some factor other than histamine is probably the cause.

**TABLE 18-1** **Nine plant families particularly significant in containing species that produce allergic contact dermatitis. Of these, the Anacardiaceae is the most important.**

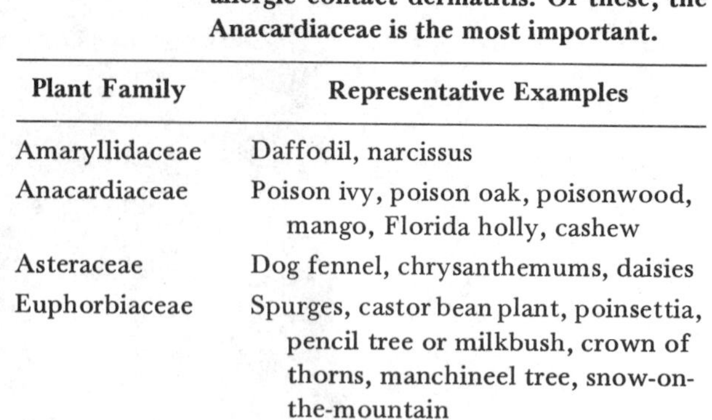

| Plant Family | Representative Examples |
|---|---|
| Amaryllidaceae | Daffodil, narcissus |
| Anacardiaceae | Poison ivy, poison oak, poisonwood, mango, Florida holly, cashew |
| Asteraceae | Dog fennel, chrysanthemums, daisies |
| Euphorbiaceae | Spurges, castor bean plant, poinsettia, pencil tree or milkbush, crown of thorns, manchineel tree, snow-on-the-mountain |
| Poaceae | Grasses, cereal grains |
| Liliaceae | Tulip, hyacinth, garlic, onions |
| Primulaceae | Primrose |
| Rutaceae | Citrus fruits |
| Apiaceae | Carrot, celery |

(a)

**FIGURE 18-6**
The inconspicuous flowers of (a) common ragweed and (b) giant ragweed produce pollen that is the major cause of hay fever in the eastern half of the United States.

(b)

TABLE 18-2 **Some pollens that cause hay fever.**

**Early spring**

Mainly tree pollens, such as box elder (*Acer negundo*), poplar (*Populus*), elm (*Ulmus*), birch (*Betula*), oak (*Quercus*), hickory (*Carya*), and ash (*Fraxinus*).

**Late spring and early summer**

Mainly grass pollens, such as timothy grass (*Phleum pratense*), Bermuda grass (*Cynodon dactylon*), Kentucky bluegrass (*Poa pratensis*), and redtop (*Agrostis gigantea*); also some broad-leaved plants.

**Late summer and autumn (especially August and September)**

Mainly ragweeds (*Ambrosia* spp.), especially in the Midwest and East, and saltbush (*Atriplex* spp.) and Russian thistle (*Salsola kali*) in the West.

Atopic allergies are often treated successfully by gradual desensitization. Sterile allergens (e.g., specific pollen extracts) are injected into a person in increasingly higher concentrations. The allergic reactions to the injections gradually decrease in intensity until the person is desensitized.

## "TRUE" POISONING

There are three main ways in which humans become intoxicated (medieval Latin, *intoxicare*, to drug or poison) by plant substances. The first is self-administration of toxic plant material for medicinal or experimental purposes. Many of the victims die; most incidents, however, are not fatal. The symptoms of intoxication are extremely unpleasant and usually include gastrointestinal pain and trauma, with or without such nervous disorders as loss of coordination, hallucinations, convulsions, respiratory abnormalities, depression of life functions, and coma.

Deaths from homemade remedies have lessened with increased dependence on standardized substances obtained through trained and licensed pharmacists. That this is a relatively new development can be shown by perusal of some of the home medicine manuals produced around the turn of the century. These manuals include recipes and dosage recommendations for use of such toxic plant materials as belladonna alkaloids, morphine, and opium, plus such toxic inorganic substances as arsenic and mercury (Fig. 18-7). Still, folk medicine remains "alive and well" in parts of the United States and there has been increased interest in it in recent years.

Certain symptoms of the disease may also require treatment, among them the cough, diarrhea, and night sweats. The cough is inevitable, so soon at least as the softening and breaking down of the lung tissue begins; hence the only object in treating the cough is to diminish the irritation and consequent exhaustion of the patient. Hence such remedies as **squills** and **ipecac** are out of place, and may even do harm; but some sedative mixture may be of service, such as the following:

Dilute hydrocyanic acid, Half a drachm.
Sulphate of **morphia,** Half a grain.
Syrup of **tolu,** water, Each one ounce.

Mix, and take half a teaspoonful every hour.

Another formula, much used in this country, is as follows:

Chloroform, A teaspoonful and a half.
Tincture of **opium,** " "
Spirits of **camphor,** " "
Aromatic spirit of ammonia, " "
**Creosote,** Three drops.
Oil of **cinnamon,** Eight drops.
**Brandy,** Two drachms.

Mix. Dissolve a teaspoonful of this in a wineglassful of ice-water, and give two teaspoonfuls out of this glass *every five minutes,* followed each time by a lump of ice.

...in certain cases of paralysis, are **strychnine** and phosphorus. The former may be given in the following prescription:

Sulphate of **strychnia,** Half a grain.
Reduced iron, Thirty grains.
Extract of **belladonna,** Eight grains.

Mix and make thirty pills. Take one morning and night.

Phosphorus can be best given dissolved in **almond oil;** one-fourth of a grain of phosphorus may be dissolved in two ounces of the oil, and a teaspoonful of this may be taken morning and night.

[Nervous exhaustion] designates a condition which is known by physicians as *neurasthenia.* It may be defined in short as a lack of nervous force.

Sugar of lead, Two drachms.
Tincture of **opium,** One ounce.
Water, Eight ounces.

The patient should also take the *tincture of the chloride of iron* internally. The following prescription may be given:

**Quinine,** Half a drachm.
Tincture of the chloride of iron, One ounce.
One ounce.
Water, Three ounces.

Mix, and take a teaspoonful in water every four hours.

FIGURE 18-7 Home-medicine recipes from an 1886 family medical book. Plant derivatives are indicated in boldface type.

A second common source of human intoxications is foraging for edible wild plants. Ingestion may be strictly experimental or the result of misidentification. There are many sensational case histories in this category.

The third, and most common, cause of human poisoning by plant substances (both wild and cultivated) is ingestion of toxic plant parts by children. The tragedy of this type of poisoning is that children tend to pick and eat almost any kind of attractive fruit or nut, plus have "tea parties," using whatever foliage is at hand for making the tea. Children are also subject to the "dares" of their playmates. Ingestion usually occurs without parental knowledge so that when such symptoms as bellyache and/or vomiting develop, parents may anticipate a bout of "flu" before considering the possibility of poisoning. In addition, the child may respond negatively to direct questioning about eating or drinking something for fear of being punished. In some cases, the intoxication may proceed through extreme, even irreversible, stages before it is recognized. By then no physical evidence of the plant material ingested may remain and the child may not be in a condition to provide any type of identification. From among the many examples of this category two commonly ingested fruits can be mentioned here—both of which have caused fatalities.

*Solanum dulcamara* (chapter opening art), climbing nightshade or European bittersweet, is an introduced member of the potato family (Solanaceae) that has poisonous foliage. The clusters of attractive orange-red berries also contain the toxic alkaloid **solanine**. This plant is common on disturbed sites, such as near railroad tracks, fences, docks, banks, roadsides, and buildings. Among other places, we have seen it rambling freely along the woven wire fencing at the children's section of a zoo. The berries may persist into late fall and early winter after the leaves have withered.

Even more than *S. dulcamara*, however, American children are attracted to and eat the seeds of *Aesculus hippocastanum*, the horse chestnut, a European native planted ornamentally in the United States. The toxic ingredient is **saponin**, present also in our native buckeyes (*Aesculus glabra*) (Fig. 18-8). Saponin is not readily absorbed through healthy intestinal lining (mucosa) but causes irritation and tissue erosion. Thus most cases only result in gastrointestinal symptoms and are not considered too serious as long as the patient is not allowed to become dehydrated. Severe systemic effects, however, can result when the nuts are eaten repeatedly, for saponin absorption is facilitated through the eroded surfaces.

**FIGURE 18-8** The shiny brown seeds that form within the fruit husk of buckeye (*Aesculus*) and its relatives are dangerously toxic.

As mentioned, the most common method of true poisoning is by ingestion of plant parts, or extracts, although poisoning may also occur by direct injection of extracts into body tissues. The most common immediate effects of ingestion are often gastric and intestinal distress—pain, vomiting, diarrhea. Following absorption, other systems may become involved; so they are discussed separately after the descriptions of poison types.

## Kinds of Medicinal/Poisonous Principles

### Nitrogen-containing Compounds

Plants produce several types of toxic nitrogen-containing substances. Most are alkaloids, phytotoxins (plant poisons), or toxic amines. In addition, many species accumulate toxic amounts of nitrates from the soil and a few species produce the enzyme thiaminase.

**Alkaloids** have received much attention since the 1960s because many have psychoactive properties. Some are valuable medicines. More than 5000 have been named and they usually exist in plants as bitter-tasting, soluble, organic acid-alkaloid salts. The alkaloids themselves are soluble in organic solvents but not in water. Table 18-3 summarizes the distribution of certain alkaloids.

**TABLE 18-3 Some types of alkaloids and poisonous plants that contain them.**

| Types of Alkaloids and Examples | Plant Examples | |
|---|---|---|
| | Genus Name | Common Name(s) |
| **Tropane** | | |
| Belladonna alkaloids atropine, hyoscyamine, scopolamine | *Atropa* | Belladonna, deadly nightshade |
| | *Datura* | Jimsonweed |
| | *Hyoscyamus* | Henbane |
| **Pyrrolizidine** | | |
| | *Crotalaria* | Rattlebox, crotalaria |
| | *Echium* | Viper's bugloss |
| | *Heliotropium* | Heliotrope |
| | *Senecio* | Groundsel, senecio |
| **Pyridine** | | |
| Coniine | *Conium* | Hemlock poison hemlock |
| | *Lobelia* | Indian tobacco, lobelia |
| | *Nicotiana* | Tobacco |
| **Isoquinoline** | | |
| Sanguinarine, berberine, protopine | *Argemone* | Prickly poppy |
| | *Chelidonium* | Celandine |
| | *Corydalis* | Fitweed |
| | *Dicentra* | Dutchman's-breeches |
| | *Papaver* | Poppies |
| | *Sanguinaria* | Bloodroot |
| **Indole** | | |
| Ergotamine, ergotoxine, and ergometrine groups | *Claviceps* | Ergot |
| | *Gelsemium* | Carolina jessamine |
| | *Hippomane* | Manchineel |
| | *Peganum* | African rue |
| **Quinolizidine** | | |
| Cytisine | *Baptisia* | False indigo |
| | *Cytisus* | Scotch broom, Scot's broom |
| | *Laburnum* | Goldenchain, laburnum |
| | *Lupinus* | Lupine, bluebonnet |
| | *Sophora* | Mescal bean, frijolito |
| **Steroid alkaloids** | | |
| (a) Solanum type—"solanidine" (see also steroid glycosides); solanine | *Lycopersicon* | Tomato |
| | *Solanum* | Potato, nightshades |
| (b) Veratrum type—"veratramine"; veratridine, veratrine | *Amianthium* | Stagger grass, fly poison |
| | *Veratrum* | False hellebore |
| | *Zigadenus* | Death camas |
| **Polycyclic diterpenes—"delphinine"** | | |
| Ajacine, ajaconine, delphinine, aconitine | *Aconitum* | Monkshood, wolfbane |
| | *Delphinium* | Larkspur |
| **Uncharacterized or incompletely characterized alkaloids (excluding algae and fungi)** | | |
| | *Allium* | Onion, garlic |
| | *Buxus* | Box, boxwood |
| | *Ervatamia* | Crape jasmine |
| | *Festuca* | Fescue |
| | *Fritillaria* | Fritillary |
| | *Gloriosa* | Glory lily |
| | *Ornithogalum* | Star-of-Bethlehem |
| Taxine | *Taxus* | Yew |

Adapted from J.M. Kingsbury, *Poisonous Plants of the United States and Canada.* Englewood Cliffs, N.J.: Prentice-Hall, Inc., 1964.

There is a strong tendency for plants in certain families to contain alkaloids (e.g., the nightshade, pea, and amaryllis families). Related plants may contain similar alkaloids, although some alkaloids (e.g., nicotine) are found in totally unrelated species. In general, alkaloid content is not subject to ecological variation and alkaloids are usually distributed throughout the body of the plant, making all parts toxic.

Although a few alkaloids elicit no reactions, most cause strong physiological reactions. Alkaloids have been known to cause severe liver damage and some cause fetal damage in sheep. The most typical effects, however, are on the nervous system, as described later.

Ingestion of one or two seeds from the large, pealike pods of the woody vine *Wisteria,* causes severe stomach pains and repeated vomiting. The bulbs, flowers, and foliage of narcissus and daffodil (*Narcissus*), snowdrop (*Galanthus*), *Amaryllis,* tuberose (*Polianthes*), spider lily (*Hymenocallis*), atamasco lily (*Zephyranthes*), snowflake (*Leucojum*), *Clivia,* and *Crinum* also cause severe vomiting and shivering. In addition to digestive symptoms, all the foregoing plants may also cause death if enough is eaten (one large bulb).

Yews (*Taxus*) produce **taxine**, a composite term for some or all of the at least ten alkaloids that are present in the foliage and seeds. Many human and livestock fatalities are attributed to yew. Human intoxications usually result from children making "tea" or women taking a decoction in an attempt to induce abortion.

Autumn crocus (*Colchicum*) and gloriosa or climbing lilies (*Gloriosa*) cause kidney damage and numerous painful gastrointestinal symptoms, which are usually delayed. The toxic alkaloid is **colchicine,** which is used in genetic and crop improvement research and cancer treatment. Colchicine arrests mitosis by interfering with development of the mitotic spindle, preventing separation of daughter chromosomes (Chapter 5). So it is useful in blocking the reproduction of cancerous cells. A side effect of colchicine poisoning, however, is damage to the generative cells of the hair follicles, causing hair loss from some or all body surfaces. A recent medical concern is about colchicine poisoning among marijuana users. Some growers have used colchicine to induce greater vegetative growth of the marijuana plants. When the plants are dried, the dangerously toxic colchicine is present and can be assimilated by persons smoking the marijuana.

**Phytotoxins** (toxalbumins) are large, complex protein molecules that act as antigens, causing allergic responses. Their toxicity is thought to be due to an ability to destroy certain critical natural proteins. Phytotoxins are absorbed in and cause damage to the digestive tract. Like other proteins, they can be destroyed by sufficient cooking. It is possible to develop an immunity to them.

Fortunately, few of our native plants manufacture phytotoxins, the black locust (*Robinia*) being one. Significantly, however, various introduced plants that do produce phytotoxins have seeds that are attractive to children—the tung nut tree (*Aleurites*), Barbados nut (*Jatropha*), and castor bean (*Ricinus*) [Fig. 18-9 (a) and (b)]; the red and black seeds of the precatory bean or rosary pea (*Abrus*) (Fig. 18-10) are used for necklaces! The precatory bean and castor bean (frequently planted as an ornamental) are the most virulent. One well-chewed precatory bean, or five or six castor beans, is a fatal dose for a child.

All parts of the black locust tree but the flowers are poisonous even though use of the seeds as a coffee substitute is occasionally recommended in foraging books. If such use is intended, the seeds must be heated sufficiently.

So far a very small number of **polypeptides** and **amines** (both proteins) produced by plants are known to be poisonous to animals. A few elicit responses alone, but some occur with and are probably synergistic with other toxic compounds. Both alkaloids and amines, for example, are present in the body of the fungus, ergot (*Claviceps*), discussed subsequently. Some plants known to produce toxic polypeptides or amines are *Microcystis,* a blue green alga of polluted waters, several poisonous mushrooms, mistletoe (*Phoradendron*), sweet pea and some wild peas (*Lathyrus*), and akee (*Blighia*).

Some authors refer to **thiaminase** as a "wayward enzyme." As its name implies, thiaminase breaks down the B vitamin called thiamine. Thiaminase is found in horsetails (*Equisetum*) and in bracken fern, the young shoots of which are often collected as food.

**Nitrates** (and nitrites) have poisoned large numbers of domestic livestock and are directly or indirectly toxic to humans. The problem has burgeoned since the late nineteenth century with increasing commercial uses of nitrate compounds.

The most obvious source of nitrate injection into the environment is nitrate fertilizer. Nitrates are absorbed and accumulated by many plants subsequently eaten by livestock or humans. Livestock also ingest nitrates as preservatives in livestock feed and in naturally high-nitrate water.

Both innate and environmental factors affect

(a)

(b)

FIGURE 18-9 The castor bean plant (*Ricinus*) is a popular ornamental (a) that produces deadly seeds (b). [Photo (b) courtesy of Marcus J. Fay.]

FIGURE 18-10 Precatory bean or rosary pea seeds are sometimes strung as beads, but are deadly if eaten. (Photo courtesy of Marcus J. Fay.)

plant nitrate accumulation. Ironically pigweeds (*Amaranthus*), ragweeds (*Ambrosia*), and Jimsonweed (*Datura*) are not normally palatable to livestock, but when treated for eradication with the herbicide 2,4-D, they not only become palatable but also begin to accumulate toxic levels of nitrate! Oat hay, corn, sorghum, and vegetables and weeds in the amaranth, goosefoot, mustard, composite, and nightshade families are known to accumulate toxic nitrate levels. Persons foraging for edible wild plants should avoid collecting where soil nitrates are unusually abundant.

Most literature discussing nitrate and nitrite poisoning has concentrated on their effects on domestic animals. Lately, however, attention has focused on the effects on humans of nitrates that accumulate in the bodies of food plants and animals. Specifically, nitrates and nitrites are added to processed meat products to retard bacterial growth. As of this writing, most weiners, bologna, bacon, sausages, and similar products still contain nitrites but not nitrates. Concern is centered on the possibility of poisoning when nitrates and nitrites are altered by human metabolism and also on the potential of these food additives for causing cancer.

### Glycosides

The presence of sugars in a compound is generally represented by use of the prefix glyco-. A **glycoside** is constructed of one or more sugars bonded to one or more nonsugars or aglycones ("a-glyco," without sugar). The glycoside molecule is broken down by enzymes, releasing the sugars from the nonsugar substances. Many aglycones are nontoxic, but others are dangerous. Glycosides are more widely distributed within the plant kingdom than alkaloids, although the latter are more numerous.

Many plants contain **cyanogenic (cyanide-producing) glycosides.** At least one person has died from eating apple seeds in quantity and livestock are commonly poisoned by foliage of wild and cultivated cherries, plums, and their relatives. The seed kernels of peaches, plums, apricots, cherries, and bitter almond contain the cyanogenic glycoside amygdalin. Amygdalin is the source of the controversial drug laetrile, reputed by some to be effective in treating certain types of cancer. Human poisoning has also occurred at least once from a person eating *Hydrangea* buds in a salad and is common when the tubers of certain tapioca or cassava plants (*Manihot*) are eaten. Considerable processing is necessary to convert toxic fresh cassava pulp into edible tapioca products.

Poisoning by cyanogenic glycosides occurs when enzymes release the small, biologically active, readily absorbed, directly toxic hydrocyanic acid molecule (HCN), commonly called cyanide. Cyanide is released within plant tissues that have been damaged by frost, wilting, or abnormal growth (stunting); or it may be released in the digestive tracts of animals. Within the animal body, HCN molecules are absorbed and transported rapidly throughout the body. Poisoned cells literally asphyxiate from lack of oxygen because cyanide interferes with cellular respiration. The liver detoxifies HCN rapidly and HCN is also excreted in the breath and urine. So within certain limitations the body is able to handle cyanide poisoning. Tolerable and lethal blood HCN levels, however, are not far apart.

Certain naturally occurring substances, including two glycosides, are **goitrogenic**. They interfere with utilization of iodine, which is essential to thyroid hormone production, causing a medical condition called goiter to develop. Goitrogenic glycosides are found in many members of the mustard family, including chard, kohlrabi, kale, broccoli, cabbage, Brussels sprouts, rutabaga, Chinese cabbage, turnip root, and the seeds of white and black mustard and rape. In humans adequate iodine in the diet ensures enjoyment of the preceding vegetables without negative effects.

**Irritant oils** are another category of poisonous glycosides. For example, mustard oils, which can blister animal tissues, are present in members of the mustard family, many of which are eaten and enjoyed for their "hotness." In small quantities, such as in radish roots and watercress leaves, they add flavor and "zing." Seeds usually contain the highest concentrations of mustard oils. The "superhot" oriental-style mustard is made by mixing powdered yellow mustard seeds with water. The more it is mixed, the hotter it gets as the enzymatic breakdown of the glycoside is facilitated. An exception is horseradish (*Armoracia*), which has a high oil content in the root.

Protoanemonin is a more dangerous irritant oil, released when the glycoside ranunculin breaks down. Protoanemonin-containing plants known to have caused human intoxications are buttercups and crowfoots (*Ranunculus*), white, red, and black baneberries (*Actaea*), anemone, windflower, and pasqueflower (*Anemone*), virgin's bower (*Clematis*), and the popular spring potherb, marsh marigold (*Caltha*). Dangerous when eaten, protoanemonin can also cause dermatitis. So the practice of treating arthritis and frostbite by applying protoanemonin-containing vegetation can have a painful result.

**Coumarin-containing glycosides** are found in various poisonous plants, including horse chestnut and buckeye (*Aesculus*) and *Daphne,* but may not be responsible for the toxic effects of these plants. Coumarins, however, have been definitely implicated in the hemorrhagic disease experienced by cattle that eat spoiled sweetclover (*Melilotus*) hay. During decomposition of the hay the coumarin is transformed into **dicoumarin**, an anticoagulant. Dicoumarin is valuable in the treatment of certain human heart and circulatory diseases and is also an ingredient in the rodent killer, Warfarin, developed by Wisconsin Alumni Research Foundation (WARF) scientists. Susceptible rodents ingesting Warfarin die from internal hemorrhaging.

The many **cardiac glycosides** come mainly from the figwort, lily, and dogbane families and have been known and used from earliest recorded times as heart stimulants. Of plants containing cardiac glycosides, the foxglove (*Digitalis*) (Fig. 18-11) is probably best known, for it is the source of the heart medication **digitalis**. Others are lily of the valley (*Convallaria*), oleander (*Nerium*), dogbane (*Apocynum*), pheasant's-eye (*Narcissus poetica*), and squill (*Scilla*).

**Saponins** are sometimes called "natural soaps" because they form a nonalkaline lather when agitated with water. Saponins are gastrointestinal irritants and can destroy red blood cells if sufficient quantities are absorbed. Some saponin-containing plants are listed in Table 18-4.

**FIGURE 18-11** The heart medication **digitalis** is a toxic derivative of the foxglove (*Digitalis*). (Photo courtesy of Fred L. Rose.)

**TABLE 18-4** **Poisonous plants containing saponin.**

| Genus | Common Name |
|---|---|
| *Aesculus* | Horse chestnut, buckeye |
| *Caulophyllum* | Blue cohosh |
| *Duranta* | Golden dewdrop, pigeonberry, skyflower |
| *Hedera* | English ivy |
| *Pachyrhizus* | Yam bean |
| *Poncirus* | Hardy-orange or mock orange (not the native southern mock orange, which is *Prunus caroliniana*) |
| *Pongamia* | Pongam |
| *Phytolacca* | Pokeweed |
| *Samanea* | Rain tree |
| *Sapindus* | Soapberry |
| *Saponaria* | Soapweed, bouncing Bet |
| *Agrostemma* | Cow cockle |
| *Momordica* | Balsam pear or bitter gourd |

#### Nonglycoside Irritant Oils

Certain nonglycoside oils are toxic but very useful medicines because of their irritant properties. Oil of wintergreen (methyl salicylate), for example, is found in certain members of the heath and birch families. It is used in ointments to soothe muscle aches and arthritis. Oil of chenopodium or wormseed oil is present in the wormseed or Mexican tea plant (*Chenopodium ambrosioides*) and has been used to expel intestinal worms.

#### Oxalates

Oxalic acid occurs in many plants, usually as a salt or oxalate. If oxalic acid combines with sodium or potassium, it forms a soluble oxalate. Insoluble forms—acid oxalates and calcium oxalate—are also common.

**Acid oxalates** give such plants as docks, sorrels, and rhubarb (*Rheum rhaponticum*) petioles their acid (sour) taste. When ingested, insoluble acid oxalates are normally eliminated from the body.

A number of plants, however, contain **soluble oxalates**, including a few in which they are present in dangerous amounts. Oxalate poisoning occurs when soluble oxalates are absorbed into the blood where they combine with blood calcium to form insoluble calcium oxalate crystals, thereby lowering the blood calcium level. Calcium oxalate crystals continue to enlarge in the bloodstream and eventually lodge in the kidney tubules. Severe blockage may cause kidney insufficiency or failure.

Rhubarb leaf blades are the most notorious cause of oxalic acid poisoning. They caused deaths in England during World War II when food was in short supply and someone misguidedly promoted the large leaf blades as potherbs. Garden sorrel or green sorrel (*Rumex acetosa*) has also caused intoxications, as have berries of Virginia creeper or American ivy (*Parthenocissus*).

Human food plants high in oxalates include spinach, New Zealand spinach, Swiss chard, beet tops, lambsquarters, and pokeweed shoots The last two, although not domesticated, are commonly eaten wild plants. Moderate amounts of oxalates, in combination with normal or enhanced calcium intake, are not apparently harmful. Increased fluid intake enables the kidneys to eliminate more oxalic acid before it can combine with calcium, thus reducing calcium oxalate formation.

Mouth and throat pain occur when plants containing bundles of needlelike **calcium oxalate** crystals (Chapter 7) are eaten. The burning pain is usually immediate so that the material is quickly expectorated, although the damage occurs so quickly that swelling of the throat, tongue, palate, oral lining, and even lips and face may result and persist over a week or so. Swallowing and breathing can be impaired and suffocation may occur.

Many arum family members (Table 18-5) contain calcium oxalate crystals. All plant parts are usually toxic and all plants in this family should be considered toxic. Several books on edible wild plants recommend both skunk cabbage and wild calla. Although their starchy underground parts can be made edible, they must be cooked in several changes of water to remove all the irritating crystals. Safer and more easily prepared wild foods are usually available.

**TABLE 18-5 Some plants that are dangerous due to the presence of calcium oxalate crystals. All are members of the arum family (Araceae).**

| Scientific Name | Common Name |
|---|---|
| **North American wild plants** | |
| *Arisaema triphyllum* | Jack-in-the pulpit |
| *Arisaema dracontium* | Green dragon, dragon root |
| *Calla palustris* | Wild calla, water arum |
| *Lysichiton americanum* | Western skunk cabbage |
| *Symplocarpus foetidus* | Eastern skunk cabbage |
| **Ornamental plants** | |
| *Alocasia* spp. | Alocasias |
| *Arum maculatum* | Cuckoopint, lords-and-ladies |
| *Caladium* spp. | Caladium |
| *Colocasia* spp. | Elephant's ear; dasheen (taro) |
| *Dieffenbachia* spp. | Dumbcane |
| *Monstera deliciosa* | Ceriman |
| *Philodendron* spp. | Philodendrons |
| *Scindapsus* spp. | Devil's ivy, pothos |
| *Xanthosoma* spp. | Malanga |
| *Zantedeschia aethiopica* | Calla lily |

#### Resinoids (Resins)

After extraction, resins or resinoids appear physically similar to the dried resin or pitch on pine or fir tree bark. Like pitch, resinoids are at least semisolid; they melt and burn easily and do not dissolve in water. In addition, they do not contain nitrogen. These toxic principles are thought to act directly on muscle and nervous tissue. Agonizing intoxications and many deaths have been attributed to plants containing toxic resinoids, especially water hemlock (*Cicuta*), discussed with other plants that affect the nervous system. At least one group, the milkweeds (*Asclepias*), includes some edible species, which, however, must be cooked to dispel danger of poisoning. Aromatic resins in the leaves of Labrador tea (*Ledum*) have endeared it as a wild tea in which obviously small quantities of the toxic principle are ingested.

The attractive ornamental shrub *Daphne* is virulently toxic and all plant parts are apparently toxic; the sap can be absorbed even through a skin wound. The berries are the most commonly ingested part, with about 30% mortality, and one or two berries can seriously poison a child. Cuckoopint, or lords-and-ladies (*Arum maculatum*), has also caused fatalities; all parts of the plant contain poisonous resins. Table 18-6 cites other toxic resin-containing plants.

#### "Mother Earth"

Some plant species growing in soils particularly high in specific minerals tend to accumulate one or more of these minerals in levels toxic to animals.

Certain western soils are high in available **selenium**, which is accumulated in the poison vetches and locoweeds (*Astragalus* and *Oxytropis*) and some other flowering plants. Prolonged grazing by horses and cattle causes the nervous system disease called "loco" and two other syndromes, "blind staggers" and "alkalai disease." Bees have even been killed working flowers for nectar and poisonous honey has been produced.

Some species grow only where selenium is present and so are utilized as "indicator species" to identify soils with high selenium levels. Tissue concentrations of such plants have been measured up to 15,000 ppm (parts per million). A potentially toxic level is considered 5 ppm.

For humans, potential hazards exist where cereal crops and forage are grown on seleniferous soils. More serious, however, is the economic problem, for livestock losses have even caused farms to be abandoned.

A few scattered soils in the United States and Canada have dangerous levels of **molybdenum**, including parts of California's San Joaquin Valley, some valleys on the Nevada side of the Sierras, and

**TABLE 18-6 Resin-containing plants considered toxic to humans and livestock.**

| Family | Scientific Name | Common Name |
|---|---|---|
| Asclepidaceae | *Asclepias* spp. | Milkweeds |
| Cannabinaceae | *Cannabis sativa* | Marijuana, hemp, hashish |
| Ericaceae | *Kalmia* spp. | Lambkill (sheep laurel), mountain laurel, bog kalmia |
| | *Ledum* spp. | Labrador teas |
| | *Leucothoe* spp. | Leucothoes, dog laurel |
| | *Menziesia ferruginea* | Rustyleaf, mock azalea, false huckleberry |
| | *Pieris japonica* | Japanese pieris, andromeda |
| | *Rhododendron* spp. | Rhododendrons, azaleas |
| Meliaceae | *Melia azedarach* | Chinaberry tree, Indian lilac |
| Apiaceae | *Cicuta* spp. | Water hemlock |

Florida's Everglades. Molybdenum poisoning is chiefly a livestock problem. Members of the pea family (legumes) have a greater tendency to accumulate molybdenum than do nonlegumes.

**Copper** poisoning is relatively rare; when it occurs, it is due more to human activities than to naturally occurring soil copper levels. Orchard areas where fungicidal Bordeaux mixture (containing copper sulfate) has been regularly applied are abnormally high in copper. Grazing livestock under such orchards is potentially hazardous, as is gardening for human consumption. A few soils in the world are naturally high in copper.

Like copper, most other mineral poisonings are due to the human addition of abnormal concentrations by one means or another. The **lead** used as an antiknock ingredient in leaded gasolines is deposited literally "by the ton" along roads. Susceptible plants are poisoned. Other species accumulate lead at levels that are toxic to animals, including humans. Consequently, it may be unwise to collect edible wild plants and fruits along heavily traveled highways.

**Cadmium** poisoning can result from ingestion of plants grown on soils treated with high levels of commercial superphosphate fertilizers. The cadmium is present as an impurity in such fertilizers. Various other substances, including fluorine, may be accumulated by plants.

#### Substances Causing Photosensitization

**Photosensitization** differs from photodermatitis discussed earlier. It is the increased sensitivity to sunlight following ingestion of certain plant pigments or pigment byproducts. After exposure to sunlight with the toxic substances present, exposed skin reddens, blisters, and may even die and slough off.

### Plants Affecting the Central Nervous System

Central nervous system (CNS) refers to the brain and spinal cord. All activities and life-supporting functions are mediated through interaction of the CNS with the rest of the body via elaborate neural and hormonal control and feedback systems. There are many CNS responses to poisoning, including perception of pain, depression of CNS functions, convulsions, and altered states of consciousness. The medical value of many plant substances is based on CNS effects.

#### CNS Depression

If CNS function is depressed, such symptoms as dizziness, headaches, an inability to concentrate, and undue excitement occur. CNS depression may lead to delirium, drowsiness, stupor, coma, and eventually death by respiratory paralysis in fatal cases. Temperature regulation and a host of other functions may be disturbed. CNS symptoms usually constitute only a portion of the total syndrome for each type of poisoning; in other words, toxins that depress the CNS also have other effects. CNS depression may also occur secondarily to the primary locus of intoxication.

### Convulsions

**Convulsions** (violent muscle spasms) are mediated through the CNS and may be tonic (sustained muscle spasms) or clonic (the muscles alternately contract and relax, causing violent movements, twitching). Convulsive periods are usually interspersed with depressive periods. Case-history descriptions of poisoning by convulsive toxins are exceptionally grim and it must be one of the most violent and painful types of poisoning.

One of the worst convulsant plants is the ubiquitous water hemlock (*Cicuta maculatum*) [Fig. 18-12 (a) and (b)]. Because it is eaten fairly

(a)

(b)

FIGURE 18-12
All parts of the ubiquitous water hemlock (*Cicuta*) contain a dangerous convulsant. The plant has caused many intoxications as well as deaths. (a) It resembles other members of the carrot family, having a compound inflorescence like that of dill. (b) The leaves are pinnately compound; some of the leaflets are also subdivided. (Photos courtesy of Marcus J. Fay.)

often, it ranks among the top three causes of plant intoxication in the United States. A member of the carrot and parsley family, water hemlock usually has a cluster of fleshy roots that smell like parsnips or carrots and it is sometimes collected by individuals believing it to be edible. For some reason, children sometimes trick other children into eating it. The hollow stems have been used to make whistles, with intoxications resulting. Most of the toxin (a resinoid called **cicutoxin**) is concentrated in the roots and stem. An adult person can die from ingestion of a single mouthful of root. Symptoms begin 15 minutes to an hour after ingestion, moving through nausea, tremors, and violent, painful seizures, with frothing at the mouth and intermittent quiescent periods. Death may occur by exhaustion and asphyxia. Curiously, individuals who survive frequently have no memory of the incident.

Alkaloids in some plants cause convulsions, as in the native plants Carolina allspice (*Calycanthus*), pink root (*Spigelia marilandica*), and moonseed (*Menispermum canadense*).

The nux vomica plant (*Strychnos nux-vomica*) is a source of the convulsant **strychnine**, a long-known poison. Decoctions of nux vomica were used medically in the past, with occasional fatal results. Some other cultivated plants can cause convulsions.

A final category of convulsants are those with so-called **extractable principles**, including members of the cypress family (arbor vitae, junipers, cedars) and a member of the daisy family, tansy (*Tanacetum vulgare*). Single doses of decoctions are seldom serious, but chronic poisoning may cause convulsions and death. Most intoxications and deaths are caused by deliberate administration of decoctions in an attempt to induce abortion.

### Altered States of Consciousness

The toxic ingredients of some plants cause altered states of consciousness and are usually called **psychomimetic** or **psychoactive**. Psychoactive drugs produce lassitude, euphoria, elation, hallucinations, distortion of time and space relationships, delusions, drowsiness, delirium, and insensitivity to pain. These and other effects are nearly always due to intentional use. Chronic use may cause physical and psychological dependence, tissue damage, and changes in behavior and motivation. Specific effects are the result of interference with normal metabolic activities. A person involved in a psychoactive episode is aware only of changes in sensations and conscious processes, but the additional metabolic and cytological effects go unnoticed at the time. Table 18-7 summarizes some of the effects and uses of psychoactive substances that are controlled by federal law; most are derived from plants or are synthetics resembling natural substances.

Nutmeg (*Myristica fragrans*), one of the earliest euphoria-producing plants, was a significant factor in the spice trade that provided the catalyst for early world exploration (Fig. 18-13). Like other "spices," it was sought not so much for its flavor as for its narcotic properties, as was true of many reputed narcotic, medicinal, or aphrodisiac plants. Nutmeg is still used as a euphoric and mild hallucinogen; however, it also causes gastritis, excessive thirst, dizziness, double vision, rapid heart rate, and anxiety. The borderline between an "effective" dose and a "toxic" dose with undesirable effects is abrupt.

The mescal bean (*Sophora secundiflora*), also called frijolitos or red bean, has a long history of use as a psychomimetic and medicinal plant among natives of the Americas. It contains the toxin **cystine**, which causes vomiting and convulsions, and a small dose of which can prove fatal. It was used as an adulterant in mezcal, or mescal (a fermented beverage made from *Agave*), to increase the intoxication, thus earning its common name. The vomiting effect was significant in certain purging ceremonies. The psychoactive peyote cactus (*Lophophora williamsii*) is mistakenly called mescal in the United States. Application of the term mescal to peyote was apparently due to lack of distinction by non-Indian observers between separate agents used simultaneously in Indian religious ceremonies.

An insidiously dangerous example of physiological damage is the relatively recent use of Madagascar periwinkle (*Catharanthus roseus* or *Vinca rosea*) as a hallucinogen. *C. roseus* contains potent **cytotoxins** and its reputation as a hallucinogen evolved from its use in the treatment of certain types of cancer tumors. One of the side effects is alteration of consciousness. *Catharanthus* chemicals, whether taken medically or for a "high," kill cells in many parts of the body. Specifically, white-blood-cell count is immediately reduced, causing increased susceptibility to infections. Prolonged use causes loss of hair, tingling of the skin, a burning sensation, and muscle deterioration.

**TABLE 18-7 Controlled substances of both natural and synthetic origin.**

| | Drugs | Often-Prescribed Brand Names | Medical Uses | Dependence Potential: Physical |
|---|---|---|---|---|
| Narcotics | Opium | Dover's Powder, Paregoric | Analgesic, antidiarrheal | High |
| | Morphine | Morphine | Analgesic | High |
| | Codeine | Codeine | Analgesic, antitussive | Moderate |
| | Heroin | None | None | High |
| | Meperidine (Pethidine) | Demerol, Pethadol | Analgesic | High |
| | Methadone | Dolophine, Methadone, Methadose | Analgesic, heroin substitute | High |
| | Other Narcotics | Dilaudid, Leritine, Numorphan, Percodan | Analgesic, antidiarrheal, antitussive | High |
| Depressants | Chloral Hydrate | Noctec, Somnos | Hypnotic | Moderate |
| | Barbiturates | Amytal, Butisol, Nembutal, Phenobarbital, Seconal, Tuinal | Anesthetic, anti-convulsant, sedation, sleep | High |
| | Glutethimide | Doriden | Sedation, sleep | High |
| | Methaqualone | Optimil, Parest, Quaalude, Somnafac, Sopor | Sedation, sleep | High |
| | Tranquilizers | Equanil, Librium, Miltown Serax, Tranxene, Valium | Anti-anxiety, muscle relaxant, sedation | Moderate |
| | Other Depressants | Clonopin, Dalmane, Dormate, Noludar, Placydil, Valmid | Anti-anxiety, sedation, sleep | Possible |
| Stimulants | Cocaine[a] | Cocaine | Local anesthetic | Possible |
| | Amphetamines | Benzedrine, Biphetamine, Desoxyn, Dexedrine | Hyperkinesis, narcolepsy, weight control | Possible |
| | Phenmetrazine | Preludin | Weight control | Possible |
| | Methylphenidate | Ritalin | Hyperkinesis | Possible |
| | Other Stimulants | Bacarate, Cylert, Didrex, Ionamin, Plegine, Pondimin, Pre-Sate, Sanorex, Voranil | Weight control | Possible |
| Hallucinogens | LSD | None | None | None |
| | Mescaline | None | None | None |
| | Psilocybin-Psilocyn | None | None | None |
| | MDA | None | None | None |
| | PCP[b] | Sernylan | Veterinary anesthetic | None |
| | Other Hallucinogens | None | None | None |
| Cannabis | Marihuana<br>Hashish<br>Hashish Oil | None | None | Degree unknown |

[a]Designated a narcotic under the Controlled Substances Act.

[b]Designated a depressant under the Controlled Substances Act.

Adapted from *Drugs of Abuse,* 3rd ed., U.S. Department of Justice, Drug Enforcement Administration.

| Dependence Potential: Psychological | Tolerance | Duration of Effects (in hours) | Usual Methods of Administration | Possible Effects | Effects of Overdose | Withdrawal Syndrome |
|---|---|---|---|---|---|---|
| High | Yes | 3–6 | Oral, smoked | Euphoria, drowsiness, respiratory depression, constricted pupils, nausea | Slow and shallow breathing, clammy skin, convulsions, coma, possible death | Watery eyes, runny nose, yawning, loss of appetite, irritability, tremors, panic, chills and sweating, cramps, nausea |
| High | Yes | 3–6 | Injected, smoked | | | |
| Moderate | Yes | 3–6 | Oral, injected | | | |
| High | Yes | 3–6 | Injected, sniffed | | | |
| High | Yes | 3–6 | Oral, injected | | | |
| High | Yes | 12–24 | Oral, injected | | | |
| High | Yes | 3–6 | Oral, injected | | | |
| Moderate | Probable | 5–8 | Oral | Slurred speech, disorientation, drunken behavior without odor of alcohol | Shallow respiration, cold and clammy skin, dilated pupils, weak and rapid pulse, coma, possible death | Anxiety, insomnia, tremors, delirium, convulsions, possible death |
| High | Yes | 1–16 | Oral, injected | | | |
| High | Yes | 4–8 | Oral | | | |
| High | Yes | 4–8 | Oral | | | |
| Moderate | Yes | 4–8 | Oral | | | |
| Possible | Yes | 4–8 | Oral | | | |
| High | Yes | 2 | Injected, sniffed | Increased alertness, excitation, euphoria, dilated pupils, increased pulse rate and blood pressure, insomnia, loss of appetite | Agitation, increase in body temperature, hallucinations, convulsions, possible death | Apathy, long periods of sleep, irritability, depression, disorientation |
| High | Yes | 2–4 | Oral, injected | | | |
| High | Yes | 2–4 | Oral | | | |
| High | Yes | 2–4 | Oral | | | |
| Possible | Yes | 2–4 | Oral | | | |
| Degree unknown | Yes | Variable | Oral | Illusions and hallucinations (with exception of MDA); poor perception of time and distance | Longer, more intense "trip" episodes, psychosis, possible death | Withdrawal syndrome not reported |
| Degree unknown | Yes | Variable | Oral, injected | | | |
| Degree unknown | Yes | Variable | Oral | | | |
| Degree unknown | Yes | Variable | Oral, injected, sniffed | | | |
| Degree unknown | Yes | Variable | Oral, injected, smoked | | | |
| Degree unknown | Yes | Variable | Oral, injected, sniffed | | | |
| Moderate | Yes | 2–4 | Oral, smoked | Euphoria, relaxed inhibitions, increased appetite, disoriented behavior | Fatigue, paranoia, possible psychosis | Insomnia, hyperactivity, and decreased appetite reported in a limited number of individuals |

FIGURE 18-13 The fruit of nutmeg is the source of two spices. Mace comes from the reticulate, fleshy red enclosure around the seed. The seed itself is the spice nutmeg, valued in past times as a drug. (Photo courtesy of Fred Rickson.)

The marijuana plant (*Cannabis sativa*) has been known for over 5000 years [Figs. 18-14 (a) and (b)]. Its greatest use in the Western world ostensibly has been as a source of hemp fibers. "Oldtimers" used a decoction of the "medicine weed" for medicinal purposes, but it is currently a popular social drug, smoked in cigarettes or pipes. Its possession and use carry criminal penalties in most of the United States.

The active ingredient, found in all aerial parts of the plant to varying degrees, is a resinoid called **tetrahydrocannabinol (THC)**. Its highest concentration is in female flowering tops. Illicit retail forms of marijuana are often "cut" with grass, alfalfa, or other vegetative matter (even manure) to increase the profits of dealers, for marijuana is sold by weight or per cigarette. Hashish is a concentrated extractive from marijuana plants and is usually smoked in pipes, ranging from a simple makeshift form to elaborate water pipes.

In 1966 tetrahydrocannabinol was first synthesized in a laboratory and has joined the natural forms in use. Drops of the pure chemical are often placed on cigarette tobacco or other dried leaves, such as parsley. Sometimes the chemical is used to increase the potency of lower-quality dried marijuana. THC has been administered to patients undergoing chemotherapy for cancer, for it relieves the undesirable side effects, including nausea, caused by the toxic injections.

Table 18-8 itemizes some of the hallucinogenic psychoactive plants in current use.

Of the drugs that have been used (and abused) by mankind for centuries, the earliest and most enduring are narcotics and stimulants derived from plants. The term **narcotics** is often used in the larger sense of referring to all drugs that induce an altered state of consciousness. The term is derived from the Greek word, *narkōtikos,* meaning to benumb. In current usage, the term refers to natural derivatives of the opium poppy (*Papaver somniferum*) or the related *P. bracteatum* or synthetic substitutes that *produce psychological and physical tolerance and dependence.*

Natural narcotics derived from *P. somniferum* include **opium, morphine**, and **codeine.** Raw opium is the dried milky sap from the unripe seed pod. At least *25* alkaloids can be extracted from opium, each with distinctive effects. Morphine is the most effective analgesic (pain reliever) known. Codeine is an effective cough suppressant and mild analgesic. In contrast, **papaverine** and **noscapine** do not

(a) (b)

FIGURE 18-14 Hemp or marijuana (*Cannabis sativa*) grows freely on disturbed soils, such as roadsides (a) and abandoned fields. (b) It can be recognized by the pointed, serrated, palmately lobed leaves.

**TABLE 18-8 Some hallucinogenic plants.**

| Plant Family | Scientific Name | Common Name | Hallucinogenic Substance and Type | Plant Part(s) |
|---|---|---|---|---|
| Apocynaceae | *Catharanthus roseus* (*Vinca rosea*) | Madagascar periwinkle | See legend[b] | Leaves |
| Cactaceae | *Lophophora williamsii* | Peyote, peyotl (plus a number of "street names)[a] | Mescaline (alkaloid) | Dried crowns or "buttons" |
| Convolvulaceae | *Ipomoea violacea* | Heavenly Blue morning glory | LSD[c] | Seeds |
| | *I. v.* var. *alba* | Pearly Gates morning glory | LSD | Seeds |
| | *Rivea (Turbina) corymbosa* | Ololiuqui, Piule, or Yerba de la Virga | LSD | Seeds |
| Myristicaceae | *Myristica fragrans* | Nutmeg | Myristicin & others | Seeds |
| | | Mace | | Fruit husk |
| Urticaceae | *Cannabis sativa* | Marijuana, hemp, pot, grass, bhang, *ad infinitum* | Tetrahydrocannabinols (resinoids) | Leaves and female flowers |
| ----- | Various species | "Magic mushrooms" | See text section on Mushroom Toxins | |

[a]Commonly but inaccurately known as "mescal" (see text).

[b]The leaves contain alkaloids, which may be responsible for the "high." However, physiological side effects caused by various cytotoxins include serious organ and tissue damage (see text).

[c]LSD is lysergic acid diethylamide.

significantly influence the central nervous system. Papaverine is used as an intestinal relaxant and noscapine as a cough suppressant.

Some exciting research has suggested the reason for the intense central nervous system effects of some opium derivatives. The human brain produces hormones called **endorphins**, which attach to specific receptor sites within the central nervous system, acting as analgesics. When a person is subjected to pain, as during childbirth, higher quantities of endorphins are produced, thus "loading" the receptor sites and decreasing the sensitivity to pain. It is now known that opiates like morphine also fit these receptor sites.

Another important natural opiate, **thebaine**, is derived from *Papaver bracteatum*. Although chemically similar to codeine and morphine, it produces stimulant rather than depressant effects.

There are also several significant semisynthetic narcotics, substances derived by modification of natural opiates. Of them, **heroin**, first synthesized from morphine in 1874, has had the greatest human impact. Like opium, heroin was initially sold and used with no serious attention paid to its potential for addiction. It first came under federal regulation in the United States in 1914.

**Hydromorphone**, the second oldest derivative of morphine, is two to eight times stronger than morphine as an analgesic, is more sedative, less euphoriant, and shorter acting. Like morphine and heroin, it is a drug of abuse.

Oxycodone, etorphine, and diprenorphine are all synthesized from thebaine. **Oxycodone** is similar to but more potent than codeine. **Etorphine** is easily a thousand times more powerful than morphine as an analgesic, sedative, and respiratory depressant. Its potency makes it dangerous for human use, but it is used by veterinarians to immobilize large animals. **Diprenorphine** counteracts the effects of etorphine.

Some entirely synthetic compounds that are chemically similar in structure or effects to the opium alkaloids are **meperidine** (pethidine), **methadone, levorphan, phenazocine, alphaprodine,** and **anileridine.** Meperidine (e.g., brand name Demerol) is second to morphine in its use to relieve intense pain and methadone has been used widely in detoxification of heroin addicts. The other substances are used mainly as analgesics. Like natural opiates, these synthetic narcotics have addiction potential.

**Cocaine** is the strongest natural *stimulant* of the central nervous system. It is extracted from the leaves of the South American coca plant (*Erythroxylon coca*). After cocaine is extracted, the leaves are used to produce flavoring extracts for cola soft drinks. The earliest patented cola drink even had a stimulative effect until the recipe was altered. For centuries, native Indians have chewed coca leaves to increase endurance and reduce fatigue. Cocaine is decreasingly used as a local anesthetic; however, its illegal use as an inhaled powder and, to a lesser degree, as an intravenous injection is increasing. Cocaine ("coke") produces intense euphoria and hence has a strong potential for psychic dependency. It also increases rate of heartbeat, blood pressure, and body temperature. Prolonged "snorting" of cocaine damages the mucous membranes of the nasal chambers and can even cause deterioration of the nasal septum.

## Nicotinelike Effects

Nicotine and several other toxins affect the digestive and nervous systems. **Nicotine** is an alkaloid that occurs in the tissues of cultivated and wild tobaccos (*Nicotiana*). Infrequent human poisoning occurs when the plants are ingested, such as for salad greens. A related genus, *Lobelia,* includes the red and blue cardinal flowers and Indian tobacco, all of which contain an alkaloid similar in effects to nicotine, **lobeline.** Lobeline poisoning generally occurs from an overdose or from medicine.

Several members of the pea family (Table 18-9) contain **cytisine** and other alkaloids, present in varying concentrations throughout the plant but in highest concentrations in the seeds. Because the pealike pods are attractive and resemble garden peas or beans (Fig. 18-15), children are usually the victims. These alkaloids are absorbed rapidly all along the digestive tract so that immediate vomiting

**TABLE 18-9 Some members of the pea family (Fabaceae) containing toxic alkaloids that have nicotinelike effects on the digestive and nervous systems.**

| Common Name | Scientific Name |
|---|---|
| **Ornamental plants** | |
| Golden chain tree | *Laburnum anagyroides* |
| Scot's broom | *Cytisus scoparius* |
| **Native plants** | |
| Kentucky coffee tree | *Gymnocladus dioca* |
| Mescal bean | *Sophora secundiflora* |
| Necklace-pod sophora | *Sophora tomentosa* |

**FIGURE 18-15**
Scot's broom (*Cytisus*), a leguminous shrub having toxic seeds.

is vital. The longer the alkaloid-containing material remains in the body, the more likelihood there is of fatality. Confusion, hallucination, paralysis, and coma may preceed death by respiratory failure.

**Coniine** and related alkaloids make poison hemlock (*Conium maculatum*) a deadly plant, producing intoxication symptoms similar to those of cytisine. It is reputed to be the plant used to execute Socrates, who described feeling the cold and stiffness creeping up his extremities as the poison affected his nervous and muscular tissues. Although native to Europe, poison hemlock now grows wild in the northern United States and Canada. All parts of the plant are dangerous; and it is usually misidentified as parsley or its seeds as wild anise. It is also mistaken easily for Queen Anne's lace, from which it can be distinguished by its nonhairy stems and the flatter shape of the mature inflorescence. Another naturalized introduced species containing lower concentrations of coniine alkaloids is lesser hemlock or dog parsley (*Aethusa cynapium*). It is sometimes mistaken for wild parsley.

## Solanaceous Alkaloids

Members of the nightshade and potato family (Solanaceae) contain alkaloids that have been of tremendous medicinal value but that are dangerously poisonous.

Medically, the three most significant solanaceous alkaloids are the belladonna alkaloids—**atropine, hyoscyamine,** and **scopolamine.** Deadly nightshade or belladonna (*Atropa belladonna*) (Fig. 18-16) characterizes plants that contain these

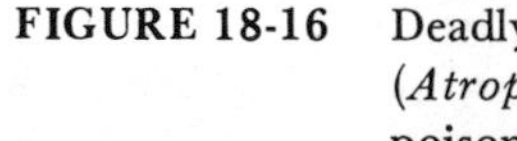

**FIGURE 18-16** Deadly nightshade or belladonna (*Atropa*) is a potent source of the poisonous alkaloid **atropine,** which has some important medical uses.

substances. The deadly aspect of the plant is indicated by its genus name—Atropos, the unbending one, was the Greek Fate who terminated life by cutting life's thread. The species name refers to "beautiful ladies" (bella donna) who used extracts of the plant to enhance their beauty by dilating their pupils. Belladonna drops—or a synthetic substitute—are still used for pupillary dilation during medical examination of the eyes.

The primary medical value of belladonna alkaloids is their capacity to block nerve functions, particularly of the CNS. The overall effect is tranquilizing. In the past, they were used to relieve the pain of childbirth and are currently used to soothe temporarily symptoms of Parkinson's disease and several types of poisoning. They are also used during surgery and in many other ways. Many nonprescription medications (motion sickness and cold-decongestant medications) list one or all of the belladonna alkaloids as active ingredients. The alkaloids, especially atropine, cause dryness of mucous membranes; therefore they help dry up a runny nose (also usually causing a dry throat). Because of the sedative effect, package instructions on preparations containing belladonna alkaloids warn against operation of a motor vehicle during use. And because of complex effects, it can be dangerous to mix medications with each other and with alcohol.

Atropine poisoning produces several consistent symptoms: pupillary dilation, hot, dry skin, and rapid heartbeat. As pupillary dilation continues, the patient becomes unable to see due to excess brightness; the mouth becomes extremely dry so that there is difficulty in swallowing and speaking. Fever and extreme nervous system symptoms develop. The classic description of atropine poisoning is "hot as a hare, blind as a bat, dry as a bone, red as a beet, and mad as a wet hen." With treatment, recovery is likely, especially if ingested material is removed from the stomach as quickly as possible.

As the name suggests, atropine is the primary alkaloid in *Atropa* as well as in Jimsonweed or thorn apple (*Datura stramonium*), henbane (*Hyoscyamis niger*), and the ancient medicinal plant, mandrake (*Mandragora offinarum*). Scopolamine (hyoscine) is also found in these and other plants. Hyoscyamine is found in henbane, *Datura* spp., and others but not in *Atropa*. Other plants containing belladonna alkaloids are *Cestrum* spp. (dayblooming, night-blooming, and green jessamine— but not yellow jessamine, *Gelsemium*), angel's-trumpet and related ornamental trees and bushes in the genus *Datura*, and matrimony vine or box thorn (*Lycium*).

Members of the nightshade family also contain various glycoalkaloids collectively called **solanine**. Solanine is found mainly in green parts and berries, particularly in the genus *Solanum* (Fig. 18-17). Several wild plants have caused human intoxication, but solanine poisoning and fatalities have also resulted from the ingestion of spoiled or green Irish potato (*Solanum tuberosum*) tubers and sprouts and from ill-considered attempts to use potato and tomato foliage as food. The cultivated potato is one of the most dangerous nightshades, with solanine present in all the tissues. Ingestion of the green fruits can be fatal. The white, edible tuber also contains solanine but at nontoxic levels; cooking also removes some. If the tuber is exposed to sunlight, chlorenchyma cells develop under the skin. Before use, all green areas must be trimmed away because they develop higher concentrations

**FIGURE 18-17**
Black nightshade (*Solanum*) berries resemble miniature black tomatoes and contain poisonous alkaloids.

of solanine. You can taste the bitterness of solanine in higher concentrations.

Most Solanaceae have poisonous fruits, but a few are nontoxic—tomatoes, eggplant (*Solanum melongana*), garden huckleberry (*Solanum intrusum*), and ground cherries (*Physalis* spp.). The foliage of all, however, is toxic. Other toxic *Solanum* species are the climbing, black, and graceful nightshades, devil's-apple or soda apple, and the cultivated Jerusalem cherry, a popular Christmas plant (Fig. 18-18). Night-blooming jessamine (*Cestrum*) and chalice vine (*Solandra*) are also dangerous.

### Plants Affecting the Circulatory System

Most circulatory system poisons affect the heart, altering such functions as rate, rhythm, and strength of contraction. These effects may be combined with inhibition of nervous reflexes within the heart and blood vessels; for example, lowered blood pressure may be enhanced by failure of the peripheral blood vessels to constrict normally. Digestive disorders, tingling of the face and skin, and various nervous system effects are common accompanying symptoms.

Two native *Aconitum* species—monkshood or aconite (sometimes called American wolfbane after the European *Aconitum*), and western monkshood—plus apparently all larkspurs (*Delphinium*) are extremely dangerous if ingested. Death from eating only 2 to 4 g of root can follow within minutes (usually within 6 hours) of coordinated twitching of heart muscle fibers (fibrillation). Intoxications are usually the result of home medicinal treatment. Tingling of the skin is the most significant first symptom.

FIGURE 18-18 The foliage and fruits of Jerusalem cherry (*Solanum*) are toxic.

The drug **digitalis** has prolonged the lives of many cardiac patients but is poisonous in overdose and has killed. The victims have been mainly children. Digitalis-containing plants are especially dangerous because they are widely cultivated, nonnative ornamentals. Lily of the valley (*Convallaria*) (Fig. 18-19) and foxglove (*Digitalis*) are particularly common to the northern United States and oleanders (*Nerium*) are widely grown in the southern regions.

Hellebores (*Veratrum*) contain **veratrum,** which severely slows heart rate and lowers blood pressure. Veratrum poisoning is usually not fatal because vomiting occurs quickly and the substance is not readily absorbed by the intestine.

Blue-flowered camas bulbs (*Camassia*) have long been used for food, particularly by the western Indians, but should only be collected in the spring when they are flowering in order to avoid accidental collection of potentially fatal bulbs of white-flowered death camas (*Zygadenus*) with which they often grow. Death camas contains **veratrine alkaloids.**

There are many poisonous shrubs in the heath family (Ericaceae), including two widespread native *Kalmia* species, mountain laurel (calico bush, sheepkill) and lambkill (sheep laurel), whose common names attest to their effects on livestock. Human death has been caused by mountain laurel, which is sometimes also used in folk medicine. Foliage, flowers, and even nectar of various other native and cultivated shrubs of this family are poisonous, including rhododendrons and azaleas (*Rhododendron*), fetterbush (*Pieris*), leucothoe (*Leucothoe*), malebush, staggerbush or fetterbush (*Lyonia*), and the introduced *Pernettya.* Their toxin, **andromedotoxin,** is even present in toxic amounts in honey produced from these plants.

### Plants Affecting the Intestine

Many poisonous plants have an initial effect on the digestive tract, causing pain and vomiting, with or without diarrhea. Some species have already

FIGURE 18-19 Lily of the valley (*Convallaria*).

been mentioned as examples of types of toxic principles. In addition, probably over a thousand plants can cause intestinal damage, with a variety of symptoms, and in many the toxic principles are unidentified. The primary danger of intestinal poisoning is severe loss of fluids and biologically important ions (electrolytes) because of the drastic catharsis that occurs.

Severe and painful, but not usually fatal, intoxication can result from ingestion of pokeweed root (*Phytolacca*). Pokeweed sprouts (poke sallatt) are much sought as a spring edible but must be cooked (changing waters) before eaten; the root must *never* be eaten. One child has died from drinking a fresh pokeberry juice beverage.

Ripe mayapples are the edible fruit of the American mandrake (*Podophyllum peltatum*), but unripe fruit and other parts of the plant are poisonous. Overdoses of home-remedy preparations are common and one fatality has been recorded. Other plants that cause initial intestinal symptoms are *Iris,* wicopy or leatherwood (*Dirca*), and rouge plant or bloodberry (*Rivinia*).

Nuts of the tung tree (*Aleurites fordii*), grown for the tung nut oil used in finishing furniture, are sometimes mistaken for pecans or Brazil nuts where grown and are eaten by children. Reportedly good tasting, they have caused death. Fatalities have also occurred as a result of intoxications by several ornamental plants, including celandine (*Chelidonium*) and the berries of English holly (*Ilex*), privet (*Ligustrum*), and honeysuckle (*Lonicera*) (Fig. 18-20). The foliage of English ivy (*Hedera*) and privet are also toxic.

### Carcinogenic Plants

Research suggests that perhaps 80 to 90% of all cancers are caused by substances contacted as a part of an animal's environment. They may be in water, air, food, or adhering to surfaces.

Several plants are known to contain **carcinogenic** principles, substances that have stimulated growth of malignant tumors at least in test animals, including the cycads, bracken fern (*Pteridium aquilinum*), and the mold *Aspergillis flavus.* Cycads (Fig. 18-21) are generally considered poisonous and hence inedible, but the seeds and roots are used as food in some countries after the toxic

FIGURE 18-20 Honeysuckle (*Lonicera*) berries.

FIGURE 18-21 A cycad. (Photo by Mark R. Fay.)

alkaloid has been thoroughly rinsed out. Bracken fern has been eaten for centuries in Europe and Asia, and many people consider the immature coiled fronds (crosiers or fiddleheads) to be one of the true delicacies of spring.

*Aspergillus flavus* grows on many grains and seeds that can be eaten by people or fed to domestic animals. During growth the mold produces metabolic substances called **aflotoxins.** In the 1960s aflotoxins were found to be a contaminant in peanut butter. In addition, peanut meal from which the oil had been expressed was used as fish food in hatcheries and fish farms. Significant numbers of fish developed tumors; the cause was found to be aflotoxins.

## Mushroom Poisoning

Mushroom poisoning is one of the trickiest types of plant poisoning to diagnose and treat. First, there may be difficulty in identifying the problem as mushroom poisoning unless the patient is conscious and able to provide this information. Secondly, the most deadly mushrooms cause delayed symptoms that neither the victim nor the doctor may initially connect with mushroom ingestion. The third critical problem is in establishing the exact identity of the offending fungus so that specific treatment and antidote (if available) can be applied. Identification requires microscopic examination by an expert mycologist of fresh material, if available, or tissue fragments recovered from the vomitus or feces of the victim.

A fourth difficulty is in the nature and distribution of mushroom poisons—they are heterogenous and each toxin produces its own set of symptoms. Within a given species, many toxins may be present. Thus symptoms may be a composite of separate effects. Fifth, total and relative concentrations of toxic substances within species vary with seasonal and environmental factors as well as with chemical races within a species. Sixth, mushroom toxins may interact with other substances present in the body, such as alcohol or drugs. **Synergism** may then occur in which the interaction of the separate entities produces greater, and usually different, effects than the sum of both working individually.

Finally, the physician (and victim) is faced not only with the unusual difficulties of diagnosis and treatment but (assuming that the victim survives) must also deal with the anticipated recovery sequence. Milder types of mushroom poisoning, for instance, allow the victim to recover a day or two after the episode. Some intoxications, however, extend recovery over weeks and even months.

Mushroom toxins vary chemically and, although a number are well characterized, many remain unidentified. For a few, specific antidotes have been formulated. A mushroom toxin may attack the body in one of several main ways:

1. by causing gastrointestinal irritation.
2. by destroying cells, either specifically, as in the liver or kidneys, or generally, wherever it is transported in the bloodstream.
3. by blocking or otherwise altering functions of the central nervous system.

Some mushroom species contain substances that attack the body in more than one way. So there is justification in listing poisonous mushrooms one species at a time and enumerating their specific syndromes as opposed to attempting to categorize them by ingredients.

A very realistic way of presenting information about mushroom poisoning is by emergency medical diagnosis, for recognizing the symptoms is essential to correct treatment. Table 18-10 is a composite

**TABLE 18-10 Classification of some of the more commonly encountered poisonous mushrooms.**

**Rapid onset**

**Group I.** Produce nausea, vomiting, and diarrhea within 2 hours after ingestion. Gastroenteritis usually only manifestation except as noted below for *Rhodophyllus* and *Scleroderma.*

*Agaricus arvensis* var. *palustris, A. hondensis, A. placomyces:* toxin(s) unidentified; beware if oozes yellow when injured and/or smells like phenol, especially when cooking

*Boletus miniato-olivaceus* var. *sensibilis, B. luridus, B. satanus*

*Cantharellus floccosus:* probably nor-caperatic acid.

*Chlorophyllum molybdites* (Lepiota molybdites, L. morgani):* frequent reports

*Lactarius glaucescens*, L. rufus, L. torminosus**: boiling usually renders safe

*Naematoloma fasciculare* (Hypholoma fasciculare):* very toxic

*Paxillus involutus:* cooking renders safe

*Phaeolepiota aurea (Togaria aurea):* edible, one report of mild gastrenteritis

*Rhodophyllus sinuatus* (Entoloma sinuatum, E. lividum):* also central nervous system, liver symptoms

*Russula emetica*

*Scleroderma cepa, S. aurantium:* also tingling, muscular rigidity (tetany)

*Tricholoma pardinum, T. venenatum*

**Group II.** Produce gastroenteritis rapidly, such as within 15 minutes after ingestion. Other symptoms are excess perspiration, salivation, visual disturbances. Toxin is muscarine, a parasympathetic stimulant. Slowed heart rate, lowered blood pressure; constriction of small breathing tubes in lungs, causing asthmatic breathing.

*Clitocybe dealbata* (C. morbifera):* about 8% mortality reported

*Inocybe napipes, I. mixtilis, I. griseolilacina, I. lacera, I. decipientoides*

*Omphalotus olearius (Clitocybe illudens)*

**Group III.** Produce gastroenteritis, vomiting, and diarrhea; drowsiness usually within 15 minutes; then lassitude and frequently a light sleep lasting about 2 hours. Excitement, euphoria, and visual disturbances follow, including compulsive speaking and shouting, elation, excessive motion, visual and/or auditory hallucinations. Deaths rare, preceded by delirium, convulsions, coma. Ibotenic acid and muscimol (pantherine) cause the central nervous system effects; gastrointestinal irritant unidentified.

*Amanita muscaria*:* fly amanita, fly agaric; in Germany about 1.5% fatality

*Amanita pantherina*:* panther amanita; in Germany about 5.5% fatality

*A. flavivolva, A. cothurnata*

**Group IV.** Produce temporary excitement, hallucinations, hilarity, pupillary dilation. Usually no gastrointestinal symptoms. Effects usually occur within 3 hours and also include drowsiness, inability to concentrate, dizziness, smothering sensation, loss of coordination, and sometimes muscular weakness, rapid breathing, and a tingling sensation. Psychoactive compounds psilocybin and psilocin are responsible toxins in *Conocybe* and *Psilocybe*; that of *Pholiota* unidentified. Main dangers are destructive behavior during intoxication and the possibility of long-term psychoses emerging. Fatalities rare but have been ascribed to *Psilocybe* and *Pholiota.*

*Conocybe cyanopus, C. smithii*

*Psilocybe caerulescens, P. mexicana, P. pelliculosa, P. cyanescens, P. baeocystis, P. cubensis*

*Pholiota spectabilis*

**TABLE 18-10 (continued)**

**Delayed onset**

**Group V.** Produce severe, frequently bloody gastroenteritis with vomiting and diarrhea and excruciating abdominal pain 6 to 12 hours after ingestion, during which time cellular damage is occurring. Subsequent symptoms are jaundice (yellowing of the skin) due to liver damage, total suppression of urine formation due to kidney damage, and drowsiness, leading to coma due to nervous system and other damages. Death occurs in more than 50% of cases; ingestion of a very small portion of the cap (or spores) is sufficient. 90% of the mushroom poisoning in the United States is by mushrooms in this group. Because the cytotoxins damage cells everywhere in the body, a wide variety of symptoms occur, indicating direct damages, such as to the liver, kidneys, or indirectly as a result of random destruction within the brain. There is no definite sequence or combination of symptoms. No specific antidote for *Amanita* poisoning; treatment by hemodialysis has saved victims.

Eight to 12 toxic proteins are classified collectively as "Amanita toxins," the primary actions probably due to *a*-Amanita. They are not affected by either cooking or drying. May or June poisoning probably due to *Helvella* sp. (Group VI); July or later, *Amanita phalloides* and others.

*Amanita phalloides**: death cup or death cap; extremely grave

*A. verna** (spring Amanita), *A. virosa** (the Destroying Angel)

*A. bisporigera**: the most toxic American mushroom

(Possibly: *A. brunnescens, A. tenuifolia, A. vernifornis, A. virosiformis:* these require verification)

*Galerina marginata**, *G. autumnalis**, *G. venenata**

**Group VI.** Produce vomiting but not usually diarrhea 6 to 12 hours after ingestion, plus jaundice.

*Helvella esculenta** *(Gyromitra esculenta, Physomitra esculenta):* false morel, lorel, lorchel

Poisonous principle is gyromitrin (formerly called helvellic acid), which is lost on adequate cooking (including parboiling) or long drying. Therefore *H. esculenta* is frequently collected for eating. Also, the western chemical race (Rocky Mountains to the Pacific) does not seem to be toxic and is collected and eaten along with related species, with no intoxications having been reported.

**Secondary intoxication**

**Group VII.** Produce temporary, severe nausea and diarrhea within 2 hours after ingestion when alcohol is present in the system. Individual is maximally sensitized to alcohol after 12 to 24 hours so that ingestion of alcoholic beverages during this time will cause reaction. Face usually becomes flushed (purplish red) within 5 to 10 minutes; flushing spreads gradually over the entire body; then subsides. There is considerable discomfort as additional symptoms of headache, vomiting, sweating, thirst, weakness, blurred vision, confusion, respiratory difficulties, and others develop. Blood pressure may drop to shock level. Symptoms usually last 30 minutes to a few hours but may recur if alcohol is again ingested during the sensitized period. Effects are due to unidentified substance that acts in a way similar to disulfiram (brand name Antabuse), used in the treatment of alcoholism.

*Coprinus atramentarius:* inky cap

---

*The asterisk indicates one or more fatalities recorded.

Adapted from K.F. Lampe and R. Fagerström, *Plant Toxicity and Dermatitis.* Baltimore: The Williams & Wilkins Company, 1968.

of information about species known to have caused poisoning in North America. They are arranged into groups I–VII by sequence and nature of the initial symptoms.

Most mushroom poisoning incidents are the result of imprudent collection for eating and, in most cases, the result is temporary gastroenteritis. Therefore the precautionary remarks about fungi in the foraging chapter cannot be overstressed.

Sometimes, however, mushrooms are deliberately ingested to produce altered states of consciousness. This is true of the Group IV species. **Psilocybin**- and **psilocin**-containing mushrooms are important to the ceremonies of many groups of Indians throughout the Americas and have been part of their social–religious–cultural heritage for centuries. They are the "magic mushrooms." On the other hand, some of the species in Group III have also been used in this manner even though uncomfortable-to-painful gastrointestinal distress precedes the nervous system effects. (Note also that Group III members have caused fatalities.) Medically speaking, experimentation with any group of mushrooms could be disastrous because of direct cellular and systemic effects, because of the possibility that personality instabilities may be adversely affected, and because long-range psychoses, such as paranoia, can result.

## PREVENTION OF POISONING BY PLANTS

Education is the key to prevention, as is becoming familiar with hazardous plants that are deliberately incorporated into the human environment or that are nearly ubiquitous in the wild. A knowledge, in general, about the potential hazards of touching or eating parts of unknown plants is especially significant and investment in a good book prepared for the nonexpert (e.g., James; see Suggested Readings) could be as important as having a fire extinguisher. Because children are so often the victims of plant poisoning, extension of the labeling laws to include species (container plants, cut flowers, seeds) that are potentially hazardous would be helpful in warning adults to treat the plant materials appropriately. Many communities have Poison Control Centers that can be called in case of suspected poisoning by plants. They are also sources of preventative information and "Mr. Yuk" stickers to identify toxic hazards to children.

## SUMMARY

A **poisonous plant** is one that causes adverse reactions when small or moderate amounts are ingested. There are several hazards other than true poisoning, including **direct physical injury, primary chemical irritation, photodermatitis,** and **allergic reactions.**

Allergic responses differ from primary chemical responses in that there must be sensitization and there is a latent period between the time of contact and the development of symptoms. Allergic reactions are side effects of immunity.

Poison oak and poison ivy cause allergic **contact dermatitis.** People having **atopic allergies** are hereditarily inclined toward abnormal immunity reactions to common antigens, such as pollen and animal dander. Urticaria ("hives"), hay fever, and asthma are atopic allergies. Some people respond to desensitization treatments for atopic allergies.

True poisoning from plant substances occurs in three main ways: using toxic plant parts or extracts for medicinal or experimental purposes; ingestion during foraging; and, most commonly, ingestion by children of attractive fruits or other plant parts. Initial reactions are usually gastrointestinal pain, vomiting, and diarrhea. After toxin absorption, other organ systems become involved.

Toxic **nitrogen-containing compounds** include alkaloids, phytotoxins, and amines. Some plants accumulate nitrates in toxic amounts and are subsequently eaten by livestock or humans.

**Glycoside** molecules contain sugars bonded to nonsugars, many of which are toxic. Some glycosides are those that are cyanogenic, are goitrogenic, that release irritant oils, that release substances that affect heart function, coumarin, and saponins.

Some plants produce **nonglycoside irritant oils** that have medicinal value and toxic potential. **Oxalates** of several types occur in plant tissues. Toxic **resins and resinoids** in various plants have caused violent intoxications, leading to death.

Many naturally occurring soil **minerals and metals** are potentially toxic, including selenium, copper, lead, cadmium, and others.

**Photosensitization** can result from ingestion of certain plant pigments.

**Central nervous system** responses to poisoning are of wide systemic consequence, including perception of pain, depression of central nervous system functions, convulsions, and altered states of consciousness.

**Nicotine** and several other alkaloids affect the digestive and nervous systems.

**Solanaceous alkaloids** are often of medicinal value but poisonous in overdoses.

Poisonous plants affecting the **circulatory system** alter heart rate, rhythm, and contractions.

Many poisonous plants have an initial effect on the **digestive system,** causing pain, vomiting, and/or diarrhea, potentially leading to severe fluid loss and loss of essential electrolytes.

**Mushroom poisoning** is difficult to diagnose and sometimes impossible to treat. The most deadly mushrooms cause delayed symptoms so that irreversible damage is done before the intoxication becomes obvious. Each toxin produces its own symptoms and mushroom toxins can interact with other substances to produce different effects than would be produced by any alone.

Prevention and prompt treatment are important factors related to plant intoxications.

## SOME SUGGESTED READINGS


Hardin, J.W., and J.M. Arena. *Human Poisoning from Native and Cultivated Plants,* 2nd ed. Durham, N.C.: Duke University Press, 1974. Includes useful descriptions and lists of species that are known to have caused human intoxications; also notes on treatment.

James, W.R. *Know Your Poisonous Plants.* Healdsburg, CA.: Naturegraph Publishers, 1973. Inexpensive paperback jampacked with line drawings and brief descriptions; a very helpful handbook.

Kingsbury, J.M. *Deadly Harvest: A Guide to Common Poisonous Plants.* New York: Holt, Rinehart and Winston, 1965. A popular treatment covering the more common poisonous plants.

—. *Poisonous Plants of the United States and Canada.* Englewood Cliffs, N.J.: Prentice-Hall Inc., 1964. A comprehensive book that includes much information on livestock poisoning in addition to details about the nature of hazardous plants.

Lampe, K.F., and R. Fagerström. *Plant Toxicity and Dermatitis.* Baltimore: The Williams and Wilkins Co., 1968. A biomedical text emphasizing human case histories (symptoms, etiology, long-range effects) and treatment; particularly useful for medical practitioners.

Muenscher, W.C. *Poisonous Plants of the United States,* 2nd ed. New York: The Macmillan Co., 1951. The classic text on poisonous plants of the United States. It includes descriptions of plants, symptoms, and some control measures; much information on livestock poisoning, a major agricultural problem.

# chapter nineteen ENVIRONMENTAL PRINCIPLES IN PRACTICE

## INDOOR PLANTS

## INTRODUCTION

Imagine the native environment of humanity—circumscribed by landscape and vegetation instead of walls and windows. It was an environment in which the human species coped for survival and improvement, as did other species.

Most of us retain an innate aesthetic appreciation for vegetation, including the desire to bring part of the outdoor environment indoors. Whether it's a nosegay on the kitchen table or a row of potted plants on the window sill, vegetation relates to the human psyche as a means of naturalizing the internal environment.

The popularity of container plants is at least as old as civilization. The Sumerians, Egyptians, Babylonians, and Hebrews all cultured them. The Hanging Gardens of Babylonia were one of the seven ancient wonders of the world. Greeks, Romans, Moors, and Mongols incorporated vegetation into their surroundings and decorative motifs.

During the seventeenth and eighteenth centuries European fleets sailed the far corners of the world for spices and other wealth. As a result, many exotic plants were collected. One estimate is that during the eighteenth century approximately 1500 exotic species were introduced to Europe. Specialized gardens for maintaining container plants through chilly winter months eventually led to use of glass houses or as we now call them, greenhouses.

In the United States, as in Europe, the Victorian period (second half of nineteenth century) was an apex of container-plant popularity. By the nineteenth century many low-light-intensity-tolerant tropical and subtropical species had been collected. Even the darkest corners of a parlor, hotel lobby, or saloon could support *Aspidistra,* known as the "cast-iron plant" for the environmental abuse it could tolerate. Victorian homes appear to have teemed with vegetation amassed on stands, in wicker trays, or in hanging baskets.

In the twentieth century plants became popular for offices. Interior design now customarily includes accent plants and plant groupings for decorative and dramatic effect as well as for visual screening and softening of functional architectural lines. Sometimes spectacular indoor settings are created (Fig. 19-1).

Nonwestern cultures have long histories of aesthetic appreciation and manipulation of landscape and vegetation. Japanese culture, for example, involves an almost symbiotic relationship of indoor living to at least the illusion of outdoor expanse. The best examples of container-grown bonsai are those that most clearly express a sense of scaled-down but authentic age and natural character.

**FIGURE 19-1** Outdoors or indoors? Natural springs from a limestone outcropping were enhanced and incorporated into the interior design of the Crown Center in Kansas City, Missouri, to create this dramatic indoor landscape.

During the 1970s there was a great upsurge of interest in container gardening. Rediscovery of the ancient Arabian art of decorative knot tying, macramé, was incorporated into this green wave and cascades of hanging plants (Fig. 19-2) still replace curtains in many windows.

So the reasons for bringing plants indoors vary from personal interaction with nature to appreciation of natural decorative potential. Attractive and healthy plants supplement furnishings. A chairless corner becomes full with a plant and a window can be dressed with "living draperies." For some individuals, growing plants is a substitute for

**FIGURE 19-2** Swedish ivy grows as a vine, but it is not a true ivy.

having a pet. A busy schedule with prolonged or frequent absences may preclude the responsibilities of caring for an animal whereas carefully selected plants manage quite well.

Many species of container plants are readily available from greenhouses, florists, and department stores. Unfortunately, the proud introduction of a flourishing sprout into the home environment can be the beginning of a discouraging sequence of events. Before the owner's anxious eyes the plant yellows, drops leaves, wilts, or otherwise demonstrates ill health, dying a prolonged death or being disposed of by a shamefaced would-be horticulturist.

Usually the difference between success and failure is not a green or brown thumb, as is often claimed. Initial specimen selection may have been at fault. Perhaps the plant had already begun to decline, as is common with plants maintained in other than greenhouse conditions prior to sale. Insects or mites may have been present, as could have been determined by a careful inspection of all leaf and stem surfaces. Perhaps the plant selected was not adapted to the particular home environment to which it was taken. On the other hand, perhaps improper maintenance procedures were responsible for the plant's ill health.

These two last problems are the central focus of this chapter.

1. What criteria are applied in selecting an appropriate, healthy specimen?
2. What is necessary to keep it alive and well in the home?

## Applied Ecology and Applied Physiology

Most of our lives are spent contained within buildings. Our primary environment is indoors—artificial. To indoor plants, this environment is total so that it is significant to look at the indoor environment in ecological terms, relating the genetic makeup of species to their environments.

Like the outdoors (Chapter 2), the indoor environment consists of two major components: the physical aspects of light, water, temperature, nutrients, atmosphere, substrate; and the biotic or biological aspects of other plants and animals (including people).

An environmental approach to indoor plant culture involves physiological ecology, relating environmental variables to their interactions with plant functions, including growth, maintenance, and reproduction. In addition, we should consider historical ecology. The origin and native environmental regime of each species determine its growth requirements, tolerances, and seasonal patterns.

Many popular houseplants are components of semitropical and tropical forest understories. Adapted to survive in moderate- to low-light conditions, they are excellent indoor plants. The various *Philodendron* species (Fig. 19-3), *Dieffenbachia, Syngonium, Spathiphyllum* (Fig. 19-4), and the cast-iron plant are examples. Species of humid climates, such as piggyback plant (*Tolmeia*) (Fig. 19-5) and African violet (*Saintpaulia*), require more uniform soil moisture than do species native to moderate-to-dry habitats.

**FIGURE 19-3** *Philodendron cordatum* is one of the hardiest plants for the home environment. The size of new leaves that are produced is an indication of the favorability (larger leaves) or the unfavorability (smaller leaves) of growing conditions.

**FIGURE 19-5** The piggyback plant is one of few popular houseplants that is a native to North America. It grows in the cool, humid climate of the Pacific Northwest.

**FIGURE 19-4**
*Spathiphyllum* also can tolerate moderate-to-low light levels.

The native soil should also be considered. In general, forest floor and epiphytic species do best in soil with a higher humus content. African violets and orchids (Fig. 19-6) will thrive in soils that are composed almost entirely of humus, but most houseplants require a balanced loam soil. (See the container soil recipes in the Substrate section that follows.)

Even within a single family ecological adaptations vary widely. Cacti are a good example. The spiny desert cacti of North America (Fig. 19-7) are adapted to direct sunlight and low-humus, hard-textured soil that dries out completely between infrequent rains. In contrast, flat-stemmed "holiday" cacti [Fig. 19-8 (a) and (b)] are semitropical forest epiphytes and do best with bright but indirect light, a balanced loam soil, and regular watering.

In summary, learn all you can about plant species that you are considering for your own abode. Many popular books are available—a few are listed at the end of this chapter along with the greatly detailed books by A.B. Graf. The rest of this chap-

**FIGURE 19-6** Some orchids, such as this *Phalaenopsis*, produce long-lasting blossoms even when grown under normal house or apartment conditions!

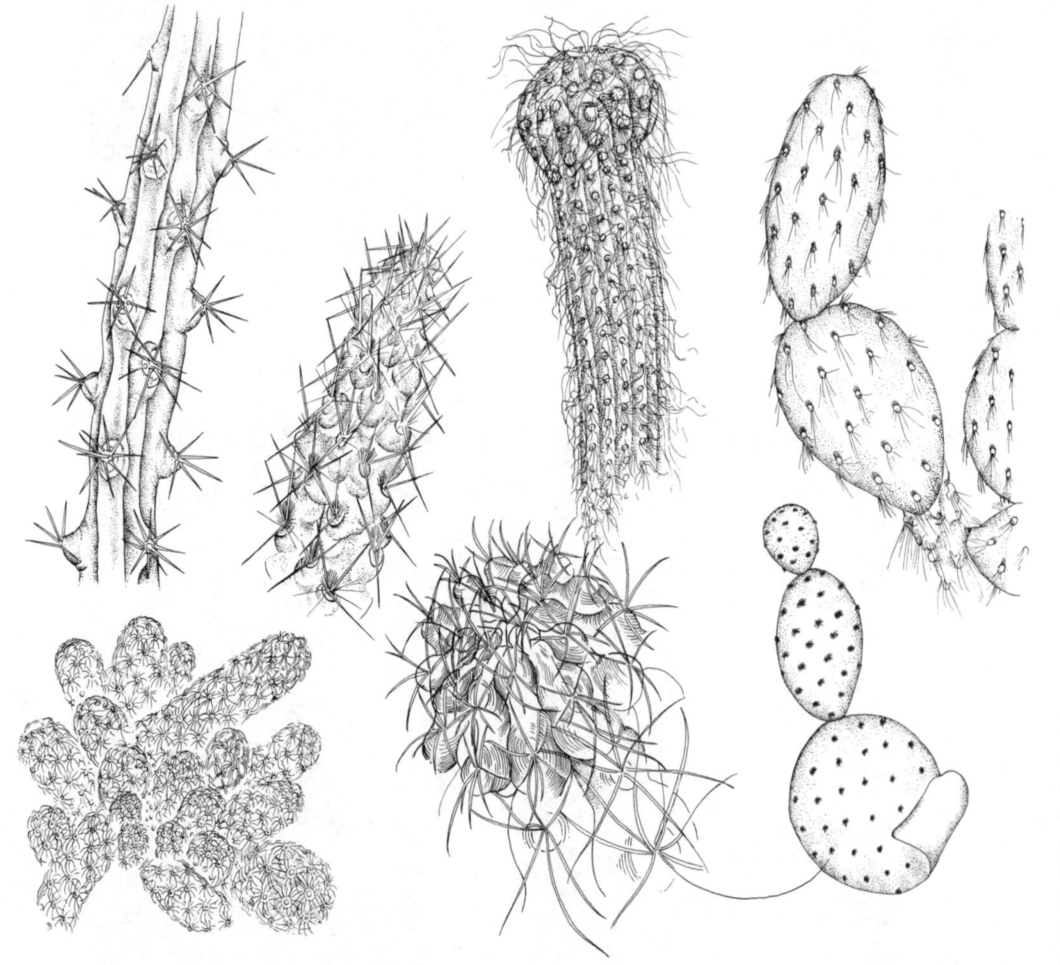

**FIGURE 19-7**
A wide variety of spiny desert cacti are available. Smaller specimens are usually propagated in nurseries. Large specimens are frequently "rustled" from protected wild desert areas, causing severe depletion of native populations.

(a)

(b)

FIGURE 19-8
Holiday cacti (Thanksgiving, Christmas, Easter) are stimulated to bloom easily, requiring a combination of shortening days and chilly nights as summer gives way to autumn. Many varieties are available. (a) A pink-blossomed species. (b) A red-blossomed species.

ter describes certain methods for matching your home conditions to prospective plant species as well as for maintaining them.

### Indoor Landscaping

Landscaping indoors? Consider the concept. The most aesthetic and functional outdoor landscaping is accomplished through careful consideration of the physical, vegetational, and use-related characteristics of a site (Chapter 20). Similar principles also apply indoors. An accumulation of potted plants usually occurs quite casually—a gift or special purchase, a cutting here and there. Before long all the windowsills and floor spaces are filled, the wall is studded with brackets and the ceiling with hooks. Whether it occurs gradually or all at once, the success or failure of this accumulation depends on attention to physiological and historical ecology. Therefore it is advisable to formulate an advance plan, practicing indoor landscaping.

Analyze **site characteristics** of each room—light (main limiting factor), temperature, air circulation, humidity, and relationships to stationary fixtures and furnishings. Decide how plants should interact with **functional aspects**, such as resting, eating, entertaining, family composition, pets, allergies. Then combine physical and functional site characteristics with **ecological characteristics**, including requirements of plant species. For almost every possible lighted spot, there is a plant that will thrive or at least survive in a respectable manner.

## BASIC REQUIREMENTS AND CARE

### Light

Natural light **intensity** indoors varies with exposure—windows facing north, south, east, or west. Intensity also varies with cloud cover or other filtering factors, such as suspended dust, smog, curtains, and shading factors, such as adjacent buildings and outside plantings. Consider the number of cloudy days per year. In the Midwest frigid winter days with bright clear sunshine are numerous compared to the coastal climates of the Northwest and Northeast, where winter days are often cloudy. Plants may need to be moved to the brightest window during winter. Artificial light may be needed, especially for blooming species. Lack of light causes etiolation. Red pigments may be lost altogether, for sunlight is required for their accumulation (Fig. 19-9).

**Quality** of light refers to the wavelengths that are present. Natural daylight contains all wavelengths needed for healthy plant growth. Window glass filters out many ultraviolet wavelengths but does not interfere significantly with lightwaves essential for photosynthesis. Artificial light varies widely in quality. Incandescent light bulbs tend to be "warmer" (higher in oranges, reds, and infrareds). Regular fluorescent lights are "cooler" (lower in oranges, reds, and infrareds), but some of the missing red bands are the ones that plants use for photosynthesis. Incandescent light, therefore, is more complete for plant growth than ordinary fluorescent light; however, sufficient incandescent illumination may produce unfavorably high temperatures! As a result, special fluorescent plant-growing bulbs have been developed that provide complete lighting at cooler temperatures. They are relatively expensive but last for many hours.

The direction from which light strikes a plant stimulates directional growth, a process called phototropism (Chapter 9). Plants should be turned regularly to maintain symmetrical growth indoors.

FIGURE 19-9 Different patterns of leaf markings and margins distinguish the decorative *Coleus blumei* varieties. To maintain bushy growth and red pigments, they must be kept in bright light and should be pinched back.

## Water

Overwatering is the most common mistake in houseplant culture. Most plants tolerate and actually seem to thrive on "benign neglect." If the top 2 to 3 cm of soil is dry to the touch when you poke your finger into it, water. If the soil feels moist, don't water. High-humus soils should not be allowed to dry out completely, for they will shrink into a nonabsorbent mass requiring patient rehydration. With soft-tissued plants, the onset of a slight wilt, together with a dry soil surface, is a good sign that water is needed. Most plants recover rapidly from slight wilting with no long-term ill effects.

Adequate drainage is essential. If a plant wilts in spite of moist soil, overwatering may have created a health problem for which a cure is not possible. Constantly wet soil encourages fungi and bacteria that attack roots and stem bases. Bacterial decomposition also increases soil acidity. The lower pH is beneficial for some plants but harmful to most. Soggy soil is also low in oxygen, which is required for healthy roots. Overwatering is a common problem with plants grown in very large containers because the soil surface may be dry even though the deeper soil is wet.

The cure for a plant with "wet feet" is reestablishment of a healthy root system and proper potting. First, propagate the plant from healthy cuttings (Chapter 5) and discard the old plant. Wash, dry, and rest the old container. It may be sterilized with chemicals, by sunlight, or in an oven.

Although apparently clear, tap water contains dissolved substances that affect plant growth. Naturally low-salt and low-mineral water is good for plants; such water, like rainwater, is said to be soft. **Hard** water (high in dissolved minerals, especially calcium salts), however, is not good for container plants. Water molecules are constantly removed from the soil, but excess minerals are left behind. So in time an extremely concentrated mineral content results, which upsets the root–soil solution osmotic balance. Some minerals may be directly toxic in quantity. Many plants develop such symptoms of salt damage as browning of leaf tips from mineral buildup. A white crust on the soil surface or on the exposed inner sides of the container indicates salt buildup and requires remedy. (See Precautions section of this chapter.)

Water from a water softener is undesirable because the softener removes calcium salts (lime) but adds sodium. Municipal water is usually chemically treated and should be left standing in an open container for a day before it is used for watering to allow most of the chlorine to diffuse out.

The most natural source of water for plants is precipitation. Many people collect rain and melted snow, but pollutants may be a problem in urban and industrial areas. An ingenious friend in Arizona collects the essentially pure water that condenses on the coils of her air conditioner. Ice or frost on refrigerator coils or freezer compartments is also condensed pure water. On-the-faucet filters and portable deionizers also provide cleaner water. Bottled distilled water can be purchased but is too expensive considering that other sources are free.

Because some plants are temperature sensitive, use room-temperature or lukewarm water for watering.

## Atmosphere

### Carbon Dioxide

Photosynthesis utilizes carbon dioxide, supplied by Earth's atmosphere and continually renewed by organic respiration and other oxidative processes, such as combustion. There is an old belief that it is unsafe to sleep in a room with plants. Perhaps it is related to the fact that plant respiration continues 24 hours per day, but at night it is not offset by continued photosynthetic production of oxygen. Suffocation by a group of houseplants, however, is highly unlikely!

### Humidity and Drafts

Houseplants are sensitive to atmospheric humidity, which affects the rate of evaporation from plant, soil, and pot surfaces. **Absolute humidity** refers to the total content of water vapor in a given volume of air. A more useful expression of humidity (and one used by weather reporters), however, is as relative humidity. **Relative humidity** is the ratio of water vapor present to the maximum amount potentially present at a given temperature.

Indoor humidity approximates the outdoor environment. During winter heating or summer air conditioning, however, household humidities may drop to desert levels! In subfreezing winter climates the air becomes especially dry and homes must be equipped with humidifiers to provide human comfort and prevent static electricity and dehydration of furniture. A relative humidity of 40 to 60% at about 20°C is comfortable and beneficial both for humans and for many popular houseplants.

Air movement increases transpiration, compounding a low-humidity problem with homes that have forced-air circulation. To maintain higher humidity for plants, mass plants together, mist plant surfaces regularly, set pots in or over water-filled trays lined with pebbles or sphagnum (Fig. 19-10), set containers of water among potted plants, or move humidity-sensitive plants to a warm, humid environment, such as the bathroom. It may be necessary to tent especially sensitive plants with plastic wrap or bags. Extremely sensitive plants might need planting in a terrarium.

**FIGURE 19-10** Low atmospheric humidity, especially during the winter heating season, is detrimental to most container-grown plants. The use of a pebble-filled evaporation tray and misting help compensate for this problem. (Photo courtesy of David Koranski.)

### Effects of Pollution

Such air-borne pollutants as fuel fumes, products of incomplete combustion, smoke, sulfur dioxide and other noxious chemicals, and a variety of suspended particles affect plants. Specific pollutants produce specific effects, although the composite effect may be generalized lack of vigor or death.

Even when outside air is relatively pollution free, there is indoor pollution. Plant surfaces accumulate dust. Dust filters light so that a very dusty leaf is actually existing in dense shade. Because light is the critical factor that is generally in short supply indoors, the additional reduction is significant. Dust also blocks stomates on upper leaf surfaces. For ordinary dust, wipe leaves with a damp cloth—a complete lukewarm shower bath is ideal (Fig. 19-11).

The other main indoor pollutant is grease from cooking. Especially during frying and broiling, minute oil droplets become suspended in the air and eventually are deposited on exposed surfaces, including plant leaves. Oil clogs stomates, limiting gas exchange, and becomes increasingly sticky as it ages, thus causing even greater dust accumulation. Greasy plants may not pose a cleaning problem in that they may simply die! If you use hardy, smooth-leaved species (Fig. 19-12) near cooking areas (as opposed to hairy-leaved specimens), however, you can wipe the leaves with a lukewarm, mild-detergent solution. Then follow with a shower or other rinse. During washing protect the soil and pot with a plastic bag to retain soil and keep out the detergent.

**FIGURE 19-12** Many *Peperomia* varieties have succulent, waxy-surfaced leaves that can tolerate kitchen "air pollution" and are easy to wash off.

**FIGURE 19-11** A good bath removes light-filtering dust from plant leaves. Note the plastic bag around the pot to keep the soil from spilling out. (Photo courtesy of David Koranski.)

Commercial products are available for cleaning and polishing leaves and milk and mayonnaise are sometimes used. All contain oils, fats, or waxes, which may attract dust and clog stomates on upper leaf surfaces. They can also cause a light-filtering, greasy buildup. Oily spray-on foliage shiners can even cause tissue damage. Some people prefer very shiny leaves; we prefer the nonplastic, natural semigloss or matte leaf finish. Foreign substances should never be applied to the undersides of the leaves because most of the stomates are there.

Hairy-leaved plants present a special problem in cleaning. They are likely to accumulate dirt more easily than smooth-leaved plants. Most houseplants can withstand a good shower. Dust on sturdy plants can be lightly (!) vacuumed or blown off. Recommendations for tropical hairy-leaved plants, such as African violets (Fig. 19-13), vary. Some

FIGURE 19-13 In contrast, African violets (like those arranged in this pyramid) have tender, hairy leaves that are sensitive to airborne oils and even to water droplets. (Photo by Mark R. Fay.)

references say to wash the leaves; water droplets should not remain, however, because the leaves are susceptible to rotting. Gentle brushing with an artists' paintbrush might suffice and African violet fanciers commonly remove an older leaf and use it as a brush to "groom" other leaves.

Aside from dust, dirt, and oils, various gases inhibit plant growth. In homes the most likely culprit is escaping fuel gas or the incompletely oxidized byproducts of fuel gas or oil combustion. In general, a kitchen is not the best place for plants.

#### Temperature

Most species grown as houseplants are from tropical and semitropical regions of the world. They are quite adaptable to usual home-temperature regimes (near 20°C), can be put outside during summer, but cannot tolerate freezing or near-freezing temperatures. Many species seem to do better when nighttime temperature is about 5 to 10° lower. Some prefer cooler locations during dormant periods.

Seasonal temperature variation is a problem mainly for window plants. In winter they are subject to chilling. If the outside air is freezing, the inside air near the window may also drop to or near freezing, causing cold damage to susceptible plants. Drafts occur due to convection currents in which the cold air near the window interacts with the warmer interior air. Drafts increase transpiration, which is also a cooling phenomenon (like sweating). The combination of chilling and draft is deleterious to many plants. So leave only hardier plants near windows in winter and move or insulate others.

High temperature (with or without air currents) is a problem in several situations. Artificial lighting may increase temperatures. Location in a south-facing window, especially during summer, may cause excessively high temperatures. Location near the hot, dry breeze from a heat duct produces an arid microclimate. In a west-facing window the afternoon sun builds up high temperatures near the window even in winter.

### Substrate

The medium within or upon which an organism lives is its substrate. The substrate indoors may be water or soil. If the plant is an epiphyte (Fig. 19-14), its substrate may be the outside of a

FIGURE 19-14 The staghorn fern (*Platycerium*) is an epiphyte; so it is grown on the outside of a pot, not on the inside! (Photo by Mark R. Fay.)

pot, a mesh-wrapped ball of moss, or a slab of bark. Even in soil, plant roots exist in a semiaquatic environment, with rootlets and water interacting among soil particles. So a discussion of substrate also includes water and dissolved substances within the substrate. Plants depend on soil for water, minerals, air, and anchorage. A good soil must be able to retain moisture while providing adequate aeration and root penetration. Chapter 10 provides a more complete discussion of soil characteristics in relation to plant requirements.

Texture is the most apparent feature of any type of soil. Clay soil is fine textured and sticks together when wet. Silty soil is fine textured but tends to be runny or oozy when wet. Sandy soil is relatively coarse textured so that it drains and yet holds water, and the particles remain independent. Good potting soil is generally a **loam** type, a combination of fine and coarse particle sizes, which drains well, holds water, and provides secure root anchorage. Such organic materials as compost and milled or sifted peat provide soil mixtures with humus for water holding and aeration. In addition, their continued decomposition releases nutrients slowly to the soil.

Soil pH for most houseplants should be nearly neutral, 6.7 to 7.5, although some forest and bog species (Fig. 19-15) prefer acidic soil, simulating their natural substrates. Monitor soil pH occasionally, for it is made more acid by plant metabolism and soil organisms; fertilizers may tend to make the soil more alkaline. The pH of water affects soil pH. Soil pH also affects the availability of minerals to plants.

The most common houseplant pH problem is excess acidity. It can be treated by "Great-Grandma's" method of adding finely crushed eggshells to the pot now and then. Bone meal also helps neutralize acids as they are formed in the soil. You can monitor soil and water pH easily and cheaply with pH test paper purchased from a pharmacist. pH test papers are available in various pH ranges, such as 4.5 to 7 and 6.8 to 8.0. For more money, you can buy a pH meter with probes that can be used for other purposes besides soil testing. You can purchase a soil test kit for precise analysis of pH and other soil factors. Table 19-1 lists application rates for the preparation of pH-adjusted potting media.

#### Potting Soil Composition

There are many specific soil mixes for container plants. Some of the commercially available potting mixtures are balanced in texture and nutrients and are suitable for nearly all popular houseplants. Many, however, are very high in humus and must be adjusted for loam-preferring plants. Specific mixtures are sold for African violets and their kin, and others for cacti. The average houseplant

**FIGURE 19-15**
The sundews (*Drosera*) and Venus flytrap (*Dionaea muscipula*) are bog plants that enhance their nitrogen uptake by capturing and digesting small insects. (Photo courtesy of Carolina Biological Supply Co.)

TABLE 19-1 **Rates of application of substances to adjust soil pH for container plants.[a]**

| Additive Materials | Addition to bench media, as in greenhouses, having a soil depth of about 4 in. (Pounds of additive per 100 sq ft of bench area) | Addition to potting media (Pounds of additive per cubic yard) |
|---|---|---|
| To increase acidity | | |
| Finely ground sulfur | $\frac{1}{2}$ lb | $\frac{1}{4}$ lb |
| Aluminum sulfate | 3 lb | $1\frac{1}{2}$ lb |
| Iron sulfate | 3 lb | $1\frac{1}{2}$ lb |
| To decrease acidity | | |
| Crushed limestone or dolomite | 5 lb | $2\frac{1}{2}$ lb |

[a]These adjustments are appropriate to the preparation of new potting media. If pH adjustment of an existing container plant is necessary, repot the plant in fresh medium. One cubic yard (3 by 3 by 3 ft) = 0.76 $m^3$. One pound = 453.6 g (0.454 kg).

*Source:* Oregon State University Extension Service.

owner is wise to begin with a *bona fide* potting soil, particularly if the alternative is to go outside and dig up some soil. Chances are great that even fertile "black dirt" will not produce good results; it is too fine textured or "heavy" and also probably harbors undesirable microorganisms and weed seeds.

If the expense of a commercial mix is unappealing, you can mix a good potting soil at home in a bucket, plastic or metal tub, large trash bag, or on a sheet of plastic or newspapers. A general mix requires three basic ingredients—sand, milled peat, and garden loam. By adjusting proportions and adding other ingredients, you can concoct mixes for individual needs. Descriptions of a variety of useful container-mix ingredients (Fig. 19-16) follow.

**FIGURE 19-16** Container soil mixtures contain various ingredients to provide adequate anchorage, nutrition, and drainage. (Photo courtesy of David Koranski.)

*Sand.* Sand increases soil porosity and drainage. Medium-to-coarse builder's (sharp) sand is the type utilized in mortar. Sand should be salt free (therefore ocean sand is not appropriate) and have sharp edges, produced by crushing, as opposed to the smooth, rounded contours of river sand.

*Peat Moss and Peat.* *Sphagnum* (peat moss) is sold in two basic forms. Peat moss or natural sphagnum appears much as it did when it was collected, consisting of whole plants—dried in a mat or sheet or wad—and is used to line hanging baskets, cachepots, terraria, and in air layering. Peat, milled peat, or sifted peat, however, is brown and has been milled into a light, fluffy-textured form. Because the native habitat of *Sphagnum* is boglike, unwashed peat moss may become acidic in solution. Properly milled peat, on the other hand, is de-acidified (acid neutralized). Peat in a mix contributes moisture-holding properties and is active in the soil colloid system (Chapter 10). In addition, its continuing decomposition releases some nutrients and contributes to soil acidity. High-peat mixes should not dry completely because dried peat resists wetting. Water simply runs down the pot sides and out; careful soaking is necessary to restore soil moisture.

*Loam.* Loam is garden soil having a balanced proportion of clay, silt, and sand particles. A good sandy loam is approximately 2/5 sand, 2/5 silt, and 1/5 clay. (See also Chapter 10.) High-loam mixtures provide firm anchorage and retain moisture longer and loam soil can dry between waterings and still remain absorbent.

Loam usually contains many nutrients but also harbors soil microorganisms—some beneficial, some harmful. Sterilize moistened loam by baking it in a kitchen oven 1 to 2 hours at 180°F on a tray. Alternately, the moistened soil and a medium-sized potato may be placed in an oven cooking bag at a 300°F oven setting. Potato and soil will be "done" at the same time. You may be tempted to try a microwave oven; it is not advisable because metals in the soil could damage the magnetron unit. Home methods are not as thorough as steam sterilization (autoclaving) but are usually adequate to kill potentially pathogenic bacteria and fungi, insect eggs, and larvae.

*Humus.* Humus is the organic content of a soil. In general, fine-rooted plants adapted to living on the forest floor and epiphytes adapted to living on the bark or in the crotches of tree branches do best with a high-humus soil. Humic soil also tends to be somewhat more acidic than soil with a higher proportion of loam. There are many sources of humus for container-soil mixes.

Compost adds humus to soil. Sifted compost from a compost pile or from such natural composts as decomposing forest leaf litter (woods soil) and leaves piled against the side of a fence may be used. Oak and conifer leaf composts should not be utilized alone, however, unless their acidity is counterbalanced by the addition of alkaline material. Leaf mold is available commercially.

Well-rotted barnyard or feed lot manure, including decomposed animal feces, straw, and hay, is a good source of humus. In its well-rotted state excess nitrogen has been leached away, as opposed to the high-nitrogen state of fresh manure and urine-soaked bedding.

Osmunda fiber consists of the shredded fibrous rhizomes of cinnamon fern and is used to grow some epiphytes, particularly orchids.

The western timber industry is the source of valuable lumber byproducts, including shredded fir bark, which is used in high-humus mixes for such acid-loving plants as azaleas and many epiphytes. Redwood bark shavings and sawdust are used in potting mixes in the Pacific states, although like other conifer bark products they do not provide nitrogen. Any coarse organic matter can be shredded and incorporated into potting mixes to contribute to soil texture and nutrient value, especially after a period of composting.

*Vermiculite.* Vermiculite is heat-expanded mica. It is totally inorganic, ionically neutral, sterile, light, spongy, and has good water-holding capacity. Because of these qualities, it is valued for starting seeds and for mixing into potting soils. It is especially helpful in making lightweight soils for hanging baskets and pots.

*Perlite.* Perlite is volcanic rock that has been expanded by heat and is a lightweight substitute for sand. Perlite is white, grainy, comes in a variety of particle sizes, permits good drainage, and is commonly used for seed germination, rooting cuttings, and as an ingredient in soil mixes.

*Other Ingredients.* Powdered limestone rock and dolomite are commonly added to help maintain near-neutral pH. Sterilized bone meal is an effective high-phosphorus fertilizer, especially beneficial to bulb plants. Charcoal particles are used in epiphyte and terrarium soil mixes to help counteract acidity that results from bacterial metabolism.

#### Container-Mix Recipes

Recipes for potting soils are as variable and personal as recipes for fruitcake! Recipes have been selected for Table 19-2 that reflect a consensus from many sources, without becoming esoteric. The lightweight soil mixes [4 (a) and (b)] are especially helpful for suspended containers and have other advantages, such as good drainage and light texture. They lack intrinsic nutrients, however. Additionally, they dry out quickly and do not provide a secure foundation for larger plants. A stone placed in a hanging container lessens swaying in the wind when lightweight mixes are used.

### Nutrients

Maintaining adequate nutrient levels is easy because so many products are available. Plants require about 16 nutrients for complete growth, some as macronutrients and others as micronutrients (Chapter 10). If a balanced potting soil is used, enough micronutrients are present to supply the plant for a long time so that regular addition of a macronutrient fertilizer (mainly nitrogen, phosphorus, and potassium) will suffice.

Underfertilization causes less vigorous growth and prolonged lack of specific nutrients can cause

TABLE 19-2 **Container soil mix recipes.**

| | |
|---|---|
| **Basic mixes for nearly all container plants** | |
| 1. Loam-type potting soil: | 1 or 2 parts loam<br>1 part peat or compost<br>1 part sharp sand or perlite<br>Some vermiculite if desired |
| 2. Humus-type potting soil: | 1 part loam<br>2 parts peat or compost<br>1 part sharp sand or perlite<br>Some vermiculite if desired |
| 3. Soilless mix for general use: | 1 part *fine* sand (0.05 to 0.5 mm)<br>1 part peat |
| 4. Lightweight soilless mixes for hanging plants, window boxes: | |
| (a) basic lightweight mix | 1 part peat<br>1 part vermiculite |
| (b) slightly heavier but more rapidly drying | 1 part sharp sand or perlite<br>1 part peat<br>1 part vermiculite |
| **Special mixes for some specified groups** | |
| 5. Cacti and similar succulents: | 1 part loam<br>1 part peat or compost<br>2 parts sharp sand or perlite, including $\frac{1}{2}$ part pea-sized pebbles or crushed clay pot pieces |
| 6. Epiphytes: | Basic humus mix plus $\frac{1}{2}$ part pea-sized pebbles or crushed clay pot pieces |
| 7. Acid lovers, such as azaleas, gardenias, citrus trees, tomatoes: | Basic humus mix plus occasional iron chelate feedings |
| 8. Pelargoniums (geraniums): | 4 parts loam<br>1 part peat or compost<br>1 part sharp sand or perlite<br>Some vermiculite if desired |

deficiency symptoms. Overfertilization can damage roots and cause mineral buildup in the soil.

To select an appropriate fertilizer, remember that nitrogen encourages foliage growth but that flowering plants require a lower proportion of nitrogen to phosphorus to encourage flowering and fruiting.

#### Organic Fertilizers

Fish remains are a time-honored fertilizer (recall fish planted under corn!). Fish meal, seaweed emulsion, and fish emulsion are popular organic fertilizers. As you might suspect, even the deodorized versions are a bit odorous; so they should not be used on a day that guests are expected. They are good fertilizers in that both macro- and micronutrients are present. Also, because the nutrients are less concentrated than in crystalline fertilizers, there is less danger of fertilizer burn. Rotted animal manure is a valuable fertilizer and humus ingredient in potting soil.

#### Inorganic Fertilizers

Inorganic fertilizers are highly concentrated and must be used with caution. In addition, most inorganic fertilizers contain only nitrogen, phosphorus, and potassium. When you select a houseplant fertilizer, read the labels carefully and attempt to find a balanced "recipe," one that includes micronutrients and chelated iron.

In addition to powder and liquid forms, houseplant fertilizers also come in time-release pellets and stakes, evening out nutrient application over a period of weeks. These forms are usually a little more expensive than soluble inorganic and organic crystals and liquids.

### Precautions

Do not fertilize a dry pot and do not exceed the recommended fertilizer concentration. Both errors create drastic osmotic stress on tender young absorptive roots. A sensitive plant can be killed by such treatment.

Do not overfertilize by frequent applications, which also kills tender roots. A plant may react to overfertilization by developing a wilted, water-soaked appearance.

Do not allow salt and mineral residues to build up in the container, as evidenced by whitish deposits on the soil surface, inside the container rim, or on the outside of a clay pot. To leach minerals from a container (must have a drainage hole), set the plant and its container in a sinkfull of tepid water. Let it remain there until the soil is thoroughly soaked; then allow it to drain. Repeat the process several times to ensure a thorough washing. If your water contains chlorine, allow it to sit for an hour or so before leaching the container soil. If your water is "hard" (containing a high mineral concentration) or is chemically "softened" (contains salt residues), it might be better to collect rainwater or melt snow for leaching.

Be aware of the changing nutrient requirements of each plant. Seedlings and new transplants require proportionately more phosphorus. Subsequent rapid juvenile growth requires ample quantities of all essential nutrients. As plants prepare to flower, proportionally more phosphorus than nitrogen is required.

During cooler, darker weather plants do not grow rapidly. They should not be fertilized in slow growth periods or during dormancy. Fertilizing during inactive periods simply increases soil nutrient concentration, possibly to toxic levels.

## Container Selection and Potting

Style is the most obvious criterion for container selection, but several general principles should be observed. Match the size and shape of the container to the growth habit of each plant—erect and narrow, spreading, moundlike, climbing, or trailing. As a rule of thumb, the visible portion of the plant should be predominant over the visible part of the container by about 2:1. Rapid growth of some plants should be considered; however, it is not desirable to overanticipate by using too large a container. It is better to trim or divide plants periodically to maintain a size or gradually to move up in pot size.

Drainage and aeration are critical considerations and, generally speaking, containers that have drainage holes are preferable to those that do not. Overwatering is probably the most common killer of houseplants, especially with drainless containers. If such a container is large enough, the surface may feel dry even though the bottom third of the container is soggy and the roots are rotting. If you have a drainless container, you may drill a hole into the bottom. If this is not feasible, provide a substantial bottom layer of gravel (for drainage) and activated charcoal (to reduce souring of the soil by bacterial metabolism) (Fig. 19-17).

Consider the composition of container material. Unglazed ceramic or pottery containers allow gas exchange through their sides and reduce the danger of overwatering. Because they do allow evaporation, more frequent watering is necessary but easier to control. For drainage and aeration, it is difficult to improve on the basic tapered clay (terra cotta) pot (Fig. 19-18). The pot "breathes," its natural appearance is aesthetic in its own way, and it may be set within a cosmetic container, or **cachepot,** if a more decorative effect is desired. Place a layer of gravel in the cachepot to allow drainage. Cachepots permit flexibility in that they may be selected for a permanent accent, but the plants themselves can be rotated to different sites without complete replanting. Also, a clay pot in a cachepot loses moisture more slowly, especially if peat moss (sphagnum) is stuffed between the walls of the two containers.

Container size is an important factor. In total biomass the root system of a plant is approximately equal to the aerial portion. When the roots have completely exploited the available growth space, both root and shoot growth rates decrease. Such a plant is said to be **rootbound** or potbound. Theoretically a potbound plant has limited potential for further significant vegetative growth. If adequate nutrients are supplied, however, continued growth can be maintained. Eventually lack of space for new root formation becomes a problem. You may recall that as roots undergo secondary growth, they lose their ability for absorption. As a result, the effective absorptive area is gradually diminished and the plant becomes senescent—surviving but no longer growing. To avoid this unhealthy situation, you should periodically repot a specimen, trimming

FIGURE 19-17
A layer of pebbles helps to keep roots from being continually waterlogged in a drainless container. (Photo courtesy of David Koranski.)

FIGURE 19-18 Tapered, porous terra cotta (clay) pots are ideal containers for most indoor plants and can be set within a decorative cachepot to achieve a different aesthetic effect. (Photo by Mark R. Fay.)

the roots if necessary. Some plants will die unless rejuvenated by the opportunity to produce new growth.

Some blooming plants and runner-producing plants (Fig. 19-19) develop maximum vegetative growth before initiating flowering or producing runners. From the plant's perspective, it is taking maximum advantage of the environment to develop the food-producing stage of life before entering the reproductive stage.

If a plant is potbound, it may "tell" you so in several ways. The roots may grow out through the drainage hole or may be forced upward and be visible around the soil surface. In extreme cases, the crowded root system may break its pot (Fig. 19-20). Increase pot size gradually through successive repottings. As when buying children's shoes, going from size 2 to size 6 is not a good idea. Root aeration is poor when a small root system occupies too large a pot. In addition, the roots are unable to make full use of all fertilizer nutrients, thereby permitting excessive mineral accumulation.

Container weight is an obvious factor with hanging plants. Free-standing pots should be heavy enough to resist tipping.

Container strength is a factor with vigorously growing and larger plants. Large specimens need containers strong enough to resist root growth pressures and handling of the pot. If a plant is too heavy to lift, consider buying or building a small platform on casters (dolly) to put under the plant

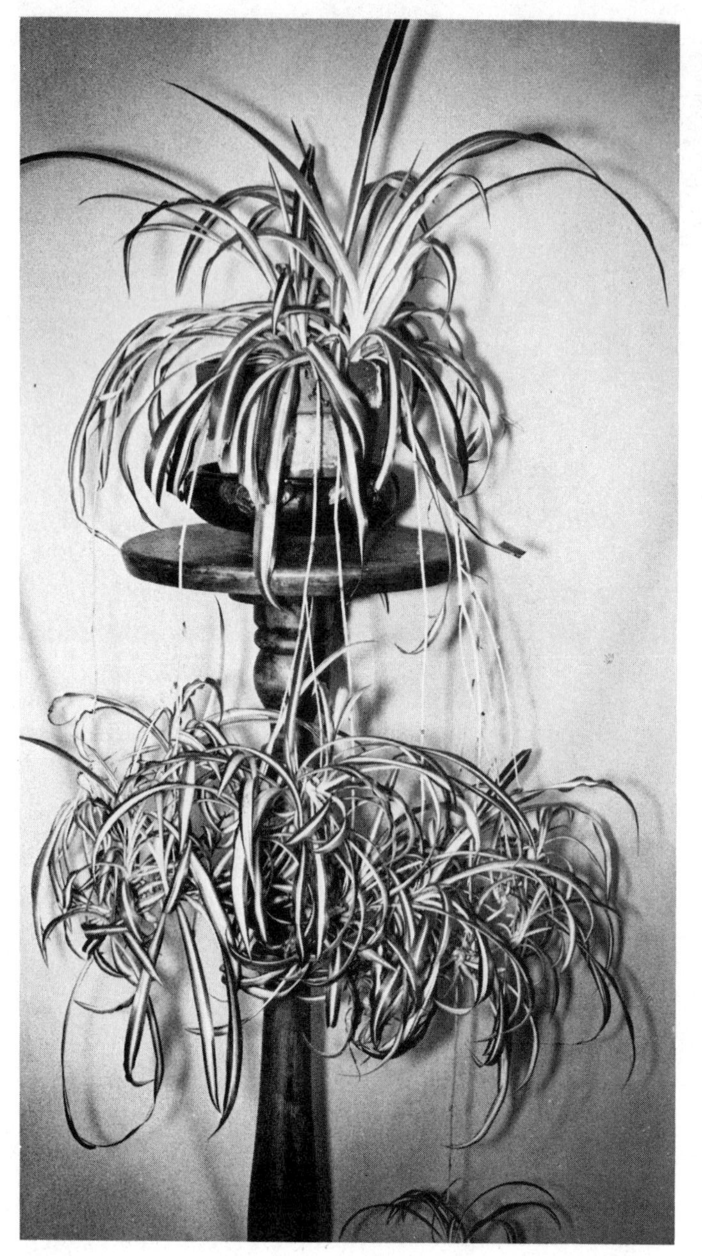

**FIGURE 19-19**
Because they produce runners (stolons), spider plants (*Chlorophytum*) make dramatic pedestal and hanging plants.

**FIGURE 19-20**
Houseplants are rarely permitted to become so rootbound that they demolish their containers, as happened here! Periodic root trimming and repotting are essential for several reasons, as detailed in the text.

so it can be moved for housecleaning or room rearrangement.

## Plant Size and Shape

Size and shape are primarily determined by genetics. Most plants, however, will not achieve their maximum potential growth in the artificial indoor environment. Within natural genetic limitations, it is possible to manipulate size and form. The secret to success is an understanding of the basic principles of plant growth.

### Control via Primary Growth

You will recall that primary growth is the elongation of shoots and roots by addition and maturation of new cells at shoot and root tips. Primary growth is influenced by apical dominance, hormonal control exerted by the apical shoot over the lateral shoots (Chapter 9).

Many houseplants grow long main stems with few side branches whereas others are naturally bushier. You can induce more compact and fuller growth form by pinching out the terminal buds or shoots. This step removes apical dominance, thus encouraging growth of lateral shoots.

### Control via Secondary Growth

The products of secondary growth are wood and bark, formed by the lateral meristems (vascular and cork cambium) (Chapter 8). Woody container plants, such as citrus, olive, myrtle, azaleas, boxwood and bonsai (pronounced "*bone* sigh") specimens, can be managed by pruning, wiring, and grafting. Woody stems are part of the permanent body of the plant and these methods allow rearrangement of existing branch patterns. Figure 19-21 (a) to (k) presents a series of stages in the transformation of a nursery juniper into a bonsai specimen.

## Dormancy and Resting Periods

Many houseplant species have dormant periods—some cease growing; others die back to the ground (Fig. 19-22). During dormancy plants require less water, less light, cooler temperatures, and no fertilizer. Fulfillment of these environmental requirements will ensure vigorous growth following dormancy. Some flowering plants require the gradual onset of dormancy-producing conditions to induce floral bud formation, including decreasing daylength, water availability, and temperature. Some species even require chilling.

## Common Diseases and Pests

Common unhealthy symptoms on indoor plants are discolored or necrotic lesions (dots or streaks); wilting even though the soil is moist; mushy, brown, rotten areas, such as in the crown or on leaves; cobwebby matter on the foliage; curled or distorted foliage; leaf or flower drop; loss of normal color; general lack of vigor; falling over due to rotting off at soil line; and death. (See also Chapter 12.)

Streaks or rings on foliage may indicate viral infection and the plant should be discarded before infection spreads to other plants. Viral infections are spread by feeding insects or mites, which act as vectors.

Fungal and bacterial diseases are frequently related to overwatering and cause rotting of crowns, petioles, and roots. Damping off of seedlings is a fungus problem.

*Botrytis* (gray mold) produces gray mold filaments and mushy, rotten plant parts (Fig. 5-43). It begins on dead plant parts and may then spread to the living. Crowding and excess humidity, excess water left on the leaves, and high nitrogen concentration in the soil are environmental conditions favorable for gray mold. Sanitation is an obvious preventative control.

Fungal leaf spot and bacterial rot are enhanced by water on the leaves, high humidity, and poor air circulation, such as might occur with overcrowding. Powdery mildew infection is shown by grayish-white powdery areas on the leaves and stunted growth; and if enough leaves are infected, the plant may die. Cold drafts, sudden temperature changes, and excess humidity are said to contribute to powdery mildew infection. Dusting powders are available for its control.

Most fungal and bacterial diseases, especially the rots, cannot be cured, but they often can be prevented by obtaining noninfected plants, by using sterile soil and containers (pretreatment with fungicide or heat), and by following good watering practices. If rotting occurs, you can propagate cuttings from noninfected parts of the plant (except seedlings); discard the infected parent plant or seedlings; discard or sterilize the soil and container in which the infected plant grew; and pot the propagated individuals in fresh, sterilized soil and containers.

(a)

(b)

(c)

(d)

(e)

FIGURE 19-21
This series of photographs demonstrates how pruning of shoots and roots, wiring, and size management contribute to the successful conversion of a typical bushy nursery-grown juniper plant to a **bonsai.** (a) Preparing the container. (b) Adequate drainage is essential. Wires will anchor the plant securely in the container. (c) and (d) Beginning at their bases, branches are wrapped with flexible but sturdy copper wire, then are bent into the desired shapes. (e) and (f) Two views of the shoot system of the specimen after trimming and wiring. In accordance with certain bonsai principles, the outline of the plant forms an asymmetrical triangle. (g) Soil is carefully removed from the root ball. (h) The roots are then trimmed, bringing the root system into balance with the shoot system. (i) The specimen is now situated in the container. (j) Soil is gently pushed into the root system to eliminate air holes, establishing good root-soil contact. (k) The finished juniper. Now time and patient tending are required to achieve growth and an "aged" appearance—both before and after the wires are removed. (Photos courtesy of Kevin Oshima.)

(f)

(h)

(i)

(g)

(j)

(k)

**FIGURE 19-22** Some *Oxalis* species grow from bulbs and naturally tend to die down yearly for a period of **dormancy.** (Photo by Mark R. Fay.)

Insects and their eight-legged relatives, the mites, attack shoots. Multilegged arthropods, such as sow bugs, centipedes, and symphilids, nibble on roots. Other root feeders are root mealybugs, the larvae (maggots) of some gnats, and soil nematodes.

Sucking insects in the order Homoptera are among the most common and destructive insects on houseplants. **Mealybugs** (Fig. 19-23) usually localize in leaf axils. They appear as masses of small, pinkish, woolly, soft-bodied, slow-moving individuals. **Aphids** or plant lice localize in succulent areas, such as flower and leaf buds and young growth. They produce a sticky shiny "honey-dew." **Scale insects** are naked when young but develop a scalelike cover at maturity (Fig. 19-24). They may be camouflaged on the bark of young twigs. **Whitefly** adults are motile and infested plants are easily identified by the cloud of small, white flies released when the plant is shaken. Also harmful, the larvae resemble soft scale insects. Tiny, elongated, black, fringe-winged **thrips** (order Thysanoptera) feed mainly on tender-tissued flowers and terminal buds, both as larvae and adults. They have combination rasping–sucking mouthparts and are usually more of a problem outdoors. **Black gnats** (order Diptera) live in the upper soil and may be seen flying about. Their chubby white larvae (maggots) eat plant tissues, such as succulent stem bases.

Closely related to spiders, the tiny but visible **spider mite** is a serious houseplant pest. Spider mites appear as moving red, yellow, green, or black dots and produce extensive cobwebs on plant shoots (Fig. 19-25). Leaves develop minute chlorotic or necrotic dots where mites have fed. Heavily infested leaves become stippled, gray, and curled and are likely to die. Warm, dry conditions are especially conducive to mites. Mites spread rapidly to adjacent plants. They are commonly brought home on potted mums. Microscopic **cyclamen mites** are considered impossible to eradicate. They localize in shoot tips, causing distorted growth.

**FIGURE 19-23** **Mealybugs** are sucking insects that commonly infest indoor plants. Mealybugs and egg clusters often are covered with woolly-looking secretions.

FIGURE 19-24
**Scale insects** are sucking insects that are related to mealybugs. They are a particular problem when there is low atmospheric humidity. The dark cone-shaped structures hide adults that have produced egg clusters. The lighter conical structures are actively feeding insects of various ages.

**Prevention** is the most important means of pest control. Use sterilized soil and clean pots. Quarantine newly acquired plants or cuttings before placing them among established healthy plants. Plants purchased from a reputable greenhouse or florist are more likely to be pest free than those purchased from a department or grocery store, where they are usually neglected in all important aspects—light, temperature, humidity, watering, and disease.

**Sanitation** is a second major factor. Examine unhealthy parts to diagnose the problem; then remove and discard them. Isolate an infested plant during treatment until it is fully decontaminated.

Finally, once infestation is diagnosed, **control** methods should begin immediately.

1. Physically remove egg masses, larvae, and adult insects and spider mites. The simplest method is washing with lukewarm water over a sink, taking care not to wash the pests down into the pot, where they can crawl back up. Use a cotton swab dipped in alcohol or acetone (fingernail polish remover) to rub away scale insects and mealybugs, taking care not to leave the solvent on plant surfaces.
2. A safe treatment for nearly all smooth-leaved plants is washing with a plain soap (not detergent) solution. Use one tablespoon of soap flakes or shavings to a gallon of lukewarm or tepid water. Spray, swab, or dip shoots into the solution, allow to dry, and then rinse

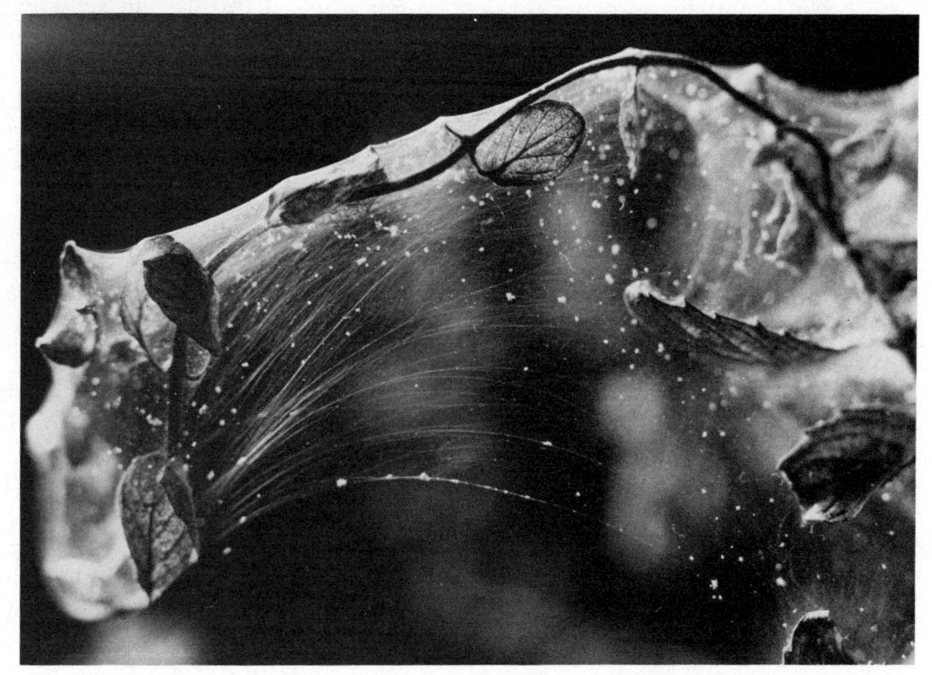

FIGURE 19-25
This is a severe infestation of spider mites, the many small dots crawling on the cobwebs that they have produced. Note also the skeletonized leaves. The problem should never be allowed to develop to this stage, for there is great danger of the mites spreading to other plants.

about an hour later. Do not get washing solution in the potting medium. The soapy treatment is especially effective against flightless sucking insects, such as adult aphids, scale insects, and mealybugs, plus all larvae.

3. The toxic alkaloid, nicotine, is poisonous to insects. Smoldering nicotine tablets are used to fumigate greenhouses. Fumigate individual plants with high-nicotine cigarette smoke in a plastic-bag enclosure at home. Or prepare a nicotine solution by soaking cigarette or cigar butts (the butts have the highest nicotine concentration) in water for washing as in treatment (2).
4. Commercially available sprays for houseplant pests contain synthetics, such as malathion and parathion, or the botanical materials pyrethrum and rotenone. Test the spray on a single leaf first to be sure that it will not damage plant tissues. Follow label instructions carefully. Protect aquaria to avoid poisoning fish with any kind of spray. (Rotenone is used as a fish poison.)
5. Some commercially available insecticides are "systemic"—that is, when applied to the soil, they are taken up by the plant and poison insects that feed on the plant.
6. Insecticide-impregnanted plastic strips are powerful and highly concentrated and can be placed among plants to control flying insects. Because the insecticide is potent, plants should not be closely confined with it or exposed to it for a prolonged time. Similarly, flea-tick pet collars may be effective against minor mite infestations.
7. Some authors suggest spraying against mites with a parakeet mite product. Test a leaf first.

## ANTHROPOMORPHISMS

Talk to your plants, sing to your plants, play them special music, take a plant to lunch, avoid aggressive or otherwise unpleasant conversation, behavior, or even thoughts when in the presence of plants!!! Such advice is an example of **anthropomorphism**, the assignment of human qualities to something that is not human. For centuries, anthropomorphic behavior has been extended to domesticated animals—and now to plants (Fig. 19-26). The "scientific" basis for such behavior toward plants lies mainly in the work of an author who wired plants to a polygraph (lie-detector) machine and then subjected them to a variety of experiences. His results indicated that plants perceive and react not only to physical conditions but also to emotional interactions, such as kindness, violence, and pain. Nonscientific magazine articles have extrapolated from these reports to various degrees. The polygraph experiments, however, have not been successfully repeated, one of the prerequisites to establishment of a scientific hypothesis as fact.

**FIGURE 19-26** Indoor plants are not human, but they do contribute to our enjoyment of life. (Drawing by Roy Massoth, Univ. of Wisconsin-Eau Claire News and Publications.)

On the positive scientific side—research *has* established a correlation between physical stress (such as from constant wind buffeting) and the formation of stouter supportive tissues in plants. Thus outdoor plants growing at the perimeter of a plant cluster react to the greater stress by developing more physical strength than more sheltered individuals. Daily shaking of some container plants has been shown to induce sturdier growth. Perhaps the vibrations of music also can contribute to plant vigor, although formulas for the intensity and duration necessary for any contribution to growth have not been established (or seriously investigated?).

As for talking to your plants, perhaps a slight increase in $CO_2$ content very close to the leaves, such as exhaled during speech, is beneficial? What concentration? For how long? Can the plant utilize any additional $CO_2$ than is already present? Such questions must occur to the organized scientific mind.

Overall the real benefit of anthropomorphisms in regard to houseplants must be an increased awareness of plant health. A person presenting an intimate monologue to a plant is likely to notice changes in vigor or the development of unhealthy symptoms. The person is probably attentive to watering and feeding procedures. A container-grown plant that is carefully and scientifically tended is more likely to be a vigorous, *healthy* plant. The owner of such a specimen is likely to be *happy*.

## SUMMARY

Bringing vegetation indoors helps naturalize and make the internal environment more pleasing.

Initial specimen selection is the most important aspect of growing houseplants. The plant must be healthy to begin with and must be able to tolerate the specific environmental conditions of the home.

Effective indoor landscaping, like that for outdoors, involves both site and functional analyses.

Indoor **light intensity** varies with exposure, cloud cover, shading, and filters, such as smog and curtains. **Light quality** refers to wavelengths present. Artificial light varies in wavelengths needed for plants whereas natural daylight is complete. Plant response to light direction is called **phototropism.**

Plants survive underwatering more easily than overwatering. A constantly wet soil encourages fungal and bacterial rot. Container plants do better with water naturally low in minerals and salts.

Most plants require a **relative humidity** of 40 to 60%.

Air-borne **pollutants,** such as dust and cooking grease, affect plants indoors. Simple cleaning processes are recommended.

Most houseplants cannot tolerate near-freezing temperatures and should be kept from winter windows and drafts.

A good **potting soil** drains well, holds water, and provides root anchorage. Soil pH can be monitored inexpensively with pH paper or, more precisely, with a soil test kit. If commercial potting soils are unavailable or are too expensive in quantity, you can concoct a suitable mix from recipes and ingredients described in this chapter.

Plants need about 16 **nutrients** for complete growth. All the micronutrients are usually found in a balanced potting soil. Rapidly used nutrients, such as nitrogen, phosphorus, and potash, should be added regularly.

Plant **containers** should reflect the plant's growth habit and requirements. Adequate drainage is important. An unglazed pot aerates well and can be placed in a more decorative cachepot.

A rootbound plant should be repotted as described.

Plant **size and shape** are determined by both genetics and environment. Pinching out terminal shoot buds encourages a bushier plant. Pruning, wiring, and grafting alter form via secondary growth.

Many plants have a natural period of **dormancy.** When dormant, houseplants require less water and light, cooler temperatures, and no fertilizers.

**Diseases and pests** may cause discolored areas, wilting even though the soil is moist, mushy, brown areas, cobwebby matter, distorted foliage, leaf or flower drop, loss of normal color, general lack of vigor, and even death.

Although it is entertaining to think of plants in anthropomorphic terms, the actual benefit of close attention is that plants are more likely to receive good care.

## SOME SUGGESTED READINGS

Ball, V. (ed.) *The Ball Red Book,* 12th ed. Chicago: Published by Geo. Ball, 1972. A manual for the commercial florist with over half the text devoted to cultural notes for specific flowering plants.

Crockett, J.U. *Foliage House Plants,* rev. ed. New York: Time-Life Books, 1976. Complete but concise cultural suggestions for over 300 popular species; well illustrated.

——. *Flowering House Plants.* New York: Time-Life Books, 1972. Like Crockett's preceding title, a well-illustrated and useful book.

Eaton, J.A. *Gardening under Glass: An Illustrated Guide to the Greenhouse.* New York: The Macmillan Co., 1973. One of several manuals for the amateur greenhouse gardener, with many useful hints for houseplants.

Graf, A.B. *Exotica,* 3rd ed. East Rutherford, N.J.: Roehrs Co., 1970. An enormous compilation of information and coded culturing information for nearly, if not all, plant species that have been collected and grown indoors from all over the world; "the Bible" of its genre.

—. *Exotica Plant Manual.* East Rutherford, N.J.: Roehrs Co., 1970. An abridged version of the larger work, covering fewer species but still including more than most greenhouse and houseplant books.

Kramer, J. *Cacti and Other Succulents.* New York: Harry N. Abrams, Publishers, 1977. Strikingly artistic color and monochrome photographs by Don Worth combined with line drawings and concise, useful information.

Laurie, A., D.C. Kiplinger, and K.S. Nelson. *Commercial Flower Forcing,* 7th ed. New York: McGraw-Hill Book Co., 1969. A manual for all phases of the florist industry, with special sections on specific flowering, foliage, and bedding plants.

Northern, H.T., and R.T. Northern. *Greenhouse Gardening,* 2nd ed. New York: Ronald Press Co., 1973. One of the better books for the amateur plantgrower. Although written for a greenhouse owner, most methods can be adapted to houseplants.

Sunset Books. *How to Grow House Plants.* Menlo Park, CA.: Lane Publishing Co., 1968. A typically well-done Sunset book.

# chapter twenty ENVIRONMENTAL PRINCIPLES IN PRACTICE

## LANDSCAPE DESIGN

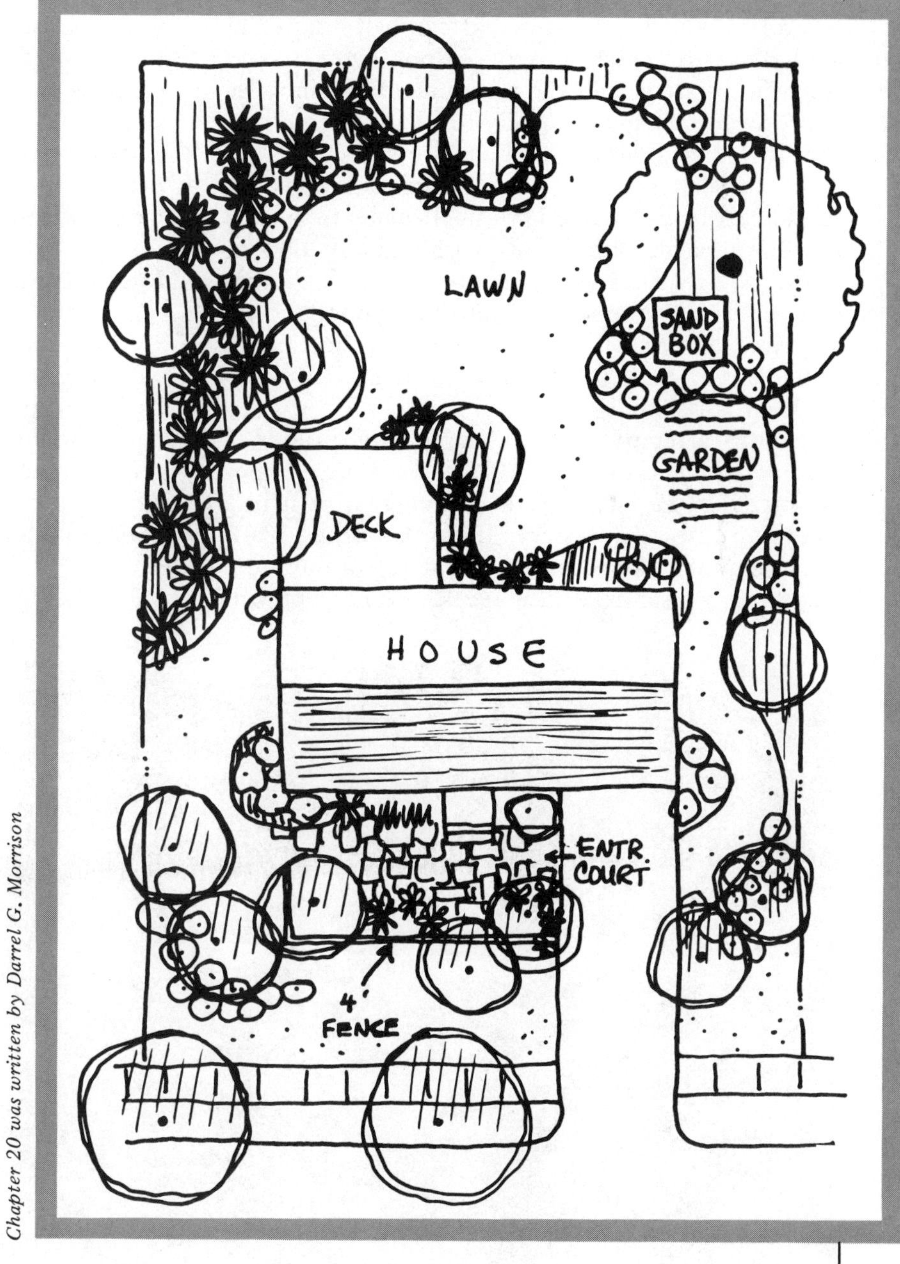


*Chapter 20 was written by Darrel G. Morrison*

## INTRODUCTION

Your piece of land, whether a 20-by 40-ft townhouse, a courtyard, or an acreage in the country, can be made more usable and enjoyable if some principles of landscape design are applied to its planning. In this chapter some of the procedures and materials employed by professional landscape architects are introduced as a first step toward making the most of the bit of landscape that you occupy. It does *not* contain a "bag of landscape tricks" for you to apply. It *does* outline some basic approaches that might be followed by a landscape architect and will provide some guidance in the art of creating outdoor spaces that can be both useful and attractive extensions of indoor living space. Keep in mind that all the recommended principles and procedures apply equally to a bare lot and one that is well established, even overgrown.

Too often we see cute gimmicks applied in residential landscape design—such elements as beds of white marble chips strewn with unlikely looking boulders. Or we see a hodgepodge of individual plants (usually one each of multitudinous species) scattered about a lawn, used purely as decorative specimens instead of serving a purpose like space enclosure or shading.

Such attempts to have "something different" are almost always disastrous, both as visual compositions (Fig. 20-1) and as environments for living. They are also unnecessary. If the landscape planning genuinely takes into account (a) site uniqueness and (b) user needs, the resulting landscape will be honestly different and will have a degree of integrity that can never be achieved by purely decorative or cosmetic techniques. In other words, a landscape that grows out of these two factors will be a happy marriage reflecting both the unique site characteristics of your piece of land and your own lifestyle and personality.

## PROCEDURES: WHERE DO YOU BEGIN?

That sounds very well, but how do you go about achieving this happy meshing of site characteristics with human needs? There are two initial steps to take. The first is to organize information on the physical characteristics of the site itself. The second step, which may progress simultaneously with the site analysis, is to state clearly the functions to be performed by the landscape—how it can accommodate the activities and experiences you would like.

### Site Analysis

Although the complexity and details may vary with the size and character of the site, certain basic information should be collected. Much of this basic site information can be plotted graphically on grid paper at a scale of 1 in. = 10 ft or 1/4 in. = 1 ft. Among the types of information that can be assembled in your **site analysis** are the following.

1. The property lines. Your city or county asses-

**FIGURE 20-1**
Clashing elements of a "bag-of-tricks" approach to landscape design create a highly artificial effect. Note the use here of harsh geometric designs accented with different colors and textures of gravel, the scattering of large rocks, the cluster of sawed-off beams standing on end, the rustic walkway, and the use of different types of retaining walls (stone wall, rough beam wall), all within one visual scope. The constructed "landscaping" elements are interspersed with randomly placed isolated shrubs and small trees. The site was a grassy slope before it was "improved." What other approach(es) might have been more effective? (Photo by Mark R. Fay.)

sor has records if you don't know the precise dimensions of your land.

2. The topography or "lay of the land." The optimum information in this regard is a topographic survey, the contour lines indicating differences in elevation. On a large or topographically complicated site, the money spent on such a survey could be a good investment; but on small lots or those with only minor topographic variation, a rudimentary topographic map can be developed with the help of a hand level and measuring stick. Starting with some assumed "base" elevation, you can plot "spot" elevations in relation to the house (Fig. 20-2).
3. Surface drainage system. Closely related to topography is the pattern followed by water after a rain. Any notable water flow should be recorded, plus springs and poorly drained depressions. They need to be recognized for later planning stages.

**FIGURE 20-2** The first step in developing a landscape design is preparation of a **topographic map** of the site with spot elevations for the major existing elements. (This and subsequent design plan sketches courtesy of Darrel G. Morrison.)

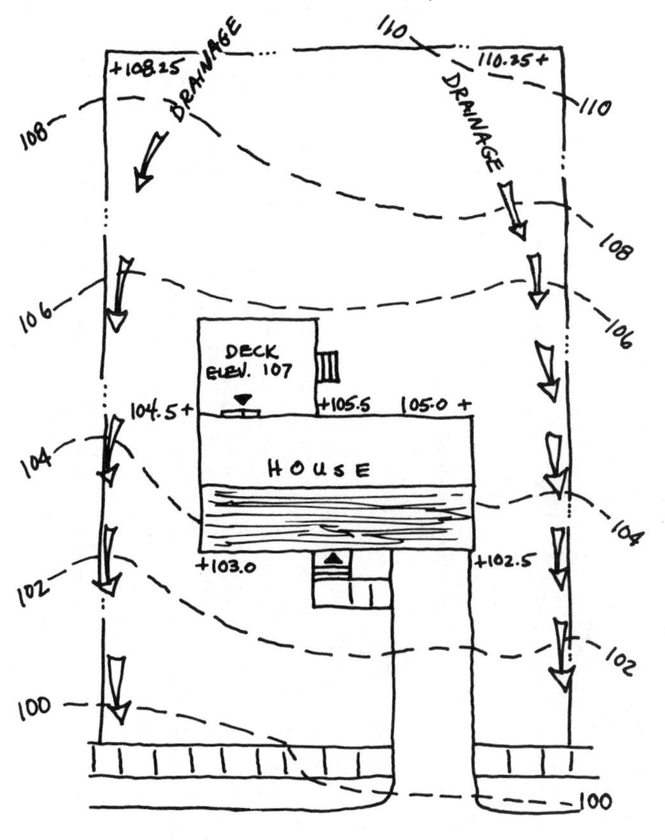


4. Soils. The amount of detail necessary is related to site size and complexity. The depth of soil as well as its texture (clay, sand, or loam) are important to note because of their influence on the potential land-use plan for your site. A vegetable garden, for instance, will do best in a deep, rich soil but cannot be expected to produce on thin, rocky soil. Sometimes the character of the soil can be determined from indicator vegetation growing on it. A more direct way is to dig into the soil or to take soil cores at various points. Unusual features, such as rock outcrops, should be located as part of the site analysis.
5. Location and size of structures. You will need to include your house and other buildings, marking location of windows and exterior doors. If there is a substantial roof overhang, it should be plotted. Existing fences, retaining walls, and such paved areas as driveway and sidewalks should also be plotted.
6. Utilities above and below ground. Power and telephone lines, well and pump head, septic system (tank, drain field, and vents), sewer, water and gas lines, and fuel storage tanks should be shown as part of the site inventory. They can affect what you place above, below, or near them.
7. Existing vegetation. Location, size, and condition of existing vegetation should be noted because of its potential incorporation into your landscape plan (Fig. 20-3). Furthermore, the presence of trees and shrubs can greatly affect site microclimate, especially sun and wind exposure. If you are fortunate enough to have an existing wooded area on your land, it may be adequate simply to delineate the outline of the woods rather than individually locate trees and shrubs. If you are analyzing a lot on which you plan to build, identify specific natural contours and patches of vegetation you wish to spare from the bulldozer. Contractors have a habit of scraping a lot "clean" and level when readying an "unimproved" site for construction.
8. Sun and wind patterns. For most residential projects, it is sufficient to plot these climatic factors somewhat generally, simply identifying areas that may represent problem spots because of too much or too little sun or prevailing wind directions (Fig. 20-4). Conveniently enough, the prevailing winds in the winter typically come from a different direction than cooling summer breezes. So you can create a

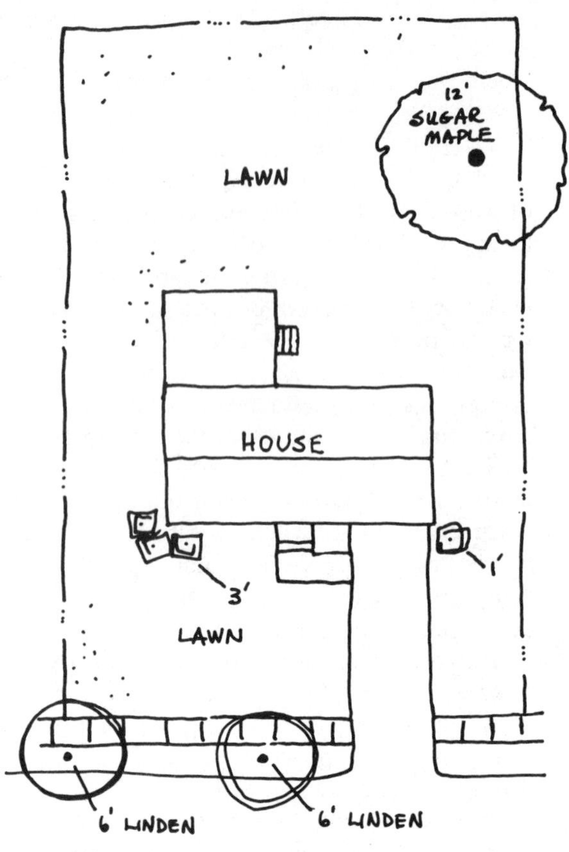


**FIGURE 20-3** Mapping the existing **vegetation.**

**FIGURE 20-4** Mapping **microclimate** factors.

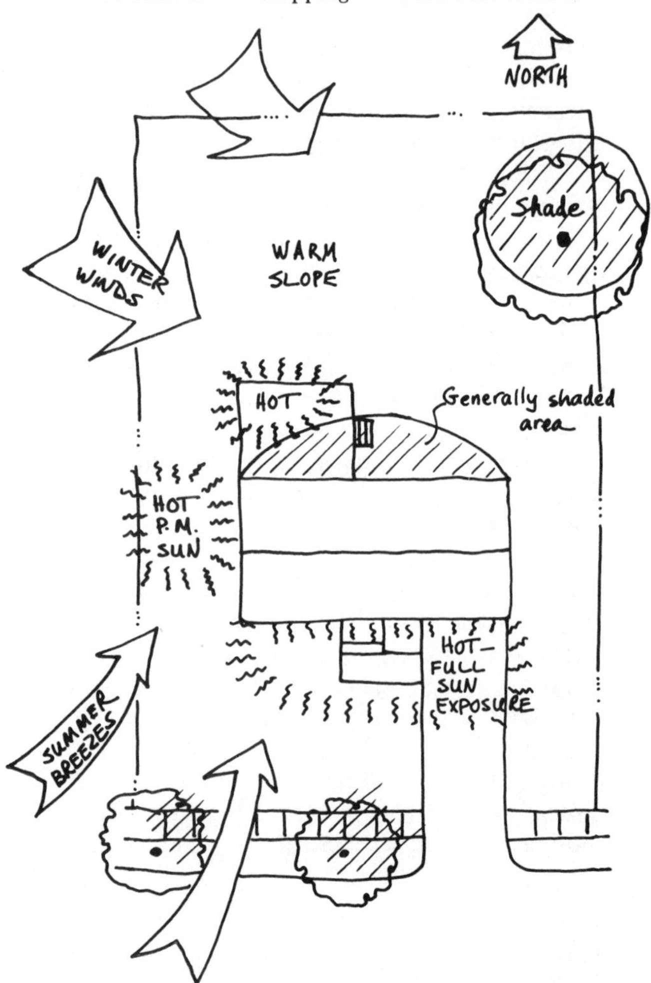


windbreak that will greatly reduce undesirable winter winds without blocking summer ones. By careful attention to sun and wind patterns in your landscape plan, you may be able to influence home energy use for summer cooling and winter heating.

9. Views, both good and bad, on and off the site. The "inside-out" views (those from inside the house to the landscapes) are especially important to consider but are surprisingly often overlooked. Perhaps this situation occurs because the landscape has been regarded merely as a setting for the house and too much effort typically is expended on framing the house as something to be exhibited. But if you think of your bit of landscape as something to *experience,* then you will create, enhance, or block certain views, depending on their character.

One approach that might be used in considering views from inside the house is to determine high-visibility vistas from important points, such as the dining area, the food preparation area, or a favorite chair in the living room. On paper, plot the **cone of vision** from that point toward the landscape (Fig. 20-5). If the view is a desirable one, it may suggest "hands off" in that area. An undesir-

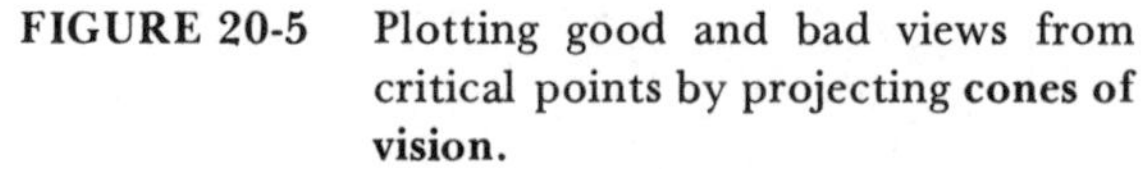

**FIGURE 20-5** Plotting good and bad views from critical points by projecting **cones of vision.**

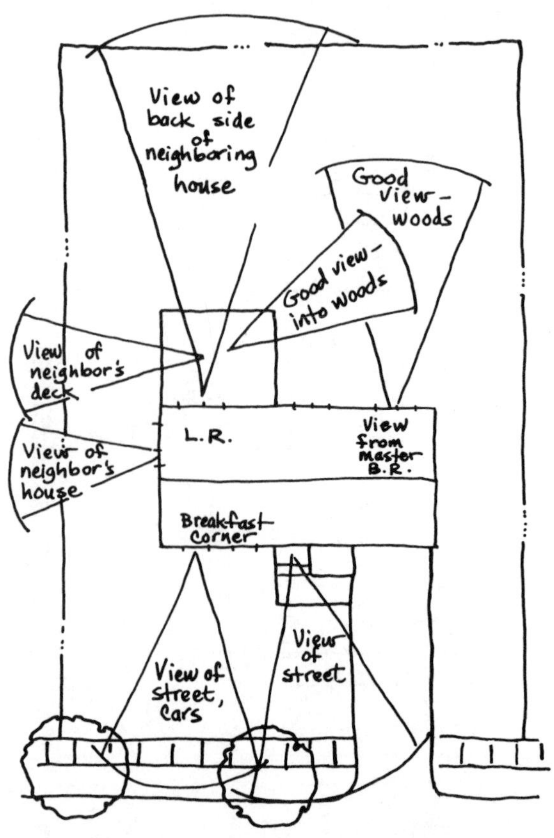

able view may suggest screening. If it is simply a nondescript view, it may provide an opportunity to create a positive view.

### Functional Analysis

In addition to analysis of site characteristics, the **functional analysis** must also be considered. The functional analysis should be a way of clearly stating not only the functions and activities that you want to accommodate in your landscape but also the quality of experience that you would like it to provide.

A good beginning toward the functional analysis can be made by developing a checklist of activities or functions you'd like to include in your bit of landscape. The items from this list that require physical space can then be incorporated graphically into a **bubble diagram** (Fig. 20-6). The bubble diagram provides a simplified way of looking at relative amounts of space required for different activities. More importantly, relationships between activities or uses can be clarified. By superimposing bubble-shaped forms over the site analysis map, you can see which activities are compatible with the physical characteristics of the site and with each other. Incompatible activities will be placed apart from each other and may even suggest physical separation in the form of structures or vegetation. Figure 20-7 shows a completed landscape plan for the site depicted in previous figures.

**FIGURE 20-6** A functional **bubble diagram** superimposes activity patterns over the site plan.

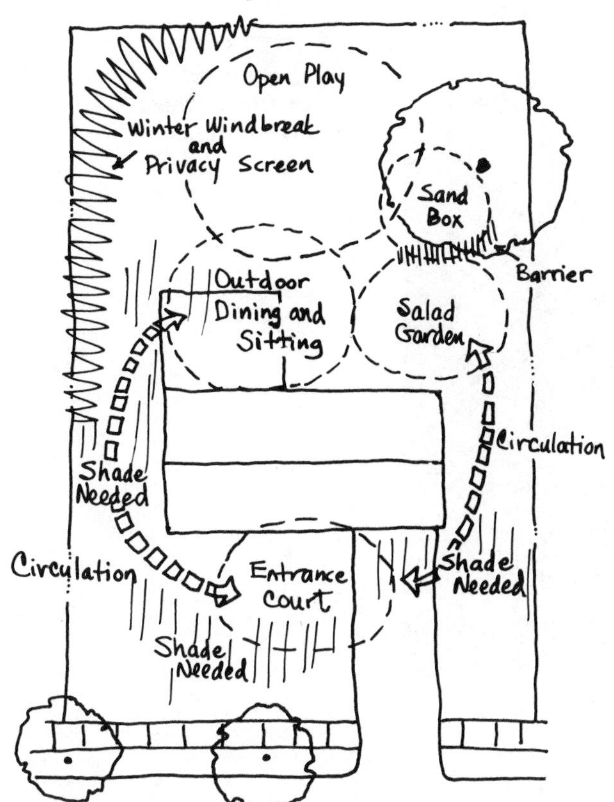


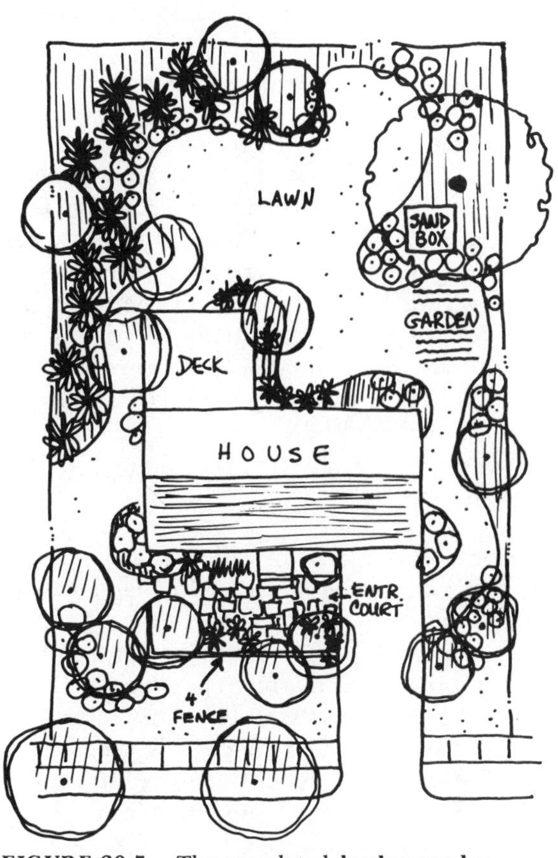


**FIGURE 20-7** The completed **landscape plan.**

## THE MATERIALS AND THEIR CHARACTERISTICS

Having completed an analysis of the site and your own needs, the next step will involve giving form to the "bubbles" representing various activities. Before continuing with this process, though, reflect on the materials or elements that are available in the design of outdoor areas: the space itself; the earth, with its individual characteristics of slope, orientation, and sometimes rock outcroppings; constructed elements, such as paving materials (stone, brick, concrete, wood), wall, and fence; vegetation, including multiple layers of growth (canopy, middle story, and ground layer); other amenities, such as sculpture, pools and fountains, and outdoor furniture.

Looking at them individually, outline the elements and materials that you have to work with.

### The Outdoor Space

Landscape architecture is actually the art of forming and manipulating outdoor space. Perhaps paradoxically, we tend not to perceive a space as a space unless it has enclosure. A landscape space can be static or it can be dynamic and have the feeling of movement. A static, dead space is likely to result if we simply enclose its perimeter with rectilinear rows of vegetation or fences. On the other hand, we can feel a dynamic, moving quality if the entire area is not visible at one time, such as when "bays" of space are enclosed by "peninsulas" or "islands" of vegetation (Fig. 20-8).

### The Earth

As the base plane on which your landscape is built, the earth is important. Most land has some natural slope, if only a slight one, that permits it to drain. Much of the character of a piece of land is due to its form, particularly on larger sites. On rural or natural land, the best policy is usually to capitalize on existing forms. In highly disturbed settings or small urban sites, some "earth sculpturing" can add interest and utility, providing, for example, privacy and sound buffering from adjacent streets. Even in such situations, it's advisable to let "made" landforms emulate gently flowing natural ones and avoid phony-looking bumps rising out of an otherwise flat surface.

### Constructed Elements

Usually they will fall into one of two categories: **floor surfaces** for outdoor activities (decks, patios, game courts, walkways) or **enclosing elements** (walls, fences, screens, and, on occasion, overhead canopies). Some guidelines and some alternatives to standard approaches are presented here.

Avoid excessively smooth or light-colored materials for floor surfaces. There are practical reasons for doing so, such as avoiding the glare that results and the hazards of slipping when the surface is wet or icy. Aesthetically, floor surfaces that are related in color and texture to the natural floor of the local landscape are the most easily absorbed into a cohesive composition.

There are many alternatives to the ubiquitous smooth concrete for sidewalks and patios. Consider these possibilities, for example: paving bricks or blocks; flagstones set in mortar or sand; rectilinear wooden blocks made from railroad ties; wood rounds, made from tree trunk or utility pole cross sections; or textured concrete, such as exposed aggregate or brushed concrete. On sloping sites, in particular, wood decks made with 2-in. decking lumber provide the potential of flat, usable surfaces without grading or building retaining walls.

For fences or privacy screens, numerous possibilities again exist, but a general guideline might be to avoid attention-demanding colors, patterns, or materials and utilize simple designs in natural materials and colors. Fences that are stained in dark, subdued colors or permitted to weather naturally provide a perfect background for plants and outdoor furniture. There may be occasions, of course,

**FIGURE 20-8**
Bays of space created by peninsulas of vegetation. (Photo courtesy of Darrel G. Morrison.)

when a fence can legitimately be considered a feature in the landscape, a design element in its own right. With architecture representing certain specific periods, for example, wrought iron fences of the appropriate design and scale could be considered. Even they should not be blatant, attention-getting fences but should blend in with the plantings and overall scheme.

### Plants and Planting

Plants and the associated butterflies, birds, and other forms of small wildlife that may find a home in them provide an exciting element of life, growth, and change in your landscape. The starting point for your planning is the existing vegetation on the site, plotted graphically as part of your site inventory and analysis. A healthy, mature tree is worth hundreds of dollars in terms of real estate value and the physical and psychological comfort it can provide. So guarding the tree against trunk and root damage during or after construction activity is well worth the effort. Similarly, shrubs and smaller trees are also of value, but they need not dictate the landscape planning for the site, for they are more easily replaced or relocated.

In selecting new planting, consider the overriding goal of providing **unity with diversity**. Too much unity can lead to monotony. On the other hand, too much diversity can lead to chaos. Therefore in selecting plants, keep in mind that the majority of them should be massed as *background,* with only a limited number of distinctive "specimen" forms or colors. In selecting plants, hardiness to local climate is a first requirement. But in addition to winter hardiness, specific microenvironmental factors should be considered, matching plant species to existing sunlight, moisture, and soil conditions.

One valid approach in selecting plant materials that is gaining acceptance is the use of **native species**—those that occur naturally in an area. There are multiple reasons for this approach, some of which are listed here.

1. Native plant materials, properly placed, are adapted to the environmental conditions of the region and typically should require little winter protection, supplementary watering, or fertilizing.
2. They can reinstate or perpetuate regional differences, serving as reminders of the vegetation that once covered the area—a prairie garden in the Midwest, a desert garden in the Southwest, or a woodland garden in the Northwest. Such landscapes can be distinctive and unique because they express a local character.
3. Native materials deserve a place in our designed landscapes for their sheer beauty alone. In a misdirected "pioneering spirit," we've too often eliminated native vegetation and replaced it with hybridized or exotic species, believing that they must be better than the natives. In reality, of course, each region has its share of truly beautiful trees, shrubs, and herbaceous plants. Consider the butterfly weed (*Asclepias tuberosa*) and bird's-foot violet (*Viola pedata*) of the Midwest prairie (Figs. 20-9 and 20-10), the lady's slipper orchids (*Cypripedium* spp.) of the forests (Fig. 20-11), the prickly pear cactus (*Opuntia* spp.) of dry, sandy areas (Fig. 20-12), and the blue wild flag (*Iris* spp.) of wet areas (Fig. 20-13). Any one can compete favorably with exotic flowers in terms of showy beauty. Then there are the more subtle beauties of native grasses, ferns, reeds, and rushes, which complement them. Native plants, growing together in the proper habitat, are aesthetically compatible and look right together.

The use of native plant species can take any of a number of forms, depending on your goal and the degree of involvement wanted. At the simplest level, native plants can be utilized like any others—for shading, for windbreaks, or for flower gardens.

**FIGURE 20-9** Butterfly weed (*Asclepias*), a showy native milkweed.

FIGURE 20-10
Bird's-foot violet (*Viola*), a native prairie perennial with narrowly dissected leaves.

FIGURE 20-11
Yellow lady's slipper orchids (*Cyrpipedium*), with wild geraniums (*Geranium*) and ferns in a shaded garden. (Photo courtesy of Darrel G. Morrison.)

FIGURE 20-12
Prickly pear cactus (*Opuntia*) of dry, sandy sites, growing in a garden setting. (Photo courtesy of Darrel G. Morrison.)

FIGURE 20-13
Wild blue flag (*Iris*), native to wet areas and adaptable to moist settings in a garden. (Photo courtesy of Darrel G. Morrison.)

Somewhat more complex but perhaps more rewarding is to group plants, including trees, shrubs, and herbaceous plants, in their natural associations. This approach might be applied to portions of the site (Fig. 20-14), with other areas treated in the more traditional lawn and garden manner.

The ultimate native landscape would eliminate any clipped lawn and imported plant species at all. This is a very real possibility in situations where a house is built in an existing natural setting. In such cases, the best approach might be to "put things back" as nearly as possible as they were before being disturbed by construction, with adequate provision made for circulation, outdoor dining, and seating.

To convert a typical manicured suburban lot into a naturalized wild yard is probably one of the most difficult approaches, partly because some of the work already done must be undone, including removing areas of established lawn to provide the right conditions for reestablishing native cover and perhaps even restoring natural contours. Also, such renaturalizing requires patience on your part and on your neighbors'. It will look somewhat barren or chaotic for the first year or two, for the very nature of this type of landscape is one of evolution toward an ever-richer and more stable composition.

Obtaining plants and seeds of native species is another stumbling block. Until recently, nursery owners and seed dealers claimed that there was no demand for natives. Believing that there was no supply, the public didn't ask for them. This situation is slowly changing. Lists of sources of native materials can be obtained from the Soil Conservation Service of the U.S. Department of Agriculture. At the beginning of the 1980s, however, many

FIGURE 20-14
Birch trees, sumac, oldfield juniper, prairie vegetation, and lawn in a Midwestern setting at a home in Madison, Wisconsin. (Photo courtesy of Darrel G. Morrison.)

catalog seed suppliers began to offer a wider variety of seeds and nursery-grown native species.

At whatever scale you're working—whether it's a perennial border of flowers or a whole community of trees, shrubs, and herbaceous plants—there are some basic guidelines to follow. One of the most important is to match the plant species with environmental factors. Know the plant in its natural habitat and plant it in a situation that is as close to it as possible in terms of light, soil, and moisture.

A second guideline in plant placement is to group several of one species together. Few plant species are found in splendid isolation in naturally evolved landscapes; instead they are among several of their own kind (Fig. 20-15). Such grouping or **massing** will contribute to a unified, harmonious landscape. Some plants, of course, are appropriate as individual **specimens** and they can be observed in the native landscape of the area.

As Fig. 20-16 suggests, a third guideline is to plan ahead, projecting your imagination to the future, based on the growth potential of your plantings.

**FIGURE 20-15** Harmony resulting from many plants of one species, here paper birches (*Betula*). (Photo courtesy of Darrel G. Morrison.)

## PUTTING THE PIECES TOGETHER

Having looked critically at your site, evaluated its physical characteristics, and listed your own requirements that the outdoor space must meet, you are now ready to utilize some or all of the preceding landscape materials in forming your own individually tailored landscape. This is the form-giving

**FIGURE 20-16** Mature boulevard tree disfigured by trimming to avoid overhead wires.

aspect of landscape design, the art of **composition.** At this point, you will work from your site analysis and abstract functional bubble diagram to allocate specific sizes and shapes to different uses and materials.

As the next step, this situation suggests the development of a **mass–space plan** in which you acknowledge more precisely the space needed for specific activities and determine just where you need architectural or vegetational space-enclosing elements. Specific spaces will require different surface treatments, such as hard surfaces for outdoor furniture and turf for lawn games. The character of enclosing elements will depend on their function (windbreaking, screening), the space available, and the theme of the landscape (whether it's highly urban, formal, or rustic).

At this stage of planning it's important to maintain a cohesive, unified design. It is easy to become so preoccupied with each part that you lose sight of the whole. To avoid this pitfall, try to make spaces flow, one into another. This doesn't mean that the whole site must be left open. In fact, partial concealment of parts of the site can create a sense of mystery and make the space seem larger than it actually is.

Once the mass–space plan is developed and you have identified paved areas, fences, walls, and other structures, plus any changes in the earth's elevation, a **planting plan** can be developed. The site development can be phased or staged in manageable parts in terms of time, energy, and finances. The open spaces in a planting plan will logically be planted with lawn or other low, ground-covering vegetation. The masses, the enclosing elements, will generally be trees and shrubs. If such mass areas are to possess some of the feeling of natural landscapes, they will include the layers found in a natural forest: a canopy level, an understory or middle layer of shrubs, saplings, and small trees, and a ground layer of wildflowers and ferns.

In such massed planting areas the time element should be considered. If shade from existing canopy trees is not available, woodland ground-layer plants that require shade and protection cannot be planted initially. In such a situation, you may be limited to sapling-sized trees that will cast sufficient shade in 5 to 10 years to shelter such plants. Meanwhile you can mulch the ground layer with woodchips, shredded bark, or fallen leaves to suppress sun-loving weeds and to create humic soil conditions similar to those in a true woods. You can also plant some ground-layer species that have a wide range of light and soil tolerance.

**FIGURE 20-17** Wild rose (*Rosa*) fruits, or "hips," are edible throughout the winter and are attractive to wildlife.

**Edges,** the places where massed vegetation meets the open spaces, are important. In selecting species for planting edges, consider the small trees and shrubs that form the rounding edges in natural settings in your own geographic region. This area of your planning affords a special opportunity to plant seed- or fruit-producing native species that will attract birds and other wildlife (Fig. 20-17).

## TIME

In any planting, accept the fact that it will take time to mature and that it will always be changing. This, in fact, is one of the exciting aspects of landscape design. It is dynamic and living and so will change yearly, weekly, even daily. Provide opportunities, particularly in the massed planting areas, for some natural reproduction and spreading of the initial planting. In 30 years, using such a scheme, there's a good chance the landscape will look like it "just happened," much as truly natural landscapes evolve. What higher compliment could you receive for your efforts at landscape design?

## SUMMARY

Thoughtful landscape design creates attractive and useful residential outdoor spaces by blending site uniqueness and user needs.

**Site analysis** includes graphed property lines, differences in elevation, drainage, soil depth and texture, buildings (location of windows, doors, roof overhangs), fences, walls, paved areas, utilities, existing vegetation, exposure to sun and wind, and views from living areas.

**Functional analysis** considers the functions and activities to be accommodated by the landscape. A **bubble diagram** of activities pinpoints which areas should be separated from others.

Static, unexciting space results from simply enclosing a perimeter with rows of vegetation or fences. A feeling of movement is created by bays, islands, and peninsulas of vegetation.

Landform determines the site's basic character and its natural tendencies should be accented. Constructed elements are generally floor surfaces and enclosing elements.

In planning vegetation, blend unity with diversity. Begin by incorporating existing vegetation into the site analysis. Native plants are becoming increasingly popular and offer several advantages. Some portions of the yard can be grouped with flowers, shrubs, and trees that normally appear together, having similar environmental requirements. Massing plants unifies a landscape.

Before composing the landscape, develop a **mass-space plan** that acknowledges exactly spaces needed for specific activities and needed space-enclosing elements and that allows the spaces to flow into one another.

Plant open spaces with low ground cover. Use trees and shrubs for enclosing elements. Layer the landscape to increase the feeling of naturalness. Plant edges with species that outline natural areas in your region.

As the years pass, a well-planned landscape will become increasingly dynamic, beautiful, and natural, as well as functional.

## SOME SUGGESTED READINGS

Brockman, C.F. *Trees of North America.* New York: Golden Press (Western Publishing Co.), 1968. A beautifully illustrated, easy-to-use field guide to all the small and large trees of the continent.

Bruce, H. *How to Grow Wildflowers and Wild Shrubs and Trees in Your Own Garden.* New York: Alfred A. Knopf, 1976. A personalized handbook combining the philosophy of preserving and appreciating native plants by home gardening, with specific information about their culture, growth forms, etc.; emphasis on eastern North America.

Crockett, J.U. *Landscape Gardening.* New York: Time-Life Books, 1971. Beautifully illustrated introduction to the basics of home landscape design.

Dietz, M.J. *Landscaping and the Small Garden.* Garden City, N.Y.: Doubleday and Co., 1973. A popular introduction to home landscaping.

Fairbrother, N. *The Nature of Landscape Design: As an Art Form, A Craft, A Social Necessity.* New York: Alfred A. Knopf, 1974. A popularly written book linking landscape design to ecology.

Malkin, R.S. *How to Landscape Your Own Home.* New York: Harper, 1955. Although old, this book has useful appendices of plants for certain conditions and special characteristics.

Reader's Digest. *Practical Guide to Home Landscaping.* Pleasantville, N.Y.: Reader's Digest Assoc., 1972. Thorough, comprehensive, and eminently useful.

Sunset Books. *Western Garden Book,* 3rd ed. Menlo Park, CA.: Lane Publishing Co., 1967. For the western states, a comprehensive guide to all aspects of gardening, including use of many native plants.

Spangler, R.L., and J. Ripperda. *Landscape Plants for the Central and Northeastern States* (including lower and eastern Canada). Minneapolis: Burgess Publishing Co., 1977. One of several textbooks of woody ornamentals for east-central North America.

Taylor, K.S., and S.F. Hamblin. *Handbook of Wild Flower Cultivation.* New York: Collier Books (Macmillan Publishing Co.), 1963. A guide to the preservation by naturalization and propagation of selected showy wild species; in paperback. Has useful lists of plants by special requirements (e.g., bog and marsh, alkaline soil, shady spots) and an extensive bibliography.

U.S. Department of Agriculture. *Landscape for Living: Yearbook of Agriculture, 1972.* A collection of articles, written for the layperson, covering all aspects of landscape design. Available from the U.S. Supt. of Documents, Washington, D.C.

Wyman, D. *Wyman's Gardening Encyclopedia.* New York: Macmillan Publishing Co., 1971. A wonderful compendium of practical and scientific information about specific plants—wild and cultivated—as well as virtually every other topic related to gardening and landscaping.

# APPENDIX

Metric Conversion Chart

| Into Metric | | | Out of Metric | | |
|---|---|---|---|---|---|
| *If You Know* | *Multiply By* | *To Get* | *If You Know* | *Multiply By* | *To Get* |
| LENGTH | | | | | |
| inches | 2.54 | centimeters | millimeters | 0.04 | inches |
| feet | 30 | centimeters | centimeters | 0.4 | inches |
| feet | 0.303 | meters | meters | 3.3 | feet |
| yards | 0.91 | meters | kilometers | 0.62 | miles |
| miles | 1.6 | kilometers | | | |
| AREA | | | | | |
| sq. inches | 6.5 | sq. centimeters | sq. centimeters | 0.16 | sq. inches |
| sq. feet | 0.09 | sq. meters | sq. meters | 1.2 | sq. yards |
| sq. yards | 0.8 | sq. meters | sq. kilometers | 0.4 | sq. miles |
| sq. miles | 2.6 | sq. kilometers | hectares | 2.47 | acres |
| acres | 0.4 | hectares | | | |
| MASS (WEIGHT) | | | | | |
| ounces | 28 | grams | grams | 0.035 | ounces |
| pounds | 0.45 | kilograms | kilograms | 2.2 | pounds |
| short ton | 0.9 | metric ton | metric tons | 1.1 | short tons |
| VOLUME | | | | | |
| teaspoons | 5 | milliliters | milliliters | 0.03 | fluid ounces |
| tablespoons | 15 | milliliters | liters | 2.1 | pints |
| fluid ounces | 30 | milliliters | liters | 1.06 | quarts |
| cups | 0.24 | liters | liters | 0.26 | gallons |
| pints | 0.47 | liters | cubic meters | 35 | cubic feet |
| quarts | 0.95 | liters | cubic meters | 1.3 | cubic yards |
| gallons | 3.8 | liters | | | |
| cubic feet | 0.03 | cubic meters | | | |
| cubic yards | 0.76 | cubic meters | | | |
| PRESSURE | | | | | |
| $lbs/in^2$ | 0.069 | bars | bars | 14.5 | $lbs/in^2$ |
| atmospheres | 1.013 | bars | bars | 0.987 | atmospheres |
| atmospheres | 1.033 | $kg/cm^2$ | $kg/cm^2$ | 0.968 | atmospheres |
| $lbs/in^2$ | 0.07 | $kg/cm^2$ | $kg/cm^2$ | 14.22 | $lbs/in^2$ |
| RATES | | | | | |
| lbs/acre | 1.12 | kg/hectare | kg/hectare | 0.892 | lbs/acre |
| tons/acre | 2.24 | metric tons/hectare | metric tons/hectare | 0.445 | tons/acre |

**Temperature Conversion Table**

**Celsius temperatures have been rounded to the nearest whole number.**

| *F* | *C* | *F* | *C* | *F* | *C* |
|---|---|---|---|---|---|
| −26 | −32 | 19 | −7 | 64 | 18 |
| −24 | −31 | 21 | −6 | 66 | 19 |
| −22 | −30 | 23 | −5 | 68 | 20 |
| −20 | −29 | 25 | −4 | 70 | 21 |
| −18 | −28 | 27 | −3 | 72 | 22 |
| −17 | −27 | 28 | −2 | 73 | 23 |
| −15 | −26 | 30 | −1 | 75 | 24 |
| −13 | −25 | 32 | 0 | 77 | 25 |
| −11 | −24 | 34 | 1 | 79 | 26 |
| −9 | −23 | 36 | 2 | 81 | 27 |
| −8 | −22 | 37 | 3 | 82 | 28 |
| −6 | −21 | 39 | 4 | 84 | 29 |
| −4 | −20 | 41 | 5 | 86 | 30 |
| −2 | −19 | 43 | 6 | 88 | 31 |
| 0 | −18 | 45 | 7 | 90 | 32 |
| 1 | −17 | 46 | 8 | 91 | 33 |
| 3 | −16 | 48 | 9 | 93 | 34 |
| 5 | −15 | 50 | 10 | 95 | 35 |
| 7 | −14 | 52 | 11 | 97 | 36 |
| 9 | −13 | 54 | 12 | 99 | 37 |
| 10 | −12 | 55 | 13 | 100 | 38 |
| 12 | −11 | 57 | 14 | 102 | 39 |
| 14 | −10 | 59 | 15 | 104 | 40 |
| 16 | −9 | 61 | 16 | 106 | 41 |
| 18 | −8 | 63 | 17 | 108 | 42 |
| | | | | 212 | 100 |

*Fahrenheit to Celsius:* Subtract 32 from the Fahrenheit figure, multiply by 5 and divide by 9.
*Celsius to Fahrenheit:* Multiply the Celsius figure by 9, divide by 5 and add 32.

# GLOSSARY

**Abscisic acid (ABA)** Powerful hormone that acts as an inhibitor, countering the effects of auxins and gibberellins; important factor in seed and tissue dormancy.

**Acclimation** Gradual physical and biochemical changes that occur in a plant, preparing it to withstand winter conditions.

**Acid** Substance that when dissolved in water produces a solution in which the $H^+$ concentration is greater than $10^{-7}$ *M.*

**Acid rain** Rain that has had its pH lowered because of contaminants from industry, such as by combination of sulfur dioxide gas with atmospheric water, forming sulfuric acid.

**Aerobic respiration** Respiration that occurs in the mitochondria and requires the presence of free oxygen.

**Active transport** Transport of a substance across a cell membrane with expenditure of energy, such as from a site of lower concentration to a site of higher concentration, against a diffusion gradient.

**Adventitious** Refers to the development of buds or roots from sites other than their usual organ locations.

**Albuminous cell** In gymnosperms, unique parenchyma cells (axial and radial) that are closely related to the functioning sieve cells.

**Aleurone** Outermost cell layer of the endosperm; high in protein and a source of enzymes for the developing seed.

**Alkaloids** Toxic organic bases found in many plants; colorless, complex in structure, bitter tasting; morphine, codeine, solanine, for example.

**Anaerobic respiration** Respiration that occurs when oxygen is absent, resulting in incomplete release of energy from glucose; most commonly alcoholic fermentation or lactic acid formation.

**Anaphase** Stage in mitosis or meiosis where the chromatids of each chromosome separate and move to opposite poles.

**Andromedotoxin** Toxic substance present in members of the heath family; affects heart and other body systems.

**Angiosperm** Flowering plant, having seed enclosed in a fruit.

**Annual** Plant that completes its life cycle in one year.

**Anthocyanin** Water-soluble pigment in the central vacuole of plant cells; generally red to blue, depending on the pH of the cell sap.

**Anthropomorphism** Interpretation of what is not human in terms of human or personal characteristics.

**Apical dominance** Suppression of lateral buds by auxins produced in and translocated downward from the apical bud.

**Apical meristems** Group of actively dividing cells at tips (apices) of roots and stems.

**Apomixis** Development of embryo from diploid parental tissue; asexual reproduction not involving meiosis or syngamy (fertilization).

**Archegonium** (*pl.,* **archegonia**) Female sex organ containing the egg in bryophytes, lower vascular plants, and gymnosperms.

**Ascus** (*pl.,* **asci**) Saclike reproductive structure of ascomycete fungi, producing ascospores.

**Atropine** Alkaloid found in belladonna (*Atropa*) and related plants; medical use to treat spasms and dilate the pupils.

**Autotroph** Organism that is capable of producing its own food, such as green plants and some bacteria; in contrast to heterotroph.

**Auxins** Plant hormones that promote cellular elongation, among other major effects.

**Axil** Site at the stem-leaf junction where a bud is located.

**Base** Substance that when dissolved in water produces a solution in which the $OH^-$ concentration is greater than $10^{-7}$ *M.*

**Basidium** (*pl.,* **basidia**) Clublike reproductive structure of basidiomycete fungi; produces basidiospores.

**Biennial** Plant that completes its life cycle in 2 years; grows vegetatively the first year; produces flowers, forms fruits and seeds the second year.

**Biological control** Use of natural biological means to control pests and diseases.

**Bioluminescence** Emission of light by living organisms.

**Biomass** Total weight of all organisms in a particular habitat or ecosystem.

**Biome** Major regional ecosystem, characterized by climate and having distinctive dominant plant and animal types.

**Bolting** Rapid development of flowers and fruit prior to normal vegetative growth, such as when certain seeds are planted at an inappropriate time during the growing season.

**Bonsai** Japanese art and culture of growing dwarfed plants; simulates interesting growth forms of full-sized plants.

**Bordered pit** Pit in the walls of tracheids; overhanging secondary walls form a chamber between the opening and the pit membrane.

**Bud** Growing point of a plant shoot; made up of the apical meristem and its derivatives.

**Bundle sheath** Specialized layer of cells surrounding a vascular bundle.

**Burl** Abnormal bulging growth on a tree; damage or disease stimulates excess cellular divisions.

**Cachepot** Ornamental container to hold and conceal a flowerpot.

**Calcium oxalate** Salt of oxalic acid making up crystalline deposits in plants; crystals of two basic types: druses (many small crystals in a spherical form) and raphides (needlelike crystals).

**Callus** Undifferentiated tissue mass, such as is formed initially at a wound site or during *in vitro* tissue culture.

**Capillary water** Water held in the soil after drainage; available to plants.

**Carbohydrate** Group of organic compounds composed of a chain of carbon atoms to which hydrogen and oxygen are attached; includes sugars, starch, cellulose, etc.; often 1C : 2H : 1O ratio.

**Carnivores** Organisms that feed on animals; in plants, refers to plants that trap animals, generally insects, for food.

**Carotene** Yellow or orange pigment in the chloroplast or chromoplast; responsible for the color of carrots, marigolds, etc.

**Casparian strip** In some plant roots, a special band of lignin and suberin encircling the radial and transverse walls of endodermis cells.

**Cell theory** Organisms are composed of cells and cell products; all cells come from preexisting cells; each cell is capable of maintaining its own existence.

**Cellulose** Carbohydrate that is the major component of the cell wall of most plants.

**Cell wall** Relatively rigid, mostly cellulose structure that surrounds each cell of plants, fungi, and bacteria.

**Cereal** Plant that produces a type of fruit called a grain or caryopsis, grass family.

**Chemical control** Application of toxic chemical substances to control pests and diseases.

**Chlorenchyma** Parenchyma cells that contain chloroplasts.

**Chlorophyll** Group of green plant pigments located in plastids, necessary for photosynthesis.

**Chloroplast** Cellular organelle with chlorophyll (green pigment); responsible for photosynthesis in eukaryotes.

**Chromoplast** Cellular organelle containing carotenoids; providing yellow to orange-red color to plant parts.

**Chromosomes** Genetic bodies in the nucleus, composed of DNA and associated substances; DNA sequences of nucleotides on chromosomes are called genes; carry species information from generation to generation.

**Climax community** Terminal or relatively stable community in a successional series.

**Clone** Population of genetically identical plants produced asexually from a single parent plant.

**Cocaine** Alkaloid that is obtained from coca leaves; stimulant.

**Codeine** Alkaloid that is taken from raw opium; used in cold medicines.

**Codon** Sequence of three nucleotides in messenger RNA that specifies a particular amino acid.

**Cohesion** Attraction or holding together of like molecules.

**Cold hardiness** Innate ability of a plant to withstand cold temperature without injury.

**Collenchyma** Elongated, living cell with irregularly thickened primary walls; functions mainly to provide support in the primary plant body.

**Commensalism** Symbiotic relationship; living together of individuals of two species in which one partner benefits and the other is neither benefited nor harmed.

**Community** In ecology, all the organisms present in a particular ecosystem.

**Companion cell** In angiosperms, specialized cell that helps maintain the metabolism of its sister cell, the sieve tube member.

**Competition** Striving of two or more organisms for such environmental factors as light, water, nutrients, and space.

**Compost** Soil additive high in humus and nutrients; developed by the natural degradation of organic refuse.

**Conifer** Cone-bearing tree; gymnosperm.

**Coniine** Deadly alkaloid found in poison hemlock (*Conium maculatum*).

**Consumers** Organisms that cannot manufacture their own food and must consume other organisms (or their remains) as a source of food.

**Convulsion** Involuntary and abnormal muscular contraction or series of contractions; can be caused by certain plant toxins.

**Cork** Protective tissue of dead cells; suberin-impregnated walls; lacking intercellular spaces; in gymnosperm and angiosperm stems and roots.

**Cork cambium** Meristematic layer of cells that produces cork.

**Cortex** Primary tissue composed mainly of parenchyma cells; in stems and roots; primarily storage tissue.

**Coumarin** Naturally occurring part of certain glycoside; converts to dicoumarin, a potent anticoagulant used to treat circulatory diseases and in such rodenticides as Warfarin.

**Crop rotation** Planting of different crops from year to year at a particular site to reduce nutrient depletion as well as disease and parasite buildup.

**Cuticle** Protective layer of cutin on the outer surface of the epidermis.

**Cutin** Lipid substance deposited on the outer epidermal cell walls of leaves and stems.

**Cutting** Portion of a plant used in vegetative propagation; usually refers to a stem cutting.

**Cytisine** Potentially deadly alkaloid found in several members of the pea family; concentrated especially in the seeds.

**Cytochrome** Iron-containing pigmented protein molecules responsible for oxidative phosphorylation in the electron transport system within the mitochondria.

**Cytokinesis** Cytoplasmic division beginning in the last stage of mitosis; in plants, a cell plate begins formation of the cell wall that separates the two daughter cells.

**Cytokinins** Plant hormones that promote cell division and influence other growth processes.

**Cytoplasm** Refers to the living protoplasm of the cell other than the nucleus.

**Day-neutral plant** Plant that is not influenced by day length to stimulate flowering.

**Deoxyribonucleic acid (DNA)** Nucleic acid that carries the genetic information of organisms and some viruses; composed of nucleotides; arranged as a double helix.

**Diatom** Minute, planktonic, single-celled photosynthetic organisms with silicified coverings; classified as algae or protists.

**Dicotyledon (dicot)** One of the two major angiosperm subgroups; term refers to the presence of two cotyledons on embryo.

**Differentiation** Increasing specialization of a cell as it approaches maturity.

**Diffusion** As a result of the random movement of molecules, the dispersal of molecules from an area of higher concentration to an area of lower concentration.

**Digitalis** Glycoside taken from foxglove (*Digitalis*) plants; an important medicine because it is a cardiac stimulant.

**Diploid** Genetic condition of having pairs (two complements) of homologous chromosomes in a cell; denoted as $2n$.

**Disease syndrome** All the reactions of the plant to a disease (wilting, color change, etc.).

**Dominant gene** Gene that exerts its phenotypic effect to the exclusion of its allele (contrasting trait).

**Dominant species** Species that controls the conditions for its existence within an ecosystem; control or dominance may come from its size or numbers.

**Dormancy** Inactive period of reduced or suspended growth in plants, such as buds, bulbs, seeds; condition associated with abscisic acid.

**Ecological succession** Sequential replacement of communities on a site in response to continuing, gradual changes in the environment.

**Ecology** Science of the interrelationships between organisms and their environments (*see* Environment).

**Ecosystem** Natural system of organisms present interacting with nonliving environmental factors in a given area.

**Edaphic factors** Environmental conditions of the soil.

**Endocytosis** Engulfing of a part of the external environment by the cell membrane, forming a vacuole; intake of liquids is called pinocytosis; intake of particles is called phagocytosis.

**Endodermis** Specialized layer of cells (innermost layer of the cortex) in the root, having casparian strips; influences water passage into root xylem.

**Endoplasmic reticulum (ER)** Double unit membrane network in the cytoplasm; compartmentalizes the cytoplasm for metabolic activities; rough ER has ribosomes attached (sites of protein synthesis).

**Endosperm** Triploid ($3n$) nutritive tissue that develops in the angiosperm ovule to provide nutrients for the embryo and seedling; formed by fusion of a sperm nucleus with two polar nuclei.

**Endotoxin** Toxin that is produced and retained within a bacterial cell.

**Enucleate** Eukaryotic cell without a nucleus, such as a sieve tube member.

**Environment** Sum of all the physical (nonliving) and biological (living) factors that affect an organism.

**Epicotyl** Portion of the embryo above the point of attachment to the cotyledon.

**Epidermis** Outer layer of cells of the primary plant body.

**Epiphyte** Plant that grows on another plant (or above-soil substrate) without being a parasite.

**Ergastic substance** Nonliving component of the cell.

**Erosion** Loss of soil, generally by natural agencies, like water and wind; a characteristic of disturbed ecosystems.

**Ethylene** Gaseous plant hormone that influences ripening, senescence, and other processes.

**Etiolation** Condition of a plant having long internodes, pale (chlorotic) leaves, and decreased vigor resulting from a lack of light.

**Etioplasts** Plastids that have developed without light.

**Eukaryotes** Cells that have membrane-bound nuclei that undergo mitotic divisions; possess other membrane-bound organelles, such as plastids and mitochondria.

**Eutrophication** Nutrient enrichment of waters resulting in rapid growth of aquatic organisms (mostly algae), often causing increased decomposition with its associated depletion of oxygen.

**Evaporation** Loss of water molecules into the atmosphere; change of water from a liquid state to a gaseous state.

**Exotoxin** Toxin that is produced and released from the cell of a bacterium.

**Exudate** Chemical substance released from the surface of a plant.

**Fascicular cambium** Vascular cambium arising between the primary xylem and primary phloem of a vascular bundle.

**Fascicle** Bundle of pine needles; or a vascular bundle.

**Fermentation** Anaerobic respiration resulting in the formation of alcohol, lactic acid, or other products.

**Fiber** Sclerenchyma cell that is elongated, generally dead at maturity, and with a thickened secondary wall; functions in support.

**Fibrous root system** A root system with many equally fine roots, such as in grasses.

**Flaccid** Opposite of turgid; tissues soft because the cells have lost water pressure, resulting in a wilted plant.

**Flower** Reproductive structure of angiosperms; evolutionarily, a modified strobilus.

**Food chain** Specific sequence of organisms, including producer, herbivore, and carnivore, through which energy and materials move within an ecosystem.

**Fossil** Preserved entire remains, impressions, or traces of a prehistoric organism.

**Fusiform initial** Spindle-shaped cambial initial that produces secondary xylem and secondary phloem cells.

**Genetic pool** All the genes available to an interbreeding population of organisms.

**Geotropism** Plant growth in relation to gravitational pull; root growth toward gravity is called positive geotropism; stem growth away from gravitational pull is called negative geotropism.

**Gibberellins** Approximately 50 hormones produced by various plants; participate in many growth activities.

**Glycolysis** Initial step in cellular respiration; sugar is broken down to pyruvic acid with the release of a small amount of energy; does not require oxygen.

**Glycoside** Compound composed of a nonsugar (aglycone) attached to a sugar; on hydrolysis yields a sugar.

**Goitrogenic** Substance that can or has a tendency to produce goiter, such as by interfering with iodine metabolism.

**Gravitational water** Water that drains off a field due to gravitational pull.

**Ground meristem** Meristematic tissue that matures into cortex and pith.

**Growth** Increase in cell number and size, often followed by differentiation and maturation.

**Growth regulators** Nonnutrient substances that affect or control plant growth; hormone or nonhormone.

**Guard cells** Specialized epidermal cells around an opening (stoma) that is opened and closed by changes in guard cell turgor.

**Guttation** Exudation of liquid water from plant leaves, such as through hydathodes.

**Gymnosperm** Plants that produce seeds not enclosed by a fruit.

**Haploid** Genetic condition of having a single set (complement) of chromosomes in a cell; denoted as *n*.

**Hard water** Water with high concentrations of calcium and magnesium ions.

**Haustorium** (*pl.*, **haustoria**) In fungi, the specialized hypha that penetrates a host cell; in parasitic vascular plants, the specialized organ that penetrates the host tissue.

**Hay fever** Allergic reaction to air-borne allergens, such as pollen and spores, resulting in coldlike symptoms of nasal area and eyes.

**Heading or heading back** Pruning of terminals of twigs.

**Heartwood** In older trees, the dark, nonliving center of the tree; used for storage.

**Herbicide** Agent used to inhibit or destroy plant growth.

**Herbivore** Organism that eats plants.

**Heroin** Strongly addictive narcotic taken from opium; derivative of morphine.

**Heterospory (heterosporous)** Condition in which plant has two kinds of spores, microspores and megaspores.

**Heterotroph** Organism that cannot produce its own food and so must rely on foods produced in other plants and animals; in contrast to autotrophs.

**Hilum** Scar left on a seed after it breaks away from the stalk attaching it to the pod; or the center of a starch grain around which layers of starch were deposited.

**Histamine** Compound responsible for the dilation and increased permeability of blood vessels; plays a major role in allergic reactions.

**Holdfast** In algae, the basal part of the thallus that attaches it to a solid substrate.

**Homeostasis** Maintenance of a stable internal physiological environment; term also can be applied to ecosystems.

**Homosporus** Plants having only one kind of spore.

**Hook** Area behind the stem tip of a germinating seed that remains bent as it breaks through the soil; appears to be an evolutionary advantage to protect the tender apex during germination.

**Hormone** Plant growth regulator that is produced and transported within a plant body; exerts its effects on growth and other functions of receptive cells when present in small amounts.

**Humus** Partially decomposed organic matter in the soil.

**Hydrolysis** Process in which water is involved in the breakdown of a complex molecule into simpler ones.

**Hydromorphone** Semisynthetic derivative of morphine.

**Hydrophyte** Plant that is adapted to growing partly or wholly in water.

**Hydroponics** Growing of plants in a nutrient solution, with or without an inert medium to provide support.

**Hygroscopic** Substance or structure that has an affinity for taking up and retaining moisture.

**Hygroscopic water** Thin sheet of water left tightly adhered to soil particles after the removal of capillary water; not available to plants.

**Hyoscyamine** Alkaloid found in belladonna and henbane plants; if ingested, causes spasms and dilates the pupils.

**Hypertonic** Environmental solution higher in solutes, so as to gain water from another solution across a membrane.

**Hyphae** (*sing.*, **hypha**) Slender, threadlike structures that compose the fungal body (mycelium).

**Hypocotyl** Portion of the embryo or seedling between the point of attachment of the cotyledons and the root.

**Hypothesis** Working explanation; postulated solution to a scientific problem that must be tested by experimentation and, if not substantiated, discarded.

**Hypotonic** Solution lower in solutes, so as to lose water across a membrane to another solution.

**Incubation period** Time interval from pathogen penetration of a host to the first appearance of symptoms.

**Inflorescence** Floral cluster.

**Inhibitor** Hormone or nonhormone growth regulator that inhibits plant growth.

**Inoculation** Exposure of plant tissues to infective parts (inoculum) of a pathogen or parasite.

**Inoculum** Infective part of the pathogen; for example, spore or hyphae.

**Intercellular spaces** Refers to spaces between adjacent cells, such as in parenchyma tissue.

**Interfascicular cambium** Vascular cambium arising between vascular bundles (fascicles); creates continuous cambium cylinder.

**Interphase** Stage between mitotic cycles; active metabolic state.

**Intracellular** Within a cell.

**Invasion** Spread of a pathogen into the healthy plant; follows inoculation and penetration.

**Isotonic** Condition when both sides of a membrane have the same solute concentration.

**Juvenile phase** Most rapid growth period, early in life.

**Krebs cycle** Second phase of aerobic respiration during which pyruvic acid is disassembled and the released energy is trapped in two intermediate compounds, ATP and $NADH_2$; also called citric acid cycle.

**Lactic acid formation** Type of anaerobic respiration in which the primary product is lactic acid.

**Lamina** Expanded flat part of a leaf; leaf blade.

**Lateral meristems** Meristems that divide to increase the girth of a plant (secondary growth); vascular cambium and cork cambium.

**Leaf mosaic** In growing plants, the tendency of leaves to arrangements most advantageous for photosynthesis.

**Leaflets** Smaller parts of a compound leaf.

**Lenticel** Specialized corky area for gas exchange, as on young woody stems before the bark forms.

**Lignin** Complex chemical compound in most secondary cell walls, providing them with durability and strength.

**Limiting factor** Environmental factor that can or does limit the growth of an individual or population.

**Limnetic zone** Open water region of a pond or lake.

**Lipid** Category of organic compounds that includes fats, oils, and waxes; has less oxygen than carbohydrates.

**Littoral zone** Shoreline zone of marine or freshwater ecosystems.

**Lobeline** Alkaloid obtained from Indian tobacco; used to stimulate respiratory activity.

**Long-day plant** Plant that flowers after exposure to day lengths that are longer than its critical light period.

**Marine** Ocean or sea environment.

**Mature phase** Adult period of growth following juvenile phase, during which sexual reproduction may occur.

**Megaspore** Spore that germinates to produce a female gametophyte.

**Megasporogenesis** Process in which the megaspore mother cell undergoes meiosis and cell division to produce megaspores; leads to egg production.

**Meiosis** Two nuclear divisions in which the chromosomes are reduced from the diploid to haploid number with the genes being distributed into four haploid cells.

**Meperidine** Synthetic narcotic drug used as an analgesic, sedative and antispasmodic; second to morphine in use to relieve pain; Demerol is one trade name.

**Mesophyll** Photosynthetic area of the leaf, between the upper and lower epidermis.

**Mesophyte** Plant adapted for growth in a terrestrial habitat having abundant moisture.

**Metabolism** Sum total of all chemical reactions occurring within a living unit.

**Metaphase** Stage in mitosis and meiosis where the chromosomes are arranged in the center of the cell across the spindle equator.

**Methadone** Synthetic narcotic drug used to relieve pain; also used as a substitute narcotic in treating heroin addicts.

**Methane** Colorless, odorless, flammable gas produced by decomposition of organic matter.

**Microfilaments** Smallest intracellular filaments (4 to 6 nm) that form a network just inside the cell membrane; apparently related to cell movements.

**Micronutrients** Nutrients required by plants in relatively minute amounts, as opposed to macronutrients, which are used by plants in much larger quantities; also called trace elements.

**Microorganism** Microscopic organism; refers to bacteria, some fungi, and protists.

**Micropyle** Opening in the ovule into which the pollen tube grows.

**Microspore** Spore that germinates to produce a male gametophyte.

**Microsporogenesis** Process in which the microspore mother cell undergoes meiosis and cell division to produce microspores.

**Microtubules** Tiny (25 nm), flexible, hollow rods within cells; many functions, including participation in cell movements and nuclear division.

**Middle lamella** Layer of pectic compounds that cements the primary cell walls together.

**Mineral** Naturally occurring chemical element or inorganic compound.

**Mineralization** Breakdown process whereby minerals are released to the soil.

**Mitochondrion** (*pl.*, **mitochondria**) Complex organelle that is the site of aerobic respiration in eukaryotic cells.

**Mitosis** Nuclear division that produces nuclei that are (normally) the same in chromosome complement as the parent cell; cytokinesis usually follows, completing cell division.

**Monera** Kingdom devised to include prokaryotic organisms.

**Monocotyledon (monocot)** In angiosperms, a plant with one cotyledon and other characteristics that distinguish it from dicots; includes grasses, lilies, orchids, and others.

**Mordant** Chemical that infiltrates a substance and that will fix a dye to the substance by forming a precipitate.

**Morphine** Highly addictive narcotic derived from opium; used to form other derivatives; an important painkiller.

**Mulch** Inorganic or organic material placed on the soil surface to reduce moisture loss; also used for weed control, clean fruit, temperature control, etc.

**Mutation** Abrupt, inheritable change in a gene or chromosome.

**Mutualism** Symbiotic relationship that is beneficial to each organism.

**Mycelium** Refers to a mass of hyphae.

**Mycoplasma** Group of organisms that are bacterialike; sometimes classified between viruses and bacteria, sometimes as bacteria.

**Mycorrhizae** Specific beneficial association of fungi with roots of vascular plants; relationship is generally considered mutualistic.

**Narcotic** Opium, its derivatives, or synthetic substitutes; used for relief of intense pain; narcotics produce physiological tolerance and dependence and are therefore highly addictive.

**Nekton** Free-swimming predators of an aquatic ecosystem, mostly fish.

**Netted venation** Leaves with veins arranged as in a net, as compared to parallel venation.

**Nicotine** Toxic alkaloid found in tobacco; used as an insecticide.

**Node** Region of the stem where a leaf or leaves are attached.

**Nodules** Enlargements on roots of plants caused by presence of nitrogen-fixing bacteria; relationship is considered mutualistic.

**Nonvascular plant** Refers to the "lower" plants, which do not possess xylem and phloem.

**Nuclear envelope** Double layer of unit membrane that surrounds the nuclear material in eukaryotes; perforated by pores to permit passage of larger molecules.

**Nuclear organizer** Special area on certain chromosomes where nucleoli are assembled.

**Nucleolus** Dense bodies within the nucleus composed of proteins and RNA granules that are ribosome precursors.

**Nucleotide** Molecule composed of a pentose sugar (ribose or deoxyribose), a nitrogen-containing base (adenine, guanine, cytosine, thymine or uracil) and a phosphoric acid unit; structural component of DNA and RNA.

**Nucleus** Largest cellular organelle in eukaryotes; contains the chromosomes and therefore controls all metabolic activities and inheritance of the cell.

**Oligotrophic** Water that is low in nutrients; thus of greater purity and lower productivity.

**Omnivore** Organism that consumes both animals and plants for food.

**Oögamous (oögamy)** Sexual reproduction where the egg is larger and nonmotile and the sperm is smaller and motile.

**Opium** Dried exudate taken from the opium poppy fruit; known for its addictive (narcotic) properties and its derivatives (morphine, codeine, and thebaine).

**Optimum** State of an environmental factor that is most favorable to an organism.

**Organ** Complex plant structure composed of different tissues; leaves, stems, roots, and reproductive structures.

**Organelle** Specialized, membrane-bounded functional structure within a cell.

**Osmosis** Diffusion of water across a differentially permeable membrane, from a site of higher concentration to lower concentration.

**Overwinter** Ability of an organism to survive through the cold winter periods to become viable again in the spring.

**Ovule** Organ located within the ovary of a seed plant; contains an egg; after fertilization, develops into a seed.

**Oxidative phosphorylation** Final stage of aerobic respiration, occurring in the electron transport system of the mitochondria; energy released during sequential transfer of $e^-$ and $H^+$ ions, resulting in formation of ATP and $H_2O$.

**Ozone** Light-blue, gaseous form of oxygen formed by electricity in the air.

**Palisade parenchyma** Columnar layer of photosynthetic cells making up the upper layer of mesophyll cells in many leaves.

**Parallel venation** Parallel arrangement of main veins in leaves; characteristic of monocots.

**Parasite** Organism that lives on or in an organism of another species, deriving nourishment from it.

**Parenchyma** Tissue made up of parenchyma cells; parenchyma cells are variable in shape, characterized by having thin walls and being alive at maturity; some remain capable of cell division.

**Parent material** Refers to the original rock from which breakdown processes eventually produce smaller and smaller pieces to finally become part of the soil.

**Pathogen** Organism that causes a disease.

**Peat** Partially decomposed sphagnum moss.

**Peat moss** *Sphagnum,* a moss that grows abundantly on wet soils, especially bogs.

**Pectin** Organic compound that acts as binding agent in the middle of primary walls of plant cells; extracted pectins are used to solidify fruit jellies.

**Pelagic region** Open waters of the sea.

**Pellicle** In many protozoans and euglenoids, the very highly modified cell membrane.

**Penetration** Initial invasion of a host by a pathogen.

**Penicillin** Antibiotic produced by molds of the genus *Penicillium;* useful against cocci bacteria.

**Perfume** Substance emitting a pleasant odor, generally derived from floral oils.

**Pericycle** Layer of cells inside the endodermis; initiates the development of lateral roots.

**Periphyton** Epiphytic aquatic plants, mostly diatoms.

**Permafrost** Permanently frozen area of the soil that underlies the tundra.

**Petiole** Stalk of a leaf.

**pH** Symbol used to identify the concentration of hydrogen ions in a solution; pH of 7 is neutral; higher pH indicates alkalinity and lower pH indicates acidity.

**Phloem** Food-conducting tissue composed of sieve cells or sieve tube elements, parenchyma cells, and sclerenchyma cells.

**Phosphorylation** Reaction in which phosphate is added to a compound.

**Photic zone** In aquatic ecosystems, the area of light penetration.

**Photoperiodism** Response of plants to the relative length of day and night.

**Photorespiration** Process occurring in $C_3$ plants that reduces the amount of carbohydrates gained by $CO_2$ fixation.

**Photosynthesis** Conversion of radiant energy (light) into chemical energy (food); complex process in which the sun's energy, with carbon dioxide and water, is transformed into sugar in the presence of chlorophyll, releasing oxygen.

**Phototropism** Bending growth movement in response to light; growth of shoot toward direction of light source is *positive* phototropism; growth of roots away from direction of light source is *negative* phototropism.

**Phylogeny** Evolutionary history of a group of organisms.

**Phytochrome** Pigment in plants associated with the absorption of far-red and red light.

**Phytotoxins** Plant poisons.

**Pioneer plant** Plant that colonizes previously uninhabited site, as lichen on a rock.

**Pistil** General term for the female reproductive

organ of a flower, composed of one or more carpels; most have a stigma, style, and ovary.

**Pith** Ground tissue making up the center of a primary plant body, around which the vascular bundles are arranged.

**Pith ray** Ground tissue between the vascular bundles; connects the pith and cortex of a stem.

**Plankton** Floating microscopic organisms in an aquatic ecosystem.

**Plasma membrane (plasmalemma)** Outermost cytoplasmic boundary, composed of a single unit membrane; limits passage of molecules into and out of the cell.

**Plasmodesma** (*pl.,* **plasmodesmata)** Membrane-lined channel through the cell wall of adjacent cells.

**Plasmodium** Phase in the life cycle of a slime mold; multinucleate mass of naked protoplasm (without cell walls).

**Plasmolysis** Shrinking of the cytoplasm from the cell wall as a result of extreme water loss from the cell.

**Plastid** Organelle bounded by a double unit membrane; of several types; chloroplast is the site of food manufacture and storage.

**Plumule** First bud development above the cotyledons, including first set of true leaves.

**Polar transport** Tendency of a substance to be transported in only one direction along axis of plant body.

**Pollen grain** Mature male gametophyte of gymnosperms and angiosperms; carries sperm nuclei to female gametophyte during pollination for eventual syngamy.

**Polypeptide** Molecular chain of amino acids.

**Polyploidy** More than two sets of chromosomes in one cell.

**Pome** General classification term for a plant that produces fruit from an inferior ovary and the receptacle tissue, such as apple and pear.

**Prickles** Sharp projections that are the outgrowth of the outer (surface) tissues of a stem.

**Primary root** First root emerging from the seed.

**Primary succession** Initial development of an ecosystem on a site not previously inhabited by organisms, such as newly exposed rocks, volcanic areas, and sandbars.

**Primary tissues** Tissues derived from the apical meristems of the root and shoot.

**Primary wall** First wall produced during cell maturation; collective term including the two adjacent walls and the middle lamella.

**Procambium** Tissue produced by the apical meristem that matures into the primary vascular tissue.

**Producers** Organisms that can manufacture their own food through direct capture of light energy; primarily green plants, via the process of photosynthesis.

**Prokaryote** More primitive organisms, not having a distinct nucleus or other membrane-bounded organelles; bacteria and blue-green "algae".

**Prophase** Earliest stage of mitosis and meiosis wherein the chromosomes condense into visible threads and the nuclear membrane and nucleolus disintegrate.

**Proplastid** Immature form of a plastid.

**Protein** Complex organic compound composed of many amino acids joined by peptide bonds.

**Protista** Major grouping (kingdom) of unicellular (acellular) eukaryotic organisms.

**Protodermis** Outer tissue produced by the apical meristem that matures into the epidermis.

**Protoplasm** Living substance, general term.

**Protoplast** Term that refers to all that is enclosed by, but not including, the cell wall.

**Psilocybin and psilocin** Psychoactive compounds found in mushroom genera *Conocybe* and *Psilocybe.*

**Radial symmetry** Organization where a structure may be bisected longitudinally in more than one plane, yielding similar halves; in contrast to bilateral symmetry, where the bisection can occur in only one plane to obtain mirror images.

**Radicle** Embryonic root.

**Ray initials** Initials in the vascular cambium dividing to produce xylem and phloem ray cells.

**Recessive** In genetics, the suppression of one phenotypic expression by a dominant allele.

**Red tide** Occasional high populations of dinoflagellates, coloring the seawater; produce sufficient toxins to poison fish; toxins absorbed into tissues of mussels and other filter feeders, making them poisonous to humans.

**Relative humidity** Ratio of the amount of water vapor in the air to the greatest amount possible at the same temperature.

**Resistance** Ability of an organism to overcome the effect of a pathogen or other stresses.

**Ribonucleic acid (RNA)** Nucleic acid formed on chromosomal DNA and involved in protein synthesis; similar in composition to DNA except that the pyrimidine uracil replaces thymine. RNA is the genetic material of many viruses. RNA has three types: messenger RNA, transfer RNA, and ribosomal RNA.

**Ribosomes** Numerous, dense, RNA-containing bodies in the cytoplasm and in chloroplasts and mitochondria; sites of protein synthesis.

**Rickettsia** Microorganisms (named after Howard T. Ricketts, an American pathologist) that cause various diseases, including typhus.

**Root cap** Protective cap of cells covering the apical meristem of a root.

**Root hairs** Tubelike protrusions of root epidermal cells at the zone of maturation.

**Rootstock** Host plant onto which a scion is grafted; used to impart disease resistance, size limitations, and other characteristics to the grafted plant.

**Sachet** Small bag containing a perfumed powder; used to scent clothes and linen.

**Sanitation** Practice of removing diseased parts and other plant debris from the growing site to reduce disease and parasite proliferation.

**Saponin** Toxic glycoside found in certain plants; produces a soaplike lather when used as a soap.

**Saprophytes** Plants that do not produce their own food but secure food by digesting nonliving organic matter.

**Sapwood** Lighter-colored outer wood that is alive and active in water transport.

**Scientific method** Organized method of gathering information, based on observation and controlled testing of hypotheses.

**Scion** Piece of twig grafted onto a rootstock.

**Sclereid** Sclerenchyma cell of varying shape but typically long with thick secondary walls and many pits.

**Sclerenchyma** Tissue specialized for support, having two basic cell types: fiber and sclereid.

**Scopolamine** Term taken from genus *Scopolia* of the nightshade family; toxic alkaloid taken from the roots and used as a truth serum and as a sedative in operations.

**Secondary cell wall** Innermost cell wall layer, mainly of cellulose reinforced and impregnated with lignin, cutin, or suberin.

**Secondary roots** Roots developing from the pericycle of the primary root.

**Secondary succession** Reestablishment of organisms after fire, flood, or some other interruption in the normal succession of an ecosystem; renewal of the damaged site.

**Seed coat** Protective covering of a seed; matures from the integument of the ovule.

**Semipermeability** Selective nature of a membrane by which some, but not all, molecules can pass through it.

**Senescence** Irreversible period of declining vigor following mature, reproductive phase.

**Shoot** General term referring to the above-ground portion of a plant, including the stem and leaves.

**Short-day plant** Plant that flowers after exposure to day lengths (photoperiods) shorter than its critical light period.

**Sieve area** Specialized area in the wall of sieve elements where pores connect adjacent protoplasts.

**Sieve cell** Long tapering cell with sieve areas; found in gymnosperms and lower vascular plants; functions in food conduction.

**Sieve element** Term including the sieve cell and sieve tube member, both involved in food conduction.

**Sieve plate** Highly specialized sieve area on the end walls of a sieve tube member.

**Sieve tube** Collective term referring to sieve tube members connected end to end.

**Slipping** Refers to a growth condition in which the vascular cambium is actively growing, thus allowing the bark to slip easily from the wood.

**Solanine** Poisonous alkaloid found in several plants in the genus *Solanum* (potatoes, tomatoes, and others) in the nightshade family.

**Species** Population of interbreeding organisms that are reproductively isolated from all other populations.

**Sphagnum** Genus of moss that grows abundantly in wet, acid areas; dried remains are used in the growing of container plants.

**Spile** Spout inserted into a tree to draw off sap.

**Spine** Hard, pointed structure developing as a modified leaf.

**Spongy parenchyma** Lower part of the leaf mesophyll; made up of loosely arranged parenchyma cells with large intercellular spaces.

**Spur** Short, woody branch where most of the flowering occurs.

**Stamen** Male organ of a flower consisting of a filament and anther; anther produces and releases pollen.

**Statoliths** Dense starch grains in root cap cells that may be involved in gravity perception.

**Stem sprouts** Vigorous shoots that arise from the lower stem (trunk) region of a woody plant; also called suckers or water sprouts.

**Stoma (stomata)** Opening and its adjacent guard cells in the epidermis of leaves and stems.

**Stone cells** Type of sclereid with thick secondary walls and many pits; brick shaped.

**Storied cambium** Regular organization of cambial initials into rows, resembling stories of a high-rise apartment house complex.

**Style** Portion of the pistil below the stigma and above the ovary.

**Suberin** Fatty substance deposited in the walls of cork cells.

**Substrate** Surface upon, within, or to which an organism is attached.

**Sulfur dioxide ($SO_2$)** Pungent gas that will produce sulfuric acid when mixed with water; a primary cause of acid rain.

**Symbiosis** Close living relationship of two or more organisms of different species, including mutualism, where relationship is beneficial to both, and commensalism, where one benefits and the other is neither harmed nor benefited; parasitism, in which one benefits and the other is harmed, is sometimes listed as a type of symbiosis.

**Synergism** Cooperative action between agents, providing a much greater reaction than the sum of the individual components.

**Syngamy** Fusion of gametes; synonym is fertilization.

**Systemic** Spreading of a chemical or pathogen throughout the tissues of a plant.

**Taproot system** Root system having a large main root, with small lateral roots, as in carrots and beets.

**Taxine** Toxic resinous substance produced in yew (*Taxus*) plants.

**Taxis (*pl.,* taxes)** Locomotion oriented in response to an environmental stimulus, such as light (phototaxis).

**Telophase** Last stage of mitosis and meiosis, when chromosomes have reached the spindle poles and begin to loosen up in organization and nuclei begin to form; cytokinesis usually begins during telophase.

**Tendril** Modified leaf that coils around other structures to support the plant shoot.

**Thebaine** Natural alkaloid found in opium.

**Thermodynamics, first and second laws** Energy cannot be created or destroyed, but it can be transferred or transformed to another state. With each successive transfer or transformation, there is a net loss of available energy.

**Thermolabile** Unstable when heated, losing properties during heating; opposite of thermostable.

**Thiaminase** Enzyme that breaks down thiamine.

**Thiamine** Also referred to as vitamin $B_1$, essential to normal metabolism.

**Thinning** Pruning process to remove lateral branches, thus encouraging elongation of selected remaining branches.

**Thorn** Hard, pointed, modified branch.

**Tissue** Group of cells specialized to perform a particular function.

**Tolerance range** Range of variation of a particular environmental factor (pH, for example) that an organism can tolerate.

**Tonoplast** Single unit membrane that surrounds the central vacuole.

**Topiary** Practice of shaping plants into unique ornamental forms.

**Tracheid** Elongated cell with thick walls and bordered pits, which functions in water conduction and support.

**Transpiration** Evaporation of water from plant surface.

**Trichomes** Epidermal outgrowths, of single or multicellular organization.

**Turgid** Firm cell condition as a result of water intake.

**Tylosis** (*pl.*, **tyloses**) Balloonlike growth from a parenchyma cell through a pit into a xylem conducting cell.

**Ultraviolet radiation** Found at the violet side of the light range outside the visible spectrum.

**Unicellular** Composed of a single cell.

**Vascular bundle** Conducting strand made up of phloem, xylem, and support tissue; meristematic tissue may be present.

**Vascular cambium** Layer of meristematic cells dividing to produce secondary xylem and phloem; increases girth of the plant.

**Vector** Animal, usually an insect, that carries the inoculum of a pathogen.

**Vein** Vascular bundle in a leaf.

**Vessel** Composite term for vessel members connected end to end to conduct water and mineral nutrients in the plant.

**Vessel element (vessel member)** Xylem cell connected end to end to form a vessel.

**Xanthophyll** Yellow chloroplast pigment; one of the carotenoid pigments.

**Xerophyte** Plant that grows in a dry habitat because of its adaptive abilities to limit water loss, store water, and otherwise endure drought.

**Xylem** Specialized tissue composed of vessels, tracheids, parenchyma, and sclerenchyma for conducting water and mineral nutrients in the plant.

**Zygote** Cell formed by syngamy.

# INDEX

## A



Numbers in boldface type represent illustration-bearing pages.

## B

## C

## D



## E

## F

## G

## H



## I

## O



## P

## T

## U



## V



## W

## X



## Y



## Z